OPEN-CHANNEL HYDRAULICS

Richard H. French

OPEN-CHANNEL HYDRAULICS

McGraw-Hill, Inc.
New York St. Louis San Francisco Auckland Bogota
Caracas Lisbon London Madrid Mexico Milan Montreal
New Delhi Paris San Juan Singapore Sydney
Tokyo Toronto

OPEN-CHANNEL HYDRAULICS
International Editions 1994

2 3 4 5 6 7 8 9 20 CMO UPE 9 8 7 6

The editors for this book were Joan Zseleczky and Rita Margolies, the designer was Elliot Epstein, and the production supervisor was Sara L. Fliess. It was set in Century Schoolbook by University Graphics, Inc.

Library of Congress Cataloging-in-Publication Data

French, Richard H.
Open-channel hydraulics.
Includes index.
1. Channels (Hydraulic engineering). I. Title.
TC175.F78 1985 627'.1 85-207
ISBN 0-07-022134-0

When ordering this title, use ISBN 0-07-113310-0

Printed in Singapore

Contents

Preface

From the twin viewpoints of quantity and quality, water-resources projects are of paramount importance to the maintenance and progress of civilization as it is known today. The knowledge of open-channel hydraulics that is essential to water resources development and the preservation of acceptable water quality has, along with other areas of engineering knowledge, exploded in the last two decades. To some degree this explosion of information is due to the advent of the high-speed digital computer which has allowed the hydraulic engineer to address and solve problems which only 20 years ago would have been computationally too large to contemplate. In addition, the concern of society with both the preservation and restoration of the aquatic environment has led the hydraulic engineer to consider open-channel flow processes which, previous to the environmental movement, would not have been considered important enough to require study. At this time there are only two well-known English-language books on the subject of open-channel hydraulics—*Open-Channel Hydraulics* by V. T. Chow, published in 1959, and *Open Channel Flow* by F. M. Henderson, published in 1966. Neither of these books provides the modern viewpoint on this important subject.

This book is designed primarily as a reference book for the practicing engineer; however, a significant number of examples are used, and it is believed that the book can serve adequately as a text in either an undergraduate or graduate course in civil or agricultural engineering. As is the case with all books, this book is a statement of what the author believes is important, with the length tempered by a very reasonable page limit suggested by the publisher. Because of the background of the author, this book approaches open-channel hydraulics from the viewpoint of presenting basic principles and demonstrating the application of these principles. This book, unlike the ones that have preceded it, emphasizes numerical methods of solving problems; like previous books, the approach to the subject is primarily one-dimensional.

The book is divided into 14 chapters. In Chapter 1 basic definitions and the equations which govern flow in open channels are introduced.

In Chapters 2 and 3 the laws of conservation of energy and momentum are introduced, and applications of these basic laws to both rectangular and non-rectangular channels are discussed.

In Chapters 4 and 5 uniform flow and its computation are discussed. In connection with the topic of uniform flow, flow resistance is also treated.

In Chapter 6 gradually and spatially varied flow is considered. Solutions to these problems that involve tabular, graphical, and numerical procedures are considered.

In Chapter 7 various methods of designing channels are presented. The types of channels considered include lined channels; stable, unlined, earthen channels; and grass-lined channels. Consideration is given to both economics and seepage losses.

In Chapter 8 methods of flow measurement are considered. Techniques included in this discussion are devices and procedures for stream gaging, weirs, flumes, and culverts.

In Chapter 9 rapidly varied flow in nonprismatic channels is discussed. Discussions of flow through bridge contractions; the control of hydraulic jumps with sharp- and broad-crested weirs, abrupt rises and drops, and stilling basins; drop-spillway structures; and the design of channel transitions are included.

In Chapter 10 the transport processes known as turbulent diffusion and dispersion are discussed. After the one- and two-dimensional governing equations have been developed, methods of estimating the vertical and transverse turbulent diffusion coefficients and the longitudinal dispersion coefficient are discussed.

In Chapter 11 the topic of turbulent, buoyant surface jets in open-channel flow is discussed.

In Chapters 12 and 13 gradually and rapidly varied unsteady flows are considered. In these chapters modern numerical methods for solving the governing equations are discussed. In the case of the dam break problem, both numerical methods and simplified methods of computation are presented.

In Chapter 14 the design, construction, and use of physical models in the study of open-channel hydraulics are discussed. Among the types of models considered are geometrically distorted and undistorted; fixed and movable bed; and ice.

It should be noted and emphasized that this book is concerned only with the flow of water in channels where the water is not transporting significant quantities of air or sediment. In subject areas where pollutant transport is the primary reason for describing the flow, specific pollutants are not discussed; rather, transport of a neutral tracer is considered. In the area of unsteady flow, tidal hydraulics are not considered since this is properly the venue of the coastal engineer.

When the author began this book in the fall of 1981, he was, from the viewpoint of 1984, not aware of the magnitude of the project he was undertaking. Because of his position as a member of the research faculty, the University of Nevada system could legitimately provide but verbal encouragement and moral support in this project. Thus, this project was completed in the evenings, on holidays, and on weekends. The author is especially indebted to Professor Vincent T. Ricca of the Ohio State University, who originally introduced the author to the subject of open-channel hydraulics, and the late Professor Hugo B. Fischer of the University of California, Berkeley, who continued the educational process. The author would also acknowledge Dr. S. C. McCutcheon of the U.S. Geological Survey and Professor J. W. Bird of the University of Nevada, Reno, who spent numerous hours of their own time reading the manuscript. The author would also acknowledge his wife, Darlene, whose assistance through the seemingly endless and late hours of work provided invaluable encouragement that was desperately needed.

Richard H. French

OPEN-CHANNEL HYDRAULICS

ONE

Concepts of Fluid Flow

SYNOPSIS

In Section 1.2, the types of flow encountered in open channels are classified with respect to time, space, viscosity, density, and gravity. In addition, the types of channels commonly encountered and their geometric properties are defined. In Section 1.3, the equations which govern flow in open channels, i.e., conservation of mass, momentum, and energy, are developed along with the equations for the energy and momentum correction coefficients. In Section 1.4, theoretical concepts such as the scaling of partial differential equations, boundary layers, and velocity distributions in both homogeneous and stratified flows are briefly discussed. In Section 1.5, the basic concepts of geometric and dynamic similarity are developed in relation to their application to the design and use of physical hydraulic models.

1.1 INTRODUCTION

By definition, an open channel is a conduit for flow which has a free surface, i.e., a boundary exposed to the atmosphere. The free surface is essentially an interface between two fluids of different density. In the case of the atmosphere, the density of air is much lower than the density for a liquid such as water. In addition the pressure is constant. In the case of the flowing fluid, the motion is usually caused by gravitational effects, and the pressure distribution within the fluid is generally hydrostatic. Open-channel flows are almost always turbulent and unaffected by surface tension; however, in many cases of practical importance, such flows are density-stratified. The interest in the mechanics of open-channel flow stems from their importance to what we have come to term civilization. As defined above, open channels include flows occurring in channels ranging from rivulets flowing across a field to gutters along residential streets and continental highways to partially filled sewers carrying waste water to irrigation channels carrying water halfway across a continent to vital rivers such as the Mississippi, Nile, Rhine, Yellow, Ganges, Amazon, and Mekong. Without exception, one of the primary requirements for the development, maintenance, and advancement of civilization is access to a plentiful and economic supply of water.

In the material which follows, it is assumed that the reader is familiar with the basic principles of modern fluid mechanics and hydraulics, calculus, numerical analysis, and computer science. It is the purpose of this chapter to review briefly a number of basic definitions, principles, and laws with the focus being on their application to the study of the mechanics of open-channel flow.

1.2 DEFINITIONS

Types of Flow

Change of Depth with Respect to Time and Space As will be demonstrated in this section, it is possible to classify the type of flow occurring in an open channel on the basis of many different criteria. One of the primary criteria of classification is the variation of the depth of flow y in time t and space x. If time is the criterion, then a flow can be classified as being either *steady*, which implies that the depth of flow does not change with time ($\partial y/\partial t = 0$), or *unsteady*, which implies that the depth does change with time ($\partial y/\partial t \neq 0$). The differentiation between steady and unsteady flows depends on the viewpoint of the observer and is a relative rather than an absolute classification. For example, consider a surge, i.e., a singular wave with a sharp front moving either up or down a channel. To a stationary observer on the bank of the channel, the flow is unsteady since he or she will note a change in the depth of flow with time. However, to an observer who moves with the wave front, the flow is steady since no variation of depth with time can be noted. If water is added or subtracted along the channel reach under consideration, which is the case with gutters and side channel spillways, then the flow may be steady, but it is nonuniform. Specifically, this type of flow is termed *spatially varied* or *discontinuous* flow.

If space is used as the classification criterion, then a flow can be classified as *uniform* if the depth of flow does not vary with distance ($\partial y/\partial x = 0$) or as *nonuniform* if the depth varies with distance ($\partial y/\partial x \neq 0$). Although conceptually an unsteady uniform flow is possible, i.e., the depth of flow varies with time but remains constant with distance, from a practical viewpoint such a flow is nearly impossible. Therefore, the terminology *uniform* or *nonuniform* usually implies that the flow is also steady. Nonuniform flow, also termed *varied flow*, is further classified as being either *rapidly varied*—the depth of flow changes rapidly over a relatively short distance such as is the case with a hydraulic jump—or *gradually varied*—the depth of flow changes rather slowly with distance such as is the case of a reservoir behind a dam.

It should be noted that from a theoretical viewpoint the classifications of steady and uniform are very restrictive. For example, the terminology *uniform flow* implies that at every point in the flow field at an arbitrary instant in time, the velocity vectors have both the same magnitude and direction. Such a strict definition is much too restrictive for practical use; therefore, the definitions given above have been extended or relaxed to a point where they are useful. For example, in practice, time and space flow classifications are commonly done on the basis of gross flow characteristics. If the spatially averaged velocity of flow $\left(\bar{u} = \iint_A u\, dA\right)$ does not vary significantly with time, then the flow is

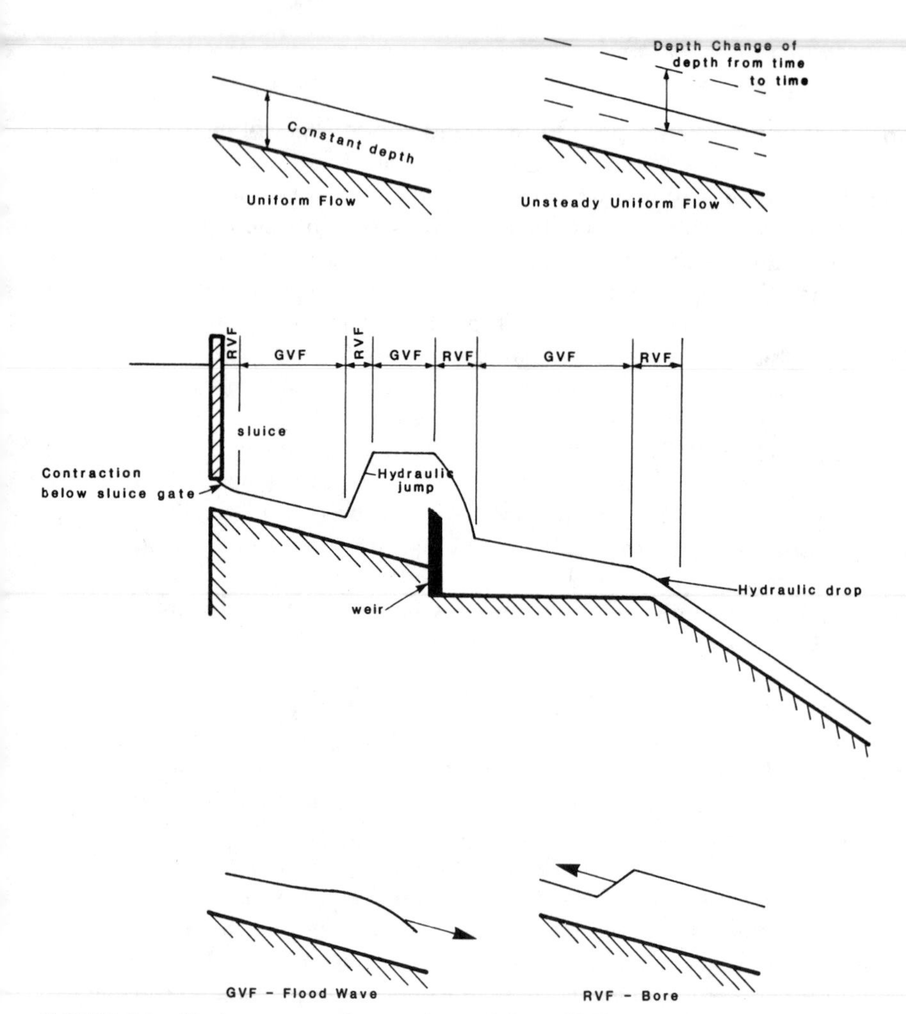

FIGURE 1.1 Various types of open-channel flow; GVF = gradually varied flow, RVF = rapidly varied flow. (*Chow, 1959.*)

classified as steady. Similarly, if the average depth of flow is constant in space, then the flow is considered uniform. Figure 1.1 schematically provides examples of the above definitions applied to field situations.

Viscosity, Density, and Gravity Recall from elementary fluid mechanics that, depending on the magnitude of the ratio of the inertial forces to the viscous forces, a flow may be classified as laminar, transitional, or turbulent. The basis for this classification is a dimensionless parameter known as the Reynolds number, or

$$\mathbf{R} = \frac{UL}{\nu} \tag{1.2.1}$$

where U = characteristic velocity of flow, often taken as the average velocity of flow
L = characteristic length
ν = kinematic viscosity

Then, a *laminar flow* is one in which the viscous forces are so large relative to the inertial forces that the flow is dominated by the viscous forces. In such a flow, the fluid particles move along definite, smooth paths in a coherent fashion. In a *turbulent flow*, the inertial forces are large relative to the viscous forces; hence, the inertial forces dominate the situation. In this type of flow, the fluid particles move in an incoherent or apparently random fashion. A *transitional flow* is one which can be classified as neither laminar nor turbulent. In open-channel flow, the characteristic length commonly used is the hydraulic radius which is the ratio of the flow area A to the wetted perimeter P. Then

$\mathbf{R} \leq 500$	laminar flow
$500 \leq \mathbf{R} \leq 12{,}500$	transitional flow
$12{,}500 \leq \mathbf{R}$	turbulent flow

The state of flow based on the ratio of inertial to viscous forces is a critical consideration when resistance to flow is considered.

Flows are classified as homogeneous or stratified on the basis of the variation of density within the flow. If in all spatial dimensions the density of flow is constant, then the flow is said to be *homogeneous*. If the density of the flow varies in any direction, then the flow is termed *stratified*. The absence of a density gradient in most natural open-channel flows demonstrates that either the velocity of flow is sufficient to completely mix the flow with respect to density or that the phenomena which tend to induce density gradients are unimportant. The importance of density stratification is that when stable density stratification exists, i.e., density increases with depth or lighter fluid overlies heavier fluid, the effectiveness of turbulence as a mixing mechanism is reduced. In two-dimensional flow of the type normally encountered in open channels, a commonly accepted measurement of the strength of the density stratification is the gradient Richardson number

$$Ri = \frac{g(\partial\rho/\partial y)}{\rho(\partial u/\partial y)^2} \tag{1.2.2}$$

where g = acceleration of gravity
ρ = fluid density

y = vertical coordinate
$\partial u/\partial y$ = gradient of velocity in vertical direction
$\partial \rho/\partial y$ = gradient of density in vertical direction

When $\partial u/\partial y$ is small relative to $\partial \rho/\partial y$, Ri is large, and the stratification is stable. When $\partial u/\partial y$ is large relative to $\partial \rho/\partial y$, Ri is small, and as $Ri \rightarrow 0$, the flow system approaches a homogeneous or neutral condition. There are a number of other parameters which are used to measure the stability of a flow, and these will be discussed, as required, in subsequent chapters of this book.

Depending on the magnitude of the ratio of inertial to gravity forces, a flow is classified as subcritical, critical, or supercritical. The parameter on which this classification is based is known as the *Froude number*

$$\mathbf{F} = \frac{U}{\sqrt{gL}} \tag{1.2.3}$$

where U = a characteristic velocity of flow and L = a characteristic length. In an open channel, the characteristic length is taken to be the *hydraulic depth,* which by definition is the flow area A divided by the width of the free surface T, or

$$D = \frac{A}{T} \tag{1.2.4}$$

If $\mathbf{F} = 1$, the flow is in a critical state with the inertial and gravitational forces in equilibrium. If $\mathbf{F} < 1$, the flow is in a subcritical state, and the gravitational

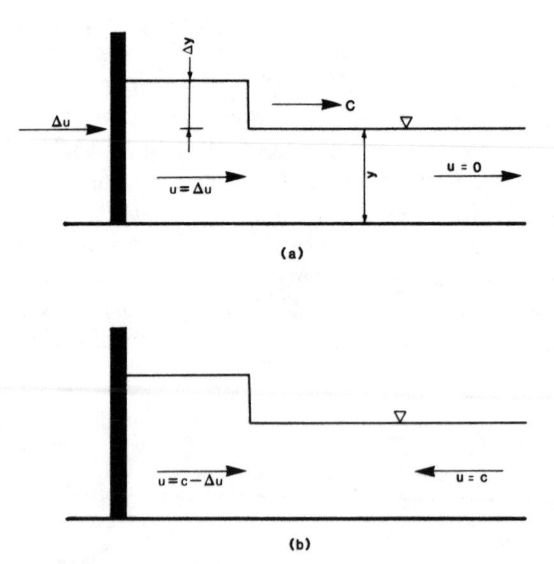

FIGURE 1.2 Propagation of an elementary wave.

forces are dominant. If **F** > 1, the flow is in a supercritical state and the inertial forces are dominant.

The denominator of the Froude number is the celerity of an elementary gravity wave in shallow water. In Fig. 1.2*a* an elementary water wave of height Δy has been created by the movement of an impermeable plate from the left to right at a velocity of Δu. The wave has a celerity c, and the velocity of flow in front of the wave is zero. The situation defined by Fig. 1.2*a* is unsteady and cannot be analyzed by elementary techniques. However, as indicated in an earlier section in this chapter, some unsteady-flow situations can be transformed to steady-flow problems (Fig. 1.2*b*). In this case, the transformation is accomplished by adapting a system of coordinates which moves at a velocity c. This is equivalent to changing the viewpoint of the observer; i.e., in Fig. 1.2*a* the observer is stationary while in Fig. 1.2*b* the observer is moving at the velocity of the wave. Application of the steady, one-dimensional equation of continuity to the situation described in Fig. 1.2*b* yields

$$cy = (y + \Delta y)(c - \Delta u)$$

and simplifying,

$$c = y\frac{\Delta u}{\Delta y} \qquad (1.2.5)$$

Application of the steady, one-dimensional momentum equation yields

$$\tfrac{1}{2}\gamma y^2 - \tfrac{1}{2}\gamma(y + \Delta y)^2 = \rho c y[(c - \Delta u) - c]$$

or

$$\frac{\Delta u}{\Delta y} = \frac{g}{c} \qquad (1.2.6)$$

Substitution of Eq. (1.2.6) in Eq. (1.2.5) yields

$$c = \frac{gy}{c}$$

or

$$c = \sqrt{gy} \qquad (1.2.7)$$

If it can be assumed that $y \simeq d$ (where d is the depth of flow), which is a valid assumption if the channel is wide, then it has been proved that the celerity of an elementary gravity wave is equal to the denominator of the Froude number. With this observation, the following interpretation can be applied to the subcritical and supercritical states of flow:

1. When the flow is subcritical, **F** < 1, the velocity of flow is less than the celerity of an elementary gravity wave. Therefore, such a wave can propagate upstream against the flow, and upstream areas are in hydraulic communication with the downstream areas.

2. When the flow is supercritical, **F** > 1, the velocity of flow is greater than the

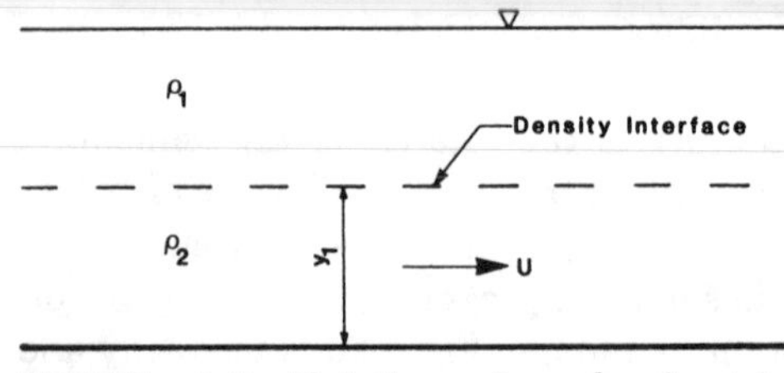

FIGURE 1.3 Notation for densimetric Froude number.

celerity of an elementary gravity wave. Therefore, such a wave cannot propagate upstream against the flow, and the upstream areas of the channel are not in hydraulic communication with the downstream areas.

Thus, the possibility of an elementary wave propagating upstream against the flow can be used as a criterion for differentiating between subcritical and supercritical flows.

In the case of density stratified open-channel flows, it is often convenient to define an overall but inverted form of the gradient Richardson number. The internal or densimetric Froude number $\mathbf{F}_D$ is defined by

$$\mathbf{F}_D = \frac{U}{\sqrt{g(\Delta\rho/\rho)L}} = \frac{U}{\sqrt{g'L}} \tag{1.2.8}$$

where, with reference to Fig. 1.3, $\Delta\rho = \rho_1 - \rho_2$, $g' = g(\rho_1 - \rho_2)/\rho$, and L = a characteristic length which is usually taken as the depth of flow in the lower layer y_1. The interpretation of $\mathbf{F}_D$ is analogous to that of $\mathbf{F}$; e.g., in an internally supercritical flow, a wave at the density interface cannot propagate upstream against the flow.

Channel Types

Open channels can be classified as either natural or artificial. The terminology *natural channel* refers to all channels which have been developed by natural processes and have not been significantly improved by humans. Within this category are creeks, rivers large and small, and tidal estuaries. The category of *artificial channels* includes all channels which have been developed by human efforts. Within this category are nagivation channels, power and irrigation canals, gutters, and drainage ditches. Although the basic principles in this book are applicable to natural channels, a comprehensive understanding of flow in natural channels is an interdisciplinary effort requiring knowledge of several technical fields, i.e., open-channel hydraulics, hydrology, geomorphology, and sediment transport. Therefore, although flow in natural channels will be discussed, an extensive treatment of the subject is outside the scope of this book.

Since many of the properties of artificial channels are controlled by design, this type of channel is much more amenable to analysis.

Within the broad category of artificial, open channels are the following subdivisions:

1. *Prismatic:* A prismatic channel has both a constant cross-sectional shape and bottom slope. Channels which do not meet this criterion are termed *nonprismatic.*
2. *Canal:* The term *canal* refers to a rather long channel of mild slope. These channels may be either unlined or lined with concrete, cement, grass, wood, bituminous materials, or an artificial membrane.
3. *Flume:* In practice, the term *flume* refers to a channel built above the ground surface to convey a flow across a depression. Flumes are usually constructed of wood, metal, masonry, or concrete. The term flume is also applied to laboratory channels constructed for basic and applied research.
4. *Chute and Drop:* A chute is a channel having a steep slope. A drop channel also has a steep slope but is much shorter than a chute.
5. *Culvert:* A culvert flowing only partially full is an open channel primarily used to convey a flow under highways, railroad embankments, or runways.

Finally, it is noted that in this book the terminology *channel section* refers to the cross section of channel taken normal to the direction of flow.

Section Elements

In this part the properties of a channel section which are wholly determined by the geometric shape of the channel and the depth of flow are defined.

1. *Depth of Flow y:* This is the vertical distance from the lowest point of a channel section to the water surface. In most cases, this terminology is used interchangeably with the terminology *depth of flow of section d,* which is the depth of flow measured perpendicular to the channel bottom. The relation between y and d is

$$y = \frac{d}{\cos \theta}$$

where θ is the slope angle of the channel bottom with a horizontal line. If θ is small

$$y \simeq d$$

Only in the case of steep channels is there a significant difference between y and d. It must be noted that in some places in this book y is also used to designate the vertical coordinate of a cartesian coordinate system. Although this dual definition of one variable can be confusing, it is unavoidable if a traditional notation system is to be used.

2. *Stage:* The stage of a flow is the elevation of the water surface relative to a datum. If the lowest point of a channel section is taken as the datum, then the stage and depth of flow are equal.

3. *Top Width T:* The top width of a channel is the width of the channel section at the water surface.

4. *Flow Area A:* The flow area is the cross-sectional area of the flow taken normal to the direction of flow.

5. *Wetted Perimeter P:* The wetted perimeter is the length of the line which is the interface between the fluid and the channel boundary.

6. *Hydraulic Radius R:* The hydraulic radius is the ratio of the flow area to the wetted perimeter or

$$R = \frac{A}{P} \tag{1.2.9}$$

7. *Hydraulic Depth D:* The hydraulic depth is the ratio of the flow area to the top width or

$$D = \frac{A}{T} \tag{1.2.10}$$

Table 1.1 summarizes the equations for the basic channel elements for the channel shapes normally encountered in practice.

Irregular channels are often encountered in practice, and in such cases values of the top width, flow area, and the location of the centroid of the flow area must be interpolated from values of these variables, which are tabulated as a function of the depth of flow. Franz (1982) has developed a rational and consistent method for performing these interpolations. Define

$$A_y = \int_0^y T_\phi \, d\phi \tag{1.2.11}$$

and

$$\bar{y}A_y = \int_0^y (y - \phi)T_\phi \, d\phi = \int_0^y A_\phi \, d\phi \tag{1.2.12}$$

where ϕ = dummy variable of integration
T_ϕ = top width of channel at distance ϕ above origin (Fig. 1.4)
A_y = flow area corresponding to depth of flow y
$\bar{y}$ = distance from origin to centroid of flow area

Franz (1982) defined consistency of interpolation to mean: (1) both the tabulated and interpolated values are consistent with Eqs. (1.2.11) and (1.2.12), and

(2) the variables vary monotonically between the tabulated values. Then, given tables of T and A as functions of y such that $y_i \leq y \leq y_{i+1}$ for $i = 0, 1, \ldots, n-1$, and $n+1 =$ the number of depth values tabulated,

$$T_y = T_i + \frac{y - y_i}{y_{i+1} - y_i}(T_{i+1} - T_i) \tag{1.2.13}$$

$$A_y = A_i + \tfrac{1}{2}(y - y_i)(T_y + T_i) \tag{1.2.14}$$

and
$$\bar{y}A_y = \bar{y}_i A_i + \tfrac{1}{2}(y - y_i)(A_y + A_i) - \tfrac{1}{6}(y - y_i)^2(T_y - T_i) \tag{1.2.15}$$

It must be noted that the accuracy of the values interpolated by Eqs. (1.2.13) to (1.2.15) hinges on the accuracy of the top width approximation because all other interpolated values are exact if the top width approximation is exact; conversely, errors in the top width propagate to all other variables.

Abstract Definitions

A *streamline* is a line constructed such that at any instant it has the direction of the velocity vector at every point; i.e., there can be no flow across a streamline. A *stream tube* is a collection of streamlines. It is noted and emphasized that neither streamlines nor stream tubes have any physical meaning; rather, they are convenient abstractions.

1.3 GOVERNING EQUATIONS

In most open-channel flows of practical importance, the Reynolds number exceeds 12,500 and the flow regime is turbulent. Therefore, for the most part laminar flow regimes are not treated in this book. The apparently random nature of turbulence has led many investigators to assume that this phenomenon can be best described in terms of statistics. On the basis of this assumption, it is convenient to define the instantaneous velocity in terms of a time-averaged velocity and a fluctuating, random component. For a cartesian coordinate system, the instantaneous velocities in the x, y, and z directions are, respectively,

$$u = \bar{u} + u'$$

$$v = \bar{v} + v' \tag{1.3.1}$$

and
$$w = \bar{w} + w'$$

Note: The average velocities used above may be determined by averaging either over time at a point in space or over a horizontal area at a point in time. In this book, the symbolism $\bar{u}_T$ indicates an average in time while $\bar{u}$ indicates an average in space. It is essential that these definitions be recalled in later

TABLE 1.1 Channel section geometric elements

Channel type	Area A	Wetted perimeter P
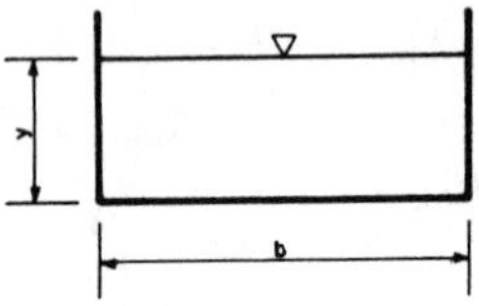 (a)	by	$b + 2y$
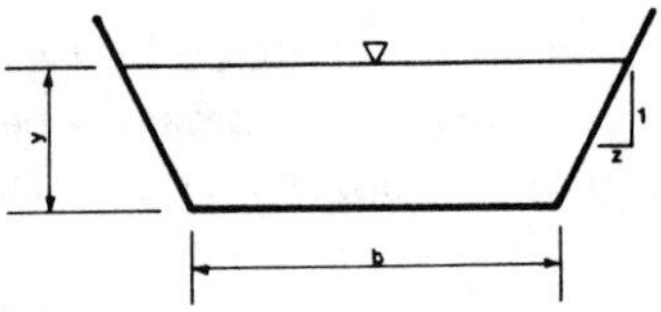 (b)	$(b + 2y)y$	$b + 2y\sqrt{1 + z^2}$
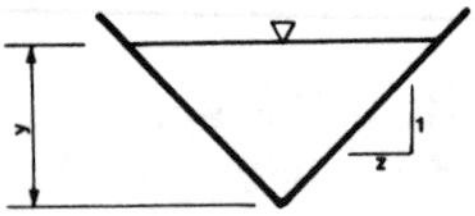 (c)	zy^2	$2y\sqrt{1 + z^2}$
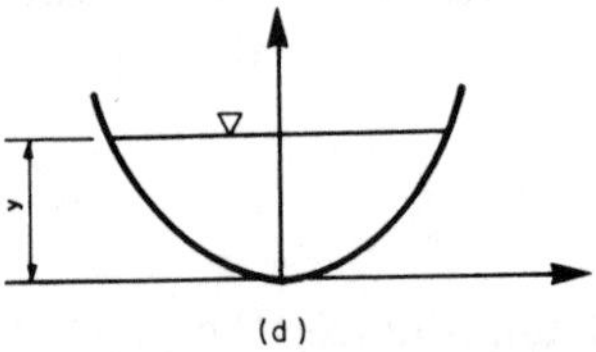 (d)	$\frac{2}{3}Ty$	$T + \frac{8}{3}\frac{y^2}{T}$*
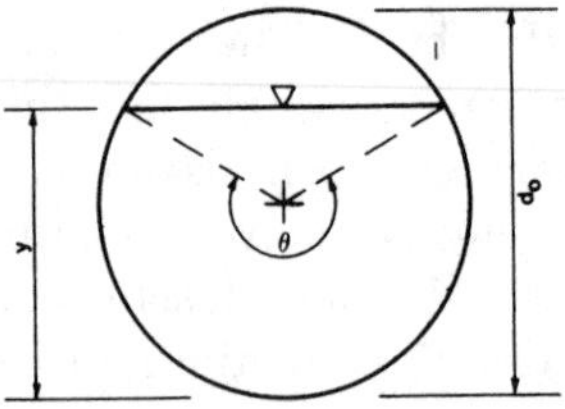 (e)	$\frac{1}{8}(\theta - \sin\theta)\, d_0^2$	$\frac{1}{2}\theta\, d_0$

*Satisfactory approximation when $0 < 4y/T \le 1$.

For $4y/T > 1$ $P = \frac{T}{2}\left[\sqrt{1 + \left(\frac{4y}{T}\right)^2} + \frac{T}{4y}\ln\left(\frac{4y}{T} + \sqrt{1 + \left(\frac{4y}{T}\right)^2}\right)\right]$

Hydraulic radius R	Top width T	Hydraulic depth D	Section factor Z
$\frac{by}{b + 2y}$	b	y	$by^{1.5}$
$\frac{(b + zy)y}{b + 2y\sqrt{1 + z^2}}$	$b + 2zy$	$\frac{(b + zy)y}{b + 2zy}$	$\frac{[(b + zy)y]^{1.5}}{\sqrt{b + 2zy}}$
$\frac{zy}{2\sqrt{1 + z^2}}$	$2zy$	$\frac{1}{2}y$	$\frac{\sqrt{2}}{2} zy^{2.5}$
$\frac{2T^2y}{3T^2 + 8y^2}$	$\frac{3}{2}\frac{A}{y}$	$\frac{2}{3}y$	$\frac{2}{9}\sqrt{6}Ty^{1.5}$
$\frac{1}{4}\left(1 - \frac{\sin\theta}{\theta}\right) d_0$	$2\sqrt{y(d_0 - y)}$	$\frac{1}{8}\left(\frac{\theta - \sin\theta}{\sin\frac{1}{2}\theta}\right) d_0$	$\frac{\sqrt{2}(\theta - \sin\theta)^{1.5}}{32\sqrt{\sin\frac{1}{2}\theta}} d_0^{2.5}$

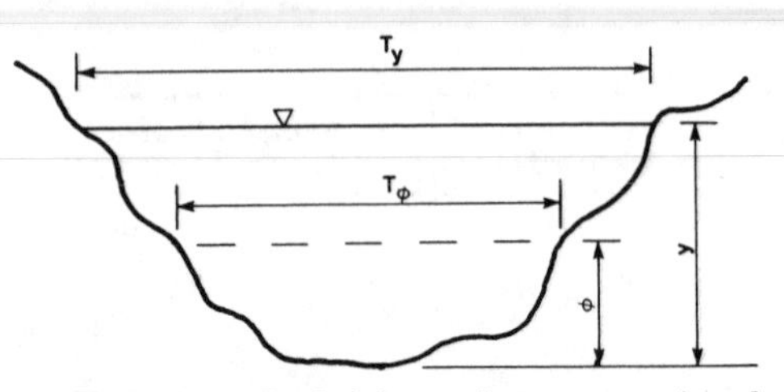

FIGURE 1.4 Definition of cross-sectional terms for interpolation.

sections. From this point forward, the pertinent statistics will be defined in the x direction only with the tacit understanding that these definitions apply to the two remaining cartesian coordinate directions. The time-averaged velocity is defined as

$$\overline{u}_T = \frac{1}{T}\int_0^T u\,dt \tag{1.3.2a}$$

where T indicates a time scale which is much longer than the time scale of the turbulence. The spatially average velocity is given by

$$\overline{u} = \frac{1}{A}\iint_A u\,dA \tag{1.3.2b}$$

Then, since the turbulent velocity fluctuations are random in terms of a time average,

$$\overline{u'} = \frac{1}{T}\int_0^T u'\,dt \tag{1.3.3}$$

Then the statistical parameters of interest are:

1. Root mean square (rms) value of the velocity fluctuations

$$\text{rms}(u') = \left[\frac{1}{T}\int_0^T (u')^2\,dt\right]^{1/2} \tag{1.3.4}$$

2. Average kinetic energy (KE) of the turbulence per unit mass

$$\frac{\text{Average KE turbulence}}{\text{Mass}} = \tfrac{1}{2}[\overline{(u')^2} + \overline{(v')^2} + \overline{(w')^2}] \tag{1.3.5}$$

3. Variable correlations measure the degree to which two variables are interdependent. In the case of the velocity fluctuations in the xy plane, the parameter

$$\overline{u'v'} = \frac{1}{T}\int_0^T u'v'\,dt \tag{1.3.6}$$

measures the correlation which exists between u' and v'. In a turbulent shear flow, $\overline{u'v'}$ is finite; therefore, it is concluded that u' and v' are correlated.

Conservation of Mass

Independent of whether a flow is laminar or turbulent, every fluid flow must satisfy the equation of conservation of mass or, as it is commonly termed, the equation of continuity. Substituting the expressions for the instantaneous velocities defined in Eq. (1.3.1) into the standard equation of continuity (density is assumed constant) (see, for example, Streeter and Wylie, 1975) yields

$$\frac{\partial(\bar{u} + u')}{\partial x} + \frac{\partial(\bar{v} + v')}{\partial y} + \frac{\partial(\bar{w} + w')}{\partial z} = 0$$

or

$$\frac{\partial \bar{u}}{\partial x} + \frac{\partial \bar{v}}{\partial y} + \frac{\partial \bar{w}}{\partial z} + \frac{\partial u'}{\partial x} + \frac{\partial v'}{\partial y} + \frac{\partial w'}{\partial z} = 0 \tag{1.3.7}$$

Taking the time or space average of each term in Eq. (1.3.7),

$$\overline{\frac{\partial \bar{u}}{\partial x}} + \overline{\frac{\partial \bar{v}}{\partial y}} + \overline{\frac{\partial \bar{w}}{\partial z}} = 0$$

and

$$\overline{\frac{\partial u'}{\partial x}} + \overline{\frac{\partial v'}{\partial y}} + \overline{\frac{\partial w'}{\partial z}} = 0$$

Therefore, Eq. (1.3.7) yields two equations of continuity, one for the mean or time-averaged motion and one for the turbulent fluctuations

$$\frac{\partial \bar{u}}{\partial x} + \frac{\partial \bar{v}}{\partial y} + \frac{\partial \bar{w}}{\partial z} = 0 \tag{1.3.8}$$

and

$$\frac{\partial u'}{\partial x} + \frac{\partial v'}{\partial y} + \frac{\partial w'}{\partial z} = 0 \tag{1.3.9}$$

In most applications of practical importance in open-channel hydraulics, Eq. (1.3.8) is the only equation of continuity which is used. At this point and for the remainder of this section, the overbar and subscript which indicate a time average will be dropped.

Other forms of the equation of continuity can be derived by specifying appropriate boundary conditions. For example, assume that the bottom boundary of the flow is given by the function

$$F = y - y_0(x,z) \tag{1.3.10}$$

where $y = y_0$ is the elevation of the bottom above a datum. Then, if it is assumed that (1) the velocity at the bottom is either zero or tangential to the surface and (2) a particle of fluid in contact with the bottom surface remains in contact with surface, the condition which must be satisfied at the bottom boundary is

$$\frac{\partial F}{\partial t} + u\frac{\partial F}{\partial x} + v\frac{\partial F}{\partial y} + w\frac{\partial F}{\partial z} = 0 \tag{1.3.11}$$

Substitution of Eq. (1.3.10) into Eq. (1.3.11) yields

$$u \frac{\partial y_0}{\partial x} + w \frac{\partial y_0}{\partial z} - v(x,y_0,z) = 0 \tag{1.3.12}$$

In an analogous fashion, the free surface is defined by the function

$$F = y - y^*(x,z,t)$$

and

$$\frac{\partial y^*}{\partial t} + u \frac{\partial y^*}{\partial x} + w \frac{\partial y^*}{\partial z} - w(x,y^*,z) = 0 \tag{1.3.13}$$

where y^* is the elevation of the free surface above a datum. Integrating Eq. (1.3.8) over the interval y_0 to y^*

$$\int_{y_0}^{y^*} \frac{\partial u}{\partial x}\, dy + \int_{y_0}^{y^*} \frac{\partial v}{\partial y}\, dy + \int_{y_0}^{y^*} \frac{\partial w}{\partial z}\, dy = 0$$

yields

$$\int_{y_0}^{y^*} \frac{\partial u}{\partial x}\, dy + \int_{y_0}^{y^*} \frac{\partial w}{\partial z}\, dy + v(x,y^*,z) - v(x,y_0,z) = 0 \tag{1.3.14}$$

Then

$$\frac{\partial}{\partial x} \int_{y_0}^{y^*} u\, dy = u(x,y^*,z) \frac{\partial y^*}{\partial x} - u(x,y_0,z) \frac{\partial y_0}{\partial x} + y^* \frac{\partial u(x,y^*,z)}{\partial x} - z_0 \frac{\partial u(x,y_0,z)}{\partial x}$$

and it is noted that

$$\int_{y_0}^{y^*} \frac{\partial u}{\partial x}\, dy = y^* \frac{\partial u(x,y^*,z)}{\partial x} - y_0 \frac{\partial u(x,y_0,z)}{\partial x}$$

Combining the above equations yields

$$\frac{\partial}{\partial x} \int_{y_0}^{y^*} u\, dy = u(x,y^*,z) \frac{\partial y^*}{\partial x} - u(x,y_0,z) \frac{\partial y_0}{\partial x} + \int_{y_0}^{y^*} \frac{\partial u}{\partial x}\, dy$$

or

$$\int_{y_0}^{y^*} \frac{\partial u}{\partial x}\, dy = \frac{\partial}{\partial x} \int_{y_0}^{y^*} u\, dy + u(x,y_0,z) \frac{\partial y_0}{\partial x} - u(x,y^*,z) \frac{\partial y^*}{\partial x} \tag{1.3.15}$$

In an analogous fashion

$$\int_{y_0}^{y^*} \frac{\partial w}{\partial z}\, dy = \frac{\partial}{\partial z} \int_{y_0}^{y^*} w\, dy + w(x,y_0,z) \frac{\partial y_0}{\partial z} - u(x,y^*,z) \frac{\partial y^*}{\partial z} \tag{1.3.16}$$

Combining Eqs. (1.3.12) to (1.3.16) yields

$$\frac{\partial}{\partial x}\int_{y_0}^{y^*} u\,dy + \frac{\partial}{\partial z}\int_{y_0}^{y^*} w\,dy + \frac{\partial y^*}{\partial t} = 0 \tag{1.3.17}$$

Equation (1.3.17) is a form of the equation of continuity from which a number of forms used in the field of open-channel hydraulics can be derived. Some examples follow.

1. Define

$$q_x = \int_{y_0}^{y^*} u\,dy$$

and

$$q_z = \int_{y_0}^{y^*} w\,dy$$

where q_x and q_z are the components of volume flux at the point (x,z) per unit time and unit widths of the vertical cross sections in the x and z directions, respectively, measured between the surface and bottom. With these definitions, Eq. (1.3.17) becomes

$$\frac{\partial q_x}{\partial x} + \frac{\partial q_y}{\partial y} + \frac{\partial y^*}{\partial t} = 0 \tag{1.3.18}$$

2. In many cases, it may be convenient to introduce values of u and w which are vertically averaged values, i.e., spatially averaged values. Define

$$\overline{u} = \frac{1}{y^* - y_0}\int_{y_0}^{y^*} u\,dy$$

and

$$\overline{w} = \frac{1}{y^* - y_0}\int_{y_0}^{y^*} w\,dy$$

Note: In this case u and w are average velocities in space rather than time. Rewriting Eq. (1.3.17)

$$\frac{\partial}{\partial x}\left[\frac{y^* - y_0}{y^* - y_0}\int_{y_0}^{y^*} u\,dy\right] + \frac{\partial}{\partial z}\left[\frac{y^* - y_0}{y^* - y_0}\int_{y_0}^{y^*} w\,dy\right] + \frac{\partial y^*}{\partial t} = 0$$

and substituting the definitions of u and w yields

$$\frac{\partial}{\partial x}[(y^* - y_0)\overline{u}] + \frac{\partial}{\partial z}[(y^* - y_0)\overline{w}] + \frac{\partial y^*}{\partial t} = 0 \tag{1.3.19}$$

3. In the case of a prismatic, rectangular channel

$$Q = \overline{u}A(t) = (y^* - y_0)b\overline{u}$$

where b = channel width. Since the channel is prismatic, the flow can be considered one-dimensional, and therefore

$$\frac{\partial}{\partial x}[(y^* - y_0)\bar{u}] + \frac{\partial y^*}{\partial t} = 0 \tag{1.3.20}$$

or

$$\frac{\partial Q}{\partial x} + b\frac{\partial y^*}{\partial t} = 0 \tag{1.3.21}$$

In this rather lengthy discussion of the equation of continuity, an attempt has been made to derive the forms of this equation which will be utilized throughout this book and which are found in the modern literature.

Conservation of Momentum

A very useful approach to the derivation of an appropriate conservation of momentum equation for open-channel flow can be obtained from a consideration of the one-dimensional control volume in Fig. 1.5. The principle of conservation of momentum states

$$\left\{\begin{array}{l}\text{Rate of accumulation}\\ \text{of momentum within}\\ \text{the control volume}\end{array}\right\} = \left\{\begin{array}{l}\text{net rate of}\\ \text{momentum}\\ \text{entering}\\ \text{control volume}\end{array}\right\} + \left\{\begin{array}{l}\text{sum of forces}\\ \text{acting on}\\ \text{control volume}\end{array}\right\} \tag{1.3.22}$$

The rate of momentum entering the control volume is the product of the mass flow rate and the velocity or

$$\text{Momentum entering} = \rho u y^*(u) - \frac{\Delta x}{2}\frac{\partial}{\partial x}(\rho u^2 y^*)$$

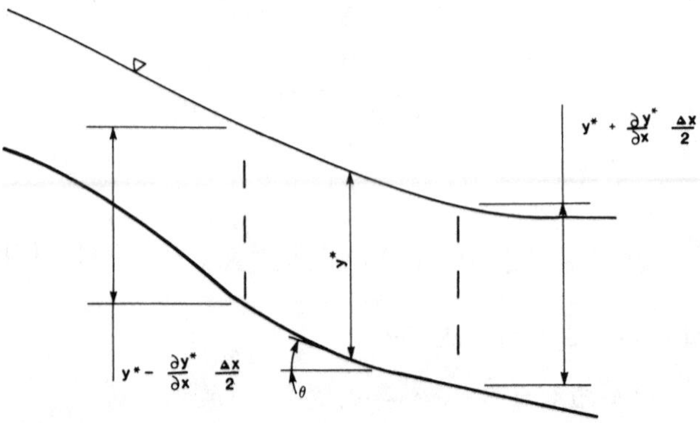

FIGURE 1.5 One-dimensional control volume.

$$\text{Momentum leaving} = \rho u y^*(u) + \frac{\Delta x}{2}\frac{\partial}{\partial x}(\rho u^2 y^*)$$

Then, the net rate at which momentum enters the control volume is

$$\rho u y^*(u) - \frac{\Delta x}{2}\frac{\partial}{\partial x}(\rho u^2 y^*) - \left[\rho u y^*(u) + \frac{\Delta x}{2}\frac{\partial}{\partial x}(\rho u^2 y^*)\right] = \Delta x \frac{\partial}{\partial x}(\rho u^2 y^*) \quad (1.3.23)$$

The forces acting on the control volume shown in Fig. 1.5 are: (1) gravity, (2) hydrostatic pressure, and (3) friction. The body force due to gravity is the weight of the fluid within the control volume acting in the direction of the x axis or

$$F_y = \rho g y^* \Delta x \sin\theta = \rho g y^* \Delta x S_x \quad (1.3.24)$$

where θ = angle the x axis makes with the bottom of the channel, and it is assumed that

$$S_x = \sin\theta$$

The pressure force on any vertical section of unit width and water depth y^* is

$$F_p = \int_0^{y^*} p\,dy = \int_0^{y^*} \rho g(y^* - y)\,dy = \tfrac{1}{2}\rho g(y^*)^2 \quad (1.3.25)$$

The frictional force which is assumed to act on the bottom and sides of the channel is given by

$$F_f = \rho g y^* \Delta x S_f \quad (1.3.26)$$

where S_f = slope of the energy grade line or the friction slope.

Combining Eqs. (1.3.23) to (1.3.26) with the law of conservation of momentum, Eq. (1.3.22), yields

$$\Delta x \frac{\partial}{\partial t}(\rho u y^*) = -\Delta x \frac{\partial}{\partial x}(\rho u^2 y^*) + g y^* \rho(S_x - S_f) - \frac{g}{2}\Delta x \frac{\partial y^{*2}}{\partial x}$$

Assuming that ρ is constant and dividing both sides of the above equation by $\rho\Delta x$,

$$\frac{\partial(u y^*)}{\partial t} + \frac{\partial}{\partial x}(u^2 y^*) + \frac{g}{2}\frac{\partial y^{*2}}{\partial x} = g y^*(S_x - S_f) \quad (1.3.27)$$

which is known as the conservation form of the momentum equation. If this equation is expanded,

$$y^* \frac{\partial u}{\partial t} + u \frac{\partial y^*}{\partial t} + u \frac{\partial}{\partial x}(u y^*) + y^* u \frac{\partial u}{\partial x} + g y^* \frac{\partial y^*}{\partial x} = g y^*(S_x - S_f) \quad (1.3.28)$$

In an analogous fashion, the law of conservation of momentum can, when required, be written in a two-dimensional form or

$$\frac{\partial}{\partial t}(uy^*) + \frac{\partial}{\partial x}(u^2y^*) + \frac{\partial}{\partial y}(uvy^*) + \frac{g}{2}\frac{\partial}{\partial x}(y^{*2}) = gy^*(S_x - S_{fx}) \quad (1.3.29)$$

and

$$\frac{\partial}{\partial t}(vy^*) + \frac{\partial}{\partial x}(uvy^*) + \frac{\partial}{\partial y}(v^2y^*) + \frac{g}{2}\frac{\partial}{\partial y}(y^*)^2 = gy^*(S_y - S_{fy}) \quad (1.3.30)$$

where S_x and S_y are the channel bottom slopes in the x and y directions, respectively, and S_{fx} and S_{fy} are the friction slopes in the x and y directions, respectively. These equations can also be expanded to yield forms similar to Eq. (1.3.28).

In a prismatic, rectangular channel where the equation of continuity is given by Eq. (1.3.20), the corresponding conservation of momentum equation is

$$\frac{\partial u}{\partial t} + u\frac{\partial u}{\partial x} + g\frac{\partial y^*}{\partial x} = g(S_x - S_f) \quad (1.3.31)$$

At this point, it is convenient to note that the conservation of momentum in a turbulent flow is governed by a set of equations known as the Reynolds equations which can be derived from the Navier-Stokes equations by the substitution of Eq. (1.3.1). The result in a general cartesian coordinate system is

$$\begin{aligned}
&\rho\left(\frac{\partial \bar{u}}{\partial t} + \bar{u}\frac{\partial \bar{u}}{\partial x} + \bar{v}\frac{\partial \bar{u}}{\partial y} + \bar{w}\frac{\partial \bar{u}}{\partial z}\right) = -\frac{\partial}{\partial x}(\bar{p} + \gamma h) \\
&\quad + \mu\left(\frac{\partial^2 \bar{u}}{\partial x^2} + \frac{\partial^2 \bar{u}}{\partial y^2} + \frac{\partial^2 \bar{u}}{\partial z^2}\right) - \rho\left(\frac{\partial \overline{u'^2}}{\partial x} + \frac{\partial \overline{u'v'}}{\partial y} + \frac{\partial \overline{u'w'}}{\partial z}\right) \\
&\rho\left(\frac{\partial \bar{v}}{\partial t} + \bar{u}\frac{\partial \bar{v}}{\partial x} + \bar{v}\frac{\partial \bar{v}}{\partial y} + \bar{w}\frac{\partial \bar{v}}{\partial z}\right) = -\frac{\partial}{\partial y}(\bar{p} + \gamma h) + \mu\left(\frac{\partial^2 \bar{v}}{\partial x^2}\right. \\
&\quad \left. + \frac{\partial^2 \bar{v}}{\partial y^2} + \frac{\partial^2 \bar{v}}{\partial z^2}\right) - \rho\left(\frac{\partial \overline{u'v'}}{\partial x} + \frac{\partial \overline{v'^2}}{\partial y} + \frac{\partial \overline{v'w'}}{\partial z}\right) \\
&\rho\left(\frac{\partial \bar{w}}{\partial t} + \bar{u}\frac{\partial \bar{w}}{\partial x} + \bar{v}\frac{\partial \bar{w}}{\partial y} + \bar{w}\frac{\partial \bar{w}}{\partial z}\right) = -\frac{\partial}{\partial z}(\bar{p} + \gamma h) + \mu\left(\frac{\partial^2 \bar{w}}{\partial x^2}\right. \\
&\quad \left. + \frac{\partial^2 \bar{w}}{\partial y^2} + \frac{\partial^2 \bar{w}}{\partial z^2}\right) - \rho\left(\frac{\partial \overline{w'u'}}{\partial x} + \frac{\partial \overline{w'v'}}{\partial y} + \frac{\partial \overline{w'^2}}{\partial z}\right)
\end{aligned} \quad (1.3.32)$$

where h = vertical distance
ρ = density
p = pressure
μ = absolute or dynamic viscosity

The above equations along with the law of conservation of mass govern all turbulent flows; but without an assumption to quantify the velocity fluctuations, a solution of this set of equations is not possible.

A popular method of quantifying the turbulent fluctuation terms in Eq. (1.3.32) is the Boussinesq assumption which defines the eddy viscosity such that the above equations become

$$\rho\left(\frac{\partial \bar{u}}{\partial t} + \bar{u}\frac{\partial \bar{u}}{\partial x} + \bar{v}\frac{\partial \bar{u}}{\partial y} + \bar{w}\frac{\partial \bar{u}}{\partial z}\right) = -\frac{\partial}{\partial x}(\bar{p} + \gamma h) + (\mu + \eta)\left(\frac{\partial^2 \bar{u}}{\partial x^2} + \frac{\partial^2 \bar{u}}{\partial y^2} + \frac{\partial^2 \bar{u}}{\partial z}\right)$$

$$\rho\left(\frac{\partial \bar{v}}{\partial t} + \bar{u}\frac{\partial \bar{v}}{\partial x} + \bar{v}\frac{\partial \bar{v}}{\partial y} + \bar{w}\frac{\partial \bar{v}}{\partial z}\right) = -\frac{\partial}{\partial y}(\bar{p} + \gamma h) + (\mu + \eta)\left(\frac{\partial^2 \bar{v}}{\partial x^2} + \frac{\partial^2 \bar{v}}{\partial y^2} + \frac{\partial^2 \bar{v}}{\partial z^2}\right) \tag{1.3.33}$$

$$\rho\left(\frac{\partial \bar{w}}{\partial t} + \bar{u}\frac{\partial \bar{w}}{\partial x} + \bar{v}\frac{\partial \bar{w}}{\partial y} + \bar{w}\frac{\partial \bar{w}}{\partial z}\right) = -\frac{\partial}{\partial z}(\bar{p} + \gamma h) + (\mu + \eta)\left(\frac{\partial^2 \bar{w}}{\partial x^2} + \frac{\partial^2 \bar{w}}{\partial y^2} + \frac{\partial^2 \bar{w}}{\partial z^2}\right)$$

where η = eddy viscosity. It must be noted that the Boussinesq assumption treats η as if it were a fluid property similar to μ; however, η is not a fluid property but is a parameter which is a function of the flow and density. In practice in turbulent flows, μ is neglected since it is much smaller than η.

Prandtl (see, for example, Schlichting, 1968), in an effort to relate the transport of momentum to the mean flow characteristics, introduced a characteristic length which is termed the *mixing length*. Prandtl claimed that

$$u' \sim v' \sim \ell\frac{d\bar{u}}{dy} \tag{1.3.34}$$

where ℓ = mixing length. Then

$$\tau = \rho\ell^2\left(\frac{d\bar{u}}{dy}\right)^2 \tag{1.3.35}$$

where τ = shear stress. A comparison of the Boussinesq and Prandtl theories of turbulence yields a relationship between the eddy viscosity and the mixing length or

$$\eta = \rho\ell^2\frac{d\bar{u}}{dy}$$

and

$$\frac{\eta}{\rho} = \ell^2\frac{d\bar{u}}{dy} = \epsilon \tag{1.3.36}$$

where η/ρ = a kinematic turbulence factor similar to the kinematic viscosity. ϵ is a direct measure of the transport or mixing capacity of a turbulent flow. In a homogeneous flow, ϵ refers to all transport processes, e.g., momentum, heat, salinity, and sediment. In a density stratified flow, the following inequality is believed to be valid:

$$\epsilon > \epsilon_M^s > \epsilon_{Ma}^s \tag{1.3.37}$$

where ϵ_M^s = stratified momentum transport coefficient and ϵ_{Ma}^s = stratified mass transport coefficient.

Energy Equation

From elementary hydraulics and fluid mechanics, recall that the total energy of a parcel of fluid traveling at a constant speed on a streamline is equal to the sum of the elevation of the parcel above a datum, the pressure head, and the velocity head. The one-dimensional equation which quantifies this statement is known as the Bernoulli energy equation or

$$H = z_A + d_A \cos\theta + \frac{u_A^2}{2g} \tag{1.3.38}$$

where subscript A = point on streamline in open-channel flow
z_A = elevation of point A above an arbitrary datum
d_A = depth of flow section
θ = slope angle of channel
u_A = velocity at point A

For small values of θ Eq. (1.3.38) reduces to

$$H = z + y + \frac{\overline{u}^2}{2g} \tag{1.3.39}$$

where $\overline{u}$ = spatially averaged velocity of flow and y = depth of flow.

Energy and Momentum Coefficients

In many open-channel problems it is both convenient and appropriate to use the continuity, momentum, and energy equations in a one-dimensional form. In such a case, the laws of conservation are:

1. *Conservation of Mass (Continuity)*

$$Q = \overline{u}A$$

2. *Conservation of Momentum*

$$\Sigma F = \rho Q(\overline{u}_2 - \overline{u}_1)$$

3. *Conservation of Energy*

$$H = z + y + \frac{\overline{u}^2}{2g}$$

Of course, no real flow is one dimensional; therefore, the true transfer of momentum through a cross section

$$\iint_A \rho u^2 \, dA$$

is not necessarily equal to the spatially averaged transfer

$$\rho Q \bar{u}$$

Thus, in situations where the velocity profile varies significantly in the vertical and/or transverse directions, it may be necessary to define a momentum correction coefficient, or

$$\beta \rho Q \bar{u} = \iint_A \rho u^2 \, dA$$

and solving

$$\beta = \frac{\iint_A \rho u^2 \, dA}{\rho Q \bar{u}} = \frac{\iint_A \rho u^2 \, dA}{\rho \bar{u}^2 A} \tag{1.3.40}$$

where β = momentum correction coefficient. In an analogous fashion, the kinetic energy correction coefficient is

$$\alpha \left(\gamma \frac{\bar{u}^3}{2g} \right) = \iint_A \gamma \frac{u^3}{2g} \, dA$$

and solving

$$\alpha = \frac{\iint_A \gamma u^3 \, dA}{\gamma \bar{u}^3 A} \tag{1.3.41}$$

where α = kinetic energy correction coefficient. The following properties of β and α are noted:

1. They are both equal to unity when the flow is uniform. In all other cases β and α must be greater than unity.
2. A comparison of Eqs. (1.3.40) and (1.3.41) demonstrates that for a given channel section and velocity distribution α is much more sensitive to the variation in velocity than β.
3. In open-channel hydraulics β and α are generally used only when the channel consists of a main channel with subchannels and/or berms and floodplains (Fig. 1.6). In such cases, the large variation in velocity from section to section effectively masks all gradual variations in velocity, and it is appropriate to consider the velocity in each of the subsections as constant. In channels of compound section, the value of α may exceed 2.

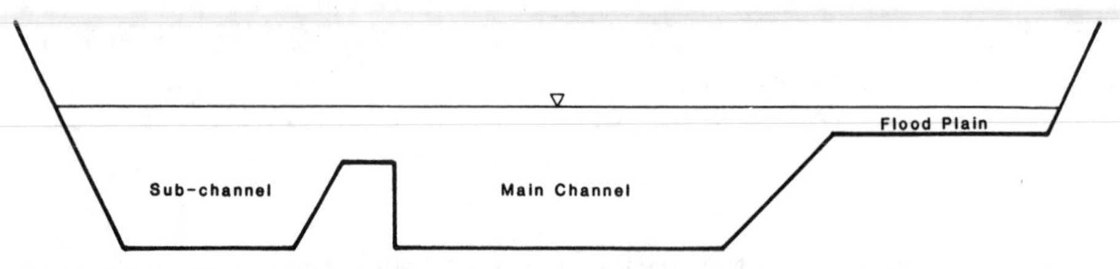

FIGURE 1.6 Main channel with subchannels and floodplains.

1.4 THEORETICAL CONCEPTS

Scaling Equations

In general, the terminology of scaling of equations refers to a process by which a complex differential equation is examined in a rational manner to determine if it can be reduced to a simpler form given a specific application. The first step in the process is to nondimensionalize the equation. As an example, consider an incompressible, laminar, constant density and viscosity flow occurring in a gravity field. The Navier-Stokes equations along with the equation of continuity govern the behavior of this flow. For simplicity, only the x component of the Navier-Stokes equations will be treated or

$$\frac{\partial u}{\partial t} + u\frac{\partial u}{\partial x} + v\frac{\partial u}{\partial y} + w\frac{\partial u}{\partial z} = -g\frac{\partial h}{\partial x} - \frac{1}{\rho}\frac{\partial p}{\partial x} + \frac{\mu}{\rho}\left(\frac{\partial^2 u}{\partial x^2} + \frac{\partial^2 u}{\partial y^2} + \frac{\partial^2 u}{\partial z^2}\right) \tag{1.4.1}$$

Define the following dimensionless quantities:

$$\hat{x} = \frac{x}{L} \qquad \hat{v} = \frac{v}{U}$$
$$\hat{y} = \frac{y}{L} \qquad \hat{w} = \frac{w}{U}$$
$$\hat{z} = \frac{z}{L} \qquad \hat{t} = \frac{tU}{L}$$
$$\hat{h} = \frac{h}{L} \qquad \hat{p} = \frac{p}{\rho U^2}$$
$$\hat{u} = \frac{u}{U} \tag{1.4.2}$$

where L and U are a characteristic length and velocity, respectively. Substituting these dimensionless variables in Eq. (1.4.1) yields

$$\frac{U^2}{L}\frac{\partial \hat{u}}{\partial \hat{t}} + \frac{U^2}{L}\hat{u}\frac{\partial \hat{u}}{\partial \hat{x}} + \frac{U^2}{L}\hat{v}\frac{\partial \hat{v}}{\partial \hat{y}} + \frac{U^2}{L}\hat{w}\frac{\partial \hat{u}}{\partial \hat{z}} = -g\frac{\partial \hat{h}}{\partial \hat{x}} - \frac{U^2}{L}\frac{\partial \hat{p}}{\partial \hat{x}} + \frac{\mu U}{\rho L^2}\left(\frac{\partial^2 \hat{u}}{\partial \hat{x}^2} + \frac{\partial^2 \hat{u}}{\partial \hat{y}^2} + \frac{\partial^2 \hat{w}}{\partial \hat{z}^2}\right) \tag{1.4.3}$$

The substitution has altered neither the form nor the units of Eq. (1.4.1). Equation (1.4.3) can be made dimensionless by dividing both sides by U^2/L with the result being

$$\frac{\partial \hat{u}}{\partial \hat{t}} + \hat{u}\frac{\partial \hat{u}}{\partial \hat{x}} + \hat{v}\frac{\partial \hat{u}}{\partial \hat{y}} + \hat{w}\frac{\partial \hat{u}}{\partial \hat{z}} = -\left(\frac{gL}{U^2}\right)\frac{\partial \hat{h}}{\partial \hat{x}} - \frac{\partial \hat{p}}{\partial \hat{x}} + \frac{\mu}{\rho UL}\left(\frac{\partial^2 \hat{u}}{\partial \hat{x}^2} + \frac{\partial^2 \hat{u}}{\partial \hat{y}^2} + \frac{\partial^2 \hat{w}}{\partial \hat{z}^2}\right) \tag{1.4.4}$$

Equation (1.4.4) is dimensionless, and the two groups of dimensionless variables which appear have been previously defined as the Froude number, i.e.,

$$\mathbf{F}^2 = \frac{U^2}{gL}$$

and the Reynolds number, i.e.,

$$\mathbf{R} = \frac{\rho UL}{\mu}$$

The dimensionless equation may be rewritten as

$$\frac{\partial \hat{u}}{\partial \hat{t}} + \hat{u}\frac{\partial \hat{u}}{\partial \hat{x}} + \hat{v}\frac{\partial \hat{u}}{\partial \hat{y}} + \hat{w}\frac{\partial \hat{u}}{\partial \hat{z}} = -\frac{1}{\mathbf{F}^2}\frac{\partial \hat{h}}{\partial \hat{x}} - \frac{\partial \hat{p}}{\partial \hat{x}} + \frac{1}{\mathbf{R}}\left(\frac{\partial^2 \hat{u}}{\partial \hat{x}^2} + \frac{\partial^2 \hat{u}}{\partial \hat{y}^2} + \frac{\partial^2 \hat{u}}{\partial \hat{x}^2}\right) \tag{1.4.5}$$

After an equation or set of equations has been scaled, e.g., Eq. (1.4.5), it is examined to determine if significant mathematical simplification can be achieved on the basis of specifics of a particular problem. For example in Eq. (1.4.5), if **R** is large, then the last group of terms on the right-hand side of the equation can be neglected.

Boundary Layers

The principle of scaling can be used to reduce the Navier-Stokes equations to a mathematically tractable set of equations known as the *boundary layer equations* (Schlichting, 1968). A well-established principle of fluid mechanics is that a particle of fluid in contact with a stationary solid boundary has no velocity. Thus, all flows over stationary boundaries exhibit velocity profiles through which the drag force caused by the boundary is transmitted outward. For example, consider a flat surface isolated in a flow. If the viscosity of the fluid is small, then the effect of the surface is confined to a thin layer of fluid in the immediate vicinity of the surface. Outside this "boundary layer" the fluid behaves as if it had no viscosity. In addition, since the boundary layer is thin compared to the typical longitudinal length scale, the pressure difference across the layer is negligible.

Consider a flat surface with U = velocity of flow outside the boundary layer. Experimental and theoretical evidence demonstrates that δ, the boundary layer thickness, depends on U, ρ, μ, and x. For a laminar boundary layer, the Blasius solution (Schlichting, 1968) yields

$$\delta = \frac{5x}{\sqrt{\mathbf{R}_x}} \qquad \text{at } \frac{u}{U} = 0.99 \tag{1.4.6}$$

where x = distance from the leading edge of the surface and $\mathbf{R}_x = \rho U x/\mu$ = a Reynolds number based on longitudinal distance. In Eq. (1.4.6) it is noted that δ is proportional to $\sqrt{x}$, and thus as x increases δ also increases. As the thickness of the laminar boundary layer increases, the boundary layer becomes unstable and transforms into a turbulent boundary layer (Fig. 1.7*a*). This transition occurs in the range

$$500{,}000 < \mathbf{R}_x < 1{,}000{,}000$$

The thickness of a turbulent boundary layer is given by

$$\delta = \frac{0.37x}{\mathbf{R}_x^{0.2}} \qquad \text{at } \frac{u}{U} = 0.99 \tag{1.4.7}$$

Even in a turbulent boundary layer there is a very thin layer near the boundary which remains for the most part laminar and is known as the *laminar sublayer.* With regard to boundary layers, especially in relation to open-channel hydraulics:

1. As mentioned, the thickness of a boundary layer, whether laminar or turbulent, is a function of U, ρ, μ, and x. In general, the relationship which exists is

$$x\uparrow, \mu\uparrow, \rho\downarrow, U\downarrow \Rightarrow \delta\uparrow$$

and

$$x\downarrow, \mu\downarrow, \rho\uparrow, U\uparrow \Rightarrow \delta\downarrow$$

where ↑ indicates an increase in value and ↓ indicates a decrease in value.

2. Boundary layers quite often grow inside other boundary layers. In most open-channel flows, the boundary layer intersects the free surface; thus the total depth of flow is the thickness of the boundary layer. A change in the shape of the channel or channel roughness may result in the formation and growth of a new boundary layer (Fig. 1.7*b*).

3. Boundary surfaces are classified as hydraulically smooth or rough on the basis of a comparison of the thickness of the laminar sublayer and the roughness height. If the boundary roughness is such that the roughness elements are covered by the laminar sublayer, then the boundary is by definition hydraulically smooth. In this case, the roughness has no effect on the flow outside the sublayer. However, if the boundary roughness elements project through the sublayer, then by definition the boundary is hydraulically rough

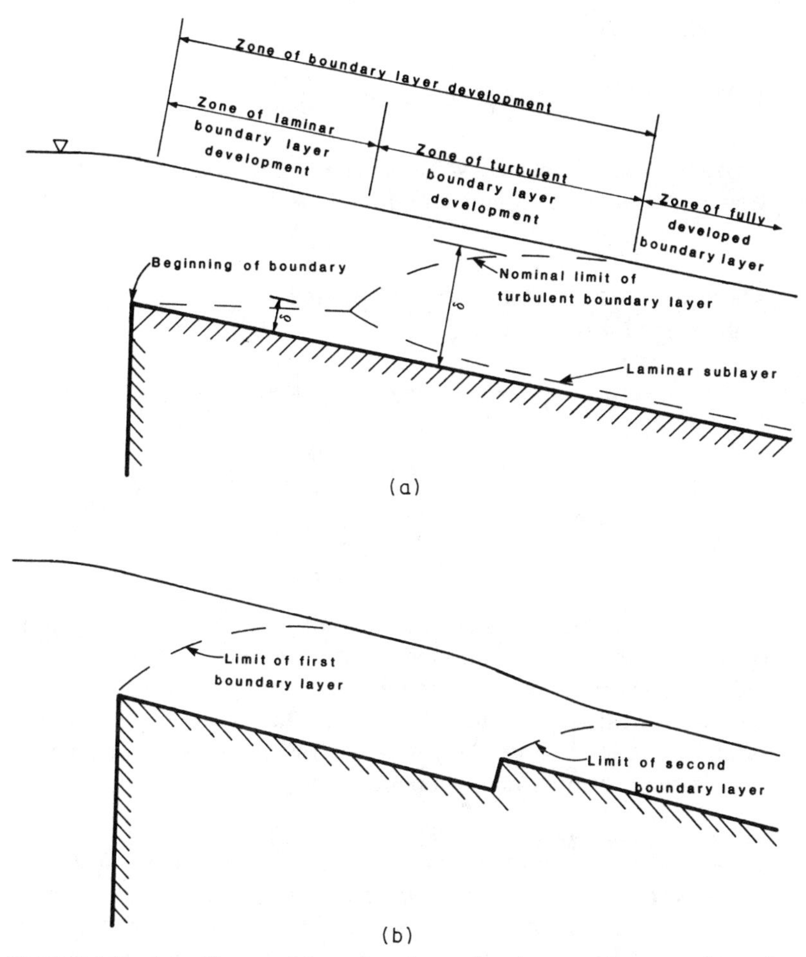

FIGURE 1.7 (*a*) Stages of boundary layer development in open channels. (*b*) Growth of boundary layers within boundary layers.

and the flow outside the sublayer is affected by the roughness. Schlichting (1968) in connection with flat surfaces and pipes experimentally determined the following criteria for classifying boundary surfaces:

Hydraulically Smooth Boundary

$$0 \le \frac{k_s u_*}{\nu} \le 5 \tag{1.4.8a}$$

Transition Boundary

$$5 \le \frac{k_s u_*}{\nu} \le 70 \tag{1.4.9a}$$

Hydraulically Rough Boundary

$$70 \leq \frac{k_s u_*}{\nu} \tag{1.4.10a}$$

where u_* = shear velocity = $\sqrt{gRS}$, and k_s = roughness height. If the Chezy resistance equation (Chapter 4) or

$$\bar{u} = C\sqrt{RS}$$

where C = Chezy resistance coefficient is used to calculate u_*, then Eqs. (1.4.8*a*) to (1.4.10*a*) become

$$0 \leq \frac{k_s \bar{u}\sqrt{g}}{C\nu} \leq 5 \tag{1.4.8b}$$

$$5 \leq \frac{k_s \bar{u}\sqrt{g}}{C\nu} \leq 70 \tag{1.4.9b}$$

$$70 \leq \frac{k_s \bar{u}\sqrt{g}}{C\nu} \tag{1.4.10b}$$

In addition, it must be emphasized that defining k_s as a roughness height conveys an improper impression. k_s is a parameter which characterizes not only the vertical size of the roughness elements but also their orientation, geometric arrangement, and spacing. For this reason, the roughness scale will rarely correspond to the height of the roughness elements. Table 1.2 summarizes some approximate values of k_s for various materials.

4. Flow around a bluff body, e.g., a sphere, usually results in what is termed separation of the boundary layer. In the case of a sphere, flow in the vicinity of the leading edge is similar to that over a flat surface. However, downstream where the surface curves sharply away from the flow the boundary

TABLE 1.2 Approximate values of k_s (Chow, 1959)

Material	k_s range, ft	k_s range, m
Brass, copper, lead, glass	0.0001–0.0030	0.00003–0.0009
Wrought iron, steel	0.0002–0.0080	0.00006–0.002
Asphalted cast iron	0.0004–0.0070	0.0001 –0.002
Galvanized iron	0.0005–0.0150	0.0002 –0.0046
Cast iron	0.0008–0.0180	0.0002 –0.0055
Wood stave	0.0006–0.0030	0.0002 –0.0009
Cement	0.0013–0.0040	0.0004 –0.001
Concrete	0.0015–0.0100	0.0005 –0.003
Drain tile	0.0020–0.0100	0.0006 –0.003
Riveted steel	0.0030–0.0300	0.0009 –0.009
Natural river bed	0.1000–3.000	0.3 –0.9

layer separates from the surface (Schlichting, 1968). Boundary layer separation results in what is known as *form drag* which may be the primary component of drag on a bluff body. Recall from basic fluid mechanics that the total drag, i.e., surface plus form drag, is given by

$$F = C_D A \frac{\rho \bar{u}^2}{2} \tag{1.4.11}$$

where F = total drag force on body
C_D = drag coefficient
A = area of body projected onto plane normal to flow

As noted above, the most common case in open-channel flow is that of a completely or fully developed boundary layer; i.e., the boundary layer fills the complete channel section. There are many cases in the chapters which follow where a basic understanding of boundary layer concepts is very important.

Velocity Distributions

Within a turbulent boundary layer, Prandtl (see, for example, Schlichting, 1968) demonstrated that the vertical velocity profile is approximately logarithmic. Equation (1.3.35) states that the shear stress at any point in the flow

$$\tau = \rho \ell^2 \left(\frac{d\bar{u}}{dy}\right)^2$$

or

$$du = \sqrt{\frac{\tau}{\rho \ell^2}}\, dy \tag{1.4.12}$$

It is usually assumed that

$$\ell = ky \tag{1.4.13}$$

where k was originally called von Karman's turbulence constant; however, it is perhaps more accurate to term k a coefficient since there is some evidence that k may vary over a range of values as a function of the Reynolds number (see, for example, Slotta, 1963, Hinze, 1964, and Vanoni 1946). If it is assumed that k can be approximated as 0.4, then substitution of Eq. (1.4.13) in Eq. (1.4.12) yields, after integration,

$$u = 2.5 \sqrt{\frac{\tau_0}{\rho}} \ln \frac{y}{y_0}$$

or

$$u = 2.5 u_* \ln \frac{y}{y_0} \tag{1.4.14}$$

where it is assumed that $\tau = \tau_0$ = shear stress on the bottom boundary, u = turbulent average velocity at a distance y above the bottom [Eq. (1.3.1)], and

y_0 = a constant of integration. Equation (1.4.14) is known as the Prandtl-von Karman universal velocity distribution law. y_0, the constant of integration, is of the same order of magnitude as the viscous sublayer thickness and is a function of whether the boundary is hydraulically smooth or rough. If the boundary is hydraulically smooth, then y_0 depends solely on the kinematic viscosity and shear velocity or

$$y_0 = \frac{m\nu}{u_*} \tag{1.4.15}$$

where m is a coefficient and equal to approximately ⅑ for smooth surfaces (Chow, 1959). Substitution of Eq. (1.4.15) in Eq. (1.4.14) yields

$$u = 2.5u_* \ln \frac{9yu_*}{\nu} \tag{1.4.16}$$

When the boundary surface is hydraulically rough, y_0 depends only on the roughness height or

$$y_0 = mk_s \tag{1.4.17}$$

where in this case m is a coefficient approximately equal to $\frac{1}{30}$ for sand grain roughness. Substitution of Eq. (1.4.17) in Eq. (1.4.14) yields

$$u = 2.5u_* \ln \frac{30y}{k_s} \tag{1.4.18}$$

Equations (1.4.16) and (1.4.18) adequately represent the vertical velocity profile which exists in a wide channel carrying an unstratified flow. In density stratified flows, the density stratification causes the velocity profile to be modified (McCutcheon, 1979, 1981, and French, 1978). Figure 1.8 illustrates the

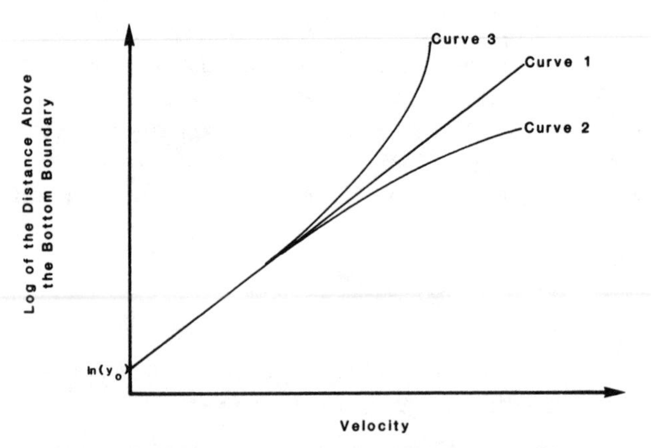

FIGURE 1.8 Types of velocity profiles expected in one-dimensional flow.

three types of vertical velocity profiles that are encountered in turbulent, fully developed, one-dimensional boundary layer flows. Curve 1 is the unstratified profile represented by Eq. (1.4.18). Curve 2 represents the expected velocity profile in a one-dimensional, stably stratified flow, while curve 3 illustrates the type of velocity profile expected in an unstably stratified, one-dimensional flow. Since unstable density stratification is only rarely encountered in open-channel flow, it will be ignored here.

The shape of the velocity profile in an unstratified flow is a function of u_* and y with viscosity being unimportant except near the boundary. The roughness scale k_s determines the magnitude of the velocity in a flow but has no effect on the shape of the profile. Monin and Obukhov (in Monin and Yaglom, 1971) introduced a parameter termed the *buoyancy flux B*, which in a density stratified flow can be used with y and u_* to describe the velocity profile. The buoyancy flux is defined by

$$B = -\frac{\overline{\rho' v'}}{g/\overline{e}} \tag{1.4.19}$$

where $\overline{\rho}$ is the average density over the depth of flow and $\overline{\rho' v'}$ is the turbulent fluctuation of density and vertical velocity averaged either over time at a point in space or over a horizontal area at a point in time. The Buckingham Π theorem can be used to combine u_*, y, and B into a single dimensionless parameter

$$\zeta = \frac{y}{L}$$

(1.4.20)

where $$L = -\frac{u_*^3}{kB} = -\frac{u_*^3\overline{\rho}}{kg\overline{\rho' v'}}$$

where k was introduced by Obukov and has been retained in all subsequent analyses (Monin and Yaglom, 1971). L is a scaling length that is related to the height of the dynamic sublayer where the shear forces of the flow dominate the effects of stratification. By convention, L is negative for stably stratified flows and positive for unstably stratified flows.

Dimensional analysis also yields an expression for the shape of the velocity profile in stratified flows or

$$\frac{\partial u}{\partial y}\frac{yk}{u_*} = \phi(\zeta) \tag{1.4.21}$$

where $\phi(\zeta)$ = a function of y/L. For stably stratified flows in the atmosphere, Monin and Obukhov, in Monin and Yaglom (1971), asserted that

$$\phi(\zeta) = 1 + \alpha_w\left(\frac{y}{L}\right) \tag{1.4.22}$$

and Webb (1970) determined that the coefficient α_w has a value of 5.2. Other investigators (e.g., Monin and Yaglom, 1971, and Panofsky, 1974) have determined values of α_w as low as 0.6 and as high as 10. McCutcheon (1979) found that α_w ranged from 4.4 to 7.8 with an average value of 5.8. Webb's use of a wide range of high-quality data is a convincing argument for assuming that α_w has a value of approximately 5. Regardless of the form $\phi(\zeta)$ and the value of any associated coefficients, $\phi(\zeta)$ must satisfy the following conditions:

$$\lim_{\zeta \to 0} \phi(\zeta) = 1 \tag{1.4.23}$$

and

$$\lim_{\zeta \to 0} \frac{\partial u}{\partial y} = \frac{u_*}{ky} \tag{1.4.24}$$

At this point, an expression for the buoyancy flux must be determined. McCutcheon (1979, 1981) demonstrated that when the stratification is slight

$$\overline{\rho' v'} = k^2 y^2 \frac{\partial \rho}{\partial y} \frac{\partial u}{\partial y} \tag{1.4.25}$$

Substituting Eqs. (1.4.25) and (1.4.22) into Eq. (1.4.21) and solving for $\partial u/\partial y$

$$\frac{\partial u}{\partial y} = \frac{u_*}{ky} \left[\frac{1}{1 + (c\alpha_w g k^2 y^2 / \bar{\rho} u_*^2)(\partial \rho / \partial y)} \right] \tag{1.4.26}$$

where c is a coefficient which measures the ratio of mass to momentum mixing lengths. c is not a constant but varies slightly; however, it is assumed here to be approximately 1. Then, with an assumption regarding the density gradient, Eq. (1.4.26) can be solved. For example, if $\partial \rho / \partial y = \beta$ (Fig. 1.8) where β is a constant, then

$$\frac{u(y)}{u_*} = \frac{1}{k} \ln \left[\frac{(y/y_0)^2 + (g\alpha_w \beta / \bar{\rho} u_*^2) k^2 y^2}{1 + (g\alpha_w \beta / \bar{\rho} u_*^2) k^2 y^2} \right]^{1/2} \tag{1.4.27}$$

If a general density gradient is assumed, that is, $\partial \rho / \partial y = f(y)$ (Fig. 1.9), then

$$\frac{u(y)}{u_*} = \frac{1}{k} \int_{y_0}^{y} \frac{dy}{y[1 + (\alpha_w g / \bar{\rho} u_*^2) k^2 y^2 f(y)]} \tag{1.4.28}$$

The vertical velocity profile equations derived previously apply to flows occurring in very wide channels. Using the Prandtl-von Karman velocity law [Eq. (1.4.14)], Keulegan (1938) derived an equation for the average velocity of flow in a channel of arbitrary shape. Following Chow's (1959) modification of the Keulegan derivation, and with reference to Fig. 1.10, the total discharge through a channel section is given by

$$Q = \bar{u} A = \int_0^d \bar{u} \, dA = \int_0^A \bar{u} \beta \, dy \tag{1.4.29}$$

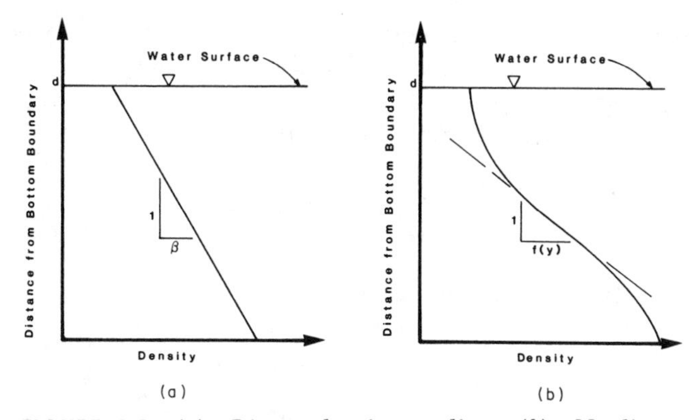

FIGURE 1.9 (*a*) Linear density gradient. (*b*) Nonlinear density gradient.

where d = depth of flow and β = length of the isovel (the curve of equal velocity). It is assumed that β is proportional to the vertical distance from the boundary or

$$\beta = P - \Gamma y$$

where Γ is a function which depends on the shape of the channel. Then

$$A = \int_0^d \beta \, dy = Pd - \frac{\Gamma d^2}{2} \qquad (1.4.30)$$

Substituting Eqs. (1.4.14) and (1.4.30) into Eq. (1.4.29), integrating, and simplifying yields

$$u = 2.5u_* \ln \left[\frac{R}{y_0} \frac{d}{R} \exp \left(-1 - \frac{\Gamma d^2}{4A} \right) \right] \qquad (1.4.31)$$

where y_0 is defined by Eqs. (1.4.15) and (1.4.17). In Eq. (1.4.31) the parameter

$$\frac{d}{R} \exp \left(-1 - \frac{\Gamma d^2}{4A} \right)$$

is a function of the shape of the channel cross section. Both Keulegan and Chow have claimed that the variation of this parameter with sections of various shapes is small, and thus this parameter can in general be represented by an overall constant β. With this assumption, Eq. (1.4.31) becomes

$$\overline{u} = u_* \left(\beta + 2.5 \ln \frac{mR}{y_0} \right)$$

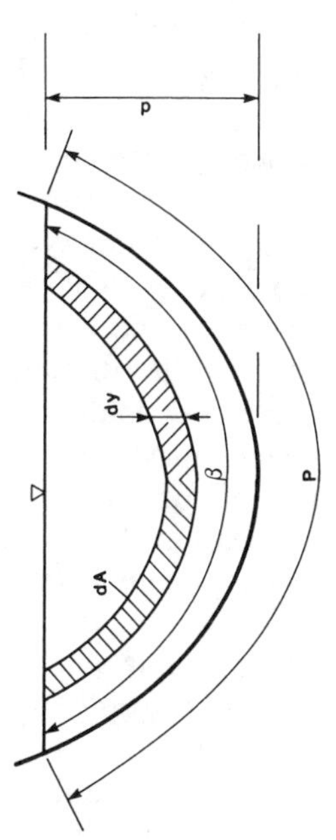

FIGURE 1.10 Definition of notation.

which is a theoretical equation for the average velocity of flow in an open channel. Keulegan (1938) arrived at the following specific conclusions:

For Hydraulically Smooth Channels

$$\bar{u} = u_* \left(3.25 + 2.5 \ln \frac{Ru_*}{\nu} \right) \tag{1.4.32}$$

For Hydraulically Rough Channels

$$\bar{u} = u_* \left(6.25 + 2.5 \ln \frac{R}{k_s} \right) \tag{1.4.33}$$

The foregoing equations for the velocity distribution in open-channel flow are one-dimensional equations; i.e., they consider only the variation of the velocity in the vertical dimension. Most channels encountered in practice exhibit velocity distributions which are usually strongly two dimensional and in many cases three dimensional. For example, the Prandtl-von Karman velocity distribution law predicts that the maximum velocity occurs at the free surface. However, both laboratory and field measurements demonstrate that the maximum velocity usually occurs below the free surface, although in wide, rapid, and shallow flows or in flows occurring in very smooth channels the maximum velocity may be at the free surface. These observations demonstrate that a one-dimensional velocity distribution law cannot completely describe flows which are two and three dimensional. In general, the velocity distribution in a channel of arbitrary shape is believed to depend on the shape of the cross section, the boundary roughness, the presence of bends, and changes in cross-sectional shape. The secondary currents caused by bends and changes in channel shape are termed *strong secondary currents*. The shape of the cross section and the distribution of the boundary roughness cause a nonuniform distribution of Reynolds stresses in the flow and result in weak secondary currents. The terminology *secondary current* or *flow* refers to a circulatory motion of the fluid around an axis which is parallel to the primary current or flow which is a translation of the fluid in the longitudinal direction. In this section, modern developments regarding weak secondary currents will be considered.

Einstein and Li (1958) attempted to identify the cause of weak secondary currents and concluded that: (1) secondary currents do not occur in either laminar or isotropic turbulent flow, (2) secondary currents in turbulent flow occur only when the isovels (the lines or contours of equal velocity) of the primary flow are not parallel to each other or the boundaries of the channel, and (3) the Reynolds stresses contribute to the secondary flow only when their distribution is not linear. Tracy (1965) directed an investigation along similar lines and came to approximately the same conclusions. The greatest progress in this area of research has been made by Liggett et al. (1965) and Chiu et al. (1967, 1971, 1976, 1978) who began by investigating weak secondary currents which occurred in corners (Liggett et al., 1965). These initial methods were later extended to flows in triangular channels (Chiu and Lee, 1971) and finally to channels of an arbitrary cross-sectional shape (Chiu et al., 1976, 1978). This method of estimating secondary currents and, more importantly, three-dimensional velocity distributions, first requires the establishment of a curvilinear coordinate system which is formed by the isovels, the ϵ curves in Fig. 1.11, and the curves η in this figure which have orthogonal trajectories to the ϵ curves. Chiu et al. (1976) then claimed that a heuristic extension of the Prandtl-von Karman velocity distribution was

$$u = \frac{u_*}{k} \ln \frac{\epsilon}{\epsilon_0} \tag{1.4.34}$$

where ϵ_0 = a constant and ϵ = the curvilinear coordinate defined above. Along the channel boundary, $\epsilon \simeq \epsilon_0$; therefore, $u = 0$. At this point, a semiempirical assumption regarding the shape of the isovels is required. In a natural channel, Chiu et al. (1976) claimed that the following equation provides an adequate estimate of the ϵ curves:

$$\epsilon = \frac{y}{d}\left(1 - \frac{|z|}{T_i}\right)^{\beta_i} \tag{1.4.35}$$

where y = vertical coordinate
d = depth of flow
z = transverse coordinate
T_i = transverse distance between y axis ($z = 0$) and left or right intersection of water surface and boundary
β_i = left and right curve coefficients (Fig. 1.11)

Combining Eqs. (1.4.34) and (1.4.35) yields

$$u = \frac{u_*}{k} \ln \left[\frac{1}{\epsilon_0}\left(\frac{y}{d}\right)\left(1 - \frac{|z|}{T_i}\right)^{\beta_i}\right] \tag{1.4.36}$$

This theory can also be extended to estimate the discharge of an irregular or natural section. From Eq. (1.4.35), the boundary of the channel is given by

$$y = f_0(z) = d\epsilon_0\left(1 - \frac{|z|}{T_i}\right)^{-\beta_i} \tag{1.4.37}$$

The cross-sectional area is

$$A_i = \int_{z=0}^{T_i} \int_{y=f_0(z)}^{d} dy\, dz = T_i d\left[1 - \frac{1}{1-\beta_i}\left(\epsilon_0 - \beta_i \epsilon_0^{-\beta_i}\right)\right] \tag{1.4.38}$$

The volumetric flow rate is then given by

$$Q = \int_{z=0}^{T_1} \int_{y=f_0(z)}^{d} u_1\, dy\, dz + \int_{z=0}^{T_2} \int_{y=f_0(z)}^{d} u_2\, dy\, dz$$

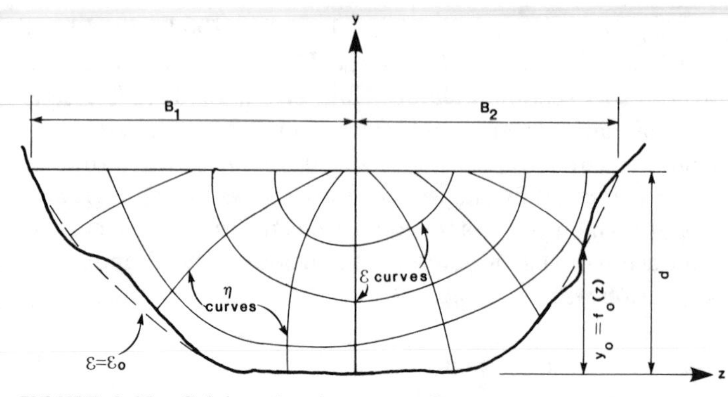

FIGURE 1.11 Schematic of ϵ, η coordinate system. (*Chiu et al., 1976.*)

or

$$Q = Td\frac{u_*}{k}\left[-1 - \ln \epsilon_0 + \epsilon_0 - \frac{1}{T}(T_1\beta_1\phi_1 + T_2\beta_2\phi_2)\right] \tag{1.4.39}$$

where

$$\phi_i = \frac{A_i}{T_i d} = 1 - \frac{1}{1-\beta_i}(\epsilon_0 - \beta_i \epsilon_0^{-\beta_i}) \tag{1.4.40}$$

In these equations, there are three parameters which must be determined: ϵ_0, β_1, and β_2. For a specific situation, values of these parameters can usually be estimated from the given information. In any commonly found situation, it is assumed that the channel geometry is known; therefore, A_i, T_i, and ϕ_i for i = 1,2 are defined. If, in addition, Q, u_*, and d are known, then ϵ_0, β_1, and β_2 are determined from the simultaneous solution of Eqs. (1.4.40) for i = 1,2 and (1.4.39). Alternatively, if only the depth of flow is known at a section, then Eqs. (1.4.39) and (1.4.40) contain four unknowns, and an auxiliary equation must be found. An examination of Eq. (1.4.35) demonstrates that on the y axis

$$\epsilon = \frac{y}{d}$$

Therefore, if two velocity measurements are available on this axis, the value of ϵ_0 can be estimated. For example, if $u(0.2)$ is the velocity at y/d = 0.2 and $u(0.8)$ is the velocity at y/d = 0.8, then

$$u(0.2) = \frac{u_*}{k}\ln\frac{0.2}{\epsilon_0}$$

$$u(0.8) = \frac{u_*}{k}\ln\frac{0.8}{\epsilon_0}$$

and solving simultaneously,

$$\epsilon_0 = \ln^{-1}\left\{\frac{0.22 - 1.61[u(0.8)/u(0.2)]}{[u(0.8)/u(0.2)] - 1}\right\} \tag{1.4.41}$$

The value of ϵ_0 estimated in Eq. (1.4.41) can then be used in Eq. (1.4.40) to estimate values of β_1 and β_2. The implication of this method is that from two measurements of the velocity at the centerline of a channel the flow in the channel can be estimated.

Equation (1.4.36) provides a method of estimating the primary velocity distribution in an open-channel flow which occurs in a channel of arbitrary shape. From Eq. (1.4.35) the orthogonal trajectories of the ϵ curves are

$$\eta = y^2 - \frac{1}{\beta_i}(\beta_i - |z|)^2 \tag{1.4.42}$$

The ϵ component of the secondary steady velocity is

$$u_\epsilon = \left(\frac{\rho}{h_\epsilon}\frac{\partial u}{\partial \epsilon}\right)^{-1}\left[-\rho u\frac{\partial u}{\partial x} - \frac{\partial}{\partial x}(p + \rho g z) + \frac{1}{h_\epsilon}\frac{\partial \tau_{\epsilon x}}{\partial \epsilon} + \frac{\tau_{\epsilon x}}{h_\epsilon h_\eta}\frac{\partial h_\eta}{\partial \epsilon}\right] \tag{1.4.43}$$

where z = mean elevation of the channel boundary and

$$\tau_{\epsilon x} = \frac{\rho k^2(\partial u/\partial \epsilon)^4}{(\partial^2 u/\partial \epsilon^2)^2} \tag{1.4.44}$$

The other velocity component is

$$u_\eta = -\frac{1}{h_\epsilon}\int_0^\eta \left(h_\epsilon h_\eta \frac{\partial u}{\partial x} + h_\eta \frac{\partial u_\epsilon}{\partial \epsilon} + u_\epsilon \frac{\partial h_\eta}{\partial \epsilon}\right) d\eta \tag{1.4.45}$$

where h_ϵ and h_η are metric coefficients given by

$$h_\epsilon = \frac{T_i d}{[1 - (|z|/T_i)]^{\beta_i - 1}\sqrt{(T_i - |z|)^2 + \beta_i^2 y^2}}$$

and

$$h_\eta = \frac{0.5T_i}{\sqrt{\beta_i^2 y^2 + (T_i - |z|)^2}}$$

The velocity components u_ϵ and u_η can then be transformed to cartesian coordinates by a standard coordinate transformation procedure.

With regard to the Chiu method of estimating primary velocities and secondary currents, the following should be noted. First, direct confirmation of the predictions yielded by Eqs. (1.4.43) and (1.4.45) is not possible in most natural channels which transport, in addition to the fluid solids, air bubbles and other impurities. The primary velocity distribution estimated by Eq. (1.4.36) has been verified with field measurements (Chiu et al., 1976). Second, the computational technique illustrated here is somewhat inadequate in that it cannot cope with the true three-dimensional irregularity of a natural channel.

1.5 SIMILARITY AND PHYSICAL MODELS

The number of important modern problems in open-channel hydraulics which can be solved satisfactorily by purely analytic techniques is severely limited.

Most problems must be solved by a combination of numerical and analytic techniques, field measurements, and physical modeling. In a physical model, two criteria must be satisfied: (1) The model and prototype must be geometrically similar, and (2) the model and prototype must be dynamically similar. The requirement of geometric similarity can be met by establishing a length scale ratio between the prototype and the model. The requirement of dynamic similarity requires that two systems with geometrically similar boundaries have geometrically similar flow patterns at corresponding instants in time. In turn, this requires that all the individual forces acting on corresponding fluid elements have the same ratios in the two systems. The primary problem in developing physical models is not meeting the requirement of geometric similarity but ensuring dynamic similarity.

One approach to developing appropriate parameters to ensure dynamic similarity is the scaling of the governing equations. For example, Eq. (1.4.5) has the same solution for two geometrically similar flow systems as long as **F** and **R** are numerically the same for the two systems. In some cases, one of these parameters may be unimportant; however, for exact dynamic similarity **F** and **R** for both the model and prototype must be equal. Equality of Froude numbers requires

$$\frac{U_M}{\sqrt{g_M L_M}} = \frac{U_P}{\sqrt{g_p L_p}} \tag{1.5.1}$$

or

$$U_R = \sqrt{g_R L_R} \tag{1.5.2}$$

where the subscript M designates the model, P the prototype, and R the ratio of the model-to-prototype variables. The requirement of equality of Reynolds numbers yields

$$U_R = \frac{\mu_R}{\rho_R L_R} \tag{1.5.3}$$

Combining Eqs. (1.5.2) and (1.5.3) and recognizing that $g_R = 1$ results in

$$L_R = \left(\frac{\mu_R}{\rho_R}\right)^{2/3} = \nu_R^{2/3} \tag{1.5.4}$$

where ν = kinematic viscosity. Thus, if an exact physical model of an open-channel flow is to be constructed, the modeler has only one degree of freedom—the choice of model fluid. Since the range of kinematic viscosity among commonly available fluids is rather limited, the requirement imposed by Eq. (1.5.4) usually results in a model almost the same size as the prototype. The conclusion is that exact physical models of open-channel flows are virtually impossible.

In many open-channel flows, the effects of viscosity are negligible relative to the effects of gravity. Thus, in most cases physical models are constructed only on the basis of Froude number equality. If only Froude number similarity is required, then

$$U_R = \sqrt{g_R L_R} \tag{1.5.5}$$

and

$$T_R = \sqrt{\frac{L_R}{g_R}} \tag{1.5.6}$$

where T_R = time scale ratio. In requiring only Froude number similarity, the modeler must be aware of two problems:

1. Care must be exercised to ensure that the model does not let viscous effects become dominant. For example, if the flow in the prototype is turbulent, then flow in the model cannot be laminar.
2. In some cases, fluid friction is important but molecular viscosity is rather unimportant, e.g., high Reynolds number flows. In such cases, the similitude of frictional effects is one of similitude of boundary roughness rather than equality of Reynolds numbers.

In the case that density stratification is an important effect, then the model should be based on the equality of the densimetric Froude number [Eq. (1.2.8)].

In a specific situation, additional dimensionless parameters must be considered. For example, in a model in which small waves are generated, surface tension must be considered. Surface tension must also be considered in situations in which small waves must be prevented. The parameter which measures the relative magnitude of the inertial and capillary forces is the Weber number or

$$\mathbf{W} = \frac{\rho U^2 L}{\sigma} \tag{1.5.7}$$

where σ = surface tension forces per unit length.

BIBLIOGRAPHY

Chiu, C.-L., "Factors Determining the Strength of Secondary Flow," *Proceedings of the American Society of Civil Engineers, Journal of the Engineering Mechanics Division,* vol. 93, no. EM4, August 1967, pp. 69–77.

Chiu, C.-L., and Lee, T.-S., "Methods of Calculating Secondary Flow," *Water Resources Research,* vol. 7, no. 4, August 1971, pp. 834–844.

Chiu, C.-L., Lin, H. C., and Mizumura, K., "Simulation of Hydraulic Processes in Open Channels," *Proceedings of the American Society of Civil Engineers, Journal of the Hydraulics Division,* vol. 102, no. HY2, February 1976, pp. 185–205.

Chiu, C.-L., Hsiung, D. E., and Lin, H., "Three-Dimensional Open Channel Flow," *Proceedings of the American Society of Civil Engineers, Journal of the Hydraulics Division,* vol. 104, no. HY8, August 1978, pp. 1119–1136.

Chow, V. T., *Open Channel Hydraulics,* McGraw-Hill Book Company, New York, 1959.

Einstein, H. A., and Li, H., "Secondary Currents in Straight Channels," *Transactions of American Geophysical Union,* vol. 39, 1958, p. 1085.

Franz, D. D., "Tabular Representation of Cross-Sectional Elements," *Proceedings of the American Society of Civil Engineers, Journal of the Hydraulics Division,* vol. 108, no. HY10, October 1982, pp. 1070–1081.

French, R. H., "Stratification and Open Channel Flow," *Proceedings of the American Society of Civil Engineers, Journal of the Hydraulics Division,* vol. 104, no. HY1, January 1978, pp. 21–31.

Hinze, J. O., "Turbulent Pipe Flow," in *The Mechanics of Turbulence,* A. Favre, ed., Gordon Breach, New York, 1964, pp. 129–165.

Keulegan, G. H., "Laws of Turbulent Flow in Open Channels," Research Paper RP 1151, *Journal of Research,* U.S. Bureau of Standards, vol. 21, December 1938, pp. 707–741.

Liggett, J. A., Chiu, C.-L., and Miao, L. S., "Secondary Currents in a Corner," *Proceedings of the American Society of Civil Engineers, Journal of the Hydraulics Division,* vol. 91, no. HY6, November 1965, pp. 99–117.

McCutcheon, S. C., "The Modification of Open Channel Vertical Velocity Profiles by Density Stratification," Ph.D. dissertation, Vanderbilt University, Nashville, TN, December 1979.

McCutcheon, S. C., "Vertical Velocity Profiles in Stratified Flows," *Proceedings of the American Society of Civil Engineers, Journal of the Hydraulics Division,* vol. 107, no. HY8, August 1981, pp. 973–988.

Monin, A. S., and Yaglom, A. M., *Statistical Fluid Mechanics: Mechanics of Turbulence,* vol. 1, J. L. Lumley, ed., MIT Press, Cambridge, MA, 1971, pp. 425–486.

Panofsky, H. A., "The Atmospheric Boundary Layer below 150 Meters," *The Annual Review of Fluid Mechanics,* M. van Dyke, ed., Annual Reviews, Palo Alto, CA, 1974, pp. 147–178.

Schlichting, H., *Boundary-Layer Theory,* 6th ed., McGraw-Hill Book Company, New York, 1968.

Slotta, L. S., "A Critical Investigation of the Universality of Karman's Constant in Turbulent Flow," in *Studies of the Effects of Variations on Boundary Conditions on the Atmospheric Boundary Layer,* H. H. Lettau, ed., University of Wisconsin Annual Report, Madison, WI, 1963, pp. 1–36.

Streeter, V. L., and Wylie, E. B., *Fluid Mechanics,* 6th ed., McGraw-Hill Book Company, New York, 1975.

Tracy, H. J., "Turbulent Flow in a Three Dimensional Channel," *Proceedings of the American Society of Civil Engineers, Journal of the Hydraulics Division,* vol. 91, no. HY6, November 1965, pp. 9–35.

Vanoni, V. A., "Transportation of Suspended Sediment by Water," *Transactions of the American Society of Civil Engineers,* 1946, vol. III, pp. 67–133.

Webb, E. K., "Profile Relationship: The Log-Linear Range, and Extension to Strong Stability," *Quarterly Journal Royal Meteorological Society,* vol. 96, 1970, pp. 67–90.

TWO

Energy Principle

SYNOPSIS

SYNOPSIS

In this chapter, the application of the law of conservation of energy to open-channel flows is discussed. Section 2.1 defines specific energy. The effect of streamline curvature on the pressure distribution within the flow field is also considered. Section 2.2 defines subcritical, critical, and supercritical flow. In this section, a number of techniques—trial and error, graphical, and explicit—for estimating critical flow in channels of various cross-sectional shapes are presented. Section 2.3 discusses upstream and downstream controls and the accessibility of various points on the *E-y* or specific energy curve. Section 2.4 introduces the reader to various applications of the energy principle in practice including the use of dimensionless *E-y* curves, computations associated with channel transitions, and problems encountered in channels of compound section.

2.1 DEFINITION OF SPECIFIC ENERGY

A central principle in any treatment of the hydraulics of open-channel flow must be the law of conservation of energy. Recall from elementary fluid mechanics that the total energy of a parcel of water traveling on a streamline is given by the Bernoulli equation or

$$H = z + \frac{p}{\gamma} + \frac{u^2}{2g} \tag{2.1.1}$$

where H = total energy
z = elevation of streamline above the datum
p = pressure
γ = fluid specific weight
p/γ = pressure head
u = streamline velocity
$u^2/2g$ = velocity head
g = local acceleration of gravity

The sum $z + (p/\gamma)$ defines the elevation of the hydraulic grade line above the datum; in general, the value of this sum varies from point to point along the streamline. To examine the variation of this sum under various circumstances, consider a particle of cross-sectional area δA, length δs, density ρ, and mass $\rho\ \delta A\ \delta s$ moving along an arbitrary streamline in the $+S$ direction (Fig. 2.1). If it is assumed that the fluid is frictionless, then there are no shear forces, and only the gravitational body force and the surface forces on the ends of the particle must be considered. The gravitational force is $\rho g\ \delta A\ \delta s$, the pressure force on the upstream face is $p\ \delta A$, and the pressure force on the downstream

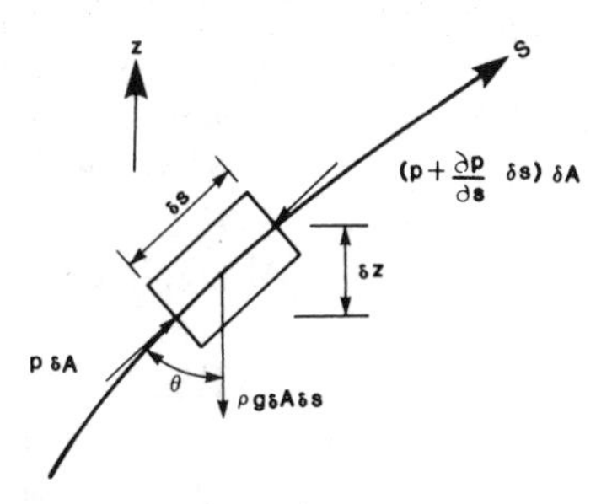

FIGURE 2.1 Component forces acting on a particle moving along a streamline.

face is $[p + (\partial p/\partial s)\ \delta s]\ \delta A$. Applying Newton's second law of motion in the direction of flow yields

$$F_s = a_s\ \delta m$$

or
$$\rho a_s\ \delta A\ \delta s = p\ \delta A - \left(p + \frac{\partial p}{\partial s}\ \delta s\right) \delta A - \rho g\ \delta A\ \delta s \cos\theta$$

where a_s = acceleration of the fluid particle along the streamline. Simplifying this equation

$$\frac{\partial p}{\partial s} + \rho g \frac{\partial z}{\partial s} + \rho a_s = 0$$

and noting that
$$\frac{\partial z}{\partial s} = \cos\theta$$

yields
$$\frac{\partial}{\partial s}(p + \gamma z) + \rho a_s = 0 \tag{2.1.2}$$

Equation (2.1.2) is known as the Euler equation for motion along a streamline. If $a_s = 0$, Eq. (2.1.2) can be integrated to yield the hydrostatic law; i.e., pressure varies in a linear fashion with depth.

The implications of Eq. (2.1.2) in open-channel flow are significant. First, if minor fluctuations due to turbulence are ignored and the streamlines have no acceleration components in the plane of the cross section, i.e., the streamlines have neither substantial curvature nor divergence, then the flow is termed parallel and a hydrostatic pressure distribution prevails. In practice, most uniform flows (Chapters 4 and 5) and gradually varied flows (Chapter 6) may be regarded as parallel flows with hydrostatic pressure distributions since the divergence and curvature of the streamlines in these cases are negligible. In a parallel flow, the sum $z + p/\gamma$ is constant and equal to the depth of flow y, if the datum is taken as the channel bottom. Then, by definition, the specific energy of an open-channel flow relative to the bottom of the channel is

$$E = y + \alpha \frac{\overline{u}^2}{2g} \tag{2.1.3}$$

where α = kinetic energy correction factor which is used to correct for the nonuniformity of the velocity profile and $\overline{u}$ = average velocity of flow ($\overline{u} = Q/A$ where A is the flow area and Q is the flow rate). The assumption inherent in Eq. (2.1.3) is that the slope of channel is small, or $\cos\theta \simeq 1$, and $y \simeq d \cos\theta$ (Fig. 2.2*a*). In general, if $\theta < 10°$ or $S < 0.018$ where S is the slope of the channel, Eq. (2.1.3) is valid.

If θ is not small, then the pressure distribution is not hydrostatic since the vertical depth of flow is significantly different from the depth measured perpendicular to the channel bed. In addition, in channels of large slope, e.g., spillways, the flow may entrain air which will change both the density of the fluid and the depth of flow. In the subsequent material, unless it is specifically stated otherwise, it is tacitly assumed that the channel slope is such that a hydrostatic pressure distribution exists.

Furthermore, if $a_s \neq 0$, then the streamlines of the flow either will have a significant amount of curvature or will diverge, and the flow is termed *curvilinear*. Such situations may occur when the bottom of the channel is curved, at sluice gates, and at free overfalls. In such cases, the pressure distribution is not hydrostatic, and a pressure correction factor must be estimated. In concave flow situations (Fig. 2.2*b*), the forces resulting from streamline curvature reinforce the gravitational forces. In the case of convex flow (Fig. 2.2*c*), the forces resulting from the curvature of the streamlines act against the gravitational forces. If a channel has a curved longitudinal profile, then the deviation of the pressure distribution from the hydrostatic condition can be estimated from an application of Newton's second law to the

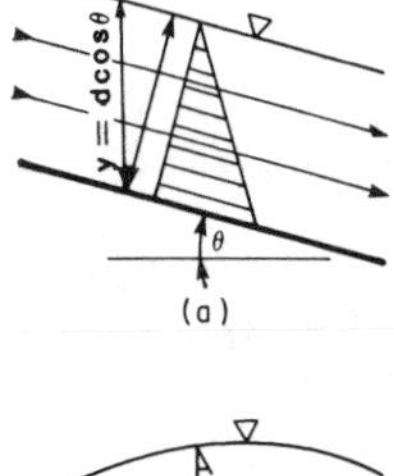

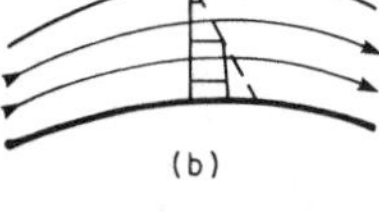

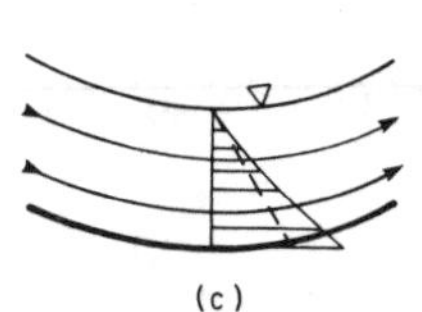

FIGURE 2.2 (*a*) Pressure distribution in parallel flow with $\cos\theta \simeq 1$. (*b*) Pressure distribution in concave flow. (*c*) Pressure distribution in convex flow.

$$c = \frac{y}{g}\frac{\overline{u}^2}{r} \tag{2.1.4}$$

where r = radius of curvature of the channel bottom. The true pressure distribution at a section is then

$$p = y \pm \frac{y}{g}\frac{\overline{u}^2}{r} \tag{2.1.5}$$

where the plus and minus signs are used with concave and convex flows, respectively. In many cases, it is convenient to define a pressure coefficient such that the pressure head in a curvilinear flow can be defined as $\alpha' y$ where α' = the pressure coefficient. It can be demonstrated that

$$\alpha' = 1 + \frac{1}{Qy}\iint_A cu\, dA \tag{2.1.6}$$

where Q = total discharge and dA = an incremental area. Then $\alpha' > 1$ for concave flow, $\alpha' = 1$ for parallel flow, and $\alpha' < 1$ for convex flow. For complex curvilinear flows, the pressure distribution can be estimated from either flow nets or model tests (see, for example, Streeter and Wylie, 1975).

2.2 SUBCRITICAL, CRITICAL, AND SUPERCRITICAL FLOW

An examination of Eq. (2.1.3) demonstrates that if $\cos\theta \approx 1$, $\alpha = 1$, and the channel section and discharge are specified, then the specific energy is only a function of the depth of flow. If y is plotted against E (Fig. 2.3), a curve with two branches results. The limb AC approaches the E axis asymptotically, and the branch AB asymptotically approaches the line $y = E$. For all the points on the E axis greater than point A, there are two possible depths of flow, known as the alternate depths of flow. Since A represents the minimum specific energy, the coordinates of this point can be found by taking the first derivative of Eq. (2.1.3) with respect to y and setting the result equal to zero or

$$E = y + \alpha\frac{\overline{u}^2}{2g} = y + \alpha\frac{Q^2}{2gA^2}$$
$$\frac{dE}{dy} = 1 - \alpha\frac{Q^2}{gA^3}\frac{dA}{dy} = 0 \tag{2.2.1}$$

where A = flow area. In Fig. 2.3, the differential water area dA near the free surface is equal to $T\,dy$ where T = the top width. By definition, the hydraulic depth D is

$$D = \frac{A}{T} \tag{2.2.2}$$

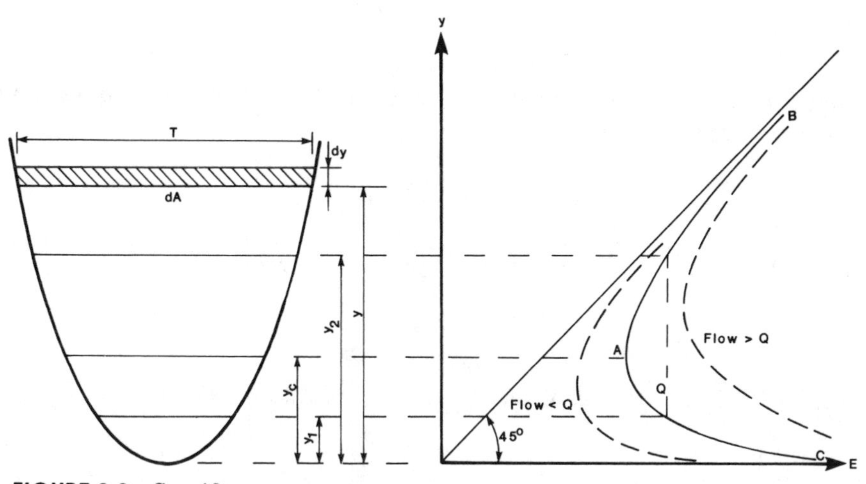

FIGURE 2.3 Specific energy curve.

Substituting $\alpha = 1$ and Eq. (2.2.2) in Eq. (2.2.1) yields

$$1 - \frac{Q^2}{gA^3}\frac{dA}{dy} = 1 - \frac{Q^2}{gA^2}\frac{T}{A} = 1 - \frac{\overline{u}^2}{gD} = 0$$

or

$$\frac{\overline{u}^2}{2g} = \frac{D}{2} \tag{2.2.3}$$

$$\frac{\overline{u}}{\sqrt{gD}} = \mathbf{F} = 1 \tag{2.2.4}$$

which is the definition of critical flow presented previously. Thus, minimum specific energy occurs at the critical hydraulic depth. With this knowledge, the branch AC can be interpreted as representing supercritical flows while the branch AB must represent subcritical flows. With regard to Fig. 2.3 and Eq. (2.2.4), the following should be considered. First, for channels of large slope angle θ and $\alpha \neq 1$, it can be easily demonstrated that the criterion for minimum specific energy is

$$\mathbf{F} = \frac{\overline{u}}{\sqrt{(gD \cos \theta/\alpha)}} \tag{2.2.5}$$

Second, in Fig. 2.3, E-y curves for flow rates greater than Q lie to the right of the curve for Q, and E-y curves for flow rates less than Q lie to the left of the curve for Q. Third, in the case of a rectangular channel of width b, Eq. (2.2.4) can be reduced to forms more suitable for computation. For example, define the flow per unit width q as

$$q = \frac{Q}{b}$$

The average velocity is then

$$\overline{u} = \frac{q}{y}$$

and for a rectangular channel, $y = D$. With these definitions, Eq. (2.2.4) can be rearranged to yield

$$y_c = \left(\frac{q^2}{g}\right)^{1/3} \tag{2.2.6}$$

Substitution of the above definitions in Eq. (2.2.3) yields

$$\frac{\overline{u}_c^2}{2g} = \frac{1}{2} y_c \tag{2.2.7}$$

And using the definition of specific energy, Eq. (2.1.3), and Eq. (2.2.7),

$$y_c = \tfrac{2}{3}E_c \tag{2.2.8}$$

where E_c = specific energy at critical depth and velocity.

At this point, it is appropriate to discuss methods of calculating the critical depth of flow in nonrectangular channels when the shape of the channel and the flow rate are specified. For simple channels, the easiest and most direct approach to this computation is an algebraic solution involving trial-and-error solutions of either Eq. (2.2.3), (2.2.4), or (2.2.5).

EXAMPLE 2.1

For a trapezoidal channel with base width $b = 6.0$ m (20 ft) and side slope $z = 2$, calculate the critical depth of flow if $Q = 17\ \text{m}^3/\text{s}$ (600 ft^3/s).

Solution

From Table 1.1

$$A = (b + zy)y = (6.0 + 2y)y$$

$$T = b + 2zy = 6 + 4y$$

$$D = \frac{A}{T} = \frac{(3 + y)y}{3 + 2y}$$

and

$$\overline{u} = \frac{Q}{A} = \frac{17}{2\,(3 + y)y}$$

Substitution of the above in Eq. (2.2.3) yields

$$\frac{[17/(6 + 2y)]^2}{g} = \frac{(3 + y)y}{3 + 2y}$$

Simplifying,

$$7.4(3 + 2y) = [(3 + y)y]^3$$

By trial and error, the critical depth is approximately

$$y_c = 0.84\ \text{m (2.8 ft)}$$

and the corresponding critical velocity is

$$u_c = \frac{17}{[6 + 2(0.84)]0.84} = 2.6\ \text{m/s (8.5 ft/s)}$$

A second method of approaching the critical depth computation problem is through semiempirical equations. A set of equations for this purpose has been developed by Straub (1982) for a number of common channel shapes. These are summarized in Table 2.1.

EXAMPLE 2.2

For a trapezoidal channel with $b = 6.0$ m (20 ft) and $z = 2$, find the critical depth of flow if $Q = 17\ \text{m}^3/\text{s}$ (600 ft^3/s).

TABLE 2.1 Semiempirical equations for the estimation of y_c (Straub, 1982)

Channel type	Equation for y_c in terms of $\psi = \alpha Q^2/g$	
Rectangular 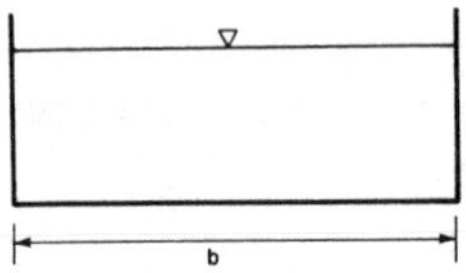	$\left(\frac{\psi}{b^2}\right)^{1/3}$	
Trapezoidal 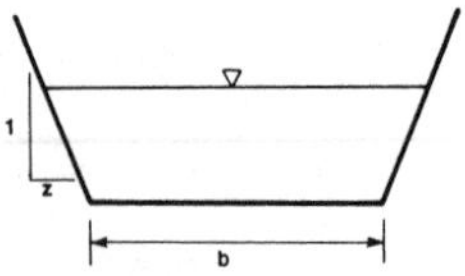	$0.81\left(\frac{\psi}{z^{0.75}b^{1.25}}\right)^{0.27} - \frac{b}{30z}$	Range of applicability $0.1 < \frac{Q}{b^{2.5}} < 0.4$ For $\frac{Q}{b^{2.5}} < 0.1$ use equation for rectangular channel
Triangular 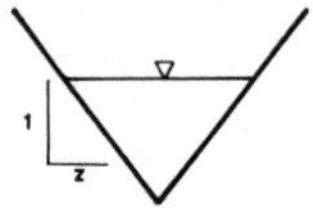	$\left(\frac{2\psi}{z^2}\right)^{0.20}$	
Parabolic 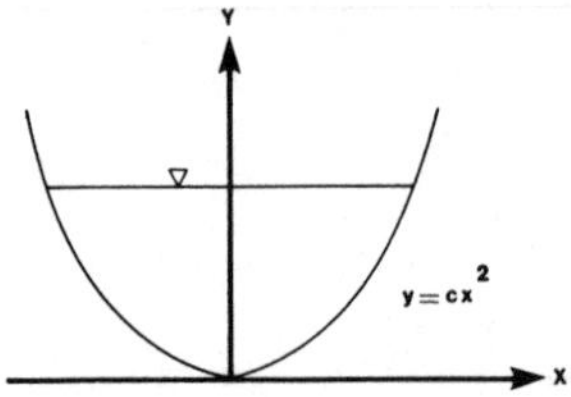	$(0.84c\psi)^{0.25}$	Perimeter equation $y = cx^2$
Circular 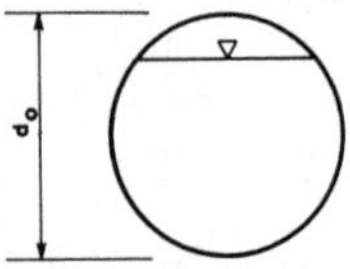	$\left(\frac{1.01}{d_0^{0.26}}\right)\psi^{0.25}$	Range of applicability $0.02 \le \frac{y_c}{d_0} \le 0.85$

TABLE 2.1 (*Continued*)

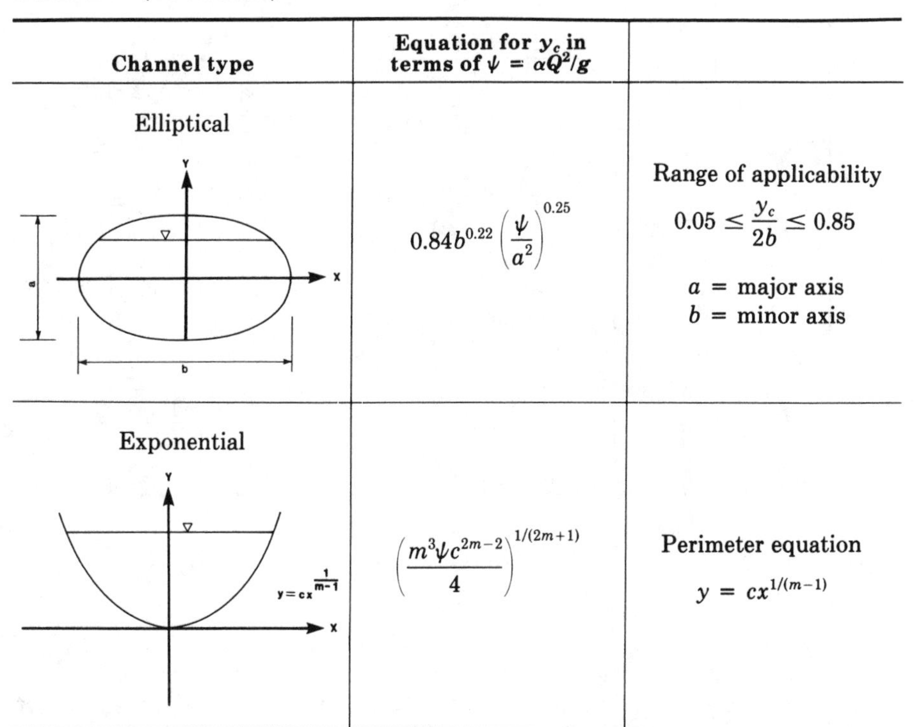

Channel type	Equation for y_c in terms of $\psi = \alpha Q^2/g$	
Elliptical	$0.84b^{0.22}\left(\dfrac{\psi}{a^2}\right)^{0.25}$	Range of applicability $0.05 \leq \dfrac{y_c}{2b} \leq 0.85$ a = major axis b = minor axis
Exponential	$\left(\dfrac{m^3\psi c^{2m-2}}{4}\right)^{1/(2m+1)}$	Perimeter equation $y = cx^{1/(m-1)}$

Solution

From Table 2.1

$$y_c = 0.81\left(\frac{\psi}{z^{0.75}b^{1.25}}\right)^{0.27} - \frac{b}{30z} \qquad \text{for } 0.1 < \frac{Q}{b^{2.5}} < 4$$

where

$$\psi = \frac{\alpha Q^2}{g}$$

As a check $Q/b^{2.5} = 17/6^{2.5} = 0.19$, confirming that the equation above from Table 2.1 applies. Substituting appropriate values,

$$\psi = \frac{1(17)^2}{9.8} = 29.5$$

$$y_c = 0.81\left(\frac{29.5}{2^{0.75}6^{1.25}}\right)^{0.27} - \frac{6}{30(2)}$$

$$= 0.86 \text{ m } (2.8 \text{ ft})$$

Comparing this result with that obtained in Example 2.1 for the same problem, one notes that the answers are very comparable and further that the method of Example 2.1 requires very little computation.

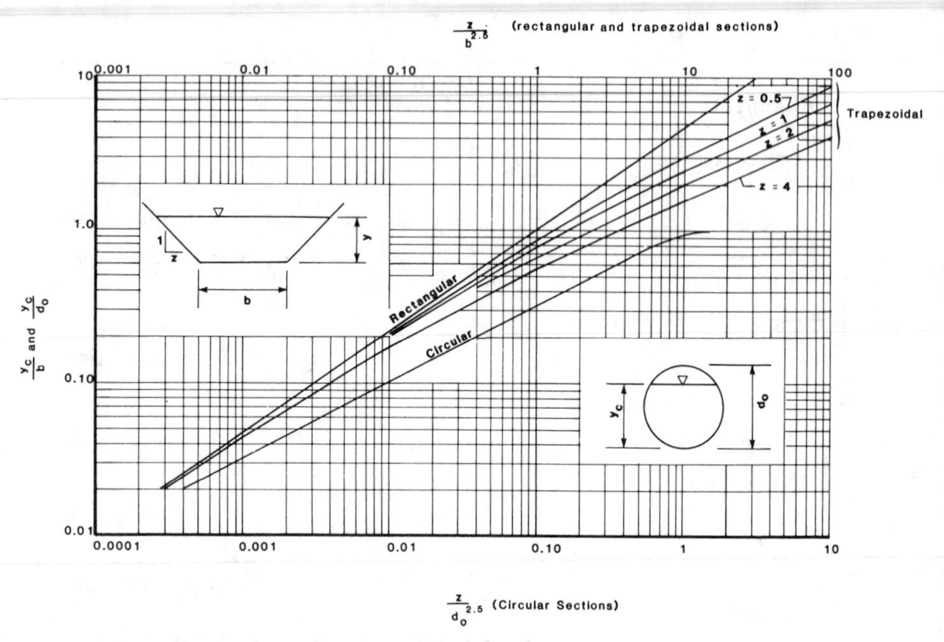

FIGURE 2.4 Curves for estimating critical depth.

A third method of determining the critical depth when the shape of the channel section and the flow rate are specified involves the use of a design chart. In developing a chart for this purpose, it is convenient to define the section factor for critical flow computation. Substituting $\overline{u} = Q/A$ in Eq. (2.2.3) yields, after simplification,

$$\frac{Q}{\sqrt{g/\alpha}} = Z = \sqrt{\frac{A^3}{T}} \tag{2.2.9}$$

The left-hand side of Eq. (2.2.9) is by definition the section factor for critical flow Z, and the right-hand side of the equation is a function of only the channel shape and the depth of flow. A design chart for the purpose of solving the critical depth problem is shown in Fig. 2.4.

EXAMPLE 2.3

A circular channel 3.0 ft (0.91 m) in diameter conveys a flow of 25 ft^3/s (0.71 m^3/s); estimate the critical depth of flow.

Solution

From Eq. (2.2.9) with $\alpha = 1.0$

$$Z = \frac{Q}{\sqrt{g/\alpha}} = \frac{25}{\sqrt{32.2}} = 4.41$$

and
$$\frac{z}{d_0^{2.5}} = \frac{4.41}{(3)^{2.5}} = 0.283$$

From Fig. 2.4

$$\frac{y_c}{d_0} = 0.54$$

$$y_c = 0.54(3) = 1.6 \text{ ft } (0.49 \text{ m})$$

For complex designed sections which cannot be treated by either the semi-empirical equations summarized in Table 2.1 or the design chart method presented in Fig. 2.4, a graphical procedure may be used. This method is also applicable to natural channels. In this procedure, a curve of y_c versus Z is constructed such that for a specified value of $Z = Q/\sqrt{g}$ the value of y_c may be estimated.

EXAMPLE 2.4

A trapezoidal channel with $b = 20$ ft (6.0 m) and $z = 1.5$ conveys a flow of 600 ft^3/s (17 m^3/s); estimate the critical depth of flow.

Solution

The first step in solving this problem is to construct a y versus Z curve (Fig. 2.5). The value of the section factor is then computed from the given data

$$Z = \frac{Q}{\sqrt{g}} = \sqrt{\frac{A^3}{T}} = \frac{600}{\sqrt{32.2}} = 105$$

Then, from Fig. 2.5, $y_c = 2.8$ ft (0.85 m).

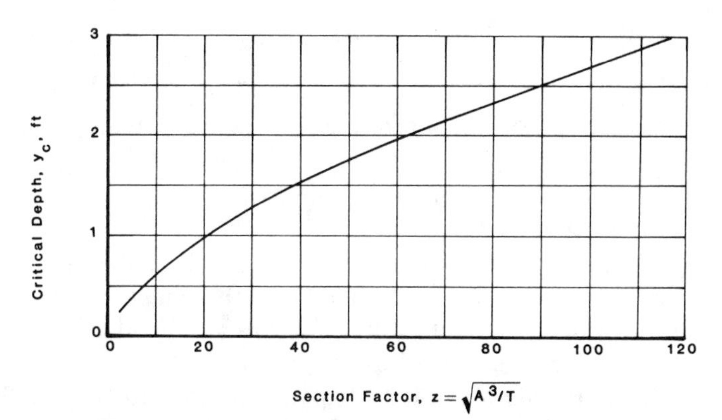

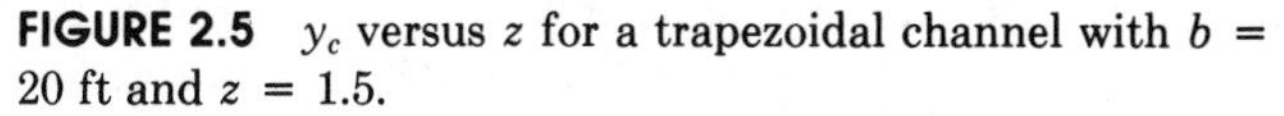

FIGURE 2.5 y_c versus z for a trapezoidal channel with b = 20 ft and z = 1.5.

2.3 ACCESSIBILITY AND CONTROLS

The introduction of the concepts of specific energy and critical flow makes it possible to discuss the reaction of the flow in a channel to changes in the shape of the channel and hydraulic structures for different steady-flow regimes.

At any cross section, the total energy is

$$H = \frac{\overline{u}^2}{2g} + y + z \tag{2.3.1}$$

where y = depth of flow, z = elevation of the channel bottom above a datum, and it is assumed that $\alpha = 1$ and $\cos\theta = 1$. Differentiating Eq. (2.3.1) with respect to the longitudinal distance x yields

$$\frac{dH}{dx} = \frac{d(\overline{u}^2/2g)}{dx} + \frac{dy}{dx} + \frac{dz}{dx} \tag{2.3.2}$$

The term dH/dx is the change of energy with longitudinal distance or the friction slope. Define

$$\frac{dH}{dx} = -S_f \tag{2.3.3}$$

The term dz/dx is the change of elevation of the bottom of the channel with respect to distance or the bottom slope. Define

$$\frac{dz}{dx} = -S_0 \tag{2.3.4}$$

For a given flow rate Q, the term

$$\frac{d(\overline{u}^2/2g)}{dx}$$

becomes

$$\frac{d(\overline{u}^2/2g)}{dx} = -\frac{Q^2}{gA^3}\frac{dA}{dy}\frac{dy}{dx} = -\frac{Q^2T}{gA^3}\frac{dy}{dx} = -\mathbf{F}^2\frac{dy}{dx} \tag{2.3.5}$$

Substituting Eqs. (2.3.3) to (2.3.5) in Eq. (2.3.2) and simplifying yields

$$\frac{dy}{dx} = \frac{S_0 - S_f}{1 - \mathbf{F}^2} \tag{2.3.6}$$

which describes the variation of the depth of flow in a channel of arbitrary shape as a function of S_0, S_f, and $\mathbf{F}^2$. In this chapter, only solutions of Eq. (2.3.6) where $S_f = 0$ (frictionless flow) will be considered. Solutions for the case where $S_f \neq 0$ will be considered in Chapter 6.

In the remainder of this chapter, solutions to Eq. (2.3.6) will be sought for the special case of $S_f = 0$, but first the behavior of the depth of flow in response to changes in cross-sectional shape will be examined from a qualitative viewpoint. For this purpose, a rectangular channel is assumed.

Case I: Constant-Width Channel

In the case of a constant-width, rectangular channel, the flow per unit width q is constant, and Eq. (2.3.6) can be rearranged to yield

$$(1 - \mathbf{F}^2)\frac{dy}{dx} + \frac{dz}{dx} = 0 \tag{2.3.7}$$

At this point a number of subcases can be considered.

1. If $dz/dx > 0$ and $\mathbf{F} < 1$, then $(1 - \mathbf{F}^2) > 0$ and dy/dx must be less than zero. Thus, the depth of flow decreases as x increases.
2. If $dz/dx > 0$ and $\mathbf{F} > 1$, then $(1 - \mathbf{F}^2) < 0$ and dy/dx must be greater than zero. Thus, under these conditions the depth of flow increases as x increases.
3. If $dz/dx < 0$ and $\mathbf{F} < 1$, then $(1 - \mathbf{F}^2) > 0$ and dy/dx must be greater than zero. Thus, under these conditions, the depth of flow increases as x increases.
4. If $dz/dx < 0$ and $\mathbf{F} > 1$, then $(1 - \mathbf{F}^2) < 0$ and dy/dx must be less than zero. Under these conditions, the depth of flow decreases as x increases.

When $dz/dx = 0$, an interesting and very useful case presents itself. Under this condition, Eq. (2.3.7) becomes

$$(1 - \mathbf{F}^2)\frac{dy}{dx} = 0$$

Then, either $dy/dx = 0$ or $\mathbf{F}^2 = 1$. This type of situation can occur at both spillways and broad-crested weirs. By physical observation, it is known that in these situations $dy/dx \neq 0$, and, therefore, $\mathbf{F}^2 = 1$. A judicious application of these results can yield an effective flow measurement method. In such cases, critical depth does not occur exactly at the free overfall but slightly before this point.

EXAMPLE 2.5

A broad-crested weir is placed in a channel of width b. If the upstream depth of flow is y_1 and the upstream velocity head of the flow and frictional losses can be neglected, develop a theoretical equation for the discharge in terms of the upstream depth of flow.

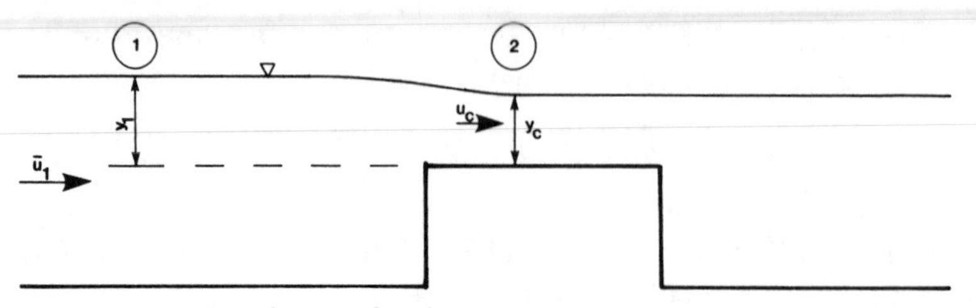

FIGURE 2.6 Broad-crested weir.

Solution

Given the situation described above and in Fig. 2.6, apply the Bernoulli energy equation between points 1 and 2 or

$$\frac{\bar{u}_1^2}{2g} + y_1 = \frac{\bar{u}_c^2}{2g} + y_c = E_c$$

For a deep, relatively slowly moving flow in the upstream section, $\bar{u}_1^2/2g \ll y_1$, and hence

$$y_1 = E_c$$

Then by Eq. (2.2.8)

$$E = \tfrac{3}{2} y_c$$

and

$$y_1 = \tfrac{3}{2} y_c$$

or

$$y_c = \tfrac{2}{3} y_1$$

Rearranging Eq. (2.2.6) yields

$$y_c^3 = \left(\frac{2}{3} y_1\right)^3 = \frac{q^2}{g}$$

or

$$\frac{q^2}{g} = \left(\tfrac{2}{3}\right)^3 y_1^3$$

and therefore

$$Q = b\sqrt{g}(\tfrac{2}{3})^{3/2} y_1^{3/2}$$

For $g = 32.2\ \text{ft/s}^2$ ($9.8\ \text{m/s}^2$)

$$Q = 3.09\ b y_1^{3/2}$$

A comparison of the coefficient in the equation above with coefficients derived empirically (see, for example, King and Brater, 1963) for this type of weir demonstrates excellent agreement and confirms that, when frictional losses are negligible and $\bar{u}_1^2/2g \ll y_1$, this type of approach is valid.

The foregoing example demonstrates an important practical application of the concepts of specific energy and critical depth; i.e., many structural means of flow measurement are based on the flow passing through critical depth since for this depth of flow there is an explicit relationship between the depth of flow and the flow rate.

Case II: Channel of Variable Width

In this case, the flow per unit width is not constant but $S_0 = 0$. Beginning with the total energy equation,

$$H = y + z + \frac{[q(x)]^2}{2gy^2}$$

where the symbolism $q(x)$ indicates that q is a function of x since the width of the channel is a function of x. Then

$$\frac{dH}{dx} = 0 = \frac{dy}{dx} + \underbrace{\frac{dz}{dx}}_{0} - \frac{[q(x)]^2}{gy^3}\frac{dy}{dx} + \frac{q(x)}{gy^2}\frac{d[q(x)]}{dx} \tag{2.3.8}$$

Since $Q = qb =$ constant

$$\frac{dQ}{dx} = 0 = b\frac{d[q(x)]}{dx} + q(x)\frac{db}{dx}$$

or

$$b\frac{d[q(x)]}{dx} = -q(x)\frac{db}{dx} \tag{2.3.9}$$

Substitution of Eq. (2.3.9) into Eq. (2.3.8) yields

$$(1 - \mathbf{F}^2)\frac{dy}{dx} - \mathbf{F}^2\frac{y}{b}\frac{db}{dx} = 0 \tag{2.3.10}$$

At this point four subcases can be considered.

1. If $db/dx > 0$ and $\mathbf{F} < 1$, then $(1 - \mathbf{F}^2) > 0$ and dy/dx must be greater than zero. Thus the depth of flow increases as x increases.
2. If $db/dx > 0$ and $\mathbf{F} > 1$, then $(1 - \mathbf{F}^2) < 0$ and dy/dx must be less than zero. Thus the depth of flow decreases as x increases.
3. If $db/dx < 0$ and $\mathbf{F} < 1$, then $(1 - \mathbf{F}^2) > 0$ and dy/dx must be less than zero. Under these conditions, the depth of flow decreases as x increases.
4. If $db/dx < 0$ and $\mathbf{F} > 1$, then $(1 - \mathbf{F}^2) < 0$ and dy/dx must be greater than zero. Under these conditions, the depth of flow increases as x increases.

The above discussion illustrates the interrelationship between the flow rate q and the depth of flow y. Then, by definition, a control is any feature which determines the depth-discharge relationship. From this definition, it follows that at any feature which acts as a control the discharge can be calculated once the depth of flow is known. This fact makes control sections attractive for flow measurement, and a critical control section, i.e., one in which the flow passes through critical depth, is especially attractive from the viewpoint of flow measurement. In Chapter 8 a number of flow measurement devices which operate by producing a critical section are discussed.

A primary application of the concepts of specific energy is the prediction of changes in the depth of flow in response to channel transitions, i.e., changes in the channel width and/or the elevation of the channel bottom. In an examination of these problems, the accessibility of various points on the E-y curve must be considered. Consider a rectangular channel of constant width b which conveys a steady flow per unit width q. In the otherwise horizontal channel bed there is a smooth upward step of height Δz (Fig. 2.7a). Given this situation, an E-y curve can be constructed (Fig. 2.7b). In this figure, the flow upstream of the step is represented by point A on the E-y curve. The point A' has the same specific energy as the point A, and the choice between the equally correct points A and A' is a function of the upstream Froude number. If the point A is chosen to represent the upstream flow, then this flow is subcritical; i.e., $\mathbf{F} < 1$. Since q is a constant, the point representing the flow downstream of the step must also lie on the same E-y curve. The location of the downstream point on the E axis can be determined by applying the Bernoulli equation between the upstream and downstream points or

$$\frac{\overline{u}_1^2}{2g} + y_1 = \frac{\overline{u}_2^2}{2g} + y_2 + \Delta z$$

$$E_2 = E_1 - \Delta z \tag{2.3.11}$$

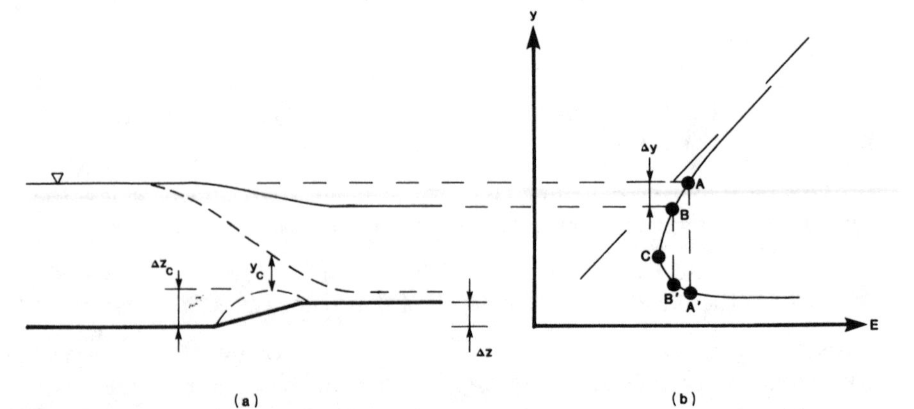

FIGURE 2.7 The accessibility problem.

where it has been tacitly assumed that the energy dissipation between these points is negligible. Having mathematically determined the value of E_2, Eq. (2.3.11) can be solved to determine the corresponding values of y_2. There are three values of y which satisfy Eq. (2.3.11). One of these solutions is negative and has no physical meaning; however, the two remaining solutions, points B and B', are equally valid solutions to the problem. The selection of the one point which is physically correct is the essence of the accessibility problem.

From the foregoing analysis of the varied-flow equation [Eq. (2.3.6)], it can be concluded that since $dz/dx > 0$ and $\mathbf{F} < 1$, the depth of flow decreases from point 1 to 2; however, since both B and B' represent depths of flow which are less than y_1, this conclusion is not useful. The E-y curve itself provides the solution to the problem. Since the width of the channel does not change, q is constant and the "flow point" can move only along the E-y curve defined by this value of q; i.e., the flow point cannot jump across the space which separates the points B and B'. Thus, the flow point must pass through point C if it is to move to point B'; however, movement to point C is possible only if the increase in the elevation of the channel is greater than the specified change Δz. This situation is represented by the dashed line in Fig. 2.7*a*. Thus, it is concluded that for the specified situation only point B is accessible from point A.

The above discussion assumes that a solution to the specified problem exists. In fact, it is quite easy to specify a problem to which there is no solution. For example, in Fig. 2.7*a* and *b* if the step height exceeds Δz_c, there is no solution. In essence, the three prescribed values, q, E, and Δz, cannot exist simultaneously. A physical interpretation of this situation is that the flow area has been sufficiently obstructed that the flow is choked. The flow will back up behind this obstruction, q will decrease because the depth of flow increases, and a new steady flow will be established on an E-y curve to the left of the one shown in Fig. 2.7*b*.

An additional observation is that near the critical condition, point C in Fig. 2.7*b*, large changes in the water surface can be effected by small changes in the bed level. Thus, flows which occur at near critical depth are inherently unstable and should be avoided.

2.4 APPLICATION OF THE ENERGY PRINCIPLE TO PRACTICE

Transition Problem

The primary application of the energy principle in practice is the solution of channel transition problems. In general, the solution of these problems can be effected by either algebraic or graphical methods. The following examples illustrate the basic solution techniques.

EXAMPLE 2.6

A rectangular channel expands smoothly from a width of 1.5 m (4.9 ft) to 3.0 m (9.8 ft). Upstream of the expansion, the depth of flow is 1.5 m

(4.9 ft) and the velocity of flow is 2.0 m/s (6.6 ft/s). Estimate the depth of flow after the expansion.

Algebraic Solution

Since there is no change in the elevation of the channel bed, the upstream specific energy E_1 is equal to the downstream specific energy E_2, or

$$E_1 = E_2$$

where $$E_1 = y_1 + \frac{\overline{u}_1^2}{2g} = 1.5 + \frac{4}{2(9.8)} = 1.7 \text{ m (5.6 ft)}$$

The velocity at the downstream station is

$$\overline{u}_2 = \frac{Q}{A_2} = \frac{2(1.5)(1.5)}{3y_2} = \frac{1.5}{y_2}$$

and therefore

$$E_2 = y_2 + \frac{\overline{u}_2^2}{2g} = y_2 + \frac{0.11}{y_2^2} = 1.7 \text{ m (5.6 ft)}$$

or $$y_2^3 - 1.7y_2^2 + 0.11 = 0$$

Solving this equation yields

$$y_2 = \begin{cases} 1.6 \text{ m (5.2 ft)} \\ 0.28 \text{ m (0.92 ft)} \end{cases}$$

A consideration of the concepts of accessibility indicates that only the subcritical depth of flow is physically possible, and hence, the correct answer is

$$y_2 = 1.6 \text{ m (5.2 ft)}$$

Graphical Solution

The graphical solution of this problem requires that an appropriate E-y curve be constructed for the downstream station. The governing equation for this curve is

$$E_2 = y_2 + \frac{0.11}{y_2^2}$$

A graph of this equation is shown in Fig. 2.8. From this graph the y values corresponding to $E_2 = 1.7$ m (5.6 ft) can be found or

$$y_2 = \begin{cases} 1.6 \text{ m (5.2 ft)} \\ 0.28 \text{ m (0.92 ft)} \end{cases}$$

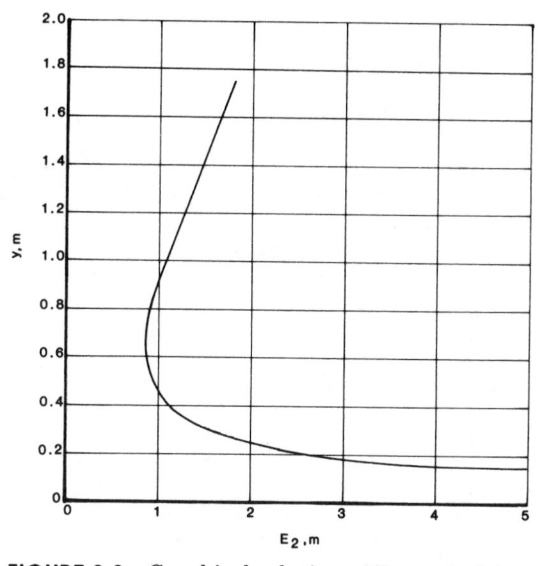

FIGURE 2.8 Graphical solution of Example 2.6.

Again, an examination of the problem statement demonstrates that the supercritical depth of flow, i.e., $y_2 = 0.28$ m (0.92 ft), is not accessible, and thus $y_2 = 1.6$ m (5.2 ft).

Although the above methods provide satisfactory answers, they require either a solution of a cubic equation or the construction of an E-y curve for each problem. For rectangular channels, these computational difficulties can be overcome by constructing a dimensionless E-y curve. Dividing both sides of the specific energy equation by the critical depth,

$$\frac{E}{y_c} = \frac{y}{y_c} + \frac{q^2}{2gy^2y_c} \tag{2.4.1}$$

Defining $E' = E/y_c$ and $y' = y/y_c$ and substituting Eq. (2.2.6) in Eq. (2.4.1) yields

$$E' = y' + \frac{1}{2(y')^2} \tag{2.4.2}$$

which is a dimensionless specific energy equation. A graph of Eq. (2.4.2) is shown in Fig. 2.9, and it is noted in this figure that the critical point occurs at the coordinates (1.5, 1.0). In practice, this graph cannot be read with sufficient precision; for this reason, Babcock (1959) reduced the dimensionless E-y graph to a tabular form (Table 2.2). All transition problems which occur in rectangular channels can be solved efficiently using either Fig. 2.9 or Table 2.2.

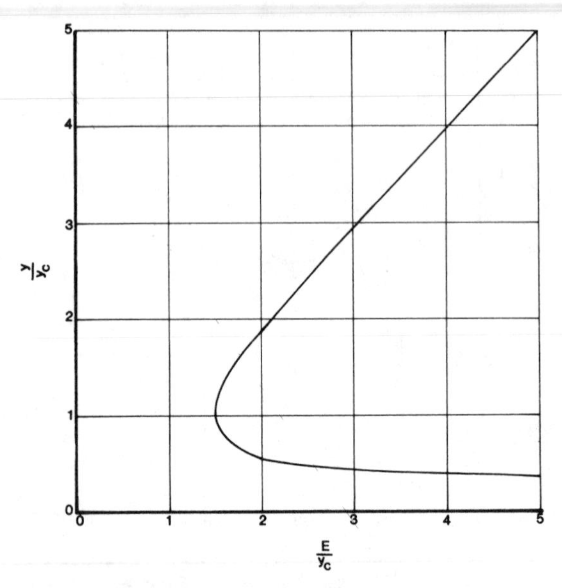

FIGURE 2.9 Dimensionless E-y curve.

TABLE 2.2 $E' = E/E_c$ **as a function of** $y' = y/y_c$ **(Babcock, 1959)**

	E' as a function of y'									
y'	0.00	0.01	0.02	0.03	0.04	0.05	0.06	0.07	0.08	0.09
0.0		5000	1250	555.6	312.5	200.1	138.9	102.1	78.21	61.82
0.1	50.10	41.43	34.84	29.71	25.65	22.37	19.69	17.47	15.61	14.04
0.2	12.70	11.55	10.55	9.682	8.920	8.250	7.656	7.129	6.658	6.235
0.3	5.856	5.513	5.203	4.921	4.665	4.432	4.218	4.022	3.843	3.677
0.4	3.525	3.384	3.254	3.134	3.023	2.919	2.823	2.734	2.650	2.572
0.5	2.500	2.432	2.369	2.310	2.255	2.203	2.154	2.109	2.066	2.026
0.6	1.989	1.954	1.921	1.890	1.861	1.833	1.808	1.784	1.761	1.740
0.7	1.720	1.702	1.684	1.668	1.653	1.639	1.626	1.613	1.602	1.591
0.8	1.581	1.572	1.564	1.556	1.549	1.542	1.536	1.531	1.526	1.521
0.9	1.517	1.514	1.511	1.508	1.506	1.504	1.502	1.501	1.501	1.500
1.0	1.500	1.500	1.501	1.501	1.502	1.504	1.505	1.507	1.509	1.511
1.1	1.513	1.516	1.519	1.522	1.525	1.528	1.532	1.535	1.539	1.543
1.2	1.547	1.552	1.556	1.560	1.565	1.570	1.575	1.580	1.585	1.590
1.3	1.596	1.601	1.607	1.613	1.618	1.624	1.630	1.636	1.642	1.649
1.4	1.655	1.662	1.668	1.674	1.681	1.688	1.695	1.701	1.708	1.715
1.5	1.722	1.729	1.736	1.744	1.751	1.758	1.766	1.773	1.780	1.788
1.6	1.795	1.803	1.810	1.818	1.826	1.834	1.841	1.849	1.857	1.865
1.7	1.873	1.881	1.889	1.897	1.905	1.913	1.921	1.930	1.938	1.946
1.8	1.954	1.963	1.971	1.979	1.988	1.996	2.004	2.013	2.022	2.030
1.9	2.038	2.047	2.056	2.064	2.073	2.082	2.090	2.099	2.108	2.116

TABLE 2.2 (*Continued*)

	E′ as a function of y′									
y′	0.00	0.01	0.02	0.03	0.04	0.05	0.06	0.07	0.08	0.09
2.0	2.125	2.134	2.142	2.151	2.160	2.169	2.178	2.187	2.196	2.204
2.1	2.213	2.222	2.231	2.241	2.249	2.258	2.267	2.276	2.285	2.294
2.2	2.303	2.312	2.322	2.330	2.340	2.349	2.358	2.367	2.376	2.385
2.3	2.394	2.404	2.413	2.422	2.431	2.440	2.450	2.459	2.468	2.478
2.4	2.487	2.496	2.505	2.515	2.524	2.533	2.543	2.552	2.561	2.571
2.5	2.580	2.589	2.599	2.608	2.618	2.627	2.636	2.646	2.655	2.665
2.6	2.674	2.683	2.693	2.702	2.712	2.721	2.731	2.740	2.750	2.759
2.7	2.769	2.778	2.788	2.797	2.807	2.816	2.826	2.835	2.845	2.854
2.8	2.864	2.873	2.883	2.892	2.902	2.912	2.921	2.931	2.940	2.950
2.9	2.960	2.969	2.979	2.988	2.998	3.008	3.017	3.027	3.036	3.046
3.0	3.056	3.064	3.075	3.084	3.094	3.104	3.113	3.123	3.133	3.142
3.1	3.152	3.162	3.171	3.181	3.191	3.200	3.210	3.220	3.229	3.239
3.2	3.249	3.258	3.268	3.278	3.288	3.297	3.307	3.317	3.326	3.336
3.3	3.346	3.356	3.365	3.375	3.385	3.395	3.404	3.414	3.424	3.434
3.4	3.443	3.453	3.463	3.472	3.482	3.492	3.502	3.512	3.521	3.531
3.5	3.541	3.551	3.560	3.570	3.580	3.590	3.600	3.609	3.619	3.629
3.6	3.639	3.648	3.658	3.668	3.678	3.688	3.697	3.707	3.717	3.727
3.7	3.737	3.746	3.756	3.766	3.776	3.786	3.795	3.805	3.815	3.825
3.8	3.835	3.844	3.854	3.864	3.874	3.884	3.894	3.903	3.913	3.923
3.9	3.933	3.943	3.953	3.962	3.972	3.982	3.992	4.002	4.012	4.021
4.0	4.031	4.041	4.051	4.061	4.071	4.080	4.090	4.100	4.110	4.120
4.1	4.130	4.140	4.150	4.159	4.169	4.179	4.189	4.199	4.209	4.218
4.2	4.228	4.238	4.248	4.258	4.268	4.278	4.288	4.297	4.307	4.317
4.3	4.327	4.337	4.347	4.357	4.366	4.376	4.386	4.396	4.406	4.416
4.4	4.426	4.436	4.446	4.456	4.465	4.475	4.485	4.495	4.505	4.515
4.5	4.525	4.535	4.544	4.554	4.564	4.574	4.584	4.594	4.604	4.614
4.6	4.624	4.634	4.643	4.653	4.663	4.673	4.683	4.693	4.703	4.713
4.7	4.723	4.732	4.742	4.752	4.762	4.772	4.782	4.792	4.802	4.812
4.8	4.822	4.832	4.842	4.851	4.861	4.871	4.881	4.891	4.901	4.911
4.9	4.921	4.931	4.941	4.951	4.960	4.970	4.980	4.990	5.000	5.010

EXAMPLE 2.7

Water flows in a rectangular channel 10 ft (3.0 m) wide at a velocity of 10 ft/s (3.0 m/s) and at a depth of 10 ft (3.0 m). There is an upward step of 2.0 ft (0.61 m). What expansion in width must take place simultaneously for this flow to be possible as specified?

Solution

First, examine the upstream flow and the step without considering the expansion. From the problem statement, the following quantities can be calculated:

$$q_1 = \frac{Q}{b_1} = \frac{10(10)(10)}{10} = 100\ (\text{ft}^3/\text{s})/\text{ft}\ [9.3\ (\text{m}^3/\text{s})/\text{m}]$$

$$y_c = \left(\frac{q_c^2}{g}\right)^{1/3} = \left(\frac{100^2}{32.2}\right)^{1/3} = 6.76\ \text{ft}\ (2.1\ \text{m})$$

$$E_1 = y_1 + \frac{\bar{u}_1^2}{2g} = 10 + \frac{10^2}{2(32.2)} = 11.6\ \text{ft}\ (3.5\ \text{m})$$

and

$$E_1' = \frac{E_1}{y_c} = \frac{11.6}{6.76} = 1.72$$

The downstream dimensionless specific energy, considering only the step, would be

$$E_2' = E_1' - \frac{\Delta z}{y_c} = 1.72 - \frac{2}{6.76} = 1.42$$

From Fig. 2.9 it can be determined that without an expansion the flow is not possible as specified since the downstream dimensionless specific energy does not lie on the E'-y' curve. If the downstream flow occurs at critical depth, then the expansion required is a minimum. The downstream condition is thus

$$E_2' = 1.5 = \frac{E_2}{y_{c2}}$$

or

$$y_{c2} = \frac{E_2}{1.5}$$

where $E_2 = E_1 - \Delta z = 9.55$ and y_{c2} = downstream critical depth. Then

$$y_{c2} = \frac{9.55}{1.5} = 6.37\ \text{ft}\ (1.9\ \text{m})$$

and

$$q = \sqrt{g y_{c2}^3} = 91.1\ (\text{ft}^3/\text{s})/\text{ft}\ [8.5\ (\text{m}^3/\text{s})/\text{m}]$$

Therefore, the downstream width is

$$b = \frac{Q}{b} = \frac{1000}{91.1} = 11\ \text{ft}\ (3.4\ \text{m})$$

or the minimum expansion required is $(11 - 10) = 1.0$ ft (0.30 m).

In the foregoing material, only the specific case of transitions in rectangular channels has been treated. It is necessary to develop an equivalent methodology for nonrectangular shapes. For a channel of arbitrary shape

$$E = y + \frac{Q^2}{2gA^2} \tag{2.4.3}$$

Dividing both sides of the equation by y_c,

$$\frac{E}{y_c} = \frac{y}{y_c} + \frac{Q^2}{2gA^2 y_c} \tag{2.4.4}$$

From Eq. (2.2.3)

$$\frac{Q^2}{g} = \frac{A_c^3}{T_c} = \frac{y_c^3 (T_c')^3}{T_c} \tag{2.4.5}$$

where A_c and T_c = flow area and channel top width when the depth of flow is y_c, respectively, and T_c' = top width of an equivalent rectangular channel, i.e., same depth of flow and flow area as the nonrectangular channel, at the critical depth of flow (Fig. 2.10).

Combining Eqs. (2.4.4) and (2.4.5) yields

$$\frac{E}{y_c} = \frac{y}{y_c} + \frac{1}{2}\left(\frac{y_c}{y}\right)^2 \left[\frac{(T_c')^3}{T_c (T')^2}\right] \tag{2.4.6}$$

For a rectangular channel, that is, $T_c' = T_c = T' = b$, Eq. (2.4.6) reduces to Eq. (2.4.2). The bracketed term in Eq. (2.4.6) is essentially a shape factor, and Silvester (1961) demonstrated that for trapezoidal, triangular, and parabolic channels this factor can be easily evaluated. In a trapezoidal channel

$$T = b + 2zy$$

and

$$T' = b + zy$$

A triangular channel is a special case of a trapezoidal channel with $b = 0$. For a parabolic channel with $T = a\sqrt{y}$ where a is a coefficient, the shape factor is

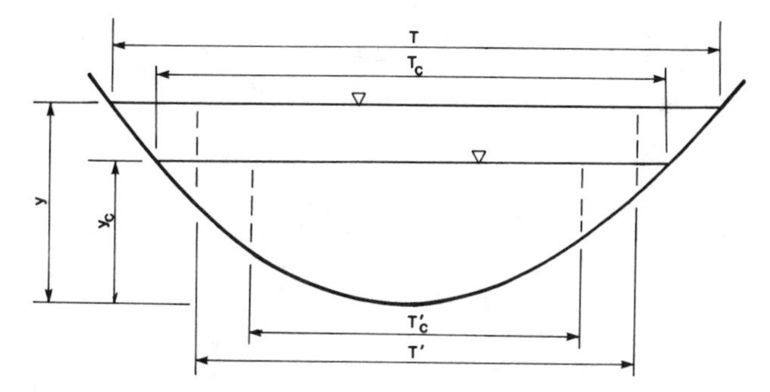

FIGURE 2.10 Definition of variables for specific energy in nonrectangular channels.

$\frac{2}{3}y_c/y$. The use of this technique is best illustrated with an example involving a trapezoidal channel.

EXAMPLE 2.8

A trapezoidal channel with $b = 20$ ft (6.1 m) and $z = 2$ carries a discharge of 4000 ft^3/s (110 m^3/s) at a depth of 12 ft (3.7 m). If a bridge spans the channel on two piers each 3.0 ft wide (0.91 m), determine the depth of flow beneath the bridge (Fig. 2.11).

Solution

From Table 2.1, for the section upstream of the bridge

$$y_c = 0.81\left(\frac{Q^2/32.2}{z^{0.75}b^{1.25}}\right)^{0.27} - \frac{b}{30z} = 8.5 \text{ ft } (2.6 \text{ m})$$

with

$$\frac{Q}{b^{2.5}} = 2.2$$

Then

$$T'_c = b + zy_c = 20 + 2(8.5) = 37 \text{ ft } (11 \text{ m})$$

$$T_c = b + 2zy_c = 20 + 2(2)(8.5) = 54 \text{ ft } (16 \text{ m})$$

$$T' = b + zy = 20 + 2(12) = 44 \text{ ft } (13 \text{ m})$$

The shape factor is

$$\left[\frac{(T'_c)^3}{T_c(T')^2}\right] = \left[\frac{(37)^3}{54(44)^2}\right] = 0.48$$

and

$$\frac{E}{y_c} = \frac{y}{y_c} + \frac{1}{2}\left(\frac{y_c}{y}\right)^2\left[\frac{(T'_c)^3}{T_c(T')^2}\right] = \frac{12}{8.5} + \frac{1}{2}\left(\frac{8.5}{12}\right)^2(0.48) = 1.5$$

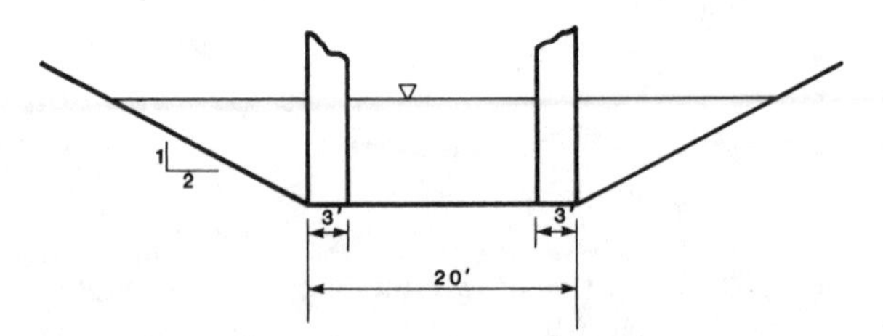

FIGURE 2.11 Schematic for Example 2.8.

The section of the channel beneath the bridge can be approximated as a trapezoidal channel with $b = 14$ ft (4.3 m). In this case

$$\frac{Q}{b^{2.5}} = 5.4$$

and, therefore, the equations in Table 2.1 cannot be used to estimate y_c. The section factor is

$$Z = \frac{Q}{\sqrt{g}} = \frac{4000}{\sqrt{32.2}} = 705$$

and

$$\frac{Z}{b^{2.5}} = \frac{705}{14^{2.5}} = 0.96$$

From Fig. 2.4

$$\frac{y_c}{b} = 0.66$$

$$y_c = 9.2 \text{ ft } (2.8 \text{ m})$$

Then

$$T_c' = b + zy_c = 14 + 2(9.2) = 32.4 \text{ ft } (9.9 \text{ m})$$

$$T_c = b + 2zy_c = 14 + 2(2)(9.2) = 50.8 \text{ ft } (15 \text{ m})$$

$$T' = b + zy = 14 + 2(12) = 38 \text{ ft } (12 \text{ m})$$

where the computation of T' is based on an estimate of the depth of flow at the bridge to avoid an implicit solution. This assumption will be subsequently checked. The shape factor is

$$\left[\frac{(T_c')^3}{T_c(T')^2}\right] = \left[\frac{(32.4)^3}{50.8(38)^2}\right] = 0.46$$

Assuming no loss of energy as the flow passes through the contraction formed by the bridge piers, the dimensionless specific energy at the bridge is

$$\frac{E_2}{y_{c2}} = E_1'\left(\frac{y_{c1}}{y_{c2}}\right) = 1.5\left(\frac{8.5}{9.2}\right) = 1.39$$

Then

$$\frac{E_2}{y_{c2}} = \frac{y_2}{y_{c2}} + \frac{1}{2}\left(\frac{y_{c2}}{y_2}\right)^2 (0.46) = 1.39$$

This equation must be solved for y_2/y_{c2} by trial and error or by some other method. By trial and error, the solution is

$$\frac{y_2}{y_{c2}} = 1.24$$

$$y_2 = 9.2(1.24) = 11.4 \text{ ft } (3.5 \text{ m})$$

Then checking the previously stated approximate computation for T',

$$T' = 14 + 2(11.4) = 36.8 \text{ ft } (11 \text{ m})$$

and the effect of this correction on the shape factor

$$\left[\frac{(T_c')^3}{T_c(T')^2}\right] = \left[\frac{(32.4)^3}{50.8(36.8)^2}\right] = 0.49$$

which is negligible modification.

Specific Energy in Channels of Compound Section

Throughout this chapter, the use of the kinetic energy correction factor has been indicated to account for the nonuniformity of the velocity distribution in open-channel flow. Recall Eq. (2.1.3), which defines specific energy in a one-dimensional channel; i.e.,

$$E = y + \frac{\alpha Q^2}{2gA^2}$$

where, from Eq. (1.3.41), α is

$$\alpha = \frac{\iint_A u^3 \, dA}{\bar{u}^3 A}$$

α becomes especially significant in channels of compound section. Traditionally (Chow, 1959, and Henderson, 1966) this factor has been computed by

$$\alpha = \frac{\sum_{i=1}^{N} (K_i^3/A_i^2)}{K^3/A^3} \tag{2.4.7}$$

where K_i = conveyance of ith channel subsection
A_i = area of ith channel subsection
K = conveyance of total section
A = total area of cross section
N = number of subsections in given section

As will be discussed in Chapter 5, the subsection conveyance is computed from the Manning equation as

$$K_i = \frac{\phi}{n_i} A_i R_i^{2/3} \tag{2.4.8}$$

where n_i = Manning resistance coefficient for ith subsection
ϕ = factor correcting for system of units used (ϕ = 1.49 for English units and ϕ = 1 for SI units)
R_i = hydraulic radius of ith subsection

The development of Eq. (2.4.7) is based on two assumptions: (1) The channel can be divided into subsections by appropriately located vertical lines which are assumed to be lines of zero shear and which do not contribute to the wetted perimeter of the subsection, and (2) the contribution of the nonuniformity of the velocity distribution within each subsection is negligible in comparison with the variation in the average velocity between the subsections. In point of fact, both of these assumptions are false. For example, Myers (1978) and Rajaratnam and Ahmadi (1979) have presented laboratory data demonstrating significant momentum transfer between channel subsections, and Blalock (1980) has shown both theoretically and experimentally the importance of the nonuniformity of the velocity within a subsection. The implications of these results in channels of compound section are crucial in both the area of specific energy calculation and the computation of gradually varied flow profiles.

The behavior of the E-y curve and the Froude number in a channel of compound section can best be examined by considering an idealized cross section (Fig. 2.12*a*) which carries a discharge of 5000 ft^3/s (140 m^3/s). This channel has the same dimensions as that used by Blalock (1980) in his original work on this topic and similar to that used by Petryk and Grant (1975). If the E-y curve for this situation is plotted, the resulting curve has two points at which the specific energy has a local minimum value (Fig. 2.12*b*). Point 1 is a minimum value for flows which occur only in the main channel, and point 2 is a minimum value for flows occurring in the compound channel section. The value of the Froude number at these minima should be 1; however, a graph of the depth of flow versus the Froude number defined by Eq. (2.2.3) or this value modified by α [Eq. (2.2.5)] yields erroneous values of the depth of flow corresponding to the minima (Fig. 2.12*c*).

Blalock (1980) and Blalock and Sturm (1981) developed a compound section Froude number $\mathbf{F}_B$, which correctly locates the points of minimum specific energy. If the kinetic energy correction factor is assumed to be a function of depth, then differentiation of Eq. (2.1.3) with respect to the depth of flow yields

$$\frac{dE}{dy} = 1 - \frac{\alpha Q^2}{gA^3}\frac{dA}{dy} + \frac{Q^2}{2gA^2}\frac{d\alpha}{dy}$$

and setting this equation equal to zero,

$$\frac{\alpha Q^2 T}{gA^3} - \frac{Q^2}{2gA^2}\frac{d\alpha}{dy} = 1 \qquad (2.4.9)$$

Then, the compound section Froude number is

$$\mathbf{F}_B = \left(\frac{\alpha Q^2 T}{gA^3} - \frac{Q^2}{2gA^2}\frac{d\alpha}{dy}\right)^{1/2} \qquad (2.4.10)$$

and $\mathbf{F}_B$ has a value of 1 at the depths of flow at which minimum specific energy occurs (Fig. 2.12*c*). In Eq. (2.4.10) the only parameter not routinely determined

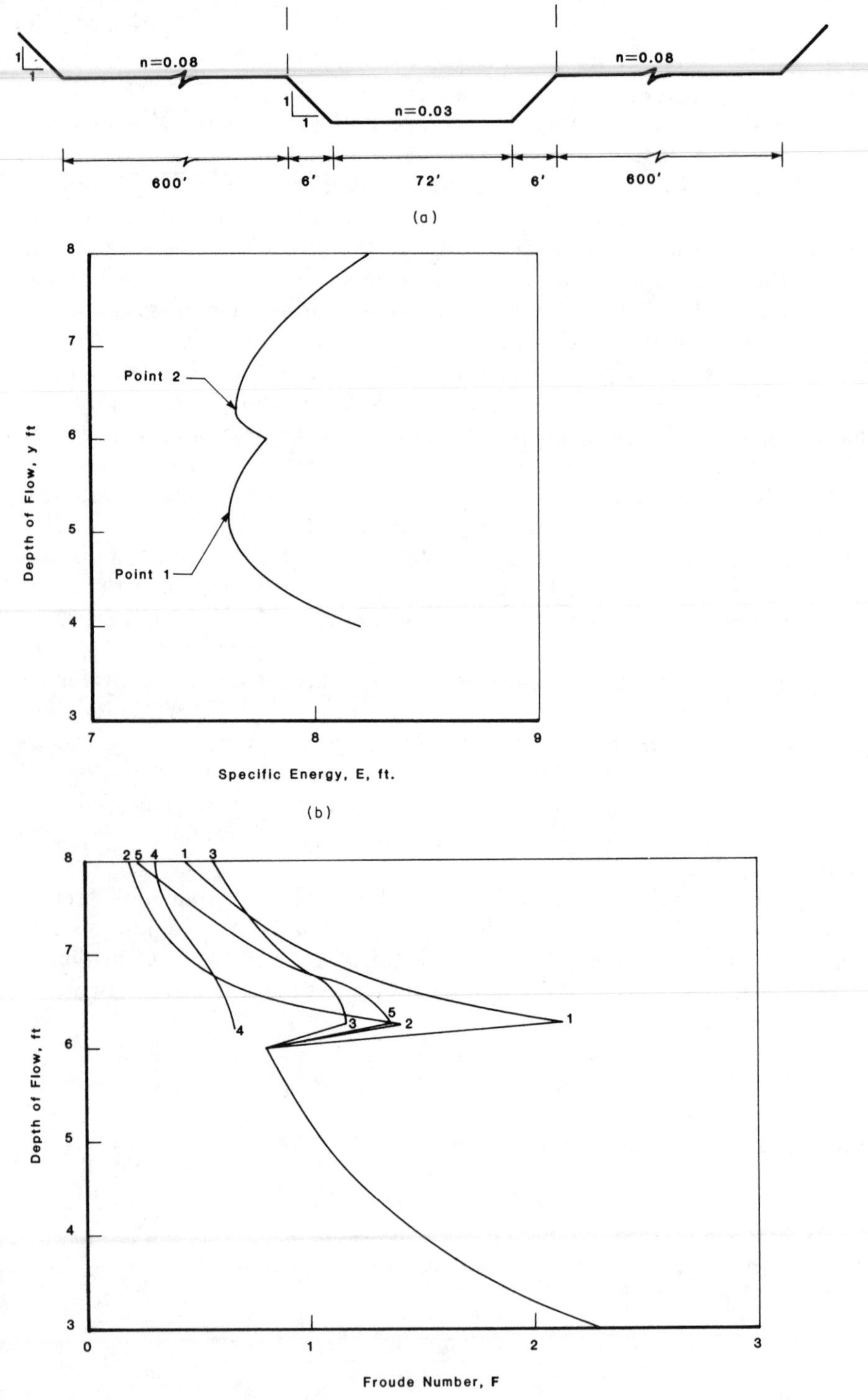

n=0.08
n=0.03
n=0.08
1
1
600'
6'
72'
6'
600'
(a)
Point 2
Point 1
Depth of Flow, y ft
Specific Energy, E, ft.
(b)
Depth of Flow, ft
Froude Number, F
(c)

is $d\alpha/dy$. Blalock (1980) and Blalock and Sturm (1981) demonstrated that

$$\frac{d\alpha}{dy} = \frac{A^2\sigma_1}{K^3} + \sigma_2\left(\frac{2AT}{K^3} - \frac{A^2\sigma_3}{K^4}\right) \tag{2.4.11}$$

where

$$\sigma_1 = \sum_{i=1}^{N}\left[\left(\frac{K_i}{A_i}\right)^3\left(3T_i - 2R_i\frac{dP_i}{dy}\right)\right]$$

$$\sigma_2 = \sum_{i=1}^{N}\left(\frac{K_i^3}{A_i^2}\right)$$

and

$$\sigma_3 = \sum_{i=1}^{N}\left[\left(\frac{K_i}{A_i}\right)\left(5T_i - 2R_i\frac{dP_i}{dy}\right)\right]$$

In these equations, T_i = top width of the ith subsection, R_i = hydraulic radius of the ith subsection, dP_i/dy = rate of change of the wetted perimeter of ith subsection with respect to the depth of flow, and N = number of subsections. Substitution of Eq. (2.4.11) into Eq. (2.4.10) yields

$$\mathbf{F}_B = \left[\frac{Q^2}{2gK^3}\left(\frac{\sigma_2\sigma_3}{K} - \sigma_1\right)\right]^{1/2} \tag{2.4.12}$$

In an analogous fashion, Konemann (1982) developed the following equation for the Froude number in a channel of compound section:

$$\mathbf{F}_K = \left[\frac{Q^2}{2g}\left(\frac{3B\,dM - M\,dB}{M^4}\right)\right]^{1/2} \tag{2.4.13}$$

where

$$B = \sum_{i=1}^{N}\left[\left(\frac{1}{n_i}R_i^{2/3}\right)^3 A_i\right]$$

$$dB = 3\sum_{i=1}^{N}\left[\left(\frac{1}{n_i}R_i^{2/3}\right)^3 T_i\right]$$

FIGURE 2.12 (a) Specific energy calculations in a channel of compound section. (b) Specific energy as a function of depth. (c) Froude number as a function of depth.

Curve 1 $\mathbf{F} = \left[\dfrac{\alpha\, Q^2\, T}{g\, A^3}\right]^{1/2}$ Curve 2 $\mathbf{F} = \left(\dfrac{Q^2\, T}{g\, A^3}\right)^{1/2}$

Curve 3 $\mathbf{F} = \left[\dfrac{Q^2}{2g\, K^3}\left(\dfrac{\sigma_2\,\sigma_3}{K} - \sigma_1\right)\right]^{1/2}$ Curve 4 $\mathbf{F} = \dfrac{K_M}{K\, A_M}\dfrac{Q}{(g\, A_M/T_M)^{1/2}}$

Curve 5 $\mathbf{F} = \left[\dfrac{Q^2}{2g}\left(\dfrac{3\, B\, \mathrm{d}M - M\, dB}{N^4}\right)\right]^{1/2}$

$$M = \sum_{i=1}^{N} \left[\left(\frac{1}{n_i} R_i^{2/3} \right) A_i \right]$$

$$dM = \frac{5}{3} \sum_{i=1}^{N} \left[\left(\frac{1}{n_i} R_i^{2/3} \right) T_i \right]$$

and n_i = Manning resistance coefficient associated with the ith subsection. Equations (2.4.12) and (2.4.13) yield results which are almost identical; however, Eq. (2.4.13) is somewhat simpler to work with from a computational viewpoint.

A fourth method of computing a compound section Froude number was given by Shearman (1976). If the discharge distribution between the subsections is assumed to be proportional to the conveyance, then the index Froude number is given by

$$\mathbf{F}_I = \frac{K_M}{KA_M} \frac{Q}{\sqrt{g(A_M/T_M)}} \tag{2.4.14}$$

where the subscript M refers to variables whose values are computed in the subsection with the largest conveyance. The results of this type of analysis are also summarized in Fig. 2.12*b*.

In general, the methods of Blalock (1980), Blalock and Sturm (1981), and Konemann (1982) yield the best representation of the variation of the Froude number with depth in a channel of compound section.

BIBLIOGRAPHY

Babcock, H. A., "Tabular Solution of Open Channel Flow Equations," *Proceedings of the American Society of Civil Engineers, Journal of the Hydraulics Division,* vol. 85, no. HY3, March 1959, pp. 17–23.

Blalock, M. E., III, "Minimum Specific Energy in Open Channels of Compound Section," thesis presented to the Georgia Institute of Technology, Atlanta, GA, in partial fulfillment of the requirements for the degree of Master of Science in Civil Engineering, June 1980.

Blalock, M. E., and Sturm, T. W., "Minimum Specific Energy in Compound Open Channel," *Proceedings of the American Society of Civil Engineers, Journal of the Hydraulics Division,* vol. 107, no. HY6, June 1981, pp. 699–717.

Chow, V. T., *Open Channel Hydraulics,* McGraw-Hill Book Company, New York, 1959.

Henderson, F. M., *Open Channel Flow,* The Macmillan Company, New York, 1966.

King, H. W., and Brater, E. F., *Handbook of Hydraulics,* 5th ed., McGraw-Hill Book Company, New York, 1963, pp. 5.1–5.51.

Konemann, N., Discussion of "Minimum Specific Energy in Compound Open Channel," *Proceedings of the American Society of Civil Engineers, Journal of the Hydraulics Division,* vol. 108, no. HY3, March 1982, pp. 462–464.

Myers, W. R. C., "Momentum Transfer in a Compound Channel," *International Association of Hydraulic Research, Journal of Hydraulic Research,* vol. 16, no. 2, 1978, pp. 139–150.

Petryk, S., and Grant, E. U., "Critical Flow in Rivers with Flood Plains," *Proceedings of the American Society of Civil Engineers, Journal of the Hydraulics Division,* vol. 101, no. HY7, July 1975, pp. 933–946.

Rajaratnam, N., and Ahmadi, R. M., "Interaction between Main Channel and Flood-Plain Flows," *Proceedings of the American Society of Civil Engineers, Journal of the Hydraulics Division,* vol. 105, no. HY5, May 1979, pp. 573–588.

Shearman, J. O., "Computer Applications of Step-Backwater and Floodway Analyses," *Open-File Report 76-499,* U.S. Geological Survey, Washington, D.C., 1976.

Silvester, R., "Specific-Energy and Force Equations in Open-Channel Flow," *Water Power,* March 1961.

Straub, W. O., Personal Communication, Civil Engineering Associate, Department of Water and Power, City of Los Angeles, Jan. 13, 1982.

Streeter, V. L., and Wylie, E. B., *Fluid Mechanics,* McGraw-Hill Book Company, New York, 1975.

THREE

The Momentum Principle

SYNOPSIS

SYNOPSIS

In this chapter, the application of the law of conservation of momentum to open-channel flow is considered. Section 3.1 defines specific momentum. In Section 3.2, the occurrence and characteristics of hydraulic jumps in rectangular and nonrectangular channels are discussed. Among the characteristics considered are the energy losses incurred in a hydraulic jump and the length of the hydraulic jump. The types of jumps treated include free jumps, submerged jumps, and jumps which occur in channels with a significant slope. In Section 3.3 the occurrence and characteristics of internal hydraulic jumps are discussed.

3.1 DEFINITION OF SPECIFIC MOMENTUM

In examining the application of Newton's second law of motion to the basic problems of steady open-channel flow, it is convenient to begin with a general-case problem such as the one shown schematically in Fig. 3.1. Within the control volume defined in this figure, there is an unknown energy loss and/or a force acting on the flow between sections 1 and 2; the result is a change in the linear momentum of the flow. In many cases, this change in momentum is accompanied by a change in the depth of flow. The application of Newton's second law, in a one-dimensional form, to this control volume yields

$$F_1' + F_3' - F_2' - \Sigma f_f' - P_f' = \frac{\gamma}{g} Q(\beta_2 \bar{u}_2' - \beta_1 \bar{u}_1') \tag{3.1.1}$$

where F_1' and F_2' = horizontal components of pressures acting at sections 1 and 2, respectively

F_3' = horizontal component of $W \sin \theta$

W = weight of fluid between sections 1 and 2

γ = specific weight of fluid

θ = channel slope angle

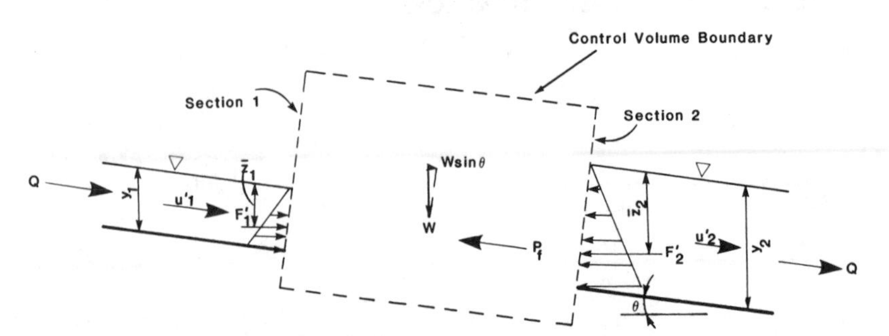

FIGURE 3.1 Schematic definition for specific momentum.

$\Sigma f_f'$ = sum of horizontal components of average velocities of flow at sections 1 and 2, respectively
P_f' = horizontal component of unknown force acting between sections 1 and 2
β_1 and β_2 = momentum correction coefficients

With the assumptions that, first, θ is small and hence $\sin\theta = 0$ and $\cos\theta = 1$, second, $\beta_1 = \beta_2 = 1$, and, third, $\Sigma f_f' = 0$, Eq. (3.1.1) becomes

$$\gamma\bar{z}_1A_1 - \gamma\bar{z}_2A_2 - P_f = \frac{\gamma}{g}Q(\bar{u}_2 - \bar{u}_1) \tag{3.1.2}$$

where $\bar{z}_1$ and $\bar{z}_2$ = distances to centroids of respective flow areas A_1 and A_2 from free surface
$F_1 = \gamma\bar{z}_1A_1$
$F_2 = \gamma\bar{z}_2A_2$

Substituting $\bar{u}_1 = Q/A_1$ and $\bar{u}_2 = Q/A_2$ into Eq. (3.1.2) and rearranging yields

$$\frac{P_f}{\gamma} = \left(\frac{Q^2}{gA_1} + \bar{z}_1A_1\right) - \left(\frac{Q^2}{gA_2} + \bar{z}_2A_2\right) \tag{3.1.3}$$

or

$$\frac{P_f}{\gamma} = M_1 - M_2 \tag{3.1.4}$$

where

$$M = \frac{Q^2}{gA} + \bar{z}A \tag{3.1.5}$$

and M is known as the specific momentum or force function.

Plotting the depth of flow y versus M produces a specific momentum curve which has two branches (Fig. 3.2). The lower limb AC asymptotically

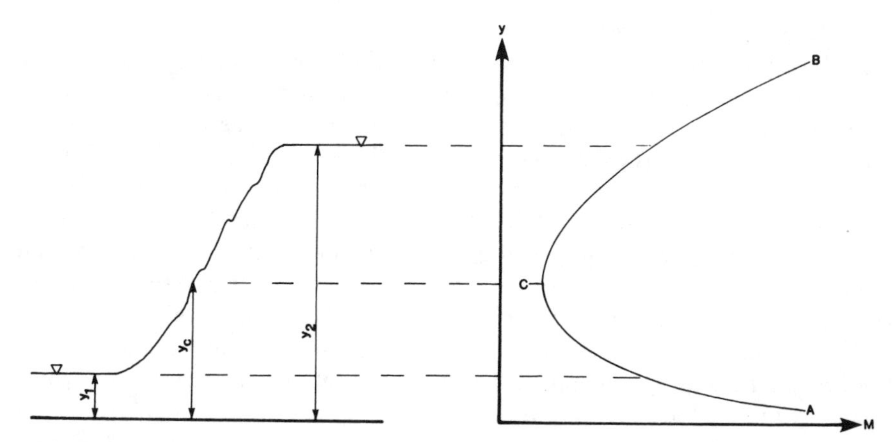

FIGURE 3.2 Specific momentum curve and sequent depths y_1 and y_2 of a hydraulic jump.

approaches the horizontal axis while the upper limb BC rises upward and extends indefinitely to the right. Thus, in analogy with the concept of specific energy, for a given value of M, the M-y curve predicts two possible depths of flow. These depths, shown in Fig. 3.2, are termed the *sequent depths of a hydraulic jump*.

The minimum value of the specific momentum function can be found under the assumptions of parallel flow and a uniform velocity distribution by taking the first derivative of M with respect to y and setting the resulting expression equal to zero or

$$\frac{dM}{dy} = -\frac{Q^2}{gA^2}\frac{dA}{dy} + \frac{d(\bar{z}A)}{dy} = 0 \tag{3.1.6}$$

and

$$-\frac{Q^2}{gA^2}\frac{dA}{dy} + A = 0 \tag{3.1.7}$$

where

$$d(\bar{z}A) = \left[A(\bar{z} + dy) + \frac{T(dy)^2}{2}\right] - \bar{z}A \simeq A dy$$

and where it is assumed that $(dy)^2 \simeq 0$. Then substituting $dA/dy = T$, $\bar{u} = Q/A$, and $D = A/T$ in Eq. (3.1.7),

$$\frac{\bar{u}^2}{2g} = \frac{D}{2} \tag{3.1.8}$$

which is the same criterion developed for the minimum value of specific energy. Therefore, for a specified discharge, minimum specific momentum occurs at minimum specific energy or critical depth.

3.2 THE HYDRAULIC JUMP

Hydraulic jumps result when there is a conflict between upstream and downstream controls which influence the same reach of channel. For example, if the upstream control causes supercritical flow while the downstream control dictates subcritical flow, then there is a conflict which can be resolved only if there is some means for the flow to pass from one flow regime to the other. Experimental evidence suggests that flow changes from a supercritical to a subcritical state can occur very abruptly through a phenomenon known as a hydraulic jump. The hydraulic jump can take place either on the free surface of a homogeneous flow or at a density interface in a stratified flow. In either case, the hydraulic jump is accompanied by significant turbulence and energy dissipation. In the field of open-channel flow, the applications of the hydraulic jump are many and include the following:

1. The dissipation of energy in flows over dams, weirs, and other hydraulic structures

2. The maintenance of high water levels in channels for water distribution purposes
3. The increase of the discharge of a sluice gate by repelling the downstream tailwater and thus increasing the effective head across the gate
4. The reduction of uplift pressure under structures by raising the water depth on the apron of the structure
5. The mixing of chemicals used for water purification or wastewater treatment
6. The aeration of flows and the dechlorination of wastewaters
7. The removal of air pockets from open-channel flows in circular channels
8. The identification of special flow conditions such as the existence of supercritical flow or the presence of a control section for the cost-effective measurement of flow

Sequent Depths

The computation of the hydraulic jump always begins with Eq. (3.1.3). If the jump occurs in a channel with a horizontal bed and $P_f = 0$, that is, the jump is not assisted by a hydraulic structure, then Eq. (3.1.4) requires that

$$M_1 = M_2 \tag{3.2.1}$$

or

$$\frac{Q^2}{gA_1} + \bar{z}_1 A_1 = \frac{Q^2}{gA_2} + \bar{z}_2 A_2 \tag{3.2.2}$$

Rectangular Sections In the case of a rectangular channel of width b, substitution of $Q = \bar{u}_1 A_1 = \bar{u}_2 A_2$, $A_1 = by_1$, $A_2 = by_2$, $\bar{z}_1 = \frac{1}{2}y_1$, and $\bar{z}_2 = \frac{1}{2}y_2$ into Eq. (3.2.2) yields

$$\frac{q^2}{g}\left(\frac{1}{y_1} - \frac{1}{y_2}\right) = \frac{1}{2}(y_2^2 - y_1^2) \tag{3.2.3}$$

where $q = Q/b$ is the flow per unit width. Equation (3.2.3) has the following solutions

$$\frac{y_2}{y_1} = \frac{1}{2}(\sqrt{1 + 8\mathbf{F}_1^2} - 1) \tag{3.2.4}$$

and

$$\frac{y_1}{y_2} = \frac{1}{2}(\sqrt{1 + 8\mathbf{F}_2^2} - 1) \tag{3.2.5}$$

Equations (3.2.4) and (3.2.5) each contain three independent variables, and two must be known before a value of the third can be estimated. It must be emphasized that the downstream depth y_2 is not the result of upstream conditions but is the result of a downstream control; i.e., if the downstream control produces the depth y_2, then a jump will form.

The use of Eqs. (3.2.4) and (3.2.5) to solve hydraulic jump problems in rectangular channels is rather obvious; however, in the case of Eq. (3.2.5) significant computational difficulties can arise. In this equation, $\mathbf{F}_2^2$ is usually small, the term $\sqrt{1 + 8\mathbf{F}_2^2}$ is close to 1, and hence the term

$$(\sqrt{1 + 8\mathbf{F}_2^2} - 1)$$

is near zero. The difficulty occurs in retaining computational precision while computing a small difference between two relatively large numbers. This difficulty can be avoided by expressing the term $\sqrt{1 + 8\mathbf{F}_2^2}$ as the binomial series expansion or

$$\sqrt{1 + 8\mathbf{F}_2^2} = 1 + 4\mathbf{F}_2^2 - 8\mathbf{F}_2^4 + 32\mathbf{F}_2^6 + \cdots \qquad (3.2.6)$$

Then, substituting Eq. (3.2.6) into Eq. (3.2.5),

$$\frac{y_1}{y_2} = 2\mathbf{F}_2^2 - 4\mathbf{F}_2^4 + 16\mathbf{F}_2^6 + \cdots \qquad (3.2.7)$$

Equation (3.2.7) should be used when $\mathbf{F}_2^2$ is very small; for example, $\mathbf{F}_2^2 \leq 0.05$.

Nonrectangular Sections In analyzing the occurrence of hydraulic jumps in nonrectangular but prismatic channels, there are no equations analogous to Eqs. (3.2.4) and (3.2.5). In such cases, Eq. (3.2.2) could be solved by trial and error, but semiempirical approximations, and other analytic techniques, are available. A few comments regarding various semiempirical and analytic solutions of Eq. (3.2.2) for circular and other common channel sections are appropriate at this point.

As noted in Chapter 2, for circular channels, a log-log plot of y_c/d versus $(Q\sqrt{\alpha}/\sqrt{g})^{2.5}$ yields a straight line in the interval $0.02 < y_c/d \leq 0.85$ (Fig. 2.4). A regression analysis of this straight line yields the equation

$$y_c = \frac{1.01}{d_0^{0.264}} \left(\frac{Q\sqrt{\alpha}}{\sqrt{g}} \right)^{0.506} \qquad (3.2.8)$$

where d_0 = pipe diameter. It should be noted that the ratio y_c/d only rarely exceeds 0.85 in practice since it is virtually impossible to maintain critical flow near the top of a circular channel. Thus, Eq. (3.2.8) applies to the region of general interest.

Straub (1978) noted that in circular conduits the upstream Froude number $\mathbf{F}_1$ can be approximated by

$$\mathbf{F}_1 = \left(\frac{y_c}{y_1} \right)^{1.93} \qquad (3.2.9)$$

where y_1 = upstream depth of flow and y_c is estimated by Eq. (3.2.8). Straub (1978) also noted that for $\mathbf{F} < 1.7$ the sequent depth y_2 can be estimated by

$$y_2 = \frac{y_c^2}{y_1} \qquad (3.2.10)$$

and for $\mathbf{F} > 1.7$

$$y_2 = \frac{y_c^{1.8}}{y_1^{0.73}} \tag{3.2.11}$$

These equations provide a practical basis for estimating hydraulic jump parameters in a channel of circular section.

EXAMPLE 3.1

A flow of 100 ft³/s (2.8 m³/s) occurs in a circular channel 6.0 ft (1.8 m) in diameter. If the upstream depth of flow is 2.0 ft (0.61 m), determine the downstream depth of flow which will cause a hydraulic jump (Fig. 3.3).

Solution by Trial and Error

At the upstream station (station 1)

$$A_1 = \tfrac{1}{8}(\theta - \sin\theta)d_0^2 = \tfrac{1}{8}[2.46 - \sin(2.46)](6)^2 \text{ (Table 1.1)}$$
$$= 8.25 \text{ ft}^2 \ (0.77 \text{ m}^2)$$

$$D_1 = \frac{d_0}{8}\frac{(\theta - \sin\theta)}{\sin\theta/2} = \frac{6}{8}\left(\frac{2.46 - \sin(2.46)}{\sin(1.23)}\right) \text{ (Table 1.1)}$$
$$= 1.46 \text{ ft } (0.44 \text{ m})$$

$$\overline{u}_1 = \frac{Q}{A_1} = \frac{100}{8.25} = 12.1 \text{ ft/s } (3.7 \text{ m/s})$$

$$\mathbf{F}_1 = \frac{\overline{u}_1}{\sqrt{g\,D_1}} = \frac{12.1}{\sqrt{32.2\,(1.46)}} = 1.76$$

For a circular open channel flowing partially full, the distance from the water surface to the centroid of the flow area must be developed from basic principles. With reference to Fig. 3.3, the flow area is given by

$$A = 2\int_r^z \sqrt{r^2 - z^2}\,dz = \left[z\sqrt{r^2 - z^2} + r^2\sin^{-1}\left(\frac{z}{r}\right)\right]_{-r}^{z}$$

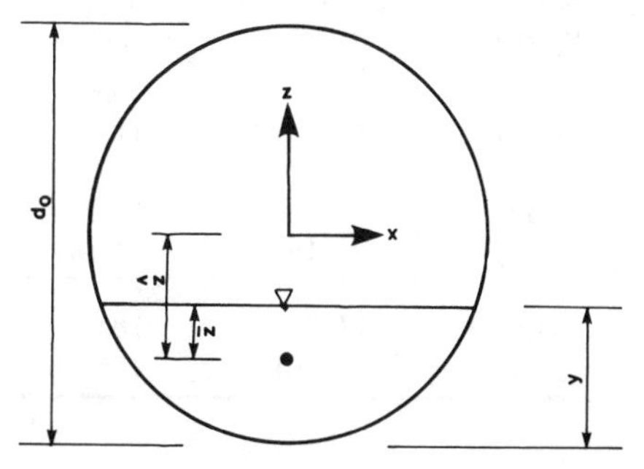

FIGURE 3.3 Circular conduit flowing partially full.

or
$$A = \frac{\pi r^2}{2} + z\sqrt{r^2 - z^2} + r^2 \sin^{-1}\left(\frac{z}{r}\right)$$

where r = radius of the circular section and $r^2 = x^2 + z^2$. The distance from the origin of the $z - x$ coordinate system to the centroid, the distance z in Fig. 3.3, is by definition

$$\hat{z}A = \int_{-r}^{z} z(2\sqrt{r^2 - z^2})\, dz = -⅔(r^2 - z^2)^{3/2}\Big|_{-r}^{z}$$

and
$$\hat{z} = -\frac{2(r^2 - z^2)^{3/2}}{3A}$$

The distance from the water surface to the centroid of the flow area

$$\bar{z} = y - (r + \hat{z})$$

Then for this example

$$\hat{z}_1 = \frac{-⅔[(r^2 - z_1^2)^3]^{1/2}}{A_1} = \frac{-⅔\{[3^2 - (-1)^2]^3\}^{1/2}}{8.25} = -1.83 \text{ ft } (0.56 \text{ m})$$

$$\bar{z} = y_1 - (r + \hat{z}_1) = 2 - (3 - 1.83) = 0.83 \text{ ft } (0.25 \text{ m}) \text{ (Fig. 3.3)}$$

$$M_1 = \frac{Q^2}{gA_1} + \bar{z}A_1 = \frac{(100)^2}{32.2(8.25)} + 0.83(8.25) = 44.5 \text{ ft}^3\ (1.26 \text{ m}^3)$$

At the downstream station (station 2)

$$M_2 = \frac{Q^2}{gA_2} + \bar{y}_2 A_2$$

The condition which must be satisfied if a hydraulic jump is to occur between stations 1 and 2 is

$$M_1 = M_2$$

The value $\bar{y}_2$ which satisfies this condition must be found by trial and error.

Trial y_2, ft (m)	$z_2 = y_2 - r$, ft (m)	A_2, ft^2 (m^2)	$\bar{z}_2$, ft (m)	$\bar{y}_2 = y_2 - (r + z_2)$ ft (m)	M_2, ft^3 (m^3)
3.40 (1.04)	0.40 (0.12)	16.5 (1.53)	−1.06 (−0.32)	1.46 (0.445)	42.9 (1.21)
3.58 (1.09)	0.58 (0.18)	17.6 (1.64)	−0.97 (−0.30)	1.55 (0.472)	44.9 (1.27)
3.65 (1.11)	0.65 (0.20)	18.0 (1.67)	−0.93 (−0.28)	1.58 (0.482)	45.7 (1.29)

Therefore, the condition for a hydraulic jump is that the downstream depth must be 3.58 ft (1.1 m).

Solution by Method of Straub

The upstream critical depth of flow estimated for $\alpha = 1$ is

$$y_c = \frac{1.01}{d_0^{0.264}}\left(\frac{Q}{\sqrt{g}}\right)^{0.506} = \frac{1.01}{(6)^{0.264}}\left(\frac{100}{\sqrt{32.2}}\right)^{0.506} = 2.69 \text{ ft } (0.820 \text{ m})$$

Since $y_c/d_0 = 0.45$, Straub's approximations can be used to estimate $\mathbf{F}_1$ and y_2 or

$$\mathbf{F}_1 = \left(\frac{y_c}{y}\right)^{1.93} = \left(\frac{2.69}{2}\right)^{1.93} = 1.77$$

Then for $\mathbf{F}_1 > 1.7$

$$y_2 = \frac{(y_c)^{1.8}}{y^{0.73}} = \frac{(2.69)^{1.8}}{(2)^{0.73}} = 3.58 \text{ ft } (1.09 \text{ m})$$

For other prismatic channel sections such as triangular, parabolic, and trapezoidal, either a graphical or trial-and-error solution of Eq. (3.2.2) is generally required. Silvester (1964, 1965) noted that for any prismatic channel the distance to the centroids of the flow areas $\bar{z}_i$ could be expressed as

$$\bar{z}_i = k_i' y_i \tag{3.2.12}$$

where the subscript i indicates the section, y is the greatest depth of flow at the specified section, and k' = a coefficient. The equation of a hydraulic jump in a horizontal channel then becomes

$$A_1 k_1' y_1 - A_2 k_2' y_2 = \frac{Q^2}{g}\left(\frac{1}{A_2} - \frac{1}{A_1}\right) \tag{3.2.13}$$

Rearrangement of Eq. (3.2.13) yields

$$k_2' \frac{A_2}{A_1}\frac{y_2}{y_1} - k_1' = \mathbf{F}_1^2 \frac{A_1}{y_1 T_1}\left(1 - \frac{A_1}{A_2}\right) \tag{3.2.14}$$

where the parameter $\mathbf{F}_1^2$ was first defined by Silvester (1964) to be $\mathbf{F}_1^2 = Q^2/gA_1^2 y_1$ and subsequently (Silvester, 1965) to be $\mathbf{F}_1^2 = Q^2/gA_1^2 D_1$ which is the standard definition of the Froude number for nonrectangular channels.

For rectangular channels, $k_1' = k_2' = ½$, $A_1/A_2 = y_1/y_2$, and $A_1/T_1 = D_1 = y_1$. With these definitions, Eq. (3.2.14) becomes

$$\left(\frac{y_2}{y_1}\right)^2 - 1 = 2\mathbf{F}_1^2\left(1 - \frac{y_1}{y_2}\right) \tag{3.2.15}$$

Although this equation can be reduced to Eq. (3.2.4), for obtaining general, graphical solutions of the sequent depth problems, the form given above is sat-

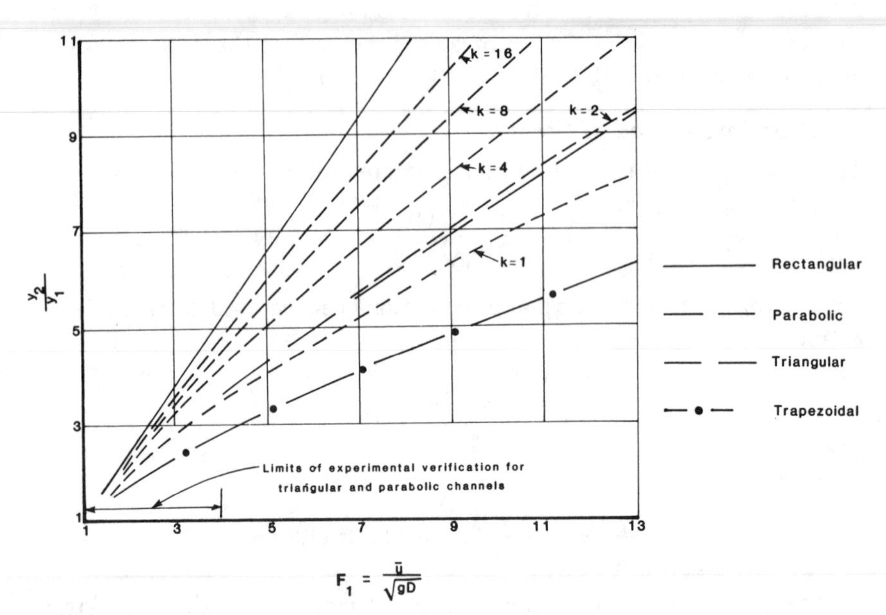

FIGURE 3.4 Analytical curves for y_2/y_1 versus $\mathbf{F}_1$.

isfactory. For comparison, the solution of this equation is plotted with the other solutions in Fig. 3.4.

For triangular channels, $k_1' = k_2' = \frac{1}{3}$, $A_1/A_2 = y_1^2/y_2^2$, and $A_1/T_1 = D_1 = y_1/2$. With these definitions, Eq. (3.2.14) becomes

$$\left(\frac{y_2}{y_1}\right)^3 - 1 = 1.5\mathbf{F}_1^2\left[1 - \left(\frac{y_1}{y_2}\right)^2\right] \tag{3.2.16}$$

For parabolic channels whose perimeters can be defined by $y = aT^2/2$, where a is a constant, $k_1' = k_2' = \frac{2}{5}$, $A_1/A_2 = (y_1/y_2)^{1.5}$, and $A_1/T_1 = D_1 = 2y_1/3$. With these definitions, Eq. (3.2.14) becomes

$$\left(\frac{y_2}{y_1}\right)^{2.5} - 1 = 1.67\mathbf{F}_1^2\left[1 - \left(\frac{y_1}{y_2}\right)^{1.5}\right] \tag{3.2.17}$$

This equation is plotted in Fig. 3.4.

For trapezoidal channels, Eq. (3.2.14) must be solved directly. Silvester (1964) defined a shape factor

$$k = \frac{b}{zy_1} \tag{3.2.18}$$

where b = bottom width of the trapezoid and z = trapezoid side slope. By use of this shape factor, Eq. (3.2.14) can be solved for this type of channel, and a family of solution curves results (Fig. 3.4).

Silvester (1964, 1965) verified the theoretical results plotted in Fig. 3.4 with laboratory data from the work of Argyropoulos (1957, 1961), Hsing (1937), Sandover and Holmes (1962), and Press (1961). In the laboratory results for triangular channels, the value of $\mathbf{F}_1$ did not exceed 4, and some scatter of the data was noted. For this case, all the experimental data were gathered in a single channel which had an included angle of 47.3° and, hence, was rather narrow and deep. For parabolic channels, the agreement between the experimental and laboratory results was generally excellent. In this case, the value of **F** also never exceeded 4. For the trapezoidal channels, there was a fair degree of scatter in the data, but Silvester (1964) ascribed the deviation of laboratory results from theory to various experimental problems and limitations. In general, the curves plotted in Fig. 3.4 provide an adequate method for estimating the value of y_2/y_1 given $\mathbf{F}_1^2$.

EXAMPLE 3.2

A flow of 100 m^3/s (3530 ft^3/s) occurs in a trapezoidal channel with side slopes of 2:1 and a base width of 5 m (16 ft). If the upstream depth of flow is 1.0 m (3.3 ft), determine the downstream depth of flow which will cause a hydraulic jump.

Solution

At the upstream station

$$A_1 = (b + zy_1)y_1 = [5 + 2(1)]1 = 7.0 \text{ m}^2 \ (75 \text{ ft}^2)$$

$$T_1 = b + 2zy_1 = 5 + 2(2)(1) = 9.0 \text{ m} \ (30 \text{ ft})$$

$$D_1 = \frac{A_1}{T_1} = \frac{7.0}{9.0} = 0.78 \text{ m} \ (2.6 \text{ ft})$$

$$\bar{u}_1 = \frac{Q}{A_1} = \frac{100}{7} = 14.3 \text{ m/s} \ (47 \text{ ft/s})$$

$$\mathbf{F}_1 = \frac{\bar{u}_1}{\sqrt{gD_1}} = \frac{14.3}{\sqrt{9.8(0.78)}} = 5.2$$

$$k = \frac{b}{zy_1} = \frac{5}{2(1)} = 2.5$$

Then, from Fig. 3.4 for $\mathbf{F}_1 = 4.6$ and $k = 2.5$

$$\frac{y_2}{y_1} = 4.7$$

$$y_2 = 4.7 \text{ m} \ (15 \text{ ft})$$

Therefore, the condition for a hydraulic jump is that the downstream depth must be 4.7 m (15 ft).

Submerged Jump

In the previous section, methodologies for determining the downstream depth which must prevail if a hydraulic jump is to form, given a specified supercritical upstream condition, were discussed. If the downstream depth is less than the sequent depth y_2, then a jump is not formed, and the supercritical flow maintains itself. If the downstream depth is greater than y_2, a submerged jump is formed (Fig. 3.5). Submerged jumps commonly form downstream of gates in irrigation systems, and the crucial unknown in such situations is the depth of submergence y_3.

Using the principles of conservation of momentum and mass, Govinda Rao (1963) demonstrated that in horizontal, rectangular channels

$$\frac{y_3}{y_1} = \left[(1 + S)^2\phi^2 - 2\mathbf{F}_1^2 + \frac{2\mathbf{F}_1^2}{(1 + S)\phi}\right]^{1/2} \tag{3.2.19}$$

where

$$S = \frac{y_4 - y_2}{y_2} \tag{3.2.20}$$

$$\phi = \frac{y_2}{y_1} = \frac{1}{2}(\sqrt{1 + 8\mathbf{F}_1^2} - 1) \tag{3.2.21}$$

and y_2 = subcritical sequent depth of the free jump corresponding to y_1 and $\mathbf{F}_1$. Govinda Rao verified Eq. (3.2.19) with data resulting from a number of laboratory experiments. For this same situation, Chow (1959) gave the following equation:

$$\frac{y_3}{y_4} = \left[1 + 2\mathbf{F}_4^2\left(1 - \frac{y_4}{y_1}\right)\right]^{1/2} \tag{3.2.22}$$

Equation (3.2.22) provides estimates of y_3 which are comparable with those obtained from Eq. (3.2.19).

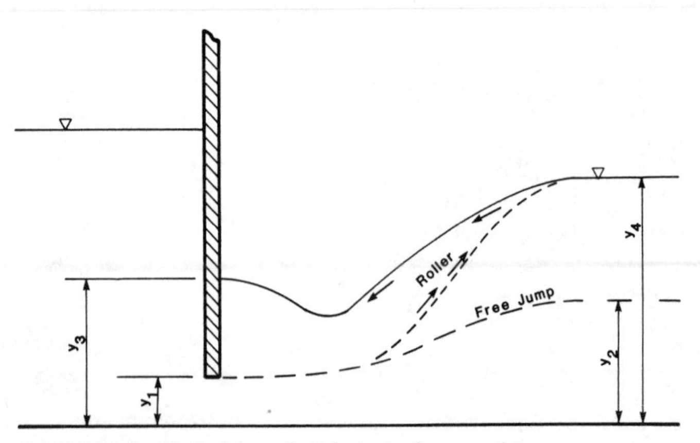

FIGURE 3.5 Definition sketch for submerged jump.

Energy Loss

In many applications the primary function of the hydraulic jump is the dissipation of energy. In a horizontal channel, the change in energy across the jump is

$$\Delta E = E_1 - E_2 \tag{3.2.23}$$

where ΔE = change in energy from section 1 to 2
E_1 = specific energy at section 1
E_2 = specific energy at section 2

Energy loss is commonly expressed as either a relative loss $\Delta E/E_1$ or as an efficiency E_2/E_1.

Rectangular Sections In the case of a horizontal, rectangular channel, it can be demonstrated that

$$\Delta E = \frac{(y_2 - y_1)^3}{4y_1y_2} \tag{3.2.24}$$

that

$$\frac{\Delta E}{E_1} = \frac{2 - 2(y_2/y_1) + \mathbf{F}_1^2[1 - (y_1/y_2)^2]}{2 + \mathbf{F}_1^2} \tag{3.2.25}$$

and that

$$\frac{E_2}{E_1} = \frac{(8\mathbf{F}_1^2 + 1)^{3/2} - 4\mathbf{F}_1^2 + 1}{8\mathbf{F}_1^2(2 + \mathbf{F}_1^2)} \tag{3.2.26}$$

Nonrectangular Sections For other prismatic channel sections, either the general specific energy equation must be solved on a case-by-case basis or a general graphical solution of this equation must be obtained. In terms of relative energy loss, the general specific energy equation is

$$\frac{\Delta E}{E_1} = \frac{y_1 + (\bar{u}_1^2/2g) - y_2 - (\bar{u}_2^2/2g)}{y_1 + (\bar{u}_1^2/2g)}$$

$$= \frac{y_1 - y_2 + (Q^2/2g)[(1/A_1^2) - (1/A_2^2)]}{y_1 + (Q^2/2gA_1^2)} \tag{3.2.27}$$

Silvester (1964, 1965) rearranged Eq. (3.2.27) to yield

$$\frac{\Delta E}{E_1} = \frac{2y_1/D_1\,[1 - (y_2/y_1)] + \mathbf{F}_1^2[1 - (A_1/A_2)^2]}{(2y_1/D_1) + \mathbf{F}_1^2} \tag{3.2.28}$$

The values of y_2/y_1 for any specified $\mathbf{F}_1$ can be obtained from Fig. 3.4, the values of the parameter $[1 - (A_1/A_2)^2]$ from Table 3.1, and the values of the hydraulic depth D from Table 1.1. With these values, a generalized graphical solution of Eq. (3.2.28) can be found (Fig. 3.6). In this figure, the relative energy loss for a hydraulic jump in a rectangular channel is plotted for comparison. Although a general graphical solution of Eq. (3.2.28) can be obtained for a circular channel,

TABLE 3.1 Formulations of the term $1 - (A_1/A_2)^2$ for use in Eq. (3.2.28)

Prismatic channel section	Depth-dependent formulation of $1 - \left(\frac{A_1}{A_2}\right)^2$
Triangular	$1 - \left(\frac{y_1}{y_2}\right)^4$
Parabolic	$1 - \left(\frac{y_1}{y_2}\right)^3$
Trapezoidal*	$1 - \left(\frac{k+1}{k+\frac{y_2}{y_1}}\right)\left(\frac{y_1}{y_2}\right)^2$

$$*k = \frac{b}{zy_1}$$

in most cases it is much simpler to solve for the energy loss on a case-by-case basis.

With regard to Fig. 3.6, it is noted that all channel shapes yield a greater relative energy loss than the rectangular channel for a specified Froude number. This result is not unexpected since channels with sloping sides provide for a

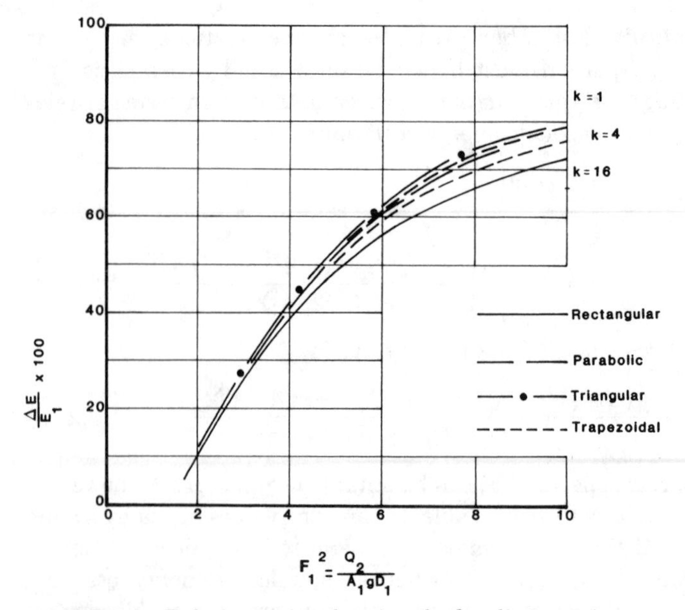

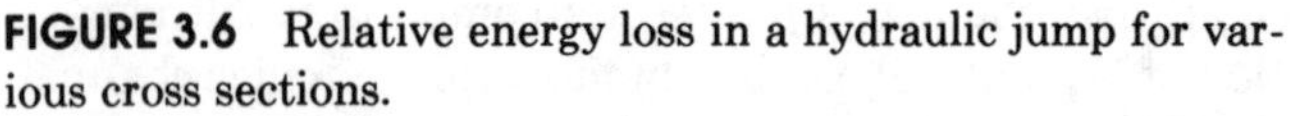

FIGURE 3.6 Relative energy loss in a hydraulic jump for various cross sections.

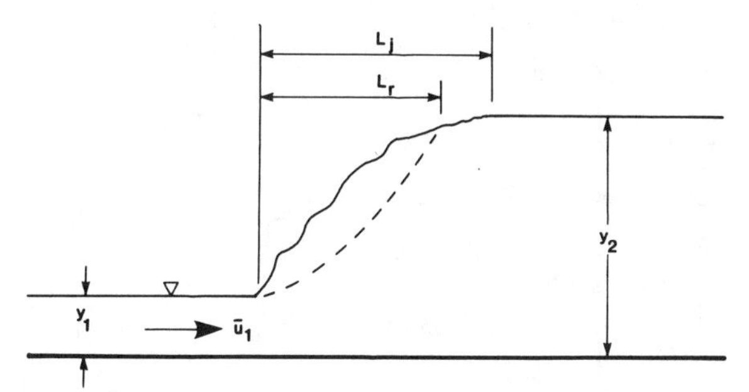

FIGURE 3.7 Schematic definition of jump and roller length.

degree of secondary circulation and, hence, increased energy dissipation. Silvester (1965) noted that for a given flow rate with upstream velocities and depths being equal, the triangular channel provides the greatest energy dissipation.

Hydraulic Jump Length

Although the length of a hydraulic jump is a crucial design parameter, in general it cannot be derived from theoretical considerations; the results of several experimental investigations (see, for example, Bakhmeteff and Matzke, 1936, Bradley and Peterka, 1957a, Peterka, 1963, and Rajaratnam, 1965) have yielded results which are to some degree contradictory. In this section, the length of the hydraulic jump L_j is defined to be the distance from the front face of the jump to a point on the surface of the flow immediately downstream of the roller associated with the jump (Fig. 3.7). In estimating this length, a number of special cases must be considered.

Rectangular Sections In the case of a classic hydraulic jump occurring in a horizontal rectangular channel, the distance L_j is usually estimated from the Bradley and Peterka (1957a) curve which is a plot of $\mathbf{F}_1$ versus L_j/y_2 (Fig. 3.8). This curve has a flat section in the range of Froude numbers yielding the best performance, and its validity is supported by the data of Rajaratnam (1965). It is noted that there is a marked difference between this curve and the one derived from the Bakhmeteff and Matzke (1936) data. Both Bradley and Peterka (1957a) and Chow (1959) ascribed this difference to the scale effect involved in the Bakhmeteff and Matzke (1936) data; i.e., the prototype behavior was not faithfully reproduced by the model used by these investigators.

The data for the length of the roller L_r result in three curves which are significantly different from each other (Fig. 3.8). From this figure, it is concluded that L_r is always less than L_j and the curve derived from the Rajaratnam (1965) data lies between the curves defined by the data of Rouse et al. (1959) and Safranez (1934).

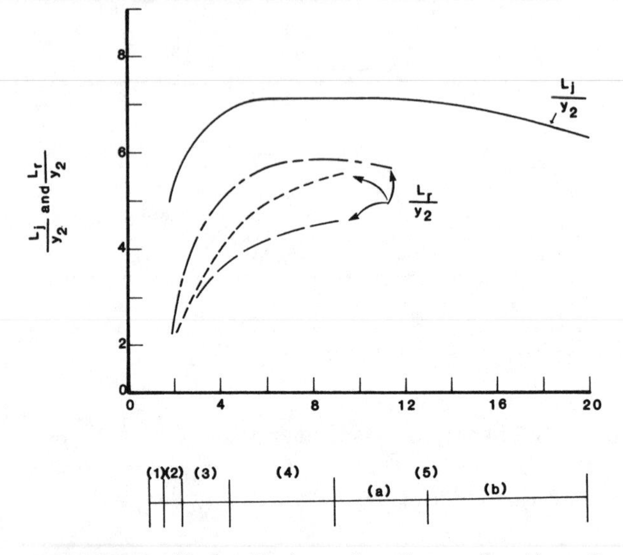

FIGURE 3.8 Hydraulic jump length as a function of sequent depth and upstream Froude number. Legend: 1, undular jump; 2, weak jump (surface tubulence only); 3, oscillating jump (wavy); 4, steady jump (best performance); 5, strong jump—(*a*) acceptable performance, (*b*) extensive silling basin and rough surface conditions. L_j = jump length; L_r = roller length. *(Bradley and Peterka, 1957a; Rouse et al., 1959; Rajaratnam, 1967; Safranez, 1934.)*

Silvester (1964) has demonstrated that for horizontal, rectangular channels the ratio L_j/y_1 is a function of the upstream supercritical Froude number.

$$\frac{L_j}{y_1} = 9.75(\mathbf{F}_1 - 1)^{1.01} \tag{3.2.29}$$

Although this functional relationship was noted in the original work of Bradley and Peterka (1957a), it was apparently not widely used because of the convenience of Fig. 3.8.

Nonrectangular Sections Silvester (1964) has hypothesized that a functional relationship exists between the ratio L_j/y_1 for prismatic channels of any shape or

$$\frac{L_j}{y_1} = \sigma(\mathbf{F}_1 - 1)^{\Gamma} \tag{3.2.30}$$

where σ and Γ are both shape factors. Although the relations presented in the material which follows have not been verified by extensive laboratory and field measurements, they provide an approximate method of estimating L_j for various channel shapes.

For triangular channels, the relationship developed by Silvester (1964) is

$$\frac{L_j}{y_1} = 4.26(\mathbf{F}_1 - 1)^{0.695} \tag{3.2.31}$$

This equation is based on data from a single channel which had an included angle of 47.3° (Argyropoulos, 1957) and probably cannot be expected to be valid in triangular channels whose side slopes are either steeper or milder.

For parabolic channels, Silvester's analysis of the Argyropoulos (1957) data yielded

$$\frac{L_j}{y_1} = 11.7(\mathbf{F}_1 - 1)^{0.832} \tag{3.2.32}$$

This equation provided a very poor fit of the available data when $\mathbf{F}_1$ exceeded 3.

In the case of trapezoidal sections, a number of different data sets were available to Silvester, e.g., Hsing (1937), Sandover and Holmes (1962), and Press (1961). In a previous section of this chapter, it was demonstrated that for trapezoidal channels the ratio y_2/y_1 is dependent not only on $\mathbf{F}_1$ but also on a shape factor k. Thus, it might be expected that for each value of k there is a unique set of values of ϕ and Γ. In the case of Press (1961) data, this was true, and the results for this data set are summarized in Table 3.2. In the case of Hsing (1937) data, the value of ϕ had no apparent effect, and these data appeared to be affected by the value of z. The Hsing data also exhibited variations in the value of $\mathbf{F}_1$ for constant values of L_j, which was probably the result of experimental difficulties in measuring the distance L_j under laboratory conditions. Sandover

TABLE 3.2 Definition of hydraulic jump length parameters σ and Γ for trapezoidal channels (after Press, 1961)

Side slope z	Shape factor k*	σ, ft	Γ
2	16	17.6	0.905
1	8	23.0	0.885
0.5	4	35.0	0.836

$*k = \dfrac{b}{zy}$

and Holmes (1962) found no correlation in their data between L_j and $\mathbf{F}_1$. Thus, the values of ϕ and Γ used to estimate L_j in the case of trapezoidal channels are based entirely on the data of Press (1961).

A salient feature of hydraulic jumps noted by both Hsing (1937) and Press (1961) was that the reverse flow of the roller caused a depth increase at the sides of the channel and a corresponding decrease in the depth of flow along the centerline of the jump. This traverse variation in the depth of flow could result in problems in determining accurately the length of the jump.

The results available for circular channels from Kindsvater (1936) were limited and not sufficient to draw useful conclusions.

Submerged Jump Length In the case of a submerged hydraulic jump (Fig. 3.5), the distance L_j is estimated by the empirical equation

$$\frac{L_j}{y_2} = 4.9\,S + 6.1 \tag{3.2.33}$$

where S = submergence factor defined by Eq. (3.2.20) (Govinda Rao, 1963). Equation (3.2.33) demonstrates that the length of a submerged jump exceeds the length of the corresponding free jump by the term 4.9 S and further that L_j is directly proportional to S. Stepanov (1959) found that the length of the roller in the case of a submerged jump can be estimated by

$$\frac{L_r}{y_c} = \frac{3.31}{[(y_4 - y_3/y_3\mathbf{F}_1)]^{0.885}} \tag{3.2.34}$$

This equation has been shown to be valid for $S \le 2$ and $1 \le \mathbf{F}_1 \le 8$ (Rajaratnam, 1967).

Hydraulic Jumps in Sloping Channels

As noted in an earlier section of this chapter, hydraulic jumps can and do occur in channels of such a slope that the gravitational forces acting on the flow must be considered. Although Eq. (3.1.1) is theoretically applicable to such problems, in practice the number of solutions available is limited. The primary difficulties in obtaining a useful solution to this problem are: (1) The term $W \sin\theta$ is generally poorly quantified because the length and shape of the jump are not well defined, (2) the specific weight of the fluid in the control volume may change significantly owing to air entrainment, and (3) the pressure terms cannot be accurately quantified.

According to Rajaratnam (1967), the earliest experiments done by Bidone on the hydraulic jump were performed in what would now be termed a sloping channel. Bazin in 1865 and Beebe and Riegel in 1917 also addressed this problem. In 1934, Yarnell began an extensive study of the hydraulic jump in sloping channels which was not completed because of his untimely death in 1937.

Kindsvater (1944), using the unpublished Yarnell data, was the first investigator to develop a rational solution to the problem. Extensive studies have also been conducted by Bradley and Peterka (1957a) and Argyropoulos (1962).

In discussing the equations and relationships available for hydraulic jumps in sloping channels, it is convenient to consider a number of cases (Fig. 3.9). It is also noted that, in the material which follows, *the end of the jump is by definition at the end of the surface roller.* The reader is cautioned that this definition differs from that used in the previous sections of this chapter. Then, with regard to the cases defined in Fig. 3.9, let y_t = tailwater depth, L_r = length of the jump measured horizontally, y_1 = the supercritical depth of flow on the slope which is assumed to be constant, y_2 = subcritical sequent depth corresponding to y_1, and y_2^* = subcritical depth given by Eq. (3.2.4) (see Fig. 3.10). If the jump begins at the end of the sloping section, then $y_2^* = y_t$ and a type A jump, governed by Eq. (3.2.4), occurs. If the end of the jump coincides with the intersection of the sloping and horizontal beds, a type C jump occurs. If y_t is less than that required for a type C jump but greater than y_2^*, the toe of the jump is on the slope and the end on the horizontal bed. This situation is termed a type B jump. If y_t is greater than that required for a type C jump, then a type D jump occurs completely on the sloping section. Type E jumps occur on sloping beds which have no break in slope, and the rare type F jump occurs only in stilling basins normally found below drop structures. Of the six types of jumps defined in Fig. 3.9, types A to D are the most common and will be discussed first.

For the type C jump, Kindsvater (1944) developed the following equation for the sequent depth:

$$\frac{y_2}{y_1} = \frac{1}{2\cos\theta}\left[\sqrt{1 + 8\mathbf{F}_1^2\left(\frac{\cos^3\theta}{1 - 2N\tan\theta}\right)} - 1\right] \tag{3.2.35}$$

where N = an empirical factor related to the length of the jump and θ = the longitudinal slope angle of the channel. If the following parameters are defined

$$\Gamma_1^2 = \frac{\cos^3\theta}{1 - 2N\tan\theta} \tag{3.2.36}$$

and

$$G_1^2 = \Gamma_1^2\,\mathbf{F}_1^2 \tag{3.2.37}$$

Eq. (3.2.37) can be written as

$$\frac{y_2}{y_1'} = \frac{1}{2}\left(\sqrt{1 + 8G_1^2} - 1\right) \tag{3.2.38}$$

where $y_1' = y_1/\cos\theta$. Bradley and Peterka (1957a) and Peterka (1963) found that N was primarily a function of θ, and Rajaratnam (1967) stated the following relationship

$$\Gamma_1 = 10^{0.027\theta} \tag{3.2.39}$$

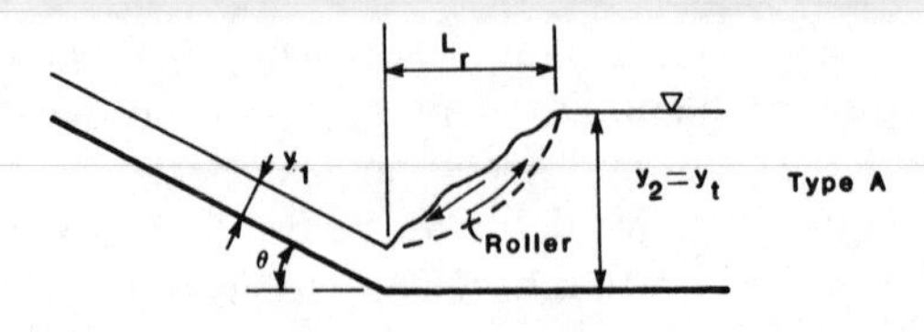

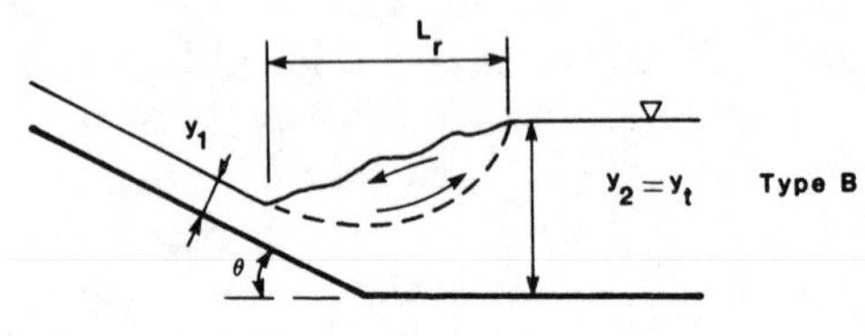

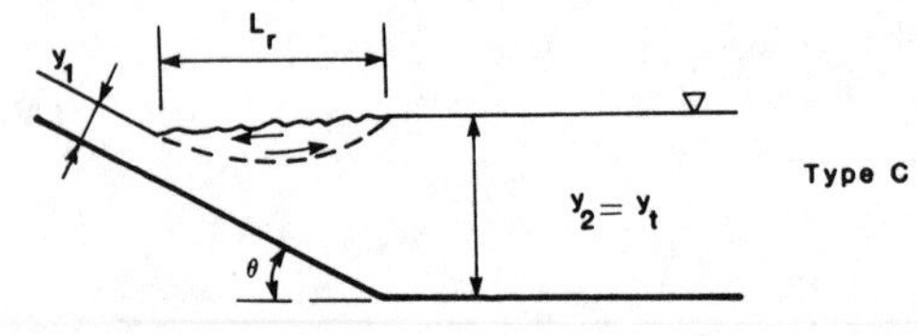

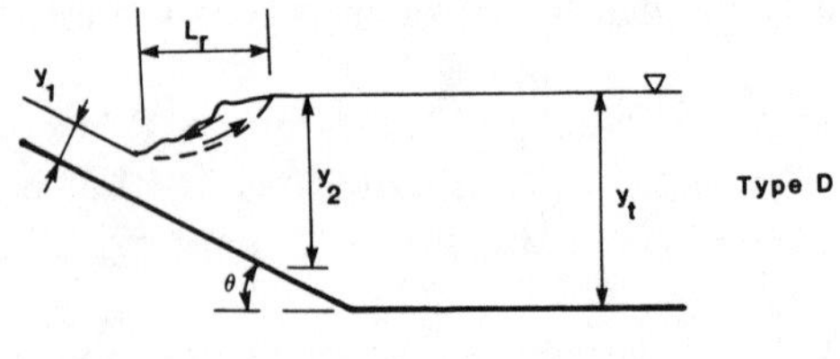

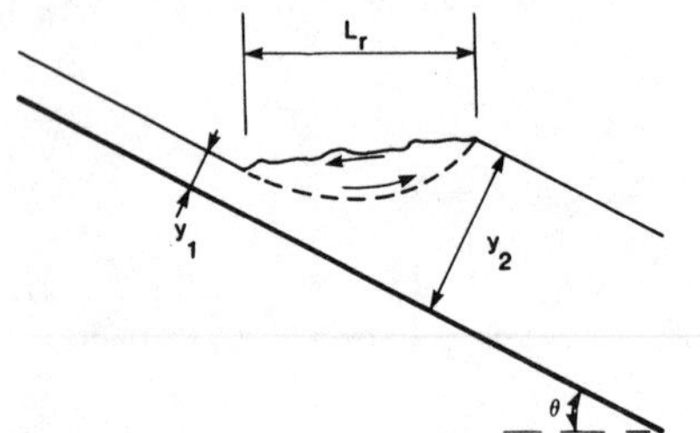

FIGURE 3.9 Definition of types of hydraulic jumps that occur in sloping channels.

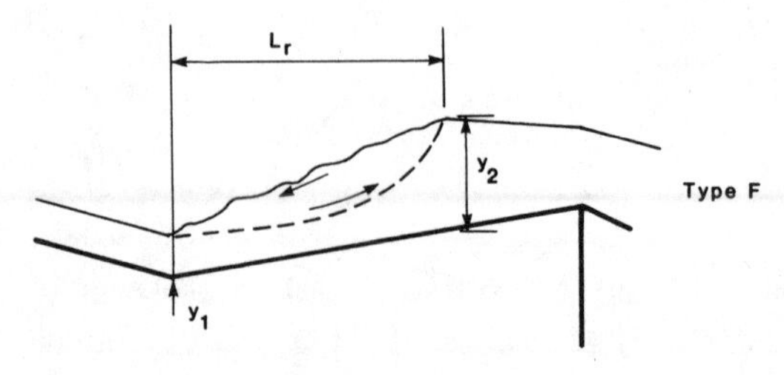

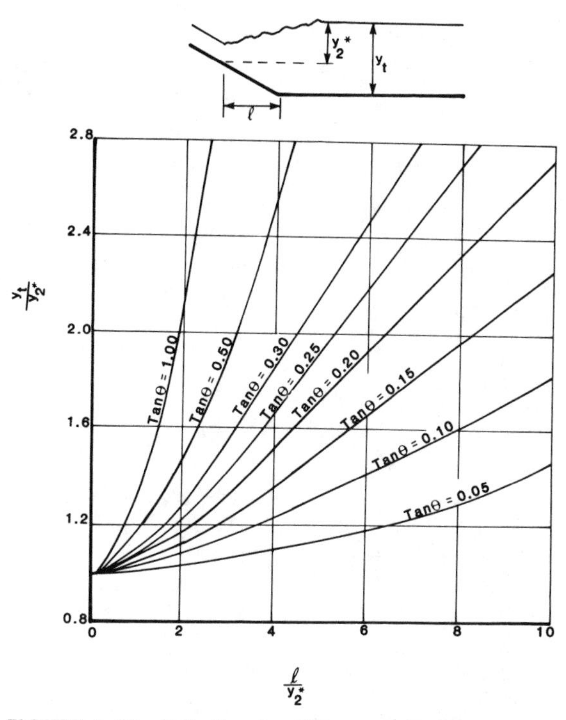

FIGURE 3.10 Solution for B jumps. *(Peterka, 1963, and Rajaratnam, 1967.)*

where θ is in degrees. In addition to the type C jump, Bradley and Peterka (1957a) and Peterka (1963) also found that Eqs. (3.2.37) to (3.2.39) could be applied to the type D jump.

Although an analytic solution for the type B jump has not yet been developed, Bradley and Peterka (1957a) and Peterka (1963) have developed a graphical solution for this type of jump based on laboratory experiments. These data are summarized in Fig. 3.10. In this figure, the original data have been supplemented with two additional curves, $\tan \theta = 0.50$ and 1.0, abstracted from Rajaratnam (1967).

The initial step in solving jump problems where the slope of the channel must be considered is to classify the jump which occurs as to type given the slope, initial supercritical flow depth, and the tailwater condition. If it is assumed that the depth of the supercritical stream is constant on the sloping bed—an assumption which closely approximates conditions which have been found to exist at the base of high dams and below sluices acting under high heads—a relatively simple jump classification procedure can be developed as shown in Fig. 3.11.

Bradley and Peterka (1957a) and Peterka (1963) also presented a number of results regarding the length of the type D jump, and these results are summarized in Fig. 3.12. This graph can also be used to estimate the lengths of type B and C jumps.

The energy loss for the type A jump can be estimated from the equations derived for a horizontal channel, e.g., Eq. (3.2.24). For jump types C and D an analytic expression can be derived. In Fig. 3.9, if the bed level at the end of the jump is taken as the datum, then the upstream energy is

$$E_1 = L_r \tan\theta + \frac{y_1}{\cos\theta} + \frac{\bar{u}_1^2}{2g} \tag{3.2.40}$$

where L_r = length of the jump and $\bar{u}_1$ = supercritical upstream velocity. The energy at the end of the jump is

$$E_2 = y_2 + \frac{\bar{u}_2^2}{2g} \tag{3.2.41}$$

where $\bar{u}_2$ = subcritical velocity at the downstream section. Then the relative energy loss is given by

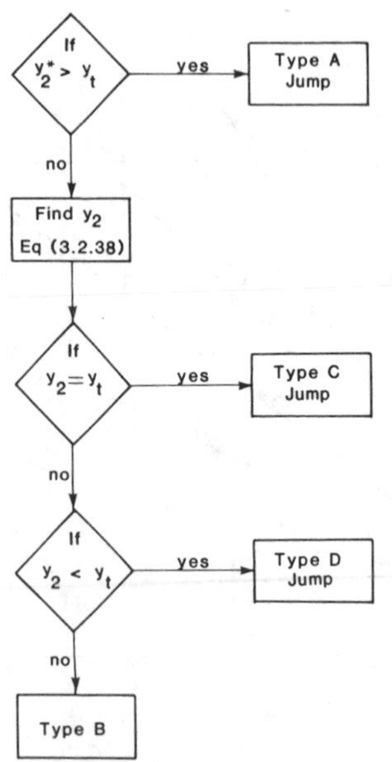

FIGURE 3.11 Determination of hyraulic jump type in sloping channels.

$$\frac{\Delta E}{E_1} = \frac{[1 - (y_2/y_1)] + \mathbf{F}_1^2/2\{1 - [1/(y_2/y_1)]^2\} + (L_r y_t/y_t y_1)\tan\theta}{1 + (\mathbf{F}_1^2/2) + (L_r y_t/y_t y_1)\tan\theta} \tag{3.2.42}$$

where it has been assumed that $y_1 \simeq y_1/\cos\theta$. In general, Eq. (3.2.42) should not be used in situations where $\mathbf{F}_1$ is less than 4 since in this range there is only very limited information regarding the ratio L_r/y_2 and in this range $\Delta E/E_1$ is sensitive to this ratio.

The computation of the energy loss for the type B jump is much more involved, and it is best illustrated with an example.

EXAMPLE 3.3

Given a rectangular channel 4.0 ft (1.2 m) wide and inclined at an angle of 3° with the horizontal, determine the jump type if Q = 5.0 ft^3/s (0.14 m^3/s), y_1 = 0.060 ft (0.018 m), and y_t = 1.4 ft (0.43 m).

Solution

At section 1

$$A_1 = by_1 = 4(0.06) = 0.24 \text{ ft}^2\ (0.022 \text{ m}^2)$$

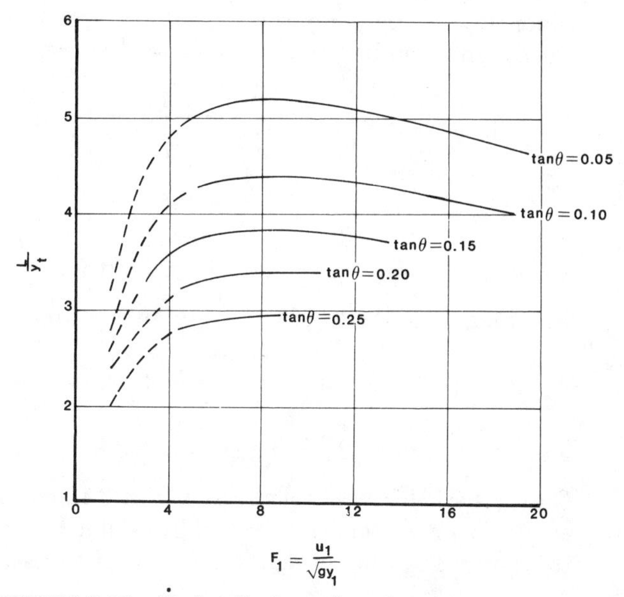

FIGURE 3.12 Hydraulic jump length in sloping channels for jump types B, C, and D. *(Peterka, 1963.)*

$$\overline{u}_1 = \frac{Q}{A_1} = \frac{5}{0.24} = 20.8 \text{ ft/s } (6.3 \text{ m/s})$$

$$\mathbf{F}_1 = \frac{\overline{u}_1}{\sqrt{gy_1}} = \frac{20.8}{\sqrt{32.2(0.06)}} = 15$$

The sequent depth in a horizontal channel y_2^* can be computed from Eq. (3.2.4).

$$y_2^* = \frac{y_1}{2}\left(\sqrt{1 + 8\mathbf{F}_1^2} - 1\right) = \frac{0.06}{2}\left[\sqrt{1 + 8(15)^2} - 1\right] = 1.2 \text{ ft } (0.37 \text{ m})$$

The jump can then be classified by the following scheme defined in Fig. 3.11. Since $y_t > y_2^*$, the depth y_2 must be calculated from Eqs. (3.2.37) to (3.2.39).

$$\Gamma_1 = 10^{0.027\theta} = 10^{0.027(3)} = 1.2$$

$$G_1^2 = \Gamma_1^2\mathbf{F}_1^2 = (1.2)^2(15)^2 = 324$$

$$y_2 = \frac{y_1}{2\cos\theta}\left(\sqrt{1 + 8G_1^2} - 1\right) = \frac{0.06}{2\cos 3}\left[\sqrt{1 + 8(324)} - 1\right]$$

$$= 1.5 \text{ ft } (0.46 \text{ m})$$

Since $y_2 > y_t$, the jump is classified as a type B jump. From Fig. 3.10, the distance can be found for $\tan\theta = 0.05$ and

$$\frac{y_t}{y_2^*} = \frac{1.4}{1.2} = 1.2$$

$$\frac{\ell}{y_2^*} = 6.5$$

and

$$\ell = 1.2(6.5) = 7.8 \text{ ft } (2.4 \text{ m})$$

From Fig. 3.12, the length of the jump is estimated

$$\frac{L_r}{y_t} = 4.9$$

and

$$L_r = 4.9\,(1.4) = 6.9 \text{ ft } (2.1 \text{ m})$$

On the basis of these calculations, Fig. 3.13 can be constructed. The energy loss can then be estimated by establishing the horizontal bed as the datum and computing the total energies at sections 1 and 2 or

$$E_1 = \ell\tan\theta + \frac{y_1}{\cos\theta} + \frac{\bar{u}_1^2}{2g} = 7.8\tan(3) + \frac{0.06}{\cos(3)} + \frac{(20.8)^2}{2(32.2)}$$

$$= 7.2 \text{ ft } (2.2 \text{ m})$$

Then at section 2

$$\bar{u}_2 = \frac{Q}{A_2} = \frac{5}{1.4(4)} = 0.89 \text{ ft/s } (0.27 \text{ m/s})$$

$$E_2 = y_2 + \frac{\bar{u}_2^2}{2g} = 1.4 + \frac{(0.89)^2}{2(32.2)} = 1.4 \text{ ft } (0.43 \text{ m})$$

and

$$\Delta E = E_1 - E_2 = 7.2 - 1.4 = 5.8 \text{ ft } (1.8 \text{ m})$$

or

$$\frac{\Delta E}{E_1} = \frac{5.8}{7.2} = 0.81$$

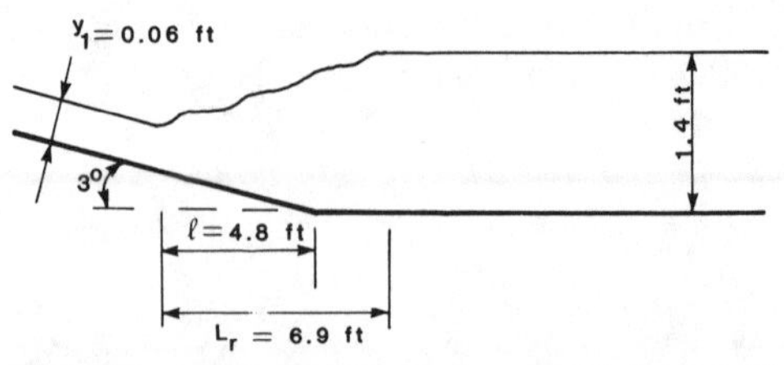

FIGURE 3.13 Results of type B hydraulic jump analysis.

These calculations indicate that the energy losses in this jump are quite high; however, these results should be used cautiously since they are subject to a number of assumptions.

EXAMPLE 3.4

Given a rectangular channel 6.1 m (20 ft) wide and inclined at an angle of 3° with the horizontal, determine the jump type if $Q = 9.0$ m^3/s (320 ft^3/s), $y_t = 2.6$ m (8.5 ft), and $y_1 = 0.09$ m (0.30 ft).

Solution

At section 1

$$A_1 = by_1 = 6.1(0.09) = 0.55 \text{ m}^2 \ (5.9 \text{ ft}^2)$$

$$\overline{u}_1 = \frac{Q}{A_1} = \frac{9.0}{0.55} = 16 \text{ m/s } (52 \text{ ft/s})$$

$$\mathbf{F}_1 = \frac{\overline{u}_1}{\sqrt{gy_1}} = \frac{16}{\sqrt{9.8(0.09)}} = 17$$

The depth y_2^* can be computed from Eq. (3.2.4).

$$y_2^* = \frac{y_1}{2}(\sqrt{1 + 8\mathbf{F}_1^2} - 1) = \frac{0.09}{2}[\sqrt{1 + 8(17)^2} - 1] = 2.1 \text{ m } (6.9 \text{ ft})$$

The jump can then be classified by following the scheme defined in Fig. 3.11. Since $y_t > y_2^*$, the depth of y_2 must be calculated from Eqs. (3.2.37) and (3.2.39).

$$\Gamma = 10^{0.027\theta} = 10^{0.027(3)} = 1.2$$

$$G_1^2 = \Gamma_1 \mathbf{F}_1^2 = (1.2)^2(17)^2 = 416$$

$$y_2 = \frac{y_1}{2 \cos \theta}(\sqrt{1 + 8G_1^2} - 1) = \frac{0.09}{2 \cos (3°)}[\sqrt{1 + 8(445)} - 1]$$

$$= 2.6 \text{ m } (8.5 \text{ ft})$$

Then since $y_2 \sim y_t$, the jump is classified as the type C jump. From Fig. 3.12, the distance L_r can be found for $\tan \theta = 0.05$ and $\mathbf{F}_1 = 17$

$$\frac{L_r}{y_t} = 4.8$$

and

$$L_r = 4.8\ (2.6) = 12 \text{ m } (39 \text{ ft}).$$

The relative energy loss is found from Eq. (3.2.42)

$$\frac{\Delta E}{E_1} = \frac{[1 - (y_2/y_1)] + \mathbf{F}_1^2/2\left[1 - \dfrac{1}{(y_2/y_1)^2}\right] + \tan\theta\,(L_r y_t/y_t y_1)}{1 + (\mathbf{F}_1^2/2) + (L_r y_t/y_t y_1)\tan\theta}$$

with

$$\frac{y_2}{y_1} = \frac{2.6}{0.09} = 29$$

and

$$\frac{L}{y_t} = 4.8$$

$$\frac{y_t}{y_1} = 29$$

Then

$$\frac{\Delta E}{E_1} = \frac{(1 - 29) + (17^2/2)(1 - 1/29^2) + \tan(3°)[4.8(29)]}{1 + 17^2/2 + 4.8(29)\tan(3°)}$$

$$= 0.76$$

The analysis of the jump designated as type E (Fig. 3.9) begins with Eq. (3.1.1). For a rectangular channel, if it is assumed that (first) $P_f = 0$, (second) $\beta_1 = \beta_2 = 1$, (third) $\Sigma f_f' = 0$, and (fourth) the weight of the water between sections 1 and 2 is given by

$$W = \frac{A_1 + A_2}{2}\frac{L_r}{\cos\theta}\gamma k'$$

where $L_r/\cos\theta$ can be approximated as

$$\frac{L_r}{\cos\theta} = L' = X(y_2 - y_1) \tag{3.2.43}$$

X = a slope factor and k' = a correction factor which results from the assumption that the jump profile is linear. Equation (3.1.1) can then be solved to yield

$$\frac{y_2}{y_1} = \frac{1}{2}(\sqrt{1 + 8M^2} - 1) \tag{3.2.44}$$

where

$$M = \frac{\mathbf{F}_1}{\sqrt{1 - [k'L'\sin\theta/(y_2 - y_1)]}}$$

Equation (3.2.44) has been derived independently by Kennison (1944), Chow (1959), and Argyropoulos (1962). Although for simplicity the value of k' can be assumed to be unity, the variation of M with X cannot be ignored except for small values of θ (Rajaratnam, 1967). Kennison (1944) suggested that X could be assumed to have a value of 3. Argyropoulos (1962) and Chow (1959) noted that X was a function of $\mathbf{F}_1$, but neither of these investigators suggests an appropriate value.

As noted previously, the type F jump is very rare, and there are few data available for this type of jump. Rajaratnam (1967) noted that in practice it is almost impossible to keep this jump completely on the reverse slope. If this type of jump is to be used in a design, its characteristics should be determined from model tests.

3.3 HYDRAULIC JUMPS AT DENSITY INTERFACES

It has been demonstrated in laboratory experiments (see, for example, Yih and Guha, 1955, Hayakawa, 1970, Stefan and Hayakawa, 1972, and Stefan et al., 1971) that hydraulic jumps can also occur at density interfaces within stratified flows. Although this phenomenon—termed an internal hydraulic jump—has not been widely observed in nature except in the atmosphere (Turner, 1973) and in tidal inlets (Gardner et al., 1980), this lack of field observations may result from the fact that this is an internal phenomenon which does not necessarily produce a noticeable effect at the air-water interface.

The types of internal hydraulic jumps which can occur are defined schematically in Fig. 3.14. In the first case (Fig. 3.14*a*), layer 1 passes from an internally supercritical state to a subcritical state by means of an internal hydraulic jump. In the second case (Fig. 3.14*b*), layer 2 passes from an internally supercritical state to an internally subcritical state by means of an internal hydraulic jump. In the discussion of internal hydraulic jumps to come, the following assumptions are required: (1) The channel in which the jump occurs is rectangular and has a width b, (2) the bed of the channel is horizontal, (3) the interfacial shear between the two layers is ignored, (4) there is no mixing between the layers, (5) all pressure distributions are hydrostatic, and (6) $\beta_1 = \beta_2 = 1$. Although assumptions (3) to (5) are crucial to the analysis which follows, the results of the analysis are only qualitatively correct because of these assumptions.

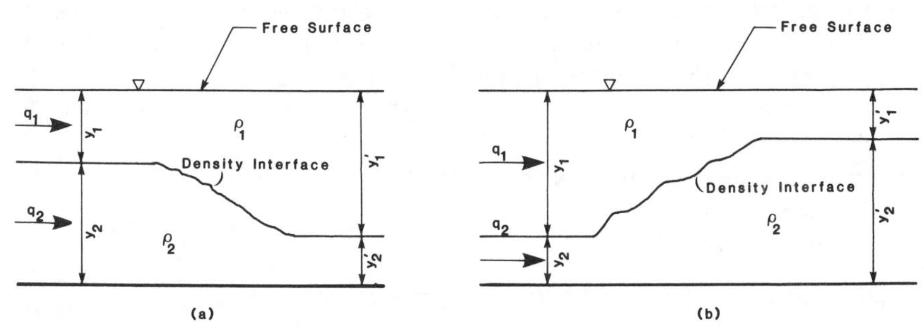

FIGURE 3.14 Internal hydraulic jump definition.

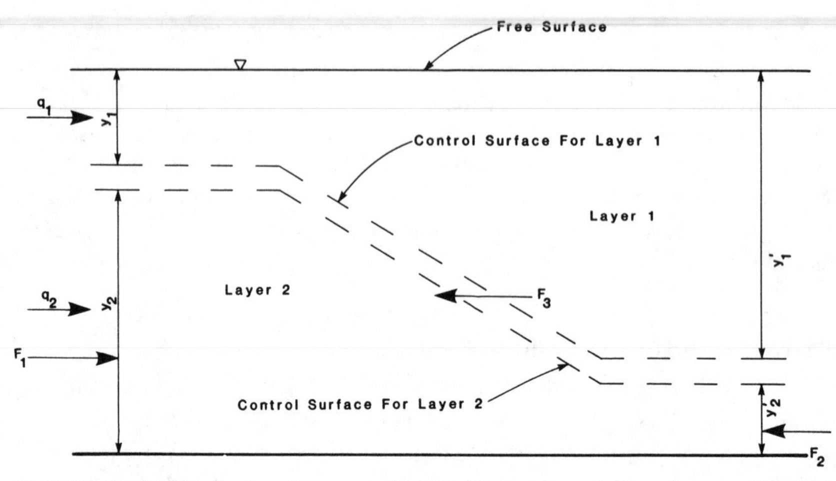

FIGURE 3.15 Control volume and definition of variables for an internal hydraulic jump.

In Fig. 3.15, control volumes of unit width are defined for layers 1 and 2. Considering only layer 2, the upstream hydrostatic force is

$$F_1 = \left[\frac{\gamma_1 y_1 + (\gamma_1 y_1 + \gamma_2 y_2)}{2}\right] y_2 = \gamma_1 y_1 y_2 + \tfrac{1}{2}\gamma_2 y_2^2 \tag{3.3.1}$$

and the corresponding downstream hydrostatic force is

$$F_2 = \gamma_1 y_1' y_2' - \tfrac{1}{2}\gamma_2 (y_2')^2 \tag{3.3.2}$$

where γ_1 and γ_2 = specific weights of fluids in layers 1 and 2, respectively. The force exerted by layer 1 on the layer 2 control volume is given by

$$F_2 = \tfrac{1}{2}(y_1 + y_1')(y_2 - y_2')\gamma_1 \tag{3.3.3}$$

where $\gamma_1[(y_1 + y_1')/2)]$ = mean pressure exerted by layer 1 on the sloping surface separating layers 1 and 2 and $y_2 - y_2'$ = projected vertical unit area of the jump. Application of the one-dimensional form of the conservation of momentum equation to the layer 2 control volume yields

$$F_1 - F_2 - F_3 = \rho_2 q_2(\overline{u}_2 - \overline{u}_1) \tag{3.3.4}$$

where q_2 = flow per unit width in layer 2
$\overline{u}_2 = q_2/y_2'$
$\overline{u}_1 = q_2/y_2$

Rearranging Eq. (3.3.4) and defining $a_2 = q_2^2/g$ and $r = \rho_1/\rho_2$,

$$2a_2(y_2 - y_1') = y_2 y_2'(y_2 + y_2')[r(y_1 - y_1') + (y_2 - y_2')] \tag{3.3.5}$$

In an analogous fashion, the application of the one-dimensional form of the conservation of momentum equation to the layer 1 control volume yields

$$2a_1(y_1 - y_1') = y_2y_2'(y_1 + y_1')[r(y_1 - y_1') + (y_2 - y_2')] \qquad (3.3.6)$$

If y_1, y_2, ρ_1, and ρ_2 are specified, then Eqs. (3.3.5) and (3.3.6) can be solved simultaneously to determine y_1' and y_2' there are at most nine solutions to this set of equations. One valid solution is $y_1' = y_1$ and $y_2' = y_2$, and of the eight remaining solutions which are conjugate to the specified state only those which are real and positive must be considered. Yih and Guha (1955) demonstrated that there can be at most three conjugate states which have a real physical interpretation, and under certain conditions a unique conjugate state exists. The conditions for uniqueness are either $\mathbf{F}_1^2 = a_1/y_1^3$ or $\mathbf{F}_2^2 = a_2/y_2^3$ must be large. The three unique solutions cited by Yih and Guha (1955) are:

1. If the downstream velocities are the same for both layers, then

$$(1 + r\lambda)\left(\frac{y_2'}{y_2}\right)^3 + r\left(\lambda - \frac{y_1}{y_2}\right)\left(\frac{y_2'}{y_2}\right)$$
$$- \left(1 + r\frac{y_1}{y_2} + 2\mathbf{F}_2^2\right)\left(\frac{y_2'}{y_2}\right) + 2\mathbf{F}_2^2 = 0 \qquad (3.3.7)$$

where $\dfrac{q_1}{q_2} = \dfrac{y_1'}{y_2'} = \lambda$. Equation (3.3.7) has two positive roots: $y_2' = y_2$ and another which corresponds to the conjugate state.

2. If $a_1 = 0_1$, layer 1 is at rest; then Eqs. (3.3.5) and (3.3.6) can be solved simultaneousy to yield

$$\frac{y_2'}{y_2} = \frac{1}{2}\left(\sqrt{1 + 8\mathbf{F}_2^2} - 1\right) \qquad (3.3.8)$$

where $\qquad \mathbf{F}_2'^2 = \mathbf{F}_2^2(1 - r)^{-1}$

The similarity of Eq. (3.3.8) with Eqs. (3.2.4) and (3.2.5) should be noted.

3. If $a_2 = 0$, that is, layer 2 is at rest, then the simultaneous solution of Eqs. (3.3.5) and (3.3.6) yields

$$\frac{y_1'}{y_1} = \frac{1}{2}\left(\sqrt{1 + 8\mathbf{F}_1'^2} - 1\right) \qquad (3.3.9)$$

where $\qquad \mathbf{F}_2'^2 = \mathbf{F}_1^2(1 - r)^{-1}$

With regard to these solutions, in the case $a_1 = 0$ and $a_2 \neq 0$, Eq. (3.3.6) reduces to

$$y_1 + y_2 = y_1' + y_2'$$

This result implies that the free surface is level and there is no evidence of the internal jump at the free surface. If $a_2 = 0$ and $a_1 \neq 0$, then from Eq. (3.3.6)

$$r(y_1' - y_1) = y_2 - y_2'$$

and the free surface does not remain level.

In the foregoing material it was explicitly assumed that fluids composing layers 1 and 2 did not mix. Stefan and Hayakawa (1972) and Stefan et al. (1971) have considered both theoretically and experimentally the case in which mixing occurs. Figure 3.16 defines the situation considered by these authors. The situation defined in Fig. 3.16 is analogous to cooling water from a thermal power plant entering a reservoir. With reference to Figure 3.16 and considering only the horizontal components of the fluxes and forces, the one-dimensional momentum equation is

$$\frac{\beta_1 \rho_1 Q_1^2}{(y_1 - w)} - \frac{\beta_0 \rho_0 Q_0^2}{y_0} + \frac{\beta_2 \rho_2 Q_2^2}{y_2}$$
$$= \tfrac{1}{2}\rho_0 g y_0^2 - \tfrac{1}{2}\rho_1 g (y_1 - w)^2 - (y_1 - w)\rho_1 y_2 - \tfrac{1}{2} g \rho_2 y_2^2 + P_s + F_s \qquad (3.3.10)$$

where the subscript 0 refers to variable values upstream of the jump, the subscript 1 refers to variable values at the end of the jump in layer 1, the subscript 2 refers to variable values at the end of the jump in the lower layer or layer 2, P_s and F_s are the horizontal components of force resulting from the normal and tangential forces, respectively, Q = volumetric flow rate, and all other variables have either been previously defined or are defined in Fig. 3.16. For the situation defined, the conservation of mass equation is

$$\rho_0 q_0 = \rho_1 q_1 - \rho_2 q_2 \qquad (3.3.11)$$

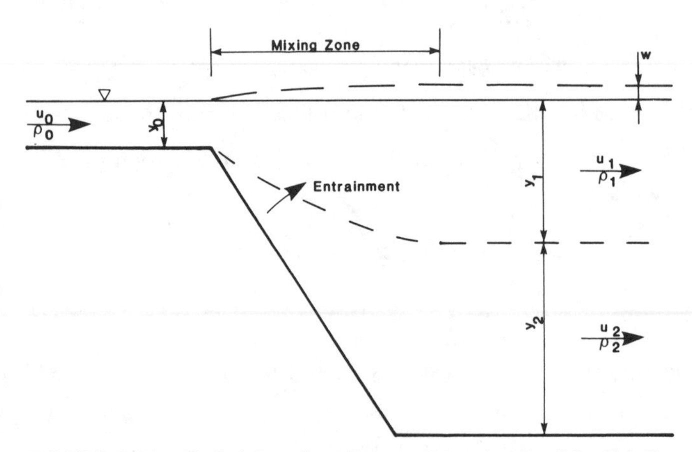

FIGURE 3.16 Definition sketch for mixing internal hydraulic jump.

and the conservation of volume equation is

$$q_0 + q_2 = q_1 \tag{3.3.12}$$

From Eqs. (3.3.11) and (3.3.12) an equation of continuity for the density excess can be derived or

$$\Delta\rho_0 q_0 = \Delta\rho_1 q_1 \tag{3.3.13}$$

where $\Delta\rho_0 = \rho_2 - \rho_0$ and $\Delta\rho_1 = \rho_2 - \rho$. Then, if the analysis is restricted to a case where $\theta = 90°$, the tangential stress at the channel bottom is neglected, and a hydrostatic pressure distribution along the vertical slope is assumed, Eq. (3.3.10) becomes

$$\frac{\beta_1\rho_1 Q_1^2}{y_1 - w} - \frac{\beta_0\rho_0 Q_0^2}{y_0} + \frac{\beta_2\rho_2 Q_2^2}{y_2} = g\rho_0 y_0\left(y_1 + y_2 - \frac{y_0}{2}\right)$$
$$+ \tfrac{1}{2}g\rho_2(y_1 + 2y_2 - y_0)(y_1 - y_0) - \tfrac{1}{2}g\rho_1(y_1 - w)(y_1 + 2y_2 - w) \tag{3.3.14}$$

Equations (3.3.12) to (3.3.14) contain four unknowns, q_1, q_2, ρ_1, and w, and an additional equation is required. Stefan and Hayakawa (1972) assumed that

$$gy_0\rho_0 + g(y_1 - y_0)\rho_2 = g(y_1 - w)\rho_1 \tag{3.3.15}$$

which is equivalent to assuming that the hydrostatic pressure in the lower layer is equal in any horizontal plane. This assumption is exactly satisfied if the internal hydraulic jump results in no mixing at the density interface or if y_2 is equal to infinity. In terms of dimensionless variables, Eq. (3.3.15) becomes

$$\frac{w}{y_0} = r\left(\frac{E - H_1}{E - r}\right) \tag{3.3.16}$$

where $E = Q_1/Q_0$, $H_1 = y_1/y_0$, and $r = \Delta\rho_0/\rho_2$. In terms of dimensionless variables Eq. (3.3.14) becomes

$$2\beta_2\mathbf{F}_0^2 = \frac{\mathrm{A}}{\mathrm{B}}$$

where $A = H_2(H_1 - r)[(H_1 - r)^2 - (1 - \mathrm{r})(E - r)]$

$B = (E - r)\ [(\beta_0/\beta_2)(1 - r)\ H_2 - (E - 1)^2]\ \{H_1 - r - [\beta_1 H_2(E - r)^2]/[\beta_0(1 - r)H_2 - \beta_2(E - 1)^2]\}$

$H_2 = y_2/y_0$ and

$$\mathbf{F}_0 = \frac{Q_0}{[(\Delta\rho_0/\rho_2)gy_0^3]^{1/2}}$$

For the case where $\beta_0 = \beta_1 = \beta_2 = 1$ and

$$0 < r < 10^{-2}$$

Stefan and Hayakawa (1972) solved Eq. (3.3.17) (Fig. 3.17). In this figure lines representing a constant downstream depth of flow H_1 and a constant down-

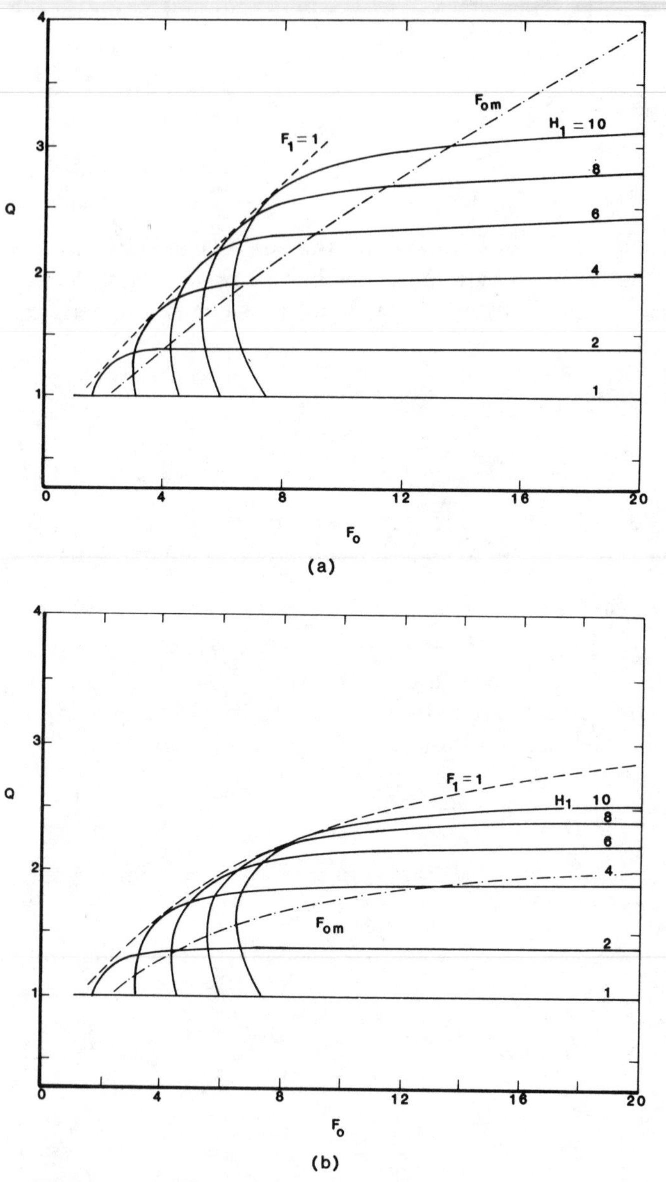

FIGURE 3.17 (*a*) Q as a function of $\mathbf{F}_0$ for $H_2 = \infty$. *(Stefan and Hayakawa, 1972.)* (*b*) Q as a function of $\mathbf{F}_0$ for $H_2 = 8$. *(Stefan and Hayakawa, 1972.)*

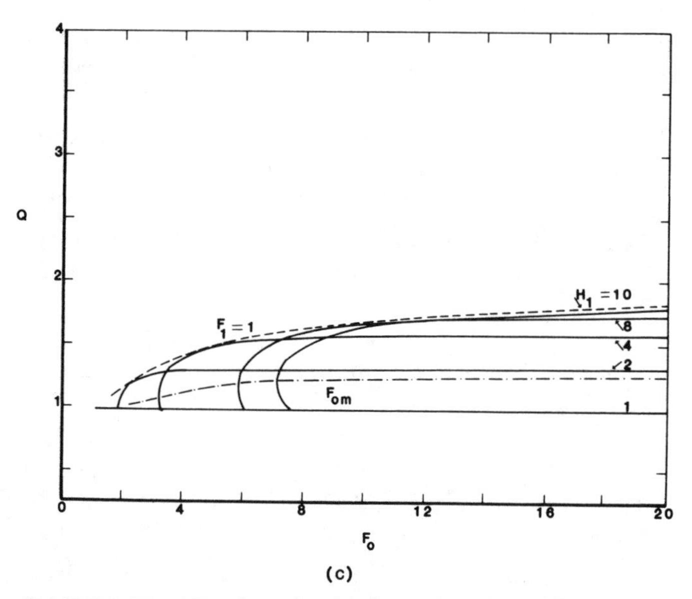

FIGURE 3.17 *(Continued)* (*c*) Q as a function of $\mathbf{F}_0$ for $H_2 = 1$. *(Stefan and Hayakawa, 1972.)*

stream densimetric Froude number $\mathbf{F}_1$ are also shown where

$$\mathbf{F}_1 = \frac{Q_1}{\sqrt{(\Delta\rho_1/\rho_2)g(y_1 - w)^3}}$$

The value of $\mathbf{F}_1$ indicates whether the flow downstream is internally subcritical or supercritical. $\mathbf{F}_1 = 1$ represents the limit of internally subcritical flows; $\mathbf{F}_{0m}$ is the minimum outlet densimetric Froude number for a given downstream dimensionless depth H_1. Figure 3.17 also demonstrates that the amount of total entrainment is limited by the thickness of the upper and lower layers. It should be noted that the theoretical results summarized in Fig. 3.17 have been confirmed with laboratory measurements.

At this point it should be noted that the study of internal hydraulic jumps is an area in which basic research is still being conducted. While it is clear that the fundamental concepts of open-channel hydraulics are applicable, further research in this area is required before the problem can be addressed in a comprehensive fashion.

BIBLIOGRAPHY

Argyropoulos, P. A., "Theoretical and Experimental Analysis of the Hydraulic Jump in a Parabolic Flume," *Proceedings of the 7th Conference, International Associaton for Hydraulic Research,* vol. 2, 1957, pp. D12.1–D12.20.

Argyropoulos, P. A., "The Hydraulic Jump and the Effect of Turbulence on Hydraulic Structures: Contribution of the Research of the Phenomena," *Proceedings of the 9th Conference, International Association for Hydraulic Research,* 1961, p. 173–183.

Argyropoulos, P. A., "A General Solution of the Hydraulic Jump in Sloping Channels," *Proceedings of the American Society of Civil Engineers, Journal of the Hydraulics Division,* vol. 88, no. HY4, July 1962, pp. 61–77.

Bakhmeteff, B. A., and Matzke, A. E., "The Hydraulic Jump in Terms of Dynamic Similarity," *Transactions of the American Society of Civil Engineers,* vol. 101, 1936, pp. 630–647.

Bradley, J. N., and Peterka, A. J., "The Hydraulic Design of Stilling Basins: Hydraulic Jumps on a Horizontal Apron (Basin I)," *Proceedings of the American Society of Civil Engineers, Journal of the Hydraulics Division,* vol. 83, no. HY5, 1957a, pp. 1–24.

Bradley, J. N., and Peterka, A. J., "The Hydraulic Design of Stilling Basins: Stilling Basin with Sloping Apron (Basin V)," *Proceedings of the American Society of Civil Engineers, Journal of the Hydraulics Division,* vol. 83, no. HY5, 1957b, p. 1–32.

Chow, V. T. *Open Channel Hydraulics,* McGraw-Hill Book Company, New York, 1959.

Gardner, G. B., Nowell, A. R. M., and Smith, J. D., "Turbulent Processes in Estuaries," *Estuarine and Wetland Processes,* P. Hamilton and K. McDonald (Eds.), Plenum Press, New York, 1980, pp. 1–34.

Govinda Rao, N. S., and Rajaratnam, N., "The Submerged Hydraulic Jump," *Proceedings of the American Society of Civil Engineers, Journal of the Hydraulics Division,* vol. 89, no. HY1, January 1963, pp. 139–162.

Hayakawa, N., "Internal Hydraulic Jump in Co-current Stratified Flow," *Proceedings of the American Society of Civil Engineers, Engineering Mechanics Division,* vol. 96, no. EM5, 1970, pp. 797–800.

Hsing, P. S., "The Hydraulic Jump in a Trapezoidal Channel," thesis presented to the State University of Iowa, Iowa City, 1937, in partial fulfillment of the requirements for the degree of Doctor of Philosophy.

Kennison, K. R., "Discussion of the Hydraulic Jump in Sloping Channels," *Transactions of the American Society of Civil Engineers,* vol. 109, 1944, pp. 1123–1125.

Kindsvater, C. E., "Hydraulic Jump in Enclosed Conduits," thesis presented to the State University of Iowa, Iowa City, 1936, in partial fulfillment of the requirements for the degree of Master of Science.

Kindsvater, C. E., "The Hydraulic Jump in Sloping Channels," *Transactions of the American Society of Civil Engineers,* vol. 109, 1944, pp. 1107–1154.

Peterka, A. J., "Hydraulic Design of Stilling Basins and Energy Dissipators," Engineering Monograph No. 25, U.S. Bureau of Reclamation, Denver, 1963.

Press, M. J., "The Hydraulic Jump," engineering honours thesis presented to the University of Western Australia, Nedlands, Australia, 1961.

Rajaratnan, N., "The Hydraulic Jump as a Wall Jet," *Proceedings of the American Society of Civil Engineers, Journal of the Hydraulics Division,* vol. 91, no. HY5, September 1965, pp. 107–132.

Rajaratnam, N., "Hydraulic Jumps," *Advances in Hydroscience,* vol. 4, Academic Press, New York, 1967, pp. 197–280.

Rouse, H., Siao, T. T., and Nagaratnam, S., "Turbulence Characteristics of the Hydraulic Jump," *Transactions of the American Society of Civil Engineers,* vol. 124, 1959, pp. 926–966.

Safranez, K., "Untersuchungen über den Wechselsprung," *Der Baningenieur,* vol. 10, no. 37, 1919, pp. 649–651. A brief summary is given in Donald P. Barnes, "Length of Hydraulic Jump Investigated at Berlin," *Civil Engineering,* vol. 4, no. 5, May 1934, pp. 262–263.

Sandover, J. A., and Holmes, P., "The Hydraulic Jump in Trapezoidal Channels," *Water Power,* vol. 14, 1962, p. 445.

Silvester, R. "Hydraulic Jump in All Shapes or Horizontal Channels," *Proceedings of the American Society of Civil Engineers, Journal of the Hydraulics Division,* vol. 90, no. HY1, January 1964, pp. 23–55.

Silvester, R., "Theory and Experiment on the Hydraulic Jump," *2nd Australiasian Conference on Hydraulics and Fluid Mechanics,* 1965, pp. A25–A39.

Stefan, H., Hayakawa, N., and Schiebe, F. R., "Surface Discharge of Heated Water," 16130 FSU 12/71, U.S. Environmental Protection Agency, Washington, D.C., December 1971.

Stefan, H., and Hayakawa, N., "Mixing Induced by an Internal Hydraulic Jump," *Water Resources Bulletin,* American Water Resources Association, vol. 8, no. 3, June 1972, pp. 531–545.

Stepanov, P. M., "The Submerged Hydraulic Jump," *Gidrotekhn i Melioratsiya,* Moscow, vol. 10, 1958. English translation by Israel Programme for Scientific Translations, 1959.

Straub, W. O., "A Quick and Easy Way to Calculate Critical and Conjugate Depths in Circular Open Channels," *Civil Engineering,* December 1978, pp. 70–71.

Turner, J. S., *"Buoyancy Effects in Fluids,"* Cambridge University Press, Cambridge, England, 1973.

Yih, C. S., and Guha, C. R., "Hydraulic Jump in a Fluid System of Two Layers," *Tellus,* vol. 7, no. 3, 1955, pp. 358–366.

FOUR

Development of Uniform Flow Concepts

SYNOPSIS

In this chapter uniform flow is defined, and the Chezy and Manning equations for uniform flow are developed. Both theoretical and applied methods of estimating the resistance coefficients used in these equations are then discussed.

4.1 ESTABLISHMENT OF UNIFORM FLOW

Although the definition of uniform flow and the assumptions required to develop the governing equations are only rarely satisfied in practice, the concept of uniform flow is central to the understanding and solution of most problems in open-channel hydraulics. By definition, uniform flow occurs when:

1. The depth, flow area, and velocity at every cross section are constant.
2. The energy grade line, water surface, and channel bottom are all parallel; that is, $S_f = S_w = S_0$ where S_f = slope of the energy grade line, S_w = slope of the water surface, and S_0 = slope of the channel bed (Fig. 4.1).

In general, uniform flow can occur only in very long, straight, prismatic channels where a terminal velocity of flow can be achieved; i.e., the head loss due to turbulent flow is exactly balanced by the reduction in potential energy due to the uniform decrease in the elevation of the bottom of the channel. In addition, although unsteady uniform flow is theoretically possible, it rarely occurs in open channels; only the case of steady, uniform flow will be treated here.

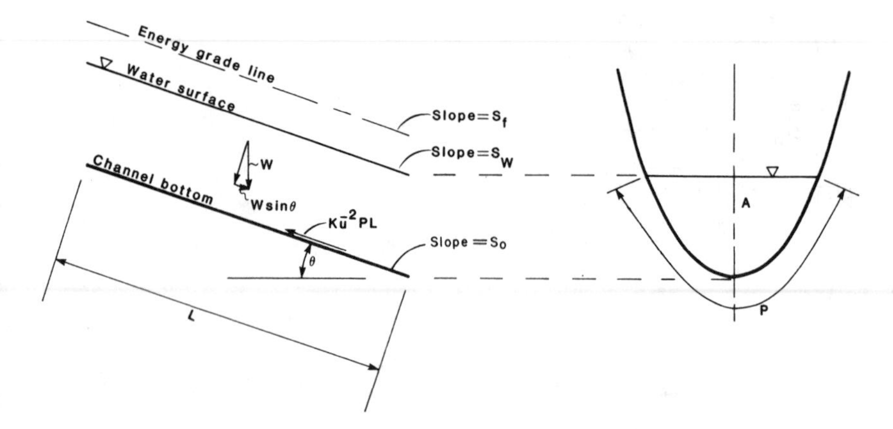

FIGURE 4.1 Schematic definition of variables for derivation of the Chezy equation.

4.2 THE CHEZY AND MANNING EQUATIONS

For computational purposes, the average velocity of a uniform flow can be computed approximately by any one of a number of semiempirical uniform flow equations. All these equations have the form

$$\overline{u} = CR^x S^y \tag{4.2.1}$$

where $\overline{u}$ = average velocity
R = hydraulic radius
S = channel longitudinal slope
C = resistance coefficient
x and y = coefficients

In this book, only the Chezy equation, developed in 1769, and the Manning equation, developed in 1889, are considered.

The Chezy equation can be derived from the definition of uniform flow with an assumption regarding the form of the flow resistance coefficient. With reference to Fig. 4.1, the definition of uniform flow requires that the forces resisting flow exactly equal the forces causing motion. The force causing motion is

$$F_m = W \sin\theta = \gamma\, AL \sin\theta \tag{4.2.2}$$

where W = weight of fluid within control volume
γ = fluid specific weight
A = flow area
L = control volume length
θ = longitudinal slope angle of channel

If θ is small, which is usually the case, then $\sin\theta \simeq S_0$. It is assumed that the force per unit area of the channel perimeter resisting motion, F_R, is proportional to the square of the average velocity, or

$$F_R \sim \overline{u}^2$$

Then, for a reach of length L with a wetted perimeter P, the force of resistance is

$$F_R = LPk\overline{u}^2 \tag{4.2.3}$$

where k = constant of proportionality. Setting the force causing motion, Eq. (4.2.2), equal to the force resisting motion, Eq. (4.2.3),

$$\gamma ALS_0 = LPk\overline{u}^2$$

or

$$\overline{u} = \left(\frac{\gamma}{k}\right)^{1/2} \sqrt{RS} \tag{4.2.4}$$

where the subscript associated with S has been dropped. For convenience, define,

$$C = \left(\frac{\gamma}{k}\right)^{1/2} \tag{4.2.5}$$

The resistance coefficient, C, defined by Eq. (4.2.5) is commonly known as the Chezy C and in practice is determined by either measurement or estimate. The coefficient of resistance defined by Eq. (4.2.5) is not dimensionless but has dimensions of acceleration; i.e., length/time^2 or LT^{-2}.

The Manning equation is the result of a curve-fitting process and is thus completely empirical in nature. In application of the Manning equation, it is essential that the system of units being used be identified and that the appropriate coefficient be used. In the SI system of units, the Manning equation is

$$\bar{u} = \frac{1}{n} R^{2/3} \sqrt{S}$$

where n = Manning resistance coefficient. As was the case with the Chezy resistance coefficient, n is not dimensionless but has dimensions of $TL^{-1/3}$ or in the specific case of the equation above s/m$^{1/3}$. The Manning equation in SI units can be converted to English units without affecting the numerical value of n by noting

$$n \text{ in } \left[\frac{\text{s}}{\text{m}^{1/3}}\right] = 0.6730\, n \text{ in } \left[\frac{\text{s}}{\text{ft}^{1/3}}\right]$$

Therefore, in English units, the Manning equation is

$$\bar{u} = \frac{1}{0.6730n} R^{2/3}\sqrt{S} = \frac{1.49}{n} R^{2/3}\sqrt{S}$$

For the sake of generality, in this book the Manning equation will be written

$$\bar{u} = \frac{\phi}{n} R^{2/3}\sqrt{S} \tag{4.2.6}$$

where ϕ = 1.00 if SI units are used and ϕ = 1.49 if English units are used.

From the viewpoint of modern fluid mechanics, the dimensions of the resistance coefficients C and n are a matter of major importance; see, for example, Simons and Senturk (1976). However, the dimensions of these coefficients appear to have been a problem of no significance to early hydraulic engineers. The reader is cautioned to be aware of the dimensions of C and n and to take these dimensions into account where appropriate.

Since the Chezy and Manning equations describe the same phenomena, the coefficients C and n must be related. Setting Eq. (4.2.4) equal to Eq. (4.2.6) yields

$$C\sqrt{RS} = \frac{\phi}{n} R^{2/3}\sqrt{S} \tag{4.2.7}$$

$$C = \frac{\phi}{n} R^{1/6}$$

In this book, the Manning equation will be used almost exclusively, even though the Chezy equation is equally appropriate from a theoretical point of view. The reader is cautioned that the above discussion applies only to channels with plane beds. In alluvial channels where bed forms are present, the analysis which applies is much more complex; see for example, Simons and Senturk (1976) or Garde and Raju (1978).

4.3 RESISTANCE COEFFICIENT ESTIMATION

Basic Theoretical Concepts

The primary difficulty in using either the Manning or Chezy equation in practice is accurately estimating an appropriate value of the resistance coefficient. In general, it is expected that n and C should depend on the Reynolds number of the flow, the boundary roughness, and the shape of the channel cross section. This is equivalent to hypothesizing that n and C behave in a manner analogous to the Darcy-Weisbach friction factor f used to assess flow resistance in pipe flow; see, for example, Streeter and Wylie (1975), or

$$S = \frac{f}{4R}\frac{\overline{u}^2}{2g} \tag{4.3.1}$$

Then

$$n = \phi R^{1/6}\sqrt{\frac{f}{8g}} \tag{4.3.2}$$

and

$$C = \sqrt{\frac{8g}{f}} \tag{4.3.3}$$

If the behavior of n and C are to be investigated from this viewpoint, then a number of basic definitions regarding the types of turbulent flow must be recalled. *Hydraulically smooth turbulent flow* refers to a flow in which the perimeter roughness elements are completely covered by the viscous sublayer. *Fully rough flow* occurs when the perimeter roughness elements project through the laminar sublayer and dominate the flow behavior. In the latter case, flow resistance is entirely due to form drag, and the resistance coefficient is independent of the Reynolds number, where $\mathbf{R} = 4R\overline{u}/\nu$ = Reynolds number. Between these two extremes, there is a transitional region. In the analysis which follows, k_s is a length parameter which characterizes the perimeter roughness and is defined as the sand grain diameter for a sand-coated surface which has the same limiting value of f. The three types of turbulent flow—smooth, transitional, and fully rough—are differentiated from each other by a Reynolds number based on k_s and the shear velocity, $u_* = \sqrt{\tau_0/\rho} = \sqrt{gRS}$ or

$$\mathbf{R}_* = \frac{u_* k_s}{\nu} \tag{4.3.4}$$

The transition region is defined by the limits

TABLE 4.1 Values of k_s for various types of concrete and masonry surfaces (Ackers, 1958, and Zegzhda, 1938, both in Anonymous, 1963b)

k_s	Surface description
0.0005	Concrete class 4 (monolithic construction, cast against oiled steel forms with no surface irregularities)
0.001	Very smooth cement-plastered surfaces, all joints and seams hand-finished flush with surface
0.0016	Concrete cast in lubricated steel molds, with carefully smoothed or pointed seams and joints
0.002	Wood-stave pipes, planed-wood flumes, and concrete class 3 (cast against steel forms, or spun-precast pipe); smooth troweled surfaces; glazed sewer pipe
0.005	Concrete class 2 (monolithic construction against rough forms or smooth-finished cement-gun surface, the latter often termed *gunite* or *shot concrete*); glazed brickwork
0.008	Short lengths of concrete pipe of small diameter without special facing of butt joints
0.01	Concrete class 1 (precast pipes with mortar squeeze at the joints); straight, uniform earth channels
0.014	Roughly made concrete conduits
0.02	Rubble masonry
0.01–0.03	Untreated gunite

$$4 \le \mathbf{R}_* \le 100 \tag{4.3.5}$$

with the lower limit defining the end of the smooth region and the upper limit defining the beginning of the fully rough region. As will be shown in a subsequent section of this chapter, the point at which fully rough flow begins can also be defined in terms of Manning's n or

$$n^6\sqrt{RS} = 1.9 \times 10^{-13} \tag{4.3.6}$$

When the boundaries are hydraulically smooth and $\mathbf{R} < 100{,}000$, the friction factor is given by the Blasius formula developed for pipe flow or

$$f = \frac{0.316}{\mathbf{R}^{0.25}} \tag{4.3.7}$$

When the Reynolds number exceeds 100,000, but the flow is still hydraulically smooth, the friction factor is given by

$$\frac{1}{\sqrt{f}} = 2.0 \log\left(\frac{\mathbf{R}\sqrt{f}}{2.51}\right) \tag{4.3.8}$$

For hydraulically rough flow, the friction factor can be estimated by

$$\frac{1}{\sqrt{f}} = 2.0 \log\left(\frac{12R}{k_s}\right) \tag{4.3.9}$$

where typical values of k_s for various materials are summarized in Table 4.1. In the transition region between hydraulically smooth and rough flows, f can be estimated by a modified Colebrook formula or

$$\frac{1}{\sqrt{f}} = -2 \log\left(\frac{k_s}{12R} + \frac{2.5}{\mathbf{R}\sqrt{f}}\right) \tag{4.3.10}$$

In Fig. 4.2, the friction factor is plotted as a function of **R** and the parameter $2R/k_s$ which is analogous to the relative roughness parameter in pipe flow.

The American Society of Civil Engineers' Task Force on Friction Factors in Open Channels concluded that for open-channel roughness similar to that encountered in pipes, a resistance diagram similar to that shown in Fig. 4.2 was adequate for estimating f and, hence, n or C (Anonymous, 1963a).

Applied Concepts

For the case of high Reynolds numbers and roughnesses of the type usually found in unlined channels, the friction factor is independent of the Reynolds number and nearly proportional to $R^{-1/3}$ (Anonymous, 1963a). If f is assumed to be given by $R^{-1/3}$ and this value is substituted in Eq. (4.3.2), it is concluded that for this situation Manning's n is a constant. Thus, Manning's equation with a constant value of n is applicable only to fully rough turbulent flow as defined by Eq. (4.3.5) or (4.3.6). In estimating an appropriate value of n for this case, a qualitative knowledge of the factors on which n depends is necessary since in most applied situations the value of n is a function of many variables.

Surface Roughness The surface roughness of a channel perimeter provides a critical point of reference in estimating n. When the perimeter material is fine, the value of n is low and is relatively unaffected by changes in the depth of flow; however, when the perimeter is composed of gravels and/or boulders, the value of n is larger and may vary significantly with the depth of flow. For example, large boulders usually collect at the bottom of such channels and result in a high value of n at low stages and relatively low values of n at high stages.

Vegetation The estimated value of n should take into account the effect of vegetation in retarding flow and increasing n. In general, the relative importance of vegetation on n is a strong function of the depth of flow and the height, density, distribution, and type of vegetation. This consideration is crucial in the design of small drainage channels, since they usually do not receive regular maintenance. Chow (1959) noted that trees from 6 to 8 in (0.15 to 0.20 m) in diameter growing on the side slopes do not impede the flow as much as brushy growths if the overhanging limbs of the trees are kept trimmed.

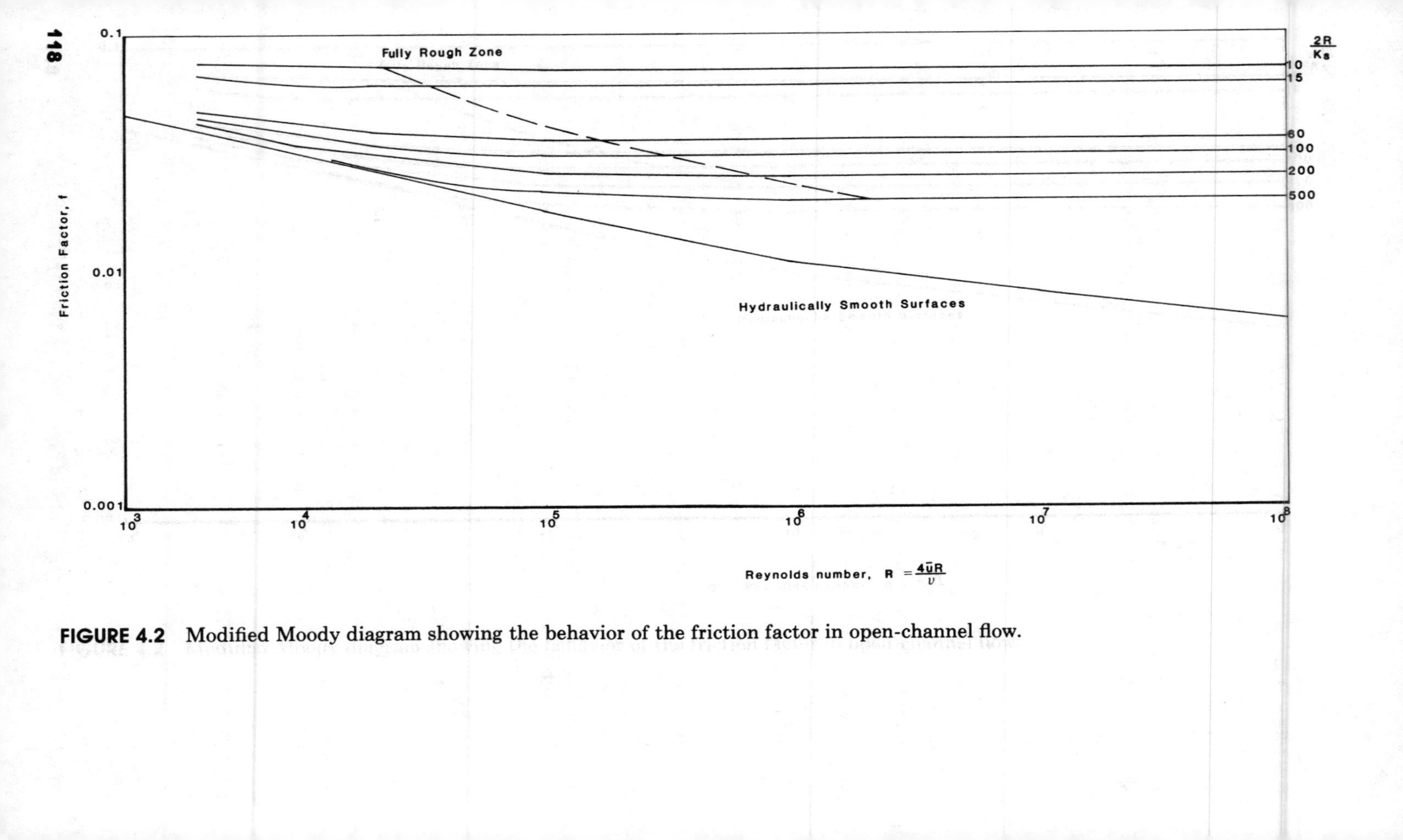

FIGURE 4.2 Modified Moody diagram showing the behavior of the friction factor in open-channel flow.

Channel Irregularity This terminology refers to variations in the channel cross section, shape, and wetted perimeter along the longitudinal axis of the channel. In natural channels, such irregularities are usually the result of deposition or scour. In general, gradual variations have a rather insignificant effect on n, while abrupt changes can result in a much higher value of n than would be expected from a consideration of only the surface roughness of the channel perimeter.

Obstruction The presence of obstructions such as fallen trees, debris flows, and log or debris jams can have a significant impact on the value of n. The degree of the effect of such obstructions depends on the number and size of the obstructions.

Channel Alignment While curves of large radius without frequent changes in the direction of curvature offer relatively little resistance, severe meandering with curves of small radius will significantly increase the value of n.

Sedimentation and Scouring In general, active sedimentation and scouring yield channel variation which results in an increased value of n. Urquhart (1975) noted that it is important to consider whether both of these processes are active and whether they are likely to remain active in the future.

Stage and Discharge The n value for most channels tends to decrease with an increase in the stage and discharge. This is the result of irregularities which have a crucial impact on the value of n at low stages when they are uncovered. However, the n value may increase with increasing stage and discharge if the banks of the channel are rough, grassy, or brush-covered, or if the stage increases sufficiently to cover the flood plain. On flood plains, the value of n usually varies with the depth of submergence. In such a case, it is necessary to compute a composite value of n—a procedure which is discussed in the next chapter.

Given these qualitative remarks regarding the variation of n with a number of factors, it is now possible to discuss several methods of estimating a value of n given a specific situation.

Soil Conservation Service Method

The Soil Conservation Service (SCS) method for estimating n involves the selection of a basic n value for a uniform, straight, and regular channel in a native material and then modifying this value by adding correction factors determined by a critical consideration of some of the factors enumerated above (Urquhart, 1975). In this process, it is critical that each factor be considered and evaluated independently. The SCS suggests that the turbulence of a flow can be used as a measure or indicator of the degree of retardance; i.e., factors which induce a greater degree of turbulence should also result in an increase in n.

TABLE 4.2 Basic n values suggested by the Soil Conservation Service (Anonymous, 1963b)

Channel character	Basic n
Channels in earth	0.02
Channels cut into rock	0.025
Channels in fine gravel	0.024
Channels in coarse gravel	0.028

TABLE 4.3 Modifying factors for vegetation (Anonymous, 1963b)

Vegetation and flow conditions comparable with:	Degree of effect on n	Range of modifying values
Dense growths of flexible turf grasses or weeds, of which Bermuda grass and blue grass are examples, where the average depth of flow is 2 to 3 times the height of vegetation Supple seedling tree switches such as willow, cottonwood, or salt cedar where the average depth of flow is 3 to 4 times the height of the vegetation	Low	0.005–0.010
Turf grasses where the average depth of flow is 1 to 2 times the height of vegetation Stemmy grasses, weeds, or tree seedlings with moderate cover where the average depth of flow is 2 to 3 times the height of vegetation Brushy growths, moderately dense, similar to willows 1 to 2 years old, dormant season, along side slopes of channel with no significant vegetation along the channel bottom, where the hydraulic radius is greater than 2 ft (0.6 m)	Medium	0.010–0.025

Step 1. *Selection of a Basic n:* In this step, a basic value for a straight, uniform, smooth channel in the native materials is selected. The channel must be visualized without vegetation, obstructions, changes in shape, and changes of alignment. The basic n values suggested by the SCS are summarized in Table 4.2.

Step 2: *Modification for Vegetation:* The retardance due to vegetation is primarily due to the flow of water around stems, trunks, limbs, and branches and only secondarily to the reduction of the flow area. In assessing the effect of vegetation on retardance, consideration must be given to the height of the vegetation in relation to the depth of flow, the capacity of the vegetation to resist bending, the degree to which

Vegetation and flow conditions comparable with:	Degree of effect on n	Range of modifying values
Dormant season, willow or cottonwood trees 8 to 10 years old, intergrown with some weeds and brush, none of the vegetation in foliage, where the hydraulic radius is greater than 2 ft (0.6 m) Growing season, bushy willows about 1-year-old intergrown with some weeds in full foliage along side slopes, no significant vegetation along channel bottom, where hydraulic radius is greater than 2 ft (0.6 m)	High	0.025–0.050
Turf grasses where the average depth of flow is less than one-half the height of vegetation Growing season, bushy willows about 1 year old, intergrown with weeds in full foliage along side slopes; dense growth of cattails along channel bottom; any value of hydraulic radius up to 10 or 15 ft (3 to 4.6 m) Growing season, trees intergrown with weeds and brush, all in full foliage; any value of hydraulic radius up to 10 or 15 ft (3 to 4.6 m)	Very high	0.050–0.100

the flow is obstructed, the transverse and longitudinal distribution of vegetation of various types, the densities and heights of vegetation in the reach being considered, and the critical season; i.e., is the vegetation dormant or growing? The SCS results regarding vegetation are summarized in Table 4.3.

Step 3: *Modification for Channel Irregularity:* In determining the modification required for channel irregularity, both changes in flow area and changes in cross-sectional shape must be considered. The effects of changes in flow area should be examined from the viewpoint of comparing the magnitude of the change with the average area. While large changes in area, if they are gradual and uniform, result in small modifying values, abrupt changes yield large modifying values. In the case of changes of channel shape, the degree to which the change causes the greatest depth of flow to migrate from side to side is critical. Shape changes which yield the largest modifying values are those which shift

TABLE 4.4 Modifying factors for changes in cross-section size and shape (Anonymous, 1963b)

Character of variations in size and shape of cross sections	Modifying value
Changes in size or shape occurring gradually	0.000
Large and small sections alternating occasionally or shape changes causing occasional shifting of main flow from side to side	0.005
Large and small sections alternating frequently or shape changes causing frequent shifting of main flow from side to side	0.010–0.015

TABLE 4.5 Modifying factors for channel surface irregularity (Anonymous, 1963b)

Degree of irregularity	Surfaces comparable with	Modifying value
Smooth	The best obtainable for the materials involved	0.000
Minor	Good dredged channels; slightly eroded or scoured side slopes of canals or drainage channels	0.005
Moderate	Fair to poor dredged channels; moderately sloughed or eroded side slopes of canals or drainage channels	0.010
Severe	Badly sloughed banks of natural channels; badly eroded or sloughed sides of canals or drainage channels; unshaped, jagged, and irregular surfaces of channels excavated in rock	0.020

the main flow from side to side in distances short enough to produce eddies and upstream currents in the shallow area. The SCS recommendations for the modifying values for this effect are summarized in Table 4.4.

The second consideration in this step is the degree of roughness or irregularity of the surface of the channel perimeter. The existing surface should be compared with the surface smoothness which can, under ideal conditions, be obtained with the native materials and with the specified depth of flow. The SCS results for this effect are summarized in Table 4.5.

Step 4: *Modification for Obstruction:* The selection of the modifying value for this factor is based on the number and characteristics of the obstructions. Obstructions considered by the SCS included debris deposits, stumps, exposed roots, boulders, and fallen and lodged logs. In assessing the relative effect of obstructions, one must give consideration to the following: (*a*) the degree to which the obstructions reduce the flow area at various depths of flow, (*b*) the shape of the obstructions (recall that angular objects produce greater turbulence than rounded objects), and (*c*) the position and spacing of the obstructions in both the transverse and longitudinal directions. The SCS recommendations for this modification are summarized in Table 4.6.

Step 5: *Modification for Channel Alignment:* The modifying value for channel alignment is found by adding the modifying values found in steps 2 to 4 to the basic value of n, step 1, to form the subtotal n'. Define ℓ_s = straight length of the reach under consideration and ℓ_m = meander length of the channel in the reach. The modifying value for alignment can then be estimated from Table 4.7 for various values of the ratio ℓ_m / ℓ_s.

Step 6: *Estimate of n:* A value of n can then be estimated by summing the results of steps 1 to 5.

The use of the SCS method in estimating n for a natural channel is best demonstrated by an example.

TABLE 4.6 Modifying factors for obstruction (Anonymous, 1963b)

Relative effect of obstructions	Modifying value
Negligible	0.000
Minor	0.010–0.015
Appreciable	0.020–0.030
Severe	0.040–0.060

TABLE 4.7 Modifying values for channel alignment (Anonymous, 1963b)

ℓ_m / ℓ_s	Degree of meandering	Modifying value
1.0–1.2	Minor	0.00
1.2–1.5	Appreciable	0.15 n'
>1.5	Severe	0.30 n'

EXAMPLE 4.1

A dredged channel has a cross section which can be approximated as a trapezoid with a bottom width of 10 ft (3.0 m) and side slopes of 1.5:1. At the maximum stage the average depth of flow is 8.5 ft (2.6 m) and the top width is 35.5 ft (10.8 m). The alignment of the channel has a moderate degree of meander. While the side slopes are fairly regular, the bottom is uneven and irregular. The variation of the cross-sectional area with longitudinal distance is moderate. The material through which the channel is cut is characterized as gray clay with the upper part of the channel being in silty clay loam. The side slopes of the channel are covered with a heavy growth of poplar trees 2 to 3 in (0.05 to 0.08 m) in diameter, large willows, and climbing vines. There is also a thick growth of water weed on the bottom of the channel. Given this description, estimate a value of n for this channel for summer conditions when the vegetation is in full foliage.

Solution

Step	Comment	Modifying value
1	Estimate basic value of n, Table 4.2	0.02
2	Description of vegetation indicates a high degree of retardance, Table 4.3	0.08
3	Description suggests an insignificant change in both channel size and shape, Table 4.4	0.00
4	Description indicates a moderate degree of irregularity, Table 4.5	0.01
5	No obstructions are indicated, Table 4.6	0.00
6	A moderate degree of alignment change is indicated; therefore, $n' = 0.02 + 0.08 + 0.01 = 0.11$ Modifying value $= 0.15(0.11) = 0.02$ Table 4.7	0.02
	Total estimated n =	0.13

Table Method of *n* Estimation A second method of estimating n for a channel involves the use of tables of values. Chow (1959) presented an extensive table of n values for various types of channels, and the information in that table is repeated here as Table 4.8. In this table, a minimum, normal, and maximum value of n are stated for each type of channel. The underlined values are those recommended for design. It is noted that the normal values assume that the channel receives regular maintenance.

TABLE 4.8 Values of the roughness of coefficient n (Chow, 1959)

Type of channel and description	Minimum	Normal	Maximum
A. Closed conduits flowing partly full			
A-1. Metal			
a. Brass, smooth	0.009	0.010	0.013
b. Steel			
1. Lockbar and welded	0.010	0.012	0.014
2. Riveted and spiral	0.013	0.016	0.017
c. Cast iron			
1. Coated	0.010	0.013	0.014
2. Uncoated	0.011	0.014	0.016
d. Wrought iron			
1. Black	0.012	0.014	0.015
2. Galvanized	0.013	0.016	0.017
e. Corrugated metal			
1. Subdrain	0.017	0.019	0.021
2. Storm drain	0.021	0.024	0.030
A-2. Nonmetal			
a. Lucite	0.008	0.009	0.010
b. Glass	0.009	0.010	0.013
c. Cement			
1. Neat, surface	0.010	0.011	0.013
2. Mortar	0.011	0.013	0.015
d. Concrete			
1. Culvert, straight and free of debris	0.010	0.011	0.013
2. Culvert with bends, connections, and some debris	0.011	0.013	0.014
3. Finished	0.011	0.012	0.014
4. Sewer and manholes, inlet, etc., straight	0.013	0.015	0.017
5. Unfinished, steel form	0.012	0.013	0.014
6. Unfinished, smooth wood form	0.012	0.014	0.016
7. Unfinished, rough wood form	0.015	0.017	0.020
e. Wood			
1. Stave	0.010	0.012	0.014
2. Laminated, treated	0.015	0.017	0.020
f. Clay			
1. Common drainage tile	0.011	0.013	0.017

TABLE 4.8 Values of the roughness of coefficient n (Chow, 1959)(*Continued*)

Type of channel and description	Minimum	Normal	Maximum
2. Vitrified sewer	0.011	0.014	0.017
3. Vitrified sewer with manholes, inlet, etc.	0.013	0.015	0.017
4. Vitrified subdrain with open joint	0.014	0.016	0.018
g. Brickwork			
1. Glazed	0.011	0.013	0.015
2. Lined with cement mortar	0.012	0.015	0.017
h. Sanitary sewers coated with sewage slimes, with bends and connections	0.012	0.013	0.016
i. Paved invert, sewer, smooth bottom	0.016	0.019	0.020
j. Rubble masonry, cemented	0.018	0.025	0.030
B. Lined or built-up channels			
B-1. Metal			
a. Smooth steel surface			
1. Unpainted	0.011	0.012	0.014
2. Painted	0.012	0.013	0.017
b. Corrugated	0.021	0.025	0.030
B-2. Nonmetal			
a. Cement			
1. Neat, surface	0.010	0.011	0.013
2. Mortar	0.011	0.013	0.015
b. Wood			
1. Planed, untreated	0.010	0.012	0.014
2. Planed, creosoted	0.011	0.012	0.015
3. Unplaned	0.011	0.013	0.015
4. Plank with battens	0.012	0.015	0.018
5. Lined with roofing paper	0.010	0.014	0.017
c. Concrete			
1. Trowel finish	0.011	0.013	0.015
2. Float finish	0.013	0.015	0.016
3. Finished, with gravel on bottom	0.015	0.017	0.020
4. Unfinished	0.014	0.017	0.020
5. Gunite, good section	0.016	0.019	0.023
6. Gunite, wavy section	0.018	0.022	0.025

Type of channel and description	Minimum	Normal	Maximum
7. On good excavated rock	0.017	0.020	
8. On irregular excavated rock	0.022	0.027	
d. Concrete bottom float finished with sides of			
1. Dressed stone in mortar	0.015	0.017	0.020
1. Random stone in mortar	0.017	0.020	0.024
3. Cement rubble masonry, plastered	0.016	0.020	0.024
4. Cement rubble masonry	0.020	0.025	0.030
5. Dry rubble or riprap	0.020	0.030	0.035
e. Gravel bottom with sides of			
1. Formed concrete	0.017	0.020	0.025
2. Random stone in mortar	0.020	0.023	0.026
3. Dry rubble or riprap	0.023	0.033	0.036
f. Brick			
1. Glazed	0.011	0.013	0.015
2. In cement mortar	0.012	0.015	0.018
g. Masonry			
1. Cemented rubble	0.017	0.025	0.030
2. Dry rubble	0.023	0.032	0.035
h. Dressed ashlar	0.013	0.015	0.017
i. Asphalt			
1. Smooth	0.013	0.013	
j. Vegetal lining	0.030		0.500
C. Excavated or dredged			
a. Earth, straight and uniform			
1. Clean, recently completed	0.016	0.018	0.020
2. Clean, after weathering	0.018	0.022	0.025
3. Gravel, uniform section, clean	0.022	0.025	0.030
4. With short grass, few weeds	0.022	0.027	0.033

TABLE 4.8 Values of the roughness of coefficient *n* (Chow, 1959)(*Continued*)

Type of channel and description	Minimum	Normal	Maximum
b. Earth winding and sluggish			
1. No vegetation	0.023	0.025	0.030
2. Grass, some weeds	0.025	0.030	0.033
3. Dense weeds or aquatic plants in deep channels	0.030	0.035	0.040
4. Earth bottom and rubble sides	0.028	0.030	0.035
5. Stony bottom and weedy banks	0.025	0.035	0.040
6. Cobble bottom and clean sides	0.030	0.040	0.050
c. Dragline-excavated or dredged			
1. No vegetation	0.025	0.028	0.033
2. Light brush on banks	0.035	0.050	0.060
d. Rock cuts			
1. Smooth and uniform	0.025	0.035	0.040
2. Jagged and irregular	0.035	0.040	0.050
e. Channels not maintained, weeds and brush uncut			
1. Dense weeds, high as flow depth	0.050	0.080	0.120
2. Clean bottom, brush on sides	0.040	0.050	0.080
3. Same, highest stage of flow	0.045	0.070	0.110
4. Dense brush, high stage	0.080	0.100	0.140
D. Natural streams			
D-1. Minor streams (top width at flood stage < 100 ft)			
a. Streams on plain			
1. Clean, straight, full stage, no rifts or deep pools	0.025	0.030	0.033
2. Same as above, but more stones and weeds	0.030	0.035	0.040

Type of channel and description	Minimum	Normal	Maximum
3. Clean, winding, some pools and shoals	0.033	0.040	0.045
4. Same as above, but some weeds and stones	0.035	0.045	0.050
5. Same as above, lower stages, more ineffective slopes and sections	0.040	0.048	0.055
6. Same as no. 4, more stones	0.045	0.050	0.060
7. Sluggish reaches, weedy, deep pools	0.050	0.070	0.080
8. Very weedy, reaches, deep pools, or floodways with heavy stand of timber and underbrush	0.075	0.100	0.150
b. Mountain streams, no vegetation in channel banks usually steep, trees and brush along banks submerged at high stages			
1. Bottom: gravels, cobbles, and few boulders	0.030	0.040	0.050
2. Bottom: cobbles with large boulders	0.040	0.050	0.070
D-2. Flood plains			
a. Pasture, no brush			
1. Short grass	0.025	0.030	0.035
2. High grass	0.030	0.035	0.050
b. Cultivated areas			
1. No crop	0.020	0.030	0.040
2. Mature row crops	0.025	0.035	0.045
3. Mature field crops	0.030	0.040	0.050
c. Brush			
1. Scattered brush, heavy weeds	0.035	0.050	0.070
2. Light brush and trees, in winter	0.035	0.050	0.060

TABLE 4.8 Values of the roughness of coefficient *n* (Chow, 1959)(*Continued*)

Type of channel and description	Minimum	Normal	Maximum
3. Light brush and trees, in summer	0.040	0.060	0.080
4. Medium to dense brush, in winter	0.045	0.070	0.110
5. Medium to dense brush, in summer	0.070	0.100	0.160
d. Trees			
1. Dense willows, summer, straight	0.110	0.150	0.200
2. Cleared land with tree stumps, no sprouts	0.030	0.040	0.050
3. Same as above, but with heavy growth of sprouts	0.050	0.060	0.080
4. Heavy stand of timber, a few down trees, little undergrowth, flood stage below branches	0.080	0.100	0.120
5. Same as above, but with flood stage reaching branches	0.100	0.120	0.160
D-3. Major streams (top width at flood stage > 100 ft); the *n* value is less than that for minor streams of similar description because banks offer less effective resistance			
a. Regular section with no boulders or brush	0.025		0.060
b. Irregular and rough section	0.035		0.100

NOTE: Values underlined are recommended for design.

Photographic Method It is the belief of the U.S. Geological Survey that photographs of channels of known resistance together with a summary of the geometric and hydraulic parameters which define the channel for a specified flow rate can be useful in estimating the resistance coefficient (Barnes, 1967.) It should also be noted that the U.S. Geological Survey maintains a program which trains engineers in the estimation of channel resistance coefficients. The results of this program indicate that trained engineers can estimate resistance coefficients with an accuracy of ± 15 percent under most conditions (Barnes, 1967).

Figures 4.3 to 4.15 are a set of photographs and tables from Barnes (1967) for channels with a wide range of resistance coefficients. Note: Most field offices of the U.S. Geological Survey have three-dimensional viewers and slides from Barnes (1967) which give a three-dimensional view of these channels. All the channels described in these figures are considered stable and meet the following criteria:

1. Sites were studied only after a major flood had occurred. Thus, the photographs represent conditions in a channel reach immediately after a flood.
2. The peak discharge in the channel reach specified was determined either by a current meter survey or from an accurate stage-discharge relation.
3. Within the reach, good high-water marks were available to define the water surface profile at peak discharge.
4. In the vicinity of the gauging station at which the peak discharge was determined, the channel was uniform.
5. The peak discharge was confined within the banks of the channel; i.e., flow did not take place in the flood plains.

The Manning resistance coefficients presented in Figs. 4.3 to 4.15 were estimated from the measured discharges, water surface profiles, and reach properties as defined by more than two cross sections in each case. For a case in which there were N cross sections

$$n = \frac{\phi}{Q}\left\{\frac{(h+h_u)_1 - (h+h_u)_N - \sum_{j=2}^{N}(k\,\Delta h_u)_{j-1,j}}{\sum_{j=2}^{N}\dfrac{L_{j-1,j}}{(AR^{2/3})_{j-1}(AR^{2/3})_j}}\right\}^{1/2} \tag{4.3.11}$$

where, with reference to Fig. 4.16, h = elevation of the water surface at a section with respect to a datum common to all sections, $(\Delta h_u)_{j-1,j}$ = change in velocity head between sections $j-1$ and j, $h_u = \alpha\bar{u}^2/2g$ = velocity head at a section, k = a coefficient accounting for the nonuniformity of the channel ($k = 0$ for a uniform reach; $k = 0.5$ for a nonuniform reach), h_f = energy loss in a reach due to boundary friction, ϕ = a coefficient which accounts for the system of units used ($\phi = 1.49$ for the English system and $\phi = 1.00$ for the SI system), and $L_{j-1,j}$ = distance between sections $j-1$ and j.

(Text continues on page 158.)

Drainage Area: 70.3 mi^2 (182 km^2)

Flood Date: May 11, 1948

Peak Discharge: 768 ft^3/s (21.7 m^3/s)

Estimated Roughness Coefficient: n = 0.026

Channel Description: Bed and banks composed of clay. Banks clear except for short grass and exposed tree roots in some places

Section	Area, ft^2 (m^2)	Top Width, ft (m)	Mean depth, ft (m)	Hydraulic radius, ft (m)	Average velocity, ft/s (m/s)	Distance between sections, ft (m)	Fall between sections, ft (m)
1	280 (26.0)	52 (16)	5.4 (1.6)	4.87 (1.48)	2.74 (0.835)	—	—
2	273 (25.4)	51 (16)	5.4 (1.6)	4.82 (1.47)	2.82 (0.860)	257 (78.3)	0.08 (0.02)
3	279 (25.9)	52 (16)	5.4 (1.6)	4.97 (1.51)	2.76 (0.841)	202 (61.6)	0.05 (0.02)

(a)

FIGURE 4.3. (*a*) Indian Fork below Atwood Dam near New Cumberland, Ohio. *(Barnes, 1967)*. (*b*) Plan sketch and cross sections, Indian Fork below Atwood Dam, near New Cumberland, Ohio. *(Barnes, 1967)*. (*c*) Section 329 from right bank, Indian Fork below Atwood Dam, near New Cumberland, Ohio. *(Barnes, 1967.)* (*d*) Section 327 from right bank, Indian Fork below Atwood Dam, near New Cumberland, Ohio. *(Barnes, 1967.)*

n=0.026

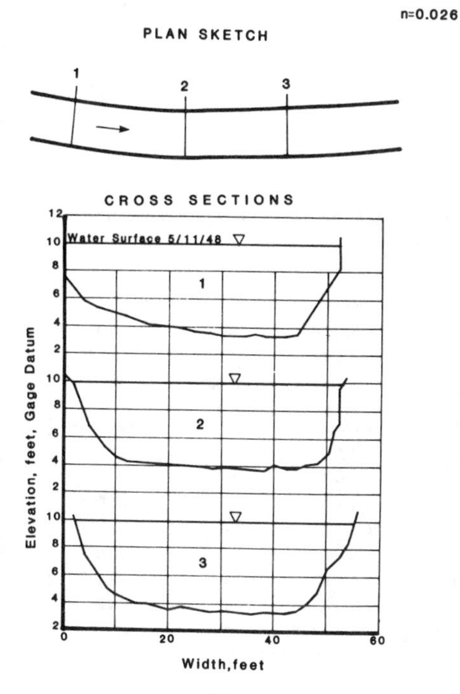
PLAN SKETCH
1
2
3
CROSS SECTIONS
Water Surface 5/11/48
Elevation, feet, Gage Datum
Width, feet
0
20
40
60

(b)

(c)

(d)

Drainage Area: 174 mi² (451 km²)

Flood Date: May 2, 1954

Peak Discharge: 1860 ft³/s (53 m³/s)

Estimated Roughness Coefficient: $n = 0.030$

Channel Description: Bed consists of sand and clay. Banks are generally smooth and free of vegetal growth during floods.

Section	Area, ft² (m²)	Top width, ft (m)	Mean depth, ft (m)	Hydraulic radius, ft (m)	Average velocity, ft/s (m/s)	Distance between sections, ft (m)	Fall between sections, ft (m)
1	528 (49.1)	72 (22)	7.4 (2.3)	6.7 (2.0)	3.46 (1.05)	—	—
2	502 (46.6)	80 (24)	6.3 (1.9)	5.9 (1.8)	3.71 (1.13)	113 (34.4)	0.09 (0.03)
3	497 (46.2)	69 (21)	7.2 (2.2)	6.6 (2.0)	3.74 (1.14)	110 (33.5)	0.06 (0.02)
4	497 (46.2)	78 (24)	6.4 (2.0)	5.9 (1.8)	3.74 (1.14)	134 (40.8)	0.05 (0.02)

(a)

FIGURE 4.4. (*a*) Salt Creek at Roca, Nebraska. *(Barnes, 1967.)* (*b*) Plan sketch and cross sections, Salt Creek at Roca, Nebraska. *(Barnes, 1967.)* (*c*) Section 824 downstream, Salt Creek at Roca, Nebraska. *(Barnes, 1967.)* (*d*) Section 825 downstream, Salt Creek at Roca, Nebraska. *(Barnes, 1967.)*

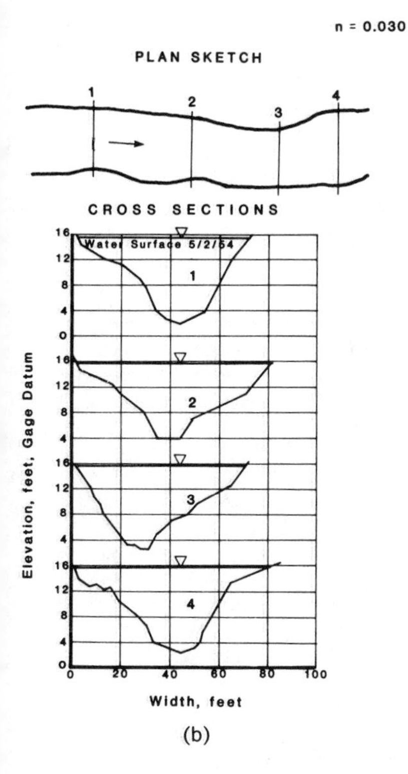
n = 0.030
PLAN SKETCH
1
2
3
4
CROSS SECTIONS
Water Surface 5/2/54
1
2
3
4
Elevation, feet, Gage Datum
0
4
8
12
16
0
20
40
60
80
100
Width, feet

(b)

(c)

(d)

Drainage Area: 3140 mi^2 (8100 km^2)

Flood Date: March 24, 1950; April 3, 1950

Peak Discharge: 1060 ft^3/s (30 m^3/s); 684 ft^3/s (19 m^3/s)

Estimated Roughness Coefficient: n = 0.032; n = 0.036

Channel Description: The bed consists of sand and gravel. The left bank is rock and the right bank is mostly gravel.

Section	Area, ft^2 (m^2)	Top width, ft (m)	Mean depth, ft (m)	Hydraulic radius, ft (m)	Average velocity, ft/s (m/s)	Distance between sections, ft (m)	Fall between sections, ft (m)
			March 24, 1950				
1	312 (29.0)	95 (29)	3.3 (1.0)	3.12 (0.951)	3.40 (1.04)	—	—
2	297 (27.6)	84 (26)	3.5 (1.1)	3.45 (1.05)	3.57 (1.09)	202 (61.6)	0.24 (0.073)
			April 3, 1950				
1	235 (21.8)	92 (28)	2.6 (0.79)	2.52 (0.768)	2.91 (0.887)	—	—
2	249 (23.1)	81 (25)	3.1 (0.94)	2.95 (0.899)	2.75 (0.838)	202 (61.6)	0.23 (0.070)

(a)

FIGURE 4.5 (*a*) Rio Chama near Chamita, New Mexico. *(Barnes, 1967.)* (*b*) Plan sketch and cross sections, Rio Chama near Chamita, New Mexico. *(Barnes, 1967.)* (*c*) Section 550 downstream from right bank, Rio Chama near Chamita, New Mexico. *(Barnes, 1967.)* (*d*) Section 547 upstream along right bank, Rio Chama near Chamita, New Mexico. *(Barnes, 1967.)*

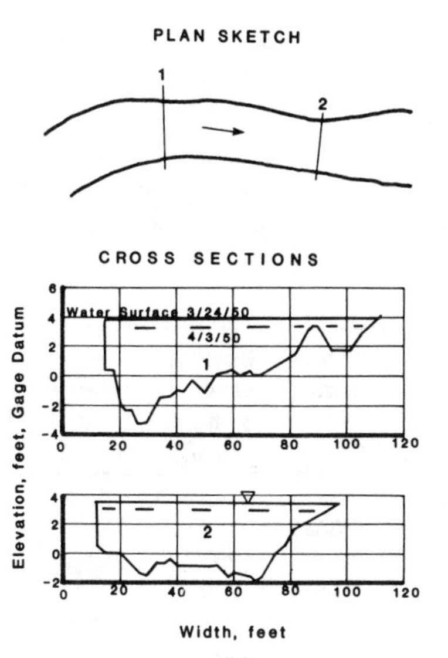
n = 0.032: 0.036
PLAN SKETCH
1
2
CROSS SECTIONS
Water Surface 3/24/50
4/3/50
1
2
Elevation, feet, Gage Datum
Width, feet
0
20
40
60
80
100
120

(b)

(c)

(d)

Drainage Area: 6230 mi^2 (16,000 km^2)

Flood Date: March 24, 1950

Peak Discharge: 1280 ft^3/s (36.2 m^3/s)

Estimated Roughness Coefficient: $n = 0.032$

Channel Description: Bed and banks consist of smooth cobbles 4–10 in (10–25 cm) in diameter, average diameter is 6 in (15 cm). A few boulders are as much as 18 in (46 cm) in diameter.

Section	Area, ft^2 (m^2)	Top width, ft (m)	Mean depth, ft (m)	Hydraulic radius, ft (m)	Average velocity, ft/s (m/s)	Distance between sections, ft (m)	Fall between sections, ft (m)
1	484 (45.0)	189 (57.6)	2.6 (0.79)	2.55 (0.777)	2.65 (0.808)	—	—
2	408 (37.9)	192 (58.5)	2.1 (0.64)	2.12 (0.646)	3.14 (0.957)	258 (78.6)	0.37 (0.11)
3	384 (35.7)	154 (46.9)	2.6 (0.79)	2.49 (0.759)	3.34 (1.02)	317 (96.6)	0.50 (0.15)
4	449 (41.7)	194 (59.1)	2.3 (0.70)	2.30 (0.701)	2.86 (0.872)	294 (89.6)	0.31 (0.094)
5	420 (39.0)	204 (62.2)	2.1 (0.64)	2.06 (0.628)	3.05 (0.930)	370 (113)	0.63 (0.19)
6	381 (35.4)	207 (63.1)	1.8 (0.55)	1.84 (0.561)	3.36 (1.02)	333 (101)	0.72 (0.22)
7	308 (28.6)	191 (58.2)	1.6 (0.49)	1.61 (0.491)	4.16 (1.27)	314 (95.7)	1.06 (0.32)

(a)

FIGURE 4.6 (*a*) Salt River below Stewart Mountain Dam, Arizona. *(Barnes, 1967.)* (*b*) Plan sketch and cross sections, Salt River below Stewart Mountain Dam, Arizona. *(Barnes, 1967.)* (*c*) Upstream along left bank, Salt River below Stewart Mountain Dam, Arizona. *(Barnes, 1967.)* (*d*) Section 818 downstream along left bank, Salt River Stewart Mountain Dam, Arizona. *(Barnes, 1967.)*

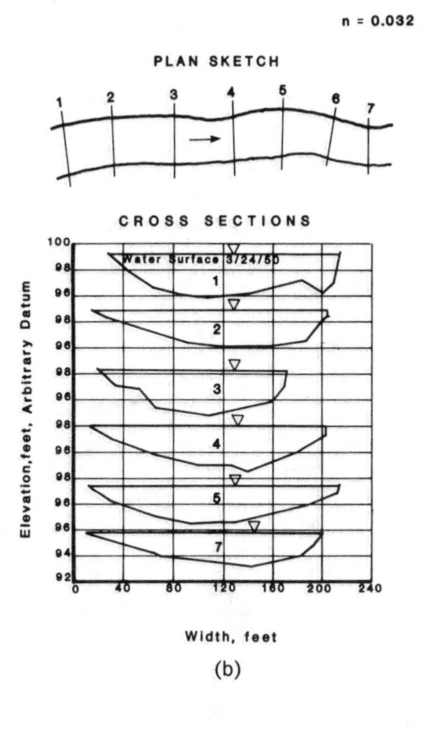
n = 0.032
PLAN SKETCH
1
2
3
4
5
6
7
CROSS SECTIONS
Water Surface 3/24/50
Elevation, feet, Arbitrary Datum
Width, feet
0
40
80
120
160
200
240

(b)

(c)

(d)

Drainage Area: 103 mi^2 (267 km^2)

Flood Date: Jan. 22, Feb. 13, Feb. 14, 1959

Peak Discharge: 2260 ft^3/s (64.0 m^3/s); 1850 ft^3/s (52.4 m^3/s); 515 ft^3/s (14.6 m^3/s)

Estimated Roughness Coefficient: n = 0.041; n = 0.039; n = 0.035

Channel Description: Bed is sand and gravel with several fallen trees in the reach. Banks are lined with overhanging trees and underbrush.

Section	Area, ft^2 (m^2)	Top width, ft (m)	Mean depth, ft (m)	Hydraulic radius, ft (m)	Average velocity, ft/s (m/s)	Distance between sections, ft (m)	Fall between sections, ft (m)
				Jan. 22, 1959			
7	618 (57.4)	60 (18)	10.3 (3.14)	8.15 (2.48)	3.66 (1.12)	—	—
8	621 (57.7)	66 (20)	9.4 (2.9)	7.77 (2.37)	3.64 (1.11)	316 (96.3)	0.22 (0.067)
9	619 (57.5)	63 (19)	9.8 (3.0)	8.18 (2.49)	3.65 (1.11)	286 (87.2)	0.18 (0.055)
10	610 (56.7)	63 (19)	9.7 (3.0)	7.77 (2.37)	3.70 (1.13)	293 (89.3)	0.14 (0.043)
11	622 (57.8	66 (20)	9.4 (2.9)	7.89 (2.40)	3.63 (1.11)	203 (61.9)	0.17 (0.052)
				Feb. 13, 1959			
7	528 (49.1)	57 (17)	9.3 (2.8)	7.37 (2.25)	3.50 (1.07)	—	—
8	531 (49.3)	64 (20)	8.3 (2.5)	6.95 (2.12)	3.48 (1.06)	316 (96.3)	0.19 (0.058)
9	532 (49.4)	59 (18)	9.0 (2.7)	7.56 (2.30)	3.48 (1.06)	286 (87.2)	0.20 (0.061)
10	525 (48.8)	61 (19)	8.6 (2.6)	7.00 (2.13)	3.52 (1.07)	293 (89.3)	0.14 (0.043)
11	530 (49.2)	62 (19)	8.6 (2.6)	7.10 (2.16)	3.49 (1.06)	203 (61.9)	0.14 (0.043)
				Feb. 14, 1959			
7	240 (22.3)	51 (16)	4.7 (1.4)	4.04 (1.23)	2.15 (0.655)	—	—
8	209 (19.4)	59 (18)	3.5 (1.1)	3.27 (0.997)	2.46 (0.750)	316 (96.3)	0.16 (0.049)
9	235 (21.8)	53 (16)	4.4 (1.3)	4.04 (1.23)	2.19 (0.668)	286 (87.2)	0.17 (0.052)
10	203 (18.6)	59 (18)	3.4 (1.0)	3.19 (0.972)	2.54 (0.774)	293 (89.3)	0.19 (0.058)
11	219 (20.3)	55 (17)	4.0 (1.2)	3.59 (1.09)	2.35 (0.716)	203 (61.9)	0.15 (0.046)

(a)

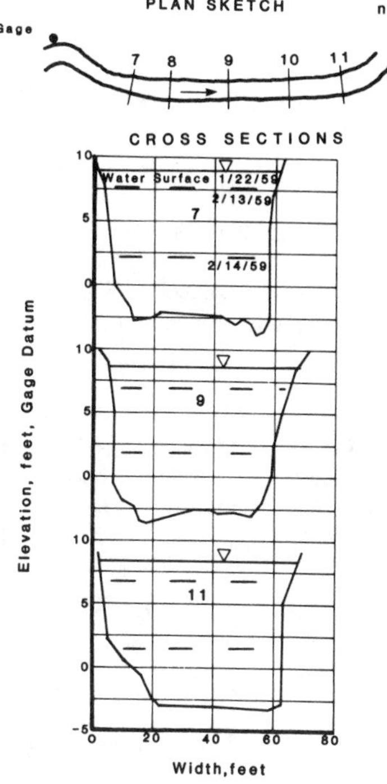

FIGURE 4.7 (*a*) Etowah river near Dawsonville, Georgia. *(Barnes, 1967.)* (*b*) Plan sketch and cross sections, Etowah river near Dawsonville, Georgia. *(Barnes, 1967.)* (*c*) Section 1167 downstream from right bank, Etowah River near Dawsonville, Georgia. *(Barnes, 1967.)* (*d*) Section 1168 upstream from right bank, Etowah River near Dawsonville, Georgia. *(Barnes, 1967.)*

(c)

(d)

Drainage Area: 388 mi^2 (1000 km^2)

Flood Date: June 1, 1948

Peak Discharge: 3220 ft^3/s (91.2 m^3/s)

Estimated Roughness Coefficient: $n = 0.041$

Channel Description: The bed is composed of sand, gravel, and angular rocks. The banks are irregular, eroded, and have a sparse cover of grass and scattered small trees. Although the channel between sections 1 and 2 is straight, the channel curves sharply above the reach and moderately below it.

Section	Area, ft^2 (m^2)	Top width, ft (m)	Mean depth, ft (m)	Hydraulic radius, ft (m)	Average velocity, ft/s (m/s)	Distance between sections, ft (m)	Fall between sections, ft (m)
1	818 (76.0)	114 (34.7)	7.2 (2.2)	6.79 (2.07)	3.94 (1.20)	—	—
2	735 (68.3)	102 (31.1)	7.2 (2.2)	6.87 (2.09)	4.38 (1.34)	315 (96.0)	0.38 (0.12)

(a)

FIGURE 4.8 (*a*) Bull Creek near Ira, Texas. *(Barnes, 1967.)* (*b*) Plan sketch and cross sections, Bull Creek near Ira, Texas. *(Barnes, 1967.)* (*c*) Section 342 downstream, Bull Creek near Ira, Texas. *(Barnes, 1967.)* (*d*) Section 345 upstream, Bull Creek near Ira, Texas. *(Barnes, 1967.)*

n = 0.041

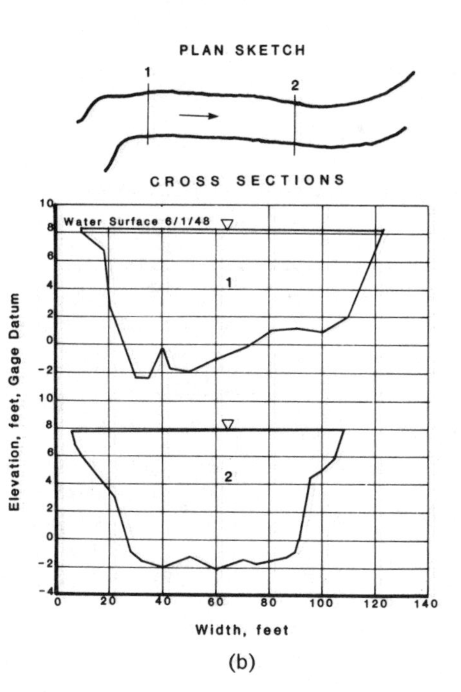
PLAN SKETCH
1
2
CROSS SECTIONS
Water Surface 6/1/48
1
2
Elevation, feet, Gage Datum
10
8
6
4
2
0
-2
10
8
6
4
2
0
-2
-4
0
20
40
60
80
100
120
140
Width, feet

(b)

(c)

(d)

Drainage Area: 24 mi^2 (62 km^2)

Flood Date: Feb. 7, 1958

Peak Discharge: 840 ft^3/s (23.8 m^3/s)

Estimated Roughness Coefficient: $n = 0.045$

Channel Description: Bed consists of sand and gravel. Both banks are lined with trees above and below the waterline.

Section	Area, ft^2 (m^2)	Top width, ft (m)	Mean depth, ft (m)	Hydraulic radius, ft (m)	Average velocity, ft/s (m/s)	Distance between sections, ft (m)	Fall between sections, ft (m)
2	215 (20.0)	45 (14)	4.8 (1.5)	4.21 (1.28)	3.93 (1.20)	—	—
3	212 (19.7)	42 (13)	5.0 (1.5)	4.20 (1.28)	3.99 (1.22)	78 (24)	0.21 (0.064)
4	227 (21.1)	50 (15)	4.5 (1.4)	4.02 (1.23)	3.73 (1.14)	124 (37.8)	0.35 (0.11)
5	202 (18.8)	43 (13)	4.7 (1.4)	4.02 (1.23)	4.19 (1.28)	92 (28)	0.21 (0.064)
6	187 (17.4)	41 (12)	4.6 (1.4)	3.74 (1.14)	4.52 (1.38)	71 (22)	0.35 (0.11)
7	200 (18.6)	48 (15)	4.2 (1.3)	3.64 (1.11)	4.23 (1.29)	88 (27)	0.11 (0.033)
8	198 (18.4)	45 (14)	4.4 (1.3)	3.94 (1.20)	4.27 (1.30)	149 (45.4)	0.38 (0.12)

(a)

FIGURE 4.9 (*a*) Murder Creek near Monticello, Georgia. *(Barnes, 1967.)* (*b*) Plan sketch and cross sections, Murder Creek near Monticello, Georgia. *(Barnes, 1967.)* (*c*) Section 1176 upstream from right bank, Murder Creek near Monticello, Georgia. *(Barnes, 1967.)* (*d*) Section 1177 downstream from left bank, Murder Creek near Monticello, Georgia. *(Barnes, 1967.)*

PLAN SKETCH
n = 0.045
2
3
4
5
6
7
8
Gage
Crest Stage Gage

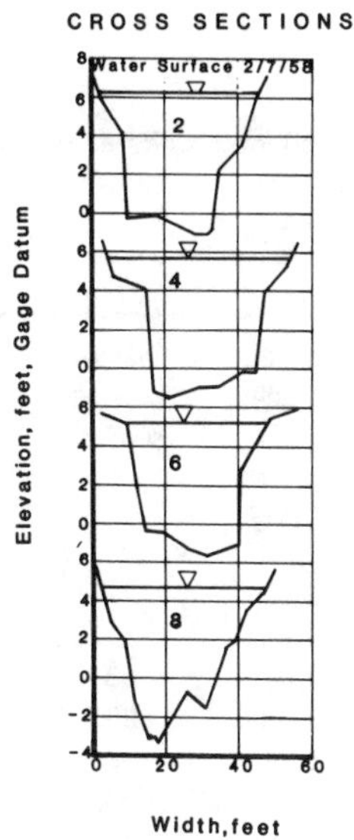
CROSS SECTIONS
Water Surface 2/7/58
Elevation, feet, Gage Datum
Width, feet

(b)

:)

(d)

Drainage Area: 233 mi² (604 km²)

Flood Date: June 13; Oct. 7, 1952

Peak Discharge: 1200 ft³/s (34.0 m³/s); 64.8 ft³/s (1.83 m³/s)

Estimated Roughness Coefficient: $n = 0.045$; $n = 0.073$

Channel Description: Bed and banks consist of smooth, rounded rocks as much as 1 ft (0.305 m) in diameter. Some undergrowth is below the water elevations of June 13, 1952.

Section	Area, ft² (m²)	Top width, ft (m)	Mean depth, ft (m)	Hydraulic radius, ft (m)	Average velocity, ft/s (m/s)	Distance between sections, ft (m)	Fall between sections, ft (m)
				June 13, 1952			
1	184 (17.1)	47 (14)	3.9 (1.2)	3.70 (1.13)	6.52 (1.99)	—	—
3	171 (15.9)	49 (15)	3.5 (1.1)	3.33 (1.02)	7.02 (2.14)	88 (26.8)	0.67 (0.20)
5	173 (16.1)	55 (17)	3.1 (0.94)	3.02 (0.920)	6.95 (2.12)	109 (33.2)	1.04 (0.317)
7	173 (16.1)	48 (15)	3.6 (1.1)	3.43 (1.05)	6.95 (2.12)	117 (35.7)	1.10 (0.335)
9	183 (17.0)	55 (17)	3.3 (1.0)	3.22 (0.982)	6.56 (2.00)	116 (35.4)	1.04 (0.317)
				Oct. 7, 1952			
1	36 (3.3)	38 (12)	1.0 (0.30)	0.95 (0.29)	1.79 (0.546)	—	—
3	38 (3.5)	34 (10)	1.1 (0.34)	1.10 (0.335)	1.70 (0.518)	88 (26.8)	0.32 (0.098)
5	34 (3.2)	32 (9.8)	1.1 (0.34)	0.82 (0.25)	1.90 (0.579)	109 (33.2)	0.84 (0.26)
7	34 (3.2)	39 (12)	0.9 (0.27)	0.86 (0.26)	1.91 (0.582)	117 (35.7)	1.28 (0.390)
9	31 (2.9)	41 (12)	0.8 (0.24)	0.76 (0.23)	2.08 (0.634)	116 (35.4)	1.12 (0.341)

(a)

FIGURE 4.10 (*a*) Provo River near Hailstone, Utah. *(Barnes, 1967.)* (*b*) Plan sketch and cross sections, Provo River near Hailstone, Utah. *(Barnes, 1967.)* (*c*) Section 770 upstream, Provo River near Hailstone, Utah. *(Barnes, 1967.)* (*d*) Section 769 downstream, Provo River near Hailstone, Utah. *(Barnes, 1967.)*

n = 0.045: 0.073

PLAN SKETCH

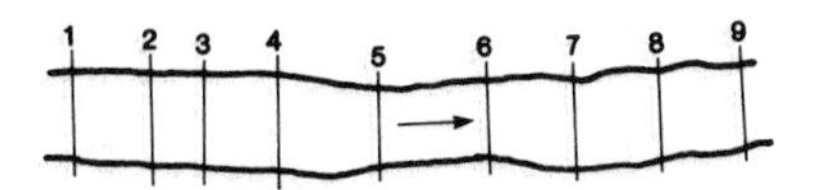

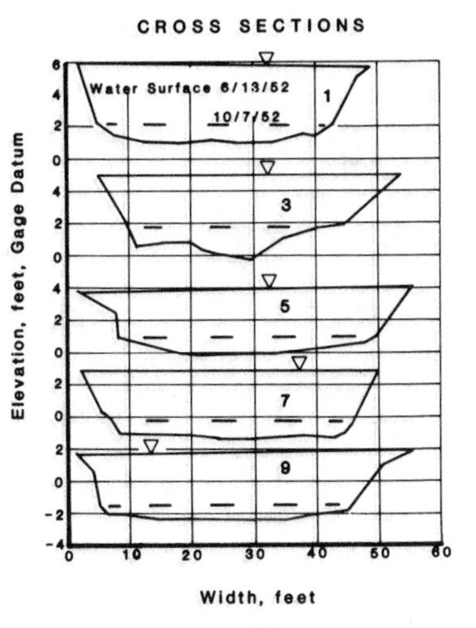

(b)

(c)

(d)

Drainage Area: 399 mi^2 (1030 km^2)

Flood Date: May 26, 1958

Peak Discharge: 1380 ft^3/s (39.1 m^3/s)

Estimated Roughness Coefficient: n = 0.050

Channel Description: Bed and banks are composed of angular boulders as much as 2.0 ft (0.61 m) in diameter.

Section	Area, ft^2 (m^2)	Top width, ft (m)	Mean depth, ft (m)	Hydraulic radius, ft (m)	Average velocity, ft/s (m/s)	Distance between sections, ft (m)	Fall between sections, ft (m)
12	200 (18.6)	43 (13)	4.6 (1.4)	4.22 (1.29)	6.90 (2.10)	—	—
13	206 (19.1)	50 (15)	4.1 (1.2)	3.83 (1.17)	6.70 (2.04)	47 (14)	0.85 (0.26)
14	183 (17.0)	52 (16)	3.5 (1.1)	3.29 (1.00)	7.54 (2.30)	39 (12)	0.65 (0.20)
15	184 (17.1)	51 (16)	3.6 (1.1)	3.36 (1.02)	7.50 (2.29)	46 (14)	0.65 (0.20)
16	184 (17.1)	55 (17)	3.3 (1.0)	2.69 (0.820)	7.50 (2.29)	32 (9.8)	0.60 (0.18)

(a)

FIGURE 4.11 (*a*) Clear Creek near Golden, Colorado. *(Barnes, 1967.)* (*b*) Plan sketch and cross sections, Clear Creek near Golden, Colorado. *(Barnes, 1967.)* (*c*) Section 1199 upstream from left bank, Clear Creek near Golden, Colorado. *(Barnes, 1967.)* (*d*) Section 1198 upstream, Clear Creek near Golden, Colorado. *(Barnes, 1967.)*

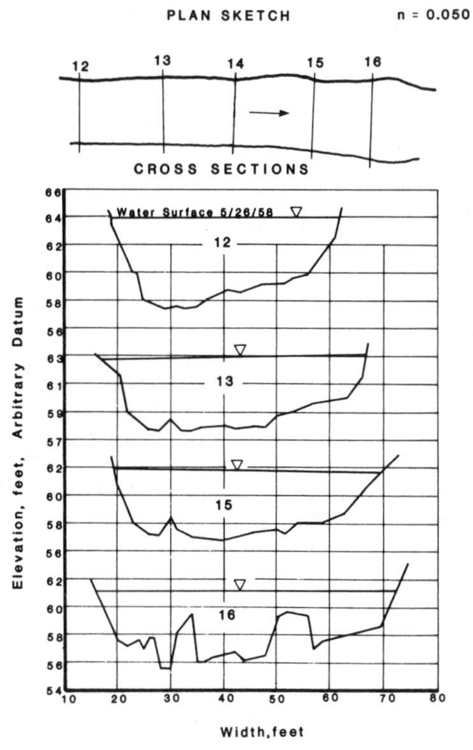
PLAN SKETCH
n = 0.050
12
13
14
15
16
CROSS SECTIONS
Water Surface 5/26/58
Elevation, feet, Arbitrary Datum
Width, feet

(b)

(c)

(d)

Drainage Area: 528 mi^2 (1370 km^2)

Flood Date: Jan. 24, Jan. 25, 1951

Peak Discharge: 3840 ft^3/s (109 m^3/s); 1830 ft^3/s (51.8 m^3/s)

Estimated Roughness Coefficient: n = 0.053; n = 0.079

Channel Description: Bed is composed of large, angular boulders. Banks consist of exposed rock, boulders, and some trees.

Section	Area, ft^2 (m^2)	Top width, ft (m)	Mean depth, ft (m)	Hydraulic radius, ft (m)	Average velocity, ft/s (m/s)	Distance between sections, ft (m)	Fall between sections, ft (m)
				Jan. 24, 1951			
1	401 (37.2)	81 (25)	5.0 (1.5)	4.27 (1.30)	9.59 (2.92)	—	—
2	384 (35.7)	65 (20)	5.9 (1.8)	5.02 (1.53)	10.00 (3.05)	102 (31.1)	1.80 (0.549)
3	295 (27.4)	45 (14)	6.6 (2.0)	5.27 (1.61)	13.02 (3.97)	62 (18.9)	3.10 (0.945)
				Jan. 25, 1951			
1	271 (25.2)	76 (23)	3.6 (1.1)	3.09 (0.942)	6.75 (2.06)	—	—
2	236 (21.9)	51 (16)	4.6 (1.4)	3.88 (1.18)	7.75 (2.36)	102 (31.1)	2.75 (0.838)
3	211 (19.6)	38 (12)	5.6 (1.7)	4.45 (1.36)	8.67 (2.64)	62 (18.9)	2.55 (0.777)

(a)

FIGURE 4.12 (*a*) Cache Creek near Lower Lake, California. *(Barnes, 1967.)* (*b*) Plan sketch and cross sections, Cache Creek near Lower Lake, California. *(Barnes, 1967.)* (*c*) Section 1191 downstream from left bank, Cache Creek near Lower Lake, California. *(Barnes, 1967.)* (*d*) Section 1193 downstream, Cache Creek near Lower Lake, California. *(Barnes, 1967.)*

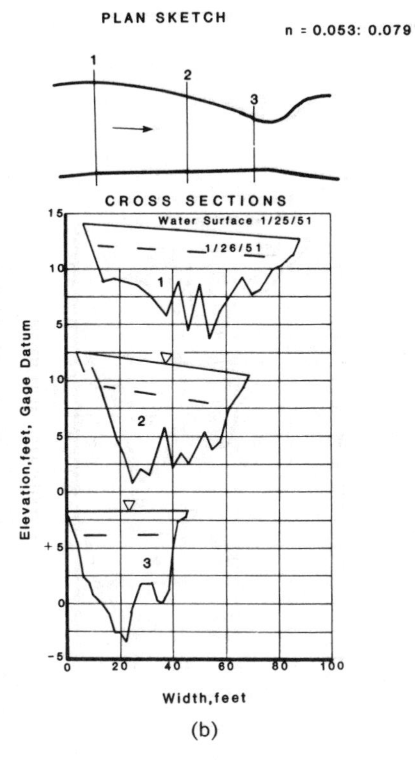
PLAN SKETCH
n = 0.053: 0.079
1
2
3
CROSS SECTIONS
Water Surface 1/25/51
1/26/51
Elevation,feet, Gage Datum
Width,feet
15
10
5
0
+5
-5
0
20
40
60
80
100

(b)

(c)

(d)

Drainage Area: 89,700 mi² (232,000 km²)

Flood Date: May 22, 1949

Peak Discharge: 406,000 ft³/s (11,500 m³/s)

Estimated Roughness Coefficient: $n = 0.024$

Channel Description: Bed consists of slime-covered cobbles and gravels. The straight and steep left bank is composed of cemented cobbles and gravel. The gently sloping right bank consists of cobbles set in gravel and is free of vegetation.

Section	Area, ft² (m²)	Top width, ft (m)	Mean depth, ft (m)	Hydraulic radius, ft (m)	Average velocity, ft/s (m/s)	Distance between sections, ft (m)	Fall between sections, ft (m)
1	47,100 (4380)	1800 (549)	26.2 (7.99)	26.16 (7.97)	8.65 (2.64)	—	—
2	49,000 (4550)	1650 (503)	29.7 (9.05)	29.56 (9.01)	8.28 (2.52)	2500 (762)	0.48 (0.15)
3	49,600 (4610)	1760 (536)	28.2 (8.60)	28.10 (8.56)	8.17 (2.49)	2500 (762)	0.49 (0.15)

(a)

FIGURE 4.13 (*a*) Columbia River at Vernita, Washington. *(Barnes, 1967.)* (*b*) Plan sketch and cross sections, Columbia River at Vernita, Washington. *(Barnes, 1967.)* (*c*) Section 66 upstream along right bank, Columbia River at Vernita, Washington. *(Barnes, 1967.)* (*d*) Section 67 upstream from top of bank, Columbia River at Vernita, Washington. *(Barnes, 1967.)*

n = 0.024

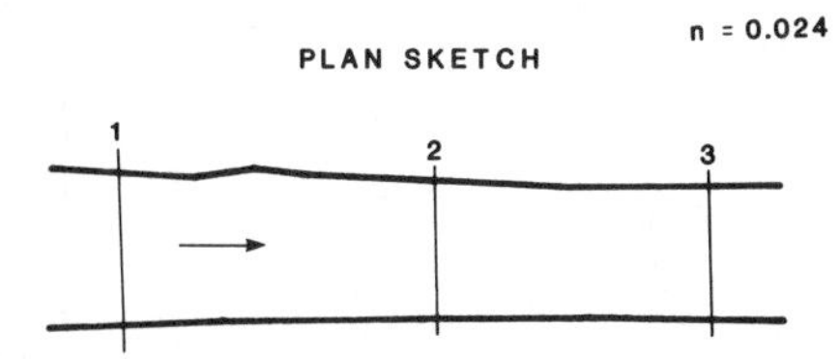
PLAN SKETCH
1
2
3

CROSS SECTIONS

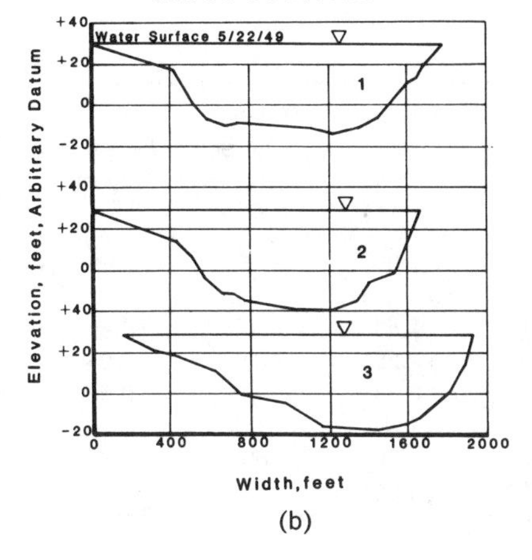
Water Surface 5/22/49
1
2
3
+40
+20
0
-20
Elevation, feet, Arbitrary Datum
0
400
800
1200
1600
2000
Width, feet

(b)

(c)

(d)

Drainage Area: 181 mi^2 (469 km^2)

Flood Date: May 17, 1950

Peak Discharge: 1950 ft^3/s (55.2 m^3/s)

Estimated Roughness Coefficient: $n = 0.065$

Channel Description: Fairly straight channel is composed of boulders with trees along top of banks.

Section	Area, ft^2 (m^2)	Top width, ft (m)	Mean depth, ft (m)	Hydraulic radius, ft (m)	Average velocity, ft/s (m/s)	Distance between sections, ft (m)	Fall between sections, ft (m)
1	308 (28.6)	78 (24)	4.0 (1.2)	3.90 (1.19)	6.33 (1.93)	—	—
2	263 (24.4)	64 (20)	4.1 (1.2)	3.98 (1.21)	7.41 (2.26)	200 (61)	3.40 (1.04)
3	309 (28.7)	63 (19)	4.9 (1.5)	4.68 (1.43)	6.31 (1.92)	40 (12)	0.50 (0.15)
4	327 (30.4)	78 (24)	4.2 (1.3)	4.09 (1.25)	5.96 (1.82)	180 (55)	1.55 (0.47)

(a)

FIGURE 4.14 (*a*) Merced River at Happy Isles Bridge, near Yosemite, California. *(Barnes, 1967.)* (*b*) Plan sketch and cross sections, Merced River at Happy Isles Bridge, near Yosemite, California. *(Barnes, 1967.)* (*c*) Section 1196 downstream, Merced River at Happy Isles Bridge, near Yosemite, California. *(Barnes, 1967.)* (*d*) Section 1197 upstream from right bank, Merced River at Happy Isles Bridge, near Yosemite, California. *(Barnes, 1967.)*

PLAN SKETCH

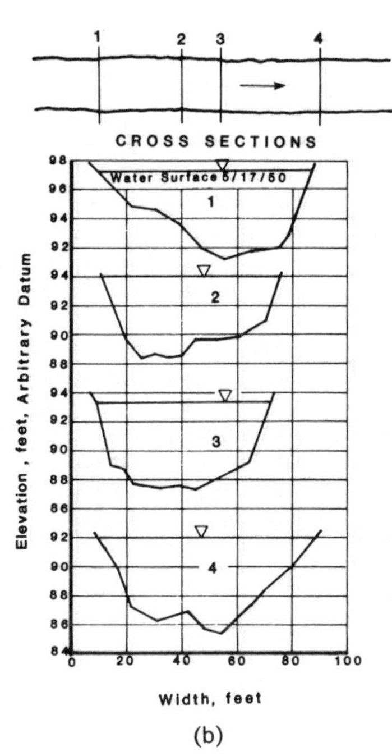
n = 0.065
1
2
3
4
CROSS SECTIONS
Water Surface 5/17/50
Elevation, feet, Arbitrary Datum
Width, feet

(b)

(c)

(d)

Drainage Area: 64.0 mi^2 (166 km^2)

Flood Date: Feb. 14, 1950

Peak Discharge: 1480 ft^3/s (41.9 m^3/s)

Estimated Roughness Coefficient: n = 0.070

Channel Description: Bed is fine sand and silt. Banks are irregular with fairly heavy growth of 2 to 8-in (5 to 20-cm) trees on the banks above low water particularly on the left bank. Reach sections 1, 2, 5, 6, 7 used to determine roughness coefficient. Bridge abutements form the constriction at section 3.

L

Section	Area, ft^2 (m^2)	Top width, ft (m)	Mean depth, ft (m)	Hydraulic radius, ft (m)	Average velocity, ft/s (m/s)	Distance between sections, ft (m)	Fall between sections, ft (m)
1	888 (82.5)	115 (35.0)	7.7 (2.3)	7.10 (2.16)	1.67 (0.509)	—	—
2	830 (77.1)	122 (37.2)	6.8 (2.1)	6.48 (1.98)	1.78 (0.543)	90 (27)	0.05 (0.015)
3	591 (54.9)	52 (16)	11.4 (3.47)	8.23 (2.51)	2.50 (0.762)	95 (29)	0.10 (0.031)
4	837 (77.8)	116 (35.4)	7.2 (2.2)	6.80 (2.07)	1.77 (0.540)	45 (14)	0.02 (0.0061)
5	818 (76.0)	94 (29)	8.7 (2.7)	8.06 (2.46)	1.81 (0.552)	79 (24)	0.05 (0.015)
6	791 (73.5)	107 (32.6)	7.4 (2.3)	7.00 (2.13)	1.87 (0.570)	156 (47.5)	0.10 (0.031)
7	854 (79.3)	115 (35.1)	7.4 (2.3)	7.00 (2.13)	1.73 (0.527)	147 (44.8)	0.07 (0.021)

(a)

FIGURE 4.15 (*a*) Pond Creek near Louisville, Kentucky. *(Barnes, 1967.)* (*b*) Plan sketch and cross sections, Pond Creek near Louisville, Kentucky. *(Barnes, 1967.)* (*c*) Section 452 downstream, Pond Creek near Louisville, Kentucky. *(Barnes, 1967.)* (*d*) Section 453 downstream from right bank, Pond Creek near Louisville, Kentucky. *(Barnes, 1967.)*

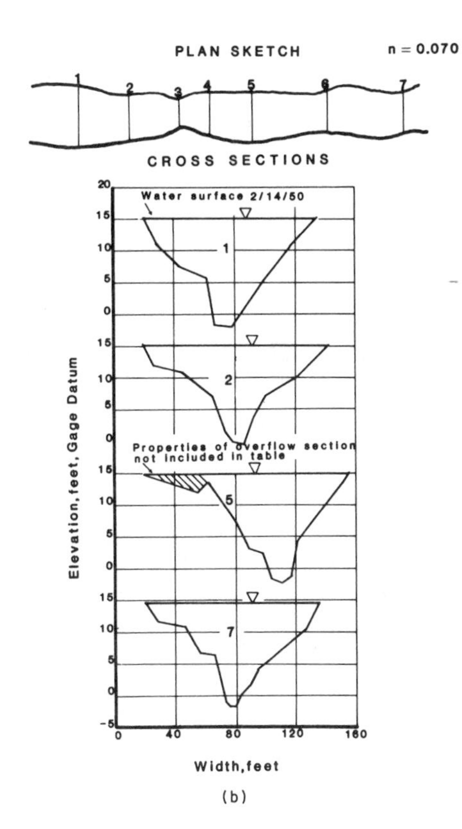
PLAN SKETCH
n = 0.070
1
2
3
4
5
6
7
CROSS SECTIONS
Water surface 2/14/50
Properties of overflow section not included in table
Elevation, feet, Gage Datum
Width, feet
20
15
10
5
0
-5
0
40
80
120
160

(b)

(c)

(d)

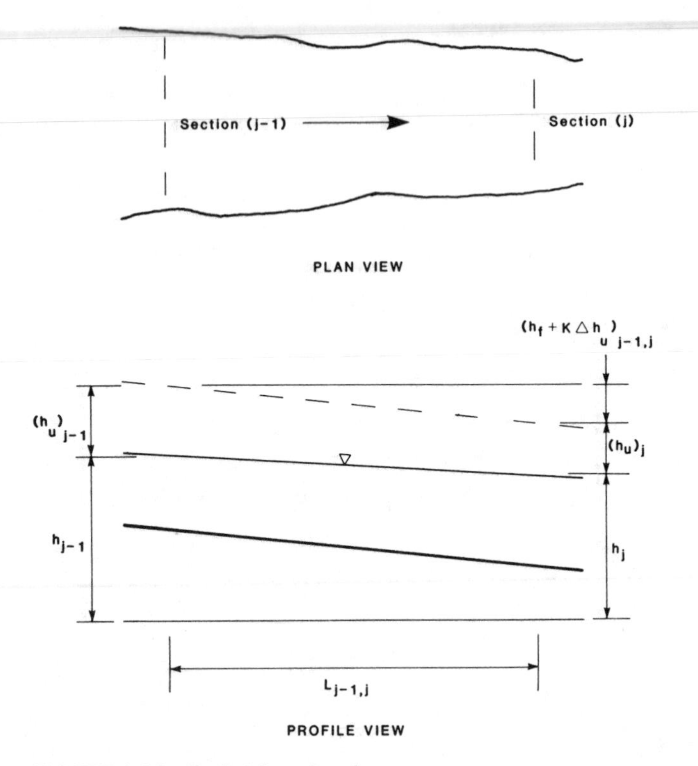

FIGURE 4.16 Definition sketch.

Velocity Measurement From a theoretical viewpoint, a value of the resistance coefficient can be estimated from velocity measurements since the velocity profile depends on the perimeter roughness height. In hydraulically rough flows, the vertical velocity distribution can be approximated by

$$u = 5.75\, u_* \log \frac{30y}{k_s} \tag{4.3.12}$$

where y = distance from the bottom boundary and k_s = roughness height. Let $u_{0.2}$ = velocity at two-tenths of the depth, i.e., at $0.8D$ above the bottom where D = depth of flow and $u_{0.8}$ = velocity at eight-tenths the depth, i.e., at $0.2D$ above the bottom. Substituting these definitions in Eq. (4.3.12),

$$u_{0.2} = 5.75\, u_* \log \frac{24D}{k_s}$$

and

$$u_{0.8} = 5.75\, u_* \log \frac{6D}{k_s}$$

Eliminating u_* from the equations above,

$$\log \frac{D}{k_s} = \frac{0.778x - 1.381}{1 - x} \tag{4.3.13}$$

where $x = u_{0.2}/u_{0.8}$. Substituting Eq. (4.3.13) in Eq. (1.4.33) where it is assumed that $R \simeq D$ yields

$$\frac{u}{u_*} = \frac{1.78(x + 0.95)}{x - 1} \tag{4.3.14}$$

Recall that

$$\frac{u}{u_*} = \frac{C}{\sqrt{g}}$$

and using this relation and Eq. (4.2.7) yields

$$\frac{u}{u_*} = \frac{D^{1/6}}{3.81} \tag{4.3.15}$$

Equating the right-hand sides of Eqs. (4.3.14) and (4.3.15) and solving for n,

$$n = \frac{(x - 1)D^{1/6}}{6.78(x + 0.95)} \tag{4.3.16}$$

Equation (4.3.16) estimates values of n for fully rough flows in a wide channel which have a logarithmic vertical velocity distribution. In the case of an actual stream, D may be taken as the mean depth. Chow (1959) hypothesized this methodology and noted that it should be used with caution since it had not been verified. Although French and McCutcheon (1977) found that Eq. (4.3.16) provided reasonable estimates of n in a study on the Cumberland River, Tennessee, additional verification of the methodology is still required.

Empirical Methods A number of empirical methods have been advanced for estimating n. Perhaps the best known of these methods is the one proposed by Strickler in 1923; see, for example, Simons and Senturk (1976), which hypothesizes that

$$n = 0.047d^{1/6} \tag{4.3.17}$$

where d = diameter in millimeters of the uniform sand pasted to the sides and bottom of the flume used by Strickler in his experiments. Simons and Senturk (1976) asserted that, because of the experimental design used by Strickler, Eq. (4.3.17) does not apply to flows over mobile beds. Note, there is significant disagreement between authors regarding the circumstances of the original Strickler experiments, the value of the coefficient in Eq. (4.3.17), the definition of the variable d, and the units associated with d. For example,

1. Henderson (1966) claimed that Strickler's research was based on streams with gravel beds, not a flume, and that d was the median size of the bed

material. The equation given by Henderson (1966) and attributed to Strickler was

$$n = 0.034d^{1/6} \tag{4.3.18}$$

where the units of d were not specified.

2. Raudkivi (1976) stated that the Strickler equation was

$$n = 0.042d^{1/6} \tag{4.3.19}$$

where d is measured in meters or

$$n = 0.013d_{65}^{1/6} \tag{4.3.20}$$

where d_{65} = diameter of the bed material in millimeters such that 65 percent of the material by weight is smaller. If the dimensions of Eq. (4.3.20) are given in feet, then

$$n = 0.034d_{65}^{1/6} \tag{4.3.21}$$

where the coefficient has the same numerical value as that given by Henderson, Eq. (4.3.18). Raudkivi (1976) further noted that Eqs. (4.3.19) to (4.3.21) are useful in selecting roughness heights in fixed bed hydraulic models.

3. Garde and Raju (1978) state that Strickler analyzed data from various streams in Switzerland which had beds formed of coarse materials and which were free from bed undulations. The equation given by these authors and attributed to Strickler was

$$n = 0.039d_{50}^{1/6} \tag{4.3.22}$$

where d_{50} = diameter of the bed material in feet such that 50 percent of the material by weight is smaller.

4. Subramanya (1982) gave the Strickler equation as

$$n = 0.047d_{50}^{1/6} \tag{4.3.23}$$

where d_{50} = diameter of the bed material in meters such that 50 percent of the material by weight is smaller. It can be easily shown that Eqs. (4.3.22) and (4.3.23) are equivalent; and therefore, there is no difference between the equations given by Garde and Raju (1978) and Subramanya (1982).

The above discussion is not intended to prove a previous author in error, rather it is presented to highlight that there are several different interpretations of an important concept published in what is now a rather obscure publication.

For mixtures of bed materials with a significant proportion of coarse grained sizes, Meyer-Peter and Muller (1948) suggested the following equation:

$$n = 0.038d_{90}^{1/6} \tag{4.3.24}$$

where d_{90} = bed size in meters such that 90 percent of material by weight is smaller. In field experiments involving canals paved with cobbles, Lane and Carlson (1953) determined that

$$n = 0.026 d_{75}^{1/6} \qquad (4.3.25)$$

where d_{75} = diameter of the bed material in inches such that 75 percent of the material is smaller by weight.

EXAMPLE 4.2

For the channel defined in Example 4.1, determine the minimum slope which is required if Manning's n is to be assumed constant.

Solution

From Example 4.1

$n = 0.13$
Depth of flow $y_N = 8.5$ ft (2.6 m)
Bottom width $b = 10$ ft (3.0 m)
Side slope 1.5:1

Then from Table 1.1

$$A = (b + zy)y = [10 + 1.5(8.5)]8.5 = 194 \text{ ft}^2 \ (18.0 \text{ m}^2)$$

$$P = b + 2y\sqrt{1 + z^2} = 10 + 2(8.5)\sqrt{3.25} = 40.6 \text{ ft} \ (12.4 \text{ m})$$

$$R = A/P = 194/40.6 = 4.8 \text{ ft} \ (1.5 \text{ m})$$

From Eq. (4.3.6)

$$S = \left(\frac{1.9 \times 10^{-13}}{n^6\sqrt{R}}\right)^2 = \left[\frac{1.9 \times 10^{-13}}{(0.13)^6\sqrt{4.8}}\right]^2 = 3.2 \times 10^{-16}$$

Therefore, if the slope of the channel described in Example 4.1 exceeds 3.2×10^{-16} the assumption of a constant value of n is valid.

BIBLIOGRAPHY

Ackers, P., "Resistance to Fluids Flowing in Channels and Pipes," Hydraulic Research Paper No. 1, H.M.S.O., London, 1958.

Anonymous, "Report of the American Society of Civil Engineers' Task Force on Friction Factors in Open Channels," *Proceedings of the American Society of Civil Engineers, Journal of the Hydraulics Division,* vol. 89, No. HY2, March 1963a, pp. 97–143.

Anonymous, "Guide for Selecting Roughness Coefficient 'n' Values for Channels," U.S. Department of Agriculture, Soil Conservation Service, Washington, December 1963b.

Barnes, H. H., Jr., "Roughness Characteristics of Natural Channels," U.S. Geological Survey Water-Supply Paper 1849, U.S. Geological Survey, Washington, 1967.

Chow, V. T., *Open Channel Hydraulics,* McGraw-Hill Book Company, New York, 1959.

French, R. H., and McCutcheon, S. C., "The Stability of a Two Layer Flow without Shear in the Presence of Boundary Generated Turbulence: Field Verification," Technical Report No. 39, Environmental and Water Resources Engineering, Vanderbilt University, Nashville, Tenn., October 1977.

Garde, R. J., and Ranga Raju, K. G., *Mechanics of Sediment Transportation and Alluvial Stream Problems,* Wiley Eastern, New Delhi, 1978.

Henderson, F. M., *Open Channel Flow,* The Macmillan Company, New York, 1966.

Lane, E. W., and Carlson, E. J., "Some Factors Affecting the Stability of Canals Constructed in Coarse Granular Materials," *Proceedings of the Minnesota International Hydraulics Convention,* September 1953.

Meyer-Peter, E., and Muller, R., "Formulas for Bed-Load Transport," *Proceedings of the 3rd Meeting of IAHR,* Stockholm, pp. 39–64, 1948.

Raudkivi, A. J., *Loose Boundary Hydraulics,* 2d ed., Pergamon Press, New York, 1976.

Simons, D. B., and Senturk, F., *Sediment Transport Technology,* Water Resources Publications, Fort Collins, Col., 1976.

Streeter, V. L., and Wylie, E. B., *Fluid Mechanics,* McGraw Hill Book Company, New York, 1975.

Subramanya, K., *Flow in Open Channels,* vol. 1, Tata McGraw-Hill Publishing Company, New Delhi, 1982.

Urquhart, W. J., "Hydraulics," *Engineering Field Manual,* U.S. Department of Agriculture, Soil Conservation Service, Washington, 1975.

Zegzhda, A. P., "Teoriia podobiia i metodika rascheta gidrotekhnicheskikh modelei" ("Theory of Similarity and Methods of Design of Models for Hydraulic Engineering"), *Gosstroiisdat,* Leningrad, 1938.

FIVE

Computation of Uniform Flow

SYNOPSIS

In this chapter primary emphasis is placed on discussing techniques of computing the normal depth of flow in open channels. The most tedious and difficult normal-flow calculation occurs when the Manning resistance equation cannot be solved explicitly for the normal depth of flow. Solution techniques including trial and error, graphical, and numerical for the solution of this implicit problem are discussed.

In the third section of this chapter, the problem of estimating composite or average boundary roughness in designed, natural, and laboratory channels is treated. Several techniques for estimating composite flow resistance coefficients in designed and natural channels are considered. In the case of laboratory flumes with artificially roughened bottoms and hydraulically smooth sidewalls, a technique of estimating the bottom boundary resistance coefficient is presented.

In the fourth section of this chapter, a number of applications of uniform flow concepts to practice are considered, e.g., slope-area peak flood flow computations, ice-covered channels, and normal discharge in channels of compound section.

5.1 CALCULATION OF NORMAL DEPTH AND VELOCITY

In the previous chapter, uniform flow, its governing equations, and various methodologies for estimating the resistance coefficient were described. Recall from that chapter that the Manning equation asserts that

$$\overline{u} = \frac{\phi}{n} R^{2/3} \sqrt{S} \tag{5.1.1}$$

and by the law of conservation of mass this equation, when multiplied by the flow area, yields an equation for the uniform flow rate or

$$Q = \overline{u}A = \frac{\phi}{n} AR^{2/3} \sqrt{S} \tag{5.1.2}$$

In Eq. (5.1.2) the parameter $AR^{2/3}$ is termed the *section factor* and

$$K = \frac{\phi}{n} AR^{2/3} \tag{5.1.3}$$

is by definition the conveyance of the channel. For a given channel where $AR^{2/3}$ always increases with increasing depth, each discharge has a corresponding unique depth at which uniform flow occurs.

An examination of Eqs. (5.1.1) and (5.1.2) demonstrates that the average uniform velocity of flow or the flow rate is a function of (1) the channel shape, (2) the resistance coefficient, (3) the longitudinal slope of the channel, and (4) the depth of flow or

$$Q = f(\Gamma, n, S, y_N) \tag{5.1.4}$$

where Γ = a channel shape factor and y_N = normal depth of flow. If four of the five variables in Eq. (5.1.4) are known, then the fifth variable can be determined.

EXAMPLE 5.1

Given a trapezoidal channel with a bottom width of 3 m (10 ft), side slopes of 1.5:1, a longitudinal slope of 0.0016, and a resistance coefficient of $n = 0.013$, determine the normal discharge if the normal depth of flow is 2.6 m (8.5 ft).

Solution

From Table 1.1

$$A = (b + zy)y = [3.0 + 1.5(2.6)]2.6 = 18 \text{ m}^2 \ (190 \text{ ft}^2)$$

$$P = b + 2y\sqrt{1 + z^2} = 3.0 + 2(2.6)\sqrt{3.25} = 12 \text{ m} \ (39 \text{ ft})$$

$$R = \frac{A}{P} = \frac{18}{12} = 1.5 \text{ m} \ (4.9 \text{ ft})$$

From Eq. (5.1.2)

$$Q = \frac{\phi}{n} AR^{2/3}\sqrt{S} = \frac{1.0}{0.013}(18)(1.5)^{2/3}\sqrt{0.0016} = 73 \text{ m}^3/\text{s} \ (2600 \text{ ft}^3/\text{s})$$

In general, the most difficult and tedious normal flow calculation occurs when Q, Γ, S, and n are known and y_N must be estimated. In such a case, an explicit solution of Eq. (5.1.2) is not possible and the problem must be solved by trial and error, design charts, or numerical methods. The initial section of this chapter will examine three methodologies which are useful in solving this type of problem.

Trial-and-Error and Graphical Solutions

If Q, Γ, S, and n are known, then Eq. (5.1.2) may be rearranged to yield

$$AR^{2/3} = \frac{nQ}{\phi\sqrt{S}} \tag{5.1.5}$$

where the right-hand side of the equation is a constant and the left-hand side is the section factor which is a function of the channel shape and the depth of flow. The most direct approach to finding y_N is a trial-and-error solution of Eq. (5.1.5).

EXAMPLE 5.2

Given a trapezoidal channel with a bottom width of 10 ft (3.0 m), side slopes of 1.5:1, a longitudinal slope of 0.0016, and estimated n of 0.13, find the normal depth of flow for a discharge of 250 ft³/s (7.1 m³/s).

Solution

$$AR^{2/3} = \frac{nQ}{\phi\sqrt{S}} = \frac{0.13(250)}{1.49\sqrt{0.0016}} = 545$$

with

$$A = (b + zy)y = (10 + 1.5y)y$$

$$P = b + 2y\sqrt{1 + z^2} = 10 + 2y\sqrt{3.25} = 10 + 3.6y$$

$$R = \frac{(10 + 1.5y)y}{10 + 3.6y}$$

Then the following table is constructed by assuming values of y and computing a corresponding value of $AR^{2/3}$. When the computed value of $AR^{2/3}$ matches the value computed from the problem statement, the correct value of y_N has been determined.

Trial y, ft	A, ft²	P, ft	R, ft	$AR^{2/3}$
8.0	176	38.8	4.54	482
9.0	212	42.4	5.00	620
8.5	193	40.6	4.75	546

From the information contained in this table, it is concluded that $y_N = 8.5$ ft (2.6 m) is the normal depth of flow for the channel and flow rate specified.

A second method of solving this problem is by the construction of a graph of the section factor versus the depth of flow.

EXAMPLE 5.3

Given a circular culvert 3.0 ft (0.91 m) in diameter with $S = 0.0016$ and $n = 0.015$, find the normal depth of flow for a discharge of 15 ft³/s (0.42 m³/s).

Solution

$$AR^{2/3} = \frac{nQ}{\phi\sqrt{S}} = \frac{0.015(15)}{1.49\sqrt{0.0016}} = 3.78$$

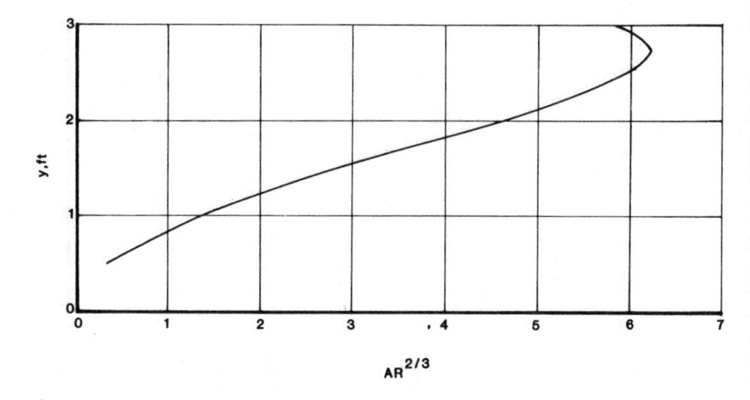

FIGURE 5.1 $AR^{2/3}$ versus y for a circular conduit 3.0 ft (0.91 m) in diameter.

A graph of $AR^{2/3}$ versus y is constructed (Fig. 5.1). For $AR^{2/3} = 3.78$, $y_N = 1.7$ ft. With regard to Fig. 5.1, it is noted that the section factor first increases with depth, and then as the full depth is approached it decreases with increasing depth. Therefore, it is possible to have two values of y_N for some values of $AR^{2/3}$.

The methods discussed above have the advantage of directness and the disadvantage of being cumbersome if a number of problems must be solved. When a number of such problems must be solved, a different approach is required.

General Design Charts

In order to simplify the computation of the normal depth for common channel shapes, dimensionless curves for the section factor as a function of the depth have been prepared for rectangular, circular, and trapezoidal channels (Fig. 5.2). Although these curves provide solutions to the problems of normal depth computation for these channel shapes in a manner similar to that used in Example 5.3, they do not provide a general method of solution.

Numerical Methods

If a computer is available and a large number of normal depth estimation problems must be solved, then a numerical trial-and-error procedure may be the best approach. A logic diagram for a numerical solution in the case of rectangular, trapezoidal, triangular, circular, and natural channels is shown in Fig. 5.3, and a listing of an interactive program which solves the normal depth problem is given in Appendix I.

In this interactive computer program, the user defines the type of channel

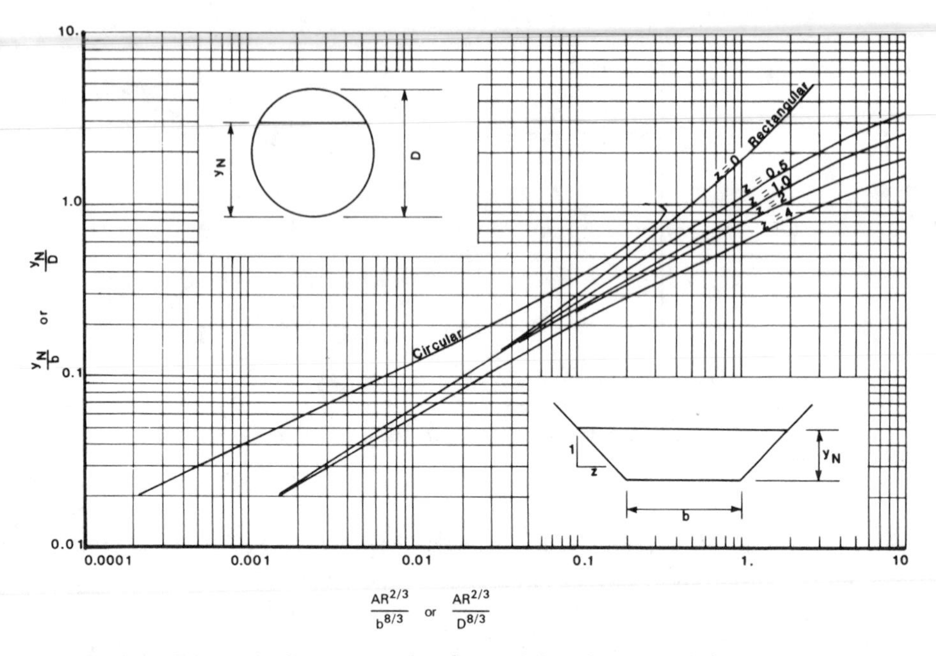

FIGURE 5.2 Dimensionless curves for determining the normal depth of flow when the section factor is known.

and specifies the value of the section factor. The subroutines RECT, TRAP, CIRC, and NATDAT create and search an array of depths of flow and their corresponding values of the section factor until an interval which includes the specified section factor is found. This interval is then subdivided and a cubic spline is fitted to the interval in subroutine SPCOEF. The value of y_M corresponding to the user-specified value of the section factor is estimated in subroutine SPLINE. With regard to this program, the following comments are noted. First, triangular channels are treated as a special case of trapezoidal channels. Second, for natural channels, the cross section must be specified in terms of the distance from one side of the channel and the height of the channel perimeter above some arbitrary datum. Third, the subroutines SPCOEF and SPLINE are standard cubic spline interpolation subroutines similar to those presented by Pine and Allen (1973).

While the numerical solution of the normal depth computation is cost-effective, if the user has a number of problems to solve and has access to a small computer, it is a methodology which can lead to serious errors when the user is not entirely familiar with the program structure. Therefore, the reader is advised to carefully review the computational scheme in Appendix I and make any adaptations required before applying it to a specific problem.

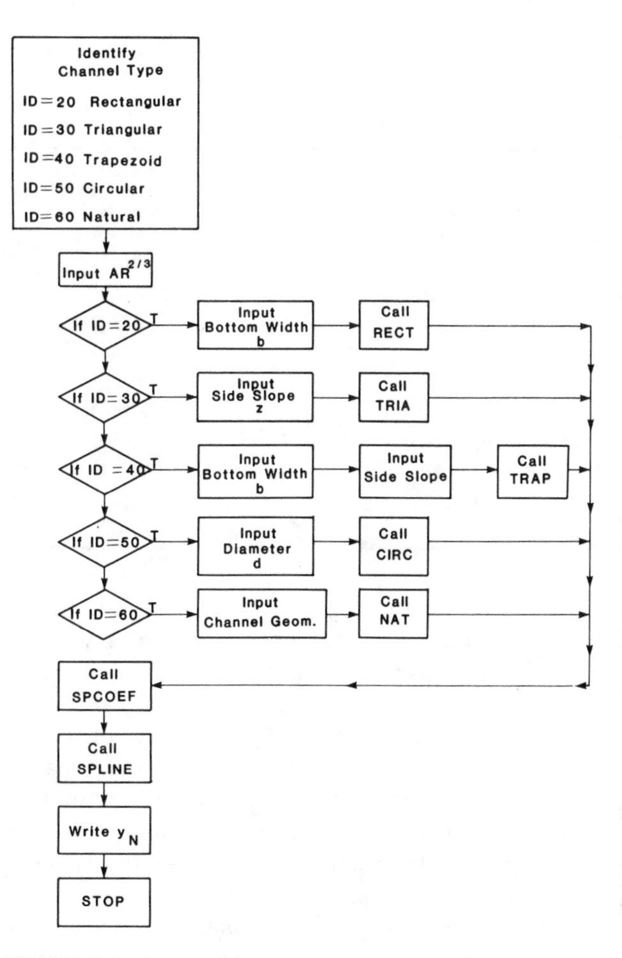

FIGURE 5.3 Logic diagram for an interactive computer solution of the normal depth of flow estimation problem.

EXAMPLE 5.4

The cross section in Fig. 5.4 is the result of a bathymetric survey of a river. If Q = 10,000 ft^3/s (283 m^3/s), n = 0.024, and S = 0.0001, estimate the normal depth of flow.

Solution

The value of the section factor is

$$AR^{2/3} = \frac{nQ}{\phi\sqrt{S}} = \frac{0.024(10{,}000)}{1.49\sqrt{0.0001}} = 16{,}107$$

The cross-sectional data required by the normal depth computer program are tabulated below.

River stage, ft (m), above an arbitrary datum	Distance to first perimeter intersection from south bank, ft (m)	Distance to second perimeter intersection from south bank, ft (m)
15 (4.6)	330(100)	330 (100)
20 (6.1)	240(73)	460 (140)
25 (7.6)	200(61)	530 (160)
30 (9.1)	170(52)	600 (180)
35 (11)	150(46)	710 (220)
40 (12)	130(40)	850 (260)
45 (14)	110(34)	1200 (365)
50 (15)	80(24)	1220 (370)
55 (17)	20(6.1)	1230 (375)

With regard to Fig. 5.4, the tabular data above, and the normal depth computation program, the following caveats should be noted. First, the channel has a single deep section. The computational scheme, although it could easily be modified, is not designed to treat a channel with two deep sections. Second, the perimeter of the channel is defined by elevations above an arbitrary datum and distances from one of the sides to the intersections with the channel boundaries. Third, the dimension statements in the program allow only 10 perimeter points to be defined. These statements could also be changed as required. The material below is the result of using the computer program in Appendix I to solve the problem. In this material, computer-generated responses are indicated by underlining, and user-provided information is indicated in bold type.

a.out

input computed value of ar**0.667

16107.

identify channel type: 20 rectangle;
30 triangle;
40 trapezoid;
50 circular;
60 natural

60

this is an interactive data program
define the channel perimeter in terms of a stage above a

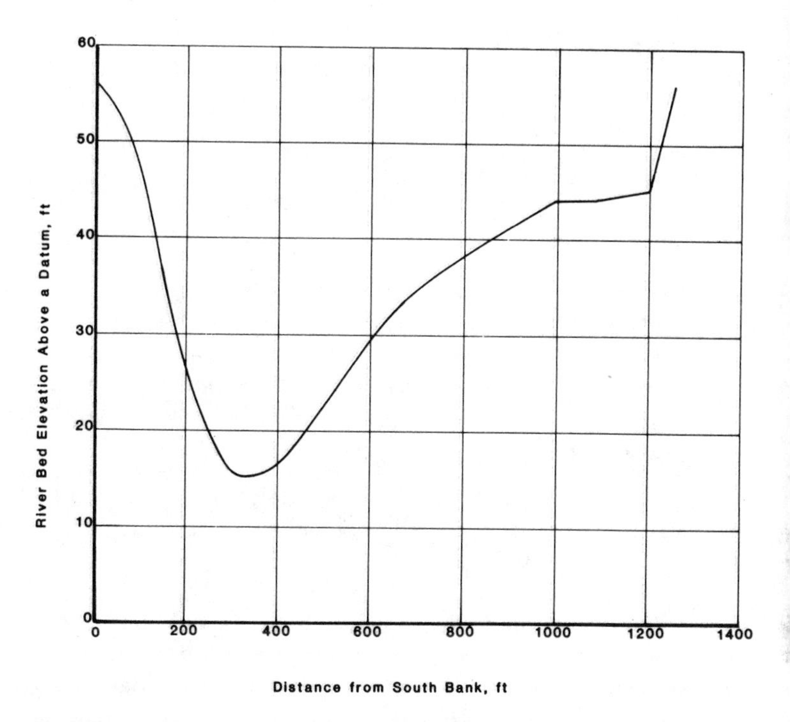

FIGURE 5.4 River bed elevation as a function of distance from the south bank.

<u>datum and two intersection points</u>
<u>input number of triplets to be read I2</u>

9

<u>input triplet number 1</u>

15. 330. 330.

<u>input triplet number 2</u>

20. 240. 460.

<u>input triplet number 3</u>

25. 200. 530.

<u>input triplet number 4</u>

30. 170. 600.

<u>input triplet number 5</u>

35. 150. 710.

<u>input triplet number 6</u>

40. 130. 850.
input triplet number 7

45. 110. 1200.
input triplet number 8

50. 80. 1220.
input triplet number 9

55. 20. 1230.
normal stage of flow = 29.86

Thus, the normal depth of flow is approximately $y_N = 30$ ft (9.1 m).

5.2 NORMAL AND CRITICAL SLOPES

If Q, n, y_N, and the channel section are defined, then Eq. (5.1.2) can be solved explicitly for the slope which allows the flow to occur as specified. By definition, this slope is a normal slope.

If the slope of a channel is varied while the discharge and roughness are held constant, it is possible to find a value of the slope such that normal flow occurs in a critical state, i.e., a slope such that normal flow occurs with $\mathbf{F} = 1$. The slope obtained in this fashion is, by definition, the critical slope but also a normal slope. The smallest critical slope for a specified channel shape, discharge, and roughness is termed the *limiting critical slope.*

On the basis of these definitions, it must be possible to define a critical slope and discharge which correspond to a specified normal depth in a given channel. This slope is known as the critical slope at the specified normal depth. The following two examples illustrate the crucial points of this discussion.

EXAMPLE 5.5

A rectangular channel has a bottom width of 6.0 m (19.7 ft) and $n = 0.020$.

a. For $y_N = 1.0$ m (3.3 ft) and $Q = 11$ m^3/s (388 ft^3/s) find the normal slope.

b. Find the limiting critical slope and normal depth of flow for $Q = 11$ m^3/s (388 ft^3/s).

c. Find the critical slope for $y_N = 1.0$ m (3.3 ft) and determine the discharge which corresponds to this depth of flow and slope.

Solution

a. For the given data

$$A = by = 6.0(1.0) = 6.0 \text{ m}^2 \text{ (65 ft}^2\text{)}$$

$$P = b + 2y = 6 + 2(1.0) = 8.0 \text{ m (26 ft)}$$

$$R = \frac{A}{P} = \frac{6}{8} = 0.75 \text{ m (2.5 ft)}$$

Rearranging Eq. (5.1.2),

$$S = \left(\frac{nQ}{\phi AR^{2/3}}\right)^2 = \left[\frac{0.02(11)}{1.0(6)(0.75)^{2/3}}\right]^2 = 0.002$$

Thus, $S = 0.002$ will maintain a uniform flow of 11 m^3/s (388 ft^3/s) in this channel at a depth of 1.0 m (3.3 ft).

b. The critical depth of flow is found by Eq. (2.2.3) or

$$\frac{\overline{u}^2}{2g} = \frac{D}{2}$$

where
$$\overline{u} = \frac{Q}{by} = \frac{11}{6y_N}$$

and since it is a rectangular channel,

$$D = y_N = \frac{Q^2}{g(by_N)^2}$$

$$y_N = \left(\frac{Q^2}{gb^2}\right)^{1/3} = \left[\frac{(11)^2}{9.8(6)^2}\right]^{1/3} = 0.70 \text{ m (2.3 ft)}$$

Then

$$A = by = 6(0.70) = 4.2 \text{ m}^2 \text{ (45 ft}^2\text{)}$$

$$P = b + 2y = 6 + 2(0.70) = 7.4 \text{ m (24 ft)}$$

$$R = \frac{4.2}{7.4} = 0.57 \text{ m (1.9 ft)}$$

Rearranging Eq. (5.1.2),

$$S_c = \left(\frac{nQ}{\phi AR^{2/3}}\right)^2 = \left[\frac{0.02(11)}{1.0(4.2)(0.57)^{2/3}}\right]^2 = 0.0058$$

where S_c = a critical slope. This is the slope which will maintain a critical and uniform flow of 11 m^3/s (388 ft^3/s) in the specified channel.

c. From part *a*, for a normal depth of 1.0 m (3.3 ft)

$$A = 6.0 \text{ m}^2 \text{ (65 ft}^2\text{)}$$

$$P = 8.0 \text{ m (26 ft)}$$

$$R = 0.75 \text{ m (2.5 ft)}$$

$$P = 8.0 \text{ m (26 ft)}$$

$$R = 0.75 \text{ m (2.5 ft)}$$

The velocity may then be found from the definition of critical flow or

$$\mathbf{F} = \frac{\overline{u}}{\sqrt{gy_N}} = 1$$

$$\overline{u} = \sqrt{9.8(1)} = 3.1 \text{ m/s (10 ft/s)}$$

Then rearranging Eq. (5.1.1),

$$S_{cN} = \left(\frac{n\overline{u}}{\phi R^{2/3}}\right)^2 = \left[\frac{0.02(3.1)}{1.0(0.75)^{2/3}}\right]^2 = 0.0056$$

where S_{cN} = slope which will maintain a critical and uniform flow at a depth of 1.0 m (3.3 ft) in the specified channel. The flow rate is given by

$$Q = \overline{u}A = 3.1(6) = 19 \text{ m}^3\text{/s (670 ft}^3\text{/s)}$$

EXAMPLE 5.6

For a trapezoidal channel with a bottom width of 20.4 ft (6.2 m), side slopes of 0.5:1, and $n = 0.02$, develop a graph of Q versus S_c, the limiting critical slope.

Solution

From Eq. (2.2.3), the condition for critical flow in a trapezoidal channel is

$$\frac{\overline{u}^2}{g} = D$$

where from Table 1.1

$$A = (b + zy)y = (20.4 + 0.5y)y$$

$$T = b + 2zy = 20.4 + y$$

$$D = \frac{A}{T} = \frac{(20.4 + 0.5y)y}{20.4 + y}$$

and

$$\overline{u} = \frac{Q}{A} = \frac{Q}{(20.4 + 0.5y)y}$$

Combining these relations yields an implicit equation for the critical depth or

$$\frac{Q^2}{g}(20.4 + y_c) = [(20.4 + 0.5y_c)y_c]^3$$

This equation, when solved, yields values of y_c which correspond to a specified flow rate. y_c is then used in the Manning equation in a manner similar to that demonstrated in Example 5.5*b* to estimate S_c. The results of this computation are summarized in Fig. 5.5. The curve in this figure separates the regions of subcritical and supercritical normal flow for the specified values of Q and S.

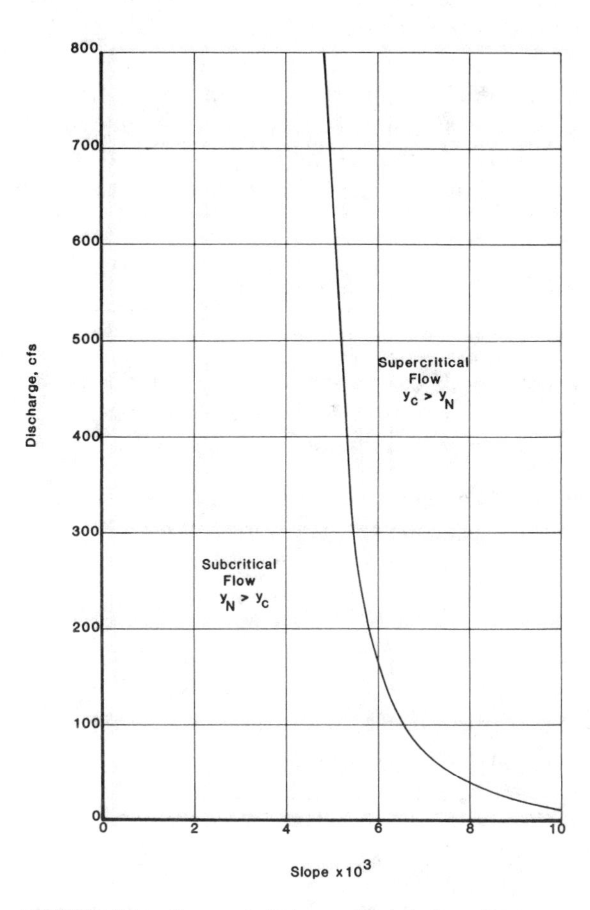

FIGURE 5.5 Curve defining subcritical and supercritical areas of normal flow for a given channel as a function of discharge and slope.

5.3 CHANNELS OF COMPOSITE ROUGHNESS

In many designed channels and most natural channels, the roughness varies along the perimeter of the channel, and in such cases it is sometimes necessary to calculate an equivalent value of the roughness coefficient for the entire perimeter. This effective roughness coefficient is then used in the normal flow calculations for the entire channel. In this section, a number of different methods of estimating the equivalent roughness coefficient for natural, designed, and laboratory channels are considered.

In a natural channel, the flow area is divided into N parts each with an associated wetted perimeter P_i and roughness coefficient n_i which are known. In these methods, the wetted perimeters, i.e., the P_i's, do not include the imaginary boundaries between the subsections. The methods of computing the equivalent roughness coefficient for this type of channel are:

1. Horton (1933) and Einstein and Banks (1950) each developed a method in which it is assumed that each of the subdivisions of the flow area is assumed to have the average velocity of the total section or $\overline{u} = \overline{u}_1 = \overline{u}_2 = \cdots = \overline{u}_N$ where $\overline{u}_i$ = the average velocity in the ith subsection. Then

$$n_e = \left[\frac{\sum_{i=1}^{N} (P_i n_i^{3/2})}{P} \right]^{2/3} \tag{5.3.1}$$

where n_e = equivalent Manning roughness coefficient
P = wetted perimeter of the complete section
N = number of subsections

2. If it is assumed that the total force-resisting motion is equal to the sum of the subsection-resisting forces, then

$$n_e = \left[\frac{\sum_{i=1}^{N} (P_i n_i^{2})}{P} \right]^{1/2} \tag{5.3.2}$$

3. If it is assumed that the total discharge of the section is equal to the sum of the subsection discharges, then

$$n_e = \frac{PR^{5/3}}{\sum_{i=1}^{N} \frac{P_i R_i^{5/3}}{n_i}} \tag{5.3.3}$$

where R_i = the hydraulic radius of the ith subsection and R = hydraulic radius of the complete section.

In a designed channel or a laboratory flume, the boundary angles are bisected and the subdivision is composed of the channel perimeter, the boundary angle bisectors, and the water surface. Although the two methods of computing n_e described below could be used for natural channels, the geometric calculations required can be quite cumbersome in such a situation.

1. In the method attributed by Cox (1973) to the Los Angeles U.S. Army Corps of Engineers District, n_e is computed by

$$n_e = \frac{\sum_{i=1}^{N} n_i A_i}{A} \tag{5.3.4}$$

where A_i = area of the ith subsection and A = total flow area.

2. The Colebatch method also described by Cox (1973) uses the following equation:

$$n_e = \left(\frac{\sum_{i=1}^{N} A_i n_i^{3/2}}{A} \right)^{2/3} \tag{5.3.5}$$

In general, any of the five methods discussed above is satisfactory for estimating equivalent values of n for natural and designed channels; however, the appropriateness and accuracy of the resulting estimates are unknown.

EXAMPLE 5.7

Given the channel described schematically in Fig. 5.6, estimate the value of n_e by the methods described above.

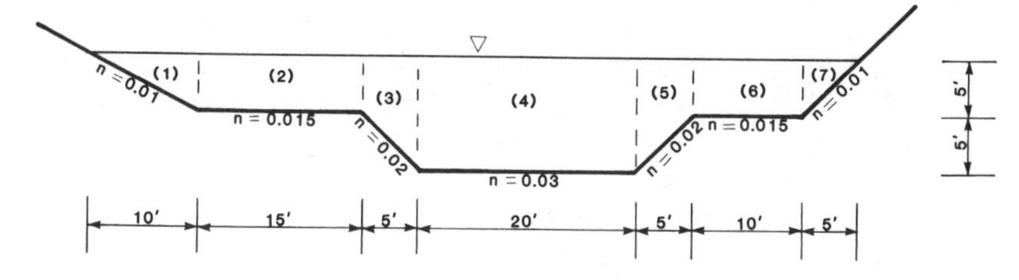

FIGURE 5.6 Channel definition for Example 5.7.

Solution

The data required for the solution of this problem are summarized in the following table:

Subdivision	Equations (5.3.1), (5.3.2), and (5.3.3)				Equations (5.3.4) and (5.3.5)
	A, ft²	P, ft	R, ft	n	A, ft²
1	25.0	11.2	2.23	0.01	27.9
2	75.0	15.0	5.00	0.015	77.2
3	37.5	7.07	5.30	0.020	53.0
4	200	20.0	10.0	0.030	159
5	37.5	7.07	5.30	0.020	53.0
6	50.0	10.0	5.00	0.015	50.0
7	12.5	7.07	1.77	0.010	17.7
Total section	437	77.4	5.65		437

Then, using

$$\text{Eq. (5.3.1)}, \ n_e = 0.019$$

$$\text{Eq. (5.3.2)}, \ n_e = 0.020$$

$$\text{Eq. (5.3.3)}, \ n_e = 0.019$$

$$\text{Eq. (5.3.4)}, \ n_e = 0.022$$

$$\text{Eq. (5.3.5)}, \ n_e = 0.023$$

It is noted that each of the methods of estimating an equivalent value of n yields approximately the same answer for this example.

The problem of compound perimeter roughness also occurs in laboratory experiments; i.e., the bottom boundary of a laboratory flume is usually roughened to duplicate natural conditions, but the walls are usually unroughened. In this case, the friction factor of the composite section is estimated from measured values of the slope of the water surface and the hydraulic radius; the problem is to estimate the friction factor or shear velocity associated with the bottom boundary roughness. The methodology described below was developed by Vanoni and Brooks (1957). In this discussion, the variables subscripted b are associated with the bottom of the channel, the variables subscripted w are associated with the walls, and the variables without a subscript are associated with the composite flow.

The following assumptions are required:

1. The channel flow area can be divided into two parts: one section in which the flow produces shear on the bottom boundary and a second section in

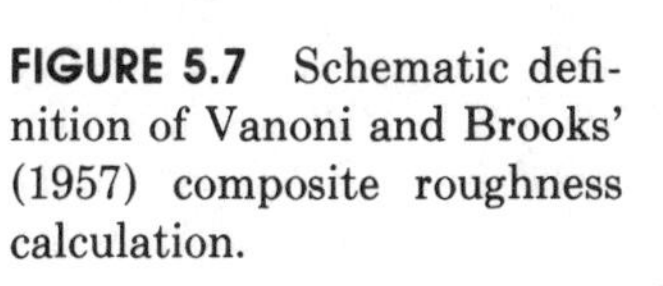

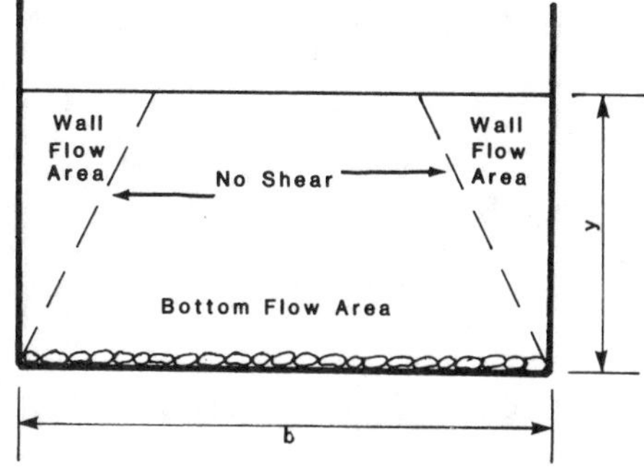

FIGURE 5.7 Schematic definition of Vanoni and Brooks' (1957) composite roughness calculation.

which the flow produces shear on the walls (Fig. 5.7). The boundaries between the bottom and wall sections are assumed to be zero shear surfaces and are not included in P_b and P_w.

2. The average velocity in the bottom section $\overline{u}_b$ equals the average velocity in the wall section $\overline{u}_w$.

3. The bottom and wall flow sections act as independent channels.

4. The bottom and wall roughnesses are homogeneous although different.

In this analysis, it is assumed that the following fundamental quantities are known:

$$R, u_*, \nu, P_w, P_b, A, F, u$$

where $f = 8(u_*/\overline{u})^2$ = Darcy-Weisbach friction factor, and ν = kinematic viscosity. The primary unknowns are

$$u_{*b}, f_b, R_b$$

It is crucial to the solution of this problem that the wall friction factor be estimated. Although the problem can be solved if the walls are not hydraulically smooth, the assumption of smooth side walls provides for a much simpler solution since under such an assumption f_w is only a function of the Reynolds number of the wall $\mathbf{R}_w$ where

$$\mathbf{R}_w = \frac{\overline{u}_w R_w}{\nu} \tag{5.3.6}$$

In this equation, $\mathbf{R}_w$ is an unknown. Equation (5.3.6) may also be written as

$$\mathbf{R}_w = \frac{4\overline{u}R}{\nu}\frac{R_w}{R} = \mathbf{R}\frac{R_w}{R} \tag{5.3.7}$$

The friction factor for the wall is given by

$$gR_wS = \frac{f_w\overline{u}^2}{8} \tag{5.3.8}$$

and the friction factor for the composite flow is

$$gRS = \frac{f\overline{u}^2}{8} \tag{5.3.9}$$

Dividing Eq. (5.3.8) by Eq. (5.3.9),

$$\frac{R_w}{R} = \frac{f_w}{f} \tag{5.3.10}$$

Substitution of this equation in Eq. (5.3.7) yields

$$\frac{\mathbf{R}_w}{f_w} = \frac{\mathbf{R}}{f} \tag{5.3.11}$$

Although the parameter $\mathbf{R}_w/f_w$ can be calculated from this equation, neither $\mathbf{R}_w$ nor f_w can be estimated alone. An auxiliary equation relating f_w and $\mathbf{R}_w$ is required. For hydraulically smooth boundaries with $\mathbf{R}_w < 100{,}000$, Eq. (4.3.7) can be used as the auxiliary equation. For $\mathbf{R}_w \geq 100{,}000$, Eq. (4.3.8) can be used. For use with Eq. (5.3.11), Eqs. (4.3.7) and (4.3.8) are plotted in Fig. 5.8. With this figure f_w can be estimated when $\mathbf{R}_w/f_w$ is known.

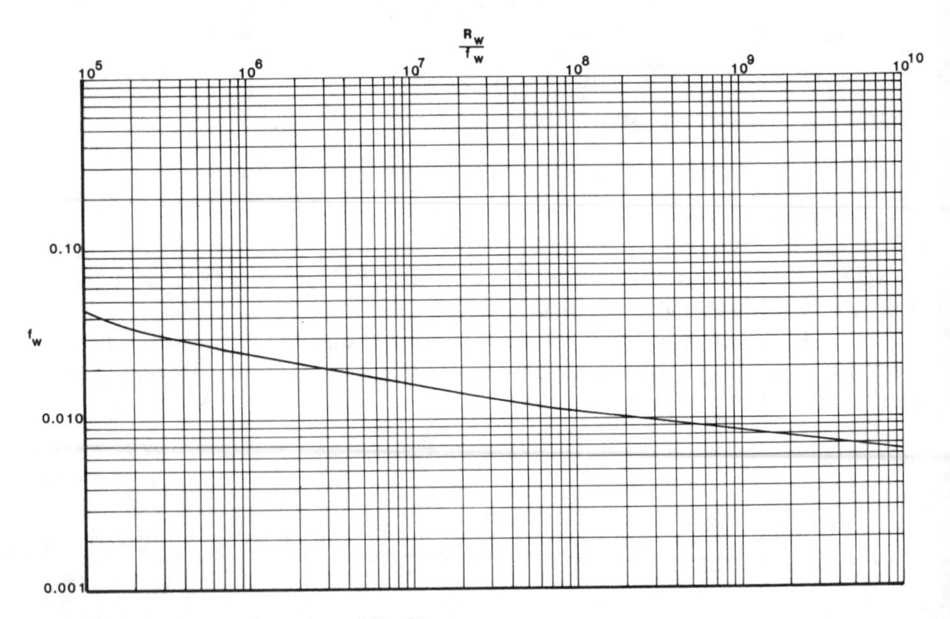

FIGURE 5.8 f_w as a function of $\mathbf{R}_w/f_w$.

Then from geometrical considerations

$$A = A_b + A_w \tag{5.3.12}$$

where with a slight rearrangement Eq. (5.3.9) yields

$$A = \frac{Pf\overline{u}^2}{8gS} \tag{5.3.13}$$

$$A_b = \frac{P_b f_b \overline{u}^2}{8gS} \tag{5.3.14}$$

and

$$A_w = \frac{P_w f_w \overline{u}^2}{8gS} \tag{5.3.15}$$

Substituting Eqs. (5.3.13) to (5.3.15) in Eq. (5.3.12) and rearranging,

$$f_b = \frac{P}{P_b} f - \frac{P_w}{P_b} f_w \tag{5.3.16}$$

For a rectangular channel, Eq. (5.3.16) becomes

$$f_b = f + \frac{2y}{b}(f - f_w) \tag{5.3.17}$$

where y = depth of flow. Also for a rectangular channel,

$$R_b = R\frac{f_b}{f} \tag{5.3.18}$$

and

$$u_{*b} = \sqrt{gR_bS} \tag{5.3.19}$$

EXAMPLE 5.8

A laboratory experiment is performed in a rectangular flume 2.75 ft (0.84 m) wide with hydraulically smooth side walls and an artificially roughened bottom. If $Q = 1.00$ ft^3/s (0.028 m^3/s), $S = 0.00278$, and $y = 0.30$ ft (0.091 m), estimate the values of f_b and u_{*_b}.

Solution

From the given data, the following variable values can be computed:

$$A = by = 2.75(0.30) = 0.825 \text{ ft}^2 \ (0.25 \text{ m}^2)$$

$$P = b + 2y = 2.75 + 2(0.30) = 3.35 \text{ ft} \ (1.0 \text{ m})$$

$$R = \frac{A}{P} = \frac{0.825}{3.35} = 0.246 \text{ ft} \ (0.075 \text{ m})$$

$$u_* = \sqrt{gRS} = \sqrt{(32.2)(0.246)(0.00278)} = 0.148 \text{ ft/s} \ (0.045 \text{ m/s})$$

$$\overline{u} = \frac{Q}{A} = \frac{1.00}{0.825} = 1.21 \text{ ft/s } (0.37 \text{ m/s})$$

$$f = 8\left(\frac{u_*}{\overline{u}}\right)^2 = 8\left(\frac{0.148}{1.21}\right)^2 = 0.120$$

Assume that $\nu = 9.55 \times 10^{-6}$ ft²/s (8.87×10^{-7} m²/s); then

$$\mathbf{R} = \frac{4\overline{u}R}{\nu} = \frac{4(1.21)(0.246)}{9.55 \times 10^{-6}} = 1.25 \times 10^5$$

and from Eq. (5.3.11)

$$\frac{\mathbf{R}_w}{f_w} = \frac{R}{f} = \frac{1.25 \times 10^5}{0.120} = 1.04 \times 10^6$$

For this value of the parameter $\mathbf{R}_w/f_w$ Fig. 5.8 yields a value of

$$f_w = 0.028$$

and $$f_b = f + \frac{2y}{b}(f - f_w) = 0.120 + \frac{2(0.30)}{2.75}(0.120 - 0.028) = 0.140$$

Using Eqs. (5.3.18) and (5.3.19),

$$R_b = R\frac{f_b}{f} = 0.246\frac{0.140}{0.120} = 0.287$$

and $$u_{*b} = \sqrt{gR_bS} = \sqrt{32.2(0.287)(0.00278)} = 0.160 \text{ ft/s } (0.049 \text{ m/s})$$

5.4 APPLICATION OF UNIFORM FLOW CONCEPTS TO PRACTICE

In this section a number of specific applications of the principles of uniform flow are considered. Although space limits the number of special cases that can be treated, it is hoped that the cases considered will indicate the manner in which the theoretical principles of uniform flow can be applied.

Slope-Area Peak Flood Flow Computations

Although flood flows are both spatially varied and unsteady, in some cases it is possible and/or necessary to analyze such flows with steady, uniform flow concepts. From a theoretical viewpoint, the use of such an approach can be justified only when the changes in flood stage and discharge are sufficiently gradual so that the friction slope reflects only the losses due to boundary friction. Also the slope-area approach is justified if the change in conveyance in the reach is less than 30 percent. From a practical viewpoint, slope-area methods can and are used when a peak flow discharge estimate is required and the data available are not sufficient to justify the use of more sophisticated techniques. For example, slope-area techniques have been used to estimate unsteady events such as flash floods in arid regions where both stream flow and precipitation records are essentially nonexistent (Glancy and Harmsen, 1975).

In applying the slope-area technique, the following data are required: (1) a measurement of the average flow area in the reach, (2) the change of elevation of the water surface through the reach, (3) the length of the reach, and (4) an estimate of the average resistance coefficient for the reach. Dalrymple and Benson (1976) provided the following guidelines for selecting a suitable reach:

1. A primary consideration in selecting a reach is the availability of high-water marks within the reach. For example, a steep-sided rock channel might have nearly perfect hydraulic characteristics, but if no high-water marks are available, it cannot be used. Benson and Dalrymple (1976) provide a number of techniques for identifying reliable high-water marks.
2. A reach with significant changes in the channel shape should be avoided, if possible, because of the uncertainties regarding the head loss in these sections. Although a straight, uniform reach is preferred, a contracting reach should be chosen over an expanding reach if there is a choice.
3. The slope-area method assumes that the total cross-sectional area of the channel is effective in transporting the flow. Therefore, conditions either upstream or downstream of the reach which cause an unbalanced flow distribution should be avoided, e.g., bridges and channel bends.
4. The slope-area technique is not applicable to reaches which include free-fall situations, e.g., waterfalls.
5. The reach should be long enough to develop a fall in the water surface whose magnitude can be determined accurately. In general, the accuracy of the slope-area method improves as the length of the reach is increased. Dalrymple and Benson (1976) recommended that one or more of the following criteria should be met in determining the reach length: (*a*) the length should be greater than or equal to 75 times the mean depth of flow, (*b*) the fall of the water surface should be equal to or greater than the velocity head, and (*c*) the fall should be equal to or greater than 0.50 ft (0.15 m).

It must be emphasized that the selection of a suitable reach is crucial to obtaining accurate estimates of the peak flow from the slope-area method.

Given the required information, the slope-area computation proceeds as follows:

1. For the specified cross-sectional areas, estimate R. Recall that the conveyance of a channel is by definition

$$K = \frac{\phi}{n} AR^{2/3} \tag{5.1.3}$$

Compute the upstream, K_u, and downstream, K_d, conveyances.

2. Compute the geometric mean conveyance of the reach or

$$\overline{K} = \sqrt{K_u K_d} \tag{5.4.1}$$

3. As a zero-order approximation of the true energy slope, let

$$S^0 = \frac{F}{L} \tag{5.4.2}$$

where S^0 = zero-order approximation of S
F = change of water surface elevation in reach
L = length of reach

4. The zero-order estimate of the peak flood flow discharge is

$$Q^0 = \overline{K}\sqrt{S^0} \tag{5.4.3}$$

5. Compute the first-order approximation of the discharge by refining the energy slope estimate; i.e.,

$$S^1 = S^0 + k\left[\frac{\alpha_u\,(\overline{u}_u^2/2g) - \alpha_d\,(\overline{u}_d^2/2g)}{L}\right] \tag{5.4.4}$$

where α = kinetic energy correction factor
$\overline{u}_u$ and $\overline{u}_d$ = upstream and downstream average velocities of flow calculated from Q^0 and the cross-sectional areas, respectively
k = constriction/expansion correction factor

If the reach is expanding, that is, $\overline{u}_u > \overline{u}_d$, then $k = 0.5$. If the reach is contracting, that is, $\overline{u}_d > \overline{u}_u$, then $k = 1.0$.

Then

$$Q^1 = \overline{K}\sqrt{S^1} \tag{5.4.5}$$

6. Step 5 is repeated until

$$Q^{n-1} \simeq Q^n$$

7. It is considered appropriate to average the discharges estimated for several reaches.

At best, the slope-area method provides a crude estimate of the peak flood flow, but in many cases the estimate obtained by this technique may be very cost-effective.

EXAMPLE 5.9

Estimate the flood discharge in a reach 4300 ft (1300 m) long if F = 7.3 ft (2.2 m), $\alpha_u = \alpha_d = 1$, $n = 0.035$, A_u = 1189 ft² (110 m²), P_u = 248 ft (76 m), A_d = 1428 ft² (133 m²), and P_d = 298 ft (91 m).

Solution

Upstream Hydraulic Radius

$$R_u = \frac{A_u}{P_u} = \frac{1189}{248} = 4.79 \text{ ft } (1.46 \text{ m})$$

Upstream Conveyance

$$K_u = \frac{1.49}{n} AR^{2/3} = \frac{1.49}{0.035}(1189)(4.79)^{2/3} = 144{,}000$$

Downstream Hydraulic Radius

$$R_d = \frac{A_d}{P_d} = \frac{1428}{298} = 4.79 \text{ ft } (1.46 \text{ m})$$

Downstream Conveyance

$$K_d = \frac{1.49}{0.035}(1428)(4.79)^{2/3} = 173{,}000$$

The percentage change in conveyance through the reach is

$$100\,\frac{K_d - K_u}{K_d} = 100\,\frac{173{,}000 - 144{,}000}{173{,}000} = 17 \text{ percent} < 30 \text{ percent}$$

The geometric average conveyance for the reach is

$$\overline{K} = \sqrt{K_u K_d} = \sqrt{(144{,}000)(173{,}000)} = 158{,}000$$

The zero-order flow estimate is

$$S^0 = \frac{F}{L} = \frac{7.3}{4300} = 0.00017$$

$$Q^0 = \overline{K}\,\sqrt{S^0} = 158{,}000\,\sqrt{0.0017} = 6514 \text{ ft}^3/\text{s } (184 \text{ m}^3/\text{s})$$

Then, since $A_u < A_d$, $k = 0.50$, and the following data apply:

Order of estimate n	Q^{n-1}, ft³/s	$\overline{u}_u$, ft/s	$\overline{u}_d$, ft/s	S^n	Q^n
1	6510	5.50	4.56	0.0017	6540
2	6540	5.50	4.58	0.0017	6540

Therefore, the discharge is estimated to be 6500 ft³/s (184 m³/s).

Ice-Covered Channels

Before a discussion of ice-covered channels can be initiated, it is essential that the reader have some knowledge of the material and the manner in which it is formed. First, as a material, ice is complex and exhibits the following characteristics:

1. Ice is nonhomogeneous and nonisotropic.
2. Under conditions of rapid loading, ice behaves elastically; however, if the load is applied slowly, viscoelastic deformation occurs.
3. Unlike most materials, ice is less dense than the water from which it is formed. Therefore, ice floats on its own melt water, at a temperature near its melting point.
4. Although the upper surface temperature of an ice cover is variable, the temperature of the lower surface which is in contact with water is always equal to the melting point temperature.
5. Within an ice cover, the ratio of the temperature of the ice to the melting point temperature, both in kelvins, is always greater than 0.80 (Sharp, 1981).
6. Although all ice cystals have a hexagonal shape, the grain structure of ice varies as a function of the mechanism of formation.

Second, two basic methods of ice formation in open channels have been identified:

1. In slow-moving or static water, nucleation occurs when the water becomes supercooled, and the result is a thin layer of skim ice on the top surface. In this stage of ice development, the crystals formed are usually needle-shaped. If the growth process of these crystals is slow, they may grow into large grains with dimensions measured in feet. However, if the growth process is rapid, the needle-shaped crystals interlock and a complex structure is formed. Since in open channels it is common for the midchannel temperature to exceed the boundary temperature, the growth of ice by this process is usually inhibited by the higher midchannel temperatures or turbulence.

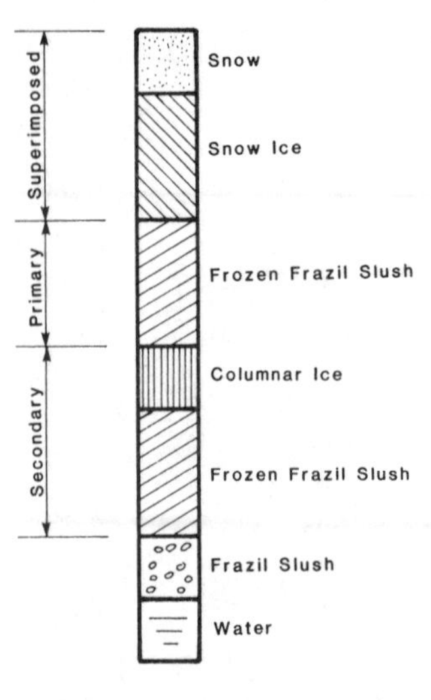

FIGURE 5.9 Typical ice profile. (*Michel and Ramseier, 1971.*)

After the development of an initial surface ice sheet, subsequent growth usually takes place in a direction parallel to the heat flow, i.e., in the vertical coordinate direction. This type of growth results in a columnar structure (Fig. 5.9). Although a columnar structure may be present over the complete ice column, in general most vertical ice column structures include some ice which is termed *snow ice*. Snow ice, which is opaque and white, is the result of snow falling on an ice cover, being wetted, and then freezing.

2. Under turbulent flow conditions, the formation of surface ice is inhibited and nucleation occurs within the water column rather than at the surface. For this process to take place, a few hundredths of a degree of supercooling is still required. Ice formed under these conditions is termed *frazil ice* and can take place throughout the entire turbulent, supercooled vertical column. Frazil ice forms by the development of small discoids which are held in suspension below the surface. Under ideal conditions, e.g., an open surface cooled by high winds, the production of frazil ice can be enormous.

Frazil ice crystals are initially very angular and adhere to one another forming floc particles. These floc particles float to the surface where they form what is termed *frazil slush*. Frazil ice floc particles also tend to adhere to any suitable object, e.g., other ice, rocks, or structures. As the frazil ice collects, a complete surface cover may be formed even under turbulent flow conditions. An excellent review of frazil ice formation was given by Osterkamp (1978).

When a channel is covered with floating ice, the wetted perimeter associated with the flow is increased; thus, for a constant bed slope, the uniform discharge of the channel must decrease. Given the significant impact that ice cover can have on channels in cold climates, it is not surprising that this problem has received attention since the early 1930s. Most of the investigations of this problem have sought to develop modifications of either the Manning or Chezy resistance coefficients which would reflect the composite nature of the channel roughness. That is, the channel is divided into two parts: one with area A_1 and hydraulic radius R_1 which is dominated by the bed roughness and the balance of the section $A_2 = A_0 - A_1$ and hydraulic radius R_2 which is dominated by the ice cover roughness, where A_0 is the total flow area. Uzuner (1975) reviewed most of the hypotheses developed between 1931 and 1969, and some of these results are summarized in Table 5.1. In this table, the subscripts 0, 1, and 2 designate parameters associated with the composite, the bed, and the ice-covered sections of the channel, respectively; $\phi = P_2/P_1$, y_j = distances from the two types of boundary to the point of maximum velocity, and $\bar{u}_j$ = mean section velocities.

The limitations of the equations summarized in Table 5.1, which are primarily for use with floating ice covers, are: (1) With the exception of Hancu's method, all the equations are based on the Manning equation and thus are applicable only to fully turbulent flow; (2) all these methods assume that the cross section can be divided into two parts with the upper part being solely affected by the ice cover and the lower part solely by the bed; (3) reliable estimates of n_2 for various types of ice cover are not available; and (4) the estimated resistance coefficient is a constant and cannot reflect the change of the ice roughness with the progression of the winter season.

TABLE 5.1 Composite resistance relationships for ice-covered channels

Developer	Equation	Comment
Pavlovskiy in Uzuner (1975)	$\frac{n_0}{n_1} = \left[\frac{1 + \phi\,(n_2/n_1)^2}{1 + a}\right]^{1/2}$	
Lotter in Uzuner (1975)	$\frac{n_0}{n_1} = \frac{\phi + 1}{1 + \phi(n_1/n_2)}$	
Belokon in Uzuner (1975)	$\frac{n_0}{n_1} = \left[1 + \phi\left(\frac{n_2}{n_1}\right)^{3/2}\right]^{2/3}$	
Sabaneev in Uzuner (1975)	$\frac{n_0}{n_1} = \frac{[1 + (n_2/n_1)^{2/2r+1}]^{(2r+1)/2}}{1 + \phi}$	
Chow (1959)	$\frac{n_0}{n_1} = \frac{1}{\sqrt{1 + \phi}}\left(\frac{n_2}{n_1}\right)\left[\phi^{3/4} + \left(\frac{n_1}{n_2}\right)^{3/2}\right]^{2/3}$	$r = 1/6$
Hancu in Uzuner (1975)	$\frac{n_0}{n_1} = \frac{1}{\sqrt{2}}\left(\frac{R_0}{R_1}\right)^{1/6}\left[\left(\frac{\bar{u}_1}{\bar{u}_0}\right)^2 + \left(\frac{\bar{u}_2}{\bar{u}_0}\right)^2\left(\frac{n_2}{n_1}\right)^2\left(\frac{R_1}{R_2}\right)^{1/3}\right]^{1/2}$	
Yu-Graf-Levine in Uzuner (1975)	$\frac{n_0}{n_1} = \left[\frac{1 + \phi^z\,(n_2/n_1)^{3/2}}{(1 + \phi)^z}\right]^{2/3}$	$z = \left(\frac{n_2}{n_1}\right)^{1/6}$
Larsen (1969)	$\frac{n_0}{n} = \frac{0.63\,[y_2/y_1 + 1]^{5/3}}{(n_1/n_2)\,(y_2/y_1)^{5/3} + 1}$	

Given the limitations of the equations in Table 5.1, Tsang (1982) began his derivation of uniform flow in an ice-covered channel with a force balance equation analogous to the one used in the previous chapter to develop the Chezy equation or

$$\rho g A_w \,\Delta h = \tau_b A_b + \tau_i A_i \tag{5.4.6}$$

where ρ = density of water
A_w = cross-sectional area of channel
A_b = surface area of channel bed
A_i = undersurface area of ice
Δh = head loss across section
τ_b = shear stress on channel bed
τ_i = shear stress on ice cover

Simplification of (5.4.6) yields

$$\rho g A_w S = \tau_b P_b \left(1 + \frac{\tau_i A_i}{\tau_b A_b}\right) \tag{5.4.7}$$

where P_b = wetted perimeter of the cross section excluding the ice cover. Let r_s represent the shear stress ratio or

$$r_s = \frac{\tau_i}{\tau_b}$$

and C_i represent the fractional surface ice coverage of the channel surface or

$$C_i = \frac{A_i}{A_b}$$

With these definitions, Eq. (5.4.7) becomes

$$\rho g R S = \tau_b(1 + r_s C_i) \tag{5.4.8}$$

It can be hypothesized that

$$\bar{u} = \phi C \sqrt{RS} \tag{5.4.9}$$

where

$$\phi = \left[\frac{1}{\psi\,(1 + rC_i)}\right]^{1/2}$$

where ψ designates an unknown function, ϕ is less than 1, and C = Chezy roughness coefficient of the channel with no ice cover. It is noted that Eq. (5.4.9) is analogous to the Chezy equation [Eq. (4.2.4)] and is applicable to any type of flow whether fully turbulent or not.

In examining winter conditions in the Beauharnois Canal between Lakes St. Francis and St. Louis on the St. Lawrence River, Tsang (1982) found that ϕ was a strong function of time measured from the day of ice formation (Fig. 5.10). With regard to this figure, which is based on 8 years of data, the following should be noted. First, following the first appearance of ice in the canal, the value of ϕ decreases significantly. Although most of the decrease takes place in the 2 days after the appearance of ice, the minimum value of ϕ is reached, on the average, in a week. Second, after the minimum value of ϕ is achieved, the value of this parameter slowly increases until it reaches a local maximum about three-fifths of the way through the ice season. Third, after reaching the local maximum, the value of ϕ decreases slowly until the day of ice breakup is reached. Fourth, following the breakup of the ice, the value of ϕ either rapidly increases or rapidly decreases and then increases depending on how the ice breaks up. This is not shown in Fig. 5.10.

It is necessary that the observations made above be correlated with a physical description of the situation. Tsang (1982) attributed the rapid decrease in the value of ϕ after the formation of ice to the presence of large amounts of frazil ice in the flow. He noted that when the concentration of frazil ice was the

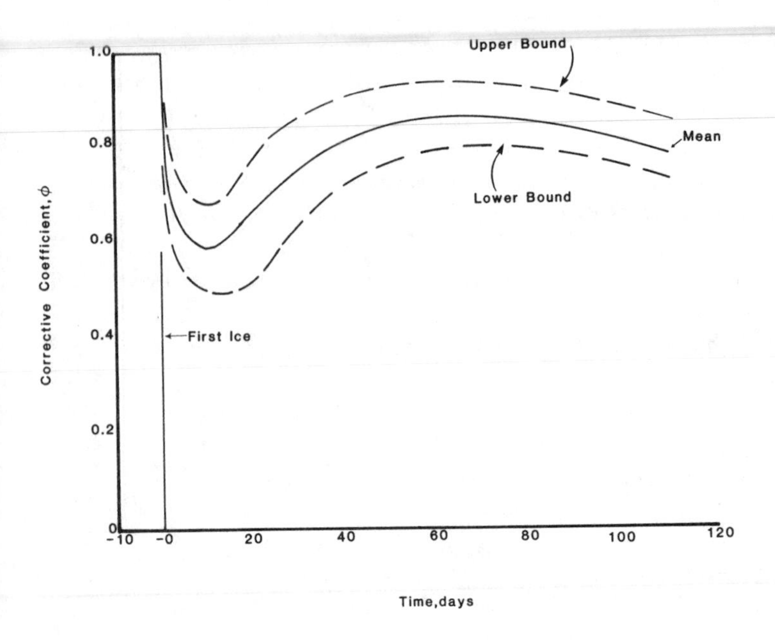

FIGURE 5.10 Variation of ϕ with time.

highest, the resistance of the canal to flow was also a maximum. The reduced velocity of flow under these conditions resulted in a consolidation of the frazil ice into a complete ice cover. As the underside of the ice cover was smoothed by the flow, the value of ϕ increased. Toward the end of winter, ϕ reaches a local maximum and begins to decrease, a phenomenon attributed to a very slight warming of the water flowing below the ice. Ashton and Kennedy (1972) have noted that an underflow only fractions of a degree above freezing can produce undulations in the underside of an ice cover which increase the roughness of the channel. Once formed, ice ripples tend to persist until the end of the winter season. Then, when the ice breaks up, ϕ should either rapidly increase to a value of 1 if the ice melts or rapidly decrease if blocks of ice are broken off and an ice jam develops.

It must be noted that in many ways these results represent a special case. For example, the flow in this channel is regulated, and during the winter months the channel is completely ice covered. In a natural river with a significant velocity of flow, the channel is seldom completely covered with ice. Thus, the results presented in Fig. 5.9 must be considered to be indicative of a pattern which may be encountered in channels which develop ice covers. The salient point is that ϕ, and therefore C or n, are a function of time or the development of the ice cover through the winter season. Thus, the assumption of a resistance coefficient whose value is stationary in time is probably inappropriate.

In a subsequent discussion of this work, Calkins and Ashton (1983) criticized the work of Tsang (1982) because it did not explicitly separate the large effect of the approximate doubling of the wetted perimeter from secondary effects such as the variation of the ice cover roughness coefficient, flow blockages due to frazil ice, and changes in the head difference associated with changes in the gradually varied flow profiles. Calkins and Ashton (1983) concluded that the methodology advocated by Tsang (1982) would appear to be inappropriate except in site-specific cases where there are extensive historical records for estimating ϕ. In reply to these comments, Tsang (1983) noted that his paper clearly demonstrated that in the Beauharnois Canal the factor which had the greatest influence on the conveyance capacity of the channel was not a solid ice cover but the presence of frazil ice in the flow. In such a case where increased flow resistance is the result of a nonsolid boundary, the methods described previously in this section are not applicable. However, until additional results are available, the methodology advocated by Tsang (1982) must be considered innovative but site-specific. *Note:* the time-dependent model of Tsang (1982) is conceptually similar to the hypothesis of Nezhikhovskiy (1964) which involved a time-dependent value of n; see, for example, Shen and Yapa (1982). Further, Shen (1983) provides a brief review regarding flow resistance due to other forms of ice such as ice jams, hanging ice dams, frazil suspensions, and moving ice covers.

Normal Discharge in Channels of Compound Section

In Chapter 2, the problem of channels composed of several distinct subsections with each subsection having its own roughness coefficient was discussed from the viewpoint of specific energy. This type of channel also requires special consideration from the viewpoint of the normal discharge computation. As the flow begins to cover the flood plain, the wetted perimeter of the section increases rapidly while the flow area increases slowly. Thus, in such a situation, the hydraulic radius, velocity, and discharge decrease with increasing flow depth, a situation which although computationally correct is physically incorrect. A number of methods have been suggested to circumvent this computational problem:

1. The whole cross-sectional area $A(abcdefgh)$, Fig. 5.11, is divided by the total wetted perimeter $P(abcdefgh)$ to obtain a mean or composite hydraulic radius. A mean velocity and discharge may then be computed. As noted above, this method yields erroneous results for shallow depths of flow in the over-bank sections.

2. The total channel is divided into a main section and two over-bank sections by the vertical lines ic and jf. The area of the main channel is $A(icdefj)$ with wetted perimeter $P(icdefj)$. The over-bank sections have the following areas and wetted perimeters: $A(abci)$, $P(abc)$, $A(jfgh)$, and $P(fgh)$. In this method, the lines ic and jf are included in the wetted perimeter of the main channel because the shear on these lines is nonzero (see, for example, Myers, 1978);

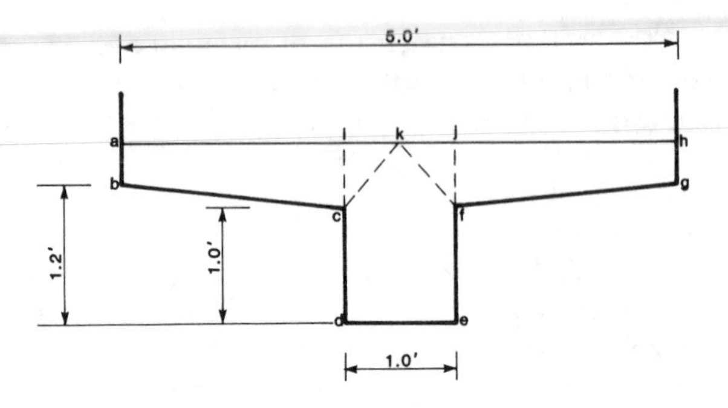

FIGURE 5.11 Cross-section definition.

however, it is assumed that the shear on these lines does not affect the over-bank flow.

3. The channel is divided as in method 2, but the lengths *ic* and *jf* are not considered part of the wetted perimeter.
4. The cross section is divided into sections by the lines *ck* and *fk* which bisect the reentrant angles. The wetted perimeter of each of the three subsections is computed including only the solid boundaries, and a weighted hydraulic radius for the whole section is computed by

$$R = \frac{\{[A(abck)]^2/P(abc)\} + \{[A(kcdef)]^2/P(cdef)\} + \{[A(kfgh)]^2/P(cgh)\}}{A(abck) + A(kcdef) + A(kfgh)}$$

Posey (1967) evaluated the accuracy of the above methods with a set of data from flume experiments performed in a flume with the dimensions shown in Fig. 5.11. In these experiments, the roughness of the main channel and the over-bank sections was uniform, and thus this evaluation primarily examined the ability of the above methods to account for a complex channel geometry. Posey (1967) concluded that: (1) method 2 is superior when the depth of flow in the over-bank sections is shallow and (2) method 1 is superior when the depth of flow in the over-bank sections is at least half as deep as the bank full main channel. It is noted that the data available by Posey were limited and additional work in this area is required.

BIBLIOGRAPHY

Ashton, G. D., and Kennedy, J. F., "Ripples on Underside of River Ice Covers," *Proceedings of the American Society of Civil Engineers, Journal of the Hydraulics Division,* vol. 98, no. HY9, September 1972, pp. 1603–1624.

Benson, M. A., and Dalrymple, T., "General Field and Office Procedures for Indirect Discharge Measurements," *Techniques of Water-Resources Investigations of the*

United States Geological Survey, book 3, chapter A1, U.S. Geological Survey, Washington, 1976.

Calkins, D. J., and Ashton, G. D., Discussion of "Resistance of Beauharnois Canal in Winter," by G. Tsang, *Journal of Hydraulic Engineering, American Society of Civil Engineers,* vol. 109, no. 4, April 1983, pp. 641–642.

Chow, V. T., *Open Channel Hydraulics,* McGraw-Hill Book Company, New York, 1959.

Cox, R. G., "Effective Hydraulic Roughness for Channels Having Bed Roughness Different from Bank Roughness," Miscellaneous paper H-73-2, U.S. Army Engineers Waterways Experiment Station, Vicksburg, MS, February 1973.

Dalrymple, T., and Benson, M. A., "Measurement of Peak Discharge by the Slope-Area Method," *Techniques of Water-Resources Investigations of the United States Geological Survey,* book 3, chapter A2, U.S. Geological Survey, Washington, 1976.

Einstein, H. A., and Banks, R. B., "Fluid Resistance of Composite Roughness," *Transactions of the American Geophysical Union,* vol. 31, no. 4, August 1950, pp. 603–610.

Glancy, P. A., and Harmsen, L., "A Hydrologic Assessment of the September 14, 1974 Flood in Eldorado Canyon, Nevada," Geological Survey Professional Paper 930, U.S. Geological Survey, Washington, 1975.

Horton, R. E., "Separate Roughness Coefficients for Channel Bottom and Sides," *Engineering News Record,* vol. III, no. 22, Nov. 30, 1933, pp. 652–653.

Larsen, P. A., "Head Losses Caused by an Ice Cover on Open Channels," *Journal of the Boston Society of Civil Engineers,* vol. 56, no. 1, 1969, pp. 45–67.

Michel, B., and Ramseier, R. O., "Classification of River and Lake Ice," *Canadian Geotechnical Journal,* vol. 8, no. 36, 1971, pp. 36–45.

Myers, W. R. C., "Momentum Transfer in a Compound Channel," *International Association for Hydraulic Research, Journal of Hydraulic Research,* vol. 16, no. 2, 1978, pp. 139–150.

Nezhikhovskiy, R. A. "Coefficients of Roughness of Bottom Surface of Slush Ice Cover," *Soviet Hydrology: Selected Papers,* no. 2, 1964, pp. 127–149.

Osterkamp, T. E., "Frazil Ice Formation: A Review," *Proceedings of the American Society of Civil Engineers, Journal of the Hydraulics Division,* vol. 104, no. HY9, September 1978, pp. 1239–1255.

Pine, L. S., and Allen, R. C., *Numerical Computing: Introduction,* Saunders, Philadelphia, 1973.

Posey, C. J., "Computation of Discharge Including Over-Bank Flow," *Civil Engineering,* American Society of Civil Engineers, April 1967, pp. 62–63.

Sharp, J. J., *Hydraulic Modelling,* London, Butterworth, 1981.

Shen, H. T., "Hydraulic Resistance of River Ice," ASCE, *Proceedings of the Conference on Frontiers in Hydraulic Engineering,* Massachusetts Institute of Technology, Cambridge, Ma., 1983, pp. 224–229.

Shen, H. T., and Yapa, P. N. D. D., "Simulation of Undersurface Roughness Coefficient of River Ice Cover," Report No. 82-6, Department of Civil and Environmental Engineering, Clarkson College, Potsdam, New York, July, 1982.

Tsang, G., "Resistance of Beauharnois Canal in Winter," *Proceedings of the American Society of Civil Engineers, Journal of the Hydraulics Division,* vol. 108, no. HY2, February 1982, pp. 167–186.

Tsang, G., Closure to "Resistance of Beauharnois Canal in Winter," *Journal of Hydraulic Engineering, American Society of Civil Engineers,* vol. 109, no. 4, April 1983, pp. 642–643.

Uzuner, M. S., "The Composite Roughness of Ice Covered Streams," *International Association for Hydraulic Research, Journal of Hydraulic Research,* vol. 13, no. 1, 1975, pp. 79–102.

Vanoni, V., and Brooks, N. H., "Laboratory Studies of the Roughness and Suspended Load of Alluvial Streams," Report No. 11, California Institute of Technology, Pasadena, 1957.

SIX

Theory and Analysis of Gradually and Spatially Varied Flow

SYNOPSIS

In this chapter the theory and analysis of gradually and spatially varied flow are considered. When the depth of flow in an open-channel flow varies with longitudinal distance, the flow is termed *gradually varied.* Such situations are found both upstream and downstream of control sections. In this chapter, tabular, digital, and graphical methods of estimating the depth of flow as a function of longitudinal distance in prismatic, nonprismatic, and natural channels are considered.

Steady, spatially varied flow is by definition a flow in which discharge varies with longitudinal distance. Such situations occur in side channel spillways, gutters collecting and conveying storm water runoff, channels with permeable boundaries, and drop structures in the bottom of channels. Tabular solutions of the differential equations governing spatially varied flow for both increasing and decreasing flow are considered.

In conclusion, a number of practical considerations are discussed including methods of computing the steady flow around islands in rivers.

6.1 BASIC ASSUMPTIONS AND THE EQUATION OF GRADUALLY VARIED FLOW

From Chapter 2, the gradual variation of the depth of flow in the longitudinal direction in an open channel is governed by

$$\frac{dy}{dx} = \frac{S_o - S_f}{1 - \mathbf{F}^2} \tag{6.1.1}$$

While Chapter 2 solutions of this equation for the special case of $S_f = 0$ were determined, in this chapter, solutions of this equation for $S_f \neq 0$ will be sought. In general, all the solution techniques which will be developed for gradually varied flow depend on the following assumptions:

1. The headloss for a specified reach is equal to the headloss in the reach for a uniform flow having the same hydraulic radius and average velocity, or in terms of the Manning equation

$$S_f = \frac{n^2\overline{u}^2}{\phi^2 R^{4/3}} \tag{6.1.2}$$

It is believed that the errors which arise from this assumption are small compared with those incurred in the normal use of the Manning equation and in the estimation of a resistance coefficient. This assumption is probably more accurate in the case of contracting flow fields than in expanding flow fields since in the former the headloss is primarily the result of frictional effects while in the latter there may be significant losses attributable to eddies.

2. The slope of the channel is small; therefore, the depth of flow is the same whether it is measured vertically or perpendicular to the bottom.
3. There is no air entrainment. If there is significant air entrainment, the problem is solved under the assumption of no air entrainment and the resulting profile is modified to account for air entrainment.
4. The velocity distribution in the channel section is fixed; therefore, α, the kinetic energy correction factor, is constant.
5. The resistance coefficient is independent of the depth of flow and constant throughout the reach under consideration.

6.2 CHARACTERISTICS AND CLASSIFICATION OF GRADUALLY VARIED FLOW PROFILES

In examining the computation of gradually varied flow profiles, it is first necessary to develop a systematic method of classifying the profiles which may occur in a specified channel reach. Recall

$$\mathbf{F}^2 = \frac{Q^2 T}{g A^3}$$

for a channel of arbitrary shape, and note that Eq. (6.1.2) can be rewritten as

$$S_f = \frac{n^2 Q^2 P^{4/3}}{\phi^2 A^{10/3}}$$

Substitution of the above equations in Eq. (6.1.1) yields

$$\frac{dy}{dx} = \frac{S_o - (n^2 Q^2 P^{4/3}/\phi^2 A^{10/3})}{1 - (Q^2 T/g A^3)} \tag{6.2.1}$$

For a specified value of Q, $\mathbf{F}$ and S_f are functions of the depth of flow y, and in a wide channel $\mathbf{F}$ and S_f will vary in much the same way with y since $P \simeq T$ and both S_f and $\mathbf{F}$ have a strong inverse dependence on the flow area. In addition, as y increases, both $\mathbf{F}$ and S_f will decrease. By definition, $S_f = S_o$ when $y = y_N$; therefore, the following set of inequalities must apply:

$$S_f \gtrless S_o \text{ according to } y \lessgtr y_N \tag{6.2.2}$$

and

$$\mathbf{F} \gtrless 1 \text{ according to } y \lessgtr y_c \tag{6.2.3}$$

These inequalities divide the channel into three sections in the vertical dimension (Fig. 6.1). By convention, these sections are labeled 1 to 3 starting at the top. For a channel of mild slope (Fig. 6.1) the following results can be stated where y is the actual depth of flow:

1. Level 1: $y > y_N > y_c$; $S_o > S_f$, $\mathbf{F} < 1$; therefore $dy/dx > 0$.

2. Level 2: $y_N > y > y_c$; $S_o < S_f$, $\mathbf{F} < 1$; therefore $dy/dx < 0$.

3. Level 3: $y_N > y_c > y$; $S_o < S_f$, $\mathbf{F} > 1$; therefore $dy/dx > 0$.

With the sign of dy/dx determined in each region, the behavior of the water surface profile can be predicted. Let us consider each of the above cases in turn.

Level 1: In this situation, at the upstream boundary, $y \to y_N$ and by definition $S_f \to S_o$; thus, $dy/dx = 0$. At the downstream boundary, $y \to \infty$ and S_f and $\mathbf{F}$ approach zero; thus $dy/dx \to S_o$ and the water surface asymptotically approaches a horizontal line. With these results, a rather clear picture of the circumstances which could cause such a profile emerges. This type of situation can occur behind a dam and is designated an M1 backwater curve where M indicates the channel slope (mild) and 1 identifies the vertical region of the channel in which the profile occurs.

Level 2: In this situation, at the upstream boundary, $y \to y_N$ and by definition $S_f \to S_o$; thus, $dy/dx \to 0$. At the downstream boundary, $y \to y_c$ and $dy/dx \to \infty$. The downstream boundary condition can never be exactly met since the water surface can never form a right angle with the bed of the channel. This type of profile can occur at either a free overfall or at a transition between a channel of mild slope and a channel with a steep slope. This water surface profile is termed an M2 drawdown curve.

Level 3: At the upstream boundary, $y \to 0$ and both S_f and $\mathbf{F}$ approach infinity, with the result being that dy/dx tends to a positive, finite limit. This result is of limited interest since a zero depth of flow can never occur. At the downstream boundary, $y \to y_c$, the derivative dy/dx is positive, and the depth of flow increases until the sequent depth is attained and a hydraulic jump forms. This profile is termed an M3 profile and can occur downstream of a sluice gate in a channel of mild slope or at the point where a channel of steep slope meets a channel of mild slope.

The information above has presented a rather detailed description of gradually varied flow profiles in channels of mild slope. In Fig. 6.1, the possibly gradually varied flow profiles for channels of various slopes are summarized schematically, and in Table 6.1 the various types of water surface profiles are summarized in a tabular form. The material which follows briefly describes the various possibilities.

S Profiles In these profiles, $S_o > S_c$ and $y_N < y_c$. The S1 profile usually begins with a jump at the upstream boundary and terminates with a profile tangent

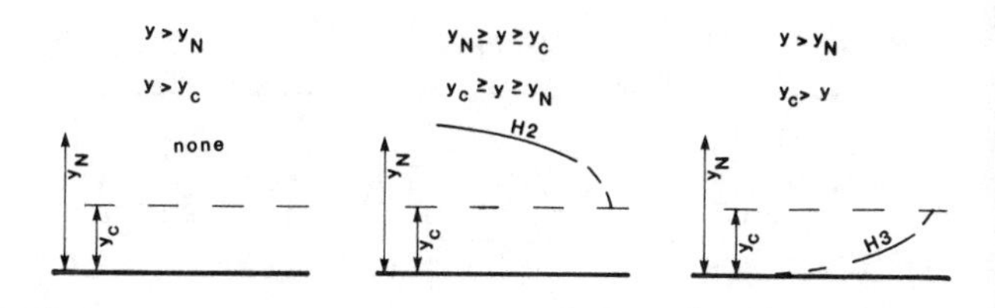

HORIZONTAL

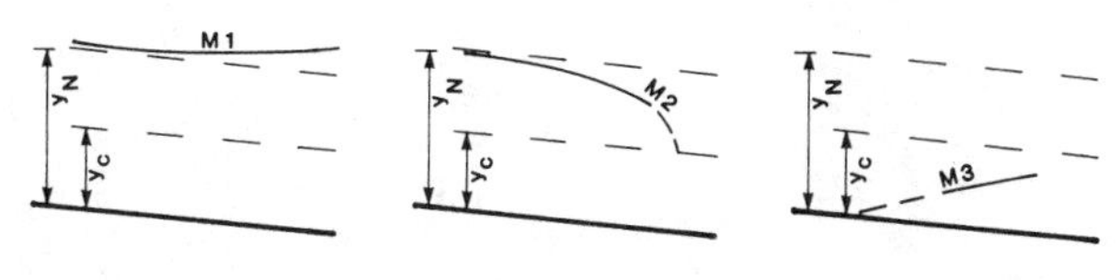

MILD

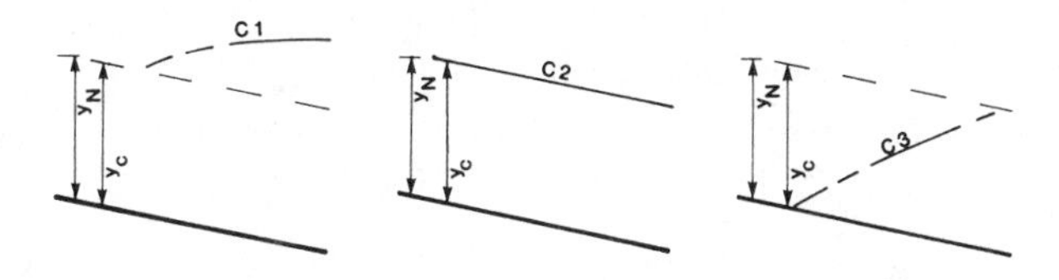

CRITICAL

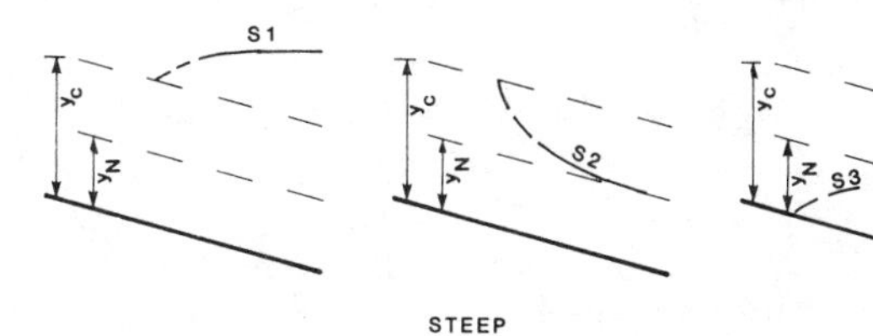

STEEP

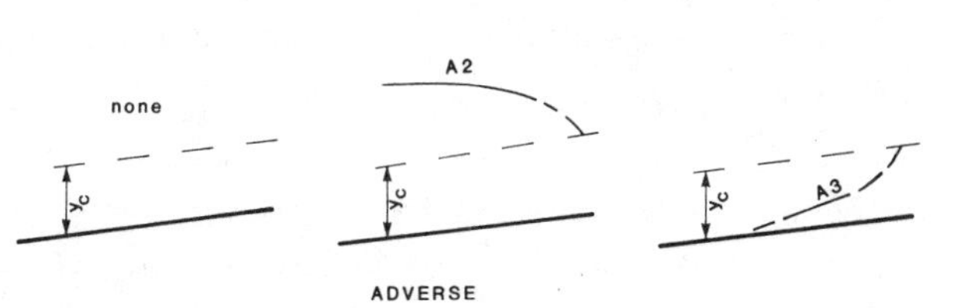

ADVERSE

FIGURE 6.1 Gradually varied flow profile classification system.

TABLE 6.1 Flow profile types

Channel slope	Profile designation Zone 1	Zone 2	Zone 3	Relative relation of y to y_N and y_c	Type of curve	Type of flow
Mild	M1			$y > y_N > y_c$	Backwater	Subcritical
$0 < S_o < S_c$		M2		$y_N > y > y_c$	Drawdown	Subcritical
			M3	$y_N > y_c > y$	Backwater	Supercritical
Critical	C1			$y > y_c = y_N$	Backwater	Subcritical
$S_o = S_c > 0$		C2		$y_c = y = y_N$	Parallel to channel bottom	Uniform-critical
			C3	$y_c = y_N > y$	Backwater	Supercritical
Steep	S1			$y > y_c > y_N$	Backwater	Subcritical
$S_o > S_c > 0$						
		S2		$y_c > y > y_N$	Drawdown	Supercritical
			S3	$y_c > y_N > y$	Backwater	Supercritical
Horizontal	None					
$S_o = 0$		H2		$y_N > y > y_c$	Drawdown	Subcritical
			H3	$y_N > y_c > y$	Backwater	Supercritical
Adverse	None					
$S_o < 0$		A2		$y_N^* > y > y_c$	Drawdown	Subcritical
			A3	$y_N^* > y_c > y$	Backwater	Supercritical

*Assumed a positive value.

to a horizontal at the downstream boundary, e.g., the flow behind a dam built in a steep channel. The S2 profile is a drawdown curve and is usually very short. At the downstream boundary, this profile is tangent to the normal depth of flow. This type of profile can occur downstream of an enlargement of a channel section and also downstream of a slope transition from steep to steeper. The S3 profile is also transitional in that it connects a supercritical flow with normal depth. Such a profile may occur at a slope transition between steep and mild or below a sluice gate on a steep slope where the water initially flows at a depth less than normal.

C Profiles In these profiles, $S_o = S_c$ and $y_N = y_c$. The C1 profile is asymptotic to a horizontal line, e.g., a profile connecting a channel of critical slope with a channel of mild slope. The C3 profile can connect a supercritical flow with a reservoir pool on a critical slope.

H Profiles In this case $S_o = 0$, and the H profiles can be considered to be a limiting case of the M profiles. The H2 drawdown profile can be found upstream of a free overfall, while the H3 profile can connect the supercritical flow below a sluice gate with a reservoir pool.

A Profiles In these profiles $S_o < 0$. In general, the A2 and A3 profiles occur only infrequently and are similar to the H2 and H3 profiles.

In summary, the following principles regarding gradually varied flow profiles can be stated:

1. The sign of dy/dx can be readily determined from Eqs. (6.2.1) through (6.2.3).
2. When the water surface profile approaches normal depth, it does so asymptotically.
3. When the water surface profile approaches critical depth, it crosses this depth at a large but finite angle.
4. If the flow is subcritical upstream but passes through critical depth, then the feature which produces critical depth determines and locates the whole water surface profile. If the upstream flow is supercritical, e.g., in the case of an M3 profile, then the control cannot come from downstream.
5. Every gradually varied flow profile exemplifies the principle that subcritical flows are controlled from downstream while supercritical flows are controlled from upstream. Gradually varied flow profiles would not exist if it were not for the upstream and downstream controls.
6. In channels with horizontal and adverse slopes, the terminology *normal depth of flow* has no meaning since the normal depth of flow is either negative or imaginary. However, in these cases, the numerator of Eq. (6.2.1) is negative and the shape of the profile can be deduced.

6.3 COMPUTATION OF GRADUALLY VARIED FLOW

In this section, a number of methods of integrating Eq. (6.1.1) to determine the variation of the depth of flow with distance are discussed. Although there are a variety of solution techniques for accomplishing this integration, given a particular situation, one method may be superior to the others. Thus, the reader is cautioned to carefully consider the problem before proceeding to a computational procedure. All solutions of the gradually varied flow equation must begin with the depth of flow at a control and proceed in the direction in which the control operates. At the upstream boundary, the gradually varied flow profile may approach a specified depth of flow asymptotically. At this point under the condition of asymptotic approach, a reasonable definition of convergence must be developed.

Uniform Channels

In the case of a uniform channel, i.e., a prismatic channel with a constant slope and resistance coefficient, a tabular solution of the gradually varied flow equation is possible. In such a channel,

$$\frac{dE}{dx} = S_o - S_f = S_o - \frac{n^2\overline{u}^2}{\phi^2 R^{4/3}} \tag{6.3.1}$$

or in finite difference form

$$\frac{\Delta E}{\Delta x} = \frac{\Delta[y + (\overline{u}^2/2g)]}{\Delta x} = S_o - \frac{n^2\overline{u}^2}{\phi^2 R^{4/3}} \tag{6.3.2}$$

In this equation, all variables, with the exception of Δx, are functions of y, the depth of flow. Thus, Eq. (6.3.2) can be solved by selecting values of y, computing $\overline{u}$ and R, and solving for the value of Δx corresponding to y or

$$\Delta x = \frac{\Delta[y + (\overline{u}^2/2g)]}{S_o - (n^2\overline{u}^2/\phi^2 R^{4/3})_m} \tag{6.3.3}$$

where the subscript m indicates a mean value over the interval being considered. This is commonly termed the *step method.* The primary difficulty with this technique is that in most cases ΔE or $\Delta[y + (\overline{u}^2/2g)]$ is small; thus, the difficulty of accurately determining a small difference between two relatively large numbers must be resolved. This computational difficulty can be overcome by using Eq. (2.2.1) in a finite difference form or

$$\Delta E = \Delta y(1 - \mathbf{F}^2) \tag{6.3.4}$$

and Eq. (6.3.3) then becomes

$$\Delta x = \frac{\Delta y(1 - \mathbf{F}^2)_m}{S_o - (n\overline{u}^2/\phi^2 R^{4/3})_m} \tag{6.3.5}$$

A secondary difficulty with this solution technique occurs in situations where $S_f \rightarrow S_o$ since in this case the difference $S_o - S_f$ is small. In general, this does not present grave difficulties from a practical viewpoint since the details of the asymptotic approach of the depth of flow to normal depth is not of interest. The details of this method of solution are best illustrated with an example.

EXAMPLE 6.1

A trapezoidal channel with $b = 20$ ft (6.1 m), $n = 0.025$, $z = 2$, and $S_o = 0.001$ carries a discharge of 1000 ft^3/s (28 m^3/s). If this channel terminates in a free overfall, determine the gradually varied flow profile by the step method.

Solution

The flow profile will be a drawdown curve, and the first step in solving the problem is to establish the depth of flow at the upstream and downstream boundaries.

1. *Upstream Boundary:* At the upstream boundary, the depth of flow will be y_N. y_N can be readily determined for the given information with the computer program contained in Appendix I with

$$AR^{2/3} = \frac{nQ}{\phi\sqrt{S}} = \frac{0.025(1000)}{1.49\sqrt{0.001}} = 531$$

and

$$y_N = 6.25 \text{ ft } (1.90 \text{ m})$$

Since at this boundary the depth of flow approaches y_N asymptotically, a reasonable definition of convergence must be established; e.g., computations will proceed until the depth of flow is $0.90y_N$ or 5.62 ft (1.71 m).

2. *Downstream Boundary:* At the downstream boundary, the depth of flow will be y_c. From Eq. (2.2.3)

$$\frac{\overline{u}^2}{2g} = \frac{D}{2} \Rightarrow \frac{Q^2}{gA^2} = \frac{A}{T}$$

or

$$\frac{Q^2}{g} = \frac{A^3}{T}$$

where

$$A = (b + zy)y = (20 + 2y)y$$

$$T = b + 2zy = 20 + 4y$$

Then

$$\frac{y_c^3(20 + 2y_c)^3}{20 + 4y_c} = \frac{(1000)^2}{32.2} = 31{,}060$$

By trial and error

$$y_c = 3.74 \text{ ft } (1.14 \text{ m})$$

Thus, the range of depths in this computation is from 3.74 ft (1.14 m) at the control to 5.62 ft at the upstream boundary. The Froude number at the upstream boundary is

$$\mathbf{F} = \frac{\overline{u}}{\sqrt{gD}} = 0.41$$

Therefore, the drawdown profile is an M2 curve. The details of the computation are contained in Table 6.2*a* and *b*, and the results are summarized in Fig. 6.2. In Table 6.2*a*, Eq. (6.3.3) was used, while in

TABLE 6.2 Example 6.1

y, ft (1)	A, ft^2 (2)	P, ft (3)	R, ft (4)	$\bar{u}$, ft/s (5)	$\frac{\bar{u}^2}{2g}$, ft (6)	E, ft (7)	$\frac{n^2\bar{u}^2}{\phi^2 R^{4/3}}$ (8)	$\left(\frac{n^2\bar{u}^2}{\phi^2 R^{4/3}}\right)_m$ (9)	$S_o - \left(\frac{n^2\bar{u}^2}{\phi^2 R^{4/3}}\right)_m$ (10)	ΔE, ft (11)	Δx, ft (12)	$x = \Sigma\Delta x$, ft (13)
(a) Solution of Example 6.1 using Eq. (6.3.3)												
3.74	103	36.7	2.81	9.71	1.46	5.20	0.00670					
								0.00601	−0.00501	+0.0400	−8.00	
4.00	112	37.9	2.95	8.93	1.24	5.24	0.00531					−8.0
								0.00478	−0.00378	+0.0700	−18.5	
4.25	121	39.0	3.10	8.26	1.06	5.31	0.00425					−27
								0.00386	−0.00286	+0.110	−38.5	
4.50	130	40.1	3.24	7.69	0.918	5.42	0.00347					−65
								0.00314	−0.00214	+0.120	−56.1	
4.75	140	41.2	3.40	7.14	0.792	5.54	0.00281					−121
								0.00256	−0.00156	+0.150	−96.1	
5.00	150	42.3	3.55	6.67	0.691	5.69	0.00231					−217
								0.00213	−0.00113	+0.170	−150	
5.25	160	43.5	3.68	6.25	0.607	5.86	0.00194					−367
								0.00179	−0.000790	+0.180	−228	
5.50	170	44.6	3.81	5.88	0.537	6.04	0.00164					−595
								0.00156	−0.000560	+0.0800	−143	
5.62	176	45.1	3.90	5.68	0.501	6.12	0.00148					−738

y, ft (1)	A, ft^2 (2)	P, ft (3)	T, ft (4)	R, ft (5)	D, ft (6)	$\bar{u}$, ft/s (7)	$\frac{n^2\bar{u}^2}{\phi^2 R^{4/3}}$ (8)	F^2 (9)	$\overline{\frac{n^2\bar{u}^2}{\phi^2 R^{4/3}}}$ (10)	$\overline{F}^2_m$ (11)	$S_o - \left(\frac{n^2\bar{u}^2}{\phi^2 R^{4/3}}\right)_m$ (12)	ΔE (13)	Δx (14)	$\Sigma\Delta x$ (15)
3.74	103	36.7	35.0	2.81	2.94	9.71	0.00670	0.996						
									0.00601	0.896	−0.00501	0.0270	−5.39	
4.00	112	37.9	36.0	2.95	3.11	8.93	0.00531	0.796						−5
									0.00478	0.722	−0.00378	0.0695	−18.4	
4.25	121	39.0	37.0	3.10	3.27	8.26	0.00425	0.648						−24
									0.0386	0.593	−0.00286	0.102	−35.7	
4.50	130	40.1	38.0	3.24	3.42	7.69	0.00347	0.537						−59
									0.00314	0.489	−0.00214	0.128	−59.8	
4.75	140	41.2	39.0	3.40	3.59	7.14	0.00281	0.441						−119
									0.00256	0.404	−0.00156	0.149	−95.5	
5.00	150	42.3	40.0	3.55	3.75	6.67	0.00231	0.368						−215
									0.00213	0.340	−0.00113	0.165	−146	
5.25	160	43.5	41.0	3.68	3.90	6.25	0.00194	0.311						−361
									0.00179	0.288	−0.000790	0.178	−225	
5.50	170	44.6	42.0	3.81	4.05	5.88	0.00164	0.265						−586
									0.00156	0.254	−0.000560	0.0895	−160	
5.62	176	45.1	42.5	3.90	4.14	5.68	0.00148	0.242						−746

(*b*) **Solution of Example 6.1 using Eqs. (6.3.4) and (6.3.5)**

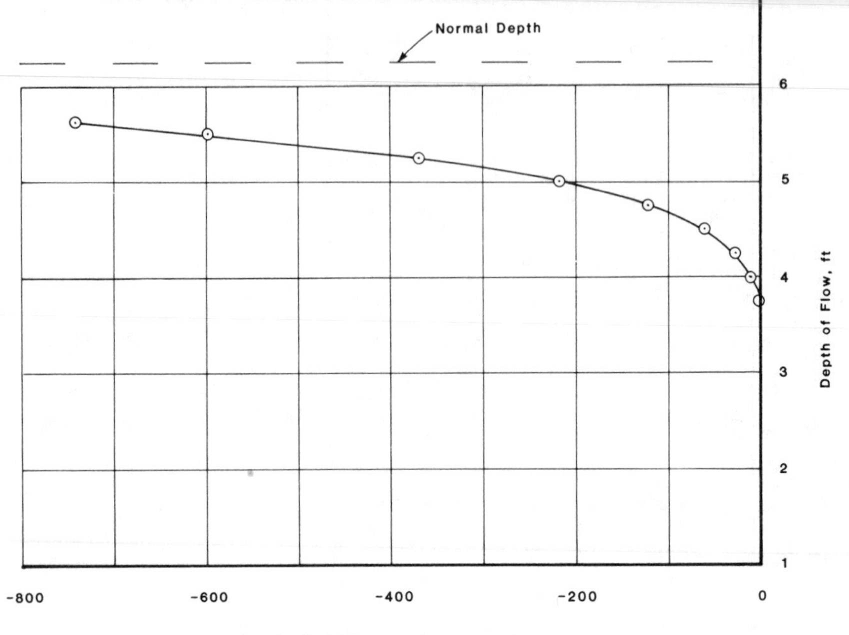

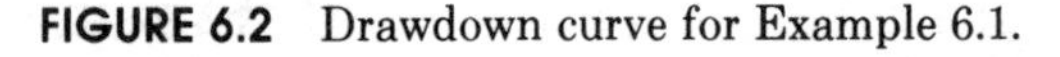

FIGURE 6.2 Drawdown curve for Example 6.1.

Table 6.2*b*, Eqs. (6.3.4) and (6.3.5) were used. The actual difference in the results at the upstream boundary is less than 2 percent.

This method is useful for artificial channels and has the distinct advantage of predicting a longitudinal distance for a specified depth of flow. However, the method has the disadvantage of not being capable of predicting a depth of flow at a specified longitudinal distance.

A second method of estimating a gradually varied flow profile in uniform channels is by direct integration of Eq. (6.1.1). In this equation, the Froude number for a channel of arbitrary shape is defined as

$$\mathbf{F}^2 = \frac{\alpha Q^2 T}{g A^3}$$

Recall that in Eq. (2.2.9) the section factor was defined as

$$Z = \sqrt{\frac{A^3}{T}}$$

Assume that a critical flow of discharge Q occurs in the subject channel; then

$$Q = Z_c \sqrt{\frac{g}{\alpha}} \tag{6.3.6}$$

where Z_c = the critical flow section factor. It is crucial that the reader differentiate between Z which is simply a numerical value computed for a specified Q and y and Z_c which is computed for a specified Q as if the flow occurred at critical depth. Substitution of the definition of Z and Eq. (6.3.6) into the Froude number definition yields

$$\mathbf{F}^2 = \frac{Z_c^2}{Z^2} \tag{6.3.7}$$

Both the slope of the channel bottom and the energy slope can be represented in terms of the conveyance defined in Eq. (5.1.3) or

$$S_o = \frac{Q^2}{K_N^2} \tag{6.3.8}$$

and

$$S_f = \frac{Q^2}{K^2} \tag{6.3.9}$$

Again it is necessary to distinguish between K which is the numerical value of the conveyance when the flow occurs at depth y and K_N which is the conveyance computed for a specified Q as if the flow occurred at normal depth. Rearranging Eq. (6.1.1) and substituting Eqs. (6.3.7) to (6.3.9) yields

$$\frac{dy}{dx} = S_o \left[\frac{1 - (K_N/K)^2}{1 - (Z_c/Z)^2} \right] \tag{6.3.10}$$

If Eq. (6.3.10) is to be solved by direct integration, it is necessary that the parameters $(K_N/K)^2$ and $(Z_c/Z)^2$ be expressed in terms of the depth of flow. Since the section factor is only a function of the channel shape and y, assume

$$Z^2 = Cy^M \tag{6.3.11}$$

where C = a coefficient and M = hydraulic exponent for critical flow. The functional relationship between M, the depth of flow, and the geometry of the channel can be established by first taking the logarithm of both sides of Eq. (6.3.11) and then differentiating that equation with respect to y or

$$\frac{d(\ln Z)}{dy} = \frac{M}{2y} \tag{6.3.12}$$

Taking the logarithm and differentiating the definition of the section factor, Eq. (2.2.9), with respect to y yields

$$\frac{d(\ln Z)}{dy} = \frac{3}{2}\frac{T}{A} - \frac{1}{2T}\frac{dT}{dy} \tag{6.3.13}$$

Combining Eqs. (6.3.12) and (6.3.13) and solving for M yields

$$M = \frac{y}{A}\left(3T - \frac{A}{T}\frac{dT}{dy}\right) \tag{6.3.14}$$

With a channel geometry specified, the value of M can be calculated for each depth of flow. For a rectangular channel, M has a value of 3. Expressions for calculating M in channels of various shapes are summarized in Table 6.3.

In an analogous fashion, it is assumed that conveyance is also a function of the depth of flow or

$$K^2 = C'y^N \tag{6.3.15}$$

where C' = a coefficient and N = hydraulic exponent for uniform flow computations. Following a procedure analogous to the technique used for M yields

$$N = \frac{2y}{3A}\left(5T - 2R\frac{dP}{dy}\right) \tag{6.3.16}$$

The value of N depends on whether the conveyance is expressed in terms of Manning's n or the Chezy C. Equation (6.3.16) is based on the Manning equation. For example, in a rectangular channel, N = 3.33 if the Manning equation is used, but N = 3 if the Chezy equation is used. Expressions for computing N in channels of various shapes are summarized in Table 6.3. It should be noted that in general N and M are functions of y and are not constants.

Substitution of Eqs. (6.3.11) and (6.3.15) into Eq. (6.3.10) yields

$$\frac{dy}{dx} = S_o\left[\frac{1 - (y_N/y)^N}{1 - (y_c/y)^M}\right] \tag{6.3.17}$$

Letting $u = y/y_N$ and rearranging yields

$$dx = \frac{y_N}{S_o}\left[1 - \frac{1}{1 - u^N} + \left(\frac{y_c}{y_N}\right)^M \frac{u^{N-M}}{1 - u^N}\right] du \tag{6.3.18}$$

If Eq. (6.3.18) is to be integrated, then N and M must be assumed to be constants. In general, it is possible to divide a channel into subsections so that N and M are essentially constant. Under this assumption

$$x = \frac{y_N}{S_o}\left[u - \int_0^u \frac{du}{1 - u^N} + \left(\frac{y_c}{y_N}\right)^M \int_0^u \frac{u^{N-M}}{1 - u^N}\, du\right] + C \tag{6.3.19}$$

where the integral $(\int_0^u)\, du/(1 - u^N) = F(u,N)$ is known as the Bakhmeteff varied flow function and C = a constant of integration. The second integral in

TABLE 6.3 Values of M and N for rectangular, trapezoidal, and triangular channels

	M	N
Rectangular	3.0	$3\frac{1}{3}$
Trapezoidal	$\dfrac{3[1 + 2z(y/b)]^2 - 2z(y/b)[1 + z(y/b)]}{[1 + 2z(y/b)][1 + z(y/b)]}$	$\dfrac{10[1 + 2z(y/b)]}{3[1 + z(y/b)]} - \dfrac{8(y/b)\sqrt{1 + z^2}}{3[1 + 2(y/b)\sqrt{1 + z^2}]}$
Triangular	5	$\frac{16}{3}$

Eq. (6.3.19) can also be expressed in the form of the varied-flow function by substituting $v = u^{N/J}$ where $J = N/(N - M + 1)$ (Chow, 1959). Then

$$\int_0^u \frac{u^{N-M}}{1-u^N}\,du = \frac{J}{N}\int_0^v \frac{dv}{1-v^J} = \frac{J}{N}F(v,J) \tag{6.3.20}$$

With this definition, Eq. (6.3.19) becomes

$$x = \frac{y_N}{S_o}\left[u - F(u,N) + \frac{J}{N}\left(\frac{y_c}{y_N}\right)^M F(v,J)\right] + C$$

The constant of integration in this equation, C, can be eliminated if Eq. (6.3.17) is integrated between two stations or

$$L = x_2 - x_1 = A\{(u_2 - u_1) - [F(u_2, N) - F(u_1, N)] + B[F(v_2, J) - F(v_1, J)]\} \tag{6.3.21}$$

where $A = y_N/S_o$ and $B = (y_c/y_N)^M J/N$. Although tables for the evaluation of $F(u,N)$ and $F(v,J)$ are available (Appendix II), it is usually convenient when using the digital computer to estimate gradually varied flow profiles to express $F(u,N)$ in the form of an infinite series or

1. For $u < 1$

$$\int_0^u \frac{du}{1-u^N} = u + \frac{1}{N+1}u^{N+1} + \cdots + \frac{1}{(P-1)N+1}u^{(P-1)N+1} + R_P \tag{6.3.22}$$

where N = hydraulic exponent
P = term number ($P = 1, 2, \ldots, \infty$)
R_P = remainder after $(P - 1)$ terms have been summed

With regard to Eq. (6.3.22), the following are noted: First,

$$R_P < \frac{u^{PN+1}}{PN+1}\left(\frac{1}{1-u^N}\right) \tag{6.3.23}$$

Second, for $u < 0.7$ the series converges rapidly, but for $u = 1$ the series diverges. Third, the smaller the value of u, the more rapidly the series converges.

2. For $u > 1$

$$\int_0^u \frac{du}{1-u^N} = \frac{1}{(N-1)u^{N-1}} + \frac{1}{(2N-1)u^{2N-1}} + \cdots + \frac{1}{(PN-1)u^{PN-1}} + R_P \tag{6.3.24}$$

For this series expansion the following are noted: First,

$$R_P < \frac{1}{(PN-1)^{PN-1}} \frac{u^N}{u^{N-1}} \tag{6.3.25}$$

Second, for $u > 1.5$ the series converges rapidly, but for $u = 1$ the series diverges. Third, the larger the value of u, the more rapid the series convergence.

At this point, it should be noted that Eq. (6.3.17) can be integrated exactly for the special case of $N = M = 3$, that is, a rectangular channel with the conveyance expressed in terms of the Chezy equation. In such a case,

$$x = \frac{1}{S_0}\left\{y - y_N\left[1 - \left(\frac{y_c}{y_N}\right)^3\right]\phi\right\} \tag{6.3.26}$$

where

$$\phi = \int \frac{du}{1-u^3} = \frac{1}{6}\log\left[\frac{u^2+u+1}{(u-1)^2}\right] - \frac{1}{\sqrt{3}}\tan^{-1}\left(\frac{\sqrt{3}}{2u+1}\right) + C$$

which is known as the *Bresse function*, and C = a constant of integration. For computational purposes, it is usually convenient to express $(y_c/y_N)^3$ as

$$\left(\frac{y_c}{y_N}\right)^3 = \frac{C^2 S_o}{g}$$

EXAMPLE 6.2

A trapezoidal change with $b = 20$ ft (6.1 m), $n = 0.025$, $z = 2$, and $S_o = 0.001$ carries a discharge of 1000 ft^3/s (28 m^3/s). If this channel terminates in a free overfall, determine the gradually varied flow profile by the method of direct integration.

Solution

In Example 6.1, for the same problem statement it was determined that

$$y_N = 6.25 \text{ ft } (1.90 \text{ m})$$

and $$y_C = 3.74 \text{ ft } (1.14 \text{ m})$$

A tabular solution of this problem by the method of direct integration is summarized in Table 6.4. This method of solution predicts that at 745 ft (227 m) from the free overfall the depth of flow is 5.62 ft (1.71 m). This result is not significantly different from the results of Example 6.1.

As with the step method discussed previously, the method of direct integration has the disadvantage of not being capable of predicting a depth of flow given a

TABLE 6.4 Example 6.2

y, ft	$\bar{u}$	N	M	J	V	$F(u,N)$	$F(v,J)$	$\bar{J}$	$\bar{N}$	$\bar{M}$	Δx	$\Sigma\Delta x$
3.74	0.598	3.63	3.39	2.93	0.529	0.620	0.552					
								2.930	3.645	3.40	−1.7	
4.00	0.640	3.66	3.41	2.93	0.573	0.670	0.607					−2
								2.950	3.67	3.425	−19	
4.25	0.680	3.68	3.44	2.97	0.620	0.721	0.664					−21
								2.975	3.69	3.45	−35	
4.50	0.720	3.70	3.46	2.98	0.665	0.775	0.725					−56
								2.99	3.71	3.47	−60	
4.75	0.760	3.72	3.48	3.00	0.712	0.834	0.794					−116
								3.01	3.73	3.49	−92	
5.00	0.800	3.74	3.50	3.02	0.759	0.899	0.871					−208
								3.025	3.75	3.51	−147	
5.25	0.840	3.76	3.52	3.03	0.805	0.974	0.958					−355
								3.04	3.77	3.53	−237	
5.50	0.880	3.78	3.54	3.05	0.853	1.066	1.065					−591
								3.055	3.785	3.545	−153	
5.62	0.899	3.79	3.55	3.06	0.876	1.118	1.130					−745

specified longitudinal distance. However, this method has the distinct advantage that each step is independent of the previous step; i.e., the total length of the gradually varied flow profile may be estimated with a single computation. For example, in Example 6.2 the distance between the points where $y = 3.74$ ft (1.14 m) and $y = 5.62$ ft (1.71 m) is given by

$$L = \frac{6.25}{0.001}\left[(0.899 - 0.598) - (1.12 - 0.620) + \frac{2.99}{3.71}\left(\frac{3.74}{6.25}\right)^{3.47}(1.13 - 0.552)\right]$$

$$= -754 \text{ ft } (-230 \text{ m})$$

where the values of M, J, and N used in this computation were obtained from the table accompanying Example 6.2 and averaged over the reach considered. This type of single-step computation is possible with the method of direct integration but cannot be done with the step method because in this method each step is dependent on the result obtained in the previous step. Unfortunately, this advantage of the method of direct integration cannot be exploited effectively since the intermediate values of x usually must be known so that a gradually varied flow profile may be plotted.

In Table 6.4, it is noted that the hydraulic exponents show very little variation throughout the entire length of channel considered. When the hydraulic exponents are a function of longitudinal distance, the Bakhmeteff-Chow methodology can be applied if the channel length is divided into subreaches so that the hydraulic exponents are constant in each subreach. In rectangular, trapezoidal, and triangular channels, the assumption of constant hydraulic exponents is usually satisfactory, while in circular channels or channels which have abrupt cross-sectional changes in shape with elevation, this is not a valid assumption.

For channels with gradually closing crowns, such as circular channels, the hydraulic exponents are variable near the crown, and while the methods described above can be applied, more accurate results can be obtained by using a numerical integration procedure such as that developed by Kiefer and Chu (1955). Let Q_o be the discharge of a circular conduit of diameter d_o when the open-channel depth of flow is d_o and the energy slope is equal to the bottom slope S_o. Then, by Eq. (5.1.3)

$$Q_o = K_o\sqrt{S_o} \tag{6.3.27}$$

For a uniform flow in this conduit of discharge Q, the same equation yields

$$Q = K_N\sqrt{S_o} \tag{6.3.28}$$

From Eqs. (6.3.27) and (6.3.28) the following relationship can be developed

$$\left(\frac{K_N}{K}\right)^2 = \left(\frac{K_N}{K_o}\right)^2\left(\frac{K_o}{K}\right)^2 = \left(\frac{Q}{Q_o}\right)^2\left(\frac{K_o}{K}\right)^2 \tag{6.3.29}$$

Since the ratio K_o/K is a function of y/d_o, Eq. (6.3.29) can be expressed as

$$\left(\frac{K_N}{K}\right)^2 = \left(\frac{Q}{Q_o}\right)^2 f_1\left(\frac{y}{d_o}\right) \tag{6.3.30}$$

In an analogous fashion, a corresponding expression for the selection factor can be developed or

$$\begin{aligned}\left(\frac{Z_c}{Z}\right)^2 &= \frac{\alpha Q^2 T}{gA^3} = \frac{\alpha Q^2}{d_o^5}\left[\frac{T/d_o}{g(A/d_o^2)^3}\right] \\ &= \frac{\alpha Q^2}{d_o^5} f_2\left(\frac{y}{d_o}\right)\end{aligned} \tag{6.3.31}$$

where the expression $(T/d_o)/g(A/d_o^2)^3$ is a function of y/d_o and represented as $f_2(y/d_o)$. Then, substituting Eqs. (6.3.30) and (6.3.31) into Eq. (6.3.10) and simplifying yields

$$dx = \frac{d_o}{S_o}\left[\frac{1-(\alpha Q^2/d_o^5)f_2(y/d_o)}{1-(Q/Q_o)^2 f_1(y/d_o)}\right] d\left(\frac{y}{d_o}\right) \tag{6.3.32}$$

Integration of Eq. (6.3.32) yields

$$x = \frac{d_o}{S_o}\left[\int_0^{y/d_o} \frac{d(y/d_o)}{1-(Q/Q_o)^2 f_1(y/d_o)} - \frac{\alpha Q^2}{d_o^5}\int^{y_o} /d_o 0 \frac{f_2(y/d_o)d(y/d_o)}{1-(Q/Q_o)^2 f_1(y/d_o)}\right] + C \tag{6.3.33}$$

where C = a constant of integration, the integrals must be evaluated by numerical methods, and $f_1(y/d_o)$ and $f_2(y/d_o)$ can be evaluated numerically. The constant of integration can be eliminated by integrating over a designated interval or

$$L = (x_2 - x_1) = A[(X_2 - X_1) - B(Y_2 - Y_1)] \tag{6.3.34}$$

where

$$X = -\int_0^{y/d_o} \frac{d(y/d_o)}{1 - (Q/Q_o)^2 f_1(y/d_o)}$$

$$Y = -\int_0^{y/d_o} \frac{d(y/d_o)}{1 - (Q/Q_o)^2 f_1(y/d_o)}$$

$$A = -\frac{d_o}{S_o}$$

and

$$B = \frac{\alpha Q^2}{d_o^5}$$

Values of the functions X and Y are summarized in Tables 6.5 and 6.6 as functions of y/d_o and Q/Q_0. In these tables, values between the tabulated values can be estimated by interpolation; however, the interpolation process cannot cross the heavy lines in these tables which identify the location of the normal depth of flow.

EXAMPLE 6.3

A circular conduit 6.0 ft (1.8 m) in diameter conveys a flow of 160 ft^3/s (4.5 m^3/s) with $S = 0.001$ and $n = 0.013$. Determine the distance between a critical depth control section and the section at which the conduit flows full.

Solution

Assume $\alpha = 1$, then from Table 2.1

$$\psi = \frac{\alpha Q^2}{g} = \frac{(1)(160)^2}{32.2} = 795$$

and

$$y_c = \frac{1.01}{d_o^{0.26}}\psi^{0.25} = \frac{1.01}{(6.0)^{0.26}}(795)^{0.25} = 3.4 \text{ ft } (1.04 \text{ m})$$

The capacity of the conduit flowing full but as an open channel is

$$Q_o = \frac{1.49}{n} AR^{2/3}\sqrt{S}$$

TABLE 6.5 The function Y as a function of y/d_o and Q/Q_o for circular channels (Kiefer and Chu, 1955)

Ratio y/d_o	Values of Q/Q_o 0.40	0.60	0.70	0.80	0.90	1.00	1.07	1.20	1.40	1.60
1.00	0	0	0	0	0		0.0245	0.1128	0.0680	0.0482
0.98	0.0003	0.0003	0.0004	0.0006	0.0009	0.0016	0.0186	0.1120	0.0677	0.0479
0.96	0.0008	0.0010	0.0012	0.0015	0.0024	0.0053	0	0.1105	0.0671	0.0477
0.94	0.0015	0.0019	0.0023	0.0028	0.0043	0.0098	0.0824	0.1079	0.0663	0.0470
0.92	0.0023	0.0029	0.0036	0.0045	0.0067	0.0150	0.1614	0.1051	0.0652	0.0466
0.90	0.0034	0.0042	0.0050	0.0063	0.0096	0.0215	0.2500	0.1028	0.0641	0.0458
0.88	0.0044	0.0055	0.0067	0.0080	0.0132	0.0303	0.1852	0.0983	0.0627	0.0452
0.86	0.0058	0.0072	0.0087	0.0112	0.0175	0.0438	0.1604	0.0950	0.0613	0.0444
0.84	0.0071	0.0090	0.0110	0.0143	0.0230	0.0667	0.1439	0.0913	0.0598	0.0435
0.82	0.0087	0.0112	0.0136	0.0180	0.0299		0.1321	0.0879	0.0583	0.0426
0.80	0.0106	0.0136	0.0166	0.0225	0.0389	0.1806	0.1230	0.0847	0.0568	0.0416
0.78	0.0128	0.0164	0.0204	0.0282	0.0524	0.1546	0.1153	0.0816	0.0552	0.0405
0.76	0.0150	0.0196	0.0247	0.0349	0.0768	0.1394	0.1088	0.0784	0.0537	0.0396
0.74	0.0177	0.0236	0.0300	0.0440	1.8457	0.1288	0.1032	0.0755	0.0521	0.0384
0.72	0.0206	0.0278	0.0361	0.0571	0.1950	0.1200	0.0980	0.0727	0.0505	0.0375
0.70	0.0240	0.0333	0.0443	0.0817	0.1650	0.1130	0.0933	0.0699	0.0489	0.0363
0.68	0.0280	0.0395	0.0556	0.1747	0.1465	0.1066	0.0891	0.0673	0.0473	0.0354
0.66	0.0324	0.0475	0.0724	0.2320	0.1352	0.1011	0.0851	0.0648	0.0458	0.0343
0.64	0.0377	0.0576	0.0978	0.1882	0.1263	0.0960	0.0813	0.0622	0.0442	0.0332
0.62	0.0438	0.0721	0.2500	0.1680	0.1189	0.0916	0.0777	0.0599	0.0426	0.0322
0.60	0.0511	0.0915	0.2780	0.1545	0.1120	0.0870	0.0743	0.0575	0.0411	0.0310
0.58	0.0600	0.1290	0.2265	0.1437	0.1054	0.0830	0.0710	0.0550	0.0396	0.0299
0.56	0.0705	0.3909	0.2006	0.1343	0.0997	0.0790	0.0678	0.0529	0.0381	0.0288
0.54	0.0840	0.3080	0.1828	0.1260	0.0944	0.0753	0.0647	0.0506	0.0365	0.0278
0.52	0.1014	0.2580	0.1686	0.1186	0.0895	0.0717	0.0618	0.0485	0.0351	0.0266
0.50	0.1253	0.2293	0.1562	0.1117	0.0850	0.0682	0.0588	0.0463	0.0336	0.0256
0.48	0.1609	0.2085	0.1452	0.1054	0.0803	0.0648	0.0561	0.0441	0.0321	0.0244
0.46	0.2235	0.1912	0.1350	0.0993	0.0761	0.0616	0.0533	0.0420	0.0306	0.0233
0.44	1.4143	0.1770	0.1256	0.0935	0.0718	0.0582	0.0505	0.0399	0.0291	0.0221
0.42	0.0500	0.1643	0.1170	0.0880	0.0680	0.0552	0.0478	0.0378	0.0276	0.0210
0.40	0.4116	0.1528	0.1090	0.0826	0.0641	0.0519	0.0451	0.0357	0.0261	0.0199
0.35	0.3090	0.1271	0.0917	0.0700	0.0546	0.0444	0.0386	0.0306	0.0224	0.0171
0.30	0.2432	0.1038	0.0755	0.0580	0.0454	0.0370	0.0321	0.0255	0.0188	0.0143
0.25	0.1903	0.0825	0.0607	0.0466	0.0366	0.0297	0.0258	0.0205	0.0151	0.0114
0.20	0.1425	0.0627	0.0462	0.0356	0.0281	0.0227	0.0197	0.0157	0.0115	0.0088
0.15	0.1000	0.0442	0.0327	0.0250	0.0197	0.0158	0.0138	0.0110	0.0081	0.0062
0	0	0	0	0	0	0	0	0	0	0

TABLE 6.6 The function X as a function of y/d_o and Q/Q_o for circular channels (Kiefer and Chu, 1955)

Ratio y/d_o	Values of Q/Q_o									
	0.40	0.60	0.70	0.80	0.90	1.00	1.07	1.20	1.40	1.60
1.00	0	0	0	0	0		1.161	0.9166	0.4075	0.2543
0.98	0.0235	0.0299	0.0368	0.0495	0.0817	0.1085	0.760	0.8567	0.3830	0.2397
0.96	0.0468	0.0592	0.0720	0.0954	0.1517	0.2777	0	0.7822	0.3555	0.2238
0.94	0.0700	0.0883	0.1067	0.1403	0.2188	0.4277	2.872	0.7016	0.3269	0.2074
0.92	0.0932	0.1173	0.1415	0.1851	0.2857	0.5763	5.052	0.6203	0.2982	0.1909
0.90	0.1165	0.1465	0.1764	0.2305	0.3541	0.7345	2.7491	0.5427	0.2701	0.1748
0.88	0.1398	0.1759	0.2119	0.2769	0.4257	0.9160	1.4754	0.4714	0.2432	0.1591
0.86	0.1632	0.2057	0.2481	0.3250	0.5023	1.1459	1.0082	0.4076	0.2178	0.1441
0.84	0.1868	0.2360	0.2852	0.3754	0.5868	1.5020	0.7486	0.3515	0.1942	0.1299
0.82	0.2105	0.2668	0.3236	0.4389	0.6835	2.5439	0.5803	0.3025	0.1724	0.1166
0.80	0.2344	0.2985	0.3637	0.4868	0.8002	0.8911	0.4619	0.2600	0.1524	0.1043
0.78	0.2585	0.3311	0.4059	0.5508	0.9552	0.6046	0.3740	0.2233	0.1343	0.0928
0.76	0.2830	0.3649	0.4509	0.6237	1.2090	0.4518	0.3065	0.1916	0.1179	0.0823
0.74	0.3077	0.4002	0.4997	0.7110	1.5967	0.3521	0.2533	0.1643	0.1031	0.0727
0.72	0.3329	0.4374	0.5536	0.8239	0.5844	0.2810	0.2105	0.1407	0.0899	0.0639
0.70	0.3585	0.4770	0.6151	0.9955	0.4016	0.2275	0.1765	0.1202	0.0781	0.0560
0.68	0.3818	0.5199	0.6885	1.5361	0.3001	0.1858	0.1465	0.1026	0.0676	0.0488
0.66	0.4117	0.5674	0.7831	0.5102	0.2329	0.1527	0.1228	0.0873	0.0583	0.0424
0.64	0.4394	0.6215	0.9263	0.3311	0.1845	0.1259	0.1023	0.0741	0.0500	0.0366
0.62	0.4683	0.6860	1.3682	0.2403	0.1479	0.1039	0.0837	0.0627	0.0428	0.0315
0.60	0.4985	0.7699	0.4083	0.1825	0.1196	0.0858	0.0713	0.0529	0.0364	0.0269
0.58	0.5304	0.8986	0.2596	0.1416	0.0970	0.0708	0.0593	0.0444	0.0308	0.0229
0.56	0.5647		0.1851	0.1113	0.0789	0.0583	0.0492	0.0371	0.0259	0.0193
0.54	0.6022	0.3114	0.1380	0.0882	0.0640	0.0478	0.0406	0.0309	0.0217	0.0163
0.52	0.6446	0.1971	0.1051	0.0701	0.0512	0.0391	0.0334	0.0255	0.0180	0.0136
0.50	0.6947	0.1385	0.0811	0.0557	0.0410	0.0318	0.0274	0.0210	0.0149	0.0112
0.48	0.7591	0.1014	0.0628	0.0442	0.0331	0.0257	0.0221	0.0171	0.0122	0.0092
0.46	0.8585	0.0756	0.0487	0.0347	0.0265	0.0206	0.0178	0.0138	0.0099	0.0075
0.44	3.3677	0.0569	0.0377	0.0273	0.0210	0.0164	0.0142	0.0110	0.0080	0.0061
0.42	0.1827	0.0430	0.0291	0.0212	0.0164	0.0130	0.0113	0.0088	0.0063	0.0048
0.40	0.1107	0.0324	0.0223	0.0165	0.0127	0.0101	0.0088	0.0069	0.0050	0.0038
0.35	0.0419	0.0155	0.0110	0.0082	0.0064	0.0052	0.0045	0.0035	0.0026	0.0020
0.30	0.0169	0.0068	0.0049	0.0038	0.0029	0.0024	0.0021	0.0016	0.0012	0.0009
0.25	0.0062	0.0027	0.0019	0.0015	0.0012	0.0009	0.0009	0.0006	0.0005	0.0004
0.20	0.0019	0.0008	0.0006	0.0005	0.0004	0.0003	0.0003	0.0002	0.0002	0.0001
0.15	0.0004	0.0002	0.0001	0.0001	0.0001	0.0001	0.0001	—	—	—
0	0	0	0	0	0	0	0	0	0	0

where

$$A = \frac{\pi d_o^2}{4} = \frac{3.14(6.0)^2}{4} = 28.3 \text{ ft}^2 \ (2.6 \text{ m}^2)$$

$$R = \frac{A}{P} = \frac{(\pi d_o^2/4)}{\pi d_o} = \frac{28.3}{3.14(6.0)} = 1.5 \text{ ft} \ (0.46 \text{ m})$$

and $$Q_o = \frac{1.49}{0.013}(28.3)(1.5)^{2/3}\sqrt{0.001} = 134 \text{ ft}^3/\text{s} \ (3.79 \text{ m}^3/\text{s})$$

Then, with reference to Eq. (6.3.34)

$$A = -\frac{d_o}{S_o} = \frac{-6}{0.001} = -6000$$

$$B = \frac{\alpha Q^2}{d_o^5} = \frac{(1)(160)^2}{(6.0)^5} = 3.29$$

and $$\frac{Q}{Q_o} = \frac{160}{134} = 1.19$$

By use of the data summarized in Tables 6.5 and 6.6, the solution proceeds as indicated in Table 6.7. From these computations, it is concluded that the conduit flows full approximately 3900 ft (1200 m) upstream of the control section.

When the Bakhmeteff-Chow methodology is applied to channels with zero and negative slopes, the techniques discussed above must be modified. In a horizontal channel, $S_o = 0$, and Eq. (6.1.1) becomes

$$\frac{dy}{dx} = -\frac{(Q/K)^2}{1-(Z_c/Z)^2} \tag{6.3.35}$$

Then, by definition, the critical slope S_c is the slope which will sustain a flow Q at critical depth, y_c; and therefore,

$$Q = K_c\sqrt{S_c} \tag{6.3.36}$$

TABLE 6.7 Tabular solution of Example 6.3

y, ft	y/d_o	X_2	X_1	Y_2	Y_1	Δx	$\Sigma\Delta x$
3.4	0.57	—	—	—	—	—	0
3.6	0.60	0.0543	0.0418	0.0588	0.0552	−3.9	−3.9
4.2	0.70	0.1245	0.0543	0.0717	0.0588	170	−170
4.8	0.80	0.2755	0.1245	0.0876	0.0717	590	−760
5.4	0.90	0.7124	0.2755	0.1141	0.0876	2100	−2900
5.9	0.98	0.8493	0.7124	0.1048	0.1141	−1000	−3900

Substituting this into Eq. (6.3.35) and letting

$$\left(\frac{K_c}{K}\right)^2 = \left(\frac{y_c}{y}\right)^N$$

$$\left(\frac{Z_c}{Z}\right) = \left(\frac{y_c}{y}\right)^M$$

and

$$P = \frac{y}{y_c}$$

yields

$$\frac{dy}{dx} = S_c \frac{P^{M-N}}{1 - P^M} \tag{6.3.37}$$

For the case where M and N are constant, Eq. (6.3.37) can be integrated to yield

$$x = \frac{y_c}{S_c}\left(\frac{P^{N-M+1}}{N - M + 1} - \frac{P^{N+1}}{N + 1}\right) + C \tag{6.3.38}$$

where C = a constant of integration.

In channels of adverse slope, the slope of the channel bottom is negative, and Eq. (6.1.1) becomes

$$\frac{dy}{dx} = \frac{-S_o - S_f}{1 + \alpha d(\overline{u}^2/2g)/dy} \tag{6.3.39}$$

It can then be shown that under these conditions with N and M constant [see, for example, Chow (1959)]

$$x = -\frac{y_N}{S_o}\left[u - \int_0^u \frac{du}{1 + u^N} - \left(\frac{y_c}{y_N}\right)^M \int_0^u \frac{u^{N-M}}{1 + u^N}\,du\right] + C \tag{6.3.40}$$

where the varied-flow functions for adverse slopes are

$$F(u, N)_{-S_o} = \int_0^u \frac{du}{1 + u^N} \tag{6.3.41}$$

$$F(v, J)_{-S_o} = \int_0^u \frac{u^{N-M}}{1 + u^N}\,du = \frac{J}{N}\int_0^v \frac{dv}{1 + v^J} \tag{6.3.42}$$

and

$$v = u^{N/J}$$

The functions $F(u,N)_{-S_o}$ and $F(u,J)_{-S_o}$ are tabulated in Appendix III.

Nonprismatic Channels

In the foregoing section, two methods of calculating gradually varied flow profiles in prismatic channels were discussed. In general, neither of these techniques is well suited to the computation of water surface profiles in nonprismatic channels where both the resistance coefficient and the channel shape are functions of longitudinal distance. In addition, because the channel properties are a function of longitudinal distance in nonprismatic channels, the method used to estimate the water surface elevation should predict the depth of flow at a specified longitudinal distance rather than vice versa which is the case in both of the methods previously discussed.

The application of the energy equation between the two stations shown in Fig. 6.3 yields

$$z_1 + \alpha_1 \frac{\overline{u}_1^2}{2g} = z_2 + \alpha_2 \frac{\overline{u}_2^2}{2g} + h_f + h_e$$

where z_1 and z_2 = elevation of water surface above a datum at stations 1 and 2, respectively
h_e = eddy loss incurred in reach
h_f = reach friction loss

The eddy losses, which may be appreciable in nonprismatic channels, are generally expressed in terms of the change in the velocity head in the reach or

$$h_e = k\,\Delta\left(\alpha \frac{\overline{u}^2}{2g}\right) \tag{6.3.43}$$

where k = a coefficient. Unfortunately, there is very little generalized information regarding k, and it is common practice for values of k to be determined on the basis of field measurements. Where this is not possible, k is assumed to have a value of 0.5 for abrupt expansions or contractions and to range between 0 and 0.2 for gradually converging or diverging reaches. The friction loss term is approximated by

$$h_f = S_f\,\Delta x = \frac{(S_{f1} + S_{f2})\,\Delta x}{2} \tag{6.3.44}$$

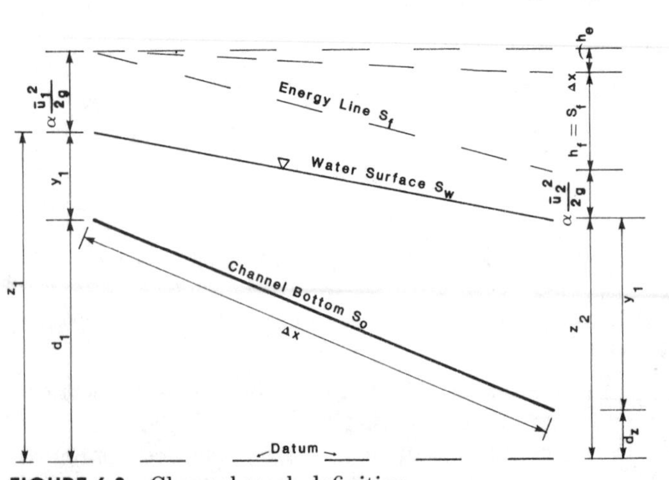

FIGURE 6.3 Channel reach definition.

where S_{f1} and S_{f2} = friction slopes at stations 1 and 2, respectively. With these approximations, the energy equation becomes

$$z_1 + \alpha_1 \frac{\overline{u}_1^2}{2g} = z_2 + \alpha_2 \frac{\overline{u}_2^2}{2g} + \frac{\Delta x(S_{f1} + S_{f2})}{2} + k\,\Delta\left(\alpha \frac{\overline{u}^2}{2g}\right) \tag{6.3.45}$$

For convenience, define

$$H_1 = z_1 + \alpha_1 \frac{\overline{u}_1^2}{2g} \tag{6.3.46}$$

and

$$H_2 = z_2 + \alpha_2 \frac{\overline{u}_2^2}{2g} \tag{6.3.47}$$

With these definitions, Eq. (6.3.45) becomes

$$H_1 = H_2 + h_f + h_e \tag{6.3.48}$$

Equation (6.3.48) is solved by trial and error; i.e., given a longitudinal distance, a water surface elevation at station 1 is assumed which allows the computation of H_1 by Eq. (6.3.46). h_f and h_e are computed, and H_1 is estimated by Eq. (6.3.48). If the two values of H_1 computed agree, then the assumed stage at station 1 is correct. In essence, the trial-and-error procedure attempts to make the difference between H_1 computed by Eq. (6.3.46) and the value estimated by Eq. (6.3.48) zero. The use of this trial-and-error step method of water surface profile computation is best illustrated with an example. Although this method was designed for use in nonprismatic channels, the initial example will be in a prismatic channel with $\alpha_1 = \alpha_2 = 1$ and $h_e = 0$.

EXAMPLE 6.4

A trapezoidal channel with $b = 20$ ft (6.1 m), $n = 0.025$, $z = 2$, and $S_o = 0.001$ carries a discharge of 1000 ft^3/s (28 m^3/s). If this channel terminates in a free overfall and there are no eddy losses, determine the gradually varied flow profile by the trial-and-error step method.

Solution

It is noted that this problem statement is identical to that of Examples 6.1 and 6.2. For comparative purposes, the elevation of the water surface will be determined at the longitudinal stations found in Example 6.2. If the elevation at the free overfall is 100 ft (30 m), then the elevation of the water surface at this point is 103.74 ft (31.62 m) (see Example 6.1). The solution of this problem is summarized in Table 6.8; with regard to this table the following points should be noted:

Column 1: Each longitudinal station is identified by its distance from the control.

TABLE 6.8 Tabular solution of Example 6.4

Station, ft (1)	z, ft (2)	y, ft (3)	A, ft^2 (4)	$\bar{u}$, ft/s (5)	$\frac{\bar{u}^2}{2g}$, ft (6)	H_1, ft (7)	R, ft (8)	S_f (9)	$\bar{S}_f$ (10)	Δx, ft (11)	h_f, ft (12)	h_e, ft (13)	H, ft (14)
0	103.74	3.74	103	9.71	1.46	105.20	2.81	0.00670	—	—	—	—	105.20
116	104.62	4.50	130	7.69	0.918	105.54	3.24	0.00347	0.00509	116	0.590	—	105.79
	105.02	4.90	146	6.85	0.729	105.75	3.48	0.00251	0.00461	116	0.535	—	105.73
355	105.56	5.20	158	6.33	0.622	106.18	3.65	0.00201	0.00226	355	0.802	—	106.55
	105.93	5.57	173	5.78	0.519	106.45	3.85	0.00156	0.00204	355	0.724	—	106.47
745	106.34	5.60	175	5.71	0.506	106.85	3.89	0.00150	0.00153	745	1.14	—	107.59
	106.96	6.21	201	4.98	0.385	107.34	4.21	0.00103	0.00130	745	0.968	—	107.42

Column 2: With the exception of the first station, this is a trial value of the water surface elevation which is either verified or rejected on the basis of subsequent computations. When a trial value is verified, it becomes the basis of the verification process for the trial value at the next longitudinal station.

Column 3: The depth of flow corresponding to the water surface elevation in col. 2. This value is computed by subtracting, from the assumed water surface elevation at a station, the elevation of the control and the product of the longitudinal distance from the control and the bottom slope of the channel. For example, in the second line of Table 6.8

$$104.62 - 100 - 0.001(116) = 4.50 \text{ ft } (1.37 \text{ m})$$

Column 4: The flow area corresponding to the depth of flow in col. 3.

Column 5: The average velocity of flow.

Column 6: The velocity head corresponding to the velocity in col. 5.

Column 7: The total head computed by Eq. (6.3.46).

Column 8: The hydraulic radius corresponding to the depth of flow in col. 3.

Column 9: The friction slope at the station estimated from

$$S_f = \frac{n^2\overline{u}^2}{2.22R^{4/3}}$$

Column 10: The average friction slope in the reach which is approximated as the average value of the friction slope in col. 9 and the corresponding value from the previous reach. Note: Other methods of computing the average friction slope are available; see, for example, Table 6.11.

Column 11: The incremental length of the reach.

Column 12: The friction loss in a reach which is estimated as the product of cols. 10 and 11.

Column 13: The estimated eddy loss for the reach.

Column 14: The total head computed from Eq. (6.3.48) or the sum of cols. 12 and 13 and the water surface elevation of the previous reach, col. 14, which should be identical to the value in col. 7. If the value obtained in this computation does not closely agree with the value in col. 7, then a new value of the water surface elevation in col. 2 must be assumed and the calculations repeated.

The crucial step in this analysis is the selection of an appropriate trial value of the water surface elevation. In Table 6.8 it is noted that no more than two trials were required to obtain the correct value of the water surface elevation in col. 7. This is because an error function equation was used to adjust the first trial value. Recall that the trial-and-error solution of Eq. (6.3.48) depends on making the difference in H_1 determined by Eq. (6.3.46) and H_1 determined by Eq. (6.3.48) zero. Let this difference be designated by e. The value of e can be modified only by changing the depth of flow y_2; therefore, the change in e with respect to changes in y_2 can be evaluated by

$$\frac{de}{dy_2} = \frac{d}{dy_2}\left(y_2 + \frac{\bar{u}_2^2}{2g} - \frac{1}{2}\Delta x\, S_{f2}\right)$$

or

$$\frac{de}{dy_2} = 1 - \mathbf{F}_2^2 + \frac{3S_{f2}\,\Delta x}{2R_2}$$

Then

$$\Delta y_2 = \frac{e}{1 - \mathbf{F}_2^2 + (3S_{f2}\,\Delta x/2R_2)} \tag{6.3.49}$$

where Δy_2 = amount by which the water level must be adjusted to force e to zero. In Table 6.5, the correction which was applied to the first trial value at x = 116 ft (35.4 m) was calculated by

$$D_2 = \frac{A_2}{T_2} = \frac{130}{38} = 3.42 \text{ ft } (1.04 \text{ m})$$

$$\mathbf{F} = \frac{\bar{u}_2}{\sqrt{gD_2}} = \frac{7.69}{\sqrt{32.2(3.42)}} = 0.73$$

Then by Eq. (6.3.49)

$$y_2 = \frac{105.54 - 105.79}{1 - (0.73)^2 + [3(0.00347)116]/2(3.24)} = -0.382 \sim -0.40 \text{ ft } (0.12 \text{ m})$$

In this case, Δy_2 must be added to the initial trial value. Properly used, Eq. (6.3.49) can significantly reduce the number of trial values needed to estimate the water surface elevation correctly.

With regard to this trial-and-error step method of calculating gradually varied flow profiles the following are noted: (1) This method yields answers similar to those estimated in Examples 6.1 and 6.2 and (2) as with all step methods the calculations are sequential; i.e., each step is dependent on the preceding step.

In most natural channels, the possibility that the channel is composed of a main section and one or more overbank sections must be considered. In such a situation, the depth of flow in the overbank section is different than the depth of flow in the main channel, the resistance coefficient of the overbank section may be significantly different from the one which characterizes the main section, and the length between sections on the river may differ. In general, the step method previously discussed with computational modifications may also be applied to this situation.

Since there is a difference in velocity heads between the main and overbank sections, the energy correction factor α must be used to define the velocity head component of the total head. From Eq. (2.4.7), the energy coefficient is given by

$$\alpha = \frac{\Sigma \bar{u}_i^3 A_i}{\bar{u}^3 \Sigma A_i} = \frac{(\Sigma A_i)^2}{(\Sigma Q_i)^3} \Sigma \left(\frac{Q_i^3}{A_i^2} \right)$$

where the subscripts indicate distinct flow subsections and the unsubscripted variables are total channel variables. If the definition of conveyance, Eq. (5.1.3), is combined with the above expression for α, then

$$\alpha = \frac{(\Sigma A_i)^2}{(\Sigma K_i)^3} \Sigma \left(\frac{K_i^3}{A_i^2} \right) \tag{6.3.50}$$

In addition, the friction slope for the reach under consideration is given by

$$S_f = \frac{Q^2}{(\Sigma K_i)^2} \tag{6.3.51}$$

When the reach under consideration is on a curve, then the length of the reach may depend on the subsection being considered. The simplest method of taking this effect into consideration is to vary the value of the resistance coefficient. For example, if the length of the overbank section is less than the main section, then the value of the resistance coefficient must be reduced to maintain S_f constant for the reach. In the case of Manning's n, the n for the overbank section is reduced in proportion to the parameter $(L_o/L_M)^{1/2}$ where L_o = overbank section length and L_M = main channel length (Henderson, 1966). The U.S. Army Corps of Engineers HEC2 model (Anonymous, 1979) and Shearman (1976) calculate a discharge-weighted reach length by the equation

$$L = \frac{\Sigma L_i K_i}{\Sigma K_i} \tag{6.3.52}$$

where L_i = subsection length. The use of the various equations and definitions in conjunction with the step method to determine a gradually varied flow profile is best demonstrated with an example.

EXAMPLE 6.5

The gradually varied flow profile for Q = 100,000 ft³/s (2800 m³/s) is to be estimated over a 1-mile reach (1600 m) whose cross section is defined in Fig. 6.4. In this reach, there is a distinct main channel and an overbank section, both of which are approximately rectangular in section. The properties of the channel are described in Table 6.9. Between river miles 10 and 10.5 the channel is straight, and it is assumed that there are no eddy losses. Between river miles 10.5 and

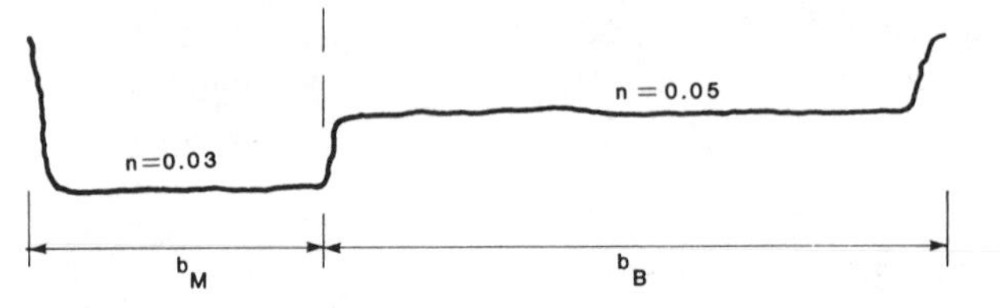

FIGURE 6.4 Channel definition for Example 6.5.

11.0, the channel is curved in such a manner that the overbank section is on the inside of the curve and has a length of 1920 ft (585 m). In this curved section, there is an eddy loss equal to 0.2 times the average velocity heads at the ends of the reach. If the stage of flow is 70 ft (21 m) at river mile 10, determine the stage at river mile 11.0.

Solution

Computationally, the solution to this problem proceeds in a manner similar to that used in Example 6.4 with minor modifications to account for the compound section. The solution is summarized in Table 6.10, and with regard to these computations, the following points should be noted:

Column 2: The notation M in this column refers to the main channel, while OB refers to the overbank section.

Column 8: This is the resistance coefficient defined in Fig. 6.4 or, in the case of the overbank section, between river miles 10.5 and 11

$$n' = \left(\frac{1920}{2640}\right)^{1/2} (0.05) = 0.04$$

TABLE 6.9 Channel properties for Example 6.5

River mile	Widths		Bed elevation	
	b_M, ft	b_B, ft	Main, ft	Berm, ft
10.00	100	600	50.0	55.0
10.25	150	550	52.5	57.5
10.50	200	500	55.0	60.0
11.00	300	400	60.0	65.0

If the Hydrologic Engineering Center (HEC) methodology were used, then the modified Δx between these stations would be

$$L = \frac{18.2 \times 10^5(2640) + 11.0 \times 10^5(1920)}{(18.2 + 11.0) \times 10^5} = 2370 \text{ ft } (722 \text{ m})$$

where the values used in this computation are from the second trial in this reach with $n_{OB} = 0.05$. The result would have been approximately the same as given; e.g., see the line in Table 6.10 designated with an asterisk.

In this solution, Eq. (6.3.49) also served as a guide to the selection of the second trial value of the stage at each station. In the reach between river miles 10.0 and 10.25, the first trial was incorrect and a second trial value was calculated from

$$R = \frac{9500}{172 + 562} = 12.9 \text{ ft } (3.93 \text{ m})$$

$$\alpha \mathbf{F}_2^2 = \frac{2}{R}\left(\alpha \frac{\overline{u}^2}{2g}\right) = \frac{2}{12.9}(2.29) = 0.36$$

$$\frac{3S_{f2}\,\Delta x}{2R} = \frac{3(0.00277)1320}{2(12.9)} = 0.43$$

and therefore,

$$\Delta y = \frac{72.7 - 74.9}{1 - 0.36 + 0.43} = -2.1 \text{ ft } (0.64 \text{ m})$$

Thus, the second trial value was 70 + 2.1 = 72.1 ft (22.0 m). It should be noted that since the first trial value was significantly in error, more than two trials were required for this reach to determine the proper stage.

Although any of the methods discussed in the foregoing material can be adapted for solution by a digital computer, some general-case computer models have been developed by agencies of the U.S. government for this purpose. In particular, there is the HEC-2 model developed by the Hydrologic Engineering Center (HEC) of the U.S. Army Corps of Engineers (Anonymous, 1979; Feldman, 1981) and a similar model developed by the U.S. Geological Survey, (Shearman, 1976, 1977). The purpose of this section is to briefly summarize the capabilities of these models and describe their theoretical basis.

The HEC-2 model was developed to calculate water surface profiles for steady, gradually varied flows in both prismatic and nonprismatic channels. Both subcritical and supercritical flow profiles can be estimated, and the effects of various obstructions such as bridges, culverts, weirs, and structures in the

TABLE 6.10 Tabular solution of Example 6.5

River mile (1)	Sub-section (2)	Stage, ft (3)	A, ft² (4)	P, ft (5)	R, ft (6)	$R^{2/3}$ (7)	n (8)	$K \times 10^{-5}$ (9)	$\frac{K^3}{A^2} \times 10^{-10}$ (10)
10	M	70.	2,000	125	16.0	6.36	0.03	6.3	6.3
	OB		9,000	615	14.6	5.99	0.05	16.0	5.1
	Total		11,000					22.3	11.4
10.25	M	70.	2,625	172	15.2	6.15	0.03	8.0	7.5
	OB		6,875	562	12.2	5.31	0.05	10.0	2.7
	Total		9,500					18.0	10.2
	M	72.1	2,940	175	16.8	6.58	0.03	9.6	10.2
	OB		8,030	565	14.2	5.88	0.05	14.0	4.3
	Total		10,970					23.6	14.5
	M	72.4	2,985	175	17.1	6.64	0.03	9.8	10.6
	OB		8,195	565	14.5	5.95	0.05	14.5	4.6
	Total		11,180					24.3	15.2
10.5	M	74.0	3,800	224	17.0	6.61	0.03	12.0	12.0
	OB		7,000	514	13.6	5.71	0.05	12.0	3.5
	Total		10,800					24.0	15.5
	M	74.5	3,900	224	17.4	6.71	0.03	13.0	14.4
	OB		7,250	514	14.1	5.84	0.05	12.6	3.8
	Total		11,150					25.6	18.2
11	M	78.	5,400	323	16.7	6.54	0.03	17.5	18.4
	OB		5,200	413	12.6	5.42	0.04	10.5	4.3
	Total		10,600					28.0	22.7
	M	78.4	5,520	323	17.1	6.64	0.03	18.2	19.8
	OB		5,360	413	13.0	5.53	0.04	11.0	4.6
	Total		10,880					29.2	24.4
*	M	78.4	5,520	323	17.1	6.64	0.03	18.2	19.8
	OB		5,360	413	13.0	5.53	0.05	8.8	2.4
	Total		10,880					27.0	22.2

overbank region are considered. The model is subject to four crucial assumptions:

1. Since the equations contain no time-dependent terms, the flow must be steady.
2. The flow must be gradually varied since the model equations assume a hydrostatic distribution of pressure.

α (11)	$\bar{u}$, ft/s (12)	$\frac{\alpha \bar{u}^2}{2g}$, ft (13)	H, ft (14)	S_f (15)	$\bar{S}_f$ (16)	Δx, ft (17)	h_f, ft (18)	h_e, ft (19)	H, ft (20)
1.24	9.09	1.59	71.6	0.00201					71.6
1.55	10.5	2.29	72.7	0.00305	0.00253	1320	3.33	—	74.9
1.33	9.12	1.72	73.8	0.00180	0.00191	1320	2.52	—	74.1
1.32	8.94	1.64	74.0	0.00169	0.00185	1320	2.44	—	74.0
1.31	9.26	1.74	75.7	0.00174	0.00171	1320	2.26	—	76.3
1.35	8.97	1.69	76.2	0.00153	0.00161	1320	2.13	—	76.1
1.17	9.43	1.60	79.6	0.00128	0.00140	2640	3.70	0.33	79.8
1.16	9.19	1.52	79.9	0.00117	0.00135	2640	3.56	0.32	80.0
1.34	9.19	1.76	80.2	0.00137	0.00145	2370	3.44	0.32	80.0

3. The flow is one-dimensional.

4. The slope of the channel is small.

The original purpose of the HEC-2 model was to determine water surface elevations for specified discharges in natural channels to aid in the U.S. Army Corps of Engineers floodplain management program. In this capacity, the model has been used to:

1. Determine areas inundated by various flood discharges for the assessment of damages
2. Study the effects of land use in floodplains from the viewpoint of flood damages
3. Study how flood damages can be mitigated by various channel improvements

The HEC-2 model can also be used to study the steady-flow, gradually varied flow profiles discussed previously in this chapter. It is noted that the HEC-2 model is written in FORTRAN IV and requires approximately 32,000 decimal words of computer memory.

Recall that the application of the energy equation between two stations on a channel yields

$$z_2 + \alpha_2 \frac{\overline{u}_2^2}{2g} = z_1 + \alpha_1 \frac{\overline{u}_1^2}{2g} + h \tag{6.3.53}$$

where

$$h = L\overline{S}_f + c \left| \alpha_2 \frac{\overline{u}_2^2}{2g} - \alpha_1 \frac{\overline{u}_1^2}{2g} \right| \tag{6.3.54}$$

where h = total energy loss between stations
$\overline{S}_f$ = representative friction slope in a reach
L = discharge-weighted reach length defined by Eq. (6.3.52)
c = expansion or contraction loss coefficient

Thus, the HEC-2 model is essentially an automated, iterative, step-method computational scheme. The model computation procedure is as follows:

1. If the computations are proceeding upstream, it is assumed that data are available to calculate all the variables in Eqs. (6.3.53) and (6.3.54) subscripted with a 1.
2. A water surface elevation at station 2 is assumed.
3. Based on the assumed water surface elevation, a corresponding total conveyance and velocity head are determined.
4. With the values obtained from step 3, $\overline{S}_f$ is determined and Eq. (6.3.54) is solved for h.
5. z_2 is computed from the values found in steps 3 and 4 with Eq. (6.3.53).
6. The value of z_2 computed in step 5 is compared with the value assumed in step 2. Steps 2 to 5 are repeated until the assumed and computed values of z_2 agree to within 0.01 ft or 0.01 m.

The method by which water surface elevations are assumed in the iterative procedure described above varies as a function of the number of previous trials.

The first trial value is estimated by projecting the previous cross section's elevation on the average friction slope of the previous two cross sections. The second trial value is estimated as the arithmetic average of the computed and assumed water surface elevations from the first trial. The third and subsequent trial values are estimated by a secant method of projecting the rate of change of the difference between the computed and assumed elevations of the previous two trials to zero. In the third and subsequent trials, the change in a single trial value is arbitrarily constrained to a maximum of ± 50 percent of the assumed water surface elevation of the previous trial. Once the correct water surface elevation has been obtained, a check is made to ascertain if the computed elevation is on the correct side of the critical depth elevation for the section; e.g., for a subcritical profile, the water surface elevation should be greater than the critical depth elevation. If the water surface elevation is on the wrong side of the critical depth elevation, then an error has been made and the user is notified. This type of error usually results from the reach lengths being too long or from a misrepresentation of the flow area at a cross section. In addition, it is noted that the technique used by the HEC-2 model to estimate the critical depth of flow may result in significant errors when it is applied to a channel of compound section. The critical depth is found by assuming water surface elevations and computing the corresponding value of specific energy. This process is continued until a minimum value of specific energy is found; however, as noted in Chapter 2, in a channel of compound section there may be more than one local minimum value of specific energy. Thus, the HEC-2 methodology does not guarantee that the correct value of the critical depth is found given a specific channel of compound section.

The HEC-2 model has a number of options and caveats regarding data which must also be noted, as follow.

1. Within the model, there are four methods of estimating the average friction slope for a reach.

a. Average conveyance

$$\overline{S}_f = \left(\frac{Q_1 + Q_2}{K_1 + K_2}\right)^2 \tag{6.3.55}$$

where K_1 and K_2 are the total conveyances at stations 1 and 2, respectively.

b. Average friction slope

$$\overline{S}_f = \frac{S_{f1} + S_{f2}}{2} \tag{6.3.56}$$

c. Geometric mean

$$\overline{S}_f = \sqrt{S_{f1} S_{f2}} \tag{6.3.57}$$

d. Harmonic mean

$$\overline{S}_f = \frac{2S_{f1}S_{f2}}{S_{f1} + S_{f2}} \tag{6.3.58}$$

In addition to the above equations, Reed and Wolfkill (1976), in their study of friction slope models, noted that the following formulations have also been used to estimate the friction slope in a reach

e.

$$\overline{S}_f = \frac{Q^2n^2}{\phi^2[2(A_1A_2/(A_1 + A_2)]^2[(R_1 + R_2)/2]^{4/3}} \tag{6.3.59}$$

f.

$$\overline{S}_f = \frac{Q^2n^2}{\phi^2[(A_1 + A_2)/2][(A_1 + A_2)/(P_1 + P_2)]^{4/3}} \tag{6.3.60}$$

g.

$$\overline{S}_f = \frac{Q^2n^2}{\phi^2[(A_1 + A_2)/2]^2[(R_1 + R_2)/2]^{4/3}} \tag{6.3.61}$$

h.

$$\overline{S}_f = \frac{Q^2n^2}{\phi^2[(A_1R_1^{2/3} + A_2R_2^{2/3})/2]} \tag{6.3.62}$$

where R_1 and R_2 = hydraulic radii at the beginning and end of the reach, respectively; A_1 and A_2 = flow areas at the beginning and end of the reach, respectively; n = Manning's resistance coefficient for the reach, and Q = the flow in the reach. Although any of the above equations will provide a satisfactory estimate of the friction slope provided that the reach is not too long, the concept of providing alternative methodologies within a model to maximize the reach length is attractive. Equation (6.3.55) is the standard HEC-2 formulation, and Eq. (6.3.57) is the formulation used by the U.S. Geological Survey in its gradually varied flow model (Shearman, 1976, 1977). The results of Reed and Wolfkill (1976) are summarized in Table 6.11 along

TABLE 6.11 Recommended friction slope models

Profile type	Friction slope model recommended by Reed and Wolfkill (1976)	Friction slope model used by HEC-2
M1	(6.3.55)	(6.3.55)
M2	(6.3.57)	(6.3.57)
M3	(6.3.58)	(6.3.56)
S1	(6.3.57)	(6.3.55)
S2	(6.3.55)	(6.3.55)
S3	(6.3.56)	(6.3.56)

with the criteria used by the HEC-2 model in selecting an optimum model of the friction slope.

2. Adequate data regarding the channel cross section are essential to the proper and accurate operation of the model. Cross sections are required along the channel where there are significant changes in either the geometry of the channel or in the hydraulic characteristics. Critical geometric changes usually involve either artificial or natural contractions or expansions, while changes in channel slope, roughness, or discharge can be regarded as significant hydraulic characteristic changes. In general, the cross sections provided should be perpendicular to the direction of flow and extend completely across the channel to the high ground on either side of the channel. In some cases where there are very significant changes in either the channel geometry or its hydraulic characteristics, a number of cross sections will be required. Care should also be exercised to ensure that only the effective flow area of the channel is included.

3. The distance between cross sections, also known as the reach length, is a function of the degree of hydraulic detail required and the human, temporal, and financial resources available. Small, nonuniform, or steep channels may require very short reach lengths, e.g., 200 ft (60 m), while in large, uniform channels with small slopes, reach lengths of up to 2 miles (3 km) may be appropriate. In addition, the type of study being performed may dictate the detail required and hence the reach length. For example, a navigation study of the Missouri River required reach lengths of 500 ft (150 m) (Feldman, 1981). Reach lengths can be optimized by adapting an appropriate friction slope equation (Table 6.11). The HEC-2 model requires reach lengths for both the right and left overbank regions of the channel, if such areas exist, and the main channel. In the main channel, this distance is measured either along the centerline in an artificial channel or along the thalweg in a natural channel, while in the overbank areas the reach length should represent the flow path length of the center of mass of the water moving in this area. Since for a given length of channel these three lengths may vary, it is sometimes necessary to calculate a hypothetical reach length by Eq. (6.3.52).

Further details regarding the operation of the HEC-2 model can be obtained from Anonymous (1979) or Feldman (1981).

In general, the U.S. Geological Survey model for computing gradually varied flow profiles, specifically computer program E431, is theoretically similar to the HEC-2 model; that is, both models solve Eq. (6.3.53) by a standard step technique. The salient differences in the models are:

1. In E431 the average friction slope of a reach is computed by

$$\overline{S}_f = \frac{Q^2}{K_1 + K_2} \tag{6.3.63}$$

where Q is the average flow in the reach, and there are no options for using other equations.

2. The iterative process used by E431 to estimate the water surface elevation z_2 is quite different from the methodology used by the HEC-2 model. The user must specify the value of four variables: (*a*) h_o, the minimum elevation of interest at each cross section; (*b*) h_M, the maximum elevation at each cross section; (*c*) Δh, an elevation increment; and (*d*) ϵ, a measure of the accuracy required for the computations. The initial estimate of z_2 is the greater of $(z_1 + 0.5\ \Delta h)$ where Δh may be either positive or negative and h_0 at station 2. If the difference in the total energies between the two cross sections is greater than ϵ, then additional trials are required. It is noted that if between two trials the difference in total energy changes sign, this demonstrates that a mathematically valid root of Eq. (6.3.53) lies in the vertical increment separating the trials. Once a vertical increment which contains a valid root of Eq. (6.3.53) has been identified, the solution is found by the method of false position. If a root is not identified in the first trial, the estimate of z_2 is increased by $0.5\ \Delta h$, and this process is then continued until the elevation h_M is reached. If a valid interval of solution has not been found by the time elevation h_M is reached, then it is assumed that the step size was too large to detect the interval of solution and a smaller step size is required. In this case, the iterative process is repeated with a vertical increment between trials of $0.25\ \Delta h$. If this series of trials also fails to identify the interval of solution, then the iterative process is again repeated with a vertical increment of $0.25\ \Delta h$ between elevations $(h_o + 0.5\ \Delta h)$ and h_M. If a valid root to Eq. (6.3.53) is still not found, then the computation is aborted and the user notified of the situation.

3. Each valid root of Eq. (6.3.53) is checked for its hydraulic validity; i.e., the user specifies a Froude number whose value cannot be exceeded, and this value is compared with a Froude number computed using the estimated water surface elevation. The crucial consideration in this computation is the method used to estimate the Froude number in a channel of compound section when the water surface elevation is specified. In E431, the Froude number is estimated in channels of compound section by

$$\mathbf{F} = \frac{K_M}{KA_M} \frac{Q}{\sqrt{g(A_M/T_M)}} \tag{6.3.64}$$

where the subscript M designates variables associated with the subsection of the channel with the largest conveyance. In general, this method of computing compound section Froude numbers is no more accurate than the methods used by HEC-2 (see, for example, Chapter 2, Fig. 2.12*b*).

Although the advent of high-speed digital mainframe computers, minicomputers, and hand-held calculators has to a large extent obviated the need for graphical solutions of the gradually varied flow equation, there are still situations in which graphical techniques may be effective. For example, in prismatic channels both the standard-step and direct-integration methods of solution predict the longitudinal distance to a specified depth of flow. In many cases, it would be much more convenient if a depth of flow could be estimated for a specified longitudinal distance.

In the case of prismatic channels, the relation between the energy available at an upstream station, station 1, and a downstream station, station 2, is

$$E_1 - E_2 = (S_o - \tfrac{1}{2}S_{f1} - \tfrac{1}{2}S_{f2})\,\Delta x \tag{6.3.65}$$

If the gradually varied flow computation is proceeding upstream, then all variables are defined at station 2, but are undefined at station 1. Rearrange Eq. (6.3.65) to yield

$$E_2 + 0.5S_{f2}\,\Delta x + (S_o - S_{f2})\,\Delta x = E_1 + \tfrac{1}{2}S_{f1}\,\Delta x \tag{6.3.66}$$

For notational convenience, define

$$U = E + 0.5S_f\,\Delta x \tag{6.3.67}$$

and Eq. (6.3.66) can then be rearranged to yield

$$U_2 - U_1 = (S_{f2} - S_o)\,\Delta x \tag{6.3.68}$$

where U is only a function of y and Δx.

Equations (6.3.66) to (6.3.68) apply to the situation in which computations are proceeding upstream. If the calculations are proceeding downstream, then conditions are defined at station 1 but are unknown at station 2. In this case, Eq. (6.3.65) is rearranged to yield

$$E_1 - 0.5S_{f1}\,\Delta x + (S_{f1} - S_o)\,\Delta x = E_2 - \tfrac{1}{2}S_{f2}\,\Delta x \tag{6.3.69}$$

Defining

$$V = E = \tfrac{1}{2}S_f\,\Delta x \tag{6.3.70}$$

and rearranging Eq. (6.3.69) yields

$$V_1 - V_2 = (S_o - S_{f1})\,\Delta x \tag{6.3.71}$$

It is noted that the difference in the parameters U and V is in essence a difference in the sign of Δx; i.e., if the calculation is proceeding upstream, Δx is negative, while if the calculation is proceeding downstream, the sign of Δx is positive. Although U and V are essentially the same parameters, it is generally convenient to regard them as different.

Equations (6.3.66) to (6.3.68) or Eqs. (6.2.69) to (6.3.71) can be used to achieve a hybrid graphical-tabular solution of the gradually varied flow problem. A suggested solution methodology for the case where the calculations are proceeding upstream is outlined in Table 6.12, and the technique is demon-

TABLE 6.12 Hybrid graphical-tabular solution of gradually varied flow problems in prismatic channels with calculations proceeding upstream

Step	Process
1	Plot $(S_f - S_o)$, E, and U, for a constant Δx, versus y, Fig. 6.5
2	Plot U versus y, Fig. 6.5
3	Obtain U_2 and $(S_f - S_o)$ at station 2 from Fig. 6.5
4	Use results of step 3 to calculate U by Eq. (6.3.67)
5	With U_1 from step 4 use Fig. 6.5 to estimate y_1

strated in the following example which, except for the solution technique, is identical to Example 6.1.

EXAMPLE 6.6

A trapezoidal channel with $b = 20$ ft (6.1 m), n = 0.025, $z = 2$, and $S_o = 0.001$ carries a discharge of 1000 ft^3/s (28 m^3/s). If this channel terminates in a free overfall, determine the gradually varied flow profile.

Solution

By use of the results of Example 6.1, the parameters $(S_f - S_o)$, E, and U for $\Delta x = 50$ and 100 ft (15 and 30 *m*, respectively) are plotted in Fig. 6.5. Two values of Δx are used so that sufficient detail regarding the shape of the profile near the free overfall can be obtained and yet the number of computations required to define the total profile is not excessive. The computations proceed as shown in Table 6.13.

This solution is not carried to completion, but a comparison of the results in Table 6.12 with those derived in Example 6.1 demonstrates that they are almost identical.

The method described in the foregoing material is a hybrid method with both graphical and tabular components. This method may be converted to a wholly graphical method by plotting another curve a distance $(S_f - S_o)\,\Delta x$ above the U-y curve for U at a specified Δx. This is shown in Fig. 6.6 for the data specified in Example 6.5 and $\Delta x = 100$ ft (30 m). With reference to Fig. 6.6, the solution proceeds as follows:

1. From the initial point on the U-y curve, a vertical line is drawn to intersect the second curve.
2. From the point found in step 1, a horizontal line is drawn until it intersects the U-y curve.
3. Step 1 is repeated from the point determined in step 2.

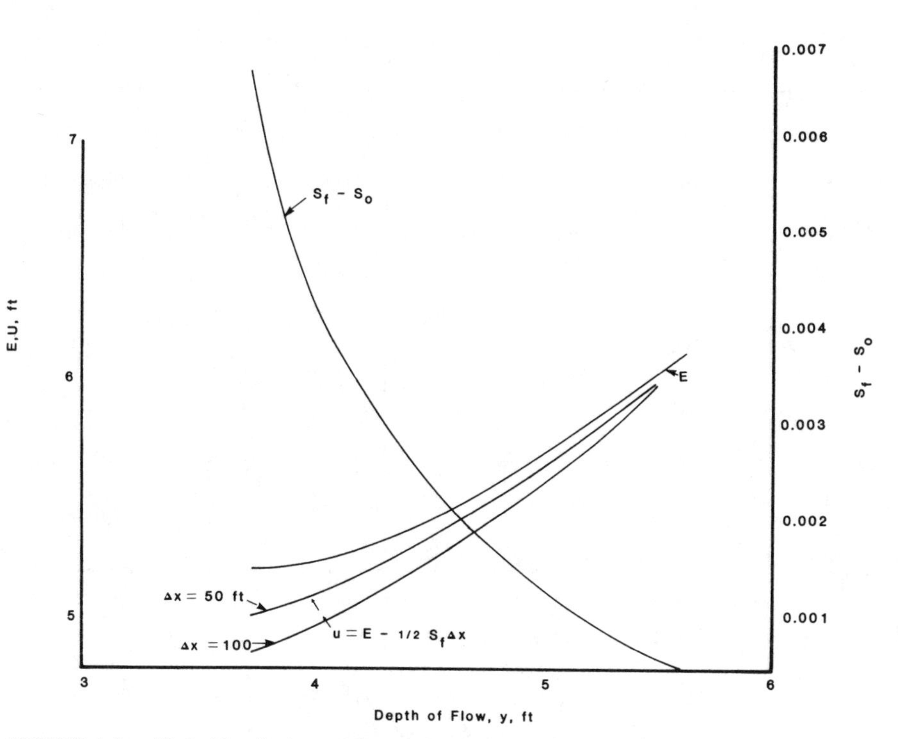

FIGURE 6.5. Hybrid solution of Example 6.6.

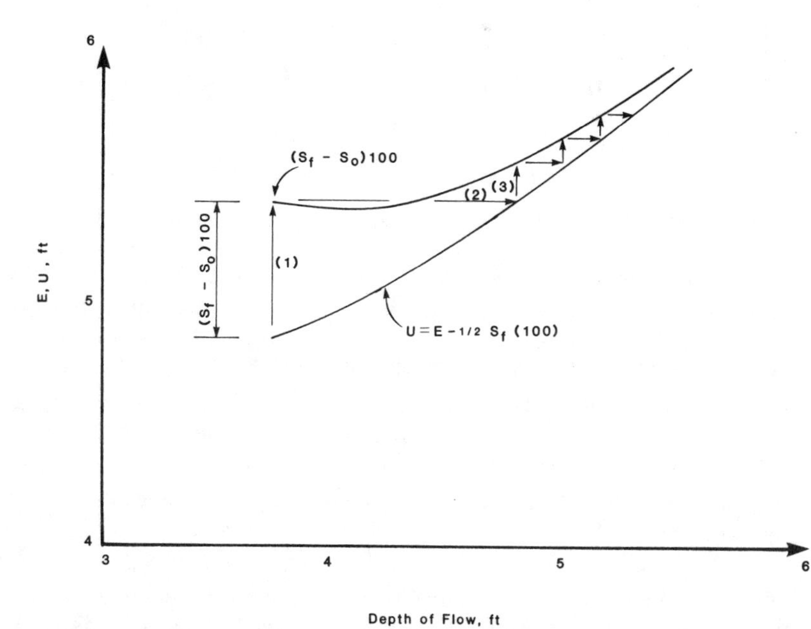

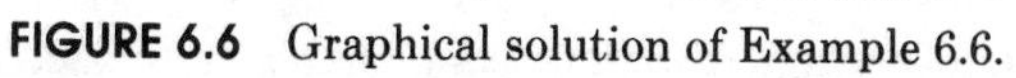

FIGURE 6.6 Graphical solution of Example 6.6.

TABLE 6.13 Hybrid solution of Example 6.6

x, ft	Δx, ft	y, ft	U	$S_f - S_o$	$\Delta U = (S_{f1} - S_o)\Delta x$
0		3.74 ①	→ 5.05 ②	→ 0.00535	③
	−50	⑤		④	→ 0.268
−50		4.46 ←	5.32 ←	⑥ 0.00260	⑦
	−50	↓	→	→	→ 0.130
−100		4.70 ← ⑨	5.45 ←	⑧	
−100		⑩ 4.70	⑪ → 5.36	⑫ → 0.00195	⑬
	−100				→ 0.195
−200		4.98	5.56	0.00135	
	−100				0.135
−300		5.18	5.70	0.00100	
	−100				0.100
−400		5.30	5.80	0.00085	
	−100				0.085
−500		5.48	5.88		

1. From control condition y_c = 3.85 ft use Fig. 6.5 to find $U(\Delta x = 50) = 5.05$.
2. For y = 3.85 ft and $U(\Delta x = 50)$ use Fig. 6.5 to find $(S_f - S_o) = 0.00535$.
3. Compute $(S_f - S_o)\ \Delta x = 50\ (0.00535) = 0.268$.
4. Compute $U + \Delta x\ (S_f - S_o) = 5.05 + 0.268 = 5.32$.
5. For $U(\Delta x = 50) - 5.32$ use Fig. 6.5 to find at $x = 50$, $y = 4.46$ ft.
6. Repeat step 2.
7. Repeat step 3.
8. Repeat step 4.
9. Repeat step 5.
10. In this step Δx is changed from 50 to 100.
11. For y = 4.70 ft find from Fig. 6.5 corresponding $U(\Delta x = 100)$.
12. Repeat step 2.
13. Repeat step 3.

Thus, this solution consists of tracing a sawtooth set of lines between the curves shown in Fig. 6.6. It is noted that at some point the two curves will intersect. This is the point at which $S_o = S_f$ and by definition uniform flow occurs. Some results from this type of graphical solution of Example 6.6 are summarized in Table 6.14. A comparison of these results with the tabular results obtained for Example 6.1 demonstrates good agreement.

While the hybrid and graphical methods described in the foregoing material for prismatic channels provide a quick and convenient method of solving the gradually varied flow equations, once the necessary curves have been constructed, the preparation of these curves requires a significant amount of effort. Thus, these methods are valuable only when a number of water surface profiles must be plotted for the same discharge. In most cases, these methods are not as cost-effective as those using digital computers or hand-held calculators.

Graphical methods can also be used to estimate gradually varied flow profiles in nonprismatic channels. The Ezra method (1954) is based on the following arrangement of Eq. (6.3.65)

$$z_1 + \alpha_1 \frac{\overline{u}_1^2}{2g} = z_2 + \alpha_2 \frac{\overline{u}_2^2}{2g} + \tfrac{1}{2} \Delta x (S_{f1} + S_{f2}) \tag{6.3.72}$$

where, as before, station 1 is upstream of station 2 and z_i = elevation of the water surface above a datum. The introduction of the water surface elevation in the energy equation results in the removal of the channel slope as a variable in Eq. (6.3.72). In applying this equation, it is usually assumed that the eddy losses can be incorporated in the resistance loss term. Equation (6.3.72) can be rewritten as

$$z_1 + F(z_1) = z_2 + F(z_2) \tag{6.3.73}$$

where

$$F(z_1) = \alpha_1 \frac{\overline{u}_1^2}{2g} - \tfrac{1}{2} S_f \, \Delta x_u \tag{6.3.74}$$

and

$$F(z_2) = \alpha_2 \frac{\overline{u}_2^2}{2g} + \tfrac{1}{2} S_f \, \Delta x_D \tag{6.3.75}$$

The functions $F(z_1)$ and $F(z_2)$ correspond to the functions U and V in the previously described graphical solution for prismatic channels. However, the Ezra method differs from the previously described technique in that the value of Δx differs from reach to reach. As will be demonstrated in the example which follows, the Ezra method is based on plotting $z_1 + F(z_1)$ and $z_2 + F(z_2)$ as a function of z. In doing this, α is calculated from Eq. (6.3.50) and S_f from Eq. (6.3.51). The value of Δx used to calculate $F(z_1)$ is the length of the next subreach upstream of the station while the Δx associated with $F(z_2)$ is the length of the next subreach downstream. The application of this graphical technique is best illustrated with the aid of an example. For convenience, the data provided with Example 6.5 and a slightly modified problem statement are used.

EXAMPLE 6.7

The gradually varied flow profile for Q = 100,000 ft^3/s (2800 m^3/s) is to be estimated over a 1-mile reach (1600 m) whose cross section and

TABLE 6.14 Summary of results from Fig. 6.6

x, ft	y, ft
0	3.74
−100	4.80
−200	5.00
−300	5.16
−400	5.30

characteristics are defined in Fig. 6.4 and Table 6.9. In this reach, there is a distinct main channel and an overbank section both of which are approximately rectangular in section. Between river miles 10 and 10.5, the channel is straight, and there are no eddy losses. Between river miles 10.5 and 11, the channel is curved in such a manner that the overbank section is on the inside of the curve and has a length of 1920 ft (585 m). For simplicity, it is assumed that there are no eddy losses in this reach. Estimate the elevation of the water surface at river mile 11 if the stage of the river is 70 ft (21 m) at river mile 10.

Solution

The values of $z_1 + F(z_1)$ and $z_2 + F(z_2)$ used in this computation are summarized in Table 6.15. With regard to these data, the following points should be noted:

1. The stage data in col. 2 were selected by examining the correct answers to Example 6.5. In the general case, these selections would be more difficult to discern.

2. In col. 3, the lengths with the asterisk were calculated from Eq. (6.3.52). Although these numbers vary in a subreach as a function of the stage, this variation is not significant over a limited range of stages.

After these data are tabulated, Fig. 6.7 can be plotted. The solution then proceeds as follows:

1. At river mile 10 the stage is 70 ft (21 m) (point A, Fig. 6.7). This point is located, and the corresponding value of $z_2 + F(z_2)$ is determined to be 72.9 ft (22.2 m).

2. The value of 72.9 ft (22.2 m) is then located on the curve $z_1 + F(z_1)$ for river mile 10.15 (point B). The stage corresponding to this value of $z_1 + F(z_1)$ is 72.4 ft (22.1 m) which compares exactly with the value found for this same location in Example 6.5.

3. Point C in Fig. 6.7 is found by extending a horizontal line from point B until it intersects the $z_2 + F(z_2)$ curve for river mile 10.25.

4. Point C corresponds to a value of $z_2 + F(z_2)$ of 75.1 ft (22.9 m).

5. The value of 75.1 ft is then located on the $z_1 + F(z_1)$ curve for river mile 10.5 (point D).

6. Then the stage at river mile 10.5 is estimated to be 74.6 ft (22.7 m), which corresponds to the value of 74.5 ft (22.7 m) in Example 6.4.

The solution continues in this fashion until it is determined that the stage at river mile 11.0 is 78 ft (23.8 m).

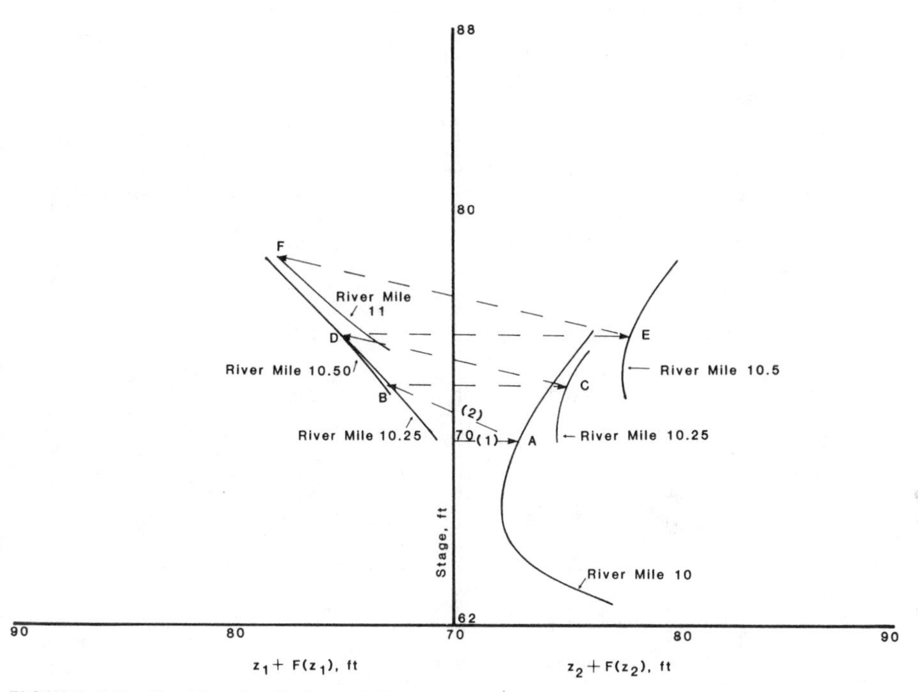

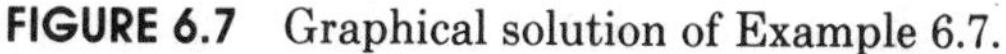

FIGURE 6.7 Graphical solution of Example 6.7.

A number of general comments regarding this methodology should be noted:

1. The z versus $z + F(z)$ curves are essentially specific energy curves, e.g., the $z_2 + F(z_2)$ curve for river mile 10 in Fig. 6.7.
2. The $z_1 + F(z_1)$ curves for river miles 10.25 and 10.50 are almost identical.
3. In this example, eddy losses were neglected; however, they could have been considered by implementing the following steps:
   a. A curve of $\alpha\bar{u}^2/2g$ against z for each subreach is constructed.
   b. The value

$$k\left(\alpha_1 \frac{\bar{u}_1^2}{2g} - \alpha_2 \frac{\bar{u}_2^2}{2g}\right)$$

is added to $z_2 + F(z_2)$ and a corrected value of $z_1 + F(Z_1)$ is found.

A distinct advantage of the Ezra method over other techniques, both digital and graphical, is that with minor modification it can be used to determine gradually varied flow profiles for a range of flow rates in a channel of specified geometry. Following Henderson (1966), define

$$k_i = \frac{Q_o^2}{Q^2} F(z_i) \tag{6.3.76}$$

TABLE 6.15 Tabular data for Example 6.7

River mile (1)	Stage, ft (2)	Δx_D Δx_u, ft (3)	A, ft^2 (4)	P, ft (5)	R, ft (6)	$R^{2/3}$ (7)	n (8)	$K \times 10^{-5}$ (9)	$\frac{K^3}{A^2} \times 10^{-10}$ (10)
10	63	—	1300	118	11.0	4.96	0.03	3.2	1.9
		1320	4800	608	7.89	3.97	0.05	5.7	0.8
			6100					8.9	2.7
	65	—	1500	120	12.5	5.39	0.03	4.0	2.9
		1320	6000	610	9.84	4.59	0.05	8.2	1.5
			7500					12.2	4.4
	67	—	1700	122	13.9	5.80	0.03	4.9	4.1
		1320	7200	612	11.8	5.18	0.05	11.0	2.6
			8900					15.9	6.7
	69	—	1900	124	15.3	6.17	0.03	5.8	5.5
		1320	8400	614	13.7	5.72	0.05	14.0	4.2
			10300					19.8	9.7
	71	—	2100	126	16.7	6.53	0.03	6.8	7.2
		1320	9600	616	15.6	6.24	0.05	18.0	6.2
			11700					24.8	13.4
	73	—	2300	128	18.0	6.87	0.03	7.8	9.1
		1320	10800	618	17.5	6.74	0.05	22.0	8.8
			13100					29.8	17.9
	75	—	2500	130	19.2	7.18	0.03	8.9	11.0
		1320	12000	620	19.4	7.22	0.05	26.0	12.0
			14500					34.9	23.0
10.25	70	1320	2625	172	15.2	6.15	0.03	8.0	7.5
		1320	6875	562	12.2	5.31	0.05	10.9	2.7
			9500					18.9	10.2
	72	1320	2925	174	16.8	6.56	0.03	9.5	10.1
		1320	7975	564	14.1	5.85	0.05	14.0	4.2
			10900					23.5	14.3
	74	1320	3225	176	18.3	6.95	0.03	11.0	13.0
		1320	9075	566	16.0	6.36	0.05	17.0	6.2
			12300					28.0	19.2
10.5	72	1320	3400	222	15.3	6.17	0.03	10.0	9.8
		2295*	6000	512	11.7	5.16	0.05	9.2	2.2
			9400					19.2	12.0
	74	1320	3800	224	17.0	6.61	0.03	12.0	12.0
		2280*	7000	514	13.6	5.71	0.05	12.0	3.4
			10800					24.0	15.4
	76	1320	4200	226	18.6	7.02	0.03	15.0	18.0
		2280*	8000	516	15.5	6.22	0.05	15.0	5.1
			12200					30.0	23.1
	78	1320	4600	228	20.2	7.41	0.03	16.9	23.0
		2255*	9000	518	17.4	6.71	0.05	18.0	7.2
			13600					34.9	30.2

α (11)	$\overline{u}$, ft/s (12)	$\frac{\alpha \overline{u}^2}{2g}$ (13)	S_f (14)	$-\frac{1}{2}S_f\,\Delta x_D$ (15)	$-\frac{1}{2}S_f\,\Delta x_u$ (16)	$F(z_1)$ (17)	$F(z_2)$ (18)	$z_1 + F(z_1)$ (19)	$z_2 + F(z_2)$ (20)
1.43	16.4	5.97	0.0126	—	8.33	—	14.3	—	77.3
1.36	13.3	3.75	0.00672	—	4.43	—	8.19	—	73.2
1.32	11.2	2.59	0.0040	—	2.61	—	5.20	—	72.2
1.32	9.71	1.93	0.00255	—	1.68	—	3.61	—	72.6
1.20	8.55	1.36	0.00163	—	1.07	—	2.43	—	73.4
1.16	7.63	1.05	0.00113	—	0.74	—	1.79	—	74.8
1.14	6.90	0.84	0.00082	—	0.54	—	1.38	—	76.4
1.36	10.5	2.33	0.00280	−1.85	1.85	0.48	4.18	70.5	74.2
1.33	9.17	1.74	0.00183	−1.20	1.20	0.54	2.94	72.5	74.9
1.30	8.13	1.33	0.00125	−0.82	0.82	0.51	2.15	74.5	76.1
1.50	10.6	2.64	0.00271	−1.79	3.11	0.85	5.75	72.8	77.8
1.30	9.26	1.73	0.00174	−1.14	1.98	0.58	3.71	74.5	77.7
1.27	8.20	1.32	0.00111	−0.73	1.27	0.59	2.59	76.6	78.6
1.31	7.35	1.10	0.00082	−0.54	0.92	0.56	2.09	78.6	80.1

TABLE 6.15 Tabular data for Example 6.7 (*Continued*)

River mile (1)	Stage, ft (2)	Δx_D Δx_u, ft (3)	A, ft^2 (4)	P, ft (5)	R, ft (6)	$R^{2/3}$ (7)	n (8)	$K \times 10^{-5}$ (9)	$\frac{K^3}{A^2} \times 10^{-10}$ (10)
11	74	2435*	4200	319	13.2	5.58	0.03	11.6	8.9
		—	3600	409	8.80	4.27	0.05	4.6	0.8
			7800					16.2	9.7
	76	2418*	4800	321	14.9	6.08	0.03	14.4	13.1
		—	4400	411	10.7	4.86	0.05	6.4	1.3
			9200					20.8	14.4
	78	2406*	5400	323	16.7	6.54	0.03	17.5	18.5
		—	5200	413	12.6	5.42	0.05	8.4	2.2
			10600					25.9	20.7
	80	2400	6000	325	18.5	6.99	0.03	20.8	25.1
		—	6000	415	14.5	5.94	0.05	10.6	3.3
			12000					31.4	28.4

*Lengths calculated from Eq. (6.3.52).

or

$$F(z_i) = \frac{Q^2}{Q_o^2} k_i = r k_i \tag{6.3.77}$$

where Q_o = a reference discharge which is chosen near the middle of the range of discharges to be considered. The k_1 and k_2 curves are plotted by calculating $F(z_1)$ and $F(z_2)$ for the flow rate Q_o. For example, in Fig. 6.8, k_1 and k_2 are plotted as functions of the stage for the channel defined in Examples 6.5 and 6.7 using the data summarized in Table 6.15. In constructing such figures, it is essential that the same scale be used for the abscissa and the ordinate. The application of this method is examined most effectively by an example.

EXAMPLE 6.8

For the data given in Example 6.7, determine the stage at river mile 11 with the curves plotted in Fig. 6.8.

Solution

The solution proceeds as follows:

1. The stage at river mile 10, that is, 70 ft (21.3 m), is located on the k_2 for this river mile, point A.

2. In this case, since $Q = Q_o$, $r = 1$ and the stage at river mile 10.25 is determined by constructing a line with a slope of -1 through point A and extending it until it intersects the k_1 curve for river mile 10.25 (point B). Thus, the stage at this location is 72.4 ft (22.1 m).

α (11)	$\bar{u}$, ft/s (12)	$\frac{\alpha \bar{u}^2}{2g}$ (13)	S_f (14)	$-\frac{1}{2}S_f\,\Delta x_D$ (15)	$-\frac{1}{2}S_f\,\Delta x_u$ (16)	$F(z_1)$ (17)	$F(z_2)$ (18)	$z_1 + F(z_1)$ (19)	$z_2 + F(z_2)$ (20)
1.39	12.8	3.55	0.00381	−4.64	—	−1.09	—	72.9	—
1.35	10.9	2.48	0.00231	−2.79	—	−0.31	—	75.7	—
1.34	9.43	1.85	0.00149	−1.79	—	0.06	—	78.1	—
1.32	8.33	1.42	0.00101	−1.22	—	0.20	—	80.2	—

3. A horizontal line is extended through point B until the k_2 curve for river mile 10.25 is intersected.

Steps 1 to 3 are repeated until the specified end point is reached, in this case point F. Thus, the stage at river mile 11 is approximately 77.6 ft (23.7 m), and this estimate compares well with the stage at this location estimated in Example 6.5.

It is concluded that Example 6.5 could have been solved by graphical techniques as described above. It is now asserted that the slope of the lines in Fig. 6.8 connecting the k_2 with the k_1 curves is determined by

$$\tan\theta = r = \frac{Q^2}{Q_o^2} \tag{6.3.78}$$

Thus, for discharges other than Q_o, the gradually varied flow profiles for this channel are determined by simply modifying the slope of the lines connecting the k_1 and k_2 curves.

EXAMPLE 6.9

For the channel defined for Examples 6.5 and 6.7, determine the stage at river mile 11 if the discharge is 132,000 ft^3/s (3740 m^3/s), the stage at river mile 10 is 71 ft (21.6 m), and eddy losses can be neglected.

Solution

The slope of the lines connecting the k_1 and k_2 curves are found from Eq. (6.3.78) or

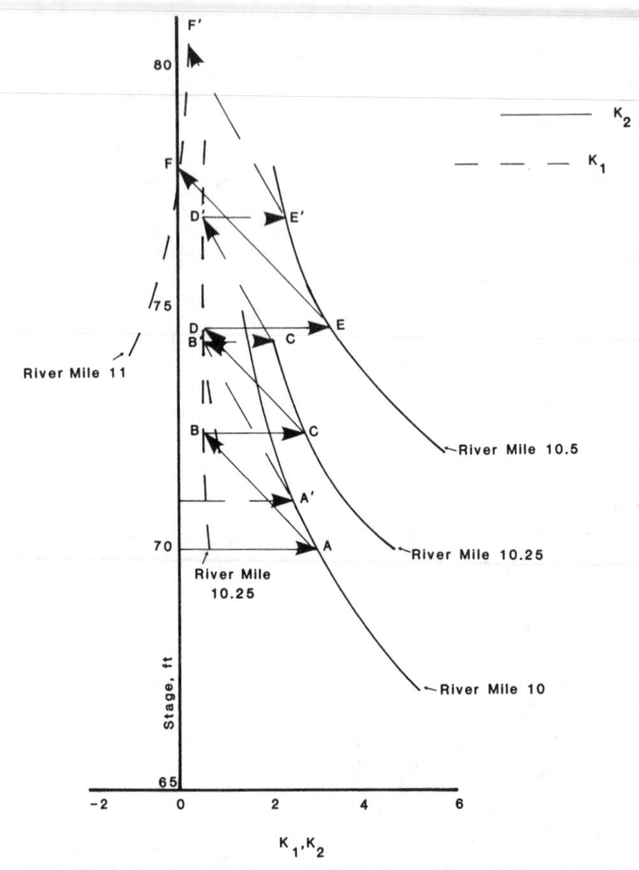

FIGURE 6.8 Gradually varied flow profile solution by the modified Ezra method.

$$\tan \theta = r = \frac{Q^2}{Q_o^2} = \frac{(132{,}000)^2}{(100{,}000)^2} = 1.74$$

$$\theta = \tan^{-1}(1.74) = 60.1°$$

Then, with reference to Fig. 6.8, the solution proceeds as follows:

1. The stage at river mile 10, that is, 71 ft, is located on the k_2 curve for this river mile, point A'.
2. A line making an angle of 60° with the horizontal is constructed through A' and extended until it intersects the k_1 curve for river mile 10.25 (point B'). The stage at this location is 74.3 ft (22.6 m).
3. A horizontal line is extended through point B' until the k_2 curve for river mile 10.25 is intersected, point C'.

Steps 1 to 3 are then repeated until the specified end point is reached, in this case point F'. The stage at river mile 11 is estimated to be 80.5 ft (24.5 m).

The methods described in the foregoing sections which are applicable to non-prismatic or natural channels all require that the cross section be completely defined at frequent intervals along the channel. In many applied situations, the engineer must estimate the gradually varied flow profile from numerical field data. A technique known as *Grimm's method* can be used to estimate the required profile in such a situation. If it is assumed that the friction slope S_f is equal to the water surface slope S_w or in effect that the velocity head is negligible, then

$$Q \sim \sqrt{S_w} \tag{6.3.79}$$

The slope of the water surface can be found for different flow rates by using the stage discharge data at two stations along the channel. Suppose that at a station on the channel, the water surface slope corresponding to a flow Q_o is S_{wo}; then, the water surface slope corresponding to a flow Q would be

$$S_w = S_{wo}\left(\frac{Q}{Q_o}\right)^2 \tag{6.3.80}$$

where S_w is the water surface slope corresponding to Q. The slope of the water surface profile for the flow rate Q is then found at a series of stations along the channel, and the profile sketched. This technique is subject to the following caveats:

1. The method is applicable only as long as the stage-discharge data remain valid. For example, if a dam is built, the method is valid; however, if the channel is improved, the method would not be valid since the stage-discharge relations would change.

2. If the velocity head of the flow is significant, the technique fails because the assumption yielding Eq. (6.3.79) is invalid.

3. The method will provide satisfactory estimates only if records exist for stages of flow which are comparable with those of the profile to be estimated. Thus, in general, the conditions causing the modification of the profile can cause only small variations in the stage.

It is often the case that the velocity head is small and that the friction slope is equal to the water surface slope, but the stages to be predicted are significantly higher than those which compose the historical record. In such a situation, Grimm's method is not applicable; however, a technique developed by Escoffier (1946) can be used. At each section along the longitudinal length of

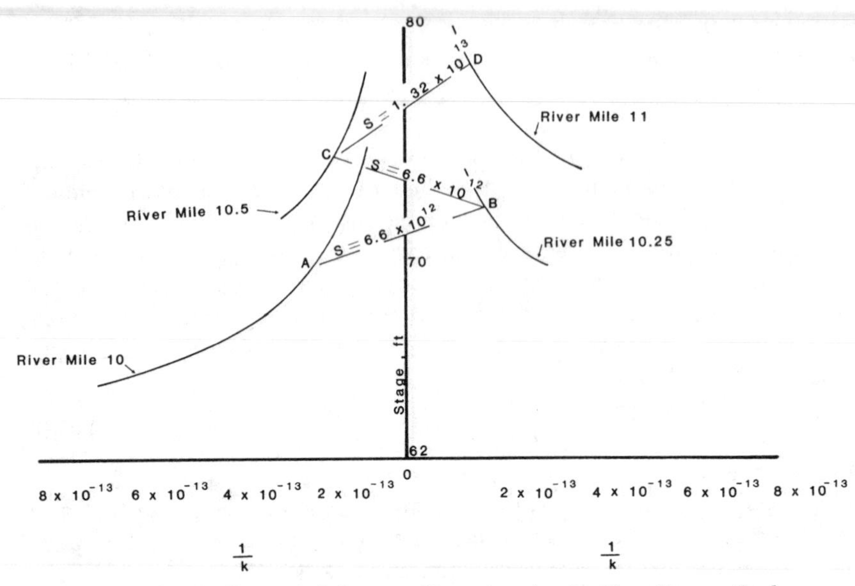

Figure 6.9 Gradually varied flow profile estimation by Escoffier method.

the channel, the parameter $1/K^2 = S_f/Q^2$ is plotted as the abscissa with the stage as the ordinate where K is the conveyance. The curves for successive channel stations are plotted alternately to the left and right of the vertical axis. For example, in Fig. 6.9 the data for the channel used in Examples 6.5, 6.7, and 6.8 are plotted. Then, for $Q = 100{,}000$ ft^3/s (2800 m^3/s) and a stage of 70 ft (21.3 m) at river mile 10 the solution proceeds as follows. The stage at river mile 10.25 is estimated by constructing a line through point A with a slope of

$$\frac{\Delta x\ Q^2}{2} = \frac{1320(100{,}000)^2}{2} = 6.6 \times 10^{12}$$

and extending this line until it intersects the curve for river mile 10.25 (point B). Δx is the distance between the two consecutive stations. The stage at river mile 10.5 is found in an analogous fashion by constructing a line with a slope of $-\Delta x\ Q^2/2$ through point B and extending it until it intersects the curve for river mile 10.5 (point C). It is noted that the answers obtained in Fig. 6.9 are in agreement with those determined in the foregoing examples; however, this method has several limitations. First, the velocity head must be negligible. Second, there is no method of accounting for eddy losses. Third, the variation in the length of the overbank and main channel sections can be accounted for, but it adds significantly to the complexity of the solution. In Fig. 6.9, the slope of the line connecting points C and D was computed by assuming that $\Delta x = 2640$ ft (800 m) even though there is a significant difference in the length of the main and overbank sections. The effect on the estimate of the stage at river mile 11 is small. The technique has the advantage of requiring only one set of curves for a wide range of discharges.

6.4 SPATIALLY VARIED FLOW

Steady, spatially varied flow is defined as flow in which the discharge varies in the direction of flow and occurs in situations involving side-channel spillways, channels with permeable boundaries, gutters for conveying storm water runoff, and drop structures in the bottom of channels. Since the salient principles governing flow in which the discharge increases with distance are different from those governing a flow in which the discharge decreases with distance, these two types of flow must be considered separately.

Increasing Discharge

In this case of spatially varied flow, which is exemplified by the side-channel spillway, a significant portion of the energy loss results from the mixing of the water entering the channel and the water flowing in the channel. Since this energy loss cannot be accurately quantified, the momentum principle has been traditionally used to develop governing equations.

Given a channel with a lateral inflow (Fig. 6.10), the continuity equation for the control volume shown is

$$\frac{dQ}{dx} = q^* \tag{6.4.1}$$

where Q = the flow rate in the channel and q^* = the lateral inflow rate per unit length. The left-hand side of Eq. (6.4.1) can be expanded to yield

$$\bar{u}\frac{dA}{dx} + A\frac{d\bar{u}}{dx} = q^* \tag{6.4.2}$$

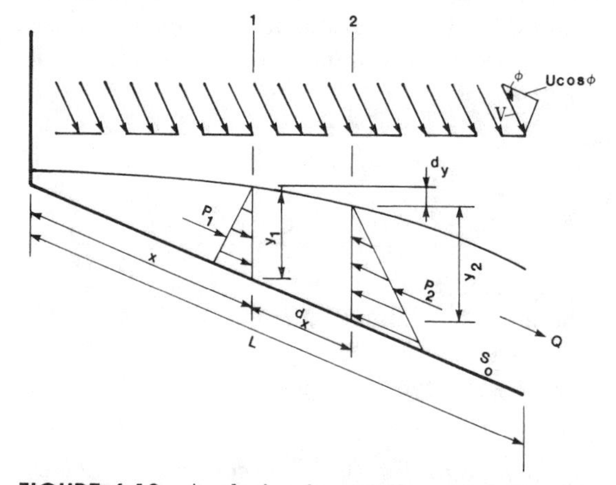

FIGURE 6.10 Analysis of spatially varied flow for increasing discharge.

where A = cross-sectional flow area and $\bar{u}$ = mean velocity of flow in the longitudinal direction. For this same control volume, the momentum flux in the x direction is

$$\Delta M = M_2 - (M_1 + M_L) \tag{6.4.3}$$

where

$$M_2 - M_1 = \iint_A \rho u_x^2 \, dA \tag{6.4.4}$$

and

$$M_L = \rho q^* \, \Delta x \; V \cos \phi \tag{6.4.5}$$

In Eqs. (6.4.4) and (6.4.5), ρ = fluid density, and V = average velocity of lateral inflow (Fig. 6.10).

If the effects of surface tension and turbulence are neglected, the forces acting on the control volume are the component of the weight of the fluid within the control volume acting in the longitudinal direction, the boundary shear resistance, and the pressure forces at stations 1 and 2. If the total change of momentum flux, Eq. (6.4.3), is equated to the sum of the forces acting, then

$$\frac{d}{dx}\iint_A \rho u_x^2 \, dA - \rho q^* V \cos\phi = \gamma S_0 A - \tau_b P - \frac{d}{dx}\iint_A p \, dA \tag{6.4.6}$$

where $S_o = \sin\theta$ = bottom slope
p = pressure
P = wetted perimeter
γ = fluid specific weight
τ_b = average boundary shear stress acting in x direction over entire cross section

Equation (6.4.6) can be simplified by introducing standard expressions for the first term on the left-hand side of the equation and the last term on the right-hand side of the equation. This result is then combined with Eq. (6.4.2) to yield an equation for the slope of the water surface (see, for example, Yen and Wenzel, 1970), or

$$\frac{dy}{dx} = \frac{S_o - S_f + (q^*/gA)(V\cos\phi - 2\beta\bar{u}) - (\bar{u}^2/g)(dB/dx) - y[d(\alpha'\cos\theta)/dx]}{\alpha'\cos\theta(1 + y/D) - \beta\bar{u}^2/gD} \tag{6.4.7}$$

where β = momentum correction factor
α' = pressure correction coefficient
D = hydraulic depth
S_f = friction slope

If θ is constant, Eq. (6.4.7) becomes

$$\frac{dy}{dx} = \frac{S_o - S_f + (q^*/gA)(V\cos\phi - 2\beta\bar{u}) - (\bar{u}^2/g)(d\beta/dx)}{\cos\theta[\alpha'(1 + y/D) + y(d\alpha'/dy)] - (\beta\bar{u}^2/gD)} \tag{6.4.8}$$

If it is assumed that the pressure distribution is hydrostatic, $V \cos \phi = 0$, $\cos \theta = 1$, and $\beta = 1$, then Eq. (6.4.8) becomes

$$\frac{dy}{dx} = \frac{S_o - S_f - (2q^* V/gA)}{1 - (\bar{u}^2/gD)} = \frac{S_o - S_f - (2Q/gA^2)(dQ/dx)}{1 - (Q^2/gA^2D)} \tag{6.4.9}$$

which is the most common form of the gradually varied flow equation with lateral inflow (see, for example, Chow, 1959, or Henderson, 1966).

Equation (6.4.9) can be solved by a number of techniques, but as was the case with the computation of the regular gradually varied flow profile, the solution procedure must begin at a control. Thus, for a given channel with a specified lateral inflow, the situation must be examined to determine if a critical flow section occurs. Following the reasoning of Henderson (1966), critical flow occurs or $dy/dx = 0$ when the numerator of Eq. (6.4.9) is zero or

$$S_o - S_f - \frac{2Q}{gA^2}\frac{dQ}{dx} = 0 \tag{6.4.10}$$

If the rate of lateral inflow is constant, as it is in most cases of practical interest,

$$\frac{dQ}{dx} = \frac{Q}{x} \tag{6.4.11}$$

where x = longitudinal distance from the beginning of the channel. Note: This formulation assumes that the flow at the beginning of the channel is zero. The friction slope can be estimated from either the Manning or Chezy equations, and for notational convenience the Chezy equation is used or

$$S_f = \frac{Q^2 P}{C^2 A^3} \tag{6.4.12}$$

Substitution of Eqs. (6.4.11) and (6.4.12) in Eq. (6.4.10) yields

$$\begin{aligned} S_o &= \frac{Q^2 P}{C^2 A^3} + \frac{2Q^2}{gA^2 x} \\ &= \mathbf{F}^2\left(\frac{gP}{C^2 T} + \frac{2A}{Tx}\right) \end{aligned} \tag{6.4.13}$$

where T = channel top width. As in Chapter 2, the possibility that $dy/dx = 0$ is now excluded, and the alternative possibility that

$$\mathbf{F}^2 = \frac{Q^2 T}{gA^3} = 1$$

is considered. Using this definition to eliminate A in Eq. (6.4.13) yields, after some rearrangement,

$$x = \frac{8Q_x^2}{gT^2[S_o - (gP/C^2T)]^3} \tag{6.4.14}$$

where $Q_x = dQ/dx$. Equation (6.4.14) may be used to estimate the location of the critical flow section, if one exists. With regard to Eq. (6.4.14), the following observations may be made. First, this type of development was first presented by Keulegan (1952). Second, if the x estimated is greater than the length of the channel, then a critical flow section does not occur. Third, if there is a critical flow section and beyond this section there is a downstream control, then it is possible that the critical section will be drowned if the downstream depth is sufficient. Fourth, as demonstrated by Keulegan (1952), explicit solutions of this equation are possible for special cases, e.g., a wide, rectangular channel. In general, Eq. (6.4.14) must be solved by trial and error since neither T nor P are known and C may not be constant. Fifth, although at this time it is not possible to justify the possibility of a critical section at which $dy/dx = 0$ and $\mathbf{F} \neq 0$, this alternative must remain an item for discussion. Sixth, in the case of a wide rectangular channel with a constant C, Eq. (6.4.14) becomes

$$x = \frac{8q_x^2}{g(S_o - g/C^2)^3} \tag{6.4.15}$$

where $q = Q_x/b$ and b = width of the channel.

Although spatially varied flow profiles can be determined by a number of trial-and-error computational procedures, the most effective solution technique is numerical integration combined with trial and error. For this procedure, Newton's second law of motion can be written as

$$\frac{\gamma}{g}[Q\,\Delta\overline{u} + (\overline{u} + \Delta\overline{u})\,\Delta Q] = -\gamma\overline{A}\,\Delta y + \gamma S_o\overline{A}\,\Delta x - \gamma S_f\overline{A}\,\Delta x$$

where $\overline{A}$ = the average flow area between stations 1 and 2 (Fig. 6.10), which are separated by a distance Δx. The average flow area may be expressed as

$$\overline{A} = \frac{Q_1 + Q_2}{\overline{u}_1 + \overline{u}_2}$$

where Q_1 and Q_2 are the flow rates at stations 1 and 2, respectively, and $\overline{u}_1$ and $\overline{u}_2$ are the average velocities of flow at these stations, respectively. Let

$$Q = Q_1$$

and

$$\overline{u}_2 = \overline{u} + \Delta\overline{u}$$

Substitution of these relations in the momentum equation yields, after some rearrangement,

$$\Delta y = -\frac{Q_1(\overline{u}_1 + \overline{u}_2)}{g(Q_1 + Q_2)}\left(\Delta\overline{u} + \frac{\overline{u}_2}{Q_1}\Delta Q\right) + S_o\,\Delta x - S_f\,\Delta x \tag{6.4.16}$$

With reference to Fig. 6.10, the drop in the water surface between stations 1 and 2 is given by

$$\Delta y' = -\Delta y + S_o\,\Delta x \tag{6.4.17}$$

Substitution of Eq. (6.4.16) in Eq. (6.4.17) yields

$$\Delta y' = \frac{Q_1(\bar{u}_1 + \bar{u}_2)}{g(Q_1 + Q_2)}\left(\Delta \bar{u} + \frac{\bar{u}_2}{Q_1}\Delta Q\right) + S_f\,\Delta x \tag{6.4.18}$$

which can be used to determine the spatially varied water surface profile for the case of increasing discharge.

EXAMPLE 6.10

A trapezoidal lateral spillway channel 400 ft (122 m) long is designed to carry a discharge which increases at a rate of 40 (ft^3/s)/ft [3.72 (m^3/s)/m]. The cross section has a bottom width of 10 ft (3.0 m) and side slopes of 0.5:1. The longitudinal slope of the channel is 0.015 and begins at an upstream bottom elevation of 73.7 ft (22.5 m). If n = 0.1505 and the velocity distribution is uniform, estimate the water surface profile of the design discharge. Note: This example was first presented by Hinds (1926).

Solution

The first step in the solution of this problem is to determine if a critical flow section exists and, if such a section exists, the second step is to determine its longitudinal position. This determination requires a trial-and-error-solution of Eq. (6.4.14) (Table 6.16). In this table the following points should be noted:

Column 1: An assumed longitudinal position.

Column 2: The total discharge at the assumed longitudinal position determined as the product of col. 1 and the lateral discharge of 40 (ft^3/s)/ft [3.72 (m^3/s)/m].

TABLE 6.16 Location of the critical flow section for Example 6.10

Trial x, ft (1)	Q, ft^3/s (2)	y_c, ft (3)	A, ft^2 (4)	P, ft (5)	T, ft (6)	R, ft (7)	C (8)	x, ft (9)
200	8000	20.8	425	56.6	30.8	7.51	139	131
131	5240	16.4	299	46.7	26.4	6.40	135	178
178	7120	19.5	385	53.7	29.5	7.17	138	142
142	5680	17.2	320	48.5	27.2	6.59	136	168
168	6720	18.9	368	52.3	28.9	7.03	137	149
149	5960	17.7	333	49.6	27.7	6.71	136	162
162	6480	18.5	356	51.4	28.5	6.92	137	153
153	6120	17.9	340	50.1	27.9	6.79	137	159
159	6360	18.3	351	51.0	28.3	6.88	137	155
155	6200	18.1	345	50.5	28.1	6.83	137	157
157	6280	18.2	348	50.8	28.2	6.85	137	156
156	6240	18.1	345	50.5	28.1	6.83	137	157

Column 3: The critical depth of flow corresponding to the assumed longitudinal distance in col. 1 and the discharge in col. 2, Eq. (2.2.3), and Table 2.1.

Columns 4 to 8: The area, wetted perimeter, top width, hydraulic radius, and Chezy resistance coefficient, respectively, corresponding to the depth of flow in col. 3 or

$$A = (b + zy) = (10 + 0.5y)y$$

$$P = b + 2y\sqrt{1 + z^2} = 10 + 2.24y$$

$$T = b + 2zy = 10 + y$$

and

$$C = \frac{\phi}{n} R^{1/6} = \frac{1.49}{0.015} R^{1/6}$$

Column 9: The calculated distance to the critical section by Eq. (6.4.14).

When the value in col. 9 of Table 6.16 agrees with the estimated value in col. 1, the calculation is completed. From Table 6.16 it is estimated that approximately 156 ft (48 m) downstream from the beginning of the channel a critical flow section occurs.

With the critical flow section located, the water surface profile both upstream and downstream of this point can be estimated from Eq. (6.4.18). The calculations for the water surface profile upstream of the critical flow section are contained in Table 6.17*a* while those for the profile downstream of this section are contained in Table 6.17*b*. In these tables the columns contain the following data.

Column 1: This is the longitudinal distance between the point of computation and the beginning of the channel.

Column 2: This is the incremental longitudinal distance between two adjacent points of computation.

Column 3: This is the elevation of the channel bottom and is obtained by substituting the product of col. 1 and the longitudinal slope of the channel from the bottom elevation at the beginning of the channel; e.g., at the critical flow section

$$z_0 = 73.7 - 156(0.1505) = 50..2 \text{ ft } (15.3 \text{ m})$$

Column 4: This is the assumed depth of flow.

Column 5: This is the elevation of the water surface and is obtained by adding cols. 3 and 4.

Column 6: This is the change in the water surface elevation and is calculated from

$$\Delta y' = -\Delta y + S_0\,\Delta x$$

Columns 7 to 13: These are, respectively, the flow area for the assumed depth of flow, the discharge, the average velocity, the sum of the flow rates, the sum of the average velocities, the difference in discharge between two adjacent stations, and the change in average velocity.

Column 14: This is the drop in the water surface due to the impact loss or

$$\Delta y'_m = \frac{Q_1(\overline{u}_1 + \overline{u}_2)}{g(Q_1 + Q_2)}\left(\Delta\overline{u} + \frac{\overline{u}_2}{Q_1}\,\Delta Q\right)$$

Column 15: This is the hydraulic radius associated with the assumed depth of flow.

Column 16: This is the head loss due to friction and is computed from

$$h_f = S_f\,\Delta x = \left(\frac{nQ}{1.49AR^{2/3}}\right)^2 \Delta x$$

Column 17: This is the drop in the water surface between two adjacent stations estimated by Eq. (6.4.18).

At each longitudinal station a trial-and-error procedure is used until the values in cols. 6 and 17 agree. This procedure is applied to the profiles above and below the critical section (Table 6.17*a* and *b*).

Decreasing Discharge

In this type of spatially varied flow which is exemplified by side weirs and bottom racks, there is no significant energy loss, and the water surface profile can be estimated from the energy principle. The total energy at a channel section relative to a datum is

$$H = z + y + \frac{Q^2}{2gA^2} \tag{6.4.19}$$

Differentiating this equation with respect to the longitudinal coordinate x

$$\frac{dH}{dx} = \frac{dz}{dx} + \frac{dy}{dx} + \frac{1}{2g}\left(\frac{2Q}{A^2}\frac{dQ}{dx} - \frac{2Q^2}{A^3}\frac{dA}{dx}\right) \tag{6.4.20}$$

TABLE 6.17 Data for Example 6.10

x, ft (1)	Δx, ft (2)	Z_o, ft (3)	y, ft (4)	Z, ft (5)	$\Delta y'$, ft (6)	A, ft^2 (7)	Q, ft^3/s (8)	$\bar{u}$, ft/s (9)	$Q_1 + Q_2$, ft^3/s (10)	$\bar{u}_1 + \bar{u}_2$, ft/s (11)	ΔQ, ft^3/s (12)	$\Delta \bar{u}$, ft/s (13)	$\Delta y'_m$, ft (14)	R, ft (15)	h_f, ft (16)	$\Delta y'$, ft (17)
(*a*) Computation of upstream subcritical water surface profile for Example 6.9																
156		50.20	18.10	68.30		345	6240	18.09								
100	56	58.70	16.00	74.70	6.33	288	4000	13.89	10,240	31.98	2240	4.20	5.56	6.28	0.09	5.65
			15.50	74.20	5.80	275	4000	14.54	10,240	32.63	2240	3.55	5.42	6.15	0.11	5.53
			15.0	73.70	5.33	262	4000	15.24	10,240	33.33	2240	2.85	5.25	6.02	0.12	5.37
50	50	66.20	12.80	79.00	5.33	210	2000	9.52	6,000	24.76	2000	5.72	5.37	5.43	0.05	5.42
			12.90	79.10	5.42	212	2000	9.42	6,000	24.66	2000	5.82	5.38	5.45	0.05	5.43
25	25	69.90	11.30	81.20	2.16	177	1000	5.65	3,000	15.07	1000	3.77	2.06	5.01	0.01	2.07
			11.20	81.10	2.06	175	1000	5.72	3,000	15.14	1000	3.70	2.06	4.98	0.01	2.07
10	15	72.20	9.90	82.10	0.96	148	400	2.70	1,400	8.42	600	3.02	0.87	4.60	0	0.87
			9.80	82.00	0.86	146	400	2.74	1,400	8.46	600	2.98	0.87	4.57	0	0.87
0	10	73.70	8.61	82.31	$\Delta y' = 2\bar{u}^2/2g$ at $x = 10$ ft (assumed).											
(*b*) Computation of downstream water surface profile for Example 6.10																
156		50.20	18.10			345	6,240	18.09								
200	44	43.60	18.60	62.20	6.12	359	8,000	22.28	14,240	40.37	1760	4.19	5.75	6.95	0.17	5.92
			18.50	62.10	6.22	356	8,000	22.46	14,240	40.55	1760	4.37	5.91	6.92	0.17	6.08
			18.40	62.00	6.32	353	8,000	22.64	14,240	40.73	1760	4.55	6.06	6.89	0.17	6.23
250	50	36.07	19.70	55.77	6.22	391	10,000	25.57	18,000	48.21	2000	2.93	6.20	7.22	0.24	6.44
			19.90	55.97	6.02	397	10,000	25.19	18,000	47.83	2000	2.55	5.84	7.27	0.23	6.07
300	50	28.55	21.00	49.55	6.42	430	12,000	27.91	22,000	53.10	2000	2.72	6.22	7.54	0.27	6.49
350	50	21.02	21.20	42.22	7.32	437	14,000	32.04	26,000	59.95	2000	4.13	8.13	7.60	0.35	8.48
			22.00	43.02	6.53	462	14,000	30.30	26,000	58.21	2000	2.39	6.21	7.79	0.30	6.51
400	50	13.50	23.00	36.50	6.53	495	16,000	32.32	30,000	62.62	2000	2.02	6.02	8.05	0.33	6.35
			22.90	36.40	6.63	491	16,000	32.57	30,000	62.87	2000	2.27	6.31	8.01	0.33	6.64

In this equation, $dH/dx = -S_f$, $dz/dx = -S_o$, and

$$\frac{dA}{dy} = \frac{dA}{dy}\frac{dy}{dx} = T\frac{dy}{dx}$$

Substituting these relations in Eq. (6.4.20) and rearranging yields

$$\frac{dy}{dx} = \frac{S_o - S_f - (Q/gA^2)(dQ/dx)}{1 - (Q^2/gA^2D)} \tag{6.4.21}$$

If a nonuniform velocity profile is present, the energy correction factor is used and Eq. (6.4.21) becomes

$$\frac{dy}{dx} = \frac{S_o - S_f - (\alpha Q/gA^2)(dQ/dx)}{1 - (\alpha Q^2/gA^2D)} \tag{6.4.22}$$

Equations (6.4.21) and (6.4.22) are known as the dynamic equations for spatially varied flow with decreasing discharge. It is noted that the only difference between Eqs. (6.4.9) and (6.4.21) is in the coefficient of the third term in the numerator.

In the case of withdrawal of water through a rack in the bottom of a rectangular channel (Fig. 6.11), an equation for the water surface profile can be derived. Assume that $\alpha = 1$ and $\theta = 0$; then the specific energy at any section of the channel is

$$E = y + \frac{Q^2}{2gb^2y^2} \tag{6.4.23}$$

where b = width of the channel. According to Chow (1959), it can be assumed that the specific energy along the channel is constant. Using this assumption and differentiating Eq. (6.4.23) with respect to longitudinal distance yields

$$\frac{dy}{dx} = -\frac{Qy(dQ/dx)}{gb^2y^3 - Q^2} \tag{6.4.24}$$

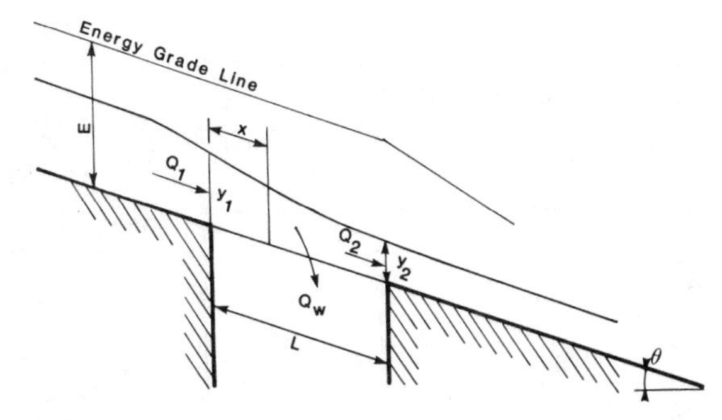

FIGURE 6.11 Analysis of spatially varied flow with bottom withdrawal.

where dQ/dx is the flow withdrawn through a length Δx of the rack. In contrast to the problem of spatially varied flow with increasing discharge, dQ/dx cannot be assumed to be constant since this term depends on the effective head on the rack.

In the case of a rack where the direction of flow through the rack openings is nearly vertical, e.g., in the case of a rack composed of vertical bars, the effective head on the rack is approximately equal to the specific energy, and the energy loss is negligible. In such a situation

$$-\frac{dQ}{dx} = \epsilon C_D b\sqrt{2gE} \tag{6.4.25}$$

where ϵ = ratio of the open area in the rack to the total rack surface area and C_D = the coefficient of discharge. According to Chow (1959), C_D varies for this situation from 0.435 for a grade of 1 on 5 to 0.497 for a horizontal channel. Then from Eq. (6.4.23), the discharge is

$$Q = by\sqrt{2g(E - y)} \tag{6.4.26}$$

Substitution of Eqs. (6.4.25) and (6.4.26) in Eq. (6.4.24) yields a first-order differential equation for the water surface profile or

$$\frac{dy}{dx} = \frac{2\epsilon C_D\sqrt{E(E - y)}}{3y - 2E}$$

and integrating this equation and evaluating the constant of integration with the condition that at $x = 0$, $y = y_1$, yields

$$x = \frac{E}{\epsilon C_D}\left(\frac{y_1}{E}\sqrt{1 - \frac{y_1}{E}} - \frac{y}{E}\sqrt{1 - \frac{y}{E}}\right) \tag{6.4.27}$$

The length of rack required to remove all the flow in the reach where the rack is located is given by

$$L = \frac{E}{\epsilon C_D}\left(\frac{y_1}{E}\sqrt{1 - \frac{y_1}{E}}\right) \tag{6.4.28}$$

since at this point $y = 0$.

In the case where the openings in the rack make an appreciable angle with the vertical, e.g., in the case of a rack composed of a perforated screen, the energy loss through the rack is not negligible, and it is assumed that the effective head on the rack is equal to the depth of flow. In this case, the discharge through the rack for an incremental distance dx is

$$-\frac{dQ}{dx} = \epsilon C_D b\sqrt{2gy} \tag{6.4.29}$$

Then in a fashion analogous to that described above

$$x = \frac{E}{\epsilon C_D}\left[0.5\cos^{-1}\left(\sqrt{\frac{y}{E}}\right) - 1.5\sqrt{\frac{y}{E}\left(1-\frac{y}{E}\right)}\right] + \Gamma \qquad (6.4.30)$$

where

$$\Gamma = \frac{E}{\epsilon C_D}\left[1.5\sqrt{\frac{y_1}{E}\left(1-\frac{y_1}{E}\right)} - 0.5\cos^{-1}\left(\frac{y_1}{E}\right)\right]$$

The length of rack required for complete flow withdrawal is

$$L = \frac{E}{\epsilon C_D}\left[1.5\sqrt{\frac{y_1}{E}\left(1-\frac{y_1}{E}\right)} - 0.25\sin^{-1}\left(1-\frac{2y_1}{E}\right) + \frac{\pi}{8}\right] \qquad (6.4.31)$$

In the case of racks composed of perforated screens, Chow (1959) noted that C_D, the discharge coefficient, ranges from 0.750 for grades of 1:5 to 0.800 for channels built on horizontal slopes.

It should be noted that the concept of flow removal by a vertical drop structure is particularly important because this situation is found in most stormwater channel systems.

6.5 APPLICATION TO PRACTICE

In the foregoing sections, the subject of gradually varied flow profile estimation has been treated from the abstract viewpoint of discussing computational procedures. In this section, a number of situations in which gradually varied flow profiles interact with other aspects of flow in open channels are considered.

Profile Behind a Dam

One of the most common gradually varied flow problems involves the estimation of backwater effects behind a dam. Although theoretically the gradually varied flow profile behind a dam on a mild slope extends in the upstream direction indefinitely, it is usually assumed to terminate at the point where the elevation of the water surface is 1 percent greater than the elevation under normal flow conditions. In most investigations, a backwater envelope is found; i.e., the terminus of the backwater curve is determined for a number of flow rates.

The backwater envelope begins at the point where the static pool of the reservoir intersects the channel floor—the case when there is zero flow. As the flow into the reservoir increases, the endpoint of the backwater curve may migrate either upstream or downstream depending on many factors. For example:

1. If the reservoir level is constant and the channel prismatic with a simple cross section, then the terminus of the backwater curve will migrate downstream as the flow rate increases.

2. An increase in the channel roughness results in the shortening of the length of the gradually varied flow profile and, hence, in a downstream migration of the backwater terminus.

3. The presence of flood plains also results in a shortening of the profile length and a downstream movement of the terminus.

In many cases, it is convenient to define the endpoint of the backwater curve in terms of the requirements of a specific problem. For example, the endpoint may be defined as the point where the rise in the water surface begins to cause damage.

Discharge Computation

When an open channel is used to connect two reservoirs, the possibility that the discharge of this channel is affected by the interaction of local features and the gradually varied flow profile must be considered. It is noted that this problem has been previously addressed by Bakhmeteff (1932), Chow (1959), and Henderson (1966), and the classification system developed by Bakhmeteff is used here. The classification system is based on the behavior of three crucial variables: (1) the depth of flow at the upstream entrance to the channel, y_1; (2) the depth of flow at the downstream outlet, y_2; and (3) the channel discharge Q. These variables are described schematically in Fig. 6.12. Based on the behavior of these variables, the following cases can be identified:

Case I (y_1 = constant)

For y_1 constant, several discharge conditions are possible. First, if $y_1 = y_N = y_2$, the flow in the channel is uniform and the discharge can be computed from the Manning equation. In Fig. 6.12*a* this situation is represented by the line *AN*, and the corresponding discharge is represented by point *N* in Fig. 6.12*b*.

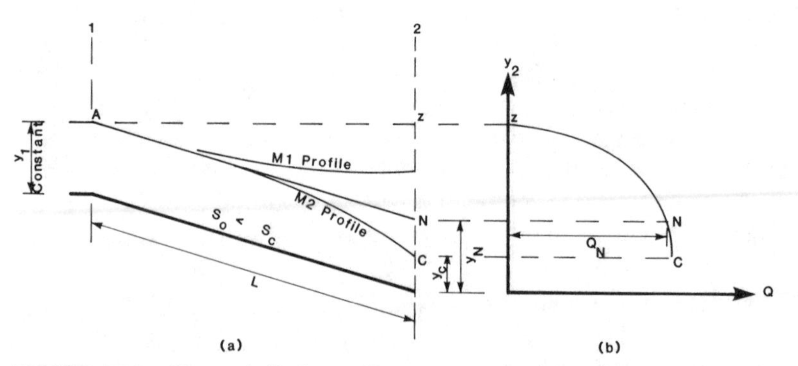

FIGURE 6.12 Channel discharge for y_1 = constant, mild slope.

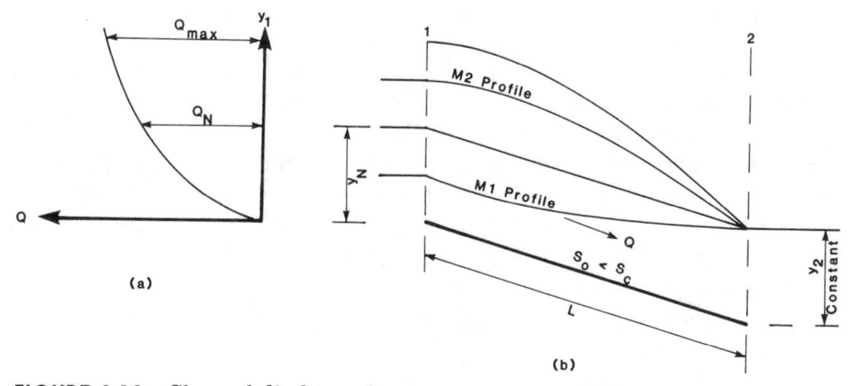

FIGURE 6.13 Channel discharge for y_2 = constant, mild slope.

Second, if $y_2 = y_c$, the discharge of the channel is the maximum attainable (line AC in Fig. 6.12a and point C in Fig. 6.12b). It is noted that if $y_2 < y_c$, then the channel terminates in a free overfall, but the discharge will not exceed the value represented by point C in Fig. 6.12b.

Third, if $y_2 > y_N$, then the flow profile in the channel is an M1 curve. For this situation to exist, the following condition must be satisfied: $y_N < y_2 < y_z$. At this point, the length of the M1 profile becomes a crucial consideration. If y_1 is constant, then the length of the M1 profile, L', cannot exceed L; however, L' may be less than L. In this latter case, the channel is termed long and the M1 curve is said to "run out"; i.e., the discharge at the upstream station is not affected by y_2 and the depth of flow in the initial channel length is normal. If $L' = L$, then the discharge to the canal is less than the normal flow.

Fourth, if $y_c < y_2 < y_N$, then an M2 gradually varied flow profile occurs in the channel and the discharge relationship can be discussed as was done for the M1 profile.

Case II (y_2 = constant)

For y_2 constant, the same types of situations discussed for case I are possible. First, if $y_1 = y_N = y_2$, flow occurs at normal depth and the discharge is the normal discharge (Fig. 6.13a and b).

Second, maximum discharge occurs when the depth y_1 corresponds to a critical discharge at section 2. It is noted that this value of y_1 is the maximum depth which can occur since any increase in y_1 would necessitate a corresponding increase in y_2.

Third, for $y_1 < y_N$ an M1 flow profile occurs, and the discharge is less than that which would occur at normal depth.

Fourth, for $y_1 > y_N$ but less than the depth which would cause critical discharge, an M2 profile occurs. In this case, the discharge would be greater than normal but less than the maximum.

Case III (Q = constant)

For a constant channel discharge, the depths y_1 and y_2 are allowed to fluctuate. This first possibility is $y_2 = y_N = y_1$, and the discharge can be found by the Manning equation, line *AN* in Fig. 6.14.

Second, for water surface positions above the line *AN*, an M1 profile occurs. The obvious upper limit occurs when $y_2 = y_1 + S_oL$. It is noted that as this limiting situation is approached, both the elevation difference between the reservoirs and the velocity of flow decrease; however, since the flow area increases, the discharge can be held constant.

Third, for water surface positions below *AB* an M2 profile occurs. The minimum depth of flow occurs when $y_2 = y_c$.

The application of these rather abstract concepts to practice is best examined through an example.

EXAMPLE 6.11

A rectangular channel 12.0 m (39 ft) wide conveys water from a reservoir whose surface elevation is 3.0 m (9.8 ft) above the channel bed at the reservoir outlet. At a distance of 1000 m (3280 ft) downstream of the reservoir, the channel contracts to a width of 7.3 m (24 ft). If n = 0.020 and S = 0.001, determine the discharge from the reservoir.

Solution

The solution of this problem requires a number of assumptions which are either accepted or rejected on the basis of calculations. Thus:

1. Assume that the contraction is not severe enough to cause a reduction in the flow. Under this assumption, the flow in the channel is uniform, and the solution is found by calculating uniform flow conditions over a range of depths until the value of the specific energy in the channel matches the value at the beginning of the channel.

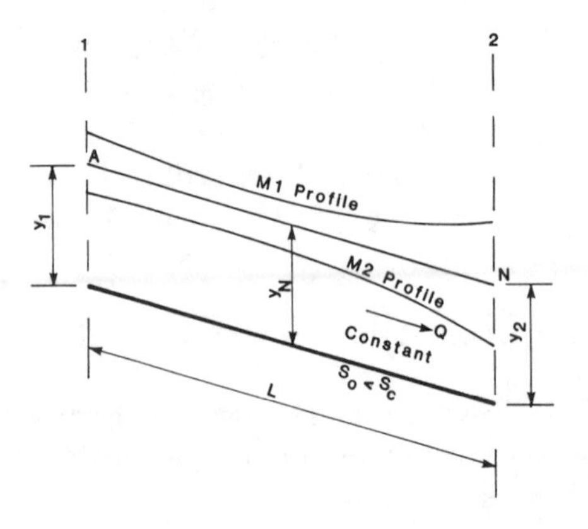

FIGURE 6.14 Channel discharge for Q = constant, mild slope.

The calculations are summarized in Table 6.18 where

$$\bar{u} = \frac{\phi\sqrt{s}}{n} R^{2/3} = \frac{\sqrt{0.001}}{0.020} R^{2/3} = 1.58R^{2/3}$$

From the results summarized in this table, it is determined that the unrestricted discharge of the reservoir is 75.8 m^3/s (2680 ft^3/s). Then within the constriction

$$q = \frac{Q}{b} = \frac{75.8}{7.3} = 10.4 \text{ (m}^3\text{/s)/m [112 (ft}^3\text{/s)/ft]}$$

and for this value of q the critical depth is, from Eq. (2.2.6),

$$y_c = \left(\frac{q^2}{g}\right)^{1/3} = \left(\frac{10.4^2}{9.8}\right)^{1/3} = 2.22 \text{ m (7.28 ft)}$$

and the critical specific energy from Eq. (2.2.8) is

$$E_c = 1.5y_c = 1.5(2.22) = 3.33 \text{ m (10.9 ft)}$$

which is greater than the available upstream specific energy if the flow is uniform. Thus, it is concluded that the contraction acts as a choke, and hence the flow in the channel cannot be uniform. The question now is if the M1 backwater profile caused by the construction influences the flow rate.

2. If it is assumed that the discharge from the reservoir is not affected by the constriction, then the depth just upstream of the constriction is found from

$$E = y + \frac{q^2}{2gy^2} = y + \frac{(75.8/12)^2}{2(9.8)y^2} = y + \frac{2.04}{y^2} = 3.33 \text{ m (10.9 ft)}$$

where it is assumed that there is no energy loss at the entrance to the constriction. Solving the above equation by trial and error yields $y = 3.15$ m (10.3 m). Then, the gradually varied upstream flow profile can be estimated from the initial depth, e.g., Example 6.1. The results of this analysis are summarized in Table 6.18*a*. This analysis demonstrates that for $Q = 75.8$ m^3/s (2680 ft^3/s) the gradually varied flow profile extends more than 1500 m (4920 ft) upstream or beyond the entrance to the channel. Therefore, it is concluded that the downstream constriction controls the flow in the channel.

3. Since the constriction 1000 m (3280 ft) downstream of the reservoir affects the flow rate, a new trial flow rate must be determined; i.e., a value of Q must be determined such that the value of E at the reservoir is 3.0 m (9.8 ft). By interpolation in Table 6.18, it is esti-

TABLE 6.18 Data for Example 6.11

Summary of calculation to determine normal depth of flow in Example 6.11

y, m (1)	A, m^2 (2)	P, m (3)	R, m (4)	$\bar{u}$, m/s (5)	F (6)	Q, m^3/s (7)	$\bar{u}^2/2g$, m (8)	E, m (9)
2	24	16	1.5	2.1	0.47	50.4	0.22	2.22
3	36	18	2.0	2.5	0.46	90.0	0.32	3.32
2.5	30	17	1.8	2.3	0.46	69.0	0.27	2.77
2.7	32	17.4	1.84	2.47	0.46	75.8	0.29	2.99

(a) Gradually varied flow profile computation, Example 6.11, $Q = 75.8$ m^3/s

y, m (1)	A, m^2 (2)	P, m (3)	R, m (4)	$\bar{u}$, m/s (5)	$\frac{\bar{u}^2}{2g}$, m (6)	E, m (7)	$\frac{n^2\bar{u}^2}{R^{4/3}}$ (8)	$\left(\frac{n^2\bar{u}^2}{R^{4/3}}\right)_m$ (9)	$S_o - \left(\frac{n^2\bar{u}^2}{R^{4/3}}\right)_m$ (10)	ΔE, m (11)	Δx, m (12)	$\Sigma\Delta x$, m (13)
3.15	37.8	18.3	2.07	2.01	0.21	3.36	0.000614					
								0.000661	0.000339	0.13	383	
3.0	36.0	18.0	2.00	2.11	0.23	3.23	0.000708					383
								0.000807	0.000193	0.21	1090	
2.75	33.0	17.5	1.89	2.30	0.27	3.02	0.000906					1470

(*b*) Gradually varied flow profile computation, Example 6.11, $Q = 72\ m^3/s$

3.0	36.0	18.0	2.00	2.00	0.20	3.20	0.000635					
								0.000667	0.000333	0.08	240	
2.9	34.8	17.8	1.96	2.07	0.22	3.12	0.000699					240
								0.000736	0.000264	0.09	341	
2.8	33.6	17.6	1.91	2.14	0.23	3.03	0.000773					581
								0.000794	0.000206	0.04	194	
2.75	33.0	17.5	1.89	2.18	0.24	2.99	0.000814					775

(*c*) Gradually varied flow profile computation, Example 6.11, $Q = 73.3\ m^3/s$

3.09	37.1	18.1	1.98	1.98	0.20	3.29	0.000614					
								0.000670	0.000330	0.16	485	
2.90	34.8	17.8	1.96	2.11	0.23	3.13	0.000726					485
								0.000765	0.000235	0.09	382	
2.80	33.6	17.6	1.91	2.18	0.24	3.04	0.000804					867
								0.000825	0.000175	0.04	224	
2.75	33.0	17.5	1.89	2.22	0.25	3.00	0.000845					1090

mated that the value of E at the reservoir is 3.12 m (10.2 ft) for $Q = 75.8\ \text{m}^3/\text{s}$ (2680 ft^3/s); thus, the correct value was missed by 0.12 m (0.39 ft). The obvious second trial value of Q is the one which produces a value of E = 0.12 m (0.39 ft) less than the previously estimated flow rate at the entrance to the constriction or

$$E_c = 3.33 - 0.12 = 3.21 \text{ m } (10.5 \text{ ft})$$

With this value of E_c

$$y_c = \tfrac{2}{3}E_c = \tfrac{2}{3}(3.21) = 2.14 \text{ m } (7.02 \text{ ft})$$

and $$u_c = \sqrt{gy_c} = \sqrt{9.8(2.14)} = 4.58 \text{ m/s } (15.0 \text{ ft})$$

Then

$$Q = u_c y_c b = 4.58(2.14)(7.3) = 71.6 \simeq 72 \text{ m}^3/\text{s } (2540 \text{ ft}^3/\text{s})$$

The depth of flow immediately upstream of the constriction is estimated by a trial-and-error solution of

$$E = y + \frac{q^2}{2gy^2} = y + \frac{1.84}{y^2} = 3.21 \text{ m } (10.5 \text{ ft})$$

or y = 3.0 m (9.8 ft). The gradually varied flow profile for Q = 72.0 m^3/s (2540 ft^3/s) is estimated in Table 6.18*b*. In this case, the profile terminates or "runs out" before the reservoir is reached. Thus, it is concluded that the correct flow rate for this situation lies between 75.8 m^3/s (2680 ft^3/s) and 72.0 m^3/s (2540 ft^3/s). Interpolation in Table 6.18*a* and *b* yields a new flow estimate of 73.3 m^3/s (2590 ft^3/s). Then

$$q = \frac{Q}{b} = \frac{73.3}{7.3} = 10.0 \ (\text{m}^3/\text{s})/\text{m} \ [108 \ (\text{ft}^3/\text{s})/\text{ft}]$$

$$y_c = \left(\frac{q^2}{g}\right)^{1/3} = \left(\frac{10.0^2}{9.8}\right)^{1/3} = 2.17 \text{ m } (7.12 \text{ ft})$$

$$E_c = 1.5y_c = 1.5(2.17) = 3.26 \text{ m } (10.7 \text{ ft})$$

and the depth of flow just upstream of the constriction is estimated from

$$E = y + \frac{q^2}{2gy^2} = y + \frac{(73.3/12)^2}{2(9.8)y^2} = y + \frac{1.90}{y^2} = 3.26 \text{ m } (10.7 \text{ ft})$$

by trial and error or y = 3.09 m (10.0 ft). The validity of this set of parameters is confirmed in Table 6.18*c*.

The discharge problem can also arise in channels of steep slope; i.e., the slope of the channel is greater than the critical slope. If the slope is not so steep that the flow becomes unsteady, then the flow rate is governed by the upstream critical section and can be easily computed. The water surface profile in such a channel is governed by the downstream water surface elevation. First, if the downstream water surface elevation is less than the depth of flow at station 2, then a free overfall occurs (Fig. 6.15). In this situation, the flow passes through critical depth at point C and approaches normal depth by means of an S2 drawdown curve. Second, if the downstream water surface elevation is greater than the depth of flow at station 2, then the S2 curve terminates in a hydraulic jump. The flow upstream of the jump is not affected by the downstream water surface elevation. Third, as the downstream water surface elevation increases, the location of the jump moves upstream until point C is reached. If the downstream water surface elevation increases beyond this elevation, the flow is affected by the downstream elevation since a drowned jump would occur.

Flow Past Islands

In natural channels, there may be reaches in which a long island divides the flow into two channels (Fig. 6.16). If the flow is normal, then Q_1 and Q_2 may be estimated from the following set of equations:

$$Q = Q_1 + Q_2$$
$$Q_1 = K_1\sqrt{S_1}$$
$$Q_2 = K_2\sqrt{S_2}$$

However, if the flow is gradually varied, then the solution procedure is more complex.

If the flow is gradually varied and subcritical, the computation begins at point B. For convenience, define E_1^A = water surface elevation at point A estimated by beginning at point B and proceeding to point A by way of channel 1 and E_2^A = water surface elevation at point A estimated by beginning at point B and proceeding to point A by way of channel 2. The discharges, Q_1 and Q_2,

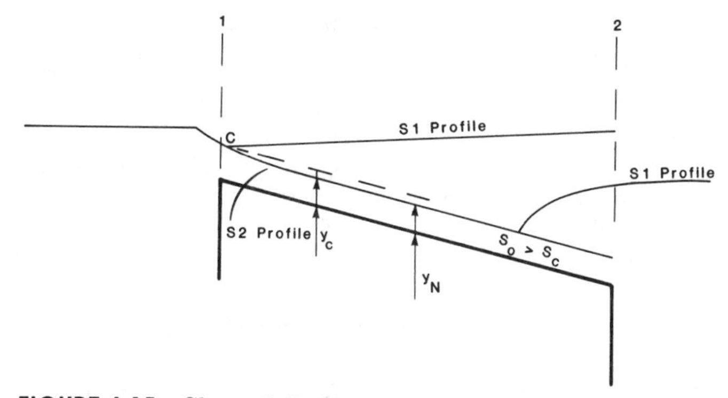

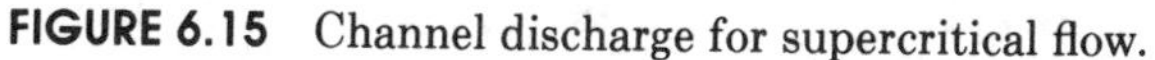
FIGURE 6.15 Channel discharge for supercritical flow.

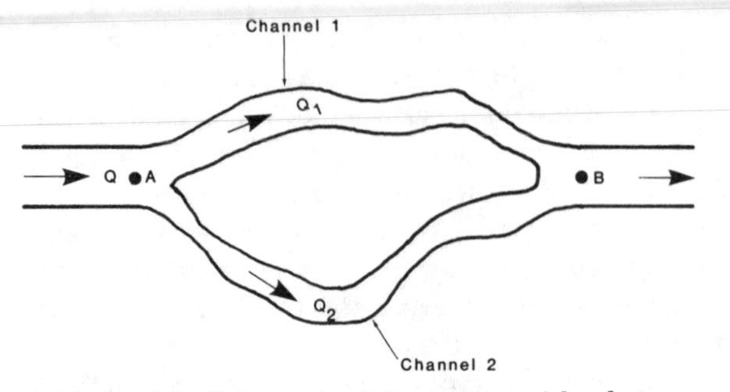

FIGURE 6.16 Schematic of flow past an island.

are assumed such that their total is Q. Then using the methods of Section 6.3, E_1^A and E_2^A are estimated for the flow rates Q_1 and Q_2, respectively. If $E_1^A = E_2^A$, as they must be if the assumed values of Q_1 and Q_2 are correct, then the solution has been determined. In general a number of trials are required. Two graphs may be used in most cases to expedite the solution (Fig. 6.17). In Fig. 6.17*a*, E_1^A is plotted as a function of E_2^A. On this figure, a 45° line is plotted, and the correct water surface elevation at point A is estimated from the intersection of this line and the E_1^A versus E_2^A curve (point C). Then, an auxiliary curve (Fig. 6.17*b*) of E_1^A versus Q is used to estimate Q_1.

If the flow in the channel is supercritical, then the solution proceeds in the same manner but begins at point A. In addition, it is noted that this procedure also serves to extend the gradually varied flow profile past the island.

Although the procedure described above provides an adequate method for determining the flow around a single island, it is not a technique which can be extended to determine the flow around a group of islands. In addition, the foregoing technique does not account for changes in the shape of the channel as it rounds the island and the minor losses associated with such changes. Wylie (1972) noted that the flow around islands can be efficiently determined by visualizing the system as a series of nodes connected by links. For example, the situation described in Fig. 6.16 is shown in node and link notation in Fig. 6.18. The solution of the node and link system rests on two principles:

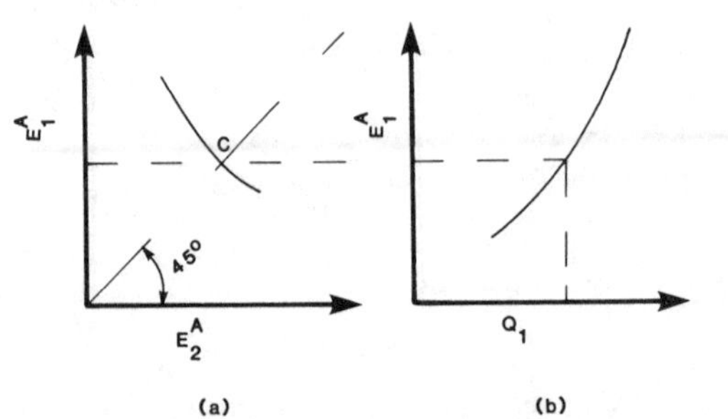

FIGURE 6.17 Graphical solution of flow passing an island.

1. Each node has a total energy associated with it which is common to each element or link terminating at the node.

2. The equation of continuity must be satisfied at each node.

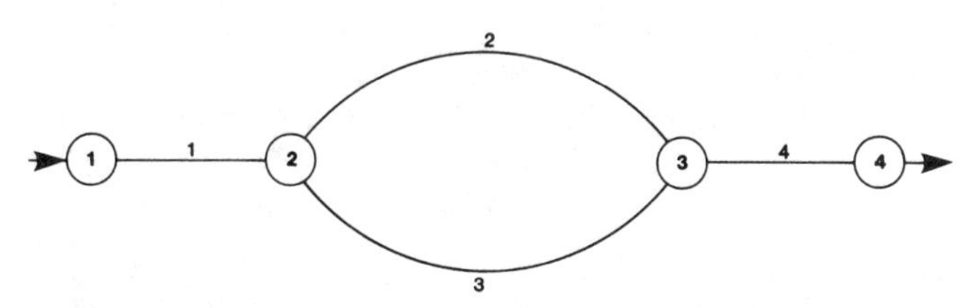

FIGURE 6.18 Link and node representation of flow past an island (see Fig. 6.16).

The most common type of link is the single channel with no minor losses. In this case, the energy equation for the link k connecting an upstream node i with a downstream node j is

$$E_i = E_j + \overline{S}_f L_k$$

where L_k = the length of the link k, $\overline{S}_f$ = average slope of the energy grade line between the nodes i and j,

$$E = z + y + \frac{Q^2}{2gA^2} \tag{6.5.1}$$

and z = elevation of the channel bottom above a datum. Although any of the previously defined expressions for $\overline{S}_f$ [Eqs. (6.3.55) to (6.3.62)] can be used, the average friction slope expression, Eq. (6.3.35), will be used here or

$$E_i = E_j + \left(\frac{S_{ki} + S_{kj}}{2}\right) L_k \tag{6.5.2}$$

The slope of the energy grade line is usually determined from the Manning equation [Eq. (4.2.6)] or

$$S = \frac{Q^2 n^2 P^{4/3}}{\phi^2 A^{10/3}}$$

Combining this expression for S and Eq. (6.5.2) yields an equation for the flow in link k or

$$Q_k = \frac{\phi}{n}\left[\frac{2(E_i - E_j)}{L_k}\right]^{1/2}\left[\frac{P_{ki}^{4/3}}{A_{ki}^{10/3}} + \frac{P_{ji}^{4/3}}{A_{ki}^{10/3}}\right]^{-1/2} \tag{6.5.3}$$

Although Eq. (6.5.3) theoretically applies only to prismatic channels, it can be applied with confidence to channels which are moderately nonprismatic. However, if there is a rapid change in the cross-sectional shape of the channel, then it may be necessary to include a link which accounts for the local or minor losses associated with the change. King and Brater (1963) suggest that the energy equation at a localized loss in subcritical flow is

$$E_i = E_j + \frac{K_k Q_k^2}{2g}\left(\frac{1}{A_{ki}^2} - \frac{1}{A_{kj}^2}\right) \tag{6.5.4}$$

where K = a loss coefficient. Rearrangement of this equation yields

$$Q_k = [2g(E_i - E_j)]^{1/2}\left[K_k\left(\frac{1}{A_{ki}^2} - \frac{1}{A_{kj}^2}\right)\right]^{-1/2} \tag{6.5.5}$$

When the element equations defined above are combined with a continuity equation written at each node, a system of nonlinear simultaneous equations which must be solved results. The continuity equation at each node is given by

$$F_i = \sum_{k=1}^{M} Q_k + Q_{Ni} = 0 \tag{6.5.6}$$

where M is the number of links connected to the node i and Q_{Ni} = flow which is added at a node. Application of Eq. (6.5.6) to node 2 in Fig. 6.18 yields

$$F_2 = Q_1 - Q_2 - Q_3 + Q_{N2} = 0 \tag{6.5.7}$$

Then, since $Q_{N2} = 0$, substitution of the link flow equations in Eq. (6.5.7) yields

$$\begin{aligned} F_2 = {} & \frac{\phi}{n_1}\left[\frac{2(E_1 - E_2)}{L_1} \Big/ \left(\frac{P_{11}^{4/3}}{A_{11}^{10/3}} + \frac{P_{12}^{4/3}}{A_{12}^{10/3}}\right)\right]^{1/2} \\ & - \frac{\phi}{n_2}\left[\frac{2(E_2 - E_3)}{L_2} \Big/ \left(\frac{P_{22}^{4/3}}{A_{22}^{10/3}} + \frac{P_{23}^{4/3}}{A_{23}^{10/3}}\right)\right]^{1/2} \\ & - \frac{\phi}{n_3}\left[\frac{2(E_2 - E_3)}{L_3} \Big/ \left(\frac{P_{33}^{4/3}}{A_{33}^{10/3}} + \frac{P_{32}^{4/3}}{A_{32}^{10/3}}\right)\right]^{1/2} \end{aligned} \tag{6.5.8}$$

At this point it should be noted that the subscripts on the variables F and E refer to a node, the subscripts on n and L to a link, and on P and A the first subscript refers to a link while the second refers to a node. For the other nodes in the system shown in Fig. 6.18, the corresponding equations are

Node 1

$$F_1 = Q_{N1} - Q_1 = 0$$

In this case, $Q_{N1} \neq 0$ and substitution of Eq. (6.5.3) yields

$$F_1 = Q_{N1} - \frac{\phi}{n_1}\left[\frac{2(E_1 - E_2)}{L_1} \Big/ \left(\frac{P_{11}^{4/3}}{A_{11}^{10/3}} + \frac{P_{12}^{4/3}}{A_{12}^{10/3}}\right)\right]^{1/2} \tag{6.5.9}$$

Node 3

$$F_3 = Q_2 + Q_3 - Q_4 + Q_{N3} = 0$$

and $Q_{N3} = 0$. Therefore,

$$F_3 = \frac{\phi}{n_2}\left[\frac{2(E_2 - E_3)}{L_2} \bigg/ \left(\frac{P_{22}^{4/3}}{A_{22}^{10/3}} + \frac{P_{23}^{4/3}}{A_{23}^{10/3}}\right)\right]^{1/2} + \frac{\phi}{n_3}\left[\frac{2(E_2 - E_3)}{L_3} \bigg/ \left(\frac{P_{33}^{4/3}}{A_{33}^{10/3}} + \frac{P_{32}^{4/3}}{A_{32}^{10/3}}\right)\right]^{1/2} - \frac{\phi}{n_4}\left[\frac{2(E_3 - E_4)}{L_4} \bigg/ \left(\frac{P_{44}^{4/3}}{A_{44}^{10/3}} + \frac{P_{43}^{4/3}}{A_{43}^{10/3}}\right)\right]^{1/2} \tag{6.5.10}$$

In a typical steady-flow problem, the unknowns would likely be Q_2, Q_1, E_1, E_2, and E_3. The depth of flow and flow rate at node 4 would generally be known and, therefore, E_4. Also, since the flow is by definition steady, $Q_{N1} = Q_{N2}$. Since the system of simultaneous equations which govern this probelm are nonlinear, a direct solution is not possible; however, a trial-and-error solution can be effected. The steps in obtaining a solution are:

1. Trial values of E_3, E_2, and E_1 are assumed. The speed at which any trial-and-error solution process will converge to the correct values of the variables depends on the accuracy of the values initially assumed. Wylie (1972) began his solutions with the assumption of constant energy throughout the system.
2. With the assumed values of E_3, E_2, and E_1, Eqs. (6.5.1) and (6.5.3) are used to determine the corresponding trial values of the depths of flow and the element flow rates.
3. Equations (6.5.8) to (6.5.10) are then used to compute values of F_1, F_2, and F_3. If the values of these functions are zero or meet a predetermined error criterion, then the solution is complete and the assumed values of E_1, E_2, and E_3 are correct. If this is not the case, then the solution is complete and the assumed values of E_1, E_2, and E_3 are correct. If this is not the case, then the trial values of E_1, E_2, and E_3 are corrected and steps 2 and 3 repeated until the error criterion is met.

Wylie (1972) used a Newton-Raphson iteration technique to achieve solutions.

It is obvious that the node-and-link method of determining the flow around a single island is not an efficient procedure compared with the method described previously unless computer programs are available to achieve the solution. However, the power of the technique is evident when more complex problems are considered. In Fig. 6.19*a* a flow system consisting of two islands, two flow sources, and a single outflow is shown. In Fig. 6.19*b*, the system is viewed as a series of links and nodes. Although many more equations will be required to describe this system, the methodology developed by Wylie is applicable. It would also be a mistake to assume that the node-and-link technique is applicable only to flow around islands. This technique can also be applied to open-channel flow in a system of storm sewers.

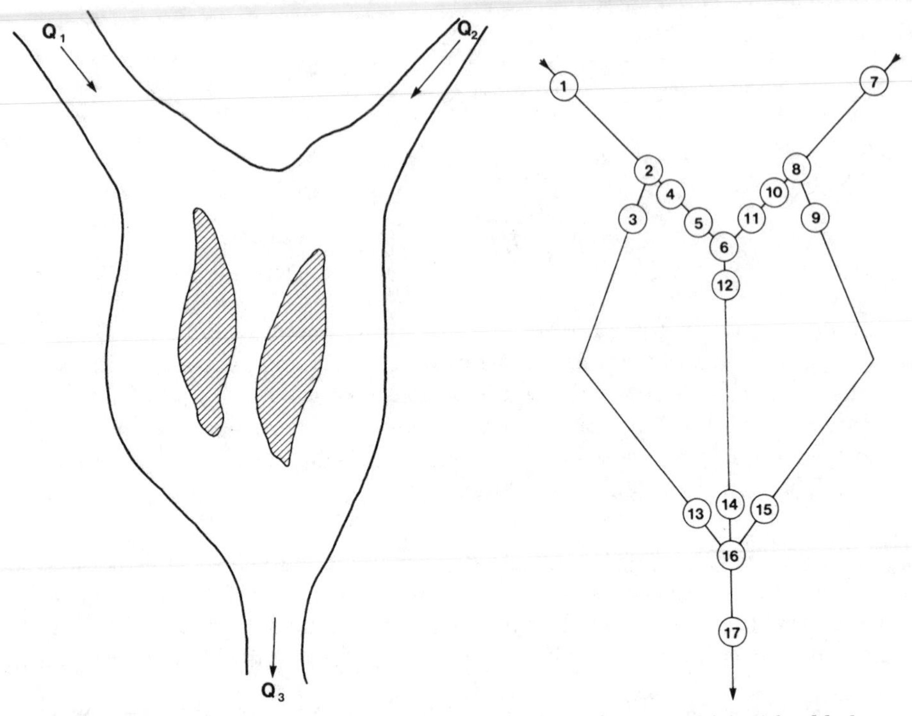

FIGURE 6.19 (*a*) Flow past two islands. (*b*) Schematic diagram of (*a*) with eddy losses at junctions included.

BIBLIOGRAPHY

Anonymous, "HEC-2, Water Surface Profiles, Computer Program 723-X6-L202A, Users Manual," U.S. Hydrologic Engineering Center, Davis, Calif., August 1979.

Bakhmeteff, B. A., *Hydraulics for Open Channels*, McGraw-Hill Book Company, New York, 1932, pp. 143–215

Chow, V. T., *Open-Channel Hydraulics*, McGraw-Hill Book Company, New York, 1959.

Escoffier, F. F., "Graphic Calculation of Backwater Eliminates Solution by Trial," *Engineering News-Record*, vol. 136, no. 26, June 27, 1946, p. 71.

Ezra, A. A., "A Direct Step Method for Computing Water Surface Profiles," *Transactions of the American Society of Civil Engineers*, vol. 119, 1954, pp. 453–462.

Feldman, A. D., "HEC Models for Water Resources Simulation: Theory and Experience," *Advances in Hydroscience*, vol. 12, 1981, pp. 297–423.

Henderson, F. M., *Open Channel Flow*, The Macmillan Company, New York, 1966.

Hinds, J., "Side Channel Spillways: Hydraulic Theory, Economic Factors, and Experimental Determination of Losses," *Transactions of the American Society of Civil Engineers*, vol. 89, 1926, pp. 881–927.

Keulegan, G. H., "Determination of Critical Depth in Spatially Variable Flow," *Proceedings of the 2d Midwestern Conference on Fluid Mechanics,* Ohio State University, Columbus, 1952.

Kiefer, C. J., and Chu, H. H., "Backwater Functions by Numerical Integration," *Transactions, American Society of Civil Engineers,* vol. 120, 1955, pp. 429–442.

King, H. W., and Brater, E. F., "Open Channels with Nonuniform Flow," *Handbook of Hydraulics,* 5th ed., McGraw-Hill Book Company, New York, 1963.

Reed, J. R., and Wolfkill, A. J., "Evaluation of Friction Slope Models," *Rivers '76 Symposium on Inland Waterways for Navigation, Flood Control, and Water Diversions,* Colorado State University, Ft. Collins, 1976, pp. 1159–1178.

Shearman, J. O., "Computer Applications for Step Backwater and Floodway Analyses," U.S. Geological Survey, Open File Report 76-499, Washington, 1976.

Shearman, J. O., "User's Guide, Step Backwater and Floodway Analyses, Computer Program J635," U.S. Geological Survey, Water Resources Division, Reston, Va., May 1977.

Wylie, E. B., "Water Surface Profiles in Divided Channels," *Journal of Hydraulic Research* (International Association for Hydraulic Research), vol. 10, no. 3, 1972, pp. 325–341.

Yen, B. C., and Wenzel, H. G., Jr., "Dynamic Equations for Steady Spatially Varied Flow," *Proceedings of the American Society of Civil Engineers, Journal of the Hydraulics Division,* vol. 96, no. HY3, March 1970, pp. 801–814.

SEVEN

Design of Channels

SYNOPSIS

SYNOPSIS

In this chapter the design of lined, unlined, and grass-lined channels is considered, and design procedures for each type of channel are discussed and demonstrated.

In the case of lined channels, a technique of minimizing lining material costs for rectangular, trapezoidal, and triangular-shaped channels is presented. Such a technique is particularly useful for long channel sections where construction procedures can be oriented to minimizing material costs or in situations in which labor costs are low relative to the lining material costs.

In the case of unlined channel design, techniques based on the principles of maximum permissible velocity and the threshold of movement are discussed, but the recommended procedure is based on the principle of tractive force. Using the principle of tractive force, the equations defining the stable hydraulic section are developed and their use demonstrated with an example. Finally, both empirical and theoretical methods of estimating the leakage from an unlined channel are considered.

In the final section of this chapter, the Soil Conservation Service method of designing grass-lined channels is discussed. In addition, a procedure for designing grass-lined channels with the center lined with gravel is presented.

7.1 INTRODUCTION

A critical topic in the area of open-channel hydraulics is the design of channels capable of transporting water between two points in a safe, cost-effective manner. Although economics, safety, and esthetics must always be considered, in this chapter only the hydraulic aspects of channel design will be examined. In addition, this discussion will be limited to the design of channels for uniform flow, and only three types of channels will be considered: (1) lined or nonerodible; (2) unlined, earthen, or erodible; and (3) grass-lined. In examining the design procedures for these types of channels, there are some basic concepts which are common to all three, and these commonalities will be discussed first.

From the Manning and Chezy equation, it is clear that the conveyance of a channel increases as the hydraulic radius increases or as the wetted perimeter decreases. Thus, from the viewpoint of hydraulics, there is among all channel cross sections of a specified geometric shape and area an optimum set of dimensions for that shape. Among all possible channel cross sections, the best hydraulic section is a semicircle since, for a given area, it has the minimum wetted perimeter. The proportions of the best hydraulic section of a specified geometric shape can be easily derived (see, for example, Streeter and Wylie,

1975; the geometric elements of these sections are summarized in Table 7.1). It should be noted that from the point of view of applications, the best hydraulic section is not necessarily the most economic section. In practice the following factors must be considered:

1. The best hydraulic section minimizes the area required to convey a specified flow; however, the area which must be excavated to achieve the flow area required by the best hydraulic section may be significantly larger if the overburden which must be removed is considered.
2. It may not be possible to construct a stable best hydraulic section in the available natural material. If the channel must be lined, the cost of the lining may be comparable with the cost of excavation.
3. The cost of excavation depends not only on the amount of material which must be removed, but also on the ease of access to the site and the cost of disposing of the material removed.
4. The slope of the channel in many cases must also be considered a variable since it is not necessarily completely defined by topographic considerations. For example, while a reduced channel slope may require a larger channel flow area to convey the specified flow, the cost of excavating the overburden may be reduced.

The terminology *minimum permissible velocity* refers to the lowest velocity which will prevent both sedimentation and vegetative growth. In general, an average velocity of 2 to 3 ft/s (0.61 to 0.91 m/s) will prevent sedimentation when the silt load of the flow is low. A velocity of 2.5 ft/s (0.76 m/s) is usually sufficient to prevent the growth of vegetation which could significantly affect the conveyance of the channel. It should be recognized that these numbers are at

TABLE 7.1 Geometric elements of best hydraulic sections

Cross section	Area A	Wetted perimeter P	Hydraulic radius R	Top width T	Hydraulic depth D
Trapezoid: half of a hexagon	$1.73\ y^2$	$3.46\ y$	$0.500\ y$	$2.31\ y$	$0.750\ y$
Rectangle: half of a square	$2\ y^2$	$4\ y$	$0.500\ y$	$2y$	y
Triangle: half a square	y^2	$2.83\ y$	$0.354\ y$	$2y$	$0.500\ y$
Semicircle	$0.500\ \pi y^2$	πy	$0.500\ y$	$2y$	$0.250\ \pi y$
Parabola: $T = 2\sqrt{2}y$	$1.89\ y^2$	$3.77\ y$	$0.500\ y$	$2.83\ y$	$0.667\ y$
Hydrostatic catenary	$1.40\ y^2$	$2.98\ y$	$0.468\ y$	$1.92\ y$	$0.728\ y$

TABLE 7.2 Suitable side slopes for channels built in various types of materials (Chow, 1959)

Material	Side slope
Rock	Nearly vertical
Muck and peat soils	¼:1
Stiff clay or earth with concrete lining	½:1 to 1:1
Earth with stone lining or earth for large channels	1:1
Firm clay or earth for small ditches	1½:1
Loose, sandy earth	2:1
Sandy loam or porous clay	3:1

best only generalized and in some cases very poor estimates of the actual minimum permissible velocity. For example, as early as 1926 Fortier and Scobey (1926) observed that canals carrying turbid waters are seldom bothered by plant growth while in channels transporting clear water some plant species flourish at velocities that are significantly in excess of the velocity which will cause erosion in the channel. The designer who obtains the highest permissible velocity that a channel will withstand from the viewpoint of hydraulic stability will have done all that can be done to prevent sedimentation and plant growth.

In most design problems, the longitudinal slope of the channel is determined by topography, the head required to carry the design flow, and the purpose of the channel. For example, in a hydroelectric power canal, a high head at the point of delivery is desirable, and a minimum longitudinal channel slope should be used.

The side slopes of a channel depend primarily on the engineering properties of the material through which the channel is excavated. From a practical viewpoint, the side slopes should be as steep as possible so that a minimum amount of land is required. In Table 7.2 side slopes for channels excavated through various types of material are suggested. These values are suitable for preliminary design purposes. In deep cuts, side slopes are often steeper above the water surface than they are below the surface. In small drainage ditches, the side slopes are steeper than they would be in an irrigation canal excavated in the same material. In many cases, side slopes are determined by the economics of construction. With regard to this subject, the following general comments are appropriate:

1. In many unlined earthen canals on federal irrigation projects, side slopes are usually 1.5:1; however, side slopes as steep as 1:1 have been used when the channel runs through cohesive materials.
2. In lined canals, the side slopes are generally steeper than in an unlined canal. If concrete is the lining material, side slopes greater than 1:1 usually require

the use of forms, and with side slopes greater than 0.75:1 the linings must be designed to withstand earth pressures. Some types of lining require side slopes as flat as those used for unlined channels.

3. Side slopes through cuts in rock can be vertical if this is desirable.

The term *freeboard* refers to the vertical distance between either the top of the channel or the top of the channel lining and the water surface which prevails when the channel is carrying the design flow at normal depth. The purpose of freeboard is to prevent the overtopping of either the lining or the top of the channel by fluctuations in the water surface caused by: (1) wind-driven waves, (2) tidal action, (3) hydraulic jumps, (4) superelevation of the water surface as the flow rounds curves at high velocities, (5) the interception of storm runoff by the channel, (6) the occurrence of greater than design depths of flow caused by canal sedimentation or an increased coefficient of friction, or (7) temporary misoperation of the canal system.

The freeboard associated with channel linings and the absolute top of the canal above the water surface can be estimated from the empirical curves in Fig. 7.1. In general, these curves apply to a channel lined with either a hard surface, a membrane, or compacted earth with a low coefficient of permeability. For unlined channels, freeboard generally ranges from 1 ft (0.30 m) for small

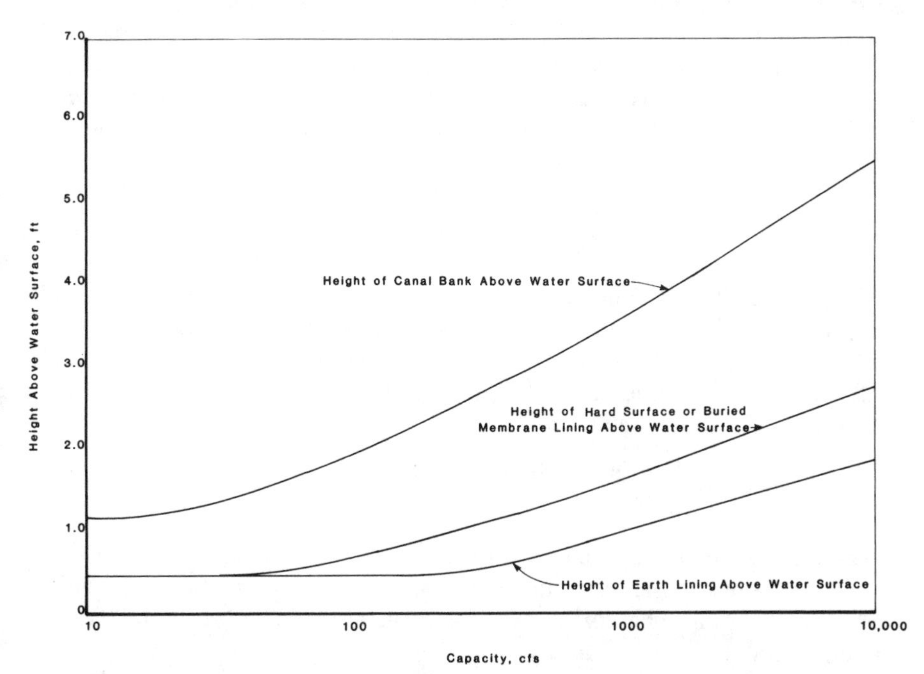

FIGURE 7.1 Freeboard for canal banks and freeboard for hard surface, buried membrane, and earth linings, U.S. Bureau of Reclamation. *(Anonymous, 1963.)*

laterals with shallow depths of flow to 4 ft (1.2 m) for channels carrying 3000 ft^3/s (85 m^3/s) at relatively large depths of flow (Chow, 1959). A preliminary estimate of freeboard for an unlined channel can be obtained from

$$F = \sqrt{Cy} \tag{7.1.1}$$

where F = freeboard, feet
y = design depth of flow, feet
C = coefficient which varies from 1.5 at Q = 20 ft^3/s (0.57 m^3/s) to 2.5 for Q = 3000 ft^3/s (85 m^3/s)

When a flow moves around a curve, a rise in the water surface occurs at the outer bank with a corresponding lowering of the water surface at the inner bank. In the design of a channel, it is important that this difference in water levels be estimated. If all the flow is assumed to move around the curve at the subcritical average velocity $\overline{u}$, then

$$\Delta h = \frac{\overline{u}^2 b}{gR} \tag{7.1.2}$$

where Δh = change in water surface elevation across channel
b = channel width
R = distance from center of curve to centerline of channel

Equation (7.1.2) always underestimates Δh because of the average velocity assumption. In some cases, this equation may underestimate by as much as 50 percent (Houk, 1956). If Newton's second law of motion is applied to each streamline of the flow as it passes around the curve, then it is possible to demonstrate that the transverse water surface profile is a logarithmic curve of the form

$$h = 2.3 \frac{\overline{u}^2}{g} \log \frac{R_o}{R_i} \tag{7.1.3}$$

where R_o and R_i are the outer and inner radii of the curve. Woodward (1920, 1941) assumed that the velocity of flow was zero at the banks and had a maximum u_M at the centerline of the curving channel. Between the sides and the center, the velocity varied according to a parabolic curve. Applying Newton's second law with these assumptions,

$$h = \frac{u_M^2}{g}\left[\frac{20}{3}\frac{R}{b} - \frac{16R^3}{b^3} + \left(\frac{4R^2}{b^2} - 1\right)^2 \ln\left(\frac{2R + b}{2R - b}\right)\right] \tag{7.1.4}$$

Of these three equations, Eq. (7.1.4) provides the best estimate of Δh. The subject of superelevation around curves will be treated in some detail in a subsequent section of this chapter, but the foregoing equations provide the basis for making initial design estimates.

There are no set rules governing the minimum radii of curvature for canals.

Shukry (1950), using laboratory data, found that the effects of curves were negligible when the ratio of the radius of curvature to the distance to the center of the canal was greater than 3 times the width of the level portion of the channel bed. In India, the minimum radii of curvature are often longer than those used in the United States. For example, some Indian engineers recommend a minimum radius of 300 ft (91 m) for canals carrying less than 10 ft^3/s (0.30 m^3/s) to 5000 ft (1500 m) for canals carrying more than 3000 ft^3/s (85 m^3/s) (Houk, 1956).

The width of the banks along a canal are usually governed by a number of considerations which include the size of the canal, the amount of excavation available for bank construction, and the need for maintenance roads. Where roads are needed, the top widths for both lined and unlined canals are 16 ft (5 m) or more. The bank top is usually graded away from the canal so that precipitation will not flow into the canal. Bank widths must also be sufficient to provide for stability against canal water pressure and keep percolating water below the ground level outside the banks.

7.2 DESIGN OF LINED CHANNELS

Lined channels are built for five primary reasons:

1. To permit the transmission of water at high velocities through areas of deep or difficult excavation in a cost-effective fashion
2. To permit the transmission of water at high velocity at a reduced construction cost
3. To decrease canal seepage, thus conserving water and reducing the waterlogging of lands adjacent to the canal
4. To reduce the annual costs of operation and maintenance
5. To ensure the stability of the channel section

The design of lined channels from the viewpoint of hydraulic engineering is a rather elementary process which generally consists of proportioning an assumed channel cross section. Some typical cross sections of lined channels used on irrigation projects in the United States are summarized in Table 7.3, and a typical lined section of the All-American Canal is shown in Fig. 7.2. Additional information regarding channel linings can be found in Willison (1958) and Anonymous (1963). A recommended procedure for proportioning a lined section is summarized in Table 7.4. In this table, it is assumed that the design flow Q_D, the longitudinal slope of the channel S, the type of channel cross section, e.g., trapezoidal, and the lining material have all been selected prior to the initiation of the channel design process.

TABLE 7.3 Typical cross sections of lined channels for selected canals in the western United States (Houk, 1956)

Canal	Project	Side slopes	Depth of flow, ft	Bottom width, ft	Ratio, width to depth	Mean velocity, ft	Discharge, ft^3/s	Freeboard, ft*
			Lined earth sections					
Contra Costa	Central Valley	1¼:1	1.84	3	1.63	2.33	26	1.16
South Branch	Yakima	1¼:1	3.80	5	1.32	5.94	220	1.00
Ridge	Yakima	1¼:1	7.46	7	0.94	4.93	600	1.04
Heart Mountain	Shoshone	1¼:1	7.51	8	1.06	7.00	914	1.49
Kittitas Main	Yakima	1¼:1	8.99	11	1.22	6.63	1320	1.25
Black Canyon	Boise	1¼:1	9.39	12	1.28	4.86	1089	1.61
Ridge	Yakima	1¼:1	11.20	14	1.25	7.02	2200	1.80
All-American	All-American	1⅜:1	10.64	22	2.07	6.62	2800	2.25
Delta-Mendota	Central Valley	1½:1	16.56*	48	2.9	—	4600	—
			Lined rock sections					
Kittitas Main	Yakima	½:1	9.00	14	1.56	7.65	1275	1.25
Main (gravity extension)	Minidoka	¼:1	13.80	20	1.45	8.20	2695	1.50
Gravity Main	Gila	¾:1	21.07	32	1.52	5.96	6000	—

*To top of lining.

TABLE 7.4 A design procedure for lined channels

Step	Process
1	Estimate n or C for specified lining material
2	Compute value of section factor $AR^{2/3} = \dfrac{nQ}{\phi\sqrt{S}}$; $\phi = 1.49$ for English units and 1 for SI units
3	Solve section factor equation for y_N given appropriate expressions for A and R (Table 1.1). See Fig. 5.2 for a section factor equation solution technique. *Note:* This step *may* require assumptions regarding side slopes, bottom widths, etc.
4	If best hydraulic section is required, compute channel parameters from Table 7.1; otherwise compute channel parameters from Table 1.1 using y_N from Step 3.
5	Check: 1. Minimum permissible velocity if water carries silt and for vegetation 2. Froude number
6	Estimate: 1. Required height of lining above water surface, Fig. 7.1 2. Required freeboard, Fig. 7.1
7	Summarize results with dimensioned sketch

EXAMPLE 7.1

A lined channel of trapezoidal cross section is to be designed to carry 400 ft^3/s (11 m^3/s). The lining of the canal is to be float-finished concrete, and the longitudinal slope of the canal is 0.0016. Determine appropriate channel proportions.

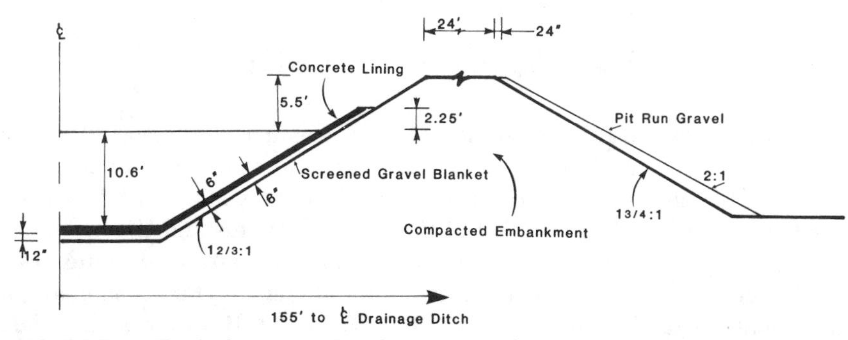

FIGURE 7.2 Typical concrete-lined section used on the All-American Canal.

Solution

Step 1 Estimate Manning's n.

$$n = 0.015 \qquad \text{(Table 4.8)}$$

Step 2 Compute the section factor.

$$Q = \frac{1.49}{n} AR^{2/3} \sqrt{S}$$

$$AR^{2/3} = \frac{nQ}{1.49\sqrt{S}} = \frac{(0.015)400}{1.49\sqrt{0.0016}} = 100.7$$

Step 3 Assume $b = 20$ ft (6.1 m) and $z = 2$. Solve the section factor equation (step 2) for y_n.

$$y_n = 2.5 \text{ ft } (0.76 \text{ m})$$

Step 4 Since the best hydraulic section is not required, this step is omitted in Table 7.4.

Step 5 Check minimum permissible velocity and Froude number.

$$A = (b + zy)y = [20 + 2(2.5)]2.5 = 62.5 \text{ ft}^2 \ (5.8 \text{ m}^2)$$

$$\bar{u} = \frac{Q}{A} = \frac{400}{62.5} = 6.4 \text{ ft/s } (2.0 \text{ m/s})$$

This velocity should prevent vegetative growth and sedimentation.

$$\mathbf{F} = \frac{\bar{u}}{\sqrt{gD}} = 0.78$$

Therefore, this is a subcritical flow.

Step 6 (a) Estimate height of the lining above water surface (Fig. 7.1) = 1.2 ft (0.37 m).
(b) Estimate height of canal bank above water surface (Fig. 7.1) = 2.9 ft (0.88 m).

The results of this design are summarized in Fig. 7.3.

A primary concern in the design of lined channels is the cost of the lining material and the development of channel dimensions which minimize this cost. The cost of the lining material is a function of the volume of lining material used which in turn is a function of the lining thickness and the magnitude of the wetted perimeter. If a channel is lined with one material of uniform thickness, then the solution which minimizes the lining cost for a channel of arbitrary shape is the best hydraulic section (Table 7.1). However, if the thickness of the lining material changes along the perimeter, then the problem becomes

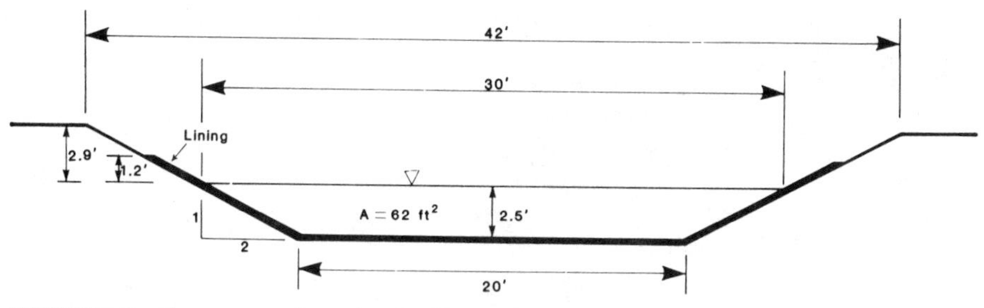

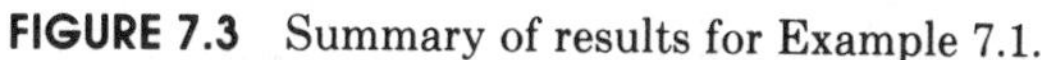

FIGURE 7.3 Summary of results for Example 7.1.

much more complex. Trout (1982) examined the problem of lining material cost minimization for trapezoidal, rectangular, and triangular channels for the special case of the material used to line the base of the channel being different from that used to line the sides. The methodology developed by Trout considers neither placement nor construction costs unless these costs can be evaluated in terms of the channel surface area. The procedure described in the material which follows is particularly useful for long channel sections where construction procedures can be oriented toward minimizing material costs or in situations in which labor costs are low relative to the lining material costs.

As noted previously in this section, the primary design parameter for designing the dimensions of a lined channel is the section factor, Eq. (5.1.5). For trapezoidal, rectangular, and triangular sections, the section factor is

$$AR^{2/3} = \frac{A^{5/3}}{P^{2/3}} = \frac{(by + zy^2)^{5/3}}{[b + 2y\sqrt{1 + z^2}]^{2/3}} = \frac{Qn}{\phi\sqrt{S}} \tag{7.2.1}$$

where z = side slope
b = bottom width
y = depth of flow
S = longitudinal channel slope
n = Manning's resistance coefficient
Q = design flow
ϕ = constant equal to 1.48 when English units are used and 1.00 when SI units are used

Equation (7.2.1) can be rearranged to yield an implicit solution for the depth of flow or

$$y = \frac{[(b/y) + z\sqrt{1 + z^2}]^{1/4}}{[(b/y) + z]^{5/8}} \left(\frac{Qn}{\phi\sqrt{S}} \right)^{3/8} \tag{7.2.2}$$

In Eq. (7.2.2) if b/y and the side slope z are specified, then the equation can be solved explicitly for y and b.

The cost of the materials used in lining a channel can usually be specified in terms of the volume of material used. Thus, if the thickness of the lining is specified, the cost of the unit length of channel is only a function of the wetted perimeter plus the freeboard and the lining material used in the corners (Fig. 7.4). With reference to Fig. 7.4

$$C_b = \mu_b \left(\frac{\text{volume}}{\text{unit length}} \right) = \mu_b t_b (b + b') = Bb + k \tag{7.2.3}$$

$$C_s = \mu_s \left(\frac{\text{volume}}{\text{unit length}} \right) = \mu_s t_s (2E + 2E') = 2\Gamma (y + F) \sqrt{1 + z^2} \tag{7.2.4}$$

and

$$C = C_b + C_s = bB + k + 2\Gamma (y + F) \sqrt{z^2 + 1} \tag{7.2.5}$$

where C = total material cost per unit length
C_b = material cost for channel base per unit length
C_s = material cost of sides per unit channel length
b' = bottom corner width
t_b = channel base lining thickness
t_s = channel side lining thickness
E = wetted side length
E' = freeboard side length
μ_b = cost of base lining material per unit volume
μ_s = cost of side lining material per unit volume
B = cost of base lining material for specified thickness per unit area
k = cost of corner materials per unit length
Γ = cost of side lining material for specified thickness per unit area
F = vertical freeboard requirement

To determine the dimensions of the minimum-cost trapezoidal, rectangular, or triangular cross section, Eq. (7.2.1) or (7.2.2) must be solved such that the cost function, Eq. (7.2.5), is minimized. As noted by Trout (1982), this open-channel cost optimization problem is analogous to the classic microeconomic problem of the minimizing production costs through input substitution. In this case, the output of the system is the hydraulic capacity; the inputs are the variables which define the channel geometry; and the production function is the

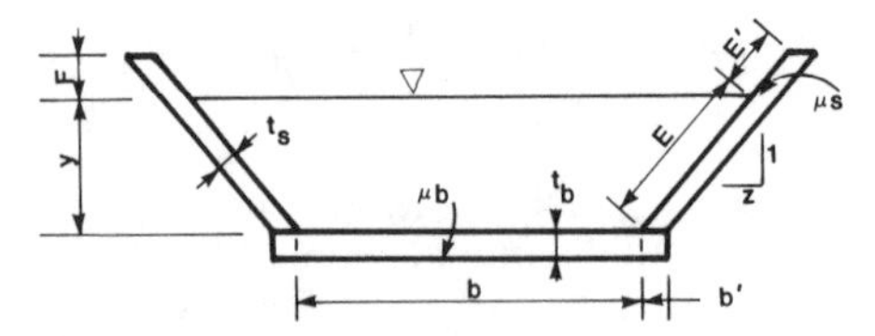

FIGURE 7.4. Definition of dimensional terms in cost optimization for trapezoidal channel. *(Trout, 1982.)*

equation for the section factor. In the terminology of business, the solution of this optimization problem requires an input mix such that the ratio of the marginal products equals the ratio of the marginal costs. If the channel side slope variable z is assumed constant, then the Lagrange multiplier method can be used to find an explicit algebraic solution to this problem.

As noted above, the dimensional combination in which the ratio of the marginal changes in the section factor are equal to the marginal changes in the costs or

$$\frac{\partial(AR^{2/3})/\partial b}{\partial(AR^{2/3})/\partial y} = \frac{\partial C/\partial b}{\partial C/\partial y} \tag{7.2.6}$$

subject to Eq. (7.2.1) represents the optimal or minimum-cost solution of the problem. If Eqs. (7.2.1) and (7.2.5) are substituted in Eq. (7.2.6) and the result simplified, then the optimal solution is

$$K_1\left(\frac{y}{b}\right)^2 + K_2\left(\frac{y}{b}\right) + K_3 = 0 \tag{7.2.7}$$

where

$$K_1 = 20(z^2 + 1) - \left[1 + 4\left(\frac{B}{\Gamma}\right)\right] 4z\sqrt{z^2 + 1} \tag{7.2.8}$$

$$K_2 = \left(1 - \frac{B}{\Gamma}\right) 6\sqrt{z^2 + 1} - 10z\left(\frac{B}{\Gamma}\right) \tag{7.2.9}$$

and

$$K_3 = -5\left(\frac{B}{\Gamma}\right) \tag{7.2.10}$$

Then the ratio of b/y for the required solution of Eq. (7.2.2) is

$$\frac{b}{y} = \frac{2K_1}{-K_2 + \left[K_2^2 + 20\left(\frac{B}{\Gamma}\right)K_1\right]^{1/2}} \tag{7.2.11}$$

which is a function of z and the ratio of the unit costs of the base to side slope material. The typical solution would proceed in the following steps:

1. Given values of S, Q, n, z, B, and Γ, values of K_1, K_2, and K_3 are determined by Eqs. (7.2.8) to (7.2.10).
2. The minimum-cost value of the ratio b/y is then estimated by Eq. (7.2.11).
3. The minimum-cost depth of flow is estimated by Eq. (7.2.2).
4. The minimum-cost bottom width can then be determined by multiplying y times the ratio b/y.

EXAMPLE 7.2

A lined trapezoidal channel is to be designed to convey 0.08 m^3/s (2.8 ft^3/s) of water on a slope of 0.001. If $n = 0.014$, $B =$ \$3.20 per square meter (\$0.30 per square foot), $\Gamma =$ \$2.00 per square meter (\$0.19 per square foot), $z = 0.5$, the vertical freeboard required is 0.15 m (0.5 ft), and the material cost for the corners is \$0.35 per meter (\$0.11 per foot), determine the depth of flow and bottom width for which the lining cost is minimized. This example was first presented by Trout (1982).

Solution

This example can be solved by both direct algebraic and graphical techniques. As will be demonstrated, the algebraic technique yields rapid results, but the graphical technique can be used to answer other important questions.

The first step in the direct algebraic technique is to determine values of K_1 and K_2 from the given data. From Eq. (7.2.8)

$$K_1 = 20(z^2 + 1) - \left[1 + 4\left(\frac{B}{\Gamma}\right)\right] 4z(z^2 + 1)^{1/2}$$

$$= 20(0.5^2 + 1) - \left[1 + 4\left(\frac{3.2}{2}\right)\right] (4)(0.5)(0.5^2 + 1)^{1/2}$$

$$= 8.45$$

and from Eq. (7.2.9)

$$K_2 = 6\left(1 - \frac{B}{\Gamma}\right)(z^2 + 1)^{1/2} - 10z\,\frac{B}{\Gamma}$$

$$= 6\left(1 - \frac{3.2}{2}\right)(0.5^2 + 1)^{1/2} - 10(0.5)\left(\frac{3.2}{2}\right)$$

$$= -12.0$$

Then, the ratio b/y for the optimal lining costs section is given by Eq. (7.2.11) or

$$\frac{b}{y} = \frac{2K_1}{-K_2 + \left[K_2^2 + 20\left(\dfrac{B}{\Gamma}\right)K_1\right]^{1/2}}$$

$$= \frac{2(8.45)}{-(-12.0) + \left[(-12.0)^2 + 20\left(\dfrac{3.2}{2}\right)(8.45)\right]^{1/2}}$$

$$= 0.522$$

With the optimal value of b/y established, Eq. (7.2.2) can be used to find the optimal depth of flow or

$$y = \frac{[(b/y) + 2(z^2 + 1)^{1/2}]^{1/4}}{[(b/y) + z]^{5/8}}\left(\frac{Qn}{\phi\sqrt{S}}\right)^{3/8}$$

$$= \frac{[0.522 + 2(0.5^2 + 1)^{1/2}]^{1/4}}{(0.522 + 0.5)^{5/8}} \left[\frac{0.08(0.014)}{1.0\sqrt{0.001}} \right]^{3/8}$$

$$= 0.36 \text{ m } (1.2 \text{ ft})$$

and therefore,

$$b = \frac{b}{y} y = 0.522(0.36) = 0.19 \text{ m } (0.62 \text{ ft})$$

The minimum lining cost per unit length of channel can then be computed from Eq. (7.2.5) or

$$C = 0.19(3.20) + 2(2)(0.36 + 0.15)(0.5^2 + 1)^{1/2} + 0.35$$
$$= \$3.24 \text{ per meter } (\$0.99 \text{ per foot})$$

Representing Eq. (7.2.2) and (7.2.11) in a graphical form is a second method of solving this problem. In Fig. 7.5 such a solution is shown for the specific case of $z = 0.5$. In this figure, the curved lines represent values of the section factor while the straight lines radiating from the origin are solutions of Eq. (7.2.11) for various values of B/Γ. With regard to this figure, it is noted, first, that a separate graph is required for each side slope value. Second, any combination of b and y which falls on the section factor line fulfills the hydraulic requirements of the design. Third, the b and y values at the point where the B/Γ line crosses the appropriate section factor line are the optimum lining cost values. Using Fig. 7.5 with

$$\frac{nQ}{\phi\sqrt{S}} = 0.035$$

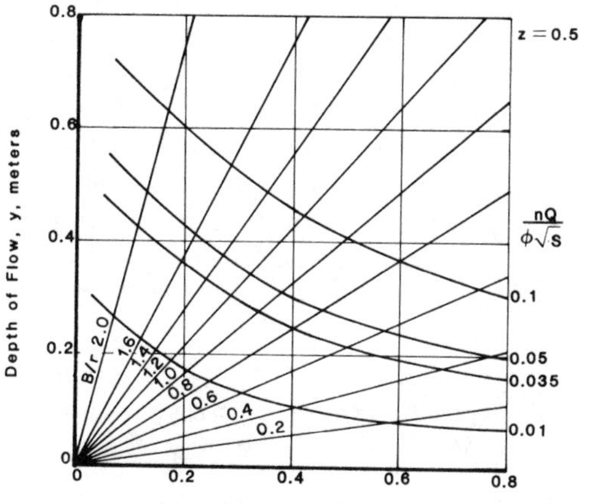

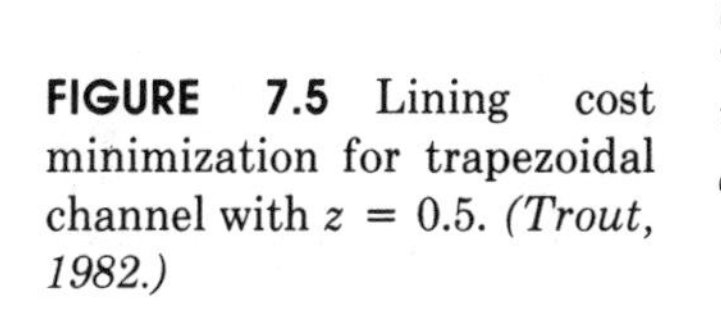
FIGURE 7.5 Lining cost minimization for trapezoidal channel with $z = 0.5$. *(Trout, 1982.)*

and
$$\frac{B}{\Gamma} = 1.6$$

it is found that

$$y = 0.36 \text{ m } (1.2 \text{ ft})$$

and
$$b = 0.19 \text{ m } (0.62 \text{ ft})$$

With these values the minimum lining cost in dollars could be calculated as before.

Once a figure such as Fig. 7.5 has been constructed for a specified side slope, other pertinent and important questions can be answered. For example, what additional cost will be incurred for the specified data if the only available construction forms have a bottom width of 0.4 m (1.3 ft)? This question can be answered by following the section factor curve until each intersects $b = 0.4$ m (1.3 ft). The depth is then determined to be $y = 0.25$ m (0.82 ft), and the cost per unit length of channel is from Eq. (7.2.5).

$$C = bB + 2\Gamma(y + F)(z^2 + 1)^{1/2} + k$$

or

$$C = 0.4(3.20) + 2(2)(0.25 + 0.15)(0.5^2 + 1)^{1/2} + 0.35$$
$$= \$3.42 \text{ per meter } (\$1.04 \text{ per foot})$$

Thus, in using the forms available, a

$$\text{Percent increase} = \frac{3.42 - 3.24}{3.24}(100) = 5.6\%$$

over the minimum lining cost is incurred.

Another question which should be examined is the sensitivity of total lining cost to variations in side slope ratios. In Fig. 7.6 total lining cost is plotted as function of the side slope ratio for the cost and flow data given in this example. Minimum lining material costs occur at $z \simeq 0.65$. At a side slope ratio of 0.92, the optimum bottom width of the channel has decreased to zero, and the minimum cost channel shape is a triangle. From Figs. 7.5 and 7.6 it can be concluded that lining material costs are rather insensitive to variations in cross-sectional dimensions if they do not deviate excessively from the optimum dimensions.

7.3 DESIGN OF STABLE, UNLINED, EARTHEN CHANNELS: A GENERAL TRACTIVE FORCE DESIGN METHODOLOGY

In comparison with the design process typically used for lined channels, the design of stable, unlined or erodible, earthen channels is a complex process involving numerous parameters, most of which cannot be accurately quantified.

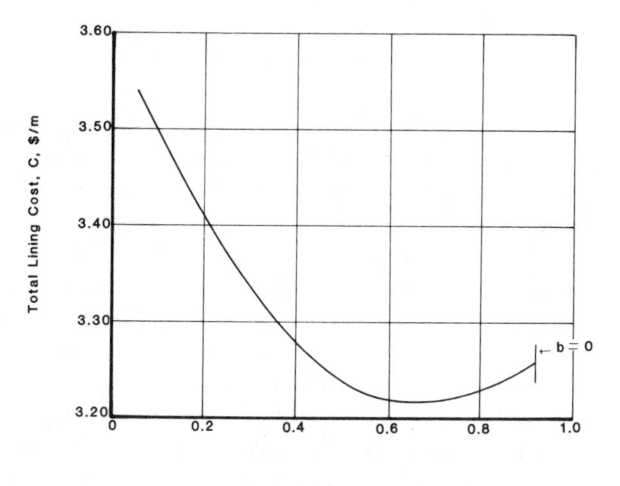

FIGURE 7.6 Total lining cost as a function of side slope ratio.

The complexity of the erodible channel design process results from the fact that in such channels stability is dependent not only on hydraulic parameters but also on the properties of the material which composes the bed and sides of the channel. In the context of this discussion, a stable channel section is one in which neither objectionable scour nor deposition occurs. In contrast, there are three types of unstable sections. In the first type, the banks and bed of the channel are scoured but no deposition occurs. This situation can occur when the channel conveys sediment-free water or water which conveys only a very small amount of sediment but with sufficient energy to erode the channel. The second type of unstable channel is one in which there is deposition but no scour. This situation can result when the water being conveyed carries a large sediment load at a velocity that permits sedimentation. The third type is the case in which both scour and deposition occur. This situation can result when the material through which the channel is excavated is susceptible to erosion and the water being conveyed carries a significant sediment load.

By the mid 1920s, it became obvious that there should exist a relationship between the flow rate or the average velocity, the engineering properties of the material composing the bed and sides of the channel, the amount and type of sediment being transported by the flow, and the stability of the channel section. The American Society of Civil Engineers' Special Committee on Irrigation Hydraulics submitted questionnaires to a number of engineers whose experience qualified them to form authoritative opinions regarding the stability of channels built in various types of material. The hypothesis of this study was that there was a relationship between the average velocity of flow, the channel perimeter material, and channel stability. The results of this survey were published in 1926 (Fortier and Scobey, 1926) and became the theoretical basis for the channel design method known as the *method of maximum permissible velocity*. The primary results of the Fortier and Scobey (1926) report are summarized in columns (1), (3), (5), (7), and (10) of Table 7.5. With regard to these data, the following caveats are noted. First, the data given are for channels with long tangents, and a 25 percent reduction in the maximum permissible velocity is recommended for canals with a sinuous alignment. Second, these data are for depths of flow less than 3.0 ft (0.91 m). For greater depths of flow, the maximum permissible velocity should be increased by 0.5 ft/s (0.15 m/s). Third, the velocity of flow in canals carrying abrasives, such as basalt ravelings, should be reduced by 0.5 ft/s (0.15 m/s). Fourth, channels diverting water from silt-laden rivers such as the Colorado River should be designed for mean design velocities 1 to 2 ft/s (0.30 to 0.61 m/s) greater than would be allowed for the same perimeter material if the water were transporting no sediment.

The pioneering work of Fortier and Scobey (1926) was the basis of channel design for many years; however, it is a design methodology based primarily on experience and observation rather than physical principles. The first step in developing a rational design process for unlined, stable, earthen channels is to examine the forces which cause scour. Scour on the perimeter of a channel occurs when the particles on the perimeter are subjected to forces of sufficient magnitude to cause particle movement. When a particle rests on the level bottom of a channel, the force acting to cause movement is the result of the flow of water past the particle. A particle resting on the sloping side of a channel is acted on not only by the flow-generated force, but also by a gravitational component which tends to make the particle roll or slide down the slope. If the resultant of these two forces is larger than the forces resisting movement, gravity, and cohesion, then erosion of the channel perimeter occurs. By definition, the tractive force is the force acting on the particles composing the perimeter of the channel and is the result of the flow of water past these particles. In practice, the tractive force is not the force acting on a single particle, but the force exerted over a certain area of the channel perimeter. This concept was apparently first stated by duBoys (1879) and restated by Lane (1955).

Recall that in uniform flow the tractive force should be approximated by the effective gravitational force acting on the water within the control volume (Fig. 4.1) and parallel to the channel bottom, or

$$F_T = \gamma ALS \tag{7.3.1}$$

where A = channel cross-sectional area
L = control volume length
S = longitudinal slope of channel

The unit tractive force is

$$\tau_o = \frac{\gamma ALS}{PL} = \gamma RS \tag{7.3.2}$$

where τ_o = average value of the tractive force per unit of wetted area or the

TABLE 7.5 Maximum permissible velocities as recommended by Fortier and Scobey (1926) for straight channels of small slope and after aging

Material (1)	n (2)	Clear water				Water transporting colloidal silts			
		$\bar{u}$, ft/s (3)	τ_o, lb/ft^2 (4)	$\bar{u}$, m/s (5)	τ_o, N/m^2 (6)	$\bar{u}$, ft/s (7)	τ_o, lb/ft^2 (8)	$\bar{u}$, m/s (9)	τ_o, N/m^2 (10)
Fine sand, noncolloidal	0.020	1.50	0.027	0.457	1.29	2.50	0.075	0.762	3.59
Sandy loam, noncolloidal	0.020	1.75	0.037	0.533	1.77	2.50	0.075	0.762	3.59
Silt loam, noncolloidal	0.020	2.00	0.048	0.610	2.30	3.00	0.11	0.914	5.27
Alluvial silts, noncolloidal	0.020	2.00	0.048	0.610	2.30	3.50	0.15	1.07	7.18
Ordinary firm loam	0.020	2.50	0.075	0.762	3.59	3.50	0.15	1.07	7.18
Volcanic ash	0.020	2.50	0.075	0.762	3.59	3.50	0.15	1.07	7.18
Stiff clay, very colloidal	0.025	3.75	0.26	1.14	12.4	5.00	0.46	1.52	22.0
Alluvial silts, colloidal	0.025	3.75	0.26	1.14	12.4	5.00	0.46	1.52	22.0
Shales and hardpans	0.025	6.00	0.67	1.83	32.1	6.00	0.67	1.83	32.1
Fine gravel	0.020	2.50	0.075	0.762	3.59	5.00	0.32	1.52	15.3
Graded loam to cobbles when noncolloidal	0.030	3.75	0.38	1.14	18.2	5.00	0.66	1.52	31.6
Graded silts to cobbles when colloidal	0.030	4.00	0.43	1.22	20.6	5.50	0.80	1.68	38.3
Coarse gravel noncolloidal	0.025	4.00	0.30	1.22	14.4	6.00	0.67	1.83	32.1
Cobbles and shingles	0.035	5.00	0.91	1.52	43.6	5.50	1.10	1.68	52.7

unit tractive force. In a wide channel, $y_N \simeq R$ and Eq. (7.3.2) becomes $\tau_o = \gamma y_N S$.

In most channels, the tractive force is not uniformly distributed over the perimeter, and, therefore, before an accurate design methodology can be developed, the distribution of the tractive force on the perimeter of the channel must be estimated.

Although many attempts to determine the distribution of the tractive force on a channel perimeter have been made using both field and laboratory data, they have not been successful (Chow, 1959). In Fig. 7.7, the maximum unit tractive force on the sides and bottoms of various channels as determined by mathematical studies are shown as a function of the ratio of the bottom width to the depth of flow. It is noted that for the trapezoidal section, which is the section generally used in unlined canals, the maximum tractive force on the bottom is approximately $\gamma y_N S$ and on the sides $0.76\gamma y_N S$ (Lane, 1955).

When a particle on the perimeter of a channel is in a state of impending motion, the forces acting to cause motion are in equilibrium with the forces resisting motion. A particle on the level bottom of a channel is subject to the tractive force $A_e \tau_L$ where τ_L = unit tractive force on a level surface and A_e = effective area. Movement is resisted by the gravitational force W_s multiplied by a coefficient of friction which is approximated by $\tan \alpha$ where W_s = submerged particle weight and α = angle of repose for the particle. When motion is incipient,

$$A_e \tau_L = W_s \tan \alpha \tag{7.3.3}$$

or

$$\tau_L = \frac{W_s}{A_e} \tan \alpha \tag{7.3.4}$$

A particle on the sloping side of a channel is subject to both a tractive force $\tau_s A_e$ and a downslope gravitational component $W_s \sin \Gamma$ where τ_s = unit tractive force on the side slopes and Γ = side slope angle. These forces and their resultant $\sqrt{(W_s \sin \Gamma)^2 + (\tau_s A_e)^2}$ are shown schematically in Fig. 7.8. The force resisting motion is the gravitational component multiplied by a coefficient of friction $W_s \cos \Gamma \tan \alpha$. Setting the forces tending to cause motion equal to those resisting motion,

$$W_s \cos \Gamma \tan \alpha = \sqrt{(W_s \tan \Gamma)^2 + (A_e \tau_s)^2} \tag{7.3.5}$$

or

$$\tau_s = \frac{W_s}{A_e} \cos \Gamma \tan \alpha \sqrt{1 - \frac{\tan^2 \Gamma}{\tan^2 \alpha}} \tag{7.3.6}$$

Equations (7.3.4) and (7.3.6) usually combine to form the tractive force ratio

$$K = \frac{\tau_s}{\tau_L} = \cos \Gamma \sqrt{1 - \frac{\tan^2 \Gamma}{\tan^2 \alpha}} = \sqrt{1 - \frac{\sin^2 \Gamma}{\sin^2 \alpha}} \tag{7.3.7}$$

where K = tractive force ratio.

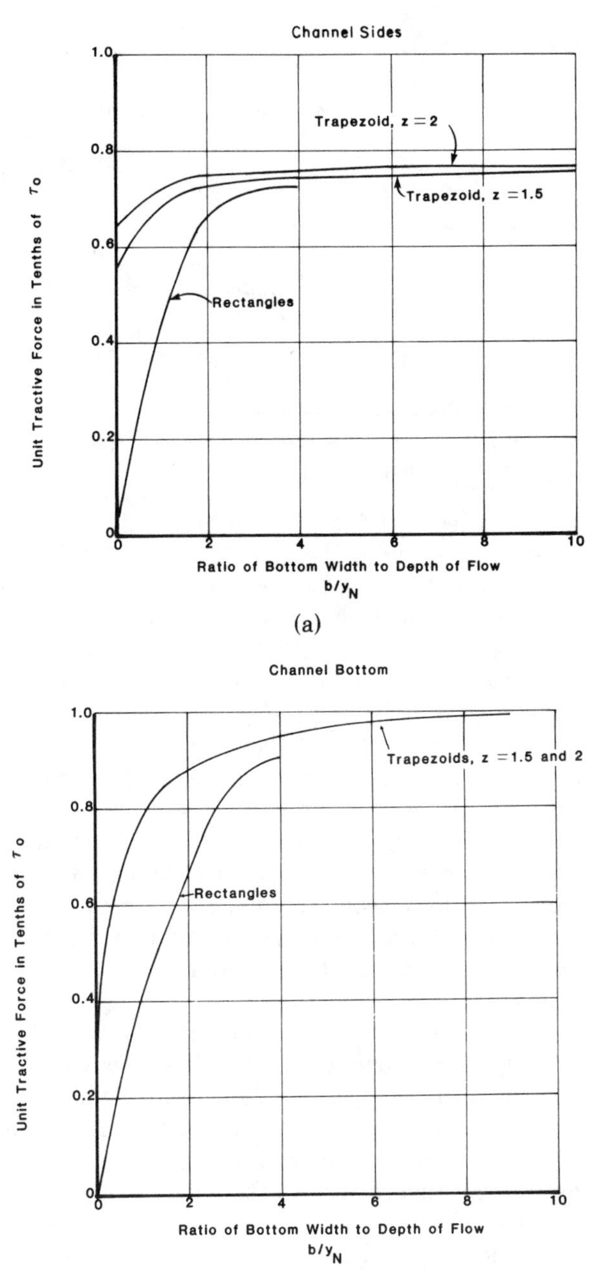

(a)

(b)

FIGURE 7.7 (*a*) Maximum unit tractive force in terms of $\gamma y_N s$ for channel sides. (*b*) Maximum unit tractive forces in terms of $\gamma y_N s$ for channel bottoms.

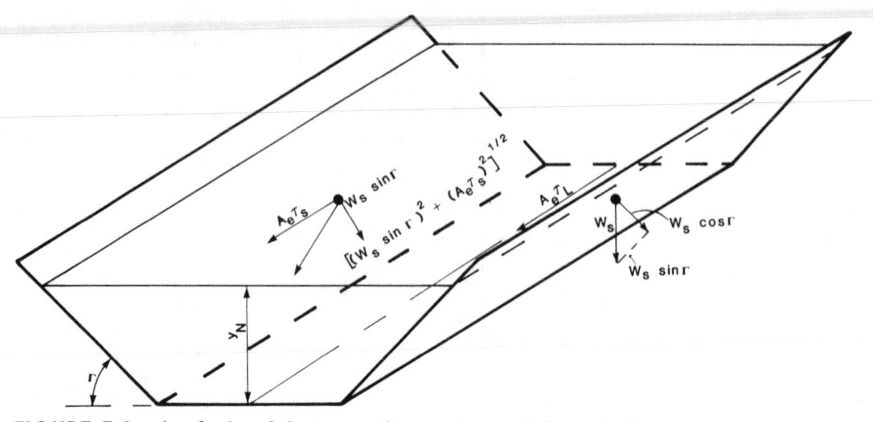

FIGURE 7.8 Analysis of forces acting on a particle resisting movement on the perimeter of a channel.

It is noted that the tractive force ratio is a function of both the side slope angle and the angle of repose of the material composing the channel perimeter. In the case of cohesive materials and fine noncohesive materials, the angle of repose is small and can be assumed to be zero; i.e., for these materials the forces of cohesion are significantly larger than the gravitational component tending to make the particles roll downslope. Lane (1955) found that, in general, the angle of repose is directly proportional to both the size and angularity of the particle. The laboratory data available to Lane (1955) are summarized in Fig. 7.9. In this figure, the particle size is the diameter of the particle of which 25 percent of all the particles, measured by weight, are larger. With regard to the data summarized in this figure, the following caveats should be noted. First, time was not available when these tests were performed to run a large number of experiments; for this reason there was a large amount of scatter in the raw data. Second, the angles of repose are limited to 41° for the very angular material and 39° for the very rounded material because of a paucity of data on the larger material.

For coarse, noncohesive materials, the laboratory data of Lane (1955) indicated that the maximum permissible tractive force in pounds per square foot was equal to 0.4 times the 25 percent diameter of the particles in inches. In recognition of the fact that actual canals can sustain tractive forces in excess of those predicted by laboratory experiment, Lane (1955) also collected data regarding actual canals. All these field data were in the form of maximum permissible velocities and thus had to be converted to tractive force data—a process which required numerous assumptions regarding the size of the canal and the depth of flow. For example, in columns (4), (6), (8), and (10) of Table 7.5 the Fortier and Scobey (1926) data have been converted to tractive force data. The results of the Lane field data collection effort are summarized in Fig. 7.10*a*. In this figure, for the *fine* noncohesive, i.e., average diameters less than 5 mm

(0.254 in), the size specified is the median size of the diameter of a particle of which 50 percent were larger by weight. Data regarding the permissible unit tractive forces for canals built in cohesive materials were presented in Chow (1959) and are summarized in Fig. 7.10*b*. The data in these tables are believed to provide conservative design information, and a factor of safety has already been built into these curves.

Lane (1955) also recognized that sinuous canals scour more easily than canals with straight alignments. To account for this observation in the tractive force design method, Lane developed the following definitions. Straight canals have straight or slightly curved alignments and are typical of canals built in flat plains. Slightly sinuous canals have a degree of curvature which is typical of canals in a slightly undulating topography. Moderately sinuous canals have a degree of curvature which is typical of moderately rolling topography. Very sinuous canals have a degree of curvature which is typical of canals in foothills or mountainous topography. Then, with these definitions, correction factors (see Table 7.6) can be defined.

Even given the limitations of the data available regarding tractive forces, this methodology is superior to the maximum permissible velocity method. In essence, at the level of basic principles, these two methods are analogous. For example, in the method of maximum permissible velocity, it is recognized that if the depth of flow is significantly different from approximately 3 ft (0.91 m), then a correction factor is required (Chow, 1959). Mehrotra (1983) demonstrated that the required corrective factor can be developed from "basic" prin-

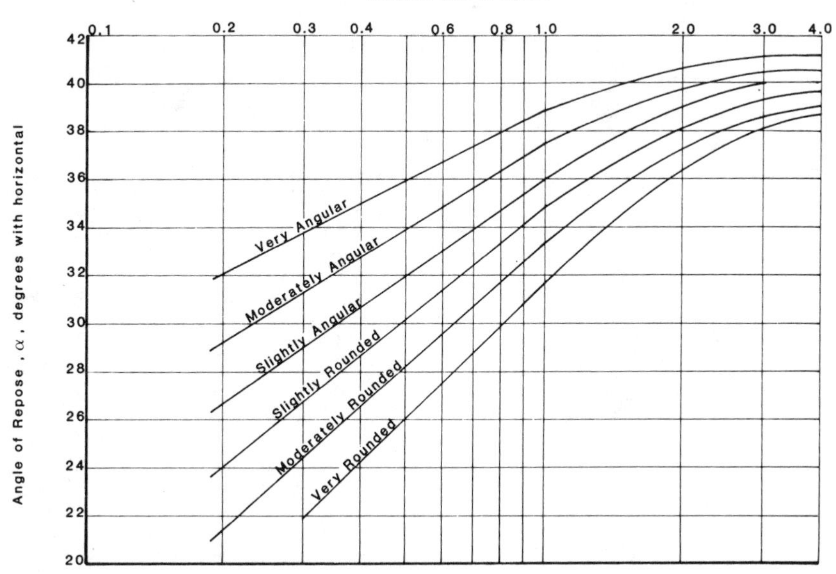

FIGURE 7.9 Angles of repose of noncohesive materials. *(Lane, 1955.)*

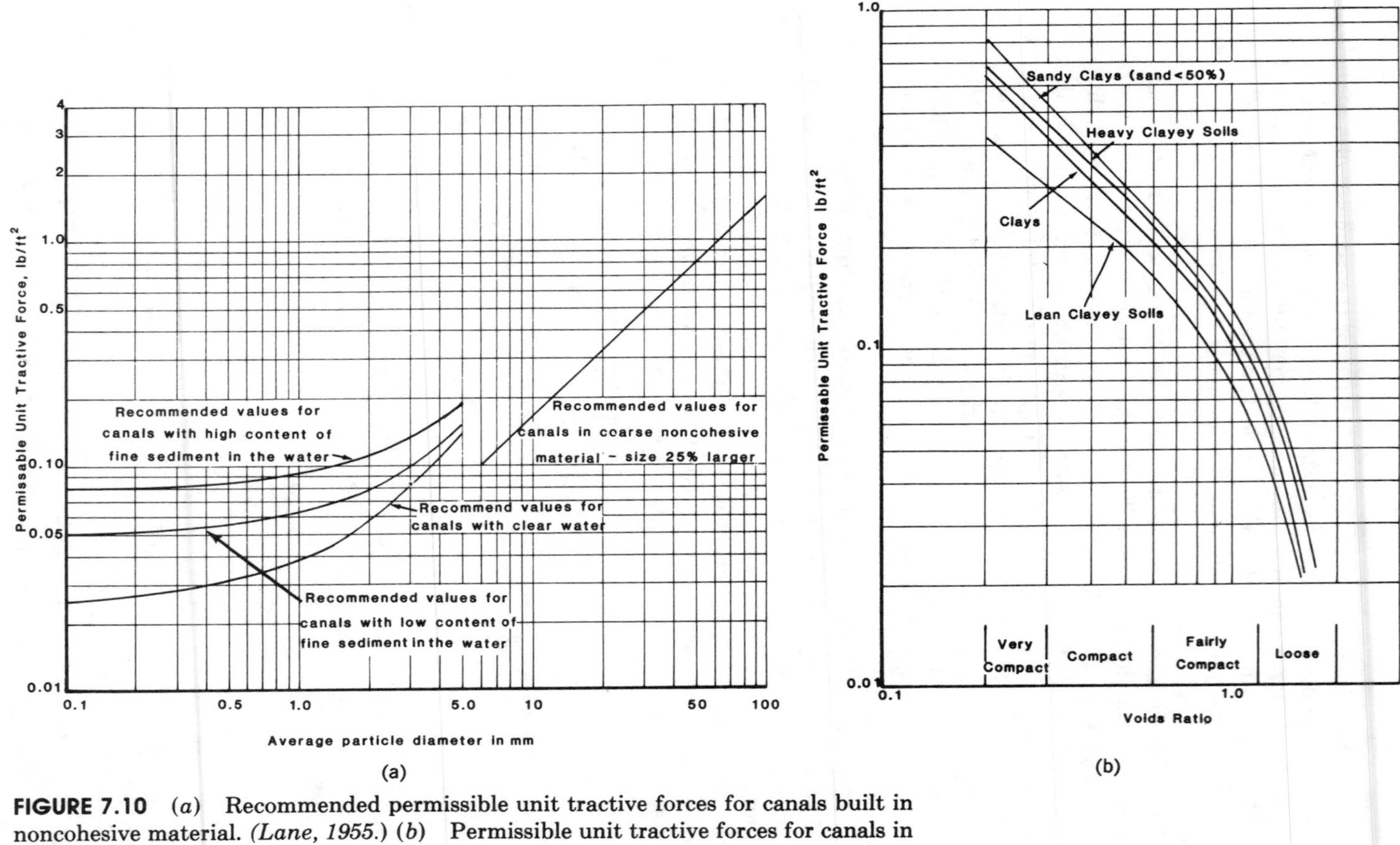

FIGURE 7.10 (*a*) Recommended permissible unit tractive forces for canals built in noncohesive material. *(Lane, 1955.)* (*b*) Permissible unit tractive forces for canals in cohesive materials. *(Chow, 1959.)*

TABLE 7.6 Comparison of maximum tractive forces for canals with varying degrees of sinuousness (Lane, 1955)

Degree of sinuousness	Relative limiting tractive force
Straight canals	1.00
Slightly sinuous canals	0.90
Moderately sinuous canals	0.75
Very sinuous canals	0.60

ciples. From Eq. (7.3.1) the critical tractive force on the channel boundary for a depth of flow y_1 is

$$\tau_c = \gamma R_1 S_1 \tag{7.3.8}$$

Where R_1 and S_1 are the hydraulic radius and slope, respectively, corresponding to the depth y_1. Then, for any other uniform depth of flow y_2 in a channel whose bed is composed of the same soil type, the critical tractive force is

$$\tau_c = \gamma R_2 S_2 \tag{7.3.9}$$

Since the same channel perimeter material is involved in both cases, the critical tractive forces must be equal, and hence

$$R_1 S_1 = R_1 S_2$$

or

$$\frac{R_1}{R_2} = \frac{S_2}{S_1} \tag{7.3.10}$$

Considering the Manning uniform flow equation for both situations, it can be demonstrated that

$$\frac{\overline{u}_2}{\overline{u}_1} = \left(\frac{R_2}{R_1}\right)^{2/3}\left(\frac{S_2}{S_1}\right)^{1/2}$$

and substituting Eq. (7.3.10),

$$\frac{\overline{u}_2}{\overline{u}_1} = \left(\frac{R_2}{R_1}\right)^{1/6} = k \tag{7.3.11}$$

The parameter k in Eq. (7.3.11) can be considered a correction factor which should be applied to the maximum permissible velocity (Table 7.5), if the uniform depth of flow y_2 is different from the depth of flow corresponding to the maximum permissible velocity. If $y_1 = 3$ ft (0.91 m) and the channel is wide, Eq. (7.3.11) becomes

$$k = \left(\frac{y_2}{3}\right)^{1/6} \tag{7.3.12}$$

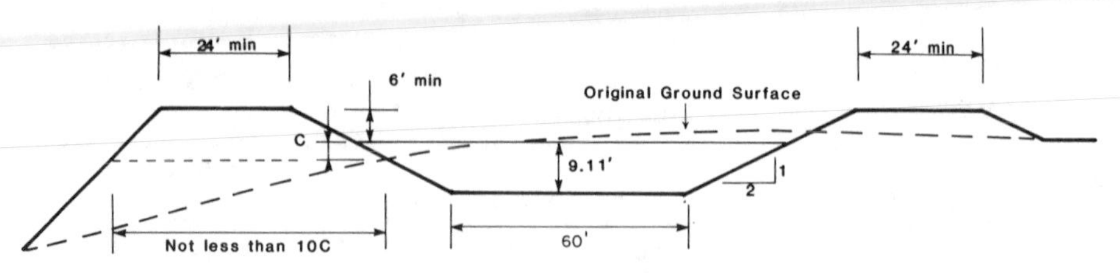

FIGURE 7.11 Typical earth section used on Coachella Canal, All-American Canal project.

However, even with these adjustments, it must be recognized that the method of maximum permissible velocity does not consider the tendency of the particles on the side slopes of a channel to roll down the slope under the influence of gravity. Many of the initial investigations of channel stability also noted these same design limitations. For example, Fortier and Scobey (1926) provided channel depth correction factors, and Houk (1926) noted that the maximum permissible velocity seemed to be directly proportional to the depth of flow.

At this point, it is appropriate to note that unlined earthen channels undergo a process known as aging; i.e., the material which composes the perimeter of the canal may be modified with the passage of time. Two examples of this process were noted by Fortier and Scobey (1926). In the Imperial Valley of California many of the large irrigation canals were initially excavated through sandy soils which would be eroded at a velocity of 2.5 ft/s (0.76 m/s). The extremely fine silt particles present in the water flowing in these canals have compacted and plastered these noncohesive sands with a coat of colloidal mud which can withstand velocities in excess of 5.0 ft/s (1.5 m/s). The coating of the original perimeter materials in this case not only reduced the scour potential but also reduced seepage losses by creating a less permeable perimeter. In the mountain valley canals of Colorado and Utah which were originally constructed in loamy gravel soils there are now very few traces of loam. The beds now consist of well-graded gravel which are compacted and arranged in such a way that a pavement is formed. It is noted that chemical interactions between the water being transported, the material composing the perimeter, and the sediment can also result in the armoring of a perimeter.

The tractive force method is the procedure recommended for designing unlined earthen canals. Table 7.7 summarizes some typical cross sections of unlined earthen canals used on irrigation projects in the United States, and a typical earthen section of the All-American Canal is shown in Fig. 7.11. A procedure for designing a stable, unlined earthen section is summarized in Table 7.8 where it is assumed that the design flow Q_D, the longitudinal slope of the channel, the relevant engineering properties of the material composing the channel perimeter, and the general shape of the section have been established prior to the initiation of the design process.

TABLE 7.7 Typical cross sections of unlined, earthen channels for selected canals in the western United States (Houk, 1956)

Canal	Project	Side slopes	Depth of flow, ft	Bottom width, ft	Ratio, width to depth	Mean velocity, ft/s	Discharge, ft^3/s	Freeboard ft*
Earth sections								
Lateral	Altus	1½:1	1.66	4	2.41	1.86	20	1.3
C Line East	Boise	1½:1	4.00	8	2.00	1.89	106	3.5
C Line East	Boise	1½:1	7.14	14	1.96	2.25	397	3.4
Altus	Altus	1½:1	6.20	20	3.23	2.48	450	3.5
Conchas	Tucumcari	1½:1	8.65	24	2.77	2.19	700	4.3
Kittitas Main	Yakima	1½:1	11.35	30	2.64	2.47	1,320	5.0
Ridge	Yakima	1½:1	9.57	40	4.18	2.50	1,300	4.0
Main (gravity extension)	Minidoka	1½:1	5.60	60	10.71	3.00	1,149	3.0
Coachella	All-American	2:1	10.33	60	5.81	3.00	2,500	6.0
Gravity Main	Gila	2:1	13.54	100	7.38	3.49	6,000	6.0
All-American	All-American	2:1	16.59	130	7.84	3.75	10,155	6.0
All-American	All-American	1¾:1	20.61	160	7.76	3.75	15,155	6.0
Rock sections								
North Unit Main	Deschutes	¼:1	5.54	20	3.61	4.63	550	2.5
Gravity Main	Gila	¾:1	21.57	33	1.53	1.79	1,900	Deep cut
Yuma	Yuma	½:1	8.46	60	7.10	3.68	2,000	Deep cut
All-American	All-American	¾:1	20.13	69	3.42	6.00	10,155	Deep cut
All-American	All-American	¾:1	22.75	94	4.13	6.00	15,155	Deep cut

*To top of bank.

TABLE 7.8 A design procedure for unlined, stable earthen channels

Step	Process
1	Estimate n or C for specified material composing the perimeter
2	Estimate angle of repose for channel perimeter material (Fig. 7.9)
3	Estimate channel sinuousness from the type of topography through which it will pass and determine the tractive force correction factor (Table 7.6)
4	Assume side slope angle (Table 7.2) and (bottom width)/(normal depth of flow)
5	Assume sides of the channel are the limiting factor in the channel design
6	Calculate the maximum permissible tractive force on sides in terms of the unit tractive force. Use the correction factor from Fig. 7.7*a* and the sinuousity correction factor (step 3)
7	Estimate tractive force ratio [Eq. (7.3.7)]
8	Estimate permissible tractive force on bottom (Fig. 7.10) and correct for sinousness (step 3)
9	Combine the results of steps 6 to 8 to determine the normal depth of flow y_N
10	Determine the bottom width with the results of steps 4 and 9
11	Compute Q and compare this value with the design flow Q_D, return to step 4, and repeat the design process with trial b/y ratios until $Q = Q_D$
12	Compare permissible tractive force on bottom (step 8) with actual tractive force given by $\gamma y_N S$ and corrected for shape (Fig. 7.7*a*)
13	Check 1. Minimum permissible velocity if the water carries silt and for vegetation 2. Froude number
14	Estimate required freeboard [Eq. (7.1.1)] or (Fig 7.1)
15	Summarize results with dimensioned sketch

EXAMPLE 7.3

A channel which is to carry 10 m³/s (350 ft³/s) through moderately rolling topography on a slope of 0.0016 is to be excavated in coarse alluvium with 25 percent of the particles being 3 cm (1.2 in) or more in diameter. The material which will compose the perimeter of this channel can be described as being moderately rounded. Assuming that the channel is to be unlined and of trapezoidal section, find suitable values of b and z.

Solution

Step 1 Estimate n from Table 4.8.

$$n = 0.025$$

Step 2 Estimate the angle of repose.

$$d_{25} = 3 \text{ cm} = \frac{3}{2.54} \text{ in} = 1.18 \text{ in}$$

From Fig. 7.9

$$\alpha = 34°$$

Step 3 Estimate channel sinuousness correction factor (Table 7.6).

$$C_s = 0.75$$

Step 4 Assume side slope 2:1 and $b/y_N = 4$.

Step 5 Assume side slopes are a limiting factor.

Step 6 Find maximum permissible tractive forces on sides (Fig. 7.7*a*).

$$\tau_s = 0.75\gamma y_N S$$

Step 7 Estimate tractive force ratio [Eq. (7.3.7)].

$$K = \frac{\tau_s}{\tau_b} = \sqrt{1 - \frac{\sin^2 \Gamma}{\sin^2 \alpha}}$$

$$\Gamma = \tan^{-1} (½) = 26.6°$$

$$K = \sqrt{1 - \frac{\sin^2 26.6°}{\sin^2 34°}} = 0.60$$

Step 8 Estimate permissible tractive force on bottom (Fig. 7.10).

$$\tau_b = 0.47 \text{ lb/ft}^2 \qquad \text{for } d_{25} = 1.18 \text{ in (30 mm)}$$

Correct for sinuousness.

$$\tau_b = C_s\tau_b = 0.75(0.47) = 0.35 \text{ lb/ft}^2 \text{ (17 N/m}^2\text{)}$$

Step 9 Estimate y_N.

$$\frac{\tau_s}{\tau_b} = K$$

$$\tau_s = K\tau_b$$

$$0.75\gamma y_N S = K\tau_b$$

$$y_N = \frac{0.60(17)}{0.75(9658)(0.0016)} = 0.88 \text{ m (2.9 ft)}$$

Step 10 $b/y_N = 4.$

$$b = 4(0.88) = 3.5 \text{ m (11 ft)}$$

Step 11 Determine Q.

$$A = (b + zy)y = [3.5 + 2(0.88)](0.88) = 4.6 \text{ m}^2 \text{ (50 ft}^2\text{)}$$

$$P = b + 2y\sqrt{1 + z^2} = 3.5 + 2(0.88)\sqrt{5} = 7.4 \text{ m (24 ft)}$$

$$R = \frac{A}{P} = \frac{4.6}{7.4} = 0.60 \text{ m (2.0 ft)}$$

$$Q = \frac{AR^{2/3}}{n}\sqrt{S} = \frac{4.6(0.60)^{2/3}}{0.025}\sqrt{0.0016} = 5.2 \text{ m}^3\text{/s (180 ft}^3\text{/s)}$$

Q is less than Q_D and, therefore, additional computations are required in which b/y_N is variable and C_s, K, permissible τ_b, and z are constant.

b/y_N	y_N, m	b, m	A, m^2	P, m	R, m	Q, m^3/s
5	0.88	4.4	5.4	8.3	0.65	6.5
8.25	0.88	7.3	8.0	11	0.71	10.2
8.15	0.88	7.2	7.9	11	0.71	10

Then, for $z = 2$ and $b/y_N = 8.15$, $y_N = 0.88$ m (2.9 ft), $b = 7.2$ m (24 ft), and $Q = 10$ m^3/s (350 ft^3/s).

Step 12 Check tractive force on bottom.

Permissible $\tau_b = C_s\tau_b = 17$ N/m^2 (see step 8)

Computed $\tau_b = 0.99\gamma y_N S$ (Fig. 7.7*b*)

$$\tau_b = 0.99(9658)(0.88)(0.0016) = 13 \text{ N/m}^2 \text{ (0.27 lb/ft}^2\text{)}$$

Since this is the actual, computed tractive force, the design is acceptable.

Step 13 Check velocity and Froude number.

$$\bar{u} = \frac{Q}{A} = \frac{10}{7.9} = 1.3 \text{ m/s (4.3 ft/s)}$$

This velocity should prevent vegetative growth and sedimentation.

$$\mathbf{F} = \frac{\overline{u}}{\sqrt{gD}} = 0.48$$

Therefore, this is a subcritical flow.

Step 14 Estimate required freeboard from Eq. (7.1.1).

Design depth of flow = 0.88 m (2.9 ft)

Design flow = 10 m^3/s (350 ft^3/s).

Estimate C in Eq. (7.1.1) as 1.6.

Then $$F = \sqrt{Cy} = \sqrt{1.6(2.9)} = 0.66 \text{ m } (2.2 \text{ ft})$$

The results of this design are summarized in Fig. 7.12.

It must be emphasized that the approach discussed in the foregoing section is not the only methodology which can be applied to the problem of designing stable channel sections. For example, the basic principles and results associated with the concept of threshold of movement can also be applied to this problem. Shields (1936) used an experimental approach to define the threshold of movement, and his results can be stated in terms of two dimensionless parameters:

$$\mathbf{R}_* = \frac{u_* d}{\nu} \tag{7.3.13}$$

and

$$F_s = \frac{\tau_o}{\gamma(S_s - 1)d} = \frac{u_*^2}{(S_s - 1)gd} \tag{7.3.14}$$

where $\mathbf{R}_*$ = a Reynolds number based on the shear velocity and the particle size (this number is also known as the particle Reynolds number)
u_* = shear velocity
ν = fluid kinematic viscosity
S_s = specific gravity of the particles composing the perimeter which is usually taken to be 2.65
d = diameter of the particles composing the perimeter of the channel

For conservative design d can usually be assumed to be the diameter of the particle of which 25 percent of all the particles, measured by weight, are larger. The results of Shields are usually summarized in a graphical form (Fig. 7.13) in

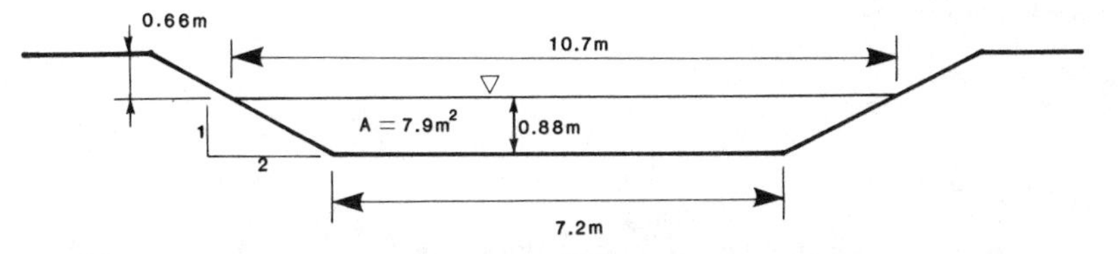

FIGURE 7.12 Summary of results for Example 7.3.

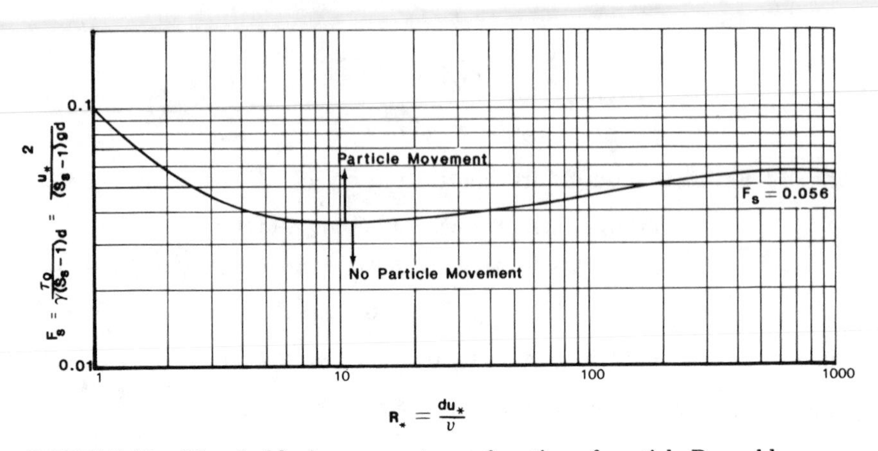

FIGURE 7.13 Threshold of movement as a function of particle Reynolds number. *(Shields, 1936.)*

which the curve delineates the threshold of movement. These results have been confirmed, in a general sense, by the theoretical results of White (1940) and the field results of Lane (1955) which are summarized in Fig. 7.10*a*.

If it is assumed that $S_s = 2.65$, then when $\mathbf{R}_*$ exceeds a value of 400, the perimeter particle size must be in excess of 0.25 in (0.006 m). In this case, the channel perimeter material can accurately be classified as coarse alluvium, and Eq. (7.3.14) becomes

$$\frac{\tau_o}{d\gamma(S_s - 1)} = 0.056 \tag{7.3.15}$$

where

$$\tau_o = \gamma RS$$

Therefore,

$$\frac{\gamma RS}{d(S_s - 1)} = \frac{RS}{d(2.65 - 1)} = 0.056$$

and

$$d \simeq 11RS \tag{7.3.16}$$

Equation (7.3.16) provides a simple method of estimating the size of the material which will remain at rest in a channel of specified R and S. Note, *for values of d less than 0.25 in (0.006 m) Eq. (7.3.16) is not valid*, but the curve in Fig. 7.9 can be used to find appropriate values of F_s and analogous relations can be developed for these sizes.

EXAMPLE 7.4

In the previous example, the specified slope was 0.0016 and 25 percent of the channel perimeter particles were 3 cm (1.2 in) or more in diameter. Use the results summarized in Fig. 7.10 to show that the design arrived at in this example represents a conservative design.

Solution

Assume $S_s = 2.65$ and that $R \simeq y_N$ for wide channels. Under these assumptions, Eq. (7.3.16) becomes for level surfaces

$$\tau = \frac{\gamma d}{11}$$

From the previous example, the tractive force ratio is

$$\frac{\tau_s}{\tau_b} = K = \sqrt{1 - \frac{\sin^2 \Gamma}{\sin^2 \alpha}} = 0.60$$

Substituting $\tau_b = C_s \left(\frac{\gamma d}{11}\right)$ and $\tau_s = 0.75\gamma y_N S$ in the above equation yields

$$\frac{0.75\gamma y_N S}{C_s(\gamma d/11)} = 0.60$$

or the maximum value of y_N for a stable channel is

$$y_N = \frac{C_s d(0.60)}{11(0.75)S} = \frac{0.75(3/100)(0.60)}{11(0.75)(0.0016)} = 1.0 \text{ m } (3.3 \text{ ft})$$

Thus, the previous computation yields a conservative result for y_N. It should be noted that this result provides a check on the validity of the solution obtained by the recommended design methodology.

The Stable Hydraulic Section

The permissible tractive force design techniques presented in the previous section yield a channel cross section in which the tractive force is equal to the permissible value on only a part of the wetted perimeter, usually the sides. It seems logical to attempt to define a channel cross section such that the condition of incipient particle motion prevails at all points of the channel perimeter. The equations defining this channel section were developed by the U.S. Bureau of Reclamation (Glover and Florey, 1951) for erodible channels carrying clear water through noncohesive materials, and this methodology yields what is known as a stable hydraulic section of maximum hydraulic efficiency.

The assumptions required to develop the equations defining the stable hydraulic section are:

1. The soil particles are held in place in the channel by the component of the submerged weight of the particle acting normal to the bed.

2. At and above the water surface, the channel side slope is the angle of repose of the noncohesive material under the action of gravity.
3. At the centerline of the channel, the side slope is zero, and the tractive force alone is sufficient to produce a state of incipient particle motion.
4. At points between the center and edge of the channel, the particles are kept at a state of incipient motion by the resultant of the tractive force and the gravity component of the submerged weight of the particles.
5. The tractive force acting on an area of the channel is equal to the component of the weight of the water above the area acting in the direction of flow. If this assumption is valid, then there is no lateral transfer of tractive force.

   It is noted that with the exception of assumptions 2 and 3, these are the same assumptions which were used in the previous section.

Consider a channel of longitudinal slope S with the side slope being defined at any point in the section (x, y) by the angle Γ (Fig. 7.14). Then, by assumption 5, the critical tractive stress acting on area AB in a unit length of channel is

$$\tau_s = \frac{\gamma y S\, dx}{\sqrt{(dx)^2 + (dy)^2}} = \gamma y S \cos \Gamma \tag{7.3.17}$$

From the previous development, the critical tractive force acting on the side of the channel is given by

$$\tau_s = K\tau_b = \gamma y_N S \cos \Gamma \sqrt{1 - \frac{\tan^2 \Gamma}{\tan^2 \alpha}} \tag{7.3.18}$$

where $\tau_b = \gamma y_N S$ is the tractive force at the centerline of the channel where the depth of flow is y_N. Combining Eqs. (7.3.17) and (7.3.18) and solving for y yields

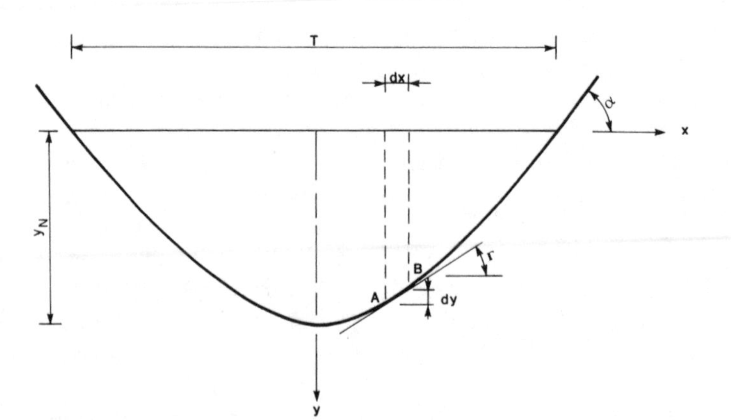

FIGURE 7.14 Schematic definition of parameters for stable hydraulic section.

$$y = \frac{y_N}{\tan \alpha} \sqrt{\tan^2 \alpha - \tan^2 \Gamma} \qquad (7.3.19)$$

Substituting $dy/dx = \tan \Gamma$ into Eq. (7.3.19) yields a differential equation that defines the shape of the cross section

$$\left(\frac{dy}{dx}\right)^2 + \left(\frac{y}{y_N}\right)^2 \tan^2 \alpha - \tan^2 \alpha = 0 \qquad (7.3.20)$$

Given the condition that at $x = 0$, $y = y_N$, the solution of Eq. (7.3.20) is

$$y = y_N \cos\left(\frac{x \tan \alpha}{y_N}\right) \qquad (7.3.21)$$

An alternate form of Eq. (7.3.21) can be obtained by noting that at $x = T/2$, $y = 0$. This condition can be satisfied only if

$$\frac{T \tan \alpha}{2y_N} = \frac{\pi}{2} \qquad (7.3.22)$$

or

$$y_N = \frac{T \tan \alpha}{\pi} \qquad (7.3.23)$$

Then Eq. (7.3.21) becomes

$$y = y_N \cos\left(\frac{\pi x}{T}\right) \qquad (7.3.24)$$

Equations (7.3.21) and (7.3.24) define the channel section which has the smallest width and the largest hydraulic radius for a given area. Thus, this channel has the greatest hydraulic efficiency of all stable, unlined, earthen channels built through noncohesive materials at a longitudinal slope S and carrying clear water.

The flow area of the channel defined by Eqs. (7.3.21) and (7.3.24) is

$$A = 2 \int_0^{T/2} y \, dx = 2y_N \int_0^{T/2} \cos\left(\frac{x \tan \alpha}{y_N}\right) dx = \frac{2Ty_N}{\pi} \qquad (7.3.25)$$

and the wetted perimeter

$$P = 2 \int_0^{T/2} \sqrt{1 + \left(\frac{dy}{dx}\right)^2} \, dx = \frac{2y_N}{\sin \alpha} E(\sin \alpha) \qquad (7.3.26)$$

where $E(\sin \alpha)$ = a complete elliptic integral of the second kind. $E(\sin \alpha)$ can be evaluated either by standard mathematical tables or by

$$E(\sin \alpha) = \frac{\pi}{2}\left[1 - \left(\frac{1}{2}\right)^2 \sin^2 \alpha - \left(\frac{1 \cdot 3}{2 \cdot 4}\right)^2 \frac{\sin^4 \alpha}{3} - \left(\frac{1 \cdot 3 \cdot 5}{2 \cdot 4 \cdot 6}\right)^2 \frac{\sin^6 \alpha}{5} - \cdots\right] \qquad (7.3.27)$$

The discharge of the channel can then be computed by the Manning equation

$$Q = \frac{2.98 y_N^{8/3} (\cos\alpha)^{2/3} \sqrt{S}}{n(\tan\alpha)[E(\sin\alpha)]^{2/3}} \tag{7.3.28}$$

The discharge Q is the flow that would be obtained in a channel designed for the greatest efficiency in a given noncohesive material at a specified longitudinal slope. If the design discharge Q_D is larger or smaller than Q, then the channel defined by Eqs. (7.3.21) and (7.3.24) must be modified. If $Q_D > Q$, then additional flow area must be provided; however, the maximum depth of flow can be no greater than the stability depth y_N since an increase in depth would result in an increased tractive force and instability. Therefore, a rectangular section is added at the center of the theoretical section (Fig. 7.15). The additional width required, T', is found by a trial-and-error solution of the Manning equation or

$$Q_D = \frac{1.49}{n}\sqrt{S}\left\{\frac{[(2y_N^2/\tan\alpha) + T'y_N]^{5/3}}{[2y_N E(\sin\alpha)/(\sin\alpha) + T']^{2/3}}\right\} \tag{7.3.29}$$

If $Q_D < Q$, then economy dictates that a portion of the theoretical section must be removed (Fig. 7.16). The width of the channel which must be removed, T'', is also found by a trial-and-error solution of the Manning equation

$$Q = \frac{1.49}{n}\sqrt{S}\,\frac{\{(2y_N^2/\tan\alpha)[\sin(T\tan\alpha/2y_N) - \sin(T''\tan\alpha/2y_N)]\}^{5/3}}{(2y_N/\tan\alpha)E[\sin\alpha, (\pi/2)(1 - T''/T)]} \tag{7.3.30}$$

where $E[\sin\alpha, (\pi/2)(1 - T''/T)]$ is an incomplete, elliptic integral of the third kind. T'' can be estimated by assuming that the mean velocity in the theoretical

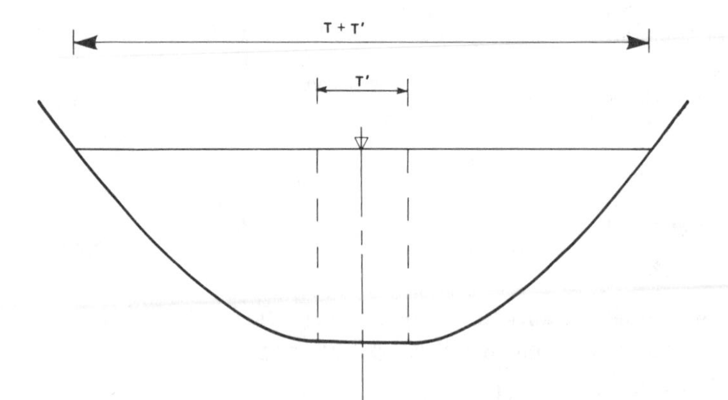

FIGURE 7.15 Definition sketch for stable hydraulic section where $Q < Q_D$.

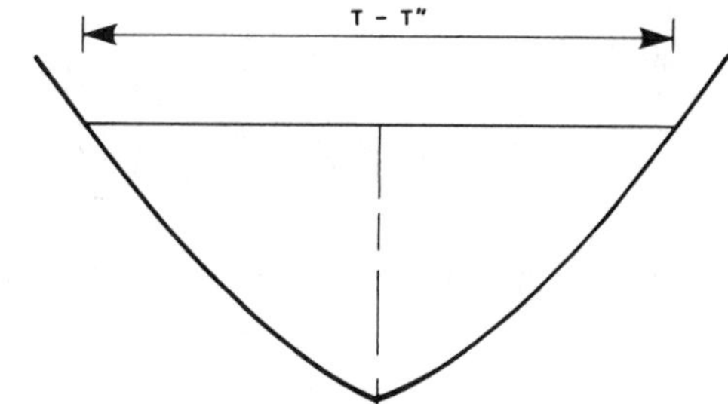

FIGURE 7.16 Definition sketch for stable hydraulic section when $Q > Q_D$.

section and the adjusted section are equal and thus that the discharges are proportional to the flow areas or

$$Q = \frac{2y_N^2}{\tan\alpha}\,\overline{u} = \frac{2T^2\tan\alpha}{\pi^2}\,\overline{u} \tag{7.3.31}$$

and

$$Q_D = \frac{2(T - T'')^2\tan\alpha}{\pi^2}\,\overline{u} \tag{7.3.32}$$

Combining Eqs. (7.3.31) and (7.3.32) results in an equation which can be used to estimate T''

$$T'' = T\left(1 - \sqrt{\frac{Q_D}{Q}}\right) \tag{7.3.33}$$

EXAMPLE 7.5

Design the stable section of greatest hydraulic efficiency for a canal which is to be constructed through a noncohesive material on a slope of 0.0004. Preliminary testing of the natural material which will compose the channel perimeter indicates that $\tau_o = 0.10$ lb/ft² (4.8 N/m²), $n = 0.02$, and $\alpha = 31°$. The design flow is 300 ft³/s (8.5 m³/s).

Solution

The depth of flow can be determined from Eq. (7.3.2) where it is assumed that $y_N = R$.

$$\tau_o = \gamma y_N S$$

$$y_N = \frac{\tau_o}{\gamma S} = \frac{0.10}{62.4(0.0004)} = 4.0 \text{ ft } (1.2 \text{ m})$$

The shape of the channel is obtained by combining Eqs. (7.3.23) and (7.3.24)

$$y_1 = y_N \cos\left(\frac{x\tan\alpha}{y_N}\right) = 4.0\cos\left(\frac{x\tan 31°}{4}\right) = 4.0\cos(0.15\,x)$$

and
$$T = \frac{\pi y_N}{\tan \alpha} = \frac{4\pi}{\tan 31°} = 21 \text{ ft } (6.4 \text{ m})$$

$$A = \frac{2Ty_N}{\pi} = \frac{2(21)(4)}{\pi} = 53 \text{ ft}^2 \ (4.9 \text{ m}^2)$$

The discharge of this channel can then be found from Eq. (7.3.28).

$$Q = \frac{2.98 y_N^{8/3} (\cos \alpha)^{2/3} \sqrt{S}}{n \tan \alpha [E(\sin \alpha)]^{2/3}} = \frac{2.98(4)^{8/3}(\cos 31°)^{2/3}\sqrt{0.0004}}{0.02(\tan 31°)(1.46)^{2/3}}$$
$$= 140 \text{ ft}^3/\text{s} \ (4.0 \text{ m}^3/\text{s})$$

Since $Q < Q_D$, a trial-and-error solution of Eq. (7.3.29) for T' is required.

$$Q_D = \frac{1.49}{n}\sqrt{S}\left\{\frac{[(2y_N^2/\tan \alpha) + T'y_N]^{5/3}}{[(2y_N/\sin \alpha)E(\sin \alpha) + T']^{2/3}}\right\}$$

$$\frac{nQ_D}{1.49\sqrt{S}} = \frac{[(2y_N^2/\tan \alpha) + T'y_N]^{5/3}}{[(2y_N/\sin \alpha)E(\sin \alpha) + T']^{2/3}}$$

$$\frac{0.02(300)}{1.49\sqrt{0.0004}} = \frac{[2(4)^2/\tan 31° + 4T']^{5/3}}{\{[2(4)(1.46)/\sin 31°] + T'\}^{2/3}}$$

$$201 = \frac{(53.3 + 4T')^{5/3}}{(22.7 + T')^{2/3}}$$

Trial T', ft	$(53.3 + 4T')^{5/3}/(22.7 + T')^{2/3}$
5	140
10	188
11.5	203

Therefore, the top width required to convey this flow is (21 + 11.5) = 32 ft (9.8 m). The results of this design are summarized in Fig. 7.17.

Channel Seepage Losses

Although a channel may require lining for many reasons (Section 7.2), one of the primary reasons for lining a channel constructed in materials which would otherwise not require lining is seepage losses. The loss of water due to seepage from an unlined channel depends on a variety of factors including, but not limited to, the dimensions of the channel, the gradation of the materials composing the perimeter, and the groundwater conditions. Although a number of attempts to theoretically estimate the seepage from a channel have been made (e.g., Bouwer, 1965), direct measurement of seepage loss is still preferred. There are basically three direct methods of measuring seepage loss:

1. In an existing channel, lined or unlined, selected reaches of the channel may be isolated by dikes to form closed basins of known volume. A mass balance will then suffice to estimate the seepage loss. Since this method usually requires that the channel be removed from service for an extended period of time, these tests are normally performed in the "off season." In such a case, care must be taken to ensure that the losses measured are typical of the season of interest.

2. If a very careful record of the inflow and outflow to a reach of channel is kept, seepage loss may be estimated from this record. In this method, the canal is not removed from service, but the accuracy of the method is less than that of the previously described ponding method.

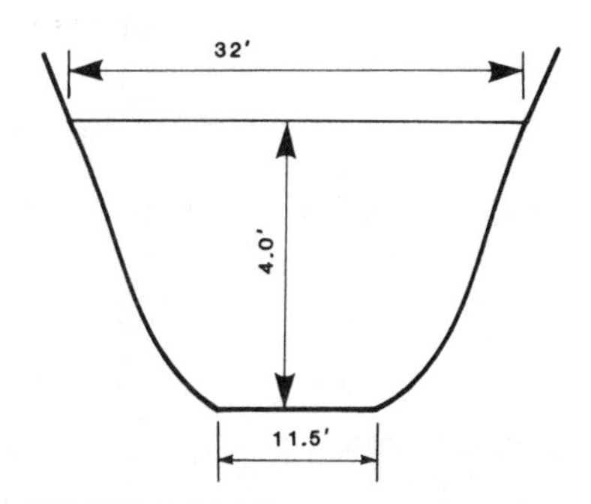

FIGURE 7.17 Summary of results for Example 7.5.

3. In the case of a proposed channel, test reaches of the channel may be constructed and the ponding method used.

A fourth method which is relatively simple but reliable is based on historical measurements. In Table 7.9, a set of values originally developed by Etcheverry and Harding in 1933 are summarized (Davis and Sorenson 1969). The values in this table are the result of many field measurements and have been found to be reasonably accurate; however, it is recommended that these values should be used only as a design guide given a specific site.

A fifth method of estimating the seepage from an unlined or partially lined channel involves the solution of the relevant porous media equations for an appropriate set of boundary conditions. Subramanya et al. (1973) examined two cases of seepage from partially lined channels. In this investigation the following assumptions were made. First, the lining was assumed to be impervious and its thickness negligible. Second, the porous material beneath the channel was assumed to be isotropic, homogeneous, and of infinite depth. Third, capillary action was assumed absent.

The first situation considered by these investigators was a channel in which the sides were lined but the bottom was unlined (Fig. 7.18*a*). In Fig. 7.19 the results are graphically summarized. In this figure, q = total seepage loss per unit length of channel, K = coefficient of permeability of material underlying the channel, β = angle made by the sides of the channel with the horizontal in

TABLE 7.9 Seepage losses for canals not affected by the groundwater table (Davis and Sorenson, 1969)

Perimeter material	Seepage loss (ft^3 of water)/(ft^2 of perimeter) for a 24-h period
Impervious clay loam	0.25–0.35
Medium clay loam underlain with hardpan at depth of not over 2–3 ft below bed	0.35–0.50
Ordinary clay loam, silt soil, or lava ash loam	0.50–0.75
Gravelly clay loam or sandy clay loam, cemented gravel, sand, and clay	0.75–1.00
Sandy loam	1.00–1.50
Loose sandy soils	1.50–1.75
Gravelly sandy soils	2.00–2.50
Porous gravelly soils	2.50–3.00
Very gravelly soils	3.00–6.00

radians, and all other variables are defined in Fig. 7.15*a*. With regard to Fig. 7.19 the following observations can be made. First, for a specified value of b/y, the effect of β on the seepage is negligible. Second, when $\beta = 0$, the solution applies to seepage from a shallow ditch.

The second situation treated by Subramanya et al. (1973) was a channel in which the sides were unlined and the bottom was lined (Fig. 7.18*b*). Depending on the relative width of the bottom of the channel, three cases can arise: (1) The relative width of the bottom is such that the two phreatic lines do not meet, (2) the relative width of the bottom is such that the two phreatic lines meet somewhere beneath the channel, and (3) the relative width of the bottom is such that the two phreatic lines meet on the bottom of the channel itself. Of these three cases, Subramanya et al. (1973) solved only the first, although they noted that the other cases could also be solved by analytic techniques. Figure 7.20 is a plot of h/y versus β/π which allows a specified seepage problem to be classified as either case 1, 2, or 3. Figure 7.21 summarizes the solution for case 1 seepage. In this figure, it is noted that the slope of the channel sides has an important and significant effect on q.

Figure 7.22 shows the variation in the percentage of seepage reduction as a function of b/y and β/π for partial channel lining. In developing this figure, Subramanya et al. (1973) calculated the unlined seepage loss q_u by the method of Vedernikov (see, for example, Palubarinova-Kochina, 1962). The dashed lines in this figure indicate the probable variation in the percentage seepage reduction for cases 2 and 3 when only a bottom lining is used. It is noted that

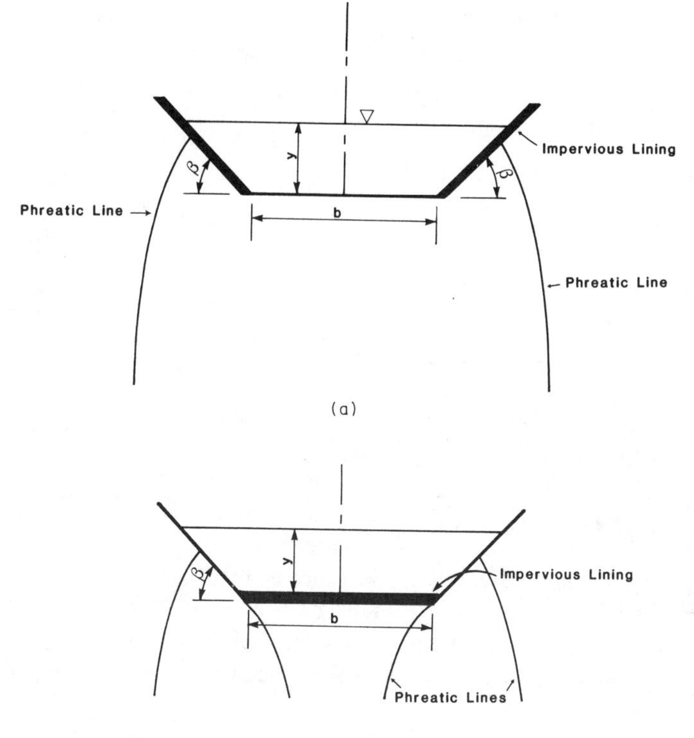

FIGURE 7.18 (*a*) Seepage from a channel with lined sides and an unlined bottom. (*b*) Seepage from a channel with a lined bottom and unlined sides.

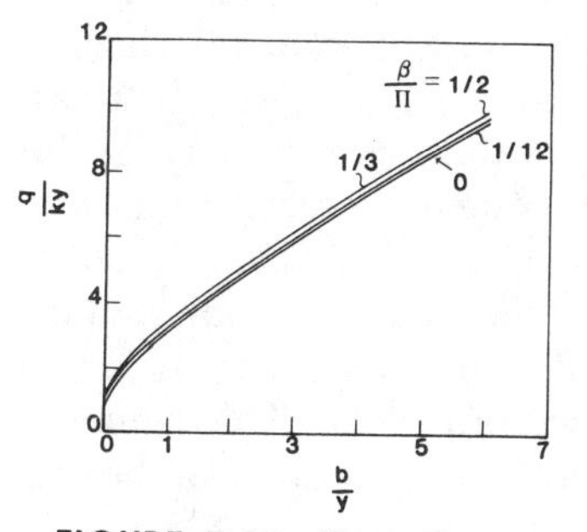

FIGURE 7.19 Variation of q/ky as a function of b/y and β/π for channels with the sides lines. *(Subramanya et al., 1973.)*

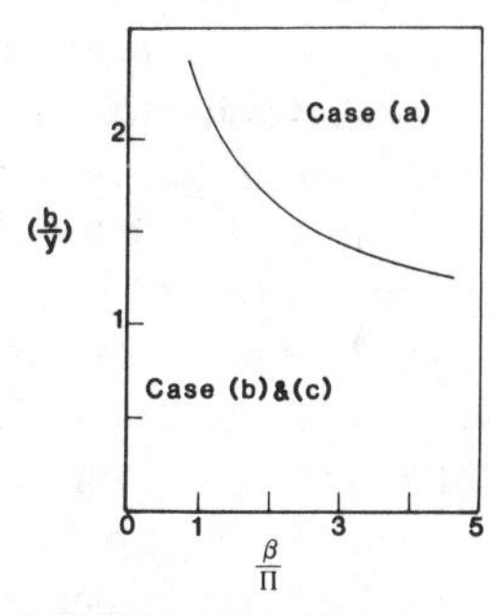

FIGURE 7.20 Graph to determine appropriate case for channel with bottom lining. *(Subramanya et al., 1973.)*

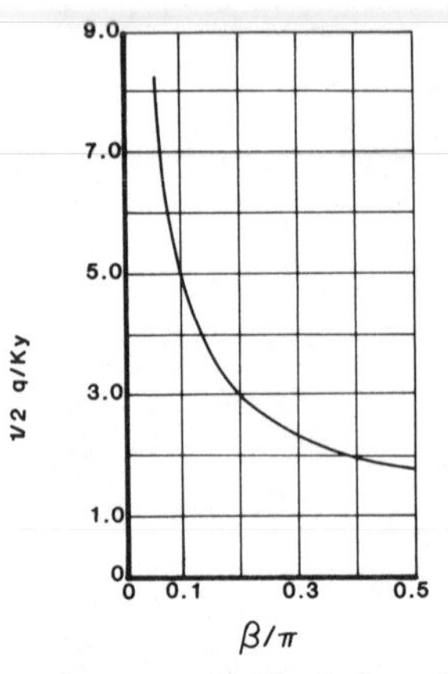

FIGURE 7.21 Variation of 1/2 (q/ky) for case 1; side lining. *(Subramanya et al., 1973.)*

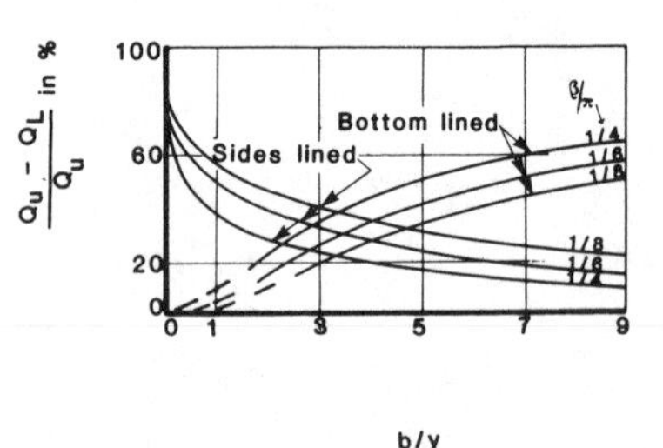

FIGURE 7.22 Variation in the percentage of seepage reduction by bottom or side lining. *(Subramanya et al., 1973.)*

for a specified value of β/π there is a value of b/y which yields equal reduction in the seepage independent of the type of partial lining. For values of b/y greater than this value, a bottom lining is more effective in reducing seepage, while for values of b/y less than this value a side lining is more effective.

7.4 DESIGN OF CHANNELS LINED WITH GRASS

The grass-lined channel is a common method of transmitting intermittent irrigation flows and controlling erosion in agricultural areas. The grass serves to stabilize the body of the channel, consolidate the soil mass of the channel perimeter, and check the movement of soil particles along the channel bottom. However, grass-lined channels cannot, in general, withstand prolonged inundation and wetness, and their design presents a number of problems which have not been encountered in the preceding sections of this chapter, e.g., the seasonal variation in the resistance coefficient due to the condition of the channel liner.

The Manning roughness coefficient, which is commonly termed the *retardance coefficient* in the literature relating to the grass-lined channel, has been found to be a function of the average velocity, the hydraulic radius, and the vegetal type (Coyle, 1975). n can be represented by a series of empirical curves of n versus $\bar{u}R$ for various degrees of retardance (Fig. 7.23). This figure must be used in conjunction with Table 7.10 which provides an estimate of the degree of retardance which various types of grass provide.

The selection of a grass for a specific application depends primarily on the climate and soil conditions which prevail. From the viewpoint of hydraulic engineering, the primary consideration must be channel stability. Table 7.11 summarizes the U.S. Soil Conservation Service recommendations regarding per-

TABLE 7.10 Classification of degree of retardance for various types of grass (Coyle, 1975)

Retardance	Cover	Condition
A	Reed canary grass	Excellent stand, tall (average 36 in)
	Yellow bluestem *Ischaemum*	Excellent stand, tall (average 36 in)
B	Smooth bromegrass	Good stand, mowed (average 12 to 15 in)
	Bermuda grass	Good stand, tall (average 12 in)
	Native grass mixture (little bluestem, blue grama, and other long and short midwest grases)	Good stand, unmowed
	Tall fescue	Good stand, unmowed (average 18 in)
	Lespedeza sericea	Good stand, not woody, tall (average 19 in)
	Grass-legume mixture—timothy, smooth	Good stand, uncut (average 20 in)
	Tall fescue, with bird's foot trefoil or lodino	Good stand, uncut (average 18 in)
	Blue grama	Good stand, uncut (average 13 in)
C	Bahia	Good stand, uncut (6 to 8 in)
	Bermuda grass	Good stand, mowed (average 6 in)
	Redtop	Good stand, headed (15 to 20 in)
	Grass-legume mixture—summer (orchard grass, redtop, Italian ryegrass, and common *Lespedeza*)	Good stand, uncut (6 to 8 in)
	Centipede grass	Very dense cover (average 6 in)
	Kentucky bluegrass	Good stand, headed (6 to 12 in)

TABLE 7.10 Classification of degree of retardance for various types of grass (Coyle, 1975) (*Continued*)

Retardance	Cover	Condition
D	Bermuda grass	Good stand, cut to 2.5-in height
	Red fescue	Good stand, headed (12 to 18 in)
	Buffalo grass	Good stand, uncut (3 to 6 in)
	Grass-legume mixture—fall, spring (orchard grass, redtop, Italian ryegrass, and common *Lespedeza*)	Good stand, uncut (4 to 5 in)
	Lespedeza sericea	After cutting to 2-in height; very good stand before cutting
E	Bermuda grass	Good stand, cut to 1.5-in height;
	Bermuda grass	burned stubble

missible velocities for various types of vegetal covers, channel slopes, and soil types. In addition, the following guidelines should be noted:

1. Where only sparse vegetal cover can be established or maintained, velocities should not exceed 3 ft/s (0.91 m/s).
2. Where the vegetation must be established by seeding, velocities in the range of 3 to 4 ft/s (0.91 to 1.2 m/s) are permitted.
3. Where dense sod can be developed quickly or where the normal flow in the channel can be diverted until a vegetal cover is established, velocities of 4 to 5 ft/s (1.2 to 1.5 m/s) can be allowed.
4. On well-established sod of good quality, velocities in the range of 5 to 6 ft/s (1.5 to 1.8 m/s) are permitted.
5. Under very special conditions, velocities as high as 6 to 7 ft/s (1.8 to 2.1 m/s) are permitted.

The design of grass-lined channels must in most cases proceed in two stages. In the first stage, a low degree of retardance, which corresponds to either dormant seasons or the period during which the vegetation is becoming established, is assumed. The second stage provides the pertinent dimensions under the assumption of a high degree of retardance. A recommended design procedure is summarized in Table 7.12 where it is assumed that the longitudinal slope of the channel, the channel shape, grass type, and the design flow have been established previous to the initiation of the design process. The channel shapes commonly used in this application are trapezoidal, triangular, and parabolic, with the latter two shapes being the most popular. Coyle (1975) has developed extensive design tables for the parabolic shape. The designer must, in designing grass-lined channels, be conscious that a primary consideration may be the ability of farm machinery to easily cross the canal during periods of no flow. This consideration may require that the side slopes of the channel be designed for this prupose rather than hydraulic efficiency or stability.

TABLE 7.11 Permissible velocities for channels lined with grass (Coyle, 1975)

Cover	Slope range,† %	Permissible velocity*	
		Erosion-reresistant soils, ft/s	Easily eroded soils, ft/s
Bermuda grass	0–5	8	6
	5–10	7	5
	Over–10	6	4
Bahia Buffalo grass Kentucky bluegrass Smooth brome Blue grama Tall fescue	0–5 5–10 Over–10	7 6 5	5 4 3
Grass mixtures Reed canary grass	0–5† 5–10	5 4	4 3
Lespedeza sericea Weeping lovegrass Yellow bluestem Redtop Alfalfa Red fescue	0–5‡	3.5	2.5
Common *Lespedeza*§,¶ Sudan grass§	0–5	3.5	2.5

*Use velocities exceeding 5 ft/s only where good covers and proper maintenance can be obtained.

†Do not use on slopes steeper than 10% except for vegetated side slopes in combination with a stone, concrete, or highly resistant vegetative center section.

‡Do not use on slopes steeper than 5% except for vegetated side slopes in combination with a stone, concrete, or highly resistant vegetative center section.

§Annuals—use on mild slopes or as temporary protection until permanent covers are established.

¶Use on slopes steeper than 5% is not recommended.

TABLE 7.12 A design procedure for channels lined with grass

Step	Process
	Stage I
1	Assume a value of n and determine $\overline{u}R$ which corresponds to this assumption (Fig. 7.23).
2	Select the permissible velocity from Table 7.11 which corresponds to the specified channel slope, lining material, and soil and compute a value of R by using the results of step 1.
3	Using the Manning equation and the assumed value of n, compute $$\overline{u}R = \frac{\phi R^{5/3}\sqrt{S}}{n}$$ where $\phi = 1.49$ if English units are used and $\phi = 1$ for SI units. The value of R found in step 2 is used in the right-hand side of this equation.
4	Repeat steps 1 through 3 until the values of $\overline{u}R$ determined in steps 1 and 3 agree
5	Determine A from the design flow and the permissible velocity (step 2).
6	Determine the channel proportions for the calculated values of R and A.
	Stage II
1	Assume a depth of flow for the channel determined in stage I and compute A and R.
2	Compute the average velocity $$\overline{u} = \frac{Q}{A}$$ for the A found in step 1.
3	Compute $\overline{u}R$ using the results of steps 1 and 2.
4	Use the results of step 3 to determine n from Fig. 7.23.
5	Use n from step 4, R from step 1, and the Manning equation to compute $\overline{u}$.
6	Compare the average velocities computed in steps 2 and 5 and repeat steps 1 through 5 until they are approximately equal.
7	Add the appropriate freeboard and check the Froude number.
8	Summarize the design in a dimensioned sketch.

EXAMPLE 7.6

Design a triangular channel which will be lined with a mixture of grasses that includes the following types: orchard grass, redtop, Italian ryegrass, and common *Lespedeza*. This channel will be built on a slope of 0.025 through a soil which can be characterized as easily eroded. An intermittent flow of 20 ft^3/s (0.57 m^3/s) is anticipated.

Solution

For a grass mixture composed of orchard grass, redtop, Italian ryegrass, and common *Lespedeza*, minimum retardance occurs during the fall, winter, and spring (curve D, Fig. 7.23). For this type of lining, the specified slope, and the soil, Table 7.11 provides an estimated maximum permissible velocity of 4 ft/s (1.2 m/s). A trial-and-error procedure must then be used to proportion the channel for stage 1 of the design.

Trial no.	Assumed n	$\bar{u}R$ from Fig. 7.23	R	$\bar{u}R = (1.49R^{5/3}\sqrt{S})/n$
1	0.04	3.0	0.75	3.65
2	0.045	2.0	0.50	1.65
3	0.043	2.5	0.62	2.5

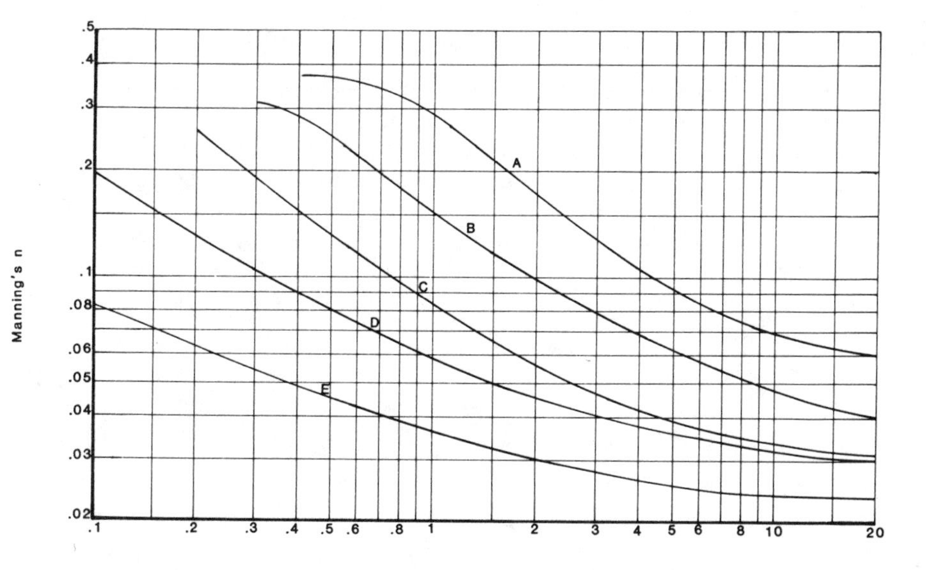

FIGURE 7.23 Manning's n as a function of velocity, hydraulic radius, and vegetal retardance. *(Coyle, 1975.)*

Therefore, $n = 0.43$ and $R = 0.62$ ft (0.19 m). The required flow area is computed by

$$A = \frac{Q}{\bar{u}} = \frac{20}{4} = 5 \text{ ft}^2 \ (0.46 \text{ m}^2)$$

Given a triangular channel, these conditions provide for an explicit solution; i.e.,

$$A = zy^2 = 5 \text{ ft}^2$$

$$R = \frac{A}{2y\sqrt{1+z^2}} = \frac{5}{2y\sqrt{1+z^2}} = 0.62$$

Solving these equations yields

$$z = 2.8$$

and

$$y_N = 1.3 \text{ ft } (0.40 \text{ m})$$

At this point the stage 1 design must be reviewed from the viewpoint of maximum retardance (curve C, Fig. 7.23). Thus, the stage II depth

Trial no.	Assumed y, ft	A, ft^2	R, ft	$\bar{u} = Q/A$, ft/s	$\bar{u}R$	n from Fig. 7.23	$\bar{u} = (1.49R^{2/3}\sqrt{S})/n$, ft/s
1	1.4	5.5	0.66	3.6	2.4	0.051	3.5
2	1.5	6.3	0.71	3.2	2.3	0.054	3.5
3	1.45	5.9	0.68	3.4	2.3	0.054	3.4

of flow is 1.45 ft (0.44 m), and the freeboard may be estimated from Eq. (7.1.1) to be 1.5 ft (0.46 m). If this is an on-farm irrigation ditch which is used intermittently, a smaller freeboard, for example, 1.0 ft (0.30 m), would be acceptable. The results are summarized in Fig. 7.24.

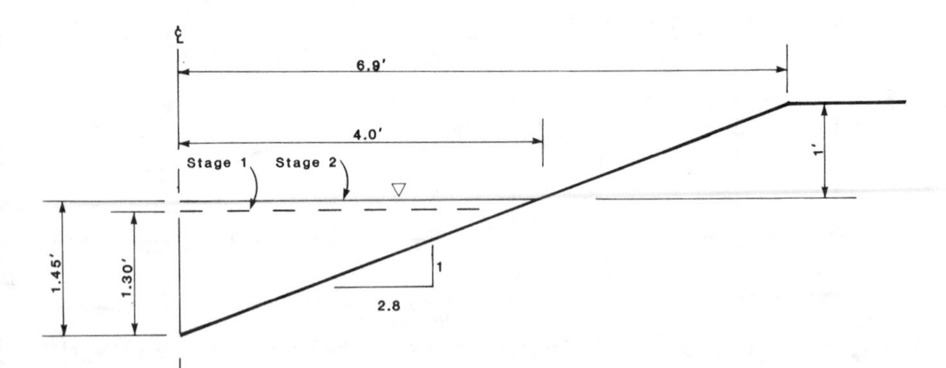

FIGURE 7.24 Summary of results for Example 7.6.

In many cases, the center of a grass-lined channel is lined with either gravel or rock to enhance the stability of the channel and/or to provide adequate drainage. Although Coyle (1975) presents a nomograph for determining the size of rock which should be used for a specified flow and channel slope, the threshold of movement concepts discussed in previous sections of this chapter can also be used.

EXAMPLE 7.7

A trapezoidal irrigation ditch 8 ft (2.4 m) wide with grass-covered sides must carry prolonged flows of 100 ft^3/s (2.8 m^3/s) on a slope of 0.01. If the side slopes are 2:1 and Manning's n is estimated to be 0.03, determine the size of gravel which should be used to line the center of this channel.

Solution

Determine the normal depth of flow.

$$Q = \frac{1.49}{n} AR^{2/3}\sqrt{S}$$

$$AR^{2/3} = \frac{nQ}{1.49\sqrt{S}} = \frac{0.03(100)}{1.49\sqrt{0.01}} = 20.1$$

$$y_N = 1.6 \text{ ft } (0.49 \text{ m})$$

Then

$$A = (b + zy)y = [8 + 2(1.6)]1.6 = 17.9 \text{ ft}^2 \ (1.7 \text{ m}^2)$$

$$P = b + 2y\sqrt{1 + z^2} = 8 + 2(1.6)\sqrt{5} = 15.2 \text{ ft } (4.6 \text{ m})$$

$$R = 1.18 \text{ ft } (0.36 \text{ m})$$

From Eq. (7.3.16)

$$d_{25} \simeq 11\ RS = 11(1.18)(0.01) = 0.13 \text{ ft} = 1.6 \text{ in } (0.04 \text{ m})$$

This is the 25 percent diameter, and since 1.6 in > 0.25 in, the use of Eq. (7.3.16) is valid. Table 4.2 also demonstrates that the assumed value of n is reasonable.

At this point it is relevant to note that there have been a number of studies regarding flow in grass-lined channels whose results should be mentioned. Kouwen et al. (1973) used flexible plastic strips in a laboratory environment to study the effects of the waving motion and bending of grass in channels on the flow characteristics. The pertinent results of this study were these:

1. Grass-lined channels are apparently characterized by two values of the local friction factor: one for the erect and waving regimes and another for the prone regimes.

2. The friction factor and, hence, Manning's n were found to be functions of the relative roughness for the erect and waving regimes but were a function of the Reynolds number or $\overline{u}R$ for the prone regime.

Phelps (1970) examined the effects of vegetative density and the depth of flow on the characteristics of the flow. The conclusions of this study were as follows:

1. The dissimilarity between the boundaries of flows over vegetation which is only partially submerged is significant and must be considered.
2. Recall that for smooth surfaces an actual Reynolds number of approximately 2000 marks the limit of laminar flow. In the case of flows over turf the critical Reynolds number varies as a function of the depth of flow.

Given the complex nature of vegetated surfaces, it is not surprising that the design of such channels is based on empiricism. It is clear that additional research in this topic area is required.

BIBLIOGRAPHY

Anonymous, "Linings for Irrigation Canals," U.S. Bureau of Reclamation, Washington, 1963.

Bouwer, H., "Theoretical Aspects of Seepage from Open Channels," *Proceedings of the American Society of Civil Engineers, Journal of the Hydraulics Division,* vol. 91, no. HY3, May 1965, pp. 37–59.

Chow, V. T., *Open Channel Hydraulics,* McGraw-Hill Book Company, New York, 1959.

Coyle, J. J., "Grassed Waterways and Outlets," *Engineering Field Manual,* U.S. Soil Conservation Service, Washington, April 1975, pp. 7-1–7-43.

Davis, C., and Sorenson, K. E., "Canals and Conduits," *Handbook of Applied Hydraulics,* 3d Ed., McGraw-Hill Book Company, New York, 1969, pp. 7-1–7-34.

duBoys, P., "The Rhone and Streams with Moveable Beds," *Annales des ponts et chaussees,* section 5, vol. 18, 1879, pp. 141–195.

Fortier, S., and Scobey, F. C., "Permissible Canal Velocities," *Transactions of the American Society of Civil Engineers,* vol. 89, 1926, pp. 940–984.

Glover, R. E., and Florey, Q. L., "Stable Channel Profiles," *Hydraulic Laboratory Report No. Hyd-325,* U.S. Bureau of Reclamation, Washington, September 1951.

Houk, I. E., "Discussion of Permissible Canal Velocities," *Transations of the American Society of Civil Engineers,* vol. 89, 1926, pp. 971–977.

Houk, I. E., *Irrigation Engineering,* vol. II, John Wiley & Sons, New York, 1956.

Kouwen, H., and Unny, T. E., "Flexible Roughness in Open Channels," *Proceedings of the American Society of Civil Engineers, Journal of the Hydraulics Division,* vol. 99, no. HY5, May 1973, pp. 713–728.

Lane, E. W., "Design of Stable Channels," *Transactions of the American Society of Civil Engineers,* vol. 120, 1955, pp. 1234–1279.

Mehrotra, S. C., "Permissible Velocity Correction Factors," *Proceedings of the American Society of Civil Engineers, Journal of Hydraulic Engineering,* vol. 109, no. 2, February 1983, pp. 305–308.

Palubarinova-Kochina, P. Y. A., *Theory of Groundwater Movement* (translated by J. M. R. DeWiest), Princeton University Press, Princeton, N.J. 1962.

Phelps, H. O., "The Friction Coefficient for Shallow Flows over a Simulated Turf Surface," *Water Resources Research,* vol. 6, no. 4, August 1970, pp. 1220–1226.

Shields, A., "Anwendung der Aehnlichkeitsmechanik und der Turbulenzforschung auf Geschiebebegung" (Application of Similarity Principles and Turbulence Research to Bed-Load Movement), *Mitteilungen der Preuss Versuchsanst für Wasserbau und Schiffbau,* No. 26, Berlin, 1936. (Available in translation by W. P. Ott and J. C. van Uchelen, Soil Conservation Service Cooperative Laboratory, California Institute of Technology, Pasadena.)

Shukry, A., "Flow around Bends in an Open Flume," *Transactions of the American Society of Civil Engineers,* vol. 115, 1950, pp. 751–788.

Streeter, V. L., and Wylie, E. B., *Fluid Mechanics,* McGraw-Hill Book Company, New York, 1975, pp. 592–595.

Subramanya, K., Madhav, M. R., and Mishra, G. C., "Studies on Seepage from Canals with Partial Lining," *Proceedings of the American Society of Civil Engineering, Journal of the Hydraulics Division,* vol. 99, no. HY12, December 1973, pp. 2333–2351.

Trout, T. J., "Channel Design to Minimize Lining Material Costs," *Proceedings of the American Society of Civil Engineers, Journal of the Irrigation and Drainage Division,* vol. 108, no. IR4, December 1982, pp. 242–249.

White, C. M., "Equilibrium of Grains on the Bed of a Stream," *Proceedings of the Royal Society of London,* Series A, vol. 174, 1940, pp. 322.

Willison, R. J., "USBR's Lower-Cost Canal Lining Program," *Proceedings of the American Society of Civil Engineers, Journal of the Irrigation and Drainage Division,* vol. 84, no. IR2, April 1958, pp. 1589-1–1589-29.

Woodward, S. M., "Hydraulics of the Miami Flood Control Project," *Miami Conservancy District Technical Report,* part VII, Dayton, Ohio, 1920.

Woodward, S. M., and Posey, C. J., *Hydraulics of Steady Flow in Open Channels,* John Wiley & Sons, New York, 1941.

EIGHT

Flow Measurement

SYNOPSIS

In this chapter, a few of the methods and devices available for estimating and/or gaging the flow in an open channel are considered. Section 8.2 discusses some of the devices available and a procedure which can be used to determine the flow rate in a channel by direct measurement of the velocity. The moving-boat method of discharge estimation is also discussed in this section. Sections 8.3 to 8.5 discuss a few of the structures which can be used for discharge measurement. Detailed consideration is given to weirs, critical depth (Parshall) flumes, and culverts.

8.1 INTRODUCTION

As has been noted in Chapter 1, water flowing in open channels serves humans in many ways, and an accurate record of discharge is crucial in each of these uses. Further, floods cause such extensive damage that records of extreme events are needed for the design of bridges, culverts, flood control reservoirs, flood plain delineation, and flood warning systems. In this chapter, a number of methodologies for measuring the discharge of an open channel either in the field or laboratory are discussed. Given the rapid advance of technology, the primary emphasis of this chapter is analytical rather than a description of available hardware.

The selection of a gaging site is usually dictated by the needs of water management personnel. Thus, in many cases, discharge records must be obtained at sites which, from a hydraulic engineering viewpoint, are far from ideal. In selecting a permanent discharge gaging site, especially in a natural channel, Carter and Davidian (1968) recommended that consideration be given to the following so that the flow can be simply related to the stage:

1. The channel should be geometrically stable. For example, a rock riffle or waterfall would indicate an ideal location since in such a situation there should be a rather stationary relationship between the depth of flow and the discharge. If a site with a movable bed must be used, as uniform a reach of channel as possible should be selected.

2. The channel should be examined from the viewpoint of establishing an artificial control section such as a critical-depth flume or a weir.

3. The possibility of the site being affected by a varied flow profile from downstream tributaries, dams, tidal backwater, or other sources must be considered. If such a site must be used, then a uniform reach should be located so that the slope of the water surface can be measured.

4. Near the gaging site there should be a cross section where good discharge measurements can be made to establish a stage-discharge relationship.

5. The possibility of the flow bypassing the gage location either in flood channels or as groundwater must be considered.
6. The availability and proximity of power and telephone lines to the site must be considered.
7. The gaging site should be accessible by roads even during floods.
8. If a permanent gaging station is to be established, then the gage site must be located properly with respect to the section at which the discharge is measured and the position of the channel which controls the stage-discharge relationship.
9. Existing structures for use in measuring extreme flow events should be available.

The above criteria define an ideal gaging site which is seldom found in nature; however, they provide a rational method of comparing alternative gaging sites in a channel reach.

Carter and Davidian (1968) also note four major considerations in establishing artificial control sections for the measurement of discharge. These considerations are:

1. The structure establishing the control section should not create undesirable flow field disturbances either above or below the control.
2. The structure should have sufficient height to eliminate the effects of variable downstream conditions.
3. The structure should be designed so that a small change in discharge at the low stage results in a measurable change in the stage. Further, the relationship between the stage and discharge should be such that it can be extended with accuracy to peak discharges.
4. The structure establishing the control should be stable, and permanence under extreme conditions ensured.

In subsequent sections of this chapter, a variety of artificial structures such as weirs, flumes, and culverts will be discussed.

Discharge measurements must also be performed in the laboratory. The primary difference between field and laboratory determinations of discharge is that in the controlled environment of a laboratory much more sophisticated measuring systems can be used. The principles used in the field determination of discharge are equally applicable to the laboratory.

8.2 DEVICES AND PROCEDURES FOR STREAM GAGING

The calibration of a functional relationship between stage and discharge requires accurate measurement of the discharge at a number of different stages.

One method of discharge measurement at a cross section involves dividing the cross section into a number of subsections, determining the average velocity in each subsection, and summarizing the products of the subsection areas and average velocities or

$$Q = \sum_{i=1}^{N} \overline{u}_i A_i \tag{8.2.1}$$

where A_i = area of subsection i
$\overline{u}_i$ = average velocity in subsection i
N = number of subsections

The most common current meter in use in the United States is the Price current meter which consists of six conical cups rotating about a vertical axis. Each complete revolution of the cups closes a battery-operated circuit and an audible click is emitted in the headphones worn by the operator. In shallow water, a similar device termed a *pygmy Price meter* can be used. In general, the Price meter overestimates the velocity; however, if the measuring station is carefully chosen, i.e., there should be a minimum of turbulence and the current should be nearly parallel to the channel axis, then the error is probably not more than 2 percent.

For the Price meter, the relation between the revolutions per unit time and the velocity is

$$u = a + bN \tag{8.2.2}$$

where u = velocity, N = number of revolutions per unit time, and a and b are calibration coefficients. If u has units of feet per second and N is in revolutions per second, then a is approximately 0.1 and b is approximately 2.2. Some variation in these constants should be expected owing to the effects of wear and manufacturing irregularities. Thus, each meter must be subjected to a calibration procedure periodically.

A second type of meter is known as a *propeller meter*. In this meter, there is a propeller which rotates around a horizontal axis. Similar to the Price meter, the rotation of the propeller causes a circuit to close and results in a click being heard by the operator. The disadvantage of the propeller meter is that the bearings of the propeller cannot be protected against the intrusion of sediment-laden water. In the case of the Price meter such protection is normally achieved by trapping air around the bearings. However, the propeller meter is not subject to rotation by vertical currents which the Price meter is.

A third type of current meter is one which is based on the electromagnetic principle. When a fluid which is conductive moves through a magnetic field at a 90° angle, an electromotive force is induced in the fluid at right angles to both the flux of the magnetic field and the velocity of the fluid. The voltage induced by the movement of the fluid through the magnetic field is proportional to the average velocity of flow. Many modern current meters are based on this principle; in general, these meters are superior in accuracy and efficiency to those described above.

Once a current meter has been selected and a cross section for discharge measurement established, the cross section is divided into a number of vertical sections (Fig. 8.1). In general, no one vertical section should include more than 10 percent of the total flow; thus 20 to 30 vertical sections are typically required (Linsley et al., 1958). In addition, the number of velocity determinations made in a cross section must be limited to those which can be made within a reasonable length of time. This is particularly critical if the stage of the flow is changing rapidly. Figure 8.1 shows a typical division of a cross section into subsections.

Although the average velocity in any one of the subsections may be determined by a variety of methods, one based on velocity measurements at 0.2 and 0.8 of the depth is in widespread use. From Chapter 1 recall that the vertical velocity profile is given by

$$u = 2.5u_* \ln \frac{y}{y_o} \tag{8.2.3}$$

where u = turbulent average longitudinal velocity
y = distance above bottom boundary
u_* = shear velocity
y_o = constant of integration related to bottom surface roughness

The average velocity in the longitudinal direction is given approximately by

$$\bar{u} = \frac{1}{D - y_o} \int_{y_o}^{D} \left(2.5u_* \ln \frac{y}{y_o} \right) dy$$

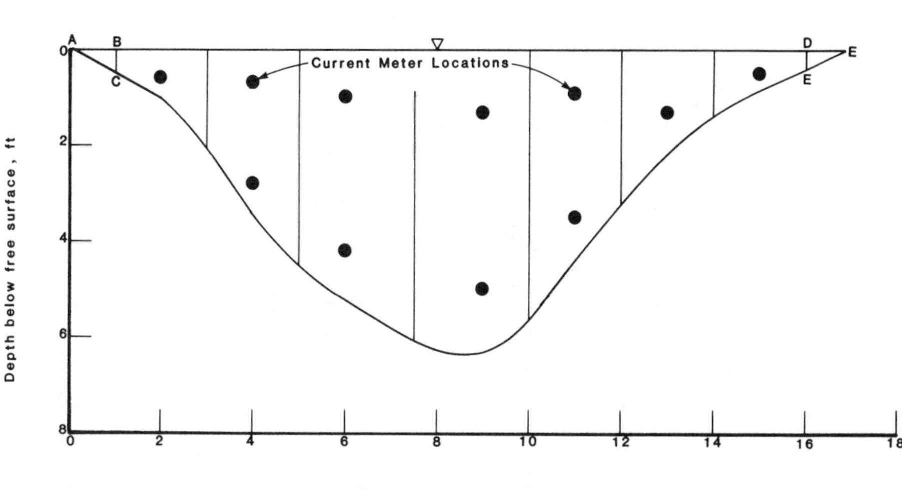

FIGURE 8.1 Discharge determination by current meter measurement.

or

$$\bar{u} = 2.5u_* \ln \frac{D}{y_o} - 2.5u_* \tag{8.2.4}$$

where D = depth of flow. From Eqs. (8.2.3) and (8.2.4) the velocity defect law is formed

$$\bar{u} - u = 2.5u_* \left[\ln \left(\frac{D}{y_o} \right) - 1 \right] - 2.5u_* \ln \frac{y}{y_o}$$

or

$$\frac{\bar{u} - u}{u_*} = 2.5 \ln \frac{D}{y} - 2.5$$

and rearranging

$$u = -2.5u_* \ln \frac{D}{y} + 2.5u_* + \bar{u} \tag{8.2.5}$$

It is then asserted that the average velocity in a vertical section is given by

$$\bar{u} = \frac{u(0.8D) + u(0.2D)}{2} \tag{8.2.6}$$

where $u(0.8D)$ = velocity measured at 0.8 the depth and $u(0.2D)$ = velocity measured at 0.2 the depth. The assertion contained in Eq. (8.2.6) can be proved by first inserting $0.8D$ for y in Eq. (8.2.5) or

$$u(0.8D) = -0.558u_* + 2.5u_* + \bar{u} \tag{8.2.7}$$

and then inserting $0.2D$ for y in Eq. (8.2.5) or

$$u(0.2D) = -4.023u_* + 2.5u_* + \bar{u} \tag{8.2.8}$$

Adding Eqs. (8.2.7) and (8.2.8) and dividing by 2 yields

$$\frac{u(0.2D) + u(0.8D)}{2} = -2.29u_* + 2.5u_* + \bar{u} \tag{8.2.9}$$

Then, if Eq. (8.2.6) is valid, from Eq. (8.2.9) we have

$$2.29u_* = 2.5u_*$$

which proves that the assertion expressed mathematically by Eq. (8.2.6) is approximately correct. In a similar fashion, it can be demonstrated that the velocity at 0.6 depth below the surface is nearly equal to the average velocity in the vertical. Thus, a procedure for determining the average velocity in a subsection of the cross section shown in Fig. 8.1 is

1. The total depth of flow is determined by sounding with the current meter cable or wading rod.

2. Raise the current meter to a position which is 0.8 the total depth and measure the velocity.
3. Raise the current meter to 0.2 the total depth and determine the velocity.
4. Determine the average velocity of this section by Eq. (8.2.6).

In shallow water, a single velocity determination at 0.6 the depth may be used.

EXAMPLE 8.1

If the data in Table 8.1 are velocity data for the cross section in Fig. 8.1, estimate the discharge.

Solution

The U.S. Geological Survey recommends estimating the discharge at a cross section by using a midsection method for computing the subsection cross-sectional areas. This method assumes that the average velocity measured in each subsection should be associated with an area which extends laterally from half the distance from the preceding current meter location to half the distance to the next current meter location and vertically from the water surface to the sounded depth. In numerical terms, this is integration using a rectangular rule. If this scheme of integration is used, then Eq. (8.2.1) becomes

$$Q = \sum_{i=1}^{N} \left(\frac{b_i - b_{i-1}}{2} + \frac{b_{i+1} - b_i}{2} \right) y_i \overline{u}_i$$

TABLE 8.1 Streamflow data for Example 8.1

Distance from bank to current meter, ft	Depth, ft	Meter depth, ft*	Velocity, ft/s
2	1.0	0.6	0.54
4	3.5	2.8	0.98
		0.7	1.6
6	5.2	4.2	1.3
		1.0	1.6
9	6.3	5.0	1.3
		1.3	1.8
11	4.4	3.5	1.5
		0.9	1.7
13	2.2	1.3	1.1
15	0.8	0.5	0.64
17	0		

*Distance from free surface to current meter.

where b_i = distance from an initial point to current meter position i. The results using this scheme of computation are summarized in Table 8.2. With regard to this solution, note that the flow in subsections ABC and DEF is assumed to be zero and that more than 10 percent of the total flow passes through a number of subsections.

A second method of obtaining a solution to this problem is to use a trapezoidal method of area integration. For example, the width of a subsection i is given by

$$w_i = \frac{b_i - b_{i-1}}{2} + \frac{b_{i+1} - b_i}{2}$$

then

$$Q = \sum_{i=1}^{N} w_i \bar{u}_i \left(\frac{d_{i-1/2} + d_{i+1/2}}{2} \right)$$

where $d_{i-1/2}$ = depth of flow at one-half the distance between current meter locations i and $i - 1$ and $d_{i+1/2}$ = depth of flow at one-half the distance between current meter locations i and $i + 1$. The results using this method are summarized in Table 8.3.

Stream-gaging procedures used by the U.S. Geological Survey are described in detail in Corbett et al. (1943).

The foregoing methods for estimating the discharge at a cross section are standardized and well known; however, in many cases of importance, e.g., large rivers and estuaries, they are impractical or involve costly and complex procedures. The U.S. Geological Survey has developed a technique termed the "moving-boat method" which overcomes these difficulties and yields results which are claimed to be within 5 percent of the results obtained by conventional methods when the width of the channel exceeds 1000 ft (300 m) (Smoot and Novak, 1969).

TABLE 8.2 Solution of Example 8.1 by midsection method

Subsection width, ft	Subsection depth, ft	Subsection average velocity, ft/s	Subsection discharge, ft^3/s	Cumulative discharge, ft^3/s
1.0	—	0	0	—
2.0	1.0	0.54	1.1	1.1
2.0	3.5	1.3	9.1	10.0
2.5	5.2	1.4	18	28
2.5	6.3	1.6	25	53
2.0	4.4	1.6	14	67
2.0	2.2	1.1	4.8	72
2.0	0.8	0.64	1.0	73
1.0	—	0	0	73

TABLE 8.3 Solution of Example 8.1 using trapezoidal rule integration

Distance from bank to current meter, ft	2	4	6	9	11	13	15
Distance from bank to subsection beginning, ft	1	3	5	7.5	10	12	14
Distance from bank to subsection end, ft	3	5	7.5	10	12	14	16
Subsection width, ft	2	2	2.5	2.5	2	2	2
Depth at beginning subsection, ft	0.5	2	4.5	6	5.5	3	1.5
Depth at end of subsection, ft	2	4.5	6	5.5	3	1.5	0.5
Subsection area, ft^2	2.5	6.5	13	14	8.5	4.5	2
Velocity at 0.8 depth, ft/s	0.54*	0.98	1.3	1.3	1.5	1.1*	0.64*
Velocity at 0.2 depth, ft/s	—	1.6	1.6	1.8	1.7	—	—
Subsection average velocity, ft/s	0.54	1.3	1.4	1.6	1.6	1.1	0.64
Subsection discharge, ft^3/s	1.3	8.4	18	22	14	5	1.3
Cumulative discharge, ft^3/s	—	9.7	28	50	64	69	70

*Velocity measured at 0.6 depth.

The moving-boat method uses a small, maneuverable boat equipped with a sonic sounder to measure depth and a horizontal axis current meter. A typical situation is shown schematically in Fig. 8.2*a*. With reference to Fig. 8.2*b*, v is the velocity measured by the current meter on the boat, v_b is the velocity of the boat in a direction perpendicular to the flow, u is the velocity of flow, and α is the angle measured by the angle indicator on the boat. Then, if α and v are specified,

$$u = v \sin \alpha \tag{8.2.10}$$

In addition,

$$L_b = \int v \cos \alpha \, dt \tag{8.2.11}$$

where L_b = the distance traveled by the boat along its true course provided that the velocity of flow is perpendicular to this course and t = time. In general, α can be assumed to be relatively constant between any two consecutive points of measurement, and under this assumption Eq. (8.2.11) becomes

$$L_b = \cos \alpha \int v \, dt \tag{8.2.12}$$

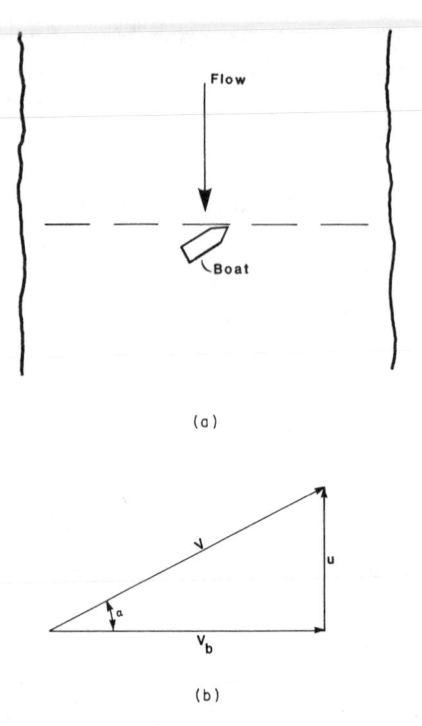

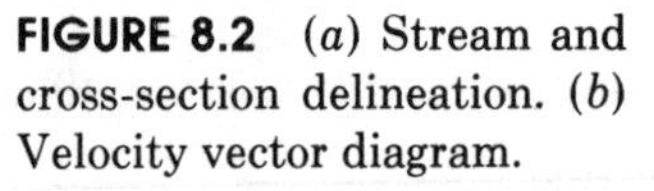

FIGURE 8.2 (*a*) Stream and cross-section delineation. (*b*) Velocity vector diagram.

The equipment developed by the U.S. Geological Survey is calibrated to determine this distance automatically.

Thus, as the boat moves across the stream, a continuous record of the bottom boundary is made, and the velocity of flow is measured at a predetermined number of points across the cross section. From these data, the total discharge can be estimated in the fashion demonstrated in Example 8.1. There are two corrections which usually have to be made when this method is used. First, Eq. (8.2.12) is based on the assumption that a right-triangle relationship exists between the velocity vectors in the analysis. If this is not the case, then the total width of the river determined by Eq. (8.2.12) will either be greater than or less than the measured width. Since in practice only the total width of the cross section is measured, only the total flow area and total flow can be corrected. Define

$$K_T = \frac{T_M}{T_C} \tag{8.2.13}$$

where K_T = width-area adjustment factor
T_M = measured top width of cross section
T_C = top width of cross section computed by Eq. (8.2.12)

K_T is then used to adjust both the total flow area and discharge. Second, the current meter mounted on the boat is set to measure the velocity at a constant depth below the water surface—usually 3 to 4 ft (0.91 to 1.2 m). Thus, the average velocity of flow in the vertical dimension is not measured. Therefore, a velocity correction coefficient must be computed. This is accomplished by developing vertical velocity curves at several locations across the cross section using conventional methods (current meters). In doing this, a velocity measurement should be made at the same depth as the setting on the meter on the boat. Define

$$K_u = \frac{\overline{u}}{u_B} \tag{8.2.14}$$

where K_u = mean in vertical velocity adjustment factor
$\overline{u}$ = mean velocity in vertical
u_B = observed velocity at depth of 3 or 4 ft (0.91 to 1.2 m)

In general, if a representative velocity correction coefficient is to be obtained, K_u must be established at several strategically placed verticals which represent a major portion of the flow. Once a representative value of K_u has been obtained at a cross section, it is not necessary to reestablish it every time the discharge is estimated. However, the validity of this coefficient should be established over a reasonable range of stages.

EXAMPLE 8.2

Using the moving-boat method of determining discharge, it is estimated that the cross section is 470 m (1540 ft) wide and has a flow area of 3900 m^2 (42,000 ft^2). The discharge is estimated as 4400 m^3/s (155,000 ft^3/s). If the actual width is 480 m (1574 ft) and K_u was previously determined to be 0.91, estimate the actual area and discharge.

Solution

The actual or corrected flow area is given by first determining the width-area adjustment factor or

$$K_T = \frac{T_M}{T_C} = \frac{480}{470} = 1.02$$

Then the adjusted flow area is

$$A = 3900\ (1.02) = 3978\ \text{m}^2\ (42{,}800\ \text{ft}^2)$$

The flow rate is also adjusted or

$$Q = 4400\ (1.02) = 4488\ \text{m}^3/\text{s}\ (158{,}000\ \text{ft}^3/\text{s})$$

The flow rate must also be multiplied by the vertical velocity adjustment factor or

$$Q = 4488\ (0.91) = 4084\ \text{m}^3/\text{s}\ (144{,}000\ \text{ft}^3/\text{s})$$

Thus, the correct flow area is 3978 m^2 (42,800 ft^2), and the correct discharge is 4084 m^3/s (144,000 ft^3/s).

8.3 STRUCTURES FOR FLOW MEASUREMENT: WEIRS

In many cases it is advantageous to emplace a structure in a channel to measure the flow rate. In some cases, it is possible to use structures placed in a channel for other purposes to estimate the flow rate. Among the structures whose primary function is the estimation of the flow rate are weirs and critical-depth flumes. Structures whose primary function is other than flow measurement, but which can be utilized for discharge estimation, are culverts, bridge piers, and dams.

In general, the dimensions of structures whose primary function is flow measurement are standardized; but the materials from which these devices are constructed may vary. The criteria on which the choice of construction material depends includes availability, cost of labor, lifetime of the structure, and prefabrication. Although the cost of construction and maintenance of a structure for the estimation of discharge is important, the primary concerns are the ease and accuracy with which discharge can be measured.

By definition, a weir is a notch of regular form through which water flows. Weirs are classified in accordance with the shape of the notch, e.g., V notch, rectangular, trapezoidal, and parabolic. The equation for the discharge over a weir cannot be derived exactly because: (1) the flow pattern of one weir differs from another of a different shape, and (2) the flow pattern varies with the discharge. In the following development, the effects of gravity are included, but the effects of viscosity, surface tension, the nature of the weir crest, the velocity distribution in the approach channel, the roughness of the weir channel, and the dimensions of the approach channel are not explicitly considered.

Broad-Crested Weirs

A broad-crested weir is by definition a structure with a horizontal crest above which the fluid pressure may be considered hydrostatic (Fig. 8.3). If such a situation is to exist, then, with reference to Fig. 8.3, the following inequality must be satisfied (Bos, 1976)

$$0.08 \leq \frac{H_1}{L} \leq 0.50 \tag{8.3.1}$$

If H_1/L is not greater than or equal to 0.08, then the energy losses over the weir crest cannot be neglected. If H_1/L is not less than or equal to 0.50, then the curvature of the streamlines over the weir block is such that the assumption of hydrostatic pressure is not valid. In Table 8.4, the discharge equations for broad-crested weirs with control sections of various shapes are summarized. These equations are derived by writing the Bernoulli equation between the

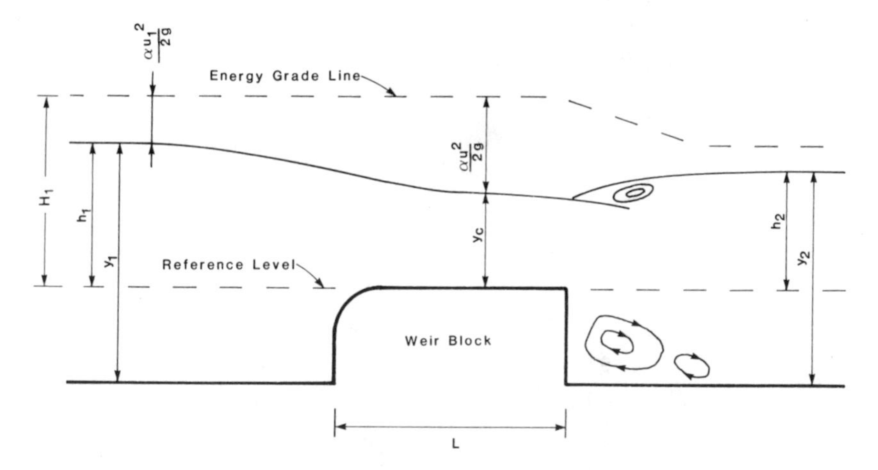

FIGURE 8.3 Definition of symbolism for broad-crested weirs.

upstream section and the control section (see Chapter 2 or Bos, 1976). In these equations, C_v is the velocity coefficient which must be used to correct for neglecting the velocity head in the approach channel. C_D is the coefficient of discharge which is introduced to account for the neglect of viscous effects, turbulence, nonuniform velocity distributions, and centripetal accelerations in the derivation of the discharge equations.

The coefficient C_v is given by

$$C_v = \left(\frac{H_1}{h_1}\right)^{\phi} \tag{8.3.2}$$

where ϕ = power of the variable h in the discharge equation. For example, for a rectangular control section $\phi = 1.5$. Since H_1 is usually not measured, the use of Eq. (8.3.2) is difficult in practice. In Fig. 8.4, C_v is plotted as a function of the parameter $C_D A^*/A_1$ where A^* = the imaginary flow area at the control section if it is assumed that the depth of flow in this section is h_1 and C_D = discharge coefficient (Bos, 1976).

Rectangular, Broad-Crested Weirs From the viewpoint of construction, the rectangular, broad-crested weir is a rather simple measuring device. In its simplest form, both the upstream and downstream weir faces are smooth, vertical planes. The weir block should be placed in a rectangular channel perpendicular to the direction of flow, and special care should be exercised to ensure that the crest surface makes a straight and sharp 90° intersection with the upstream weir face. Figure 8.5 shows schematically this type of weir with typical dimensions (Bos, 1976).

Depending on the value of the parameter H_1/L, four different flow regimes over this type of weir can be identified.

TABLE 8.4 Summary of discharge equations for broad- and sharp-crested weirs

Type	Schematic definition of control section	Discharge equation	Equation number
Rectangular	T, b, h_1	Broad-crested $Q = C_D C_v \frac{2}{3}(\frac{2}{3}g)^{1/2} T h_1^{3/2}$	(8.3.3)
		Sharp-crested $Q = C_e \frac{2}{3}(2g)^{1/2} b h_1^{3/2}$	(8.3.18)
Parabolic	h_1, f, f/2, f/2	Broad-crested $Q = C_D C_v (\frac{3}{4} f g)^{1/2} h_1^2$	(8.3.4)
		Sharp-crested $Q = C_e \frac{1}{2}\pi (fg)^{1/2} h_1^2$	(8.3.19)

Section	Equation	Eq. no.
Triangular	Broad-crested $Q = C_D C_v \frac{16}{25} (\frac{2}{5} g)^{1/2} \tan (\frac{1}{2}\Theta) h_1^{5/2}$	(8.3.5)
	Sharp-crested $Q = C_e \frac{8}{15} (2g)^{1/2} \tan (\frac{1}{2}\Theta) h_1^{5/2}$	(8.3.20)
Truncated triangular	Broad-crested $H_1 \leq 1.25 H_b$ [use Eq. (8.3.5)]	(8.3.5)
	$H_1 \geq 1.25 H_b$ $Q = C_D C_v T \frac{2}{3} (\frac{2}{3} g)^{0.5} (h_1 - \frac{1}{2} H_b)^{3/2}$	(8.3.6)
	Sharp-crested $H_1 \leq H_b$ [use Eq. (8.3.20)]	(8.3.20)
	$H_1 \geq H_b$ $Q = C_e \frac{4}{15} (2g)^{0.5} \frac{T}{H_b} [h_1^{2.5} - (h_1 - H_b)^{2.5}]$	(8.3.21)
Trapezoidal	Broad-crested $Q = C_D (T y_c + m y_c^2)[2g(H_1 - y_c)]^{1/2}$	(8.3.7)
	Sharp-crested $Q = C_e \frac{2}{3} (2g)^{0.5} (b + \frac{4}{5} h_1 \tan \frac{1}{2}\Theta) h_1^{1.5}$	(8.3.22)

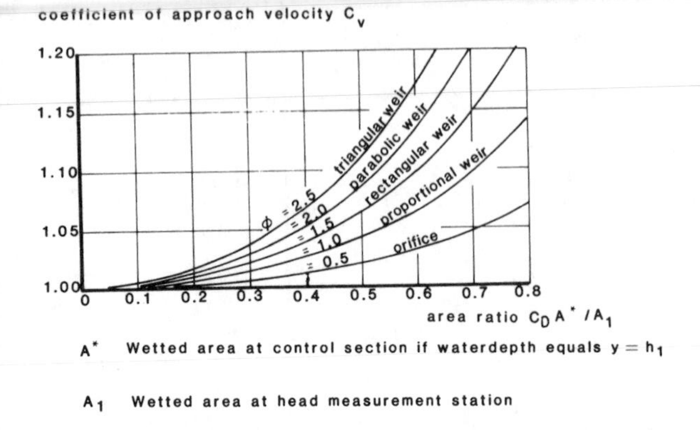

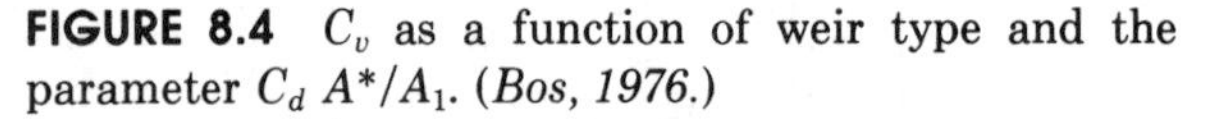

A^* Wetted area at control section if waterdepth equals $y = h_1$
A_1 Wetted area at head measurement station

FIGURE 8.4 C_v as a function of weir type and the parameter $C_d\ A^*/A_1$. (*Bos, 1976.*)

1. $H_1/L < 0.08$: In this case, the flow over the weir crest is subcritical, and the weir cannot be used to determine the discharge.
2. $0.08 \leq H_1/L \leq 0.33$: In this range, the weir can accurately be described as broad-crested since a region of parallel flow occurs in the vicinity of the mid-point of the crest. In general, C_D has a constant value in this range of H_1/L.

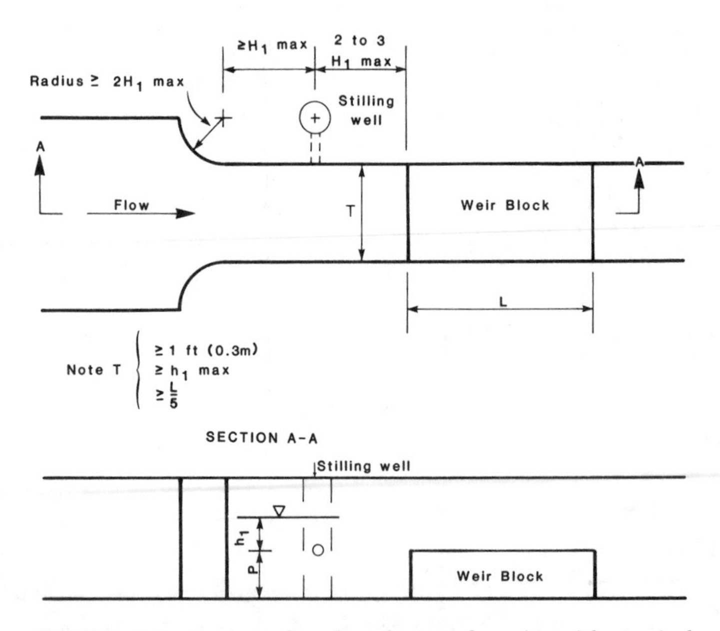

FIGURE 8.5 Rectangular broad-crested weir with typical dimensions.

3. $0.33 \leq H_1/L \leq 1.5$ to 1.8: In this range, the weir can no longer be termed broad-crested but should be classified as short-crested.

4. $1.5 \leq H_1/L$: In this range, the nappe may separate completely on the crest of the weir, and the flow pattern over the crest is unstable. For values of H_1/L greater than 3, the weir is similar to a sharp-crested weir and can be used to estimate the flow rate.

For the rectangular, broad-crested weir, Eq. (8.3.2) (Table 8.4) can be used to estimate the discharge. Experimental evidence suggests that C_D is a function of the parameters H_1/L and $H_1/(H_1 + p)$. If there is parallel flow at the control section and the approach velocity does not influence the flow over the crest of the weir, then C_D remains reasonably constant. Therefore, C_D is assumed constant if

$$0.08 \leq \frac{h_1}{L} \leq 0.33 \qquad \text{and} \qquad \frac{h_1}{h_1 + p} \leq 0.35$$

and has a value of 0.848, a value which is termed the *basic discharge coefficient*. If one of these limits is not met, then the basic discharge coefficient must be multiplied by a corrective coefficient F which can be obtained from Figs. 8.6 and 8.7. Once C_D is determined, C_v can be determined from Fig. 8.4.

A variation of the rectangular, broad-crested weir is the round-nosed, horizontal, broad-crested weir (Fig. 8.8). In this design, the upstream corner is rounded in such a manner that no flow separation occurs. Downstream of the horizontal crest there may be a vertical face, a downward sloping surface, or a rounded corner. For this type of weir, the coefficient of discharge is given by

$$C_D = \left[1 - \frac{2x(L - r)}{T}\right]\left[1 - \frac{x(L - r)}{h_1}\right]^{3/2} \tag{8.3.8}$$

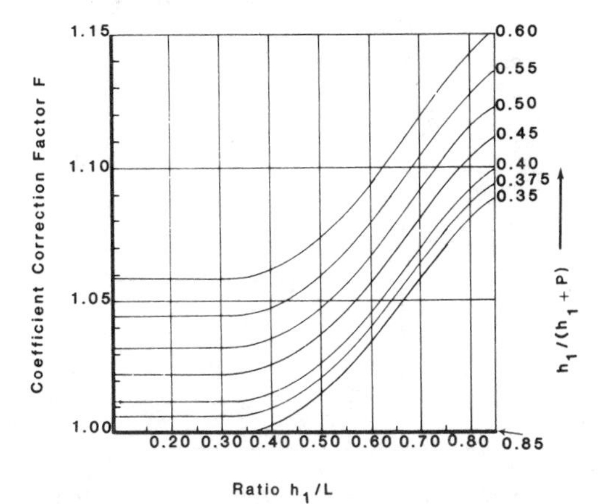

FIGURE 8.6 Correction factor F as a function of h_1/L and $h_1/(h_1 + p)$ for rectangular broad-crested weirs. *(Bos, 1976.)*

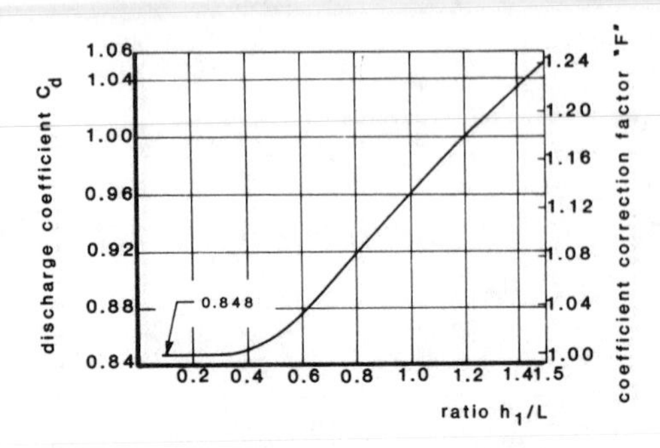

FIGURE 8.7 C_D and F as a function of h_1/L for $h_1/(h_1 + p) \leq 0.35$ for rectangular broad-crested weirs. *(Bos, 1976.)*

where all variables are defined in Fig. 8.8 and x is a parameter which accounts for boundary layer effects (Bos, 1976). For field installations with the weir made of well-finished concrete, $x \simeq 0.005$, and for laboratory installations using clean water, $x \simeq 0.003$ (Bos, 1976).

Triangular, Broad-Crested Weirs In natural streams and irrigation canals where a wide range of discharges must be measured, the triangular weir has a number of advantages which recommend its use. First, at large flows it provides a large breadth so that the backwater effect is not excessive. Second, at low flows, the width is reduced so that the sensitivity of the weir is acceptable.

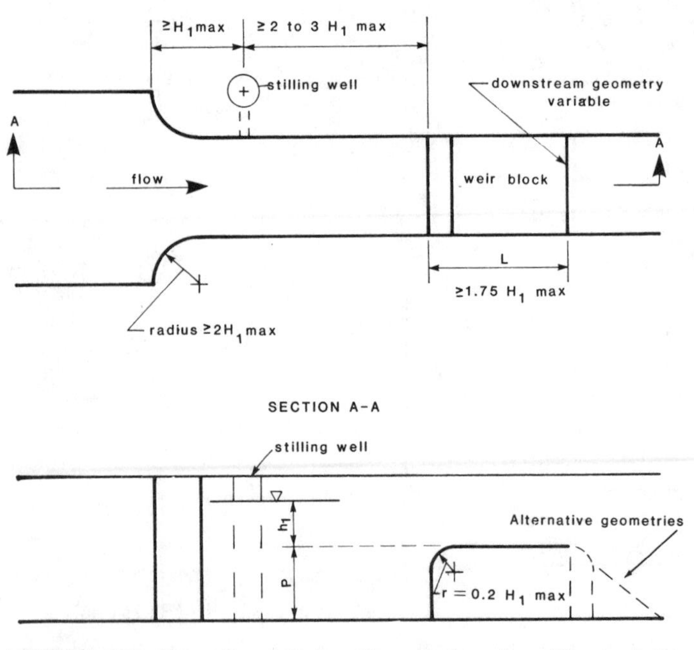

FIGURE 8.8 Round-nosed broad-crested weir with typical dimensions.

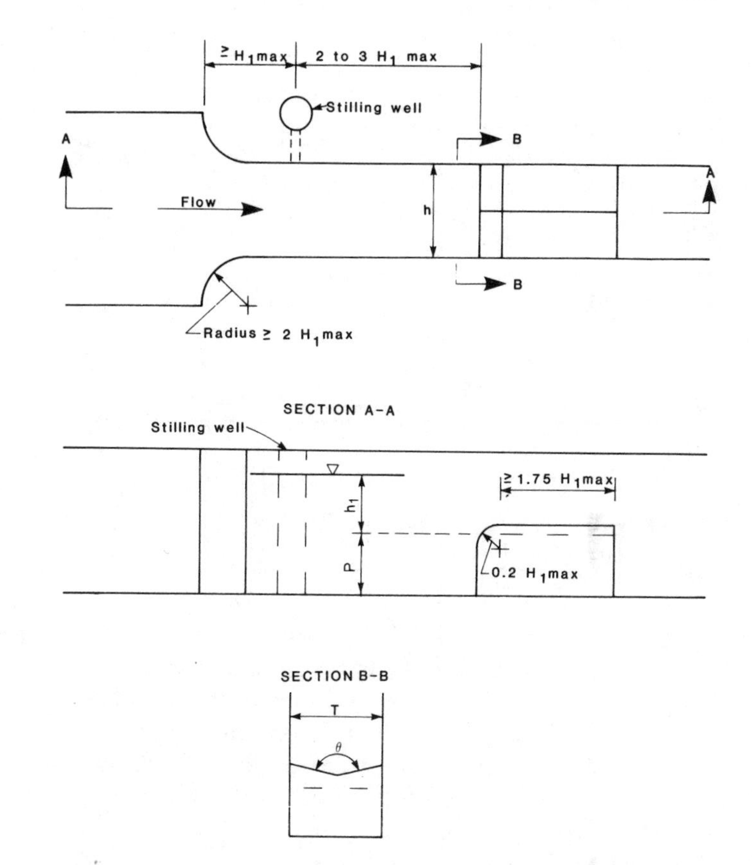

FIGURE 8.9 Triangular broad-crested weir with typical dimensions.

In Table 8.4, the triangular weir discharge equation [Eq. (8.3.5)] is a special case of the truncated triangular weir. Figure 8.9 defines typical dimensions for the truncated triangular weir. For this type of weir, the coefficient of discharge is obtained from Fig. 8.10, and the velocity coefficient can be determined from Fig. 8.4.

Other Types

At this point it should be noted that there are not sufficient published data to allow a generic estimate of the coefficients of discharge for the parabolic and trapezoidal broad-crested weirs.

Sharp-Crested Weirs

If the length of the crest of the weir in the direction of flow is such that $H_1/L > 15$ (Fig. 8.11), then the weir is termed *sharp-crested.* In practice, the crest

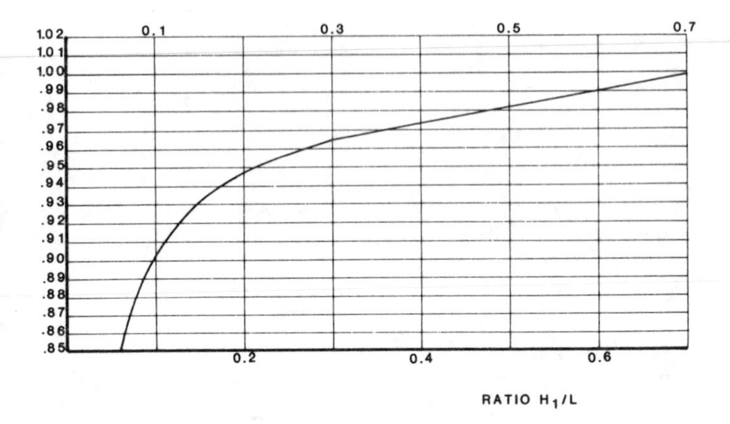

FIGURE 8.10 C_D as a function of H_1/L for triangular broad-crested weirs. *(Bos, 1976.)*

length of a sharp-crested weir is usually less than 6.6×10^{-3} ft (2.0×10^{-3} m) so that even at minimum levels of operation the flow springs clear of the weir body downstream of the weir. In this case, an air pocket is formed beneath the nappe (Fig. 8.11), from which air is continuously removed by the overflowing jet. In practice, it is necessary to design the sharp-crested weir so that the pressure in this air pocket is kept constant or the weir will have the following undesirable performance characteristics.

1. As the air pressure in the pocket decreases, the curvature of the overflowing jet increases, and the value of the coefficient of discharge will also increase.
2. If the supply of air to the pocket is irregular, then the jet will vibrate and the flow over the weir will be unsteady. If the frequency of the irregular air sup-

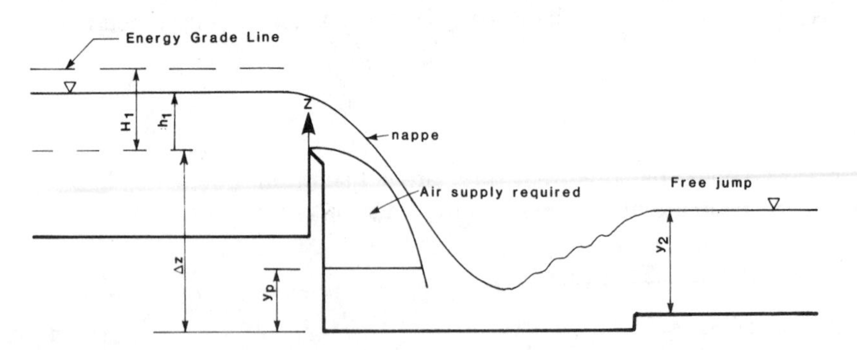

FIGURE 8.11 Schematic definition of sharp-crested weir.

ply to the pocket, the overflowing jet, and the weir structure are approximately equal, then the vibrating jet can result in failure of the structure.

Bos (1976) empirically determined that the maximum flow rate of air required to provide for full aeration per unit width of the weir crest is given by

$$q(\text{air}) = 0.1\left[\frac{q}{(y_p/h_1)^{1.5}}\right] \tag{8.3.9}$$

where $q(\text{air})$ = flow rate per unit width of weir required for full aeration
h_1 = head on weir
q = flow per unit width over weir
y_p = depth of water in pool beneath nappe

Bos (1976) further noted that if a free hydraulic jump is formed downstream of the weir, then

$$y_p = \Delta z\left(\frac{q^2}{g\,\Delta z^2}\right)^{0.22} \tag{8.3.10}$$

If a submerged jump exists downstream of the weir, then $y_p = y_2$.

EXAMPLE 8.3

If $h_1 = 0.60$ m (2.0 ft), $q = 0.86$ (m^3/s)/m [9.2(ft^3/s)/ft], and $y_p = 0.90$ m (3.0 ft), determine the total air demand for a weir 6.5 m (21 ft) wide. Further, if this air supply to the pocket beneath the nappe is to be provided by a simple steel pipe 2.5 m (8.2 ft) long with one right-angle elbow and a sharp-cornered entrance, determine the diameter of pipe required. *Note:* This example has been adopted from Bos (1976).

Solution

From Eq. (8.3.9)

$$q(\text{air}) = 0.1\left[\frac{q}{(y_p/h_1)^{1.5}}\right] = 0.1\left[\frac{0.86}{(0.90/0.60)^{1.5}}\right]$$

$$= 0.047(\text{m}^3/\text{s})/\text{m}\ [0.51(\text{ft}^3/\text{s})/\text{ft}]$$

Therefore, if the width of the weir is 6.5 m (21 ft), the total airflow rate required is

$$Q_A = bq(\text{air}) = 6.5(0.047) = 0.30\ \text{m}^3/\text{s}\ (11\ \text{ft}^3/\text{s})$$

The diameter of steel pipe required to provide this flow rate to the air pocket beneath the nappe can be determined by applying the Bernoulli energy equation with both frictional and minor losses to the situation. In practice, the flow of air through the vent pipe is facilitated by allowing a small under pressure in the pocket below the nappe. In this case, assume that the maximum allowable under pressure is 0.04 m (0.13 ft) of water. Then, with reference to Fig. 8.12, the Bernoulli equation is

$$\frac{p_1}{\gamma} = \frac{p_2}{\gamma} + \sum_{i=1}^{3} K_i \frac{u^2}{2g} + \frac{fL}{D}\frac{u^2}{2g}$$

and rearranging

$$\frac{p_1 - p_2}{\rho_a g} = \left[\sum_{i=1}^{3} K_i + \frac{fL}{D}\right]\frac{u^2}{2g}$$

$$= \left[K_e + K_b + K_{ex} + \frac{fL}{D}\right]\frac{u^2}{2g} \tag{8.3.11}$$

where ρ_a = density of air
p_1 = pressure at point 1
p_2 = pressure at point 2
K_e = entrance loss coefficient (assume K_e = 0.5)
K_b = elbow loss coefficient (assume K_b = 1.1)
K_{ex} = exit loss (assume K_{ex} = 1.)
f = Darcy Weisbach friction factor (assume f = 0.02)
u = velocity of air in pipe
L = vent pipe length
D = pipe diameter

Since the Bernoulli equation is written in terms of the fluid flowing and air, and since the under pressure in the pocket beneath the nappe is specified in terms of meters of water, a conversion is required.

$$\left(\frac{p_2}{\rho g}\right)_{air} = \left(\frac{p_2}{\rho g}\right)_{water}\frac{\rho_w}{\rho_a}$$

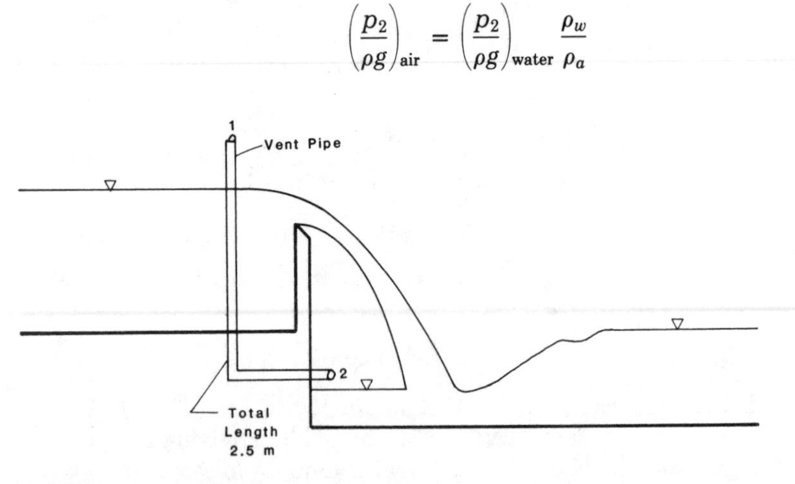

FIGURE 8.12 Schematic of vent pipe.

Also since relative rather than absolute pressures are involved, assume $p_1 = 0$. Equation (8.4.11) becomes

$$\left(\frac{p_2}{\rho g}\right)_{\text{water}} = \frac{\rho_a}{\rho_w}\left(K_e + \frac{fL}{D} + K_b + K_{ex}\right)\frac{u^2}{2g}$$

or

$$0.04 = \frac{1}{830}\left[0.5 + \frac{0.02(2.5)}{D} + 1.1 + 1\right]\frac{u^2}{2g} \qquad (8.3.12)$$

The equation of continuity yields

$$Q_{\text{air}} = 0.30 = \frac{\pi D^2}{4}u$$

or

$$u = \frac{4(0.30)}{\pi D^2} \qquad (8.3.13)$$

Substituting Eq. (8.3.13) in Eq. (8.3.12) and simplifying yields

$$0.04 = \frac{1}{830}\left[2.6 + \frac{0.05}{D}\right]\frac{(0.30)^2}{12D^4} \qquad (8.3.14)$$

Solving Eq. (8.3.14) by trial and error yields $D \simeq 0.16$ m (0.52 ft).

In the case of a sharp-crested weir, the concept of critical depth is not applicable. For this type of discharge-measuring device, the discharge equation is derived by assuming that the weir behaves as an orifice with a free water surface and that the following assumptions are valid:

1. The height of the water level above the weir crest is h_1 and there is no contraction.
2. The velocities over the weir crest are almost horizontal.
3. The approach velocity head can be neglected.

With reference to Fig. 8.11, the velocity at an arbitrary point in the control section is found from the Bernoulli equation as

$$u = [2g(h_1 - z)]^{0.5} \qquad (8.3.15)$$

The total flow over the weir is then obtained by integration or

$$Q = \sqrt{2g}\int_{z=0}^{z=h_1} b(z)(h_1 - z)^{0.5}\,dz \qquad (8.3.16)$$

where $b(z)$ = width of the weir at an elevation z above the weir crest. At this point, an effective discharge coefficient C_e is introduced to account for the assumptions made, and the resulting weir discharge equation is

$$Q = C_e\sqrt{2g}\int_{z=0}^{z=h_1} b(z)(h_1 - z)^{0.5}\,dz \tag{8.3.17}$$

Equation (8.3.17) has proved to be a satisfactory representation in practice and is widely used. The discharge equations for various control sections are summarized in Table 8.4.

Rectangular, Sharp-Crested Weirs The rectangular, sharp-crested weir is best described as a rectangular notch symmetrically located in a thin plate which is placed perpendicular to the sides and bottom of a straight, usually rectangular, open channel (Fig. 8.13). Within this category there are three subdivisions:

1. *Fully Contracted:* A fully contracted weir is one in which the sides and bottom of the channel are sufficiently distant from the weir crest that they have no effect on the contraction of the nappe.

2. *Partially Contracted:* A partially contracted weir is one in which the contraction of the nappe is not fully developed because of the proximity of the channel boundaries.

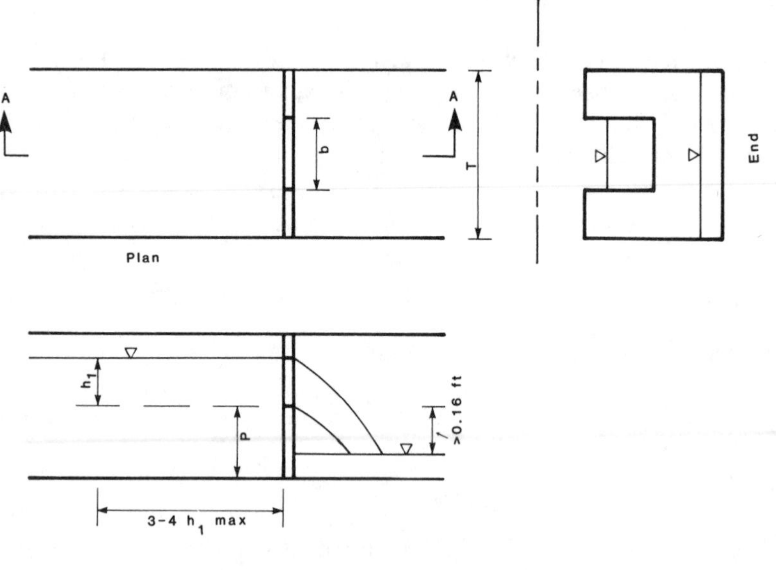

FIGURE 8.13 Rectangular sharp-crested weir.

TABLE 8.5 Definition of a fully contracted, rectangular, sharp-crested weir

$$B - b \geq 4h_1$$
$$\frac{h_1}{p} \leq 0.5$$
$$\frac{h_1}{b} \leq 0.5$$
$$0.23 \leq h_1 < 2 \text{ ft}$$
$$b \geq 1 \text{ ft}$$
$$p \geq 1 \text{ ft}$$

3. *Full Width:* A full-width weir is one in which the notch extends completely across the approach channel; i.e., with reference to Fig. 8.13, $b/T = 1$.

In Table 8.5 geometric criteria for classifying a weir as fully contracted are summarized. Weirs which are rectangular and sharp-crested, but do not meet these criteria, are not full-width weirs and must be considered to be partially contracted.

From Table 8.4, the theoretical discharge of a rectangular, sharp-crested weir is given by Eq. (8.3.18). Kindsvater and Carter (1957) modified this theoretical equation so that it would apply to all rectangular, sharp-crested weirs regardless of whether they were fully contracted, partially contracted, or full width. The equation for estimating discharge through this type of weir is

$$Q = \tfrac{2}{3}\, C_e \sqrt{2g}\; b_e h_e^{1.5} \qquad (8.3.23)$$

where b_e = effective width $= b + K_b$
h_e = effective head $= h_1 + K_n$
C_e = effective discharge coefficient

The parameters K_b and K_n represent the combined effects of the flow phenomena caused by viscosity and surface tension. K_n is generally considered to be constant, with an appropriate value of 0.003 ft (0.001 m) recommended for all values of b/T and h_1/p. K_b has been empirically determined to be a function of b/t (Fig. 8.14). Equations for C_e as a function of b/T and h_1/p are summarized in Table 8.6.

The following limitations regarding the use of this weir should be noted:

1. The recommended lower limit on h_1 is approximately 0.10 ft (0.03 m). This limit is derived from a consideration of the accuracy with which h_1 can be measured and the relative importance of viscosity and surface tension in the flow.

2. If h_1/p exceeds 5, critical depth may occur in the approach channel and

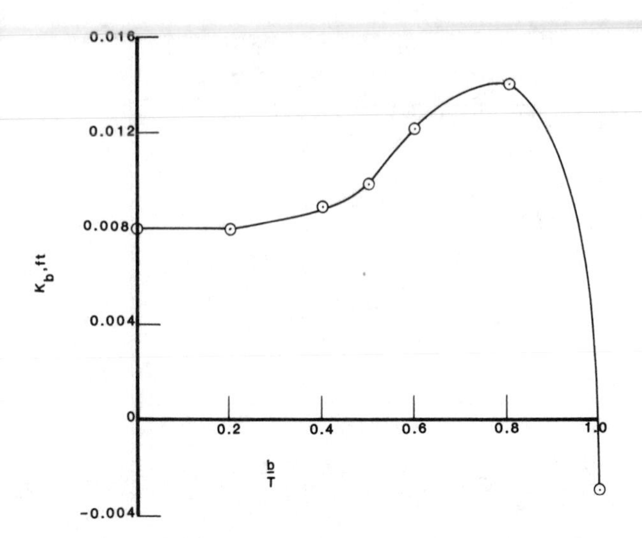

FIGURE 8.14 K_b/T as a function of b/T for rectangular, sharp-crested weirs. (*Bos, 1976.*)

invalidate the assumptions under which Eq. (8.3.18) was derived (Bos, 1976). Bos (1976) recommended that for precise measurements of discharge $h_1/p \leq 2$ and $p \geq 0.30$ ft (0.10 m).

3. The width of the weir should exceed 0.50 ft (0.15 m); that is,

$$b \geq 0.50 \text{ ft}$$

4. Because of aeration requirements, the tailwater level should be at least 0.16 ft (0.05 m) below the crest elevation.

TABLE 8.6 C_e **as a function of** b/T **and** h_1/p **to rectangular, sharp-crested weirs**

b/T	C_e
1.0	$C_e = 0.602 + 0.075\ h_1/p$
0.9	$C_e = 0.599 + 0.064\ h_1/p$
0.8	$C_e = 0.597 + 0.045\ h_1/p$
0.7	$C_e = 0.595 + 0.030\ h_1/p$
0.6	$C_e = 0.593 + 0.018\ h_1/p$
0.5	$C_e = 0.592 + 0.011\ h_1/p$
0.4	$C_e = 0.591 + 0.0058\ h_1/p$
0.3	$C_e = 0.590 + 0.0020\ h_1/p$
0.2	$C_e = 0.589 - 0.0018\ h_1/p$
0.1	$C_e = 0.588 - 0.0021\ h_1/p$
0	$C_e = 0.587 - 0.0023\ h_1/p$

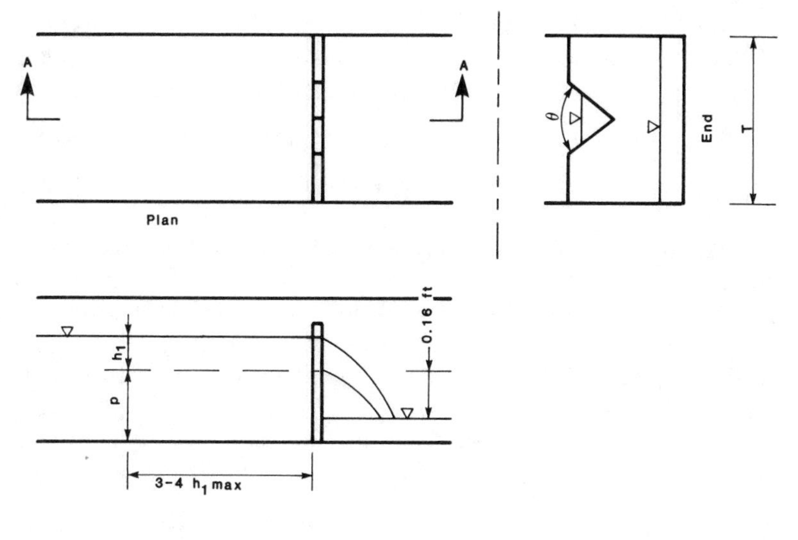

FIGURE 8.15 Triangular or V-notch sharp-crested weir.

Triangular (V-Notch), Sharp-Crested Weirs The triangular, sharp-crested weir is best described as a V-shaped notch symmetrically located in a thin plate which is placed perpendicular to the sides and bottom of an open channel (Fig. 8.15). Within this category there are two subdivisions:

1. Fully contracted
2. Partially contracted

In Table 8.7, geometric criteria for classifying this type of weir as either fully or partially contracted are specified (Bos, 1976). From these data, it appears that if h_1 is small, then the weir is fully contracted and as h_1 increases, it becomes partially contracted.

TABLE 8.7 Criteria for classifying a triangular, sharp-crested weir as fully or partially contracted

Partially contracted weir	Fully contracted weir
$h_1/p \leq 1.2$	$h_1/p \leq 0.4$
$h_1/T \leq 0.4$	$h_1/T \leq 0.2$
$0.16 < h_1 \leq 2$ ft	$0.16 < h_1 \leq 1.25$
$p \geq 0.3$ ft	$p \geq 1.5$ ft
$T \geq 2$ ft	$T \geq 3$ ft

From Table 8.4, the theoretical discharge of a triangular weir is given by Eq. (8.3.20). In order that this equation might apply to both fully and partially contracted weirs, the theoretical discharge equation is modified to yield

$$Q = \tfrac{8}{15} C_e (2g)^{0.5} (\tan \tfrac{1}{2}\theta)\, h_e^{2.5} \tag{8.3.24}$$

where h_e = effective head = $h_1 + K_h$ and C_e effective discharge coefficient. Empirically determined values of K_h as a function of the notch angle θ are summarized in Fig. 8.16. For a fully contracted weir, values of C_e can be obtained from Fig. 8.17. At the present time, sufficient data are not available to allow the estimation of C_e for other cases. Bos (1976) recommended that partially contracted triangular weirs should be located in rectangular approach channels, while fully contracted weirs of this type should be located in nonrectangular channels.

The following limitations regarding the use of this type of weir should be noted:

1. $h_1/p \le 1.2$ and $h_1/T \le 0.4$
2. $0.16 \le h_1 \le 2.0$ ft ($0.049 \le h_1 \le 0.61$ m)
3. $p \ge 0.3$ ft ($p \ge 0.09$ m)
4. $T \ge 2$ ft ($T \ge 0.61$ m)
5. The tailwater depth should remain below the vertex of the triangular notch.

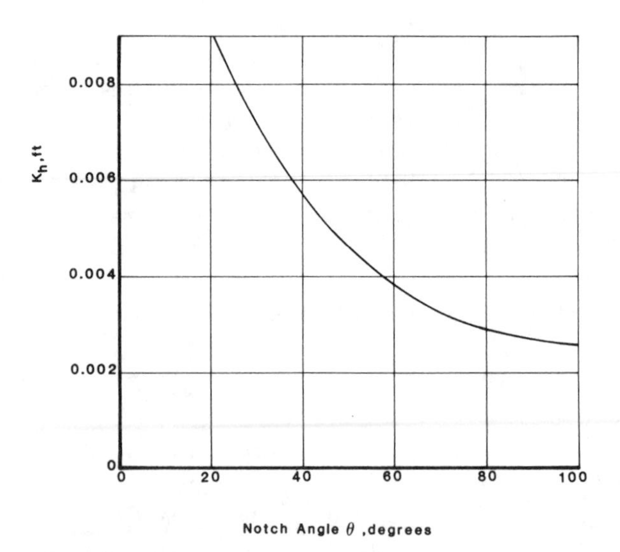

FIGURE 8.16 K_h as a function of notch angle θ. (*After Bos, 1976.*)

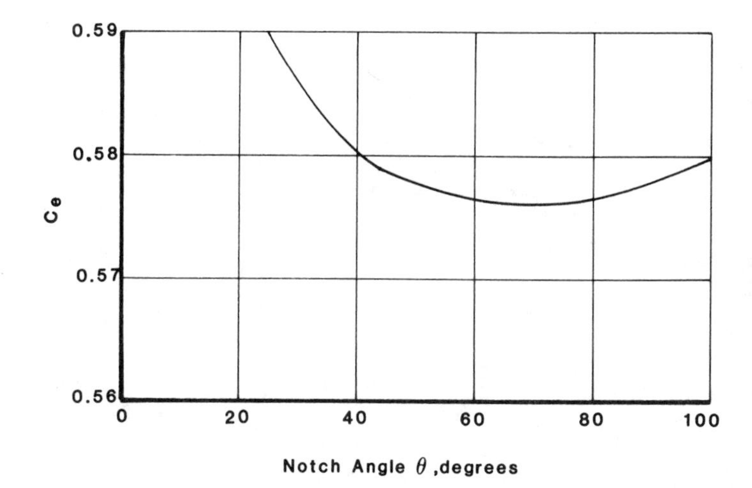

FIGURE 8.17 C_e as a function of notch angle θ for fully contracted, sharp-crested, triangular weirs. *(Bos, 1976.)*

Cipoletti Weir

The Cipoletti weir is a modification of the fully contracted, rectangular, sharp-crested weir. This weir has a trapezoidal control section (Fig. 8.18), with the crest being horizontal and the sides sloping outward with slopes of 4:1. Although the accuracy of this weir in measuring discharge is not as great as that which is obtainable with either rectangular or triangular sharp-crested weirs, it has two significant advantages. First, the discharge equation is given by (8.3.18) or

$$Q = \tfrac{2}{3} C_D C_v \sqrt{2g}\, T h_1^{1.5}$$

with $C_D \simeq 0.63$ and C_v plotted in Fig. 8.4. Second, this weir may be used in nonrectangular approach channels provided that the geometrical limitations shown in Fig. 8.18 are satisfied and

1. $h_1/b \leq 0.50$

2. $0.20 \leq h_1 \leq 2.0$ ft $(0.06 \leq h_1 \leq 0.61$ m)

Proportional or Sutro Weir

The proportional or Sutro weir is designed so that when the total head on the weir exceeds an arbitrary reference level, the discharge is linearly proportional to the total head. The Sutro weir is composed of a rectangular section joined to a curved portion (Fig. 8.19), which provides proportionality for all heads above

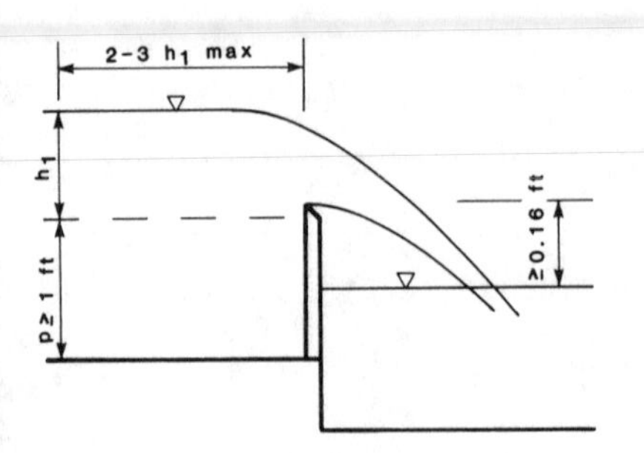

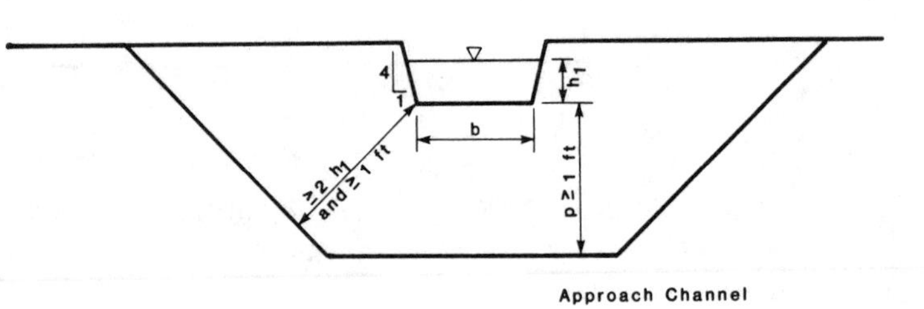

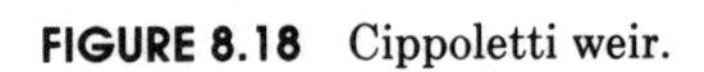

FIGURE 8.18 Cippoletti weir.

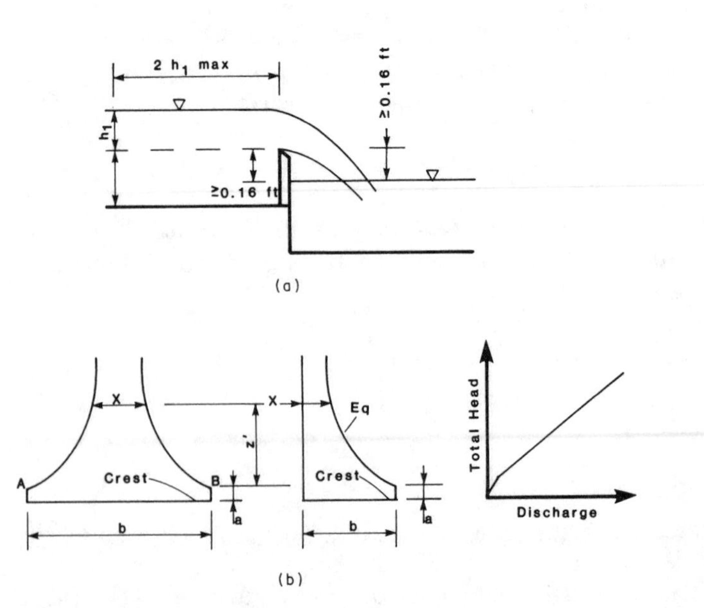

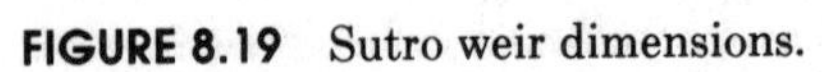

FIGURE 8.19 Sutro weir dimensions.

TABLE 8.8 Symmetrical Sutro weir discharge coefficients as a function of *a* and *b* (Soucek, Howe, Mavis, 1936)

	b, ft				
a, ft	0.50	0.75	1.0	1.25	1.50
0.02	0.608	0.613	0.617	0.6185	0.619
0.05	0.606	0.611	0.615	0.617	0.6175
0.10	0.603	0.608	0.612	0.6135	0.614
0.15	0.601	0.6055	0.610	0.6115	0.612
0.20	0.599	0.604	0.608	0.6095	0.610
0.25	0.598	0.6025	0.6065	0.608	0.6085
0.30	0.597	0.602	0.606	0.6075	0.608

the line *AB*. For this type of weir, the curved portion is defined by the equation

$$\frac{x}{b} = \left(1 - \frac{2}{\pi}\tan^{-1}\sqrt{\frac{z'}{a}}\right) \tag{8.3.25}$$

It should be noted that the Sutro weir can be either symmetrical or unsymmetrical. The head-discharge relationship for this type of weir can be shown to be (Bos, 1976):

$$Q = C_D b\sqrt{2ga}\,(h_1 - \tfrac{1}{3}a) \tag{8.3.26}$$

where the coefficient of discharge is primarily a function of the shape of the control section. Values of C_D for both symmetrical and unsymmetrical Sutro weirs are summarized in Tables 8.8 and 8.9.

Proportional weirs are particularly valuable for use in rectangular channels

TABLE 8.9 Unsymmetrical Sutro weir discharge coefficients as a function of *a* and *b* (Soucek, Howe, and Mavis, 1936)

	b, ft				
a, ft	0.50	0.75	1.0	1.25	1.50
0.02	0.614	0.619	0.623	0.6245	0.625
0.05	0.612	0.617	0.621	0.623	0.6235
0.10	0.609	0.614	0.618	0.6195	0.620
0.15	0.607	0.6115	0.616	0.6175	0.618
0.20	0.605	0.610	0.614	0.6155	0.616
0.25	0.604	0.6085	0.6125	0.614	0.6145
0.30	0.603	0.608	0.612	0.6135	0.614

to provide downstream control, e.g., as controls for float-regulated chemical dosing or as flow meters. In using these weirs, the following should be noted:

1. Ventilation of the nappe is crucial; therefore, the tailwater must always remain below the crest of the weir.
2. In most applications, the weir profile should be superimposed directly on the bottom of the channel to prevent the accumulation of sediment against the upstream face of the weir.
3. The discharge of the weir is linearly proportional to the head if

$$h_1 \geq 1.2a$$

h_1 should never be less than 0.10 ft (0.03 m) or the influences of viscosity and surface tension become dominant.

4. The width of the weir b should not be less than 0.5 ft (0.15 m) if the discharge coefficients tabulated in Tables 8.8 and 8.9 are to be used.
5. To achieve a fully contracted weir condition

$$\frac{b}{p} \geq 1$$

and

$$\frac{T}{b} \geq 3.0$$

6. Proportional weirs which do not comply with these requirements can be employed provided that a calibration procedure is used to establish the relevant values of C_D.

8.4 STRUCTURES FOR FLOW MEASUREMENT: FLUMES

Although weirs are an effective method of artificially creating a critical section at which the flow rate can be determined, a weir installation has at least two disadvantages. First, the use of a weir results in relatively high head losses. Second, most weirs create a dead zone upstream of the installation which can serve as a settling basin for sediment and other debris present in the flow. An examination of the principles developed in Chapter 2 suggests that both of these disadvantages can be overcome with an open flume having a contraction in width which is sufficient to cause the flow to pass through critical depth. Such a flume produces an effect similar to that of a Venturi meter in pipe flow (see, for example, Streeter and Wylie, 1975). Venturi flumes have been used and found undesirable because the difference in the water level between the upstream section and the critical flow section is relatively small, especially at low Froude numbers. This problem can be alleviated by designing a flume which has a constructed throat section in which critical flow occurs followed by

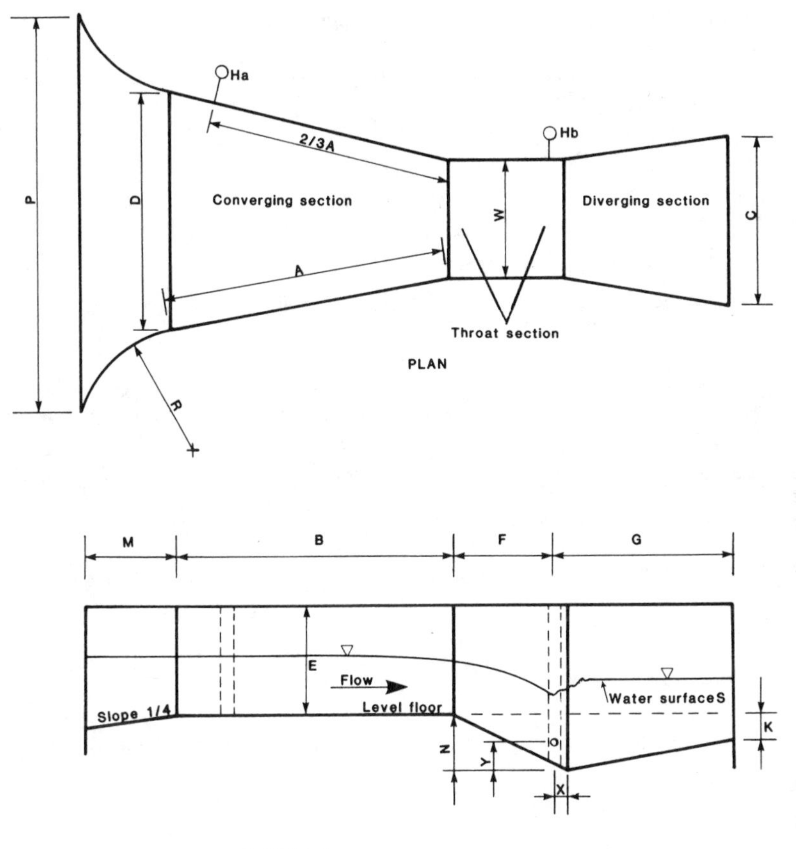

FIGURE 8.20 Plan, elevation, and dimensions of the Parshall flume.

a short length of flume in which supercritical flow occurs. At the end of the supercritical flow section, a hydraulic jump occurs (Fig. 8.20). A flume of this type was designed by R. L. Parshall in approximately 1920 (Parshall, 1938) and is widely known as the *Parshall flume*. The first practical installation of this device was in the Las Animas Consolidated Ditch near Las Animas, Colorado, in 1926 (Parshall, 1938).

Empirical depth-discharge relationships for Parshall flumes of various sizes—the size of a Parshall flume is determined by the throat width—are given in Table 8.10. In these equations, Q = free discharge in ft^3/s, W = throat width in feet, and H_a = upstream depth of flow at section A, Fig. 8.20, in feet. In Table 8.11 dimensions and capacities for Parshall flumes of various throat widths are specified. The terminology *free discharge* denotes that a free-flow condition exists in the exit section of the flume with critical depth occurring in this section and a hydraulic jump occurring at the exit section. Although it is

TABLE 8.10 Free discharge as a function of throat width

Throat width, ft	Free discharge equation, ft^3/s
0.25	$Q = 0.992\ H_a^{1.547}$
0.50	$Q = 2.06\ H_a^{1.58}$
0.67	$Q = 3.07\ H_a^{1.53}$
$1 \le W \le 8$	$Q = 4\ W\ H_a^{1.522 W^{0.026}}$
$10 \le W \le 50$	$Q = (3.6875\ W + 2.5) H_a^{1.6}$

desirable to design the Parshall flume so that free flow occurs, under some flow conditions the hydraulic jump at the exit section will be submerged, and the free-flow condition will not exist. Nonfree discharge or submerged flow occurs when

$$\frac{H_b}{H_a} \ge 0.6 \qquad \text{for } W = 0.25, 0.50, 0.75 \text{ ft } (0.076, 0.15, 0.23 \text{ m})$$

$$\frac{H_b}{H_a} \ge 0.7 \qquad \text{for } 1 \le W \le 8 \text{ ft } (0.30 \le W \le 2.4 \text{ m})$$

and

$$\frac{H_b}{H_a} \ge 0.8 \qquad \text{for } 10 \le W \le 50 \text{ ft } (0.24 \le W \le 15 \text{ m})$$

TABLE 8.11 Dimensions and capacities of the Parshall measuring flume, for

W, ft	*W*, in	*A*, ft	*A*, in	⅔*A*, ft	⅔*A*, in	*B*, ft	*B*, in	*C*, ft	*C*, in	*D*, ft	*D*, in	*E*, ft	*E*, in	*F*, ft	*F*, in
0	3	1	6⅜	1	¼	1	6	0	7	0	10³⁄₁₆	2	0	0	6
0	6	2	⁷⁄₁₆	1	4⁵⁄₁₆	2	0	1	3½	1	3⅝	2	0	1	0
0	9	2	10⅝	1	11⅛	2	10	1	3	1	10⅝	2	6	1	0
1	0	4	6	3	0	4	4⅞	2	0	2	9¼	3	0	2	0
1	6	4	9	3	2	4	7⅞	2	6	3	4⅜	3	0	2	0
2	0	5	0	3	4	4	10⅞	3	0	3	11½	3	0	2	0
3	0	5	6	3	8	5	4¾	4	0	5	1⅞	3	0	2	0
4	0	6	0	4	0	5	10⅝	5	0	6	4¼	3	0	2	0
5	0	6	6	4	4	6	4½	6	0	7	6⅝	3	0	2	0
6	0	7	0	4	8	6	10⅜	7	0	8	9	3	0	2	0
7	0	7	6	5	0	7	4¼	8	0	9	11⅜	3	0	2	0
8	0	8	0	5	4	7	10⅛	9	0	11	1¾	3	0	2	0

where H_b = depth of flow at section B (Fig. 8.20). The effect of submergence is to reduce the discharge of the flume. In such a situation, the equations summarized in Table 8.10 overestimate the flow rate, and a correction must be applied. The curves in Fig. 8.21 can be used to correct for the effects of submergence in Parshall flumes of various sizes. With regard to the data summarized in Fig. 8.21, it is noted that the correction for a 1-ft (0.30-m) flume can be made applicable to larger flumes by multiplying by the factors specified in Tables 8.12 and 8.13. Empirically derived diagrams for determining the headloss incurred in Parshall flumes of various sizes are given in Fig. 8.22.

EXAMPLE 8.3

Design a Parshall flume for measuring flows which average 20 ft^3/s (0.57 m^3/s). The channel conveying this flow has a mild slope and a normal depth of flow of 2.5 ft (0.76 m).

Solution

Discharges of this order of magnitude can be measured by flumes of several sizes; therefore, the flume dimensions determined in this solution represent only one of several possible solutions.

Assume a throat width of 4 ft (1.2 m) and a submergence ratio of 0.70. *Note:* When conditions do not permit free-flow operation, the percentage of submergence, H_b/H_a, should be kept as low as possible.

various throat widths *W* (letters refer to dimensions shown in Fig. 8.20)

G,		*K*,	*N*,	*R*,		*M*,		*P*,		*X*,	*Y*,	Free-flow capacity	
ft	in	in	in	ft	in	ft	in	ft	in	in	in	Min ft^3/s	Max ft^3/s
1	0	1	2¼	1	4	1	0	2	6¼	1	1½	0.03	1.9
2	0	3	4½	1	4	1	0	2	11½	2	3	0.05	3.9
1	6	3	4½	1	4	1	0	3	6½	2	3	0.09	8.9
3	0	3	9	1	8	1	3	4	10¾	2	3	0.11	16.1
3	0	3	9	1	8	1	3	5	6	2	3	0.15	24.6
3	0	3	9	1	8	1	3	6	1	2	3	0.42	33.1
3	0	3	9	1	8	1	3	7	3½	2	3	0.61	50.4
3	0	3	9	2	0	1	6	8	10¾	2	3	1.3	67.9
3	0	3	9	2	0	1	6	10	1¼	2	3	1.6	85.6
3	0	3	9	2	0	1	6	11	3½	2	3	2.6	103.5
3	0	3	9	2	0	1	6	12	6	2	3	3.0	121.4
3	0	3	9	2	0	1	6	13	8¼	2	3	3.5	139.5

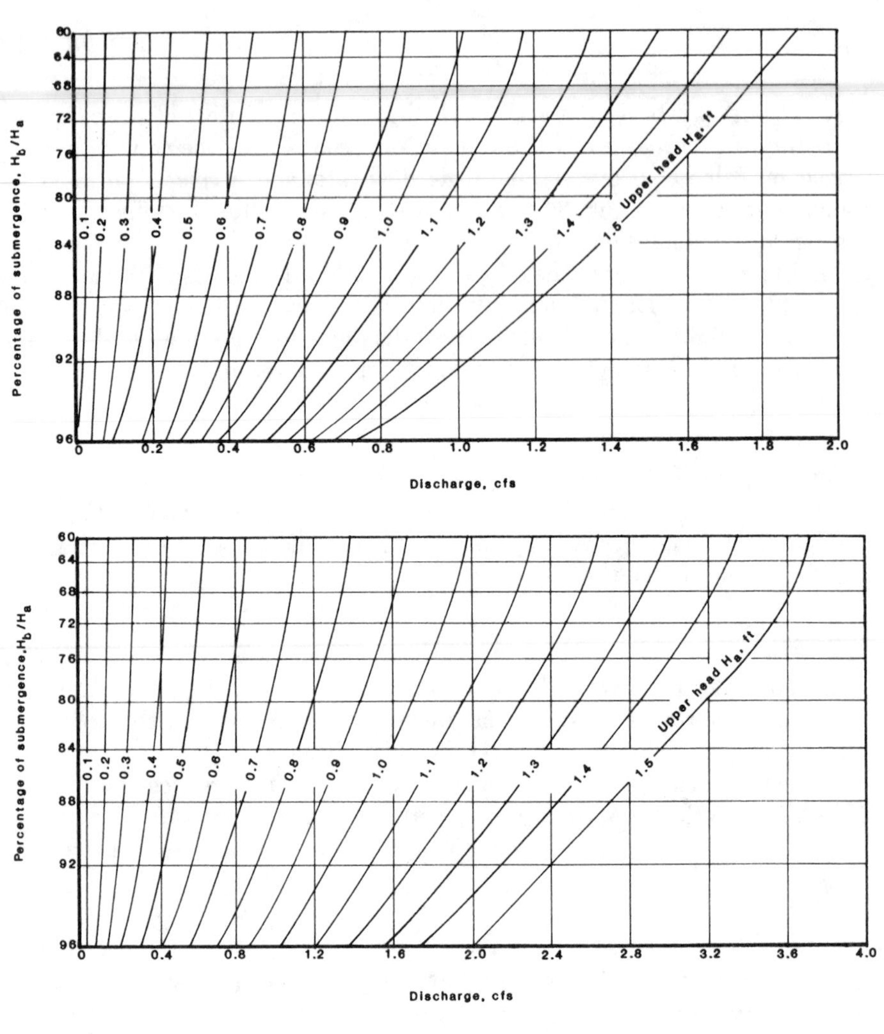
Percentage of submergence, H_b/H_a
60
64
68
72
76
80
84
88
92
96
0.1
0.2
0.3
0.4
0.5
0.6
0.7
0.8
0.9
1.0
1.1
1.2
1.3
1.4
1.5
Upper head H_a, ft
0
0.2
0.4
0.6
0.8
1.0
1.2
1.4
1.6
1.8
2.0
Discharge, cfs
Percentage of submergence, H_b/H_a
60
64
68
72
76
80
84
88
92
96
0.1
0.2
0.3
0.4
0.5
0.6
0.7
0.8
0.9
1.0
1.1
1.2
1.3
1.4
1.5
Upper head H_a, ft
0
0.4
0.8
1.2
1.6
2.0
2.4
2.8
3.2
3.6
4.0
Discharge, cfs

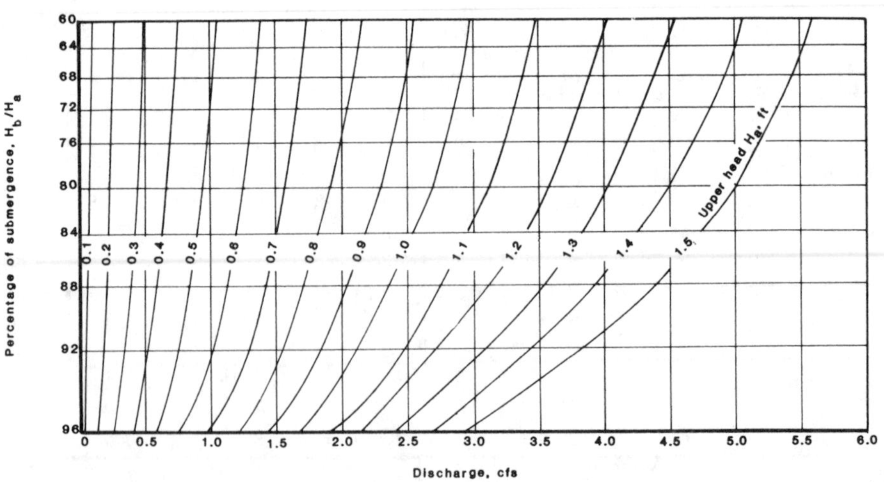
Percentage of submergence, H_b/H_a
60
64
68
72
76
80
84
88
92
96
0.1
0.2
0.3
0.4
0.5
0.6
0.7
0.8
0.9
1.0
1.1
1.2
1.3
1.4
1.5
Upper head H_a, ft
0
0.5
1.0
1.5
2.0
2.5
3.0
3.5
4.0
4.5
5.0
5.5
6.0
Discharge, cfs

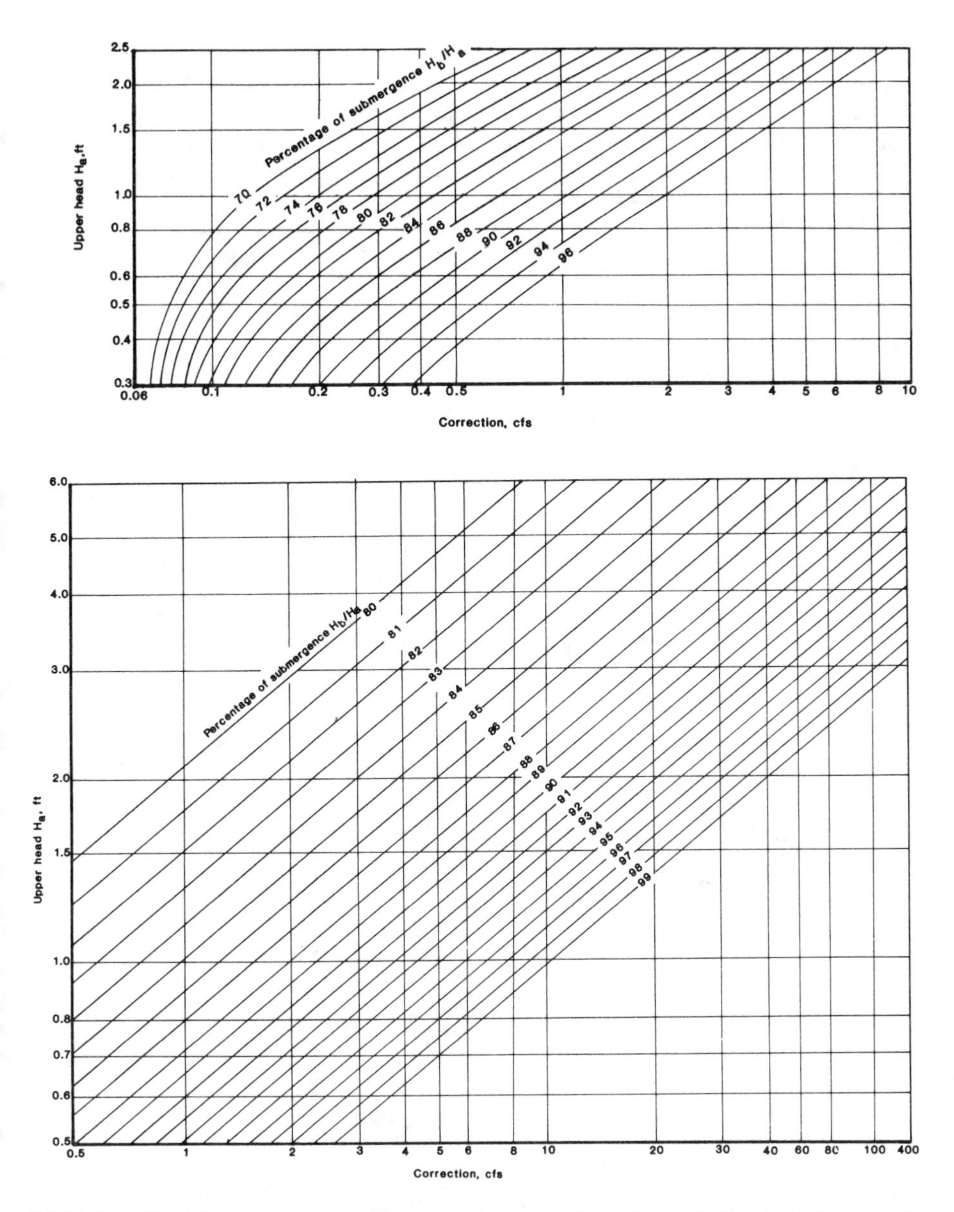

FIGURE 8.21 Diagrams for computing submerged flow through Parshall flumes of various sizes.

TABLE 8.12 Submergence correction factors for Parshall flumes with throat widths between 1 and 10 ft based on the correction for a 1-ft flume (Fig. 8.21)

Size of flume in use, B_T, ft	Correction factor
1	1.0
1.5	1.4
2	1.8
3	2.4
4	3.1
6	4.3
8	5.4

TABLE 8.13 Submergence correction factors for Parshall flumes with throat widths between 10 and 50 ft based on the correction for a 10-ft flume (Fig. 8.21)

Size of flume in use, B_T, ft	Correction factor
10	1.0
12	1.2
15	1.5
20	2.0
25	2.5
30	3.0
40	4.0
50	5.0

Discharge equations for San Dimas flume

Flume width, ft	Discharge equation, ft³/s
1.0	$Q = 6.35\, H^{1.321}$
2.0	$Q = 13.05\, H^{1.277}$
3.0	$Q = 19.90\, H^{1.245}$

When H_b/H_a exceeds 0.95, the Parshall flume does not produce reliable results. From Table 8.10,

$$Q = 4wH_a^{1.522W^{0.026}}$$

or

$$20 = 4(4)H_a^{1.522(4)^{0.026}} = 16H_a^{1.58}$$

$$H_a = (20/16)^{0.634} = 1.15 \text{ ft } (0.35 \text{ m})$$

For a submergence ratio of 0.70

$$H_b = 0.70H_a = 0.81 \text{ ft } (0.25 \text{ m})$$

Then, with reference to Fig. 8.23

$$x = 2.5 - 0.81 = 1.69 \text{ ft } (0.52 \text{ m})$$

From Fig. 8.22, the headloss through this flume is $h_L = 0.43$ ft (0.13 m); therefore, the depth of flow upstream is

$$2.5 + 0.43 = 2.9 \text{ ft } (0.88 \text{ m})$$

Similar calculations for 2- and 3-ft Parshall flumes can be made. In general, the most economical flume would be the one with the smallest dimensions.

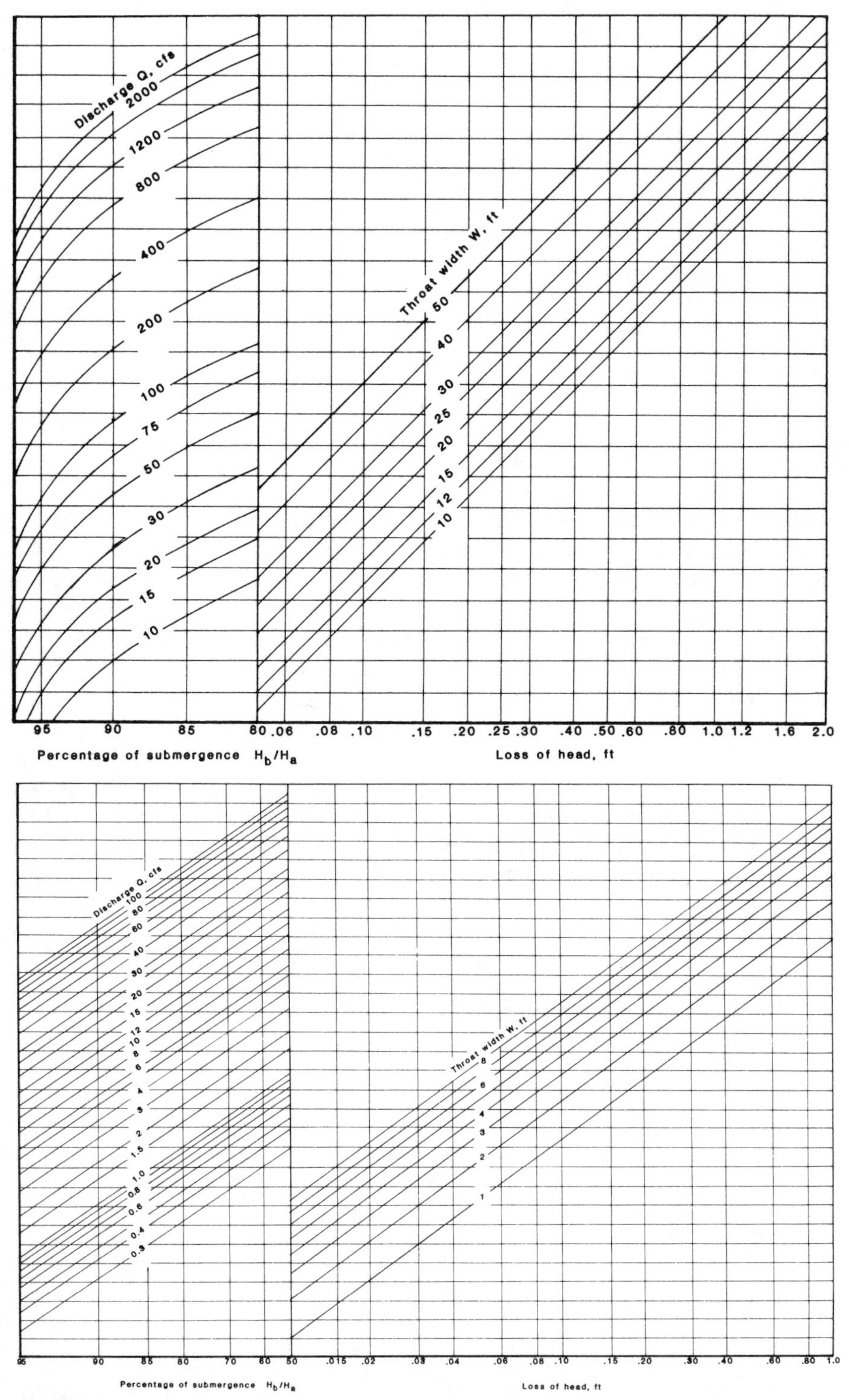

FIGURE 8.22 Diagrams for determining the loss of head through Parshall flumes of various sizes.

However, if a narrow flume is used, then, depending on the channel dimensions, moderate to long wingwalls might be required which could offset the economy achieved by using a small flume. As a rule of thumb, the throat width of Parshall flume should be one-third to one-half the width of the channel.

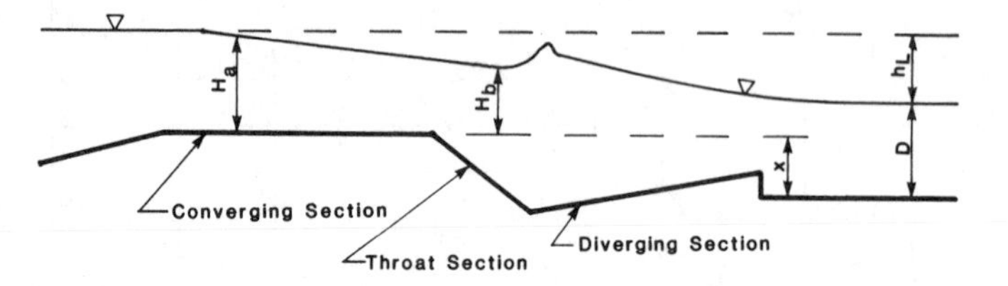

FIGURE 8.23 Section of Parshall flume for Example 8.3.

Because the Parshall flume causes critical depth to occur by contracting, the velocity of the water flowing through the flume is higher than in the natural channel. For this reason, sand and sediment particles being transported by the flow are swept through the flume; however, if significant amounts of debris are present in the flow, the Parshall flume may be rendered inoperable or inaccurate by blockage or deposition. Under such circumstances, a modified Parshall flume known as the *San Dimas flume* should be used (Wilm et al., 1938). For the purposes of this section, debris flows can be divided into (1) muddy water laden with fine silt which can be assumed to be uniformly distributed throughout the cross section, (2) water carrying a suspended load of coarse material, the bulk of which is transported in the lower part of the flow, and (3) water transporting rocks which move by rolling and slipping along the channel bed. These latter two categories of flow often occur in mountain canyons and render

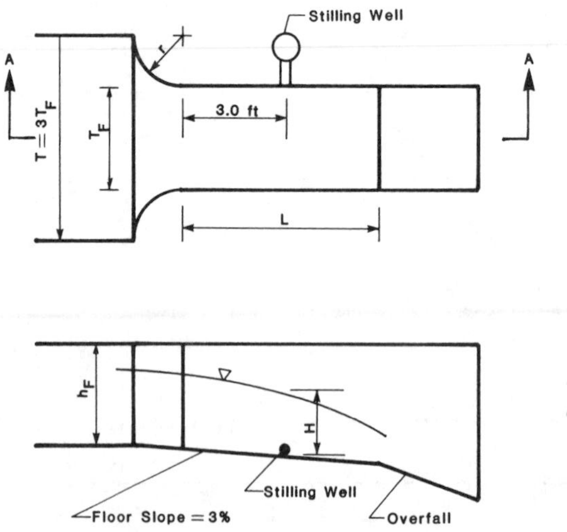

FIGURE 8.24 Schematic of San Dimas flume. *(Wilm et al., 1938.)*

discharge-measuring devices such as the Parshall flume inoperable. In response to this problem, Wilm et al. (1938) developed the San Dimas flume (Fig. 8.24), which functions as a broad-crested weir with the depth of flow being measured downstream of the point where critical depth occurs. With reference to Fig. 8.24, T = width of the approach channel, $T_F = \frac{1}{3}T$ = flume width, h_F = flume height = (maximum flow)/($5T_F$), L = flume length = $r + 2h_F$, and r = transition radius = T_F. Empirically derived discharge equations for this type of flow are summarized in Table 8.13. *Note:* Since these equations are empirical, the English system of units must be used. With regard to this type of flume, Wilm et al. (1938) claimed that it yields accurate discharge measurements and is unaffected by either the velocity in the approach channel or the presence of a bed load; however, at the time the original design was published, a number of discussions criticizing this design were also published. Therefore, caution should be exercised in using the San Dimas flume.

8.5 STRUCTURES FOR FLOW MEASUREMENT: CULVERTS

The placement of fill in a natural channel with a culvert through the resulting embankment to pass the flow yields a situation which can be used to gage the flow in the channel if other means are not available. With reference to Fig. 8.25, the following physical features of a culvert should be noted. The approach channel section is usually defined at one culvert opening width upstream of the culvert entrance. The loss of energy in the vicinity of the culvert entrance is related to the rapid contraction of the flow entering the culvert and the subsequent rapid expansion of the flow within the culvert barrel. The geometry of the culvert entrance can have a significant effect on this entrance loss. Within

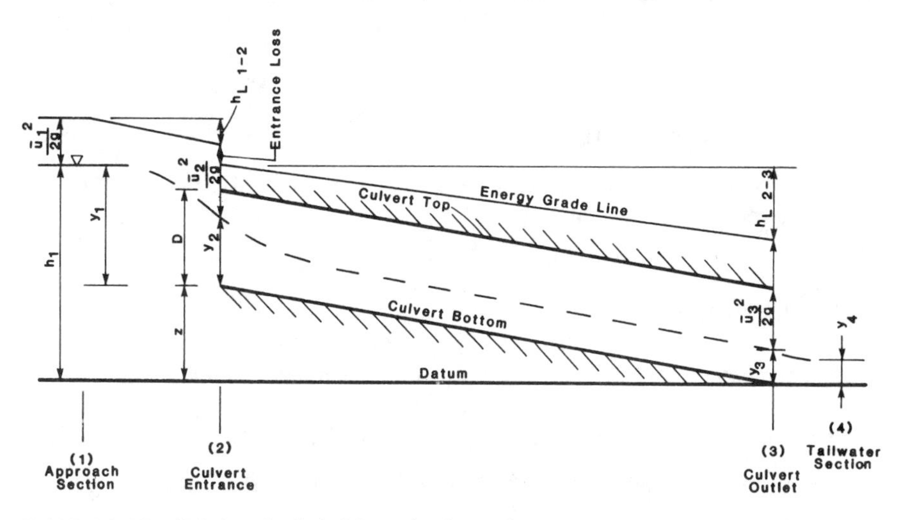

FIGURE 8.25 Schematic definition of culvert flow.

the culvert, there is an additional loss due to boundary friction. Although this friction loss is usually minor, it can be significant in long, rough culverts, or in culverts placed on a flat slope.

The discharge of a culvert is determined by the application of the continuity and energy equations between the approach section and a downstream section which is within the culvert barrel (Fig. 8.25). The location of the downstream section depends on the state of flow within the culvert.

For computational convenience, flow through culverts is divided into six categories based on the relative heights of the head and tailwaters. The six types of flow and their respective characteristics are summarized in Table 8.14. In this table, D = maximum vertical dimension of the culvert, y_1 = depth of flow in the approach section, y_c = critical depth of flow, z = elevation of the culvert entrance relative to a datum through the culvert exit, and y_4 = tailwater depth of flow. In Fig. 8.26, the discharge equations for the various types of culvert flow are summarized. In these equations, C_D = discharge coefficient, A_c = flow area at critical depth, $\bar{u}_1$ = average velocity of the approach section, α_1 = kinetic energy correction coefficient for the approach section, h_{f1-2} = $L_W Q^2/K_1 K_c$ = headloss due to friction from the approach section to the culvert entrance, L_W = distance from the approach section to the culvert entrance, K_1 = the conveyance at the approach section, K_c = the conveyance of the critical depth section, h_{f2-3} = $LQ^2/K_2 K_3$ = headloss due to friction in the culvert barrel, and L = length of the culvert barrel. With regard to these six flow classifications, the following characteristics should be noted.

Type 1 Flow

In this flow classification, critical depth occurs in the vicinity of the culvert entrance. For this type of flow to exist, the following requirements must be met:

1. The headwater-culvert diameter ratio cannot exceed 1.5.
2. The slope of the culvert barrel S_o must be greater than the critical slope S_c.
3. The tailwater elevation y_4 must be less than the elevation of the water surface at the critical section.

Type 2 Flow

In this flow classification, critical depth occurs at the culvert outlet. For this type of flow to exist, the following requirements must be met:

1. The headwater-culvert diameter ratio cannot exceed 1.5.
2. The slope of the culvert barrel S_o must be less than the critical slope S_c.
3. The tailwater elevation y_4 cannot exceed the elevation of the water surface at the critical section.

TABLE 8.14 Culvert flow characteristics (Bodhaine, 1976)

Flow type	Culvert barrel flow	Location of downstream section	Control type	Culvert slope	y_1/D	y_4/y_c	y_4/D
1	Partly full	Inlet	Critical depth	Steep	<1.5	<1.0	≤ 1.0
2	Partly full	Outlet	Critical depth	Mild	<1.5	<1.0	≤ 1.0
3	Partly full	Outlet	Backwater	Mild	<1.5	>1.0	≤ 1.0
4	Full	Outlet	Backwater	Any	>1.0	....	>1.0
5	Partly full	Inlet	Entrance geometry	Any	≥ 1.5	....	≤ 1.0
6	Full	Outlet	Entrance and barrel geometry	Any	≥ 1.5	....	≤ 1.0

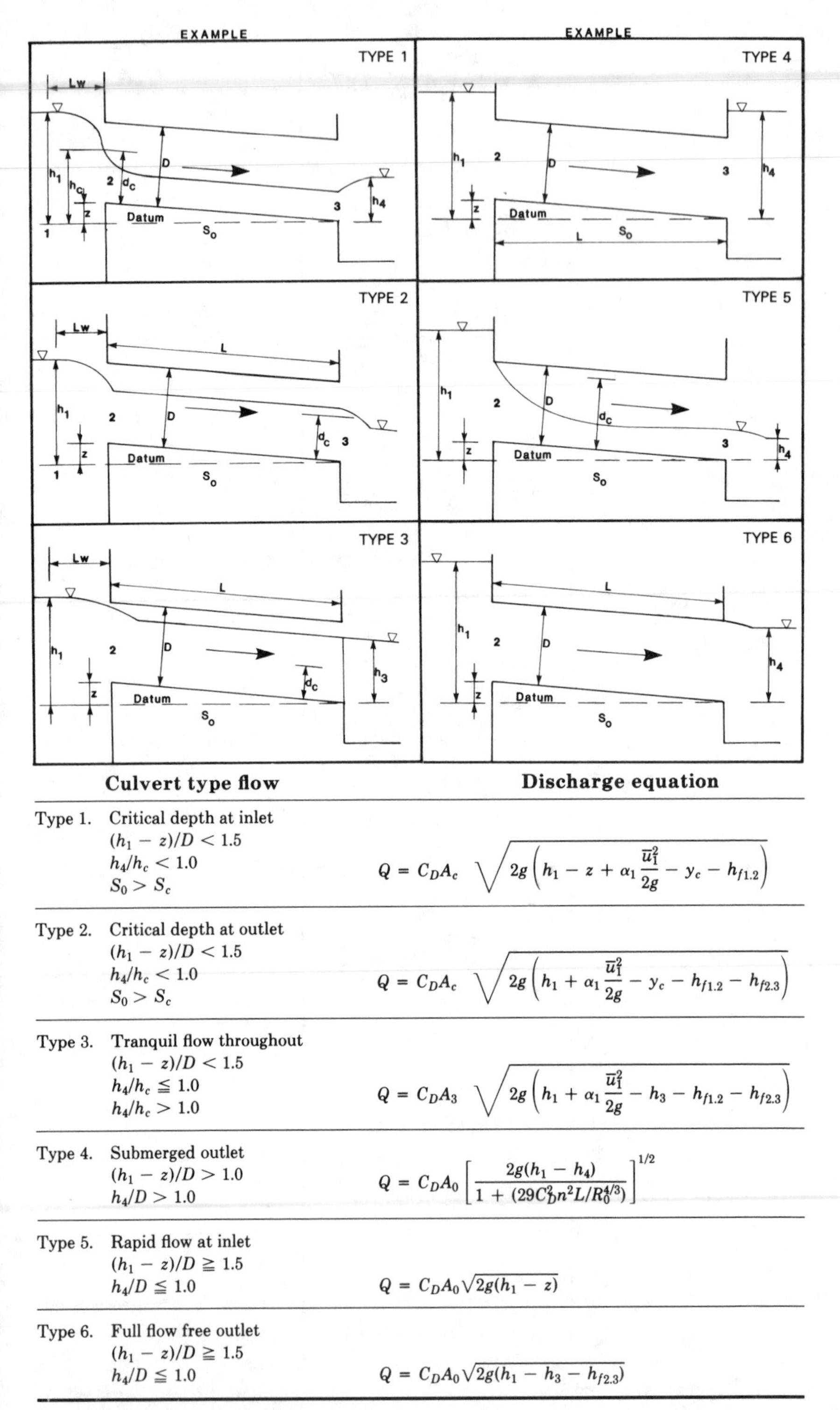

Culvert type flow		Discharge equation
Type 1.	Critical depth at inlet $(h_1 - z)/D < 1.5$ $h_4/h_c < 1.0$ $S_0 > S_c$	$Q = C_D A_c \sqrt{2g\left(h_1 - z + \alpha_1 \dfrac{\overline{u}_1^2}{2g} - y_c - h_{f1.2}\right)}$
Type 2.	Critical depth at outlet $(h_1 - z)/D < 1.5$ $h_4/h_c < 1.0$ $S_0 > S_c$	$Q = C_D A_c \sqrt{2g\left(h_1 + \alpha_1 \dfrac{\overline{u}_1^2}{2g} - y_c - h_{f1.2} - h_{f2.3}\right)}$
Type 3.	Tranquil flow throughout $(h_1 - z)/D < 1.5$ $h_4/h_c \leqq 1.0$ $h_4/h_c > 1.0$	$Q = C_D A_3 \sqrt{2g\left(h_1 + \alpha_1 \dfrac{\overline{u}_1^2}{2g} - h_3 - h_{f1.2} - h_{f2.3}\right)}$
Type 4.	Submerged outlet $(h_1 - z)/D > 1.0$ $h_4/D > 1.0$	$Q = C_D A_0 \left[\dfrac{2g(h_1 - h_4)}{1 + (29C_D^2 n^2 L/R_0^{4/3})}\right]^{1/2}$
Type 5.	Rapid flow at inlet $(h_1 - z)/D \geqq 1.5$ $h_4/D \leqq 1.0$	$Q = C_D A_0 \sqrt{2g(h_1 - z)}$
Type 6.	Full flow free outlet $(h_1 - z)/D \geqq 1.5$ $h_4/D \leqq 1.0$	$Q = C_D A_0 \sqrt{2g(h_1 - h_3 - h_{f2.3})}$

FIGURE 8.26 Classification of culvert flow.

Type 3 Flow

In this type of flow, the existence of a gradually varied flow profile is the controlling factor, critical depth cannot occur, and the upstream water surface elevation is a function of the tailwater elevation. In this flow classification, the flow is subcritical throughout the length of the culvert. For this type of flow to exist, the following requirements must be met:

1. The headwater-culvert diameter ratio must be less than 1.5.
2. The tailwater elevation is not sufficient to submerge the culvert exit; however, it exceeds the elevation of critical depth at the outlet.
3. The lower limit of the tailwater is such that (*a*) the tailwater elevation is greater than the elevation of critical depth at the culvert entrance if flow conditions are such that critical depth would result at the entrance, and (*b*) the tailwater elevation is greater than the elevation of critical depth at the culvert exit if the slope of the culvert is such that critical depth would occur at this section under free-fall conditions.

Type 4 Flow

In this flow classification, the culvert flows full, and the flow rate can be estimated directly from the energy equation. For type 4 flow, the head losses incurred between sections 1 and 2 and sections 3 and 4 can usually be neglected. The loss due to the rapid expansion of the flow field at the culvert outlet is assumed to be given by $(h_3 - h_4)$.

Type 5 Flow

In this flow classification, the flow is supercritical at the inlet to the culvert, and the headwater-culvert diameter ratio exceeds 1.5. Since the tailwater elevation is below the crown of the culvert, the culvert flows partially full.

Type 6 Flow

In this flow classification, the headwater-culvert diameter ratio exceeds 1.5, the culvert flows full, and the culvert outlet is not submerged. Table 8.15 is a decision chart which may prove useful in classifying culvert flows into one of the above categories.

Before the computations required to determine the discharge in each of the above categories are examined, it is necessary to discuss typical values of the Manning roughness coefficient for various culvert materials. The corrugated metal used to manufacture standard culverts has a 2⅔-in (0.068-m) pitch and a rise of 0.5 in (0.013 m). In laboratory tests, Neill (1962) found that for full culvert flow, n ranged from 0.0266 for a 1-ft (0.30-m) diameter culvert to 0.0224

TABLE 8.15

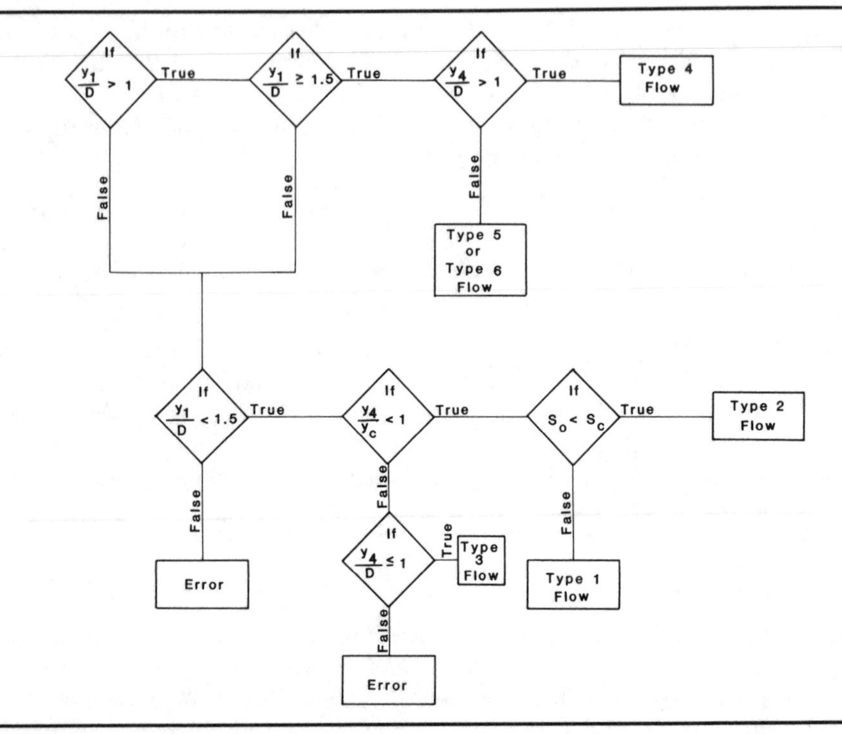

for an 8-ft (2.4-m) diameter culvert. For culverts flowing only partially full, the value of n was slightly smaller. Table 8.16 summarizes these results. In general, 0.024 is considered to be a satisfactory estimate of n for all riveted culverts of standard size.

In multiplate culverts, the corrugations are larger with a 6-in (0.15-m) pitch and 2-in (0.051-m) rise. The values of n recommended by Bodhaine (1976) for this type of culvert construction are summarized in Table 8.17. Bodhaine (1976) also noted that corrugated pipe with corrugations half the size of those used in multiplate construction is being manufactured. In the case of this type of culvert construction, Bodhaine (1976) recommended that until actual test data are available roughness coefficients lying halfway between those recommended for equal sizes of standard and multiplate sections be used; e.g., use a value of n = 0.028 for a 7-ft (2.1-m) diameter pipe.

In culverts constructed of concrete, the roughness coefficient depends on the condition of the concrete and the surface irregularities which are the result of the construction method used to build the culvert. Values of n suggested by Bodhaine (1976) for concrete culverts are summarized in Table 8.18. In concrete culverts, the value of n may be significantly increased if sections of the culvert are displaced either vertically or laterally at the joints. Bodhaine (1976)

TABLE 8.16 Manning's *n* as a function of culvert diameter for standard corrugated metal

Culvert diameter, ft	Manning's n
1	0.027
2	0.025
3–4	0.024
5–7	0.023
8	0.022

TABLE 8.17 Manning's *n* as a function of culvert diameter for multiplate culvert construction (Bodhaine, 1976)

Culvert diameter, ft	Manning's n
5–6	0.034
7–8	0.033
9–11	0.032
12–13	0.031
14–15	0.030
16–18	0.029
19–20	0.028
21–22	0.027

noted that when such displacement occurs, the increase in n is commensurate with the degree of displacement and that the degree of displacement must be significant before the value of n is significantly affected.

The coefficients of discharge, C_D, used in the equations summarized in Fig. 8.27, have been established by laboratory tests and found to vary from 0.39 to 0.98. In general, the coefficient of discharge is a function of the degree of channel contraction and the geometry of the culvert entrance. In the case of some entrance geometries, a base coefficient of discharge must be multiplied by a factor which reflects either the degree of culvert entrance rounding or beveling. Further, coefficient of discharge values are applicable to both single-barrel and multibarrel installations if the web between the barrels in a multibarrel installation is less than 0.1 of the width of a single barrel.

In examining the variation of the coefficient of discharge, it is convenient to consider three groups; i.e., flow types 1, 2, and 3 form one group; flow types 4 and 6 a second group; and flow type 5 a third group. Four classifications of entrance geometry are also considered: (1) flush setting in a vertical headwall, (2) wingwall entrance, (3) projecting entrance, (4) mitered culvert set flush with a sloping embankment.

TABLE 8.18 Manning's *n* as a function of concrete condition for concrete culverts (Bodhaine, 1976)

Concrete condition	Manning's n
Very smooth (spun pipe)	0.010
Smooth (cast or tamped pipe)	0.011–0.015
Ordinary field construction	0.012–0.015
Badly spalled	0.015–0.020

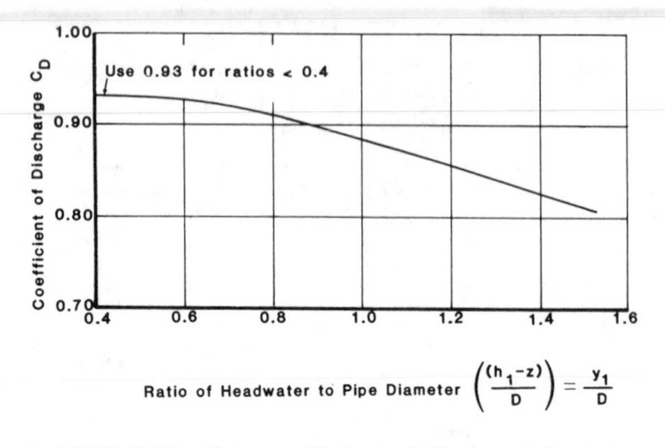

FIGURE 8.27 Base coefficient of discharge for types 1, 2, and 3 flows in pipe culverts with square entrance mounted flush with vertical headwall. *(Bodhaine, 1976.)*

Coefficient of Discharge for Flow Types 1, 2, and 3

The coefficient of discharge for square-ended culverts set flush in a vertical headwall is a function of y_1/D where the base coefficient of discharge for this situation can be determined from Fig. 8.27. If the entrance is rounded or beveled, the coefficient of discharge obtained from Fig. 8.27 must be applied by either k_r or k_w. k_r and k_w are functions of the degree of entrance rounding or beveling and are determined from Figs. 8.28 and 8.29. At this point, it is appropriate to note that tongue-and-groove reinforced concrete culverts ranging in diameter from 18 to 36 in (0.46 to 0.91 meters) have been tested, and no relation between the coefficient of discharge and y_1/D has been found. Therefore, C_D =

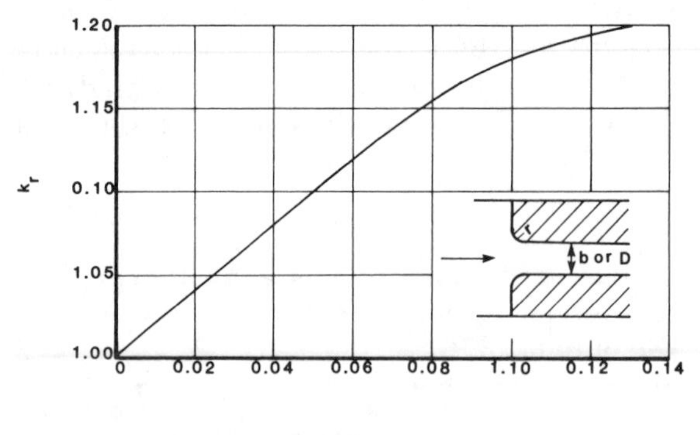

FIGURE 8.28 k_r as a function of r/b or r/D for types 1, 2, and 3 flows in box or pipe culverts set flush with a vertical headwall. *(Bodhaine, 1976.)*

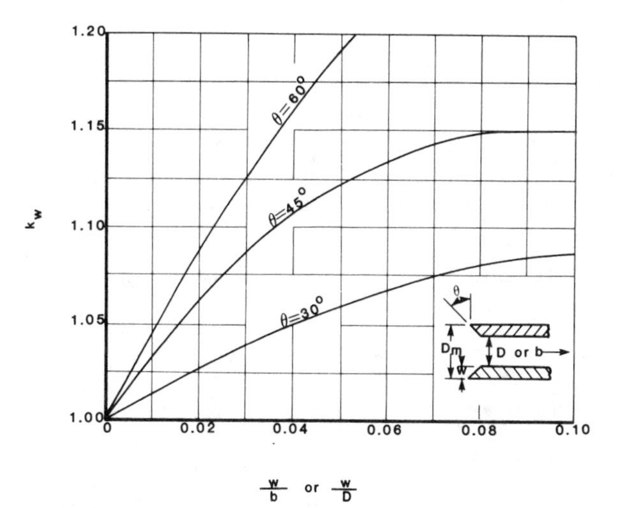

FIGURE 8.29 k_w as a function of w/b or w/D for types 1, 2, and 3 flows in box or pipe culverts set flush with a vertical headwall. *(Bodhaine, 1976.)*

0.95 is used for all sizes of culverts of this type of construction. Bell-mouthed, precast concrete pipe is also assumed to be in this category; that is, $C_D = 0.95$. Table 8.19 summarizes typical values of r/D and w/D for standard riveted corrugated metal culvert. In multiplate culvert construction, the entrance is usually considered to be beveled with w averaging 1.2 in (0.03 m). Bodhaine (1976) recommends that field measurements be used to determine r and w when possible.

The coefficient of discharge for box culverts set flush with a vertical headwall has been found to be a function of the Froude number. For flow types 1 and 2, the Froude number is 1 and $C_D = 0.95$. For type 3 flow, the Froude number should be determined at section 3 (Fig. 8.26), and C_D is read from Fig. 8.30. Figure 8.30 may be extrapolated to Froude numbers of 0.1 if needed. If

TABLE 8.19 Typical values of r/D and w/D as a function of D for standard riveted corrugated-metal culvert (Bodhaine, 1976)

D, in	*D*, m	*r/D*	*w/D*
24	0.61	0.031	0.0125
36	0.91	0.021	0.0083
48	1.2	0.016	0.0062
60	1.5	0.012	0.0050
72	1.8	0.010	0.0042

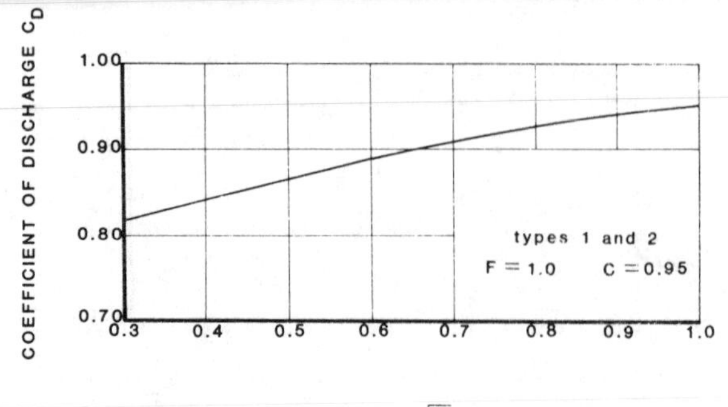

FIGURE 8.30 Base coefficient of discharge for types 1, 2, and 3 flows in box culverts with a square entrance mounted flush in a vertical headwall. *(Bodhaine, 1976.)*

the entrance of the culvert is rounded or beveled, an adjusted coefficient of discharge is obtained by multiplying the base coefficient of discharge by either k_r or k_w which is estimated from Figs. 8.28 and 8.29.

The addition of wingwalls to the entrance of pipe culverts set flush in a vertical headwall has no effect on the coefficient of discharge. In the case of a box culvert, the base coefficient of discharge obtained in Fig. 8.30 must be multiplied by k_θ, a parameter which is a function of the angle of the wingwall and is obtained from Fig. 8.31. If the entrance wingwalls are not symmetrical, then a value of C_D should be determined for each side, independently, and the results averaged.

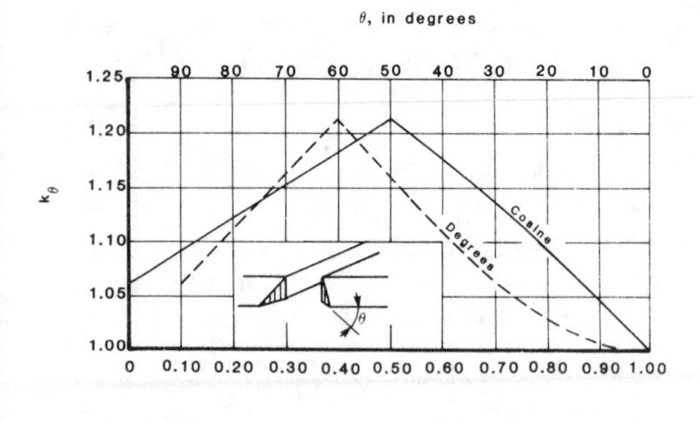

FIGURE 8.31 k_θ as a function of wingwall angle for flow types 1, 2, and 3 in box culverts with wingwalls set flush with a sloping embankment. *(Bodhaine, 1976.)*

TABLE 8.20 k_L **for protruding box culverts as a function of** L_p/D **(Bodhaine, 1976)**

L_p/D	k_L	L_p/D	k_L
0.00	1.00	0.0	1.00
0.01	0.99	0.1	0.92
0.02	0.98	0.2	0.92
0.03	0.98	0.3	0.92
0.04	0.97	0.4	0.91
0.05	0.96	0.5	0.91
0.06	0.95	0.6	0.91
0.07	0.94	0.7	0.91
0.08	0.94	0.8	0.90
0.09	0.93	0.9	0.90
0.10	0.92	≥ 1.0	0.90

For a pipe culvert which extends beyond a headwall or embankment, the coefficient of discharge is determined by following the procedure outlined for a culvert set flush with a headwall and then multiplying by the parameter k_L. In Table 8.20, values of k_L as a function of L_p/D are summarized where L_p = length which the culvert projects beyond the headwall or embankment. The coefficient of discharge to which k_L is applied must not exceed 0.95 because this is the limiting value of C_D. For a projecting concrete culvert with a beveled end, the coefficient of discharge is computed in the same manner as a concrete culvert set flush with the headwall.

For a mitered culvert set flush with a sloping bank, the coefficient of discharge is estimated from Fig. 8.32.

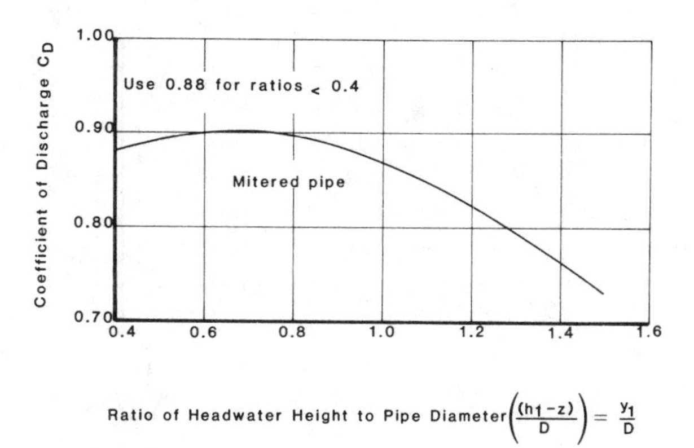

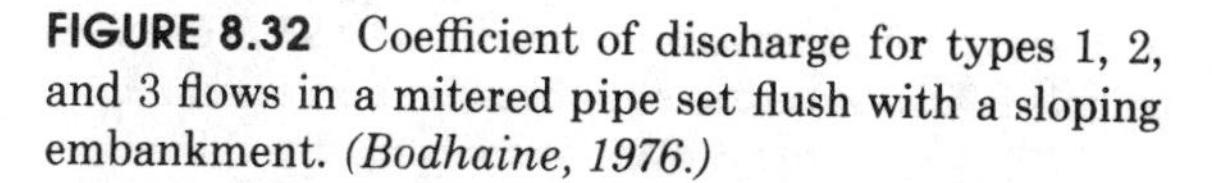

FIGURE 8.32 Coefficient of discharge for types 1, 2, and 3 flows in a mitered pipe set flush with a sloping embankment. *(Bodhaine, 1976.)*

For all culverts conveying types 1, 2, and 3 flows one final adjustment of the coefficient of discharge is required. Define

$$m = 1 - \frac{A}{A_1} \tag{8.5.1}$$

where m = ratio of channel contraction
A_1 = flow area at approach section
A = flow area at terminal section

For $m = 0.80$, an adjusted value of the coefficient of discharge can be estimated from Fig. 8.33. For values of $m < 0.80$, the adjusted coefficient can be estimated by

$$C_D' = 0.98 - \frac{(0.98 - C_D)m}{0.80} \tag{8.5.2}$$

EXAMPLE 8.4

Figure 8.34 describes schematically a flow which is known to occur through a corrugated metal pipe culvert with a 10-ft (3.0-m) diameter. If $r = 0.06$ ft (0.018 m), $n = 0.024$, $A_1 = 1000$ ft^2 (93 m^2), and $K_1 = 300{,}000$, determine the flow rate if the culvert is set flush with the headwall. This example was first presented by Bodhaine (1976).

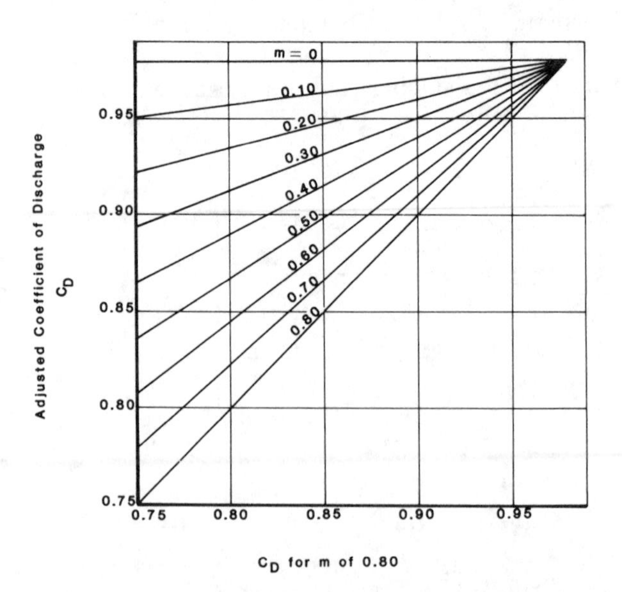

FIGURE 8.33 Adjustment of coefficient of discharge for degree of channel contraction for types 1, 2, and 3 flows. *(Bodhaine, 1976.)*

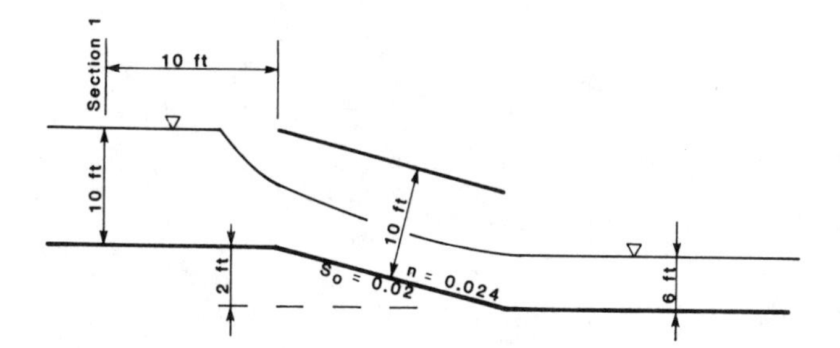

FIGURE 8.34 Schematic definition for Example 8.4.

Solution

From the problem statement and Fig. 8.34, the following data are extracted:

$$y_1 = 10 \text{ ft } (0.30 \text{ m})$$

$$S_o = 0.02$$

$$\frac{r}{D} = \frac{0.06}{10} = 0.006$$

$$y_4 = 6 \text{ ft } (1.8 \text{ m})$$

$$\frac{y_1}{D} = 1$$

From Table 8.15, it is determined that the flow must be type 1, 2, or 3 since y_1/D does not exceed 1. Therefore, from Fig. 8.27 $C_D = 0.88$, and from Fig. 8.28 for $r/D = 0.006$, $k_r = 1.01$. The adjusted coefficient of discharge is

$$C_D = k_r C_D = 1.01(0.88) = 0.89$$

The next computational step depends on classifying the flow, and the next classification step in Table 8.15 requires an estimate of the critical depth of flow. It is observed that if $y_4 < y_c$, type 3 flow is not possible. To differentiate between types 1 and 2 flows, a graph developed by Bodhaine (1976) is used (Fig. 8.35). For

$$\frac{S_o D^{1/3}}{n^2} = \frac{0.02(10)^{1/3}}{(0.024)^2} = 74.8$$

and

$$\frac{y_1}{D} = 1$$

it is concluded that this may be a type 1 flow. Then, from Fig. 8.36 with $y_1/D = 1$

$$\frac{y_c}{D} = 0.65$$

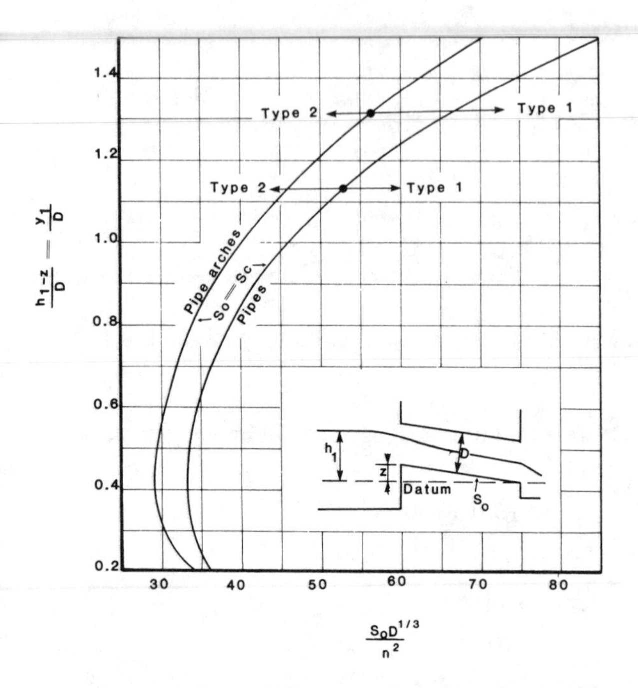

FIGURE 8.35 Critical slope as a function of approach section head for types 1 and 2 flows in pipes and pipe arches. *(Bodhaine, 1976.)*

and $\qquad y_c = 0.65\ D = 0.65\ (10) = 6.5 \text{ ft } (2.0 \text{ m})$

The flow rate is then estimated from the empirical equation specified in Table 2.1 for circular open channels or

$$y_c = \frac{1.01}{D^{0.26}}\left(\frac{Q^2}{g}\right)^{0.25}$$

where α has been assumed to be 1. Rearranging and solving this equation for an estimate of Q yields

$$Q = \left(\frac{y_c D^{0.26}}{1.01}\, g^{0.25}\right)^2 = \left[\frac{6.5(10)^{0.26}}{1.01}\,(32.2)^{0.25}\right]^2 = 780 \text{ ft}^3/\text{s } (22 \text{ m}^3/\text{s})$$

The headloss due to friction between the approach section and the culvert inlet is given by

$$h_{f1-2} = \frac{Q^2 L}{K_1 K_c}$$

where $\qquad K_c = \dfrac{1.49}{n} A_c R_c^{2/3}$

From Table 1.1

$$A_c = \tfrac{1}{8}(\theta - \sin\theta)D^2$$

and

$$R_c - \tfrac{1}{4}\left(1 - \frac{\sin\theta}{\theta}\right)D$$

For $\theta = 3.75$ rad

$$A_c = \tfrac{1}{8}[3.75 - \sin(3.75)]10^2 = 54 \text{ ft}^2\ (5.0 \text{ m}^2)$$

$$R_c = \tfrac{1}{4}\left[1 - \frac{\sin(3.75)}{3.75}\right]10 = 2.9 \text{ ft}\ (0.88 \text{ m})$$

and

$$K_c = \frac{1.49}{0.024}(54)(2.9)^{2/3} = 6820$$

$$h_{f1-2} = \frac{Q^2 L_w}{K_1 K_c} = \frac{(780)^2 10}{300{,}000(6820)} = 0.003$$

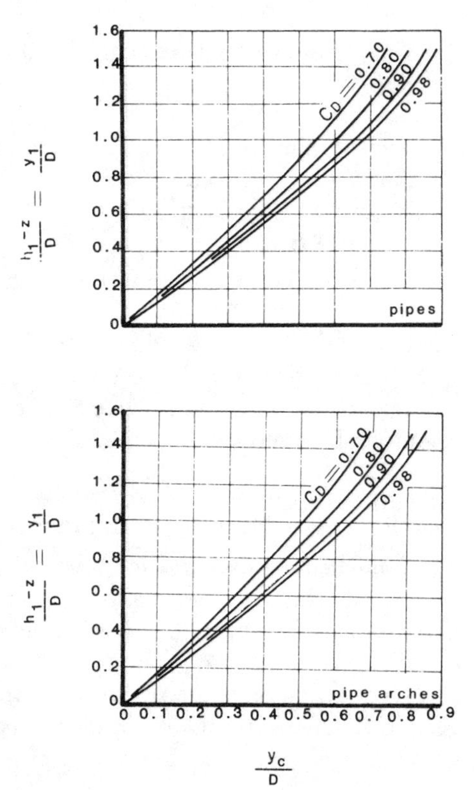

FIGURE 8.36 The relation between y_1/D, C_D, and critical depth for pipe and pipe arch culverts. *(Bodhaine, 1976.)*

The velocity head at the approach section is

$$\frac{\overline{u}_1^2}{2g} = \frac{(Q/A_1)^2}{2g} = \frac{(780/1000)^2}{2(32.2)} = 0.0094$$

At this point, the coefficient of discharge can be adjusted for the channel contraction or

$$m = 1 - \frac{A}{A_1} = 1 - \left(\frac{54}{1000}\right) = 0.946$$

and from Eq. (8.5.2)

$$C_D = 0.98 - \left(\frac{0.98 - C_D}{0.80}\right) m = 0.98 - \left(\frac{0.98 - 0.89}{0.80}\right) 0.946 = 0.87$$

From Fig. 8.24, the discharge for type 1 culvert flow is given by

$$\begin{aligned} Q &= C_D A_c \left[2g \left(h_1 + \frac{\overline{u}_1^2}{2g} - y_c - h_{f1-2} \right) \right]^{1/2} \\ &= 0.87(54)[2(32.2)(10 + 0.0094 - 6.5 - 0.003)] \\ &= 710 \text{ ft}^3/\text{s } (20 \text{ m}^3/\text{s}) \end{aligned}$$

This final answer is reasonably close to the first estimate of the flow rate.

The fact that this flow is a type 1 flow can then be confirmed by demonstrating that $S_o > S_c$ (Table 8.14).

The critical slope is given by

$$S_c = \left(\frac{Q}{K_c}\right)^2 = \left(\frac{710}{6820}\right)^2 = 0.0108$$

Therefore, this is a type 1 flow.

At this point it should be noted that if the culvert under consideration is rectangular in shape, then Fig. 8.37 should be used in place of Fig. 8.35 to differentiate between types 1 and 2 flows.

Coefficient of Discharge for Flow Types 4 and 6

The coefficient of discharge for pipe and box culverts set flush in a vertical headwall is determined from Table 8.21 for these flow types. For culverts with flared-pipe end sections, $C_D = 0.90$ regardless of the culvert diameter and value of y_1/D.

The addition of wingwalls to the entrance of a pipe culvert set flush with a vertical headwall has no effect on the coefficient of discharge. For box culverts with wingwalls and a square top entrance

$$C_D = 0.87 \qquad \text{for } 30° \leq \theta \leq 75°$$

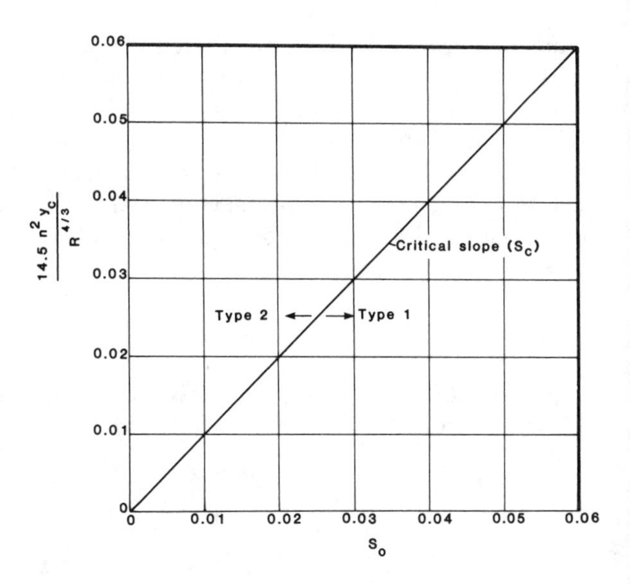

FIGURE 8.37 Critical slope for culverts of rectangular shape, types 1 and 2 flows. *(Bodhaine, 1976.)*

and $$C_D = 0.75 \qquad \text{for } \theta = 90°$$

where the angle θ is defined in Fig. 8.26. If the top entrance of the box culvert with wingwalls is beveled or rounded with $30° \le \theta \le 75°$, then C_D is selected from Table 8.21 on the basis of r/D or w/D for the top of the entrance. In such a case, $C_D = 0.87$ is the lower limiting value. For the case where $\theta = 90°$ and the top of the entrance is rounded or beveled, the base value of the coefficient of discharge is 0.75, and this value must be adjusted by either k_r or k_w, which can be obtained from either Fig. 8.28 or 8.29. For situations in which $75° \le \theta \le 90°$, values of C_D are interpolated between 0.87 and 0.75, respectively, and then adjusted by either k_r or k_w as required.

In the case of corrugated metal culverts which protrude from a headwall or

TABLE 8.21 Coefficients of discharge for box and pipe culverts, flow types 4 and 6 (Bodhaine, 1976)

r/b, w/b, w/D, or r/D	C_D
0	0.84
0.02	0.88
0.04	0.91
0.06	0.94
0.08	0.96
0.10	0.97
0.12	0.98

embankment, C_D is determined from Table 8.21 and then adjusted by the parameter k_L which is obtained from Table 8.20. For concrete culverts with beveled ends, which project from an embankment or headwall, C_D is determined from Table 8.21.

In the case of mitered-pipe culverts which are set flush with an embankment, $C_D = 0.74$. If the mitered-pipe culvert extends beyond the embankment, the coefficient of discharge is estimated by multiplying 0.74 by the parameter k_L which is estimated from Table 8.20.

EXAMPLE 8.5

Flow takes place through a level concrete culvert 4.0 ft (1.2 m) in diameter. If conditions are as shown in Fig. 8.38 and

$$w = 0.3 \text{ ft } (0.09 \text{ m})$$

$$y_1 = 7.0 \text{ ft } (2.0 \text{ m})$$

$$y_4 = 5.0 \text{ ft } (1.5 \text{ m})$$

$$z = 0$$

$$L_{2-3} = 50 \text{ ft } (15 \text{ m})$$

$$n = 0.012$$

estimate the flow rate. This example was first presented by Bodhaine (1976).

Solution

From the given date

$$\frac{y_1}{D} = \frac{7}{4} = 1.75$$

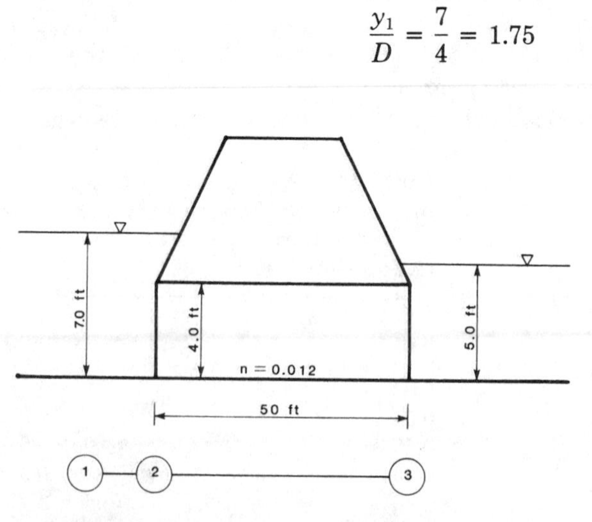

FIGURE 8.38 Schematic for Example 8.5.

and
$$\frac{y_4}{D} = \frac{5}{4} = 1.25$$

Therefore, by Tables 8.14 and 8.15, it is concluded that this is a type 4 culvert flow.

The coefficient of discharge is determined from Table 8.21 with $w/D = 0.3/4 = 0.075$ by linear interpolation or

$$C_D = 0.955$$

From Fig. 8.26

$$Q = C_D A_o \left[\frac{2g(h_1 - h_4)}{1 + (29C_D^2 n^2 L_{23}/R_o^{4/3})} \right]^{1/2}$$

where the subscript o designates a full culvert value. For a 4.0-ft (1.2-m) diameter culvert,

$$A_o = \frac{\pi D^2}{4} = \frac{\pi(4)^2}{4} = 12.6 \text{ ft}^2 \ (1.17 \text{ m}^2)$$

$$P_o = \pi D = \pi(4) = 12.6 \text{ ft} \ (3.84 \text{ m})$$

$$R_o = \frac{A_o}{P_o} = \frac{12.6}{12.6} = 1 \text{ ft} \ (0.30 \text{ m})$$

Then

$$Q = 0.955(12.6) \left\{ \frac{2(32.2)(7.0 - 5.0)}{1 + [29(0.955)^2(0.012)^2(50)/(1)^{4/3}]} \right\}^{1/2}$$
$$= 130 \text{ ft}^3/\text{s} \ (3.7 \text{ m}^3/\text{s})$$

Coefficient of Discharge for Type 5 Flow

The coefficients of discharge for pipe or box culverts set flush in a vertical headwall for type 5 flow are summarized in Table 8.22. Although type 5 flow does not normally occur in culverts with flared-pipe end sections, if $L/D < 6$ and $S_o > 0.03$, type 5 flow may occur. Even under these conditions, type 5 flow may eventually become type 6 flow. However, if type 5 flow is believed to occur, then the coefficient of discharge is a function of y_1/D (Table 8.23).

The coefficient of discharge for pipe culverts set flush with a vertical headwall is not affected by the addition of wingwalls. In the case of a box culvert with wingwalls and a square-top entrance, the coefficient of discharge is a function of y_1/D and θ (Table 8.24).

In the case of a corrugated metal culvert which projects from either a headwall or an embankment, a base coefficient of discharge is determined from Table 8.22 and modified by the parameter k_L (Table 8.20). For projecting concrete-pipe culverts with beveled ends, the coefficient of discharge is determined from Table 8.22.

TABLE 8.22 Coefficient of discharge for pipe or box culverts set flush in a vertical headwall as a function of y_1/D and r/D or w/D, type 5 flow (Bodhaine, 1976)

	r/b, w/b, r/D, or w/D						
$(h_1 - z)/D$	0	0.02	0.04	0.06	0.08	0.10	0.14
1.4	0.44	0.46	0.49	0.50	0.50	0.51	0.51
1.5	0.46	0.49	0.52	0.53	0.53	0.54	0.54
1.6	0.47	0.51	0.54	0.55	0.55	0.56	0.56
1.7	0.48	0.52	0.55	0.57	0.57	0.57	0.57
1.8	0.49	0.54	0.57	0.58	0.58	0.58	0.58
1.9	0.50	0.55	0.58	0.59	0.60	0.60	0.60
2.0	0.51	0.56	0.59	0.60	0.61	0.61	0.62
2.5	0.54	0.59	0.62	0.64	0.64	0.65	0.66
3.0	0.55	0.61	0.64	0.66	0.67	0.69	0.70
3.5	0.57	0.62	0.65	0.67	0.69	0.70	0.71
4.0	0.58	0.63	0.66	0.68	0.70	0.71	0.72
5.0	0.59	0.64	0.67	0.69	0.71	0.72	0.73

For mitered-pipe culverts set flush with an embankment, the coefficient of discharge is estimated by determining a base coefficient from Table 8.22 and multiplying this base coefficient by 0.92. If the mitered culvert is thin-walled, e.g., corrugated metal, the parameter k_L (Table 8.20) should also be used.

TABLE 8.23 Coefficient of discharge as a function of y_1/D for type 5 flow in a culvert with a flared-pipe end section (Bodhaine, 1976)

$(h_1 - z)/D$	C_D	$(h_1 - z)/D$	C_D
1.4	0.48	2.0	0.57
1.5	0.50	2.5	0.59
1.6	0.52	3.0	0.61
1.7	0.53	3.5	0.63
1.8	0.55	4.0	0.65
1.9	0.56	5.0	0.66

EXAMPLE 8.6

Flow takes place in a 4.0-ft (1.2-m) diameter corrugated-metal culvert. If the situation is as shown in Fig. 8.39 and

$$\frac{r}{D} = 0.016$$

$y_1 = 6.0$ ft (1.8 m)

$z = 2.0$ ft (0.61 m)

$L_{2-3} = 50$ ft (15 m)

$S_o = 0.04$

$y_4 = 1.0$ ft (0.30 m)

estimate the discharge. This example was first used by Bodhaine (1976).

TABLE 8.24 Coefficient of discharge for box culverts with wingwalls, type 5 flow (Bodhaine, 1976)

	Wingwall angle, θ				
$(h_1 - z)/D$	30°	45°	60°	75°	90°
1.3	0.44	0.44	0.43	0.42	0.39
1.4	0.46	0.46	0.45	0.43	0.41
1.5	0.47	0.47	0.46	0.45	0.42
1.6	0.49	0.49	0.48	0.46	0.43
1.7	0.50	0.50	0.48	0.47	0.44
1.8	0.51	0.51	0.50	0.48	0.45
1.9	0.52	0.52	0.51	0.49	0.46
2.0	0.53	0.53	0.52	0.49	0.46
2.5	0.56	0.56	0.54	0.52	0.49
3.0	0.58	0.58	0.56	0.54	0.50
3.5	0.60	0.60	0.58	0.55	0.52
4.0	0.61	0.61	0.59	0.56	0.53
5.0	0.62	0.62	0.60	0.58	0.54

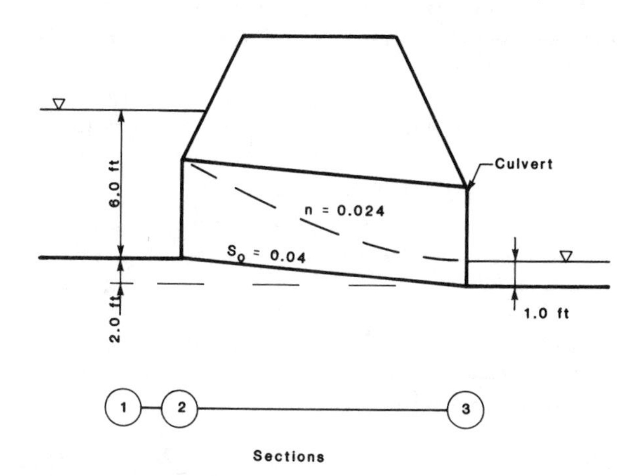

FIGURE 8.39 Schematic for Example 8.6.

Solution

From the given data

$$\frac{y_1}{D} = \frac{6}{4} = 1.5$$

and

$$\frac{y_4}{D} = \frac{1}{4} = 0.25$$

Therefore, from Tables 8.14 and 8.15, it is concluded that this is either a type 5 or 6 culvert flow. Bodhaine (1976) presented two figures for differentiating between types 5 and 6 culvert flows, Figs. 8.40 and 8.41. Using Fig. 8.40,

$$\frac{r}{D} = 0.016$$

$$\frac{L}{D} = \frac{50}{4} = 12.5$$

$$R_o = \frac{(\pi D^2/4)}{\pi D} = \frac{D}{4} = 1 \text{ ft } (0.30 \text{ m})$$

and

$$\frac{29n^2 y_1}{R_o^{4/3}} = \frac{29(0.024)^2 6}{1} = 0.10$$

Therefore, from this figure, it is concluded that this is a type 5 flow. By use of these data in Fig. 8.41, it is also concluded that this is a type 5 flow.

For a type 5 flow, the coefficient of discharge is estimated from Table 8.22; or for $r/D = 0.016$, $C_D = 0.484$. From Fig. 8.26

$$Q = C_D A_o \sqrt{2gy_1} = 0.484 \frac{\pi(4)^2}{4} [2(32.2)(6)]^{1/2}$$

$$= 120 \text{ ft}^3/\text{s } (3.4 \text{ m}^3/\text{s})$$

EXAMPLE 8.7

Flow takes place in a concrete culvert 4.0 ft (1.2 m) in diameter with a beveled entrance (Fig. 8.41). If a ponded flow condition exists upstream, and

$$\frac{w}{D} = 0.075$$

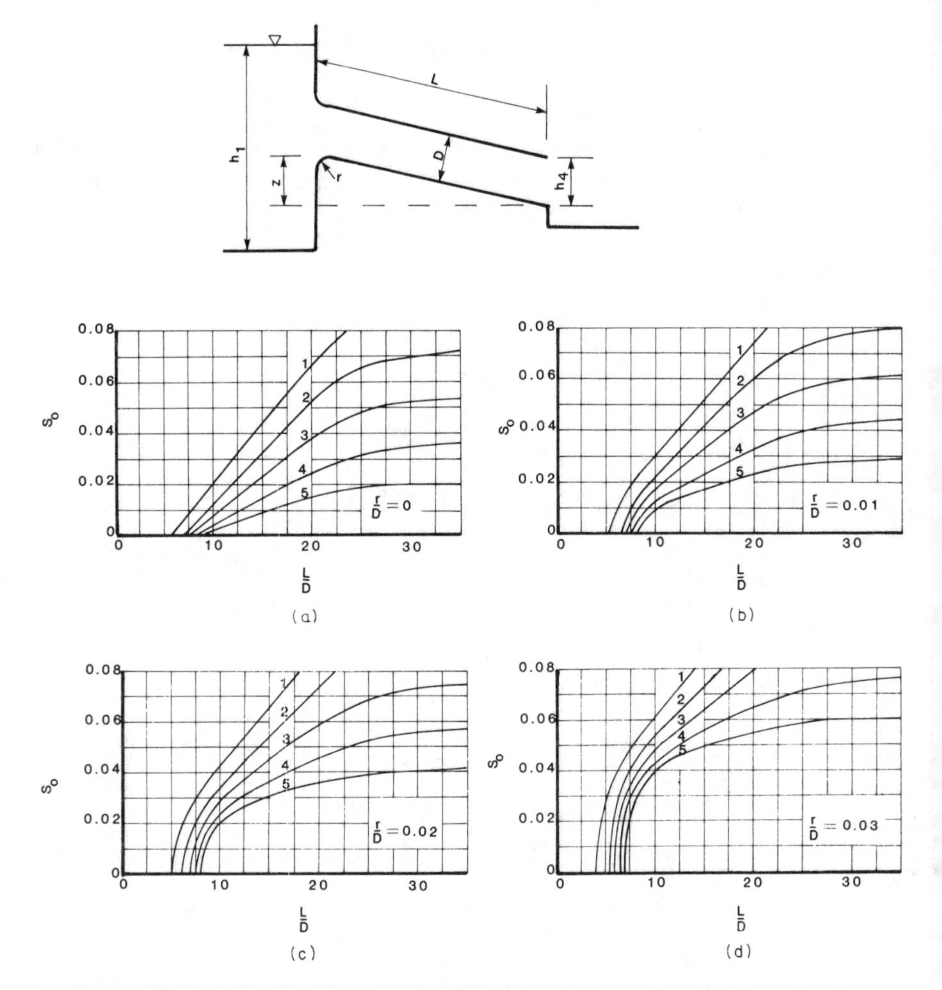

FIGURE 8.40 Criterion for differentiating between types 5 and 6 flows in pipe or box culverts with rough barrels. *(Bodhaine, 1976.)*

$$y_1 = 7.0 \text{ ft } (2.1 \text{ m})$$

$$y_4 = 1.0 \text{ ft } (0.30 \text{ m})$$

$$z = 1.0 \text{ ft } (0.30 \text{ m})$$

$$L_{2-3} = 50 \text{ ft } (15 \text{ m})$$

$$n = 0.012$$

estimate the discharge under these conditions. This example was first presented by Bodhaine (1976).

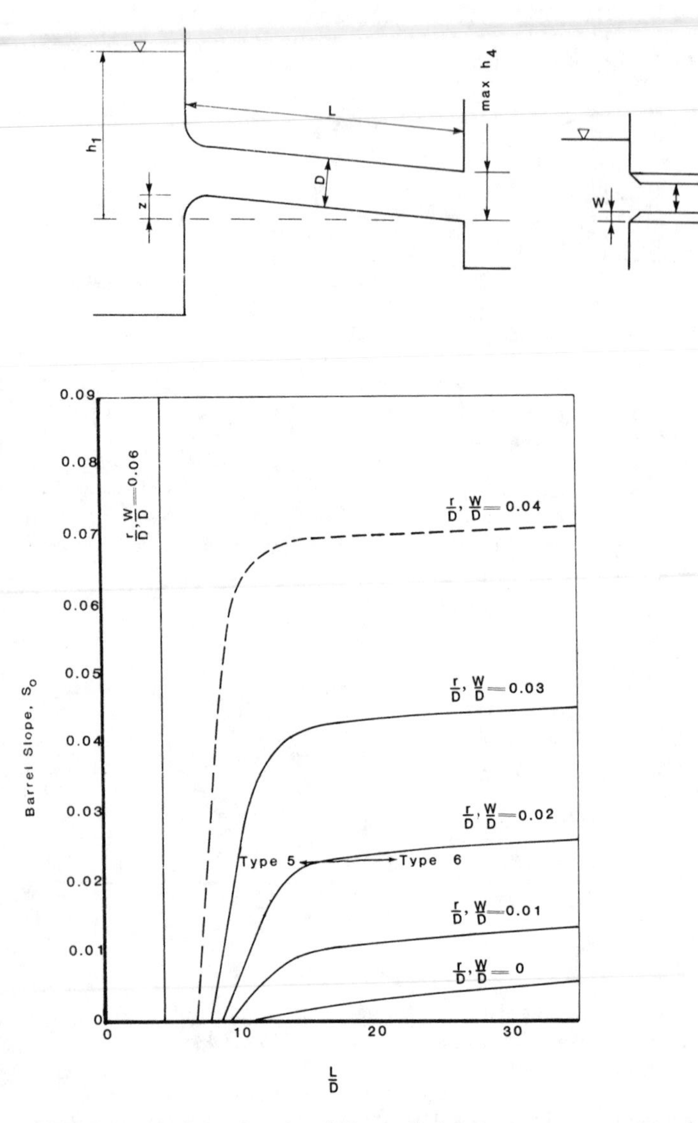

FIGURE 8.41 Criterion for differentiating between types 5 and 6 flows in pipe or box culverts with concrete barrels, square, rounded, or beveled entrances, and with or without wingwalls. *(Bodhaine, 1976.)*

Solution

For the given conditions, from Tables 8.14 and 8.15, type 5 or 6 flow is indicated since $y_1/D = 7/4 = 1.7$ and $y_4/D = 1/4 = 0.25$. From Fig. 8.40 with $L/D = 50/4 = 12.5$ and $S_o = 1/50 = 0.02$, type 6 flow is indicated.

From Fig. 8.26 the equation for estimating the discharge under these conditions is

$$Q = C_D A_o \sqrt{2g(h_1 - h_3 - h_{f2-3})}$$

The coefficient of discharge from Table 8.21 is $C_D = 0.95$, and the head-loss due to friction between sections 2 and 3 is given by

$$h_{f2-3} = \frac{Q^2 L}{K_2 K_3}$$

Since Q must be known before h_{f2-3} can be estimated, a trial-and-error solution is required. The initial estimate of Q is obtained by assuming $h_{f2-3} = 0$ or

$$Q = C_D A_o \sqrt{2g(h_1 - h_3)} = 0.95 \frac{\pi(4)^2}{4} \sqrt{2(32.2)(8 - 1)}$$

$$= 267 \text{ ft}^3\text{/s } (7.6 \text{ m}^3\text{/s})$$

Then

$$K_2 = K_3 = \frac{1.49}{n} AR^{2/3} = \frac{1.49}{0.012} \left[\frac{\pi(4)^2}{4}\right] \left\{\frac{[\pi(4)^2]/4}{4\pi}\right\}$$

$$= K_3 = 1560$$

and

$$h_{f2-3} = \frac{Q^2 L}{K_2 K_3} = \frac{(267)^2(50)}{(1560)^2} = 1.5$$

At this point a table can be constructed to continue the solution process.

Trial Q, ft³/s	h_{f2-3}	New estimates of Q, ft³/s
256	1.5	225
225	1.0	235
235	1.1	233

Therefore, for the given conditions it is estimated that

$$Q = 230 \text{ ft}^3\text{/s } (6.5 \text{ m}^3\text{/s})$$

In the foregoing paragraphs the methods of estimating the discharge through a culvert when certain variable values are specified have been discussed. Under most field conditions, the computation of culvert discharge is reasonably accurate. In low-head flow situations, i.e, where $y_1/D < 1.25$, very accurate results may be expected unless critical depth occurs between the approach section and the culvert entrance. In high-head flow situations the results may also be expected to be accurate if the flow type is definitely known and $y_1/D > 1.75$. The results for type 6 culvert flows are also accurate. In the range of transition between culvert flow types, the discharges estimated are less accurate. Bodhaine (1976) noted that in this situation better results are to be expected from circular culverts than from box culverts. In assessing the accuracy of culvert-derived discharge estimates, Bodhaine (1976) noted six factors which should be considered:

1. The accuracy with which the headwater and tailwater elevations are known
2. The stability of the approach channel
3. The closeness of the entrance conditions to standard conditions
4. The shape and condition of the culvert
5. Scour of fill in the culvert
6. The possibility of the culvert being partially plugged by debris

Bodhaine (1976) describes how stage-discharge relationships can be established for culverts, and the reader is referred to this document for additional information regarding culvert flows.

BIBLIOGRAPHY

Bodhaine, G. L., "Measurement of Peak Discharge at Culverts by Indirect Methods," *Techniques of Water Resources Investigations of the United States Geological Survey,* book 3, chapter A3, U.S. Geological Survey, Washington, 1976.

Bos, M. G. (Ed.), *Discharge Measurement Structures,* International Institute for Land Reclamation and Improvement, Wagemingen, The Netherlands, 1976.

Carter, R. W., and Davidian, J., "General Procedure for Gaging Streams," *Techniques of Water Resources Investigation of the United States Geological Survey,* book 3, chapter A6, U.S. Geological Survey, Washington, 1968.

Corbett, D. M., et al., "Stream Gaging Procedure," Water Supply Paper 888, U.S. Geological Survey, Washington, 1943.

Kindsvater, C. E., and Carter, R. W. C., "Discharge Characteristics of Rectangular Thin Plate Weirs," *Proceedings of the American Society of Civil Engineers, Journal of the Hydraulics Division,* vol. 83, no. HY6, December 1957, pp. 1453/1–1453/36.

Linsley, R. K., Jr., Kohler, M. H., and Paulhus, J. L. H., *Hydrology for Engineers,* McGraw-Hill Book Company, New York, 1958.

Neill, C. R., "Hydraulic Tests on Pipe Culverts," Alberta Highway Research Report 62-1, Edmonton Research Council, Edmonton, Canada, 1962.

Parshall, R. L., Discussion of "Measurement of Debris-Laden Stream Flow with Critical Depth Flumes," *Transactions of the American Society of Civil Engineers,* vol. 103, 1938, pp. 1254–1257.

Smoot, G. F. and Novak, C. E., "Measurement of Discharge by the Moving Boat Method," *Techniques of Water Resources Investigations of the United States Geological Survey,* book 3, chapter A11, U.S. Geological Survey, Washington, 1969.

Soucek, J., Howe, H. E., and Mavis, F. T., "Sutro Weir Investigations Furnish Discharge Coefficients," *Engineering News Record,* Nov. 12, 1936.

Streeter, V. L., and Wylie, E. B., *Fluid Mechanics,* McGraw-Hill Book Company, New York, 1975.

Wilm, H. G., Cotton, J. S., and Storey, H. C., "Measurement of Debris-Laden Stream Flow with Critical-Depth Flumes," *Transactions of American Society of Civil Engineers,* vol. 103, 1938, pp. 1237–1253.

NINE

Rapidly Varied Flow in Nonprismatic Channels

SYNOPSIS

In this chapter, rapidly varied flow through nonprismatic channel sections is considered. In particular, the chapter contains discussions of flow through bridge contractions; the control of hydraulic jumps with sharp- and broad-crested weirs, abrupt rises and drops, and stilling basins; drop spillway structures; and the design of channel transitions. It is also noted that previous chapters have treated other rapidly varied flow in nonprismatic channels such as broad- and sharp-crested weirs and critical flow flumes. The thrust of this chapter is applied rather than theoretical.

9.1 INTRODUCTION

The occurrence of rapidly varied flow through a nonprismatic section is not unusual in the study of open-channel hydraulics. For example, in previous chapters broad-crested weirs, sharp-crested weirs, and critical flow flumes have been discussed, and all these are channel appurtenances which involve rapidly varied flow through nonprismatic channels. In this chapter, several additional examples of this type of flow will be considered.

9.2 BRIDGE PIERS

Bridge piers are an obstruction to the flow; therefore, there must be an interaction between the piers and the flow. Although the force of the flow on the piers is minor in comparison with the other loads such structural elements must bear, one cannot ignore the backwater effect of the piers on the flow which results in increased depths of flow upstream of the bridge. The significance and importance of the upstream backwater effect derives from the increased economic costs which will be incurred for flood protection. Thus, it is highly desirable that the engineer be able to estimate accurately the backwater effect for a specified flow condition and a set of bridge piers.

In a schematic fashion, Fig. 9.1 defines the situation which occurs at a single pair of bridge piers. When an area constriction such as that shown in Fig. 9.1 is introduced into a friction-controlled flow, then:

1. The water in the vicinity of section 1 begins to accelerate in the center area, while in the vicinity of the boundaries it decelerates. An adequate approximation of the location of section 1 is determined by assuming that it is one opening width upstream from the center of the opening (Chow, 1959). Matthai (1976) asserted section 1 occurs at a distance equal to twice the opening width upstream from the bridge.
2. At the constriction, the flow is characterized by accelerations both parallel and normal to the streamlines and by a rapid lowering of the water surface.

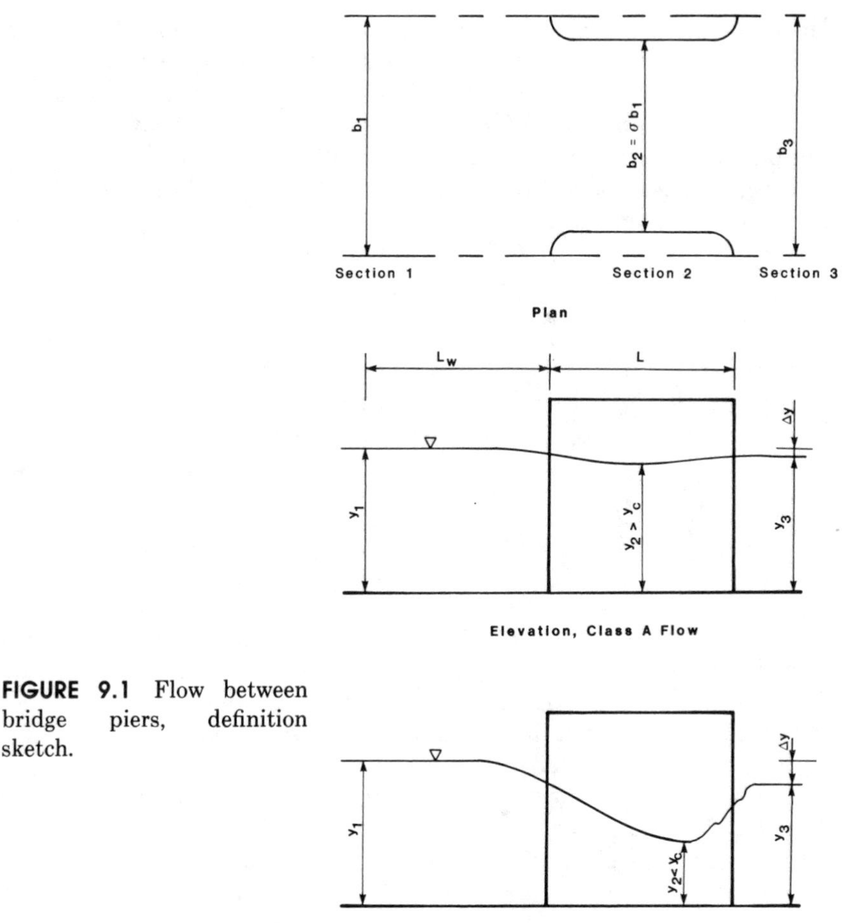

FIGURE 9.1 Flow between bridge piers, definition sketch.

3. Within the constriction, the "live stream" contracts to a width that is less than the width of the opening. The space between the live stream and the solid boundary is occupied by eddies.
4. At section 2, the live stream reaches a minimum width which is analogous to the *vena contracta* in an orifice flow.
5. At section 4 a uniform flow is reestablished. Between sections 3 and 4 the flow is gradually varied.

Given the situation described above, there are a number of possible approaches to a solution. For example, it could be assumed that

$$M_1 - M_2 = P_f \tag{9.2.1}$$

where M_1 and M_2 = specific momentum functions at sections 1 and 2, respectively, and P_f = drag force of the pier. Although Eq. (9.2.1) is a rational approach to a solution, it is in general not considered adequate because it does not take into account the shape of the piers. The accurate solution of the bridge pier problem requires recourse to experiments.

Yarnell (1934a and 1934b) examined the flow through bridge piers experimentally and determined

$$\frac{\Delta y}{y_3} = k\,\mathbf{F}_3^2\,(k + 5\mathbf{F}_3^2 - 0.6)\,(\sigma + 15\sigma^4) \tag{9.2.2}$$

where, with reference to Fig. 9.1, Δy = difference in the depth of flow between sections 1 and 3; $\mathbf{F}_3$ = Froude number at section 3; k = an empirical parameter characterizing the bridge pier shape (Table 9.1); and

$$\sigma = 1 - \Gamma \tag{9.2.3}$$

where

$$\Gamma = \frac{b_2}{b_1} \tag{9.2.4}$$

In his experiments, Yarnell (1934a, 1934b) examined situations in which Γ ranged between 0.88 and 0.50. Although these values of Γ represent contractions which are more severe than those found in modern practice, the systematic trend of the Yarnell data is such that extrapolation to modern values of σ in the range of 0.95 should be valid.

The Yarnell experiments assume that σ is not small enough to choke the flow, i.e., yield critical conditions. The limiting value of Γ, that is, the value of Γ below which the flow will be choked, can be estimated under two different assumptions. If $E_1 = E_2$, then the limiting value of Γ can be estimated by

$$\Gamma_L = \left[\frac{27\mathbf{F}_1^2}{(2 + \mathbf{F}_1^2)^3}\right]^{1/2} \tag{9.2.5}$$

TABLE 9.1 Values of *k* for Eq. (9.2.2)

Pier shape	k
Semicircular nose and tail	0.9
Lens-shaped nose and tail*	0.9
Twin-cylinder piers with connecting diaphragm	0.95
Twin-cylinder piers without diaphragm	1.05
90° triangular nose and tail	1.05
Square nose and tail	1.25

*A lens-shaped nose or tail is formed from two circular curves each having a radius of twice the pier width and each tangential to a pier face.

and if $M_2 = M_3$, then

$$\Gamma_L = \frac{\left[2 + (1/\sigma)\right]^3 \mathbf{F}_3^4}{(1 + 2\mathbf{F}_3^2)^3} \tag{9.2.6}$$

where Γ_L = limiting value of Γ and E_1 and E_2 are the specific energies at sections 1 and 2, respectively. Henderson (1966) recommended that Eq. (9.2.6) be used to estimate Γ because it does not depend on energy conservation assumptions; further, its independent variables are initially known. However, Eq. (9.2.5) was used by Yarnell to distinguish between class A and class B flow. When class B flow occurs, a hydraulic jump occurs downstream of section 2.

The contraction of a flow field at a bridge can also serve as an indirect measure of the flow. An equation for the discharge results from writing and combining the continuity and energy equations between sections 1 and 3 or

$$Q = C_D A_2 \sqrt{2g\left(\Delta y + \alpha_1 \frac{\overline{u}_1^2}{2g} - h_f\right)} \tag{9.2.7}$$

where Q = discharge
C_D = coefficient of discharge
$\overline{u}_1$ = velocity at section 1
h_f = frictional loss between sections 1 and 2
A_2 = flow area at section 2

Matthai (1976) noted that section 2 defines the minimum area on a line parallel to the contraction and is usually located between the bridge abutments. If there is a scour hole between the abutments, then a section between the abutments may not be a minimum section. In addition, when a scour hole is extremely large, the flow between the abutments may be controlled by the scour hole rather than by the contraction.

In Eq. (9.2.7) the kinetic energy correction factor is estimated by

$$\alpha = \frac{\sum_{i=1}^{N} (K_i)^3/(A_i)^2}{K_T^3/A_T^2} \tag{9.2.8}$$

where A = flow area
K = conveyance
i = subscript specifying a subsection of the total cross section
T = subscript indicating the entire cross section

The frictional loss in Eq. (9.2.7) is estimated by dividing the distance between sections 1 and 3 into two reaches: from section 1 to the upstream side of the bridge opening and from the upstream side of the bridge opening to section 3.

The variable h_f is then estimated by

$$h_f = L_w \left(\frac{Q^2}{K_1 K_3} \right) + L \left(\frac{Q}{K_3} \right)^2 \tag{9.2.9}$$

where L_w = length of approach section
L = length of bridge contraction
K_1 and K_3 = conveyances at sections 1 and 3, respectively

The distances L_w and L are defined in Fig. 9.1.

In Eq. (9.2.7) the coefficient of discharge is a combination of (1) a coefficient of contraction, (2) a coefficient for the eddy losses, and (3) a velocity head correction coefficient for the control section. In functional notation,

$$C_D = f\left(\sigma, \frac{L}{b}, \frac{r}{b}, \frac{w}{b}, \Theta, \frac{x}{b}, \frac{y_a + y_b}{2b}, \frac{t}{y_3 + \Delta y}, j, \mathbf{F}, \zeta, e, S_s, L_d, \frac{L_d}{b_d} \right) \tag{9.2.10}$$

The variables used in Eq. (9.2.10) are defined in Table 9.2 and Figs. 9.1 to 9.5. The variables σ, L/b, and $\mathbf{F}$ are common to all types of bridge openings; of these, σ is the most important and $\mathbf{F}$ the least important (Matthai, 1976).

As noted, the geometric properties of the bridge opening have a significant effect on the flow. Most bridge openings can be classified as belonging to one of four basic types:

Type 1: The type 1 bridge contraction (Fig. 9.2) has vertical embankments and abutments with or without wingwalls. The entrance rounding, wingwall angle, angularity of the contraction with respect to the flow, and the Froude number affect the discharge.

Type 2: The type 2 bridge contraction (Fig. 9.3) has sloping embankments but vertical abutments. The depth of water at the abutments and the angularity of the contraction with respect to the flow affect the discharge.

Type 3: The type 3 bridge contraction (Fig. 9.4) has both sloping embankments and sloping abutments. The entrance geometry and the angularity of the contraction with respect to the flow affect the discharge.

Type 4: The type 4 bridge contraction (Fig. 9.5) has sloping embankments, vertical abutments, and wingwalls. The wingwall angle, angularity of the contraction with respect to the flow, embankment slopes, and Froude number affect the discharge. It must be noted that the addition of wingwalls alone does not necessarily designate a type 4 contraction. If the flow passes around a vertical edge at the upstream corner of the wingwall, then the contraction is type 1; however, if the flow passes around a sloping corner of the wingwall, then the contraction is type 4.

TABLE 9.2 Definition of variables for Eq. (9.2.10)

Variable	Definition
σ	Channel contraction ratio
b	Width of bridge opening—distance between abutments (Figs. 9.1 to 9.5)
L	Length of abutment or piers, defined differently for different types of bridges (Figs. 9.1 to 9.5)
r	Radius of rounding of entrance corner of abutment for vertical-faced contractions
w	Measure of a wingwall or chamber
θ	Acute angle between a wingwall and the plane of contraction
x	Horizontal distance from the intersection of the abutment and embankment slopes to the location on the upstream embankment having the same elevation as the water surface as section 1
y_a, y_b	Depth of water at the toe of each abutment at section 3
y_3	Average depth of water in section 3 $$y_3 = \frac{A_3}{b_t}$$
A_3	Gross area of section 3
t	Vertical distance between the water level at section 1 and the lowest horizontal member of a partially submerged bridge
j	Ratio of the projected area of the submerged parts of piers or piles in the bridge opening to A_3 or $$j = \frac{A_j}{A_3}$$
A_j	Submerged cross-sectional area of piers or piles projected to section 3
$\mathbf{F}$	Froude number of the contracted section
ζ	Acute angle between the plane of the contraction and a line normal to the thread of the stream
e	Eccentricity ratio $$e = \frac{K_a}{K_b} \leq 1$$
S_s	Slope of the embankments
L_d	Length of dikes
b_d	Offset distance for straight dikes

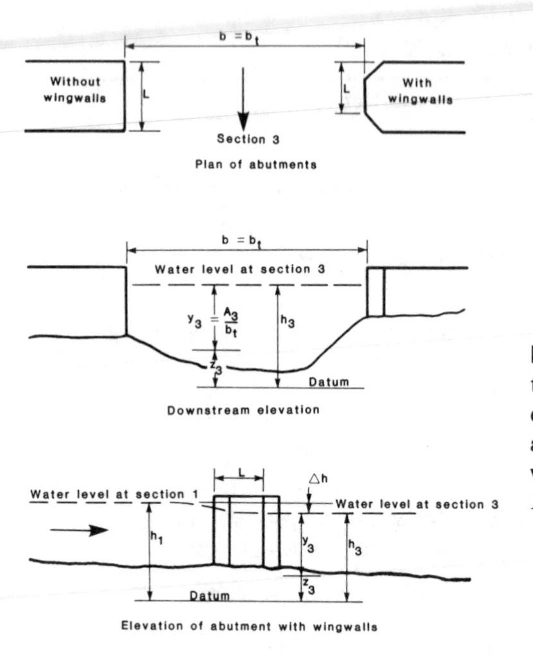

FIGURE 9.2 Definition of type 1 opening, vertical embankments, and vertical abutments with or without wingwalls. *(After Matthai, 1976.)*

In the analysis of Matthai (1976) the channel contraction ratio is defined as the proportion of the total flow that enters the contraction from the sides of the channel or

$$\sigma = \frac{Q - q'}{Q} = 1 - \frac{q'}{Q} \tag{9.2.11}$$

where Q = total discharge of the channel and q' = the discharge that would pass through the bridge opening if the contraction were not there. If it is assumed that the total discharge is distributed across the approach section in proportion to the channel subsections, then Eq. (9.2.11) becomes

$$\sigma = 1 - \frac{K_q}{K} \tag{9.2.12}$$

where K = total conveyance of the approach section, and K_q = conveyance of the subsection of the total channel occupied by the flow q' (Fig. 9.6). With reference to Fig. 9.6, the total conveyance of the approach section is given by

$$K = K_a + K_q + K_b \tag{9.2.13}$$

The determination of the coefficient of discharge specified in Eq. (9.2.7) usually requires a number of computational steps. After the type of contraction is identified, the following steps are taken:

1. With σ and L/b specified, Figs. 9.7 through 9.17 are consulted and a base coefficient of discharge C' is determined. If all the conditions specified on the

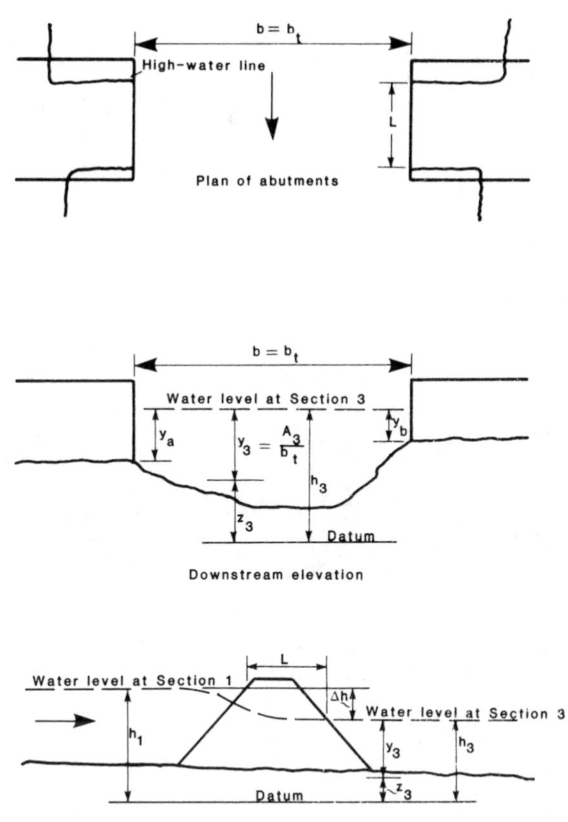

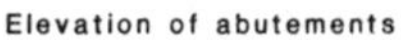

FIGURE 9.3 Definition of type 2 opening, sloping embankments, and vertical abutments. *(Matthai, 1976.)*

base curve are satisfied, then the computation is complete; if not, then additional steps are required.

2. If a bridge opening departs from the standard conditions only with respect to the Froude number and the angularity, then the final discharge coefficient is given by

$$C_D = C' k_F k_A \tag{9.2.14}$$

where C' = base coefficient of discharge for bridge type and given values of L/b

k_F = coefficient which adjusts C' for nonstandard values of $\mathbf{F}$

k_A = coefficient which adjusts C' for influence of angularity

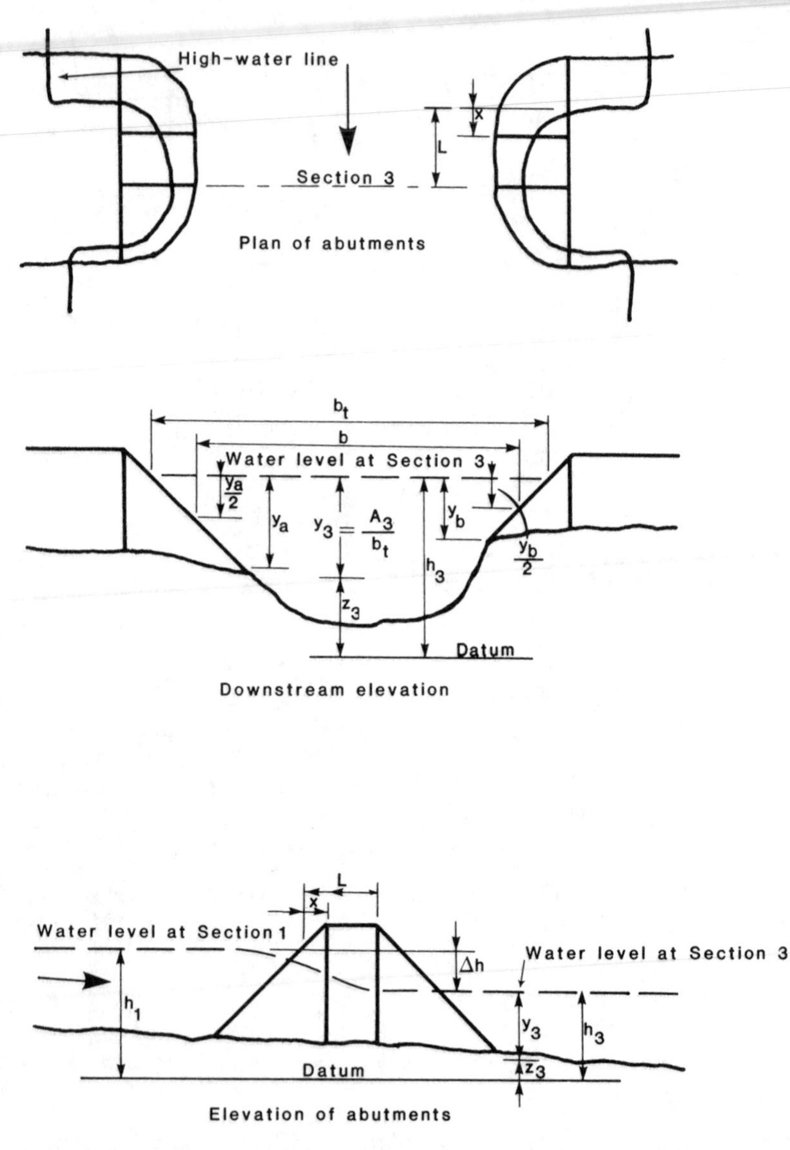

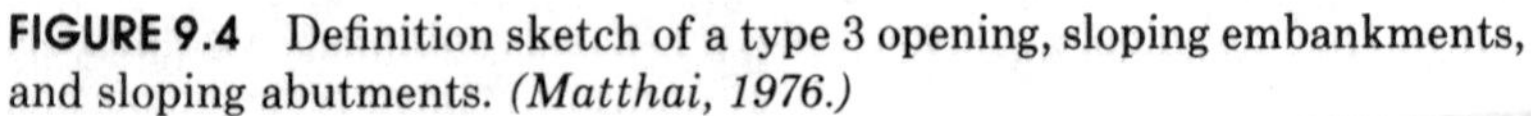

FIGURE 9.4 Definition sketch of a type 3 opening, sloping embankments, and sloping abutments. *(Matthai, 1976.)*

3. The eccentricity e of a bridge opening is defined as the ratio of the conveyances K_a and K_b (Fig. 9.6) or

$$e = \frac{K_a}{K_b} \tag{9.2.15}$$

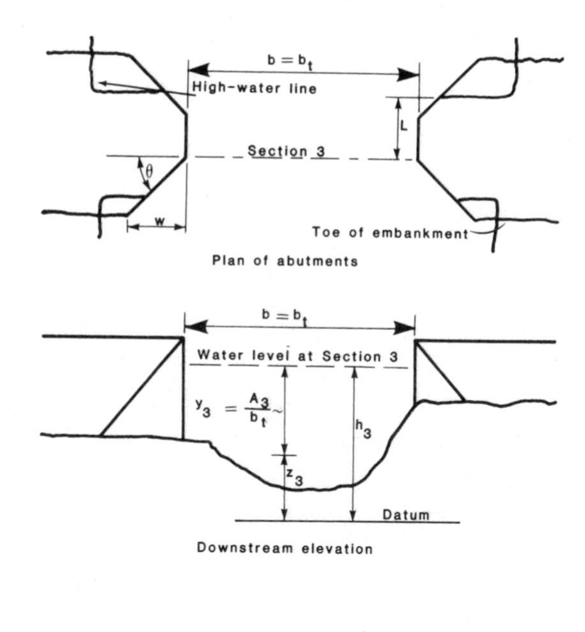

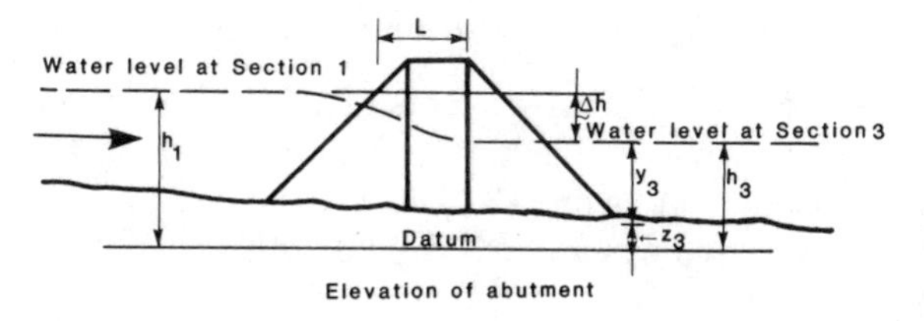

FIGURE 9.5 Definition sketch of a type 4 opening, sloping embankments, and vertical abutments with wingwalls. *(Matthai, 1976.)*

If $e > 0.12$, no correction for eccentricity is required. If $e = 0$, then the following procedure is used:

a. Section 1, the approach section, is located a distance $L = 2b$ upstream from the bridge.

b. The base coefficient of discharge is determined by using the ratio $L/2b$ and only the abutment on the contracted side.

c. The water surface elevation at section 1 is taken to be the average of the elevations at the ends of the section.

d. The water surface elevation at section 3 is taken as the elevation on the contracted side only; Δy is determined on this side also.

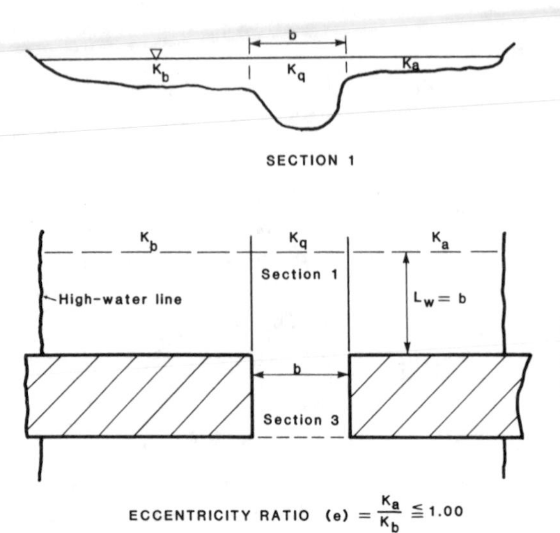

FIGURE 9.6 Definition sketch for the constriction ration σ and an eccentric contraction. *(Matthai, 1976.)*

When $e = 0$, the section is termed a *fully eccentric contraction* and is considered equal to half of a normal contraction. For this reason, the contraction width for estimating C' and L is taken as $2b$. The adjustment factor for eccentricity k_e is determined as a function of e from Table 9.3.

4. The effect of bridge piers or piles within the contraction must also be taken into account. The total submerged area of the piers or piles is projected onto a plane at section 3. The ratio of the projected pier or pile area, A_j, to the flow area at section 3 is by definition

$$J = \frac{A_j}{A_3} \tag{9.2.16}$$

The adjustment factor k_j is defined in Fig. 9.17. k_j for piers is a function of σ, while k_j for piles is a function of both σ and L/b. In Fig. 9.17, the dashed lines illustrate the determination of k_j for piles when $\sigma = 0.41$, $J = 0.04$, and $L/b = 0.69$. For these data, enter the right-hand graph at $\sigma = 0.41$; move vertically until the curve for $L/b = 0.69$ is intersected; move horizontally until the line $J = 0.10$ in the left-hand graph is intersected; vertically until the line for $J = 0.04$ is intersected; and horizontally until the value of k_j is determined; in this case $k_j = 0.967$.

5. When extreme flow events occur in the channel, the lower structural members of the bridge may be submerged, resulting in both an increased wetted perimeter in the contraction and increased obstruction of the flow. Let t be the vertical distance between the water surface at section 1 and the lowest submerged horizontal member of the bridge (Fig. 9.18). The submergence adjustment factor k_t is defined in terms of the ratio of t to the sum $(y_3 + \Delta y)$ (Fig. 9.18). Under very extreme flow conditions, the deck or floor of the

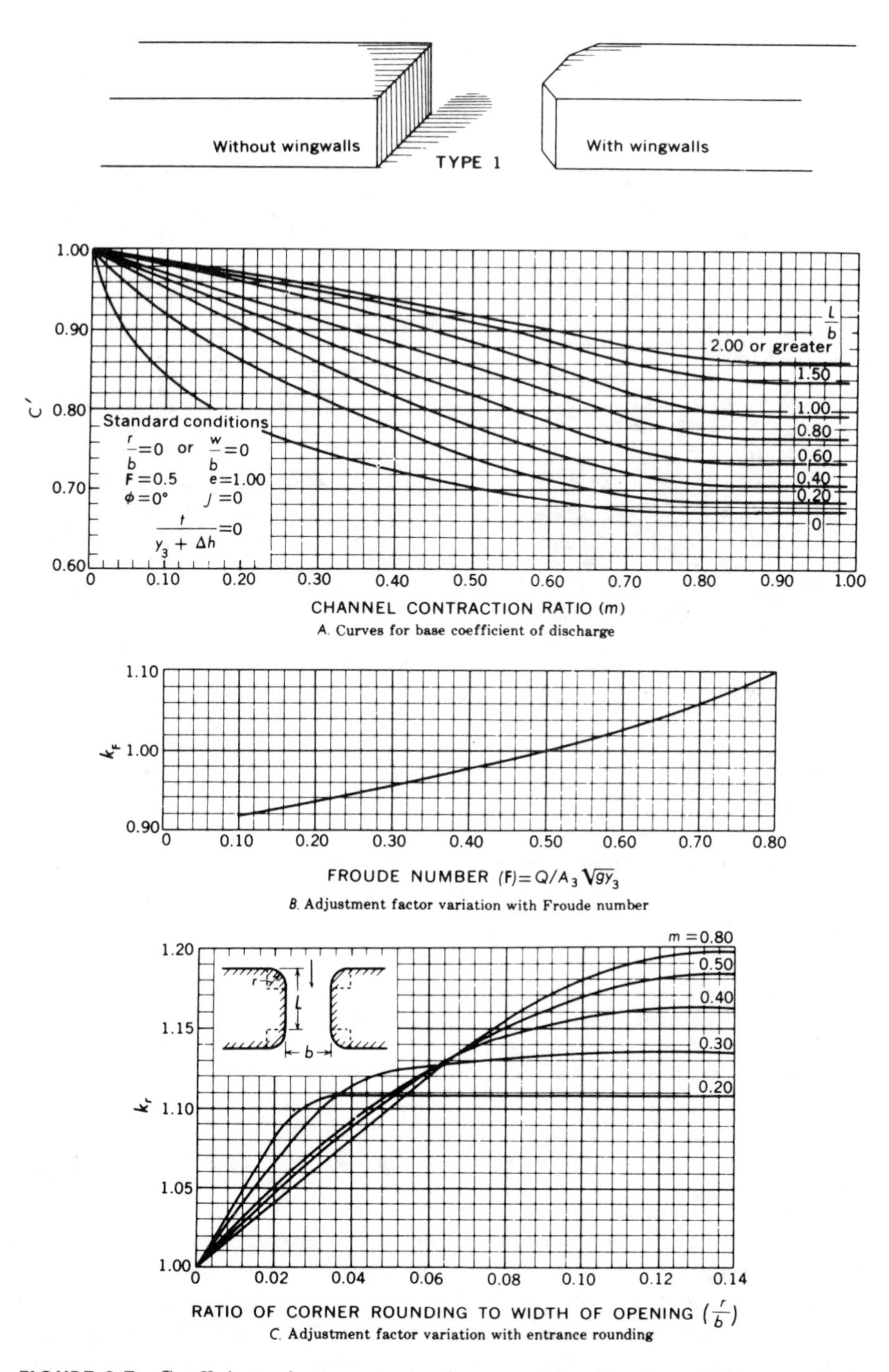

FIGURE 9.7 Coefficients for type 1 opening, vertical embankments, and vertical abutments. *(Matthai, 1976.)*

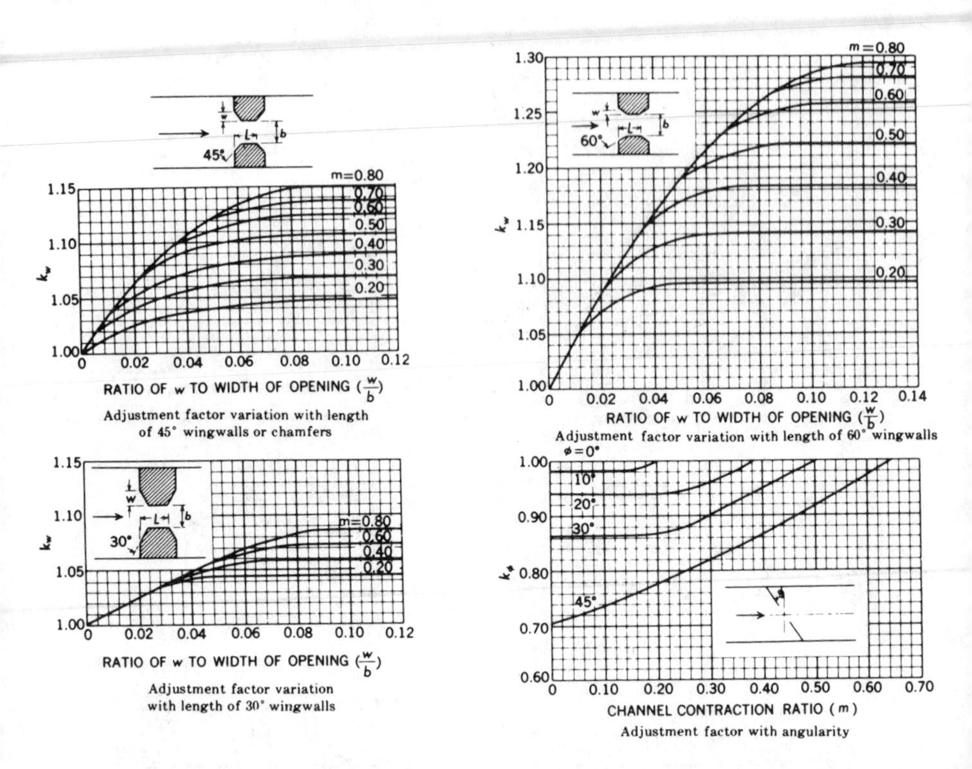

FIGURE 9.8 Adjustment factor variation for vertical embankment and abutment of type 1 opening. *(Matthai, 1976.)*

bridge may be completely submerged. If such a situation occurs, Matthai (1976) recommended that the discharge be estimated in the following manner:

a. The flow area at section 3, A_3, is estimated as the product of b and y_3 less the cross-sectional area of the submerged portion of the bridge.

b. Assume $k_t = 1.00$.

c. Estimate the discharge by Eq. (9.2.7).

d. The discharge in step c is an estimate of the total flow through the contraction.

6. The alignment of the abutments with the flow is also a crucial factor. When the contraction is at an angle ζ with the flow, the abutments may be either parallel to the flow or perpendicular to the embankments. The adjustment factor k_ζ is the same for both conditions if $\zeta \leq 20°$. If $\zeta > 20°$, then an additional adjustment factor is required; however, the data required to estimate this factor are not presently available, and no adjustment for this effect can be made.

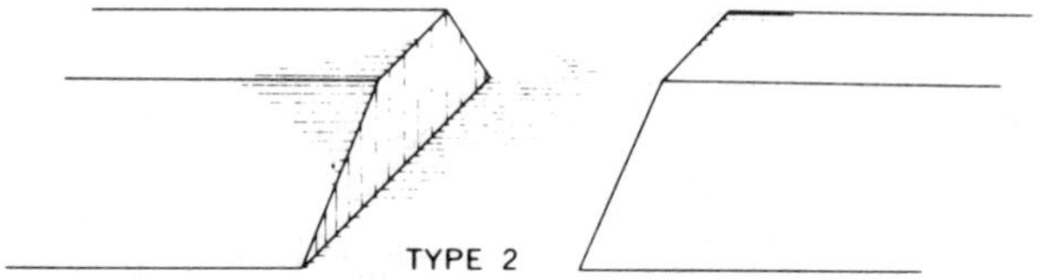

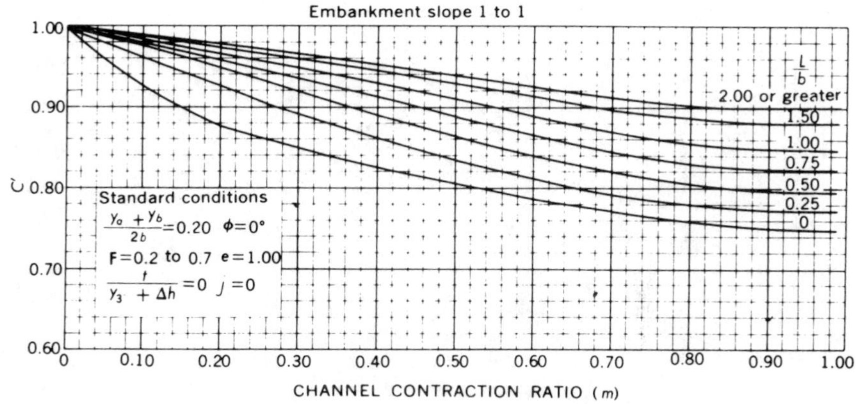

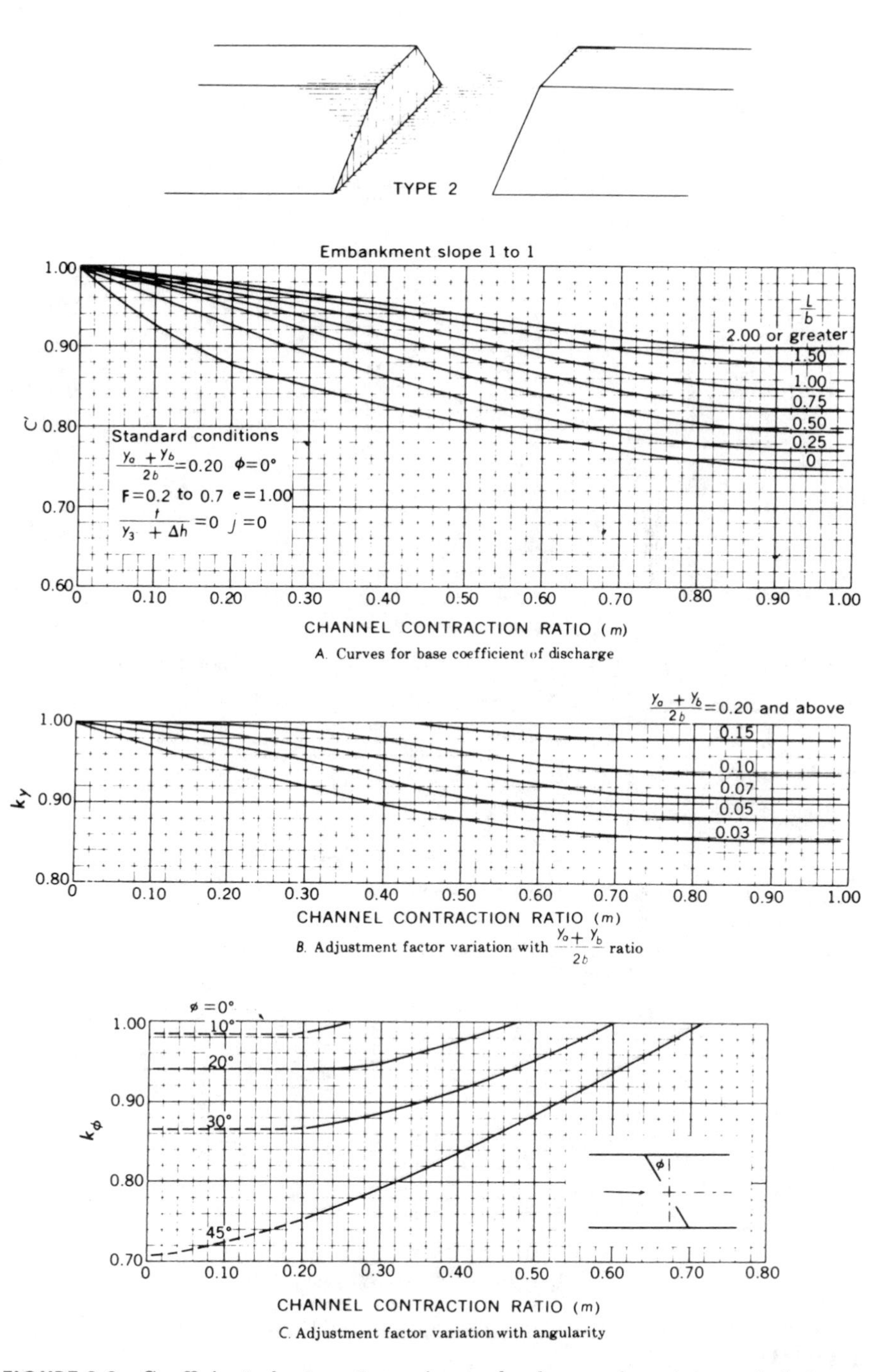

A. Curves for base coefficient of discharge

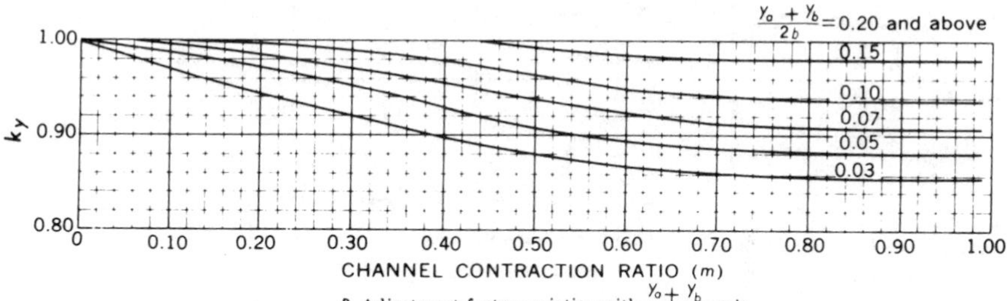

B. Adjustment factor variation with $\frac{y_a + y_b}{2b}$ ratio

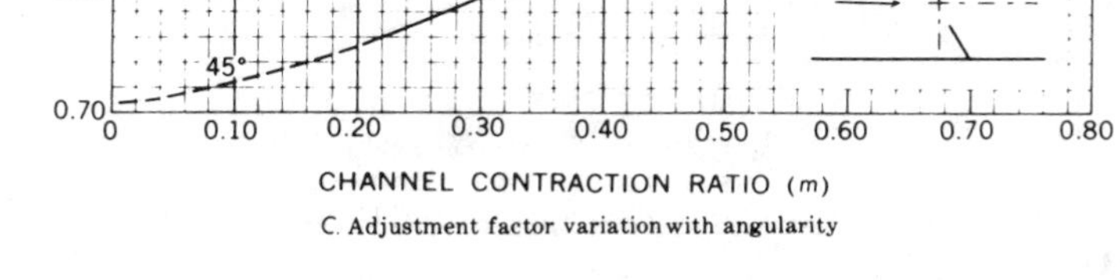

C. Adjustment factor variation with angularity

FIGURE 9.9 Coefficients for type 2 opening, embankment slope 1:1, vertical abutment. *(Matthai, 1976.)*

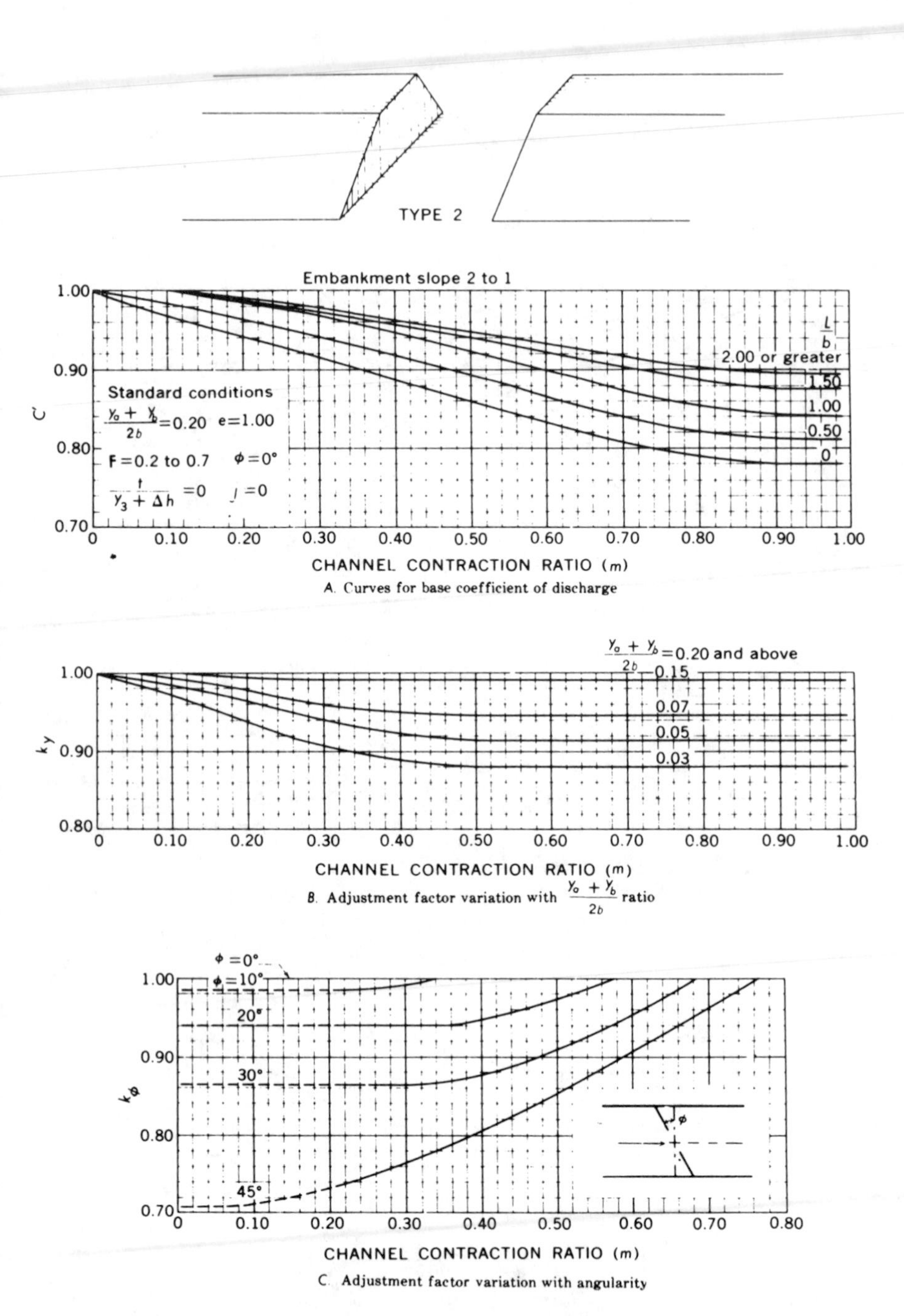

FIGURE 9.10 Coefficients for type 2 opening, embankment slope 2:1, vertical abutment. *(Matthai, 1976.)*

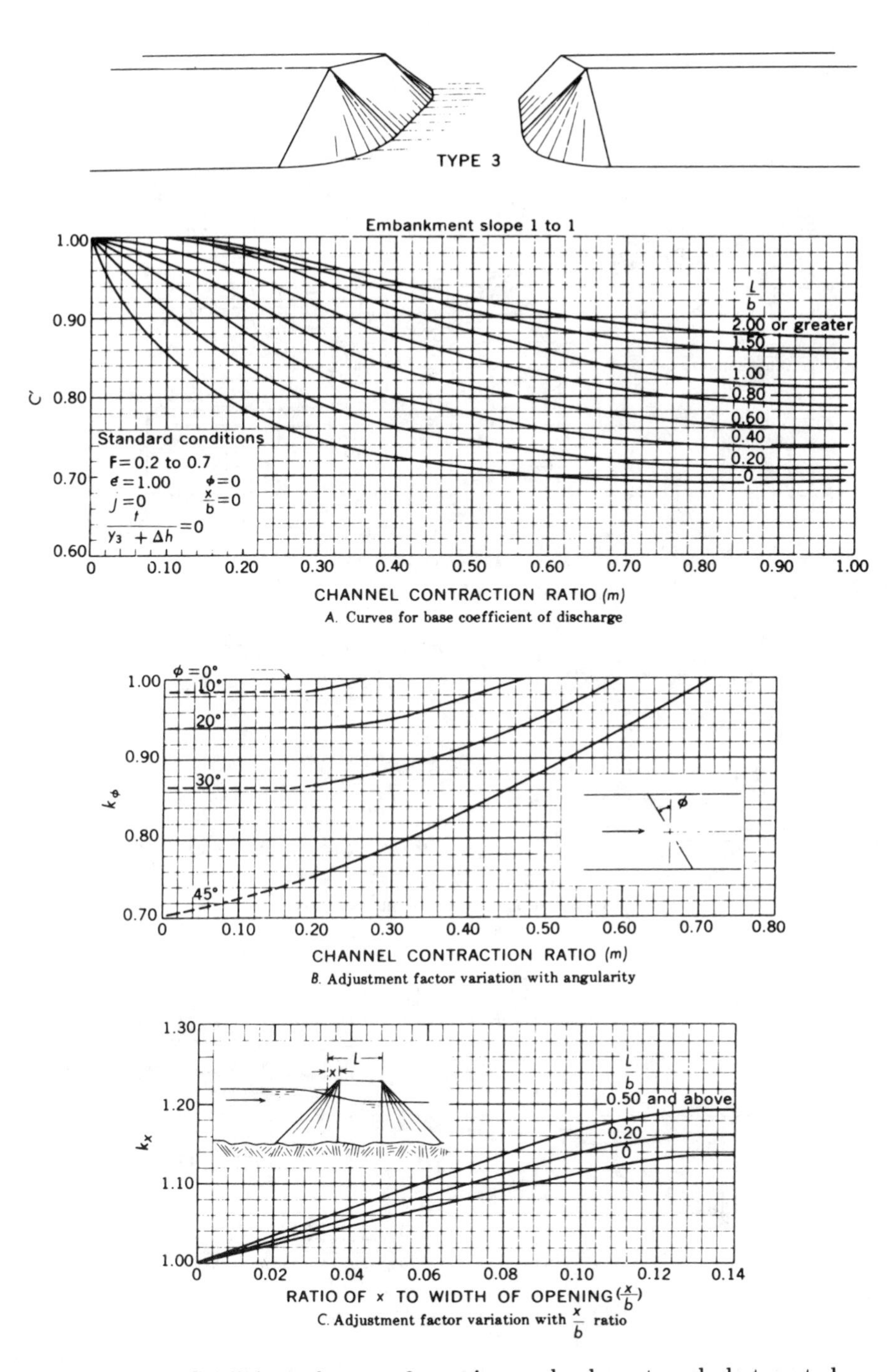

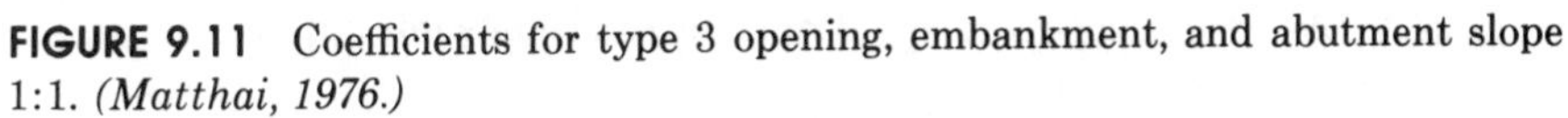

FIGURE 9.11 Coefficients for type 3 opening, embankment, and abutment slope 1:1. *(Matthai, 1976.)*

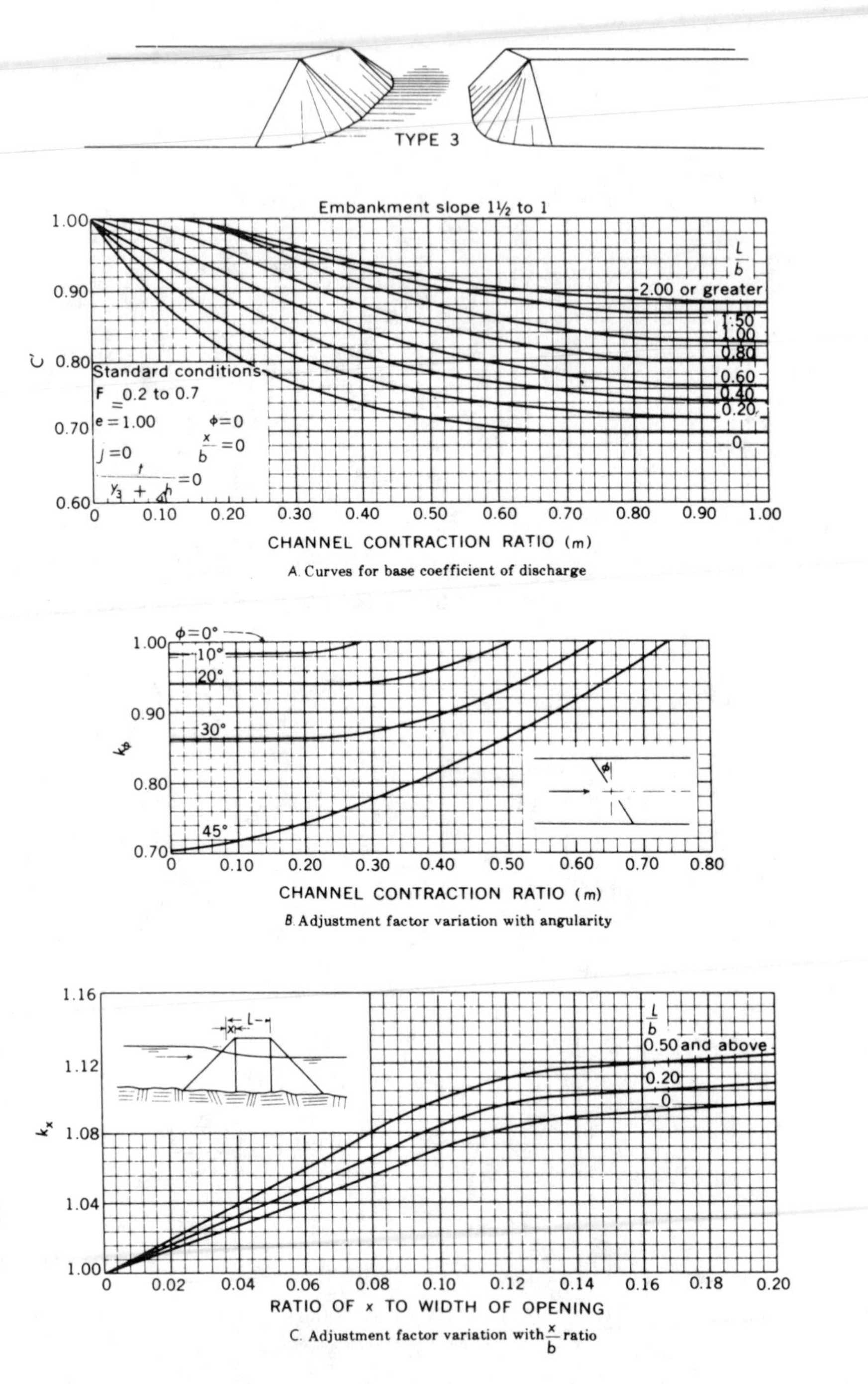

FIGURE 9.12 Coefficients for type 3 opening, embankment, and abutment slope 1½:1. *(Matthai, 1976.)*

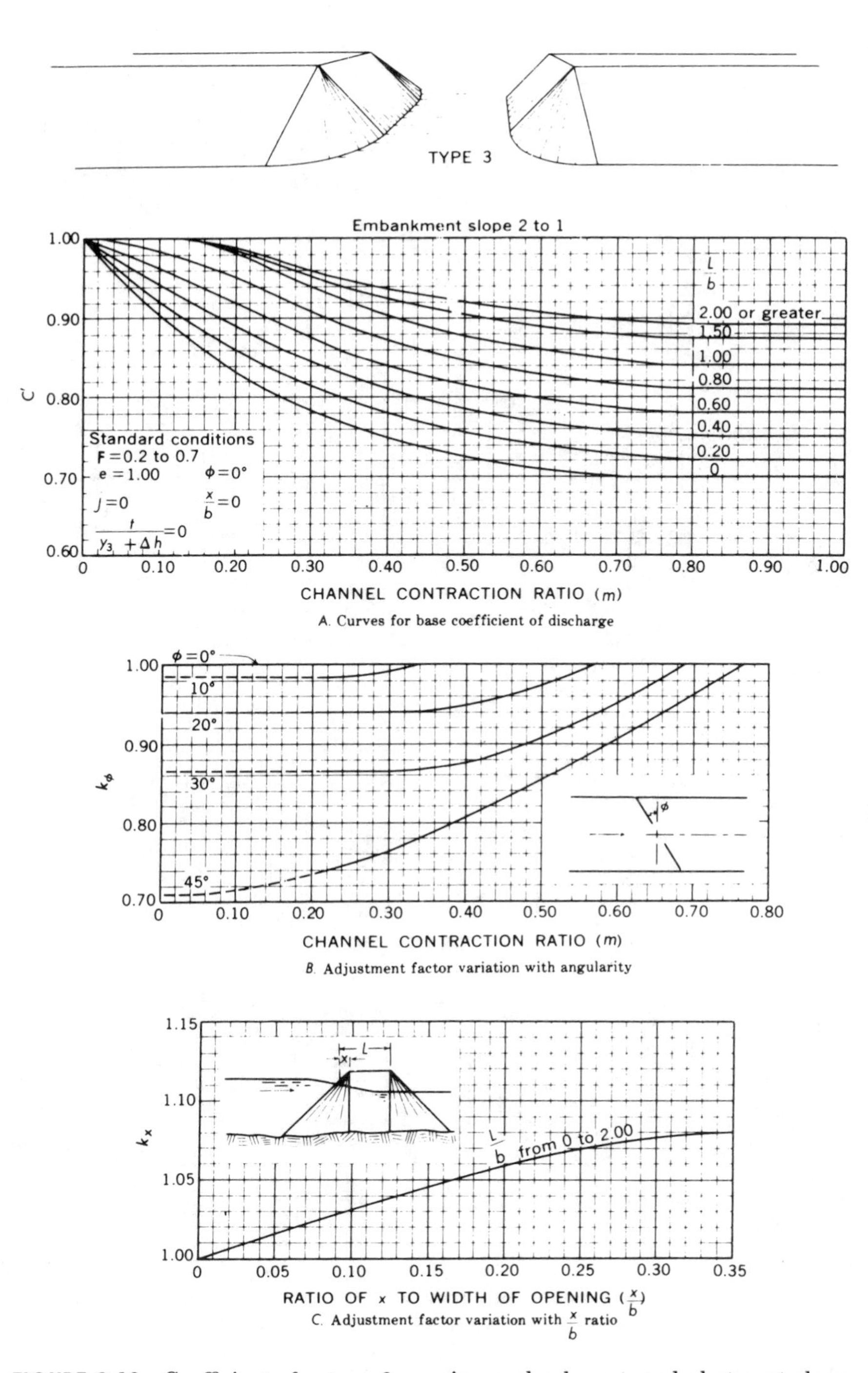

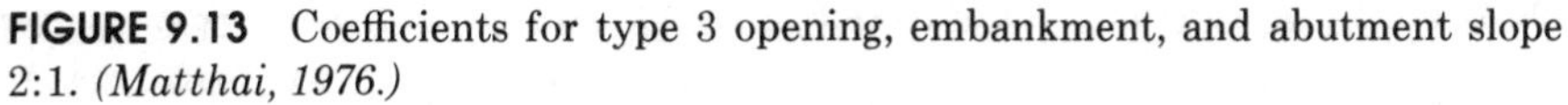

FIGURE 9.13 Coefficients for type 3 opening, embankment, and abutment slope 2:1. *(Matthai, 1976.)*

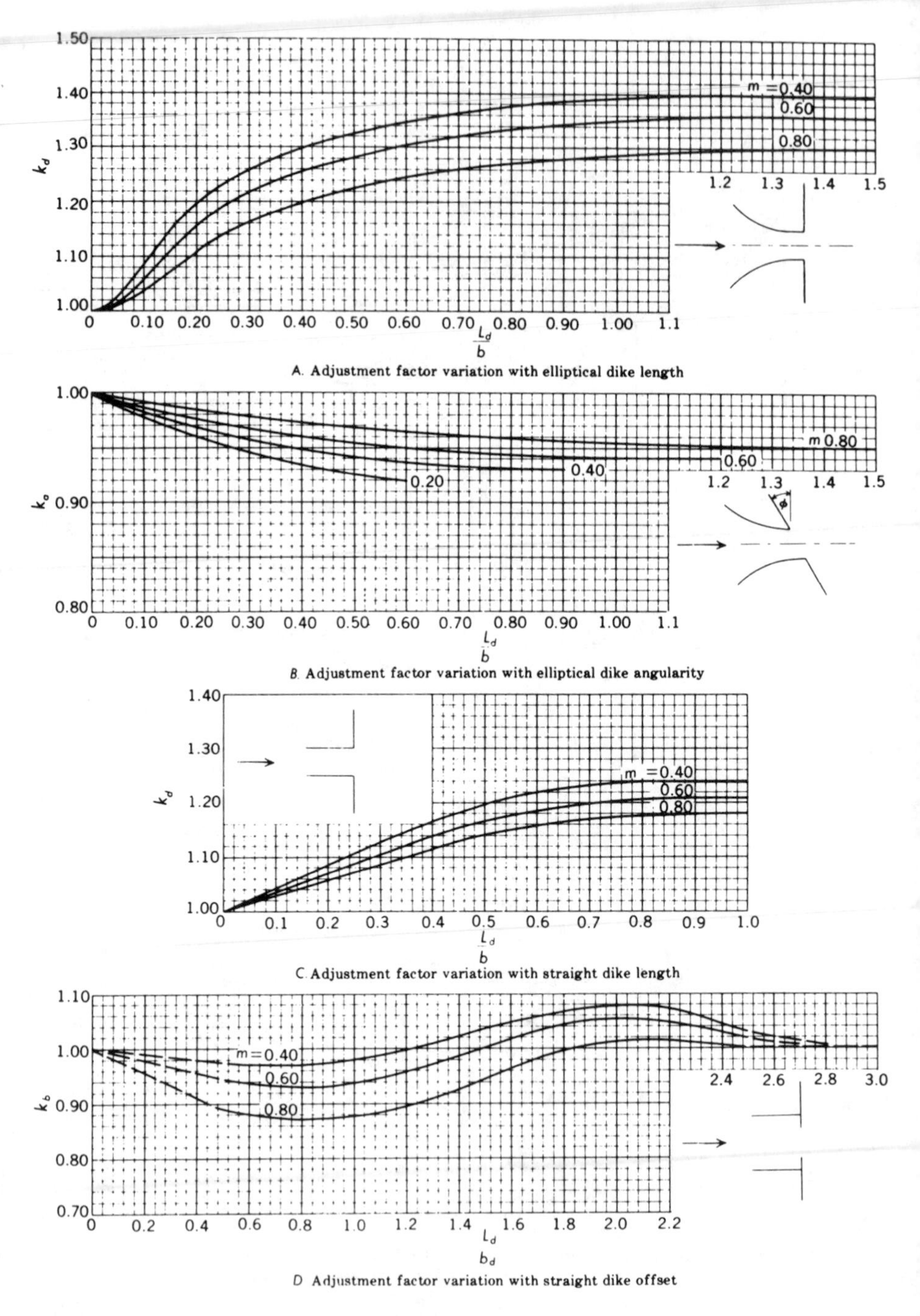

FIGURE 9.14 Adjustment factor variation for spur dikes. *(Matthai, 1976.)*

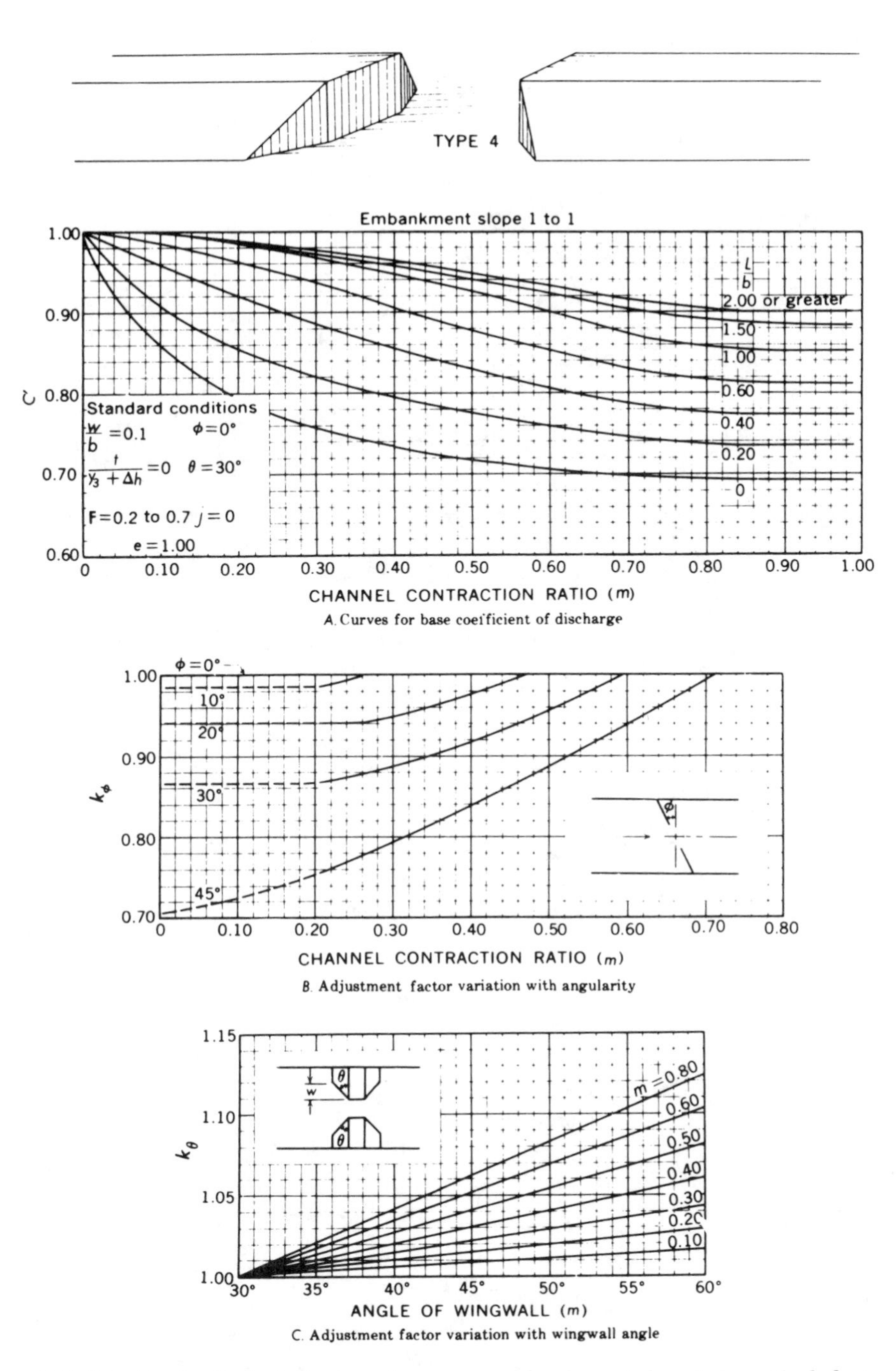

FIGURE 9.15 Coefficients for type 4 opening, embankment slope 1:1, vertical abutments with wingwalls. *(Matthai, 1976.)*

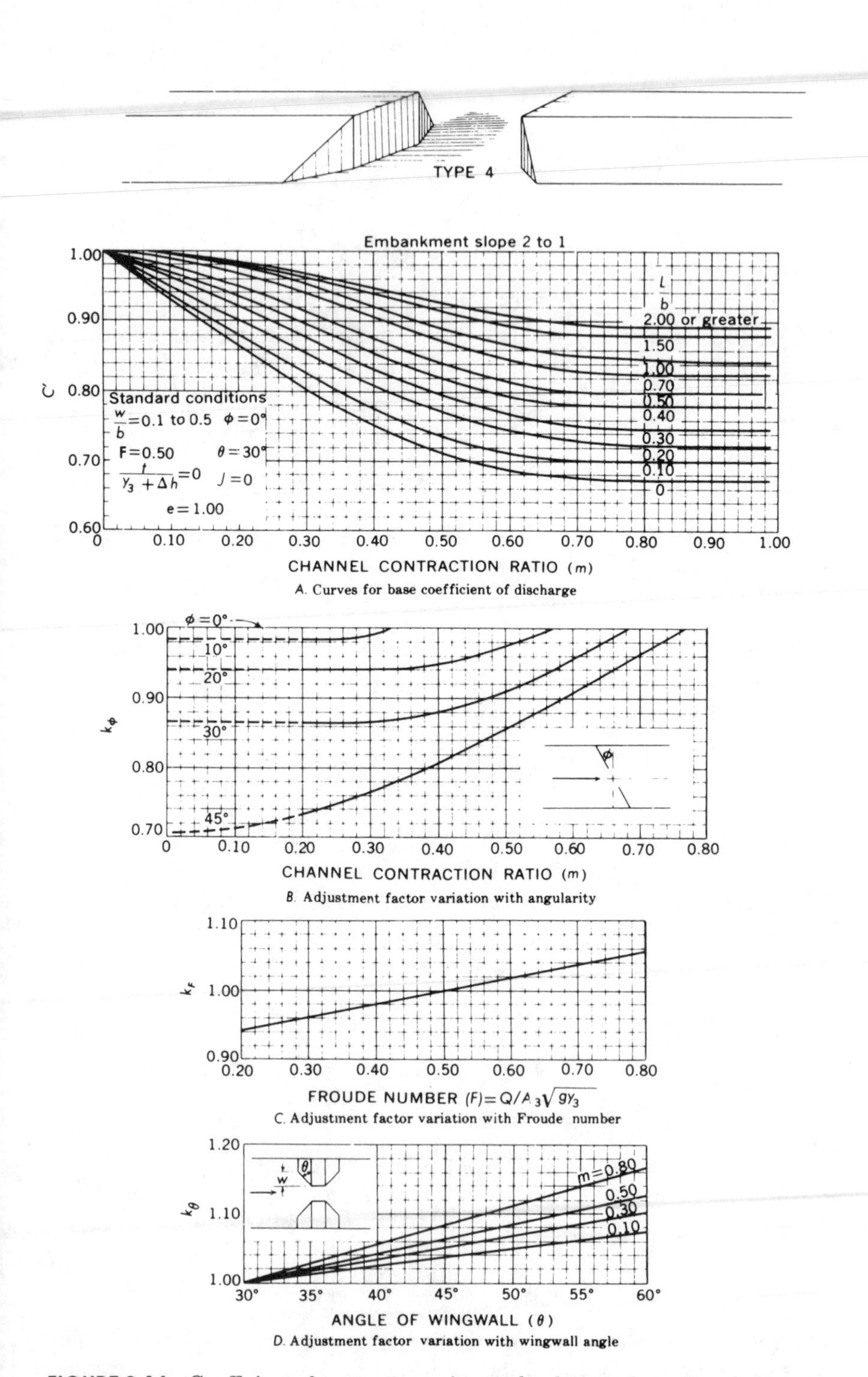

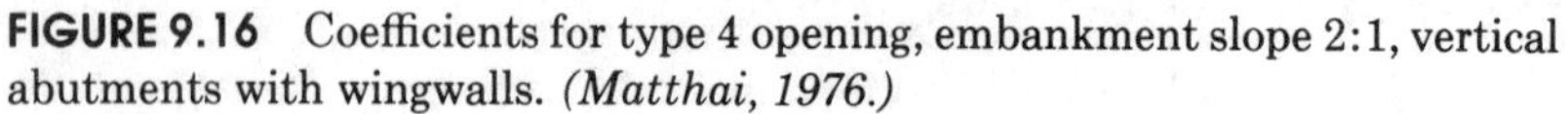

FIGURE 9.16 Coefficients for type 4 opening, embankment slope 2:1, vertical abutments with wingwalls. *(Matthai, 1976.)*

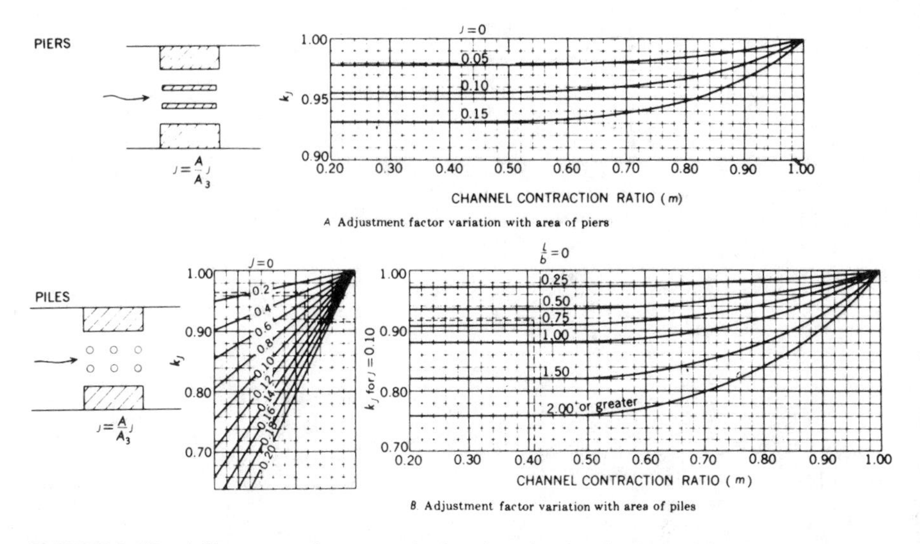

FIGURE 9.17 Adjustment factor variation for piers and piles. *(Matthai, 1976.)*

The foregoing adjustments are applicable to the four bridge types considered here, and their effect on the coefficient of discharge may in a specific situation range from minor to major.

Other coefficient of discharge adjustment factors may also be required depending on the type of contraction. The curves for estimating these factors accompany the base discharge coefficient curves for the four types of bridges considered here. With regard to these additional factors, the following comments should be noted:

1. Angularity is the relation between a skewed contraction and the flow lines for the unobstructed channel (Fig. 9.19*a* and *b*). The adjustment factor for angularity k_ζ is a function of the contraction ratio σ and the degree of angularity ζ. Angularity must not be confused with the curvature of the channel.

TABLE 9.3 Adjustment factor for eccentricity (Matthai, 1976)

e	k_e
0	0.953
0.02	0.966
0.04	0.976
0.06	0.984
0.08	0.990
0.10	0.995
0.12	1.00

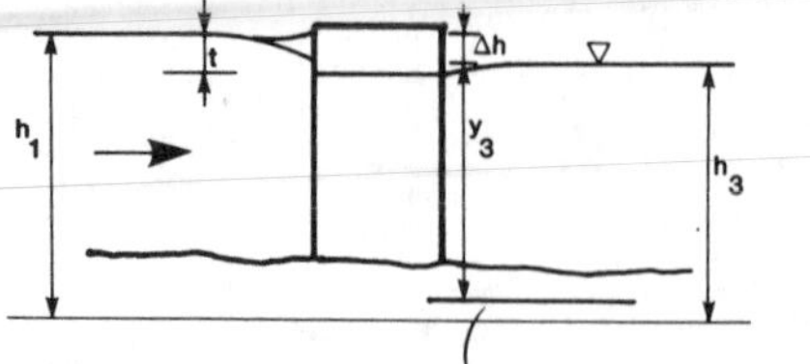

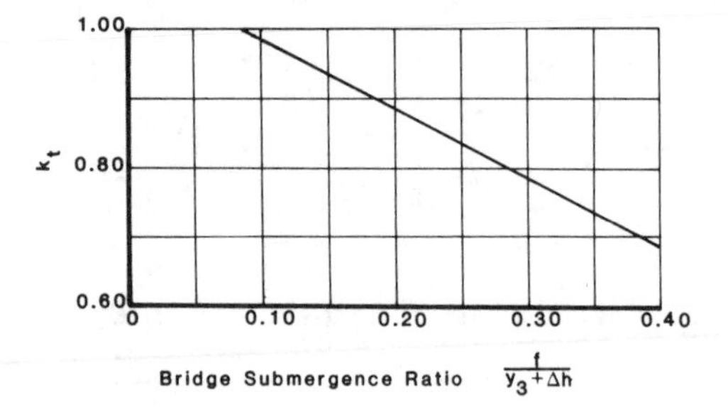

FIGURE 9.18 Adjustment factor variation with bridge submergence ratio. *(Matthai, 1976.)*

Angularity can also have a significant effect on the distance L, that is, the distance between sections 1 and 2. For other than straight channels and parallel sections, the distance L should be measured from the centroid of the approach section to the middle of section 2. In Fig. 9.19*b*, no adjustment for angularity is required, but the distance L is measured along the indicated chord.

2. In situations in which the slope of the embankment is a factor which requires adjustment of the coefficient of discharge, interpolation may be required. In general, the results presented here are for embankment slopes of 1:1 and 2:1. For other embankment slopes which are within this range of values, linear interpolation for the adjustment factor is usually satisfactory (Matthai, 1976).

3. In some cases, the two bridge abutments will be of different types. In such cases, Matthai (1976) recommended that an average coefficient of discharge be estimated by

$$C_D = \frac{C_L K_L + C_R K_R}{K_L + K_R} \tag{9.2.17}$$

where the subscripts L and R designate the left and right abutment coefficients of discharge and conveyance.

It should be kept in mind that the data and results presented here are by no means comprehensive. In some cases, the engineer will have to use judgment

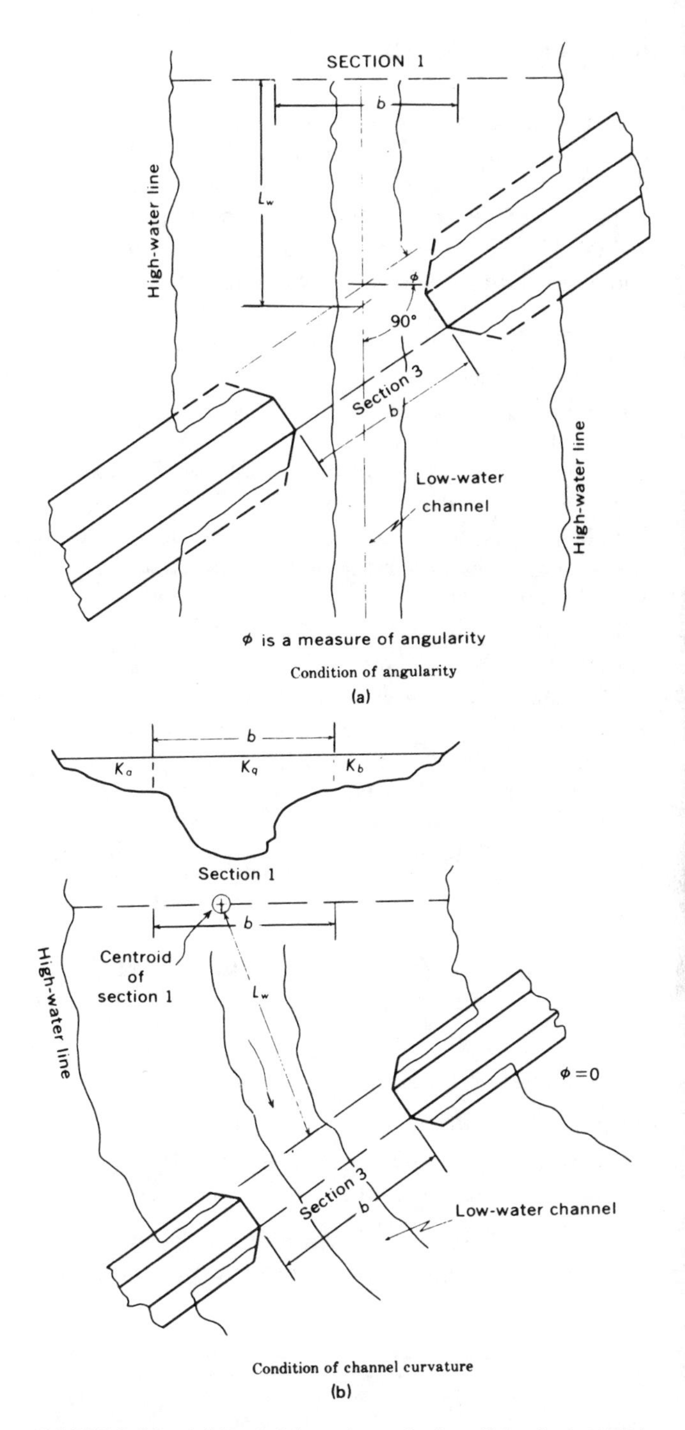

FIGURE 9.19 (*a*) Definition of angularity. *(Matthai, 1976.)* (*b*) Definition sketch comparing conditions of angularity and channel curvature. *(Matthai, 1976.)*

and the available data to estimate a reasonable coefficient of discharge. It is also appropriate to note at this point that the coefficient of discharge can never exceed 1.00. If the application of the adjustment factors to the base coefficient of discharge yields a value greater than 1.00, the value of C_D should be set equal to 1.00. Further, if both the upstream and downstream bottom chords of the bridge are submerged, then the maximum value of the coefficient of discharge is k_t.

A multiple-opening contraction is defined as a series of single-opening contractions all of which freely conduct water originating from a common approach channel. The premise of independent contractions is assumed to be valid when: (1) two or more pairs of abutments and one or more interior embankments exist or (2) piers or webs separating several openings between two abutments exist. In the case of multiple-opening bridge contractions, the flow is estimated by establishing pseudochannel boundaries and determining the discharge within these boundaries under the assumption that the flows are independent. This procedure may result in the arbitrary line defining section 1 as being a series of discontinuous lines at varying distances from the bridge instead of a single continuous line across the channel.

The general discharge equation for a bridge contraction [Eq. (9.2.7)] must be solved by trial and error. An explicit equation for the discharge can be derived if Q/A_1 is substituted for $\bar{u}_1$ and Eq. (9.2.9) is substituted for h_f in Eq. (9.2.7) and the resulting equation solved for Q or

$$Q = 8.02\, C_D A_3 \left[\frac{\Delta y}{1 - \alpha_1 C_D^2 (A_3/A_1)^2 + 2g C_D^2 (A_3/K_3)^2 (L + L_w K_3/K_1)}\right]^{1/2} \tag{9.2.18}$$

If C_D, the discharge coefficient, is a function of the Froude number at section 3, a trial-and-error solution of Eq. (9.2.18) for the discharge will still be required.

EXAMPLE 9.1

Estimate the peak flood discharge through a bridge opening in a straight, uniform channel if:

1. The approach section is located 300 ft (91 m) upstream of the bridge and has the following geometric and hydraulic characteristics:

Subsection	A, ft² (m²)	P, ft (m)	n	α
Main section	5360 (498)	225 (69)	0.035	1.10
Side section	5710 (530)	405 (123)	0.040	1.11

2. At the point where the bridge spans the channel, the side section of the channel is completely blocked by the bridge embankment.
3. The bridge contraction is a type 1 contraction (Fig. 9.2) 180 ft (55 m) wide and 30 ft (9.1 m) long.
4. The average water surface elevation at the contracted section is 5 ft (1.5 m) below that of the approach section.
5. The main and side channels are approximately rectangular in shape with the main channel being 180 ft (55 m) wide and 29.8 ft (9.1 m) deep and the side channel being 390.4 ft (119 m) wide and 14.6 ft (4.5 m) deep.

Note: This example was first presented by Chow (1959).

Solution

This problem is solved by first determining the base coefficient of discharge, modifying this coefficient as required for the geometry of the bridge contraction under consideration, and then solving Eq. (9.2.18) for the discharge.

From Eq. (9.2.12), the contraction ratio is

$$\sigma = 1 - \frac{K_q}{K}$$

where K = total conveyance of the approach section and K_q = conveyance of the subsection of the total channel occupied by the flow q'. From the given data

$$\begin{aligned} K &= \left(\frac{1.49}{n} AR^{2/3}\right)_{\text{main}} + \left(\frac{1.49}{n} AR^{2/3}\right)_{\text{side}} \\ &= \frac{1.49}{0.035}(5360)\left(\frac{5360}{225}\right)^{2/3} + \frac{1.49}{0.040}(5710)\left(\frac{5710}{405}\right)^{2/3} \\ &= 1.89 \times 10^6 + 1.24 \times 10^6 = 3.13 \times 10^6 \end{aligned}$$

and

$$K_q = \left(\frac{1.49}{n} AR^{2/3}\right)_{\text{main}} = 1.89 \times 10^6$$

Then

$$\sigma = 1 - \frac{K_q}{K} = 1 - \frac{1.89 \times 10^6}{3.13 \times 10^6} = 0.40$$

The problem statement implied that for this situation the eccentricity [Eq. (9.2.15)] is zero; therefore, the second parameter required to estimate the base coefficient of discharge from Fig. 9.7*a* is

$$m = \frac{L}{2b} = \frac{30}{2(180)} = 0.08$$

Then, from Fig. 9.7*a*

$$C' = 0.75$$

and correcting for the eccentricity (Table 9.3)

$$C_D = k_i C' = 0.953\ (0.75) = 0.71$$

From the given information, it must be assumed that there are no other relevant "geometric" adjustments to the coefficient of discharge; i.e., there is no rounding of the abutments, there are no wingwalls, there are no piers or piles, the angularity is zero, no structural members of the bridge are submerged, and the Froude number at section 3 is 0.5 (from the assumptions inherent in Fig. 9.7*a*).

At this point, a trial-and-error solution for the discharge is required since the remaining adjustment factor for the coefficient of discharge is a function of the Froude number at section 3, and the determination of the Froude number at section 3 requires an estimate of the flow rate.

The value of the kinetic energy correction factor at section 1 is given by

$$\alpha_1 = \frac{\sum_{i=1}^{N} (\alpha_i K_i^3/A_i^2)}{\left(\sum_{i=1}^{N} K_i\right)^3 / A^2} = \frac{1.10(1.89 \times 10^6)^3/5360^2 + 1.11(1.24 \times 10^6)^3/5710^2}{(1.89 \times 10^6 + 1.24 \times 10^6)^3/(5360 + 5710)^2}$$

$$= 1.29$$

And, at section 3

$$A_3 = 180(29.8 - \Delta y) = 180(29.8 - 5) = 4460 \text{ ft}^2\ (415 \text{ m}^2)$$

$$P_3 = 180 + 2(29.8 - \Delta y) = 180 + 2(29.8 - 5) = 229.6 \text{ ft}\ (70 \text{ m})$$

$$R_3 = \frac{A}{P} = \frac{4460}{229.6} = 19.4 \text{ ft}\ (5.9 \text{ m})$$

and $$K_3 = \frac{1.49}{n} A_3 R_3^{2/3} = \frac{1.49}{0.035}(4460)(19.4)^{2/3} = 1.37 \times 10^6$$

Then, the substitution of these variable values in Eq. (9.2.18) yields

$$Q = 8.02\, C_D A_3 \left[\frac{\Delta y}{1 - \alpha_1 C_D^2 (A_3/A_1)^2 + 2g\, C_D^2\, (A_3/K_3)^2\, [L + (L_w K_3/K_1)]}\right]^{1/2}$$

$$= 8.02(0.71)(4460) \Big[\, 5/\{1 - 1.29(0.71)^2(4460/11{,}070)^2$$

$$+ 2(32.2)(0.71)^2(4460/1.37 \times 10^6)^2$$

$$\times \left[30 + 300\,(1.37 \times 10^6/3.13 \times 10^6)]\}\right]^{1/2}$$
$$= 58{,}300 \text{ ft}^3/\text{s } (1600 \text{ m}^3/\text{s})$$

The initial assumption regarding the Froude number at section 3 can now be checked:

$$\bar{u}_3 = \frac{Q}{A_3} = \frac{58{,}300}{4460} = 13.1 \text{ ft/s } (4.0 \text{ m/s})$$

$$D_3 = \frac{A_3}{T_3} = \frac{4460}{180} = 24.8 \text{ ft } (7.6 \text{ m})$$

and
$$\mathbf{F}_3 = \frac{\bar{u}_3}{\sqrt{gD_3}} = \frac{13.1}{\sqrt{32.2(24.8)}} = 0.46$$

From Fig. 9.7*b* the adjustment factor for the Froude number at section 3 is

$$k_F = 0.995$$

and, hence, the new estimate of the coefficient of discharge is

$$C_D = k_e k_F C' = 0.953(0.995)(0.75) = 0.71$$

Therefore, the initial estimate of the discharge is accepted as correct.

The preceding material in this section has discussed, in some detail, techniques which can be used to determine the flow passing through a bridge opening. An equally important question involves the determination of flow profiles through various types of shapes of bridges. There are several methodologies available for estimating energy losses through bridges; see, for example, U.S. Army Hydrologic Engineering Center (Anonymous, 1979), Bureau of Public Roads (Anonymous, 1960), and the U.S. Geological Survey (Anonymous, 1964). Eichert (1970a) reviewed and compared the bridge loss subroutines used in several common numerical models for estimating gradually varied profiles. In a subsequent paper, Eichert and Peters (1970b) noted that the various techniques available do not yield comparable answers for a specified set of conditions. These authors also noted several other serious limitations to existing techniques:

1. When the free surface of the water is obstructed by the bridge, pressure flow or a combination of pressure flow and weir flow takes place. Although methodologies for estimating the loss associated with either a pressure flow or a weir flow are well established, very little information is available for bridge flow.

2. Procedures are not available for accurately determining the conditions under which a bridge will fail.

3. Procedures which take into account scour under high-flow conditions are not available.

4. Although trash and debris obstructions can cause a significant increase in the upstream water surface elevation, these conditions are not taken in account.

Thus, a great deal of additional development in this area is required.

The Hydrologic Engineering Center (HEC) technique of estimating losses incurred when a flow passes through a bridge opening under low-flow conditions is described in several publications (see, for example, Eichert and Peters, 1970b, Feldman, 1981, and Anonymous, 1979).

Note: Low-flow conditions exist when the free surface of the water is not in contact with the lower chord or deck of the bridge. In this methodology, the losses due to expansion and contraction on the upstream and downstream sides of the structures are estimated in the same fashion as losses at expansions and contractions are estimated when no bridge is present. That is, a loss coefficient is multiplied times the absolute difference between the velocity heads inside and outside the constriction. Within the bridge opening, for bridges without piers, the boundary friction along the sides of the bridge is accounted for by the boundary friction terms within the gradually varied flow model. When piers are present, the losses associated with the piers are estimated by using the principle of conservation of momentum.

Figure 9.20 defines a system of notation for the following development. Section 1 is located immediately upstream of the nose of the bridge piers, while section A is located immediately downstream of the pier nose. Applying the momentum equation between sections 1 and A,

$$\beta_1 Q\rho\bar{u}_1 + \gamma A_1\bar{y}_1 - \beta_A Q\rho\bar{u}_A - \gamma A_A\bar{y}_A = F \tag{9.2.19}$$

where

A = flow rate

ρ = fluid density

$\bar{u}_1$ and $\bar{u}_A$ = average velocities at sections 1 and A, respectively

β_1 and β_A = momentum correction coefficients at sections 1 and A, respectively

A_1 and A_A = flow areas at sections 1 and A, respectively

$\bar{y}_1$ and $\bar{y}_A$ = vertical distances from water surface to centroids of sections 1 and A, respectively

γ = specific weight of fluid

F = force exerted by piers on flow

Equation (9.2.19) is valid for a channel of arbitrary shape and is identical to the general momentum equation developed in Chapter 3 [Eq. (3.1.2)]. Equation (9.2.19) also tacitly assumes that the boundary friction forces are negligible relative to the force exerted by the bridge piers on the flow. If $\beta_1 = \beta_A = 1$, Eq. (9.2.19) can be rearranged to yield

$$\frac{Q^2}{gA_1} + A_1\bar{y}_1 - \frac{Q^2}{gA_A} - A_A\bar{y}_A = \frac{F}{\gamma} \tag{9.2.20}$$

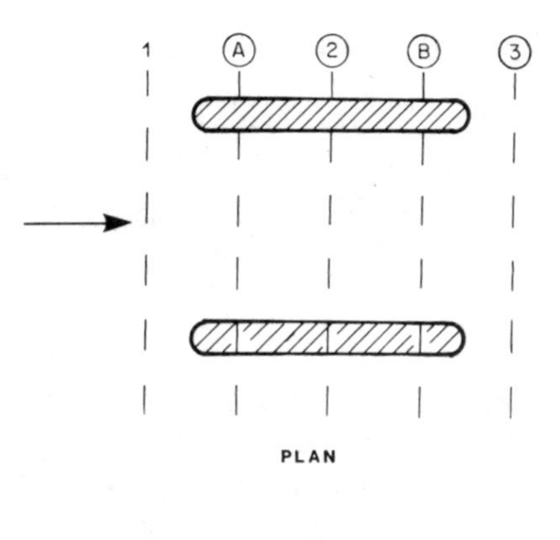

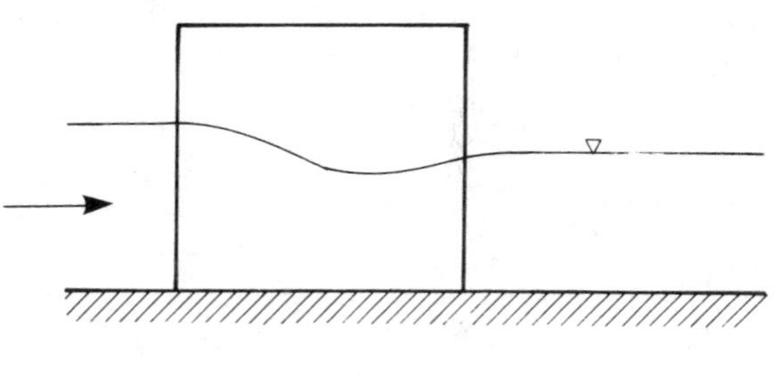

FIGURE 9.20 Plan and profile views of flow past bridge piers. *(Eichert and Peters, 1970b.)*

If it is assumed (see, for example, Eichert and Peters, 1970b) that the static and dynamic forces exerted by the piers are

$$F_s = \gamma A_{P1} \overline{y}_{P1} \tag{9.2.21}$$

and

$$F_D = \gamma \frac{C_D}{2} \left(\frac{A_{P1}}{A_1} \right) \left(\frac{Q^2}{gA_1} \right) \tag{9.2.22}$$

where F_s = static force

F_D = dynamic force

A_{P1} = projected area of the piers or piers and debris normal to the direction of flow corresponding to the depth of flow y_1

$\overline{y}_{P1}$ = vertical distance from the water surface to the centroid of A_{P1}

Equation (9.2.20) can then be rearranged to yield

$$M_A = \frac{Q^2}{gA_A} + A_A \overline{y}_A = A_1 \overline{y}_1 - A_{P1} \overline{y}_{P1} + \frac{Q^2}{gA_1^2} \left(A_1 - \frac{C_D}{2} A_{P1} \right) \tag{9.2.23}$$

where by definition M_A = momentum flux at section A. Eichert and Peters (1970b) suggested that

$$C_D = 2$$

for square-end piers and

$$C_D = 1.33$$

for piers with semicircular ends. In a similar fashion, the momentum principle can be applied between sections B and 3 or

$$M_B = \frac{Q^2}{gA_B} + A_B\bar{y}_B = A_3\bar{y}_3 - A_{P3}\bar{y}_{P3} + \frac{Q^2}{gA_3} \tag{9.2.24}$$

where A_{P3} = projected area of piers normal to the direction of flow corresponding to the depth of flow y_3, $\bar{y}_{P3}$ = vertical distance between the water surface and the centroid of A_{P3}, and there is only a static force component.

If it is assumed that the force exerted by the bridge piers is insignificant between sections A and B, then the momentum flux must be constant at these and all intermediate sections or

$$M_A = M_2 = M_B \tag{9.2.25}$$

Equations (9.2.23) to (9.2.25) can be combined to yield

$$A_1\bar{y}_1 - A_{P1}\bar{y}_{P1} + \frac{Q^2}{gA_1^2}\left(A_1 - \frac{C_D}{2}A_{P1}\right) = \frac{Q^2}{gA_2^2} + A_2\bar{y}_2$$

$$= A_3\bar{y}_3 - A_{P3}\bar{y}_{P3} + \frac{Q^2}{gA_3} \tag{9.2.26}$$

Equation (9.2.26) contains three expressions for the momentum flux within the bridge opening which are illustrated schematically in Fig. 9.21. These three expressions for the momentum flux relate the depths of flow upstream from, downstream of, and within the bridge opening. In addition, these expressions are functions of only the discharge, the channel geometry, and the bridge pier geometry.

Within the bridge opening there are six possible low-flow conditions (Fig. 9.22). If the flow would be subcritical were the bridge piers not present, then only the conditions shown in Fig. 9.22*a* to *c* are possible. For subcritical flow conditions, the gradually varied flow profile is calculated from a downstream control upstream to section 3. The depth of flow estimated at section 3 by this calculation can exist only if the momentum flux in the bridge opening calculated on the basis of the downstream depth exceeds the critical momentum flux in the constriction, designated as M_{crit}. Figure 9.21*b* is used to determine the momentum flux that would exist in the bridge opening for a specified depth at section 3. A comparison of the momentum flux in the constriction based on the specified depth of flow at section 3, M^T, with M_{crit}, determines the flow type, or:

1. If $M^T < M_{\text{crit}}$, then a hydraulic jump is formed (Fig. 9.22*c*).

2. If $M^T = M_{\text{crit}}$, then the flow is as shown in Fig. 9.22*b*.

3. If $M^T > M_{\text{crit}}$, then the flow is as shown in Fig. 9.22*a*.

For the flow types illustrated in Fig. 9.22*b* and *c*, the momentum flux that exists in the constriction is the critical momentum flux, and the upstream and downstream unknown depths of flow can be determined from Fig. 9.21*a* and *e*. For the flow type illustrated in Fig. 9.22*a*, the momentum flux that exists in the constriction can be determined from Fig. 9.21*c* on the basis of the previously specified depth at section 3. Then, the unknown depths of flow upstream from and within the bridge opening can be determined for this momentum flux from Fig. 9.21*a* and *b*, respectively.

In the case where the flow in the reach would be supercritical, reasoning analogous to that described above can be used to determine the flow conditions within the bridge opening. In such a situation the starting point for the calculation is section 1. Then, a comparison of the momentum flux in the constriction based on the specified depth of flow at section 1, M^T, with M_{crit} determines the flow types or

1. If $M^T > M_{\text{crit}}$, then the flow is as shown in Fig. 9.22*d*.

2. If $M^T = M_{\text{crit}}$, then the flow is as shown in Fig. 9.22*e*.

3. If $M^T < M_{\text{crit}}$, then the flow is as shown in Fig. 9.22*f*.

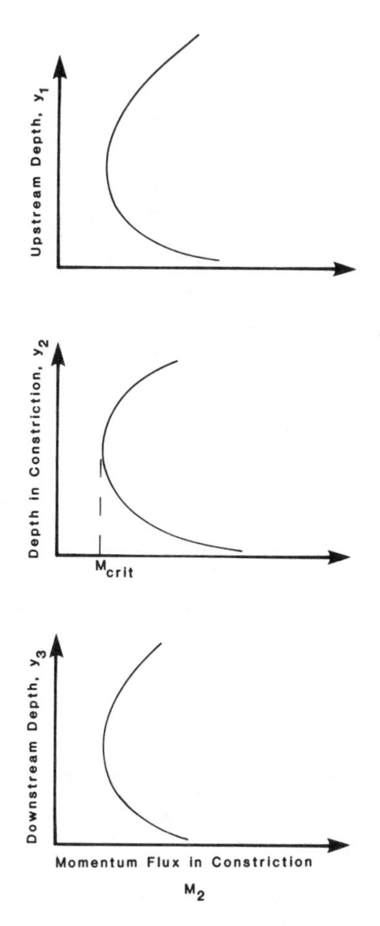

FIGURE 9.21 Relation of momentum flux in constriction to depths (*a*) upstream from, (*b*) within, and (*c*) downstream from the constriction. *(Eichert and Peters, 1970b.)*

At this point, it should be noted that in the case of the flow illustrated in Fig. 9.22*a*, Eichert and Peters (1970b) recommend using the semiempirical Yarnell equation, i.e., Eq. (9.2.2). The form of this equation used by Eichert and Peters (1970b) is

$$\Delta y = 2k\left(k + \frac{10\bar{u}_3^2}{2gy_3} - 0.6\right)\left[\frac{A_{P3}}{A_3} + 15\left(\frac{A_{P3}}{A_3}\right)^4\right]\frac{\bar{u}_3^2}{2g} \tag{9.2.27}$$

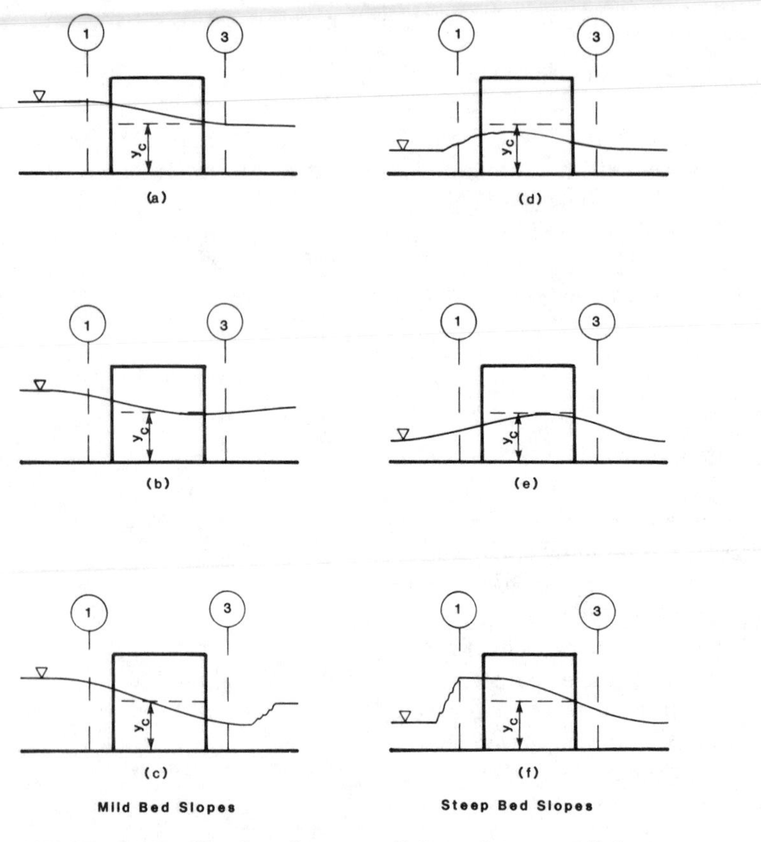

FIGURE 9.22 Six low-flow conditions in constrictions. *(Eichert and Peters, 1970b.)*

where $\Delta y = y_1 - y_3$ = difference between the upstream and downstream water surface elevations and k = a coefficient (Table 9.1).

9.3 CONTROL OF HYDRAULIC JUMPS

When a hydraulic jump occurs on a smooth, horizontal surface, there is little variation in depths upstream and downstream of the jump. In such a situation, there may be a tendency for the jump to migrate upstream or downstream unless it is held in place by some special structural feature, e.g., a sharp- or broad-crested weir, an abrupt rise or drop, or a designed stilling basin. In the material which follows, a number of special features which can be used to control the location of a hydraulic jump will be discussed. It should be noted that although in many cases the control of a hydraulic jump can be evaluated analytically, recourse to experiments is usually required if reasonably accurate quantitative results are required. In the case of hydraulic jump control by a sill, i.e., a sharp- or broad-crested weir or an abrupt rise or drop, dimensional analysis indicates that

$$\frac{\Delta z}{y_1} = \Theta\left(\mathbf{F}_1, \frac{X}{y_2}, \frac{y_3}{y_1}\right) \tag{9.3.1}$$

where with reference to Fig. 9.23, $\mathbf{F}_1$ = Froude number of the approaching flow, Δz = height of the sill, y_1 = depth of approaching flow, y_2 = depth of flow immediately upstream of the sill, y_3 = downstream depth of flow, and X = distance from the toe of the jump to the sill. The symbol Θ indicates a functional relationship which must be determined given a specific situation.

Sharp-Crested Weir

Utilizing both theoretical analysis and experimental evidence, Forster and Skrinde (1950) developed a relation between the Froude number of the

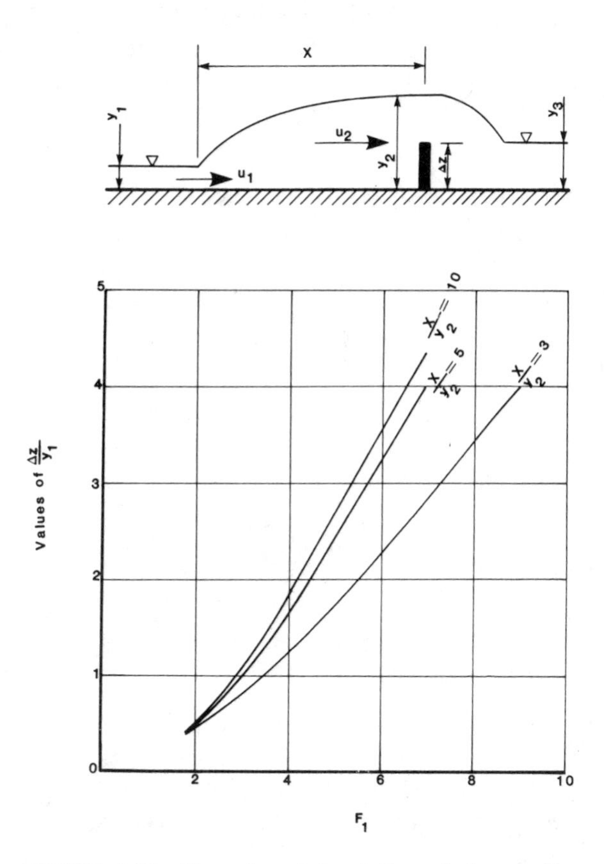

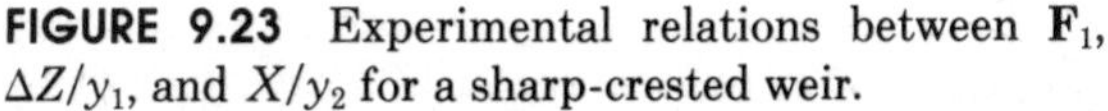

FIGURE 9.23 Experimental relations between $\mathbf{F}_1$, $\Delta Z/y_1$, and X/y_2 for a sharp-crested weir.

approaching flow, the ratio $\Delta z/y_1$, and the ratio X/y_2 for the case of the sharp-crested weir (Fig. 9.23). This figure permits the effect of a sharp-crested weir to be evaluated if the upstream flow conditions are known and

$$y_3 < y_2 - 0.75\,\Delta z \tag{9.3.2}$$

The condition specified by Eq. (9.3.2) is that the tailwater does not affect the discharge over the weir crest. For this reason, the third ratio, that is, y_3/y_1, is not considered in Fig. 9.23.

In Fig. 9.23, the coordinates ($\mathbf{F}_1$, $\Delta z/y_1$) define a unique point. If this point lies between the experimental curves, then a hydraulic jump occurs and the interpolated value of X/y_2 indicates the relative position of the jump. Points which are above the curves indicate that the jump will be forced upstream and in some cases a drowned jump will occur. Even if a drowned jump occurs, the weir will serve to deflect high-velocity bottom currents and protect the downstream channel. If the point lies below the curve $X/y_2 = 5$, a shortened, incomplete jump or spray action may occur (Forster and Skrinde, 1950). In this latter case, only a qualitative prediction of the result is possible, and this prediction depends on the distance of the point from the experimental curves. Forster and Skrinde (1950) further recommended that the curve $X/y_2 = 5$ be used for design under the conditions of maximum discharge.

Broad-Crested Weir

When a broad-crested weir is used to regulate the location of a hydraulic jump (Fig. 9.24), an analytic solution is available. If the depth of flow downstream of the weir is less than the critical depth on the top of the weir or

$$y_3 < \frac{2y_2 + \Delta z}{3} \tag{9.3.3}$$

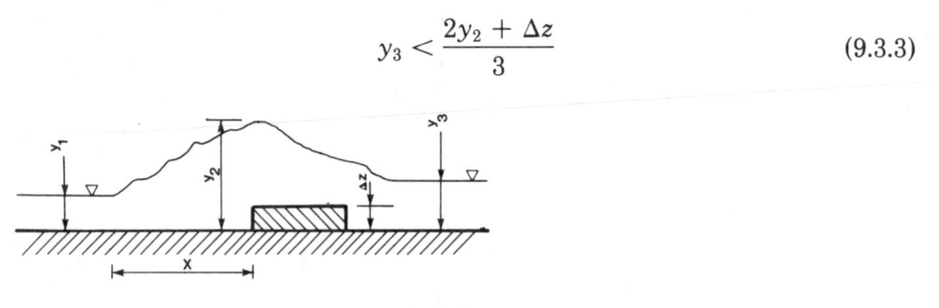

FIGURE 9.24 Hydraulic jump control with a broad-crested weir.

then the tailwater does not affect the flow over the weir. In developing the analytic relationship for this situation, Forster and Skrinde (1950) used a discharge relationship developed for flow over a broad-crested weir in a rectangular channel developed by Doeringsfeld and Barker (1941) or

$$q = 0.433\,\sqrt{2g}\left(\frac{y_1}{y_1 + \Delta z}\right)^{1/2} H^{3/2} \tag{9.3.4}$$

where q = flow per unit width and H = head on the weir or

$$H = y_2 - \Delta z \tag{9.3.5}$$

A comparison of Eq. (9.3.4) with Eq. (8.3.3), which is applicable to the same situation, exposes some differences; however, in this analysis these differences are not an issue. By use of $q = \bar{u}_1 y_1$, $\mathbf{F}_1 = \bar{u}_1/\sqrt{g y_1}$ and Eq. (9.3.5), Eq. (9.3.4) can be rearranged to yield

$$2.667\,\mathbf{F}_1^2 \left[1 + \frac{(\Delta z/y_1)}{(y_2/y_1)}\right] = \left(\frac{y_2}{y_1} - \frac{\Delta z}{y_1}\right)^3 \tag{9.3.6}$$

Then combining Eq. (3.2.4) which describes the frictional relationship between y_2/y_1 and $\mathbf{F}_1$ for a hydraulic jump and Eq. (9.3.6) yields

$$\frac{21.34\,\mathbf{F}_1^2}{\sqrt{1 + 8\mathbf{F}_1^2} - 1} = \frac{[\sqrt{1 + 8\mathbf{F}_1^2} - 1 - 2\,(\Delta z/y_1)]^3}{\sqrt{1 + 8\mathbf{F}_1^2} - 1 + 2\,(\Delta z/y_1)} \tag{9.3.7}$$

Equation (9.3.7) describes a functional relationship between the ratio $\Delta z/y_1$ and the upstream Froude number $\mathbf{F}_1$. Equation (9.3.7) is plotted in Fig. 9.25. Forster and Skrinde (1950) found that Eq. (9.3.7) described their experimental data for $y_3 = y_c$ and $X = 5\,(\Delta z + y_3)$ where y_c = critical depth. Since additional experimental data are not available, Eq. (9.3.7) can be used to guide the design of stilling basins for hydraulic jump control by a broad-crested weir. In comparison with other types of jump control, e.g., the sharp-crested weir or the

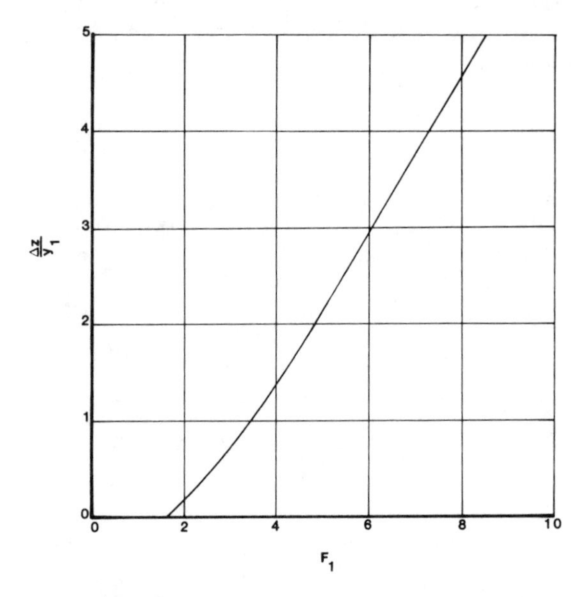

FIGURE 9.25 Analytic relation between $\mathbf{F}_1$ and $\Delta Z/y_1$ for hydraulic jump control by a broad-crested weir.

abrupt rise, the broad-crested weir has the advantages of structural stability and rather low construction costs.

Abrupt Rise If the approach section is lowered below the bed of the normal channel, an abrupt rise which may be used to control the location of the hydraulic jump results (Fig. 9.26). Forster and Skrinde (1950) demonstrated that this problem could be solved analytically by: (1) writing the momentum equation between sections 1 and 2 to estimate the force on the face of the rise; (2) writing the momentum equation between sections 2 and 3 and using the results of step 1; and (3) using the continuity equations between the necessary sections to eliminate y_2, $\overline{u}_2$, and $\overline{u}_3$ from the equation developed in step 2. The result is

$$\left(\frac{y_3}{y_1}\right)^2 = 1 + 2\mathbf{F}_1^2\left(1 - \frac{y_1}{y_3}\right) + \frac{\Delta z}{y_1}\left(\frac{\Delta z}{y_1} - \sqrt{1 + 8\mathbf{F}_1^2} + 1\right) \qquad (9.3.8)$$

Equation (9.3.8) is plotted in Fig. 9.27. In this figure two lines which pass through the origin divide the graph into three main regions. The line $\Delta z/y_1 = 0$ is defined by the equation for a hydraulic jump in a level rectangular channel and therefore represents the equality between the tailwater depth y_3 and the sequent depth of the supercritical flow. The region above this line represents situations in which $\Delta z < 0$ or situations in which a drop in the channel bottom, rather than an abrupt rise, is required to maintain a hydraulic jump. The curves of $\Delta z/y_1$ = constant all pass through a minimum value of the upstream Froude number. The line of minimum $\mathbf{F}_1$ values is found by taking the derivative of $\mathbf{F}_1$ in Eq. (9.3.8) with respect to y_3/y_1 and setting the result equal to zero. The result is

$$\frac{y_3}{y_1} = \mathbf{F}_1^{2/3} \qquad (9.3.9)$$

It can be easily shown that Eq. (9.3.9) corresponds to the condition that the tailwater depth is critical depth. In Fig. 9.27, the region below the line specified by Eq. (9.3.9) is an area in which $y_3 < y_c$. Thus, in this region a hydraulic jump

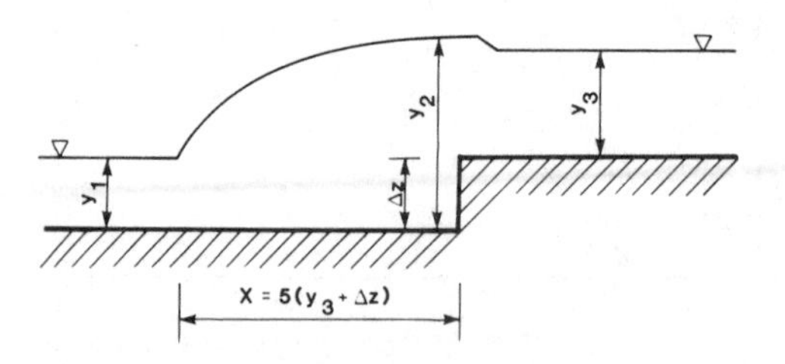

FIGURE 9.26 Hydraulic jump control with an abrupt rise.

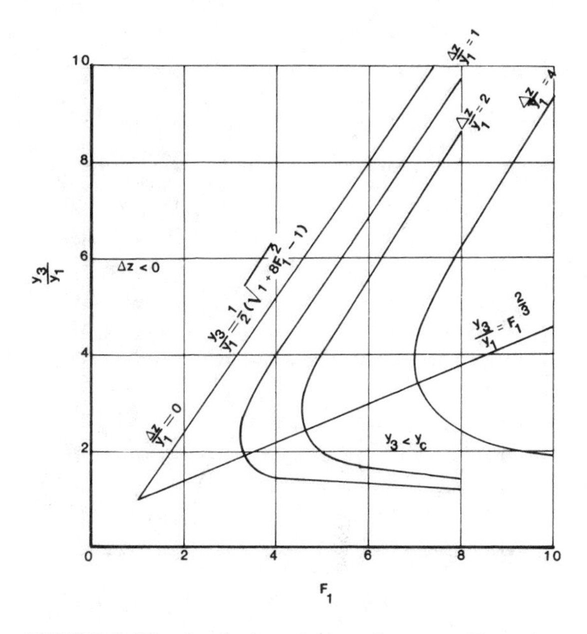

FIGURE 9.27 Analytic relations between $\mathbf{F}_1$, y_3/y_1, and $\Delta Z/y_1$ for abrupt rise.

is not formed and supercritical flow shoots over the abrupt rise. The region between the two lines defined by $\Delta z/y_1 = 0$ and Eq. (9.3.9) defines the solution of Eq. (9.3.8) in which a hydraulic jump is formed and is complete just upstream of the abrupt rise.

Figure 9.28 is a plot of the Forster and Skrinde (1950) experimental data for an abrupt rise for $X = 5\ (\Delta z + y_3)$. In this figure, it can be seen that the experimental results show a significant deviation from the theoretical results summarized in Fig. 9.27. Figure 9.28 can be used for design purposes when $\overline{u}_1$, y_1, and y_3 are known.

Abrupt Drop As noted in the previous section, if the tailwater depth is larger than the sequent depth for a normal jump, then a drop in the bed of the channel must be used to ensure that a jump forms. For a specified value of $\mathbf{F}_1$, the depth of flow may fall in any of the five regions shown in Fig. 9.29. The lower limit of region 1 is the depth of flow at which the jump will begin to travel upstream. The upper limit of region 5 is the depth of flow at which the jump will begin to move downstream. If the depth of flow downstream of the abrupt drop falls into either of these regions, the drop will not be effective in controlling the jump. The drop will be effective in producing a stable jump only if the depth of flow falls in either regions 2 or 4. Region 3 is an intermediate region in which an undular jump results.

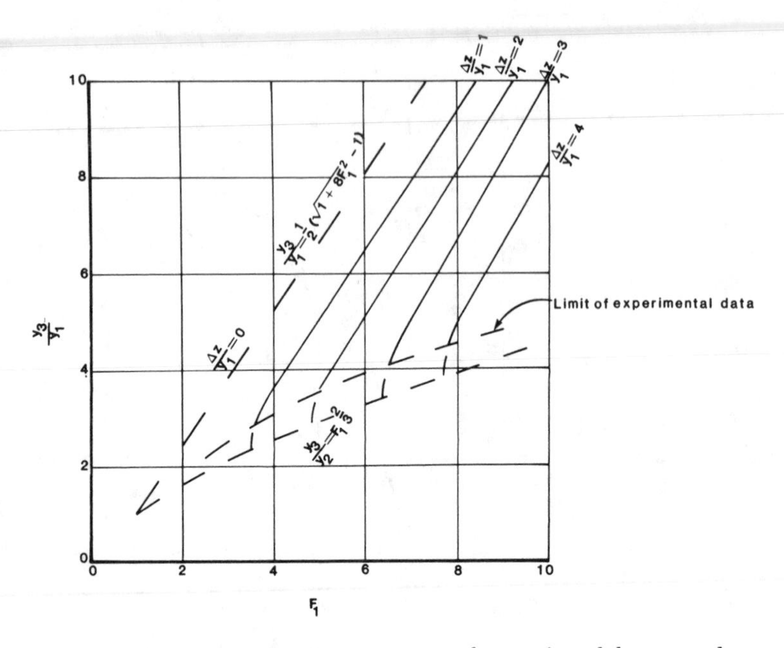

FIGURE 9.28 The hydraulic jump at an abrupt rise—laboratory data. *(Forster and Skrinde, 1950.)*

Using the continuity and momentum equations, Hsu (1950) developed the following equations:

For Region 2

$$\mathbf{F}_1^2 = \frac{y_3/y_1}{2\left[(y_3/y_1) - 1\right]}\left[-1 + \left(\frac{y_3}{y_1} - \frac{\Delta z}{y_1}\right)^2\right] \tag{9.3.10}$$

For Region 4

$$\mathbf{F}_1^2 = \frac{y_3/y_1}{2\left[(y_3/y_1) - 1\right]}\left[-\left(\frac{\Delta z}{y_1} + 1\right)^2 + \left(\frac{y_3}{y_1}\right)^2\right] \tag{9.3.11}$$

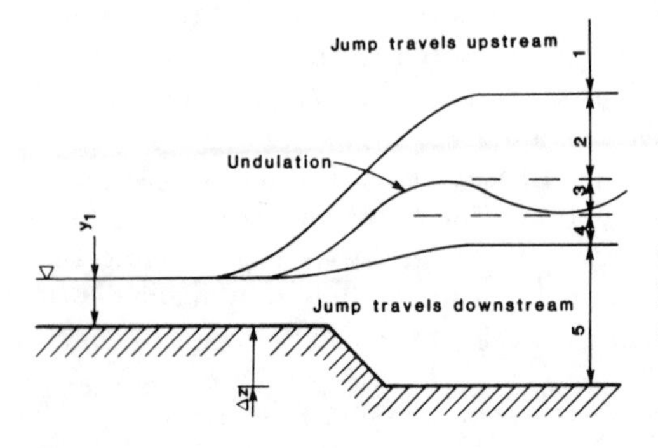

FIGURE 9.29 Forms of the hydraulic jump at an abrupt drop. *(Hsu, 1950.)*

Equations (9.3.10) and (9.3.11) are plotted in Fig. 9.30. In this figure each curve for a specified value of $\Delta z/y_1$ has two relatively straight limbs connected by a short, straight portion near the middle of the curve. The left-hand limb of each of these curves corresponds to region 2 and the right-hand side to region 4. The validity of this figure has been established with experimental data (Hsu, 1950) and can be used to determine the relative height of the drop required to stabilize a hydraulic jump.

Stilling Basins

By definition, a *stilling basin* is a short length of paved channel placed at the end of a spillway or any other source of supercritical flow. Because of the widespread use of stilling basins, a number of generalized designs have been developed based on model studies, experience, and the observation of existing stilling

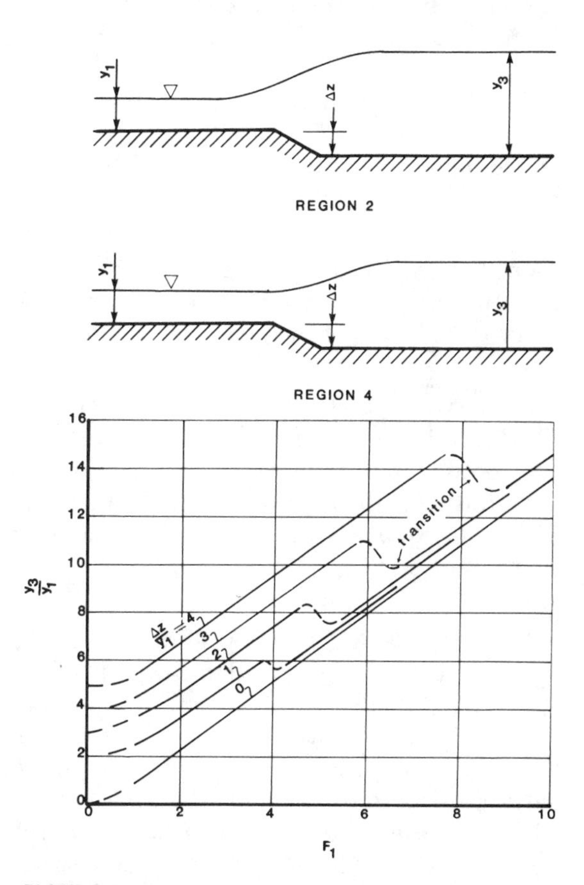

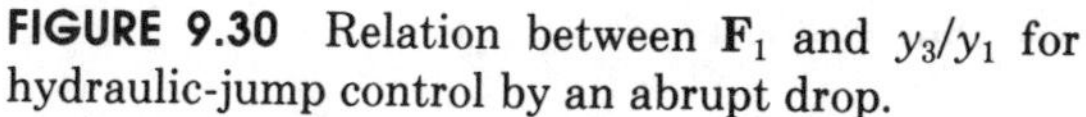

FIGURE 9.30 Relation between $\mathbf{F}_1$ and y_3/y_1 for hydraulic-jump control by an abrupt drop.

basins. In the material which follows, three generalized stilling designs will be considered:

1. *Saint Anthony Falls Basin (SAF):* This type of basin is usually used in conjunction with small spillways, outlet works, and canal structures where

$$1.7 \leq \mathbf{F}_1 \leq 17$$

   If the appurtenances designed for the basin are used, the reduction in basin length ranges from 70 to 90 percent.

2. *U.S. Bureau of Reclamation (USBR) Basin II:* This type of basin is generally used in conjunction with large structures where $\mathbf{F}_1 > 4.5$. The length of the basin can usually be reduced by 33 percent if the appurtenances appropriate to the basin are used.

3. *USBR Basin IV:* This type of basin is generally used in conjunction with canal structures and diversion structures where

$$2.5 \leq \mathbf{F}_1 \leq 4.5$$

   This basin is effective in reducing the waves associated with weak jumps.

The appurtenances usually associated with stilling basins are described below.

Chute Blocks Chute blocks are used in the approach section to channelize the flow and lift part of the flow off the bed. The purpose of these blocks is to shorten the length of the jump and stabilize it.

Sill, Dentated or Solid The dentated or solid sill occurs at the end of the basin. The function of this appurtenance is to further reduce the length of the jump and control scour. In large stilling basins, the sill is usually dentated to aid in the diffusion of the high-velocity jet that may reach the end of the basin. *Note:* The dentated sill is sometimes termed the *Rehbock sill* after Theodor Rehbock who developed it.

Baffle Piers Baffle piers are placed at intermediate locations in the basin, and their primary function is the dissipation of energy by impact. When the approach section velocities are low, baffle piers can be very effective; however, when incoming velocities are high, this type of appurtenance may be inappropriate because of the possibility of cavitation. In some situations, baffle piers must be designed to withstand the impact of debris and ice. As noted by Chow (1959), the generalized designs presented in the following material should be used cautiously since they are generic designs which may or may not be suitable to a particular situation.

The SAF stilling basin (Fig. 9.31) was developed for use on small drainage structures. Blaisdell (1948 and 1949) summarized the following design rules:

1. The length L_B of the stilling basin is given by

$$L_B = \frac{4.5y_2}{\mathbf{F}_1^{0.76}} \tag{9.3.12}$$

for

$$1.7 \leq \mathbf{F}_1 \leq 17 \tag{9.3.13}$$

2. The height of the chute and floor blocks is y_1, and the spacing and width are $0.75y_1$.

3. The distance from the upstream end of the stilling basin to the upstream face of the floor blocks is $L_B/3$.

4. A floor block should not be placed closer to a side wall than $\frac{3}{8}\, y_1$.

5. The floor blocks are placed in the openings between the chute blocks (Fig. 9.31).

6. The floor blocks will, in general, occupy 40 to 55 percent of the stilling basin width.

7. For diverging stilling basins, the widths and spacings of the floor blocks should be increased in proportion to the increase in the stilling basin width at the location of the floor blocks.

8. The height of the sill at the end of the stilling basin is given by

$$C = 0.07y_2 \tag{9.3.14}$$

where y_2 = theoretical sequent depth corresponding to y.

9. The depth of the tailwater above the stilling basin floor is given by

$$\begin{aligned} y_2' &= \left(1.10 - \frac{\mathbf{F}^2}{120}\right) y_2 \qquad \text{for } 1.7 \leq \mathbf{F}_1 \leq 5.5 \\ &= 0.85y_2 \qquad \text{for } 5.5 < \mathbf{F}_1 \leq 11 \\ &= \left(1.00 - \frac{\mathbf{F}_1^2}{800}\right) y_2 \qquad \text{for } 11 < \mathbf{F}_1 \leq 17 \end{aligned} \tag{9.3.15}$$

where y_2' = tailwater depth.

10. The height of the side wall above the maximum tailwater depth is given by

$$z = \frac{y_2}{3} \tag{9.3.16}$$

where z = height of side wall above the maximum tailwater depth.

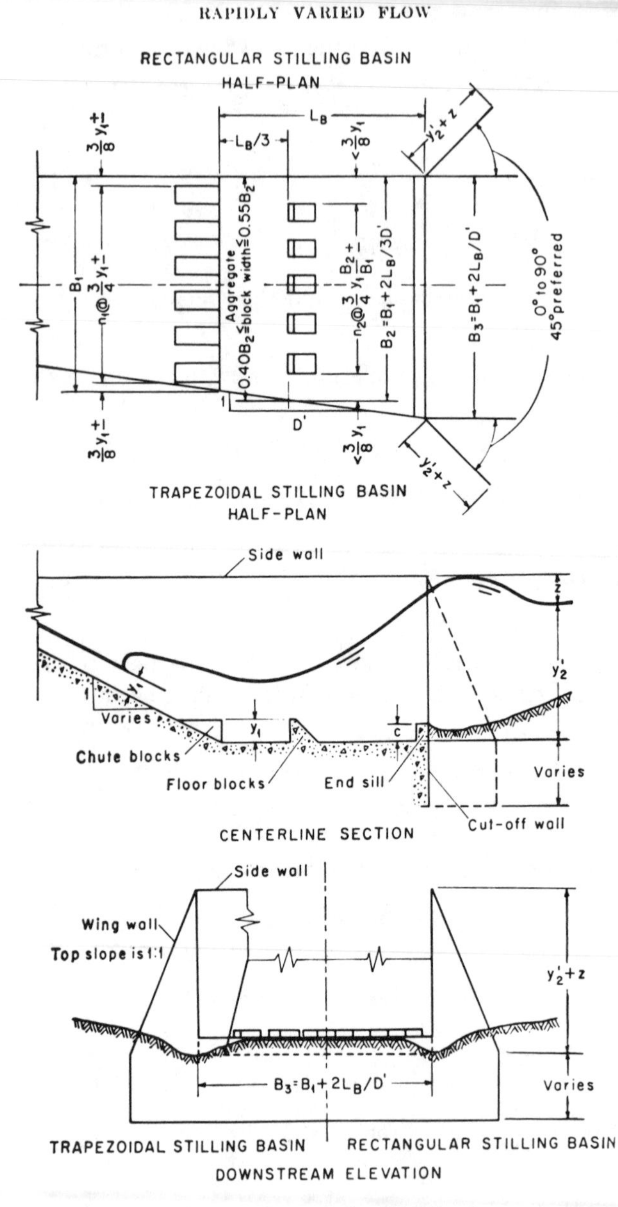

FIGURE 9.31 Proportions of the SAF basin.

11. Wingwalls (Fig. 9.31) should have a height equal to the stilling basin side walls, and the top of these walls should have a slope of 1 on 1.

12. The wingwalls should be placed on an angle of 45° to the outlet centerline.

13. The side walls of the stilling basin may either be parallel, as in the case of

the rectangular stilling basin, or they may diverge, as in the case of a trapezoidal stilling basin.

14. A cutoff wall of nominal depth should be used at the end of the stilling basin.

15. The effect of entrained air is neglected in the design process.

From intensive studies of existing structures and laboratory investigations, the U.S. Bureau of Reclamation has developed a set of generalized designs for stilling basins under many conditions. This work is summarized in a series of papers by Bradley and Peterka (1957). In this work only Basins II and IV will be considered. USBR Basin I is for the control of jumps occurring on a flat floor with no structural appurtenances. USBR Basin III is similar to the SAF basin but with a larger safety factor. USBR Basin V is used in conjunction with sloping floors.

The USBR Basin II was designed for use in conjunction with high dam and earth dam spillways or with large canal structures. Although chute blocks are used at the upstream end and a dentated sill at the lower end, floor blocks are not used because of the possibility of cavitation. With reference to Fig. 9.32, the recommended rules for design are:

1. The apron elevation should be such that the full sequent tailwater depth plus a factor of safety is utilized. In Fig. 9.32*b* the dashed lines are guides based on various ratios of the actual tailwater depth to sequent depths. Although the U.S. Bureau of Reclamation studies indicate that most basins were designed for the sequent tailwater depth or less, there is a lower limit which is labeled *minimum TW depth* in Fig. 9.32*b*. This curve defines the point at which the front of the jump moves away from the chute blocks. Additional lowering of the tailwater depth would result in the jump leaving the basin. Thus, for design purposes, the USBR Basin II should not be designed for less than sequent depth, and a minimum safety factor of at least 5 percent should be added to the sequent depth.

2. Although USBR Basin II is theoretically effective down to a Froude number of 4, its utility at lower Froude numbers should not be taken for granted. At low Froude numbers, stilling basin designs incorporating wave suppression should be considered.

3. Figure 9.32*e* can be used to estimate the length of stilling basins required.

4. The height of the chute blocks should be equal to the depth of flow in the approach section; the width and lateral spacing of these blocks should also be equal to the depth of flow in the approach section. The width and spacing of the chute block is somewhat variable, so that the need for fractional blocks is precluded. Along each side wall a spacing of $0.5y$ is recommended to reduce spray and maintain desirable pressures.

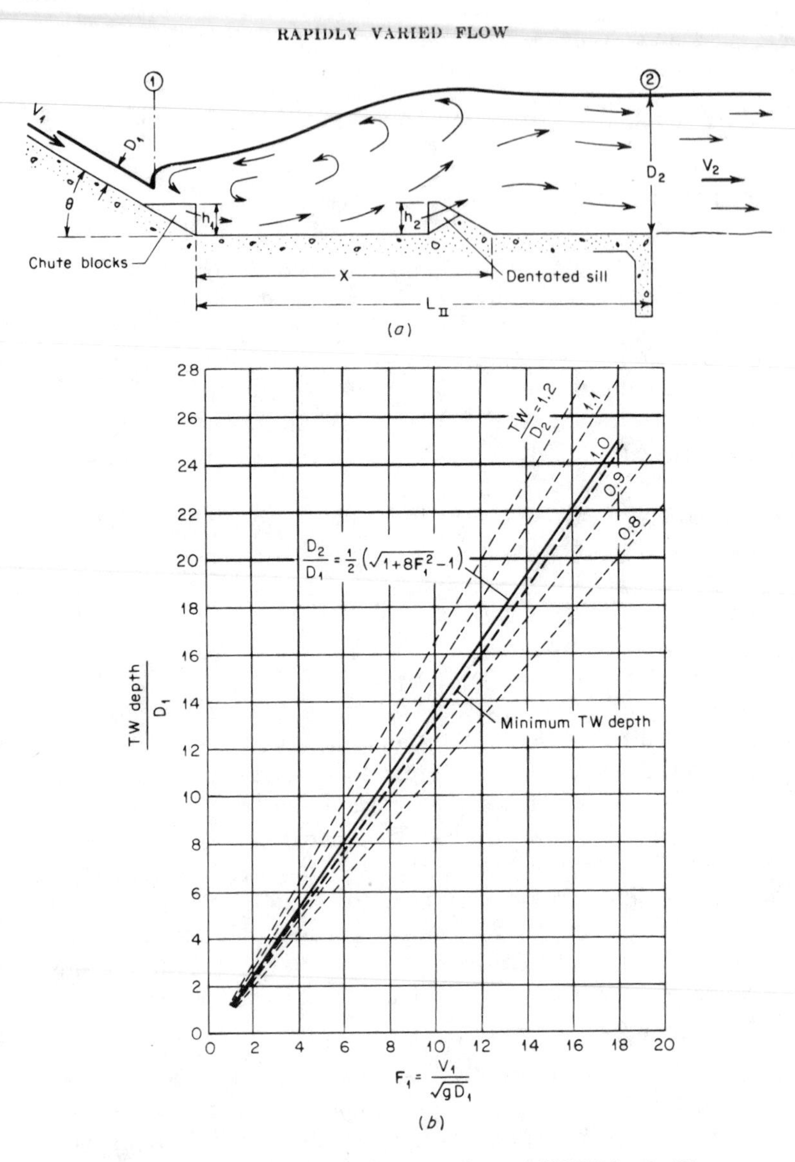

FIGURE 9.32 Design curves and proportions of USBR basin II.

5. The height of the dentated sill at the end of the basin should be equal to $0.2y_2$. The width and spacings of the dentates is $0.15y_2$. It is also recommended that a block be adjacent to each side wall. The slope of the continuous portion of the sill should be 2:1. In narrow basins, it may be advisable to reduce the width and spacing of the dentates so that the number of dentates is increased. The minimum width and spacing of the dentates in such a situation is governed only by structural considerations.

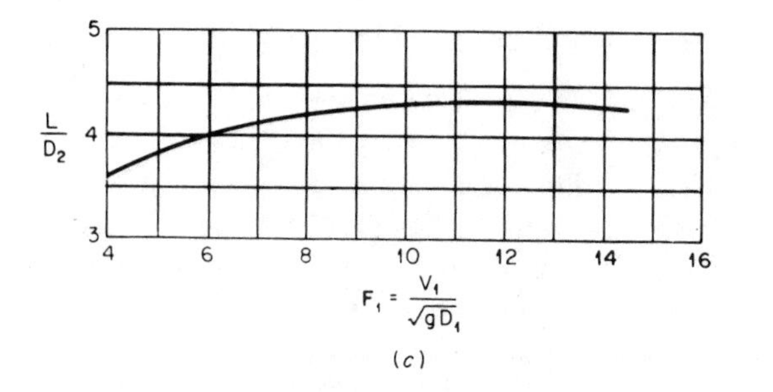

(c)

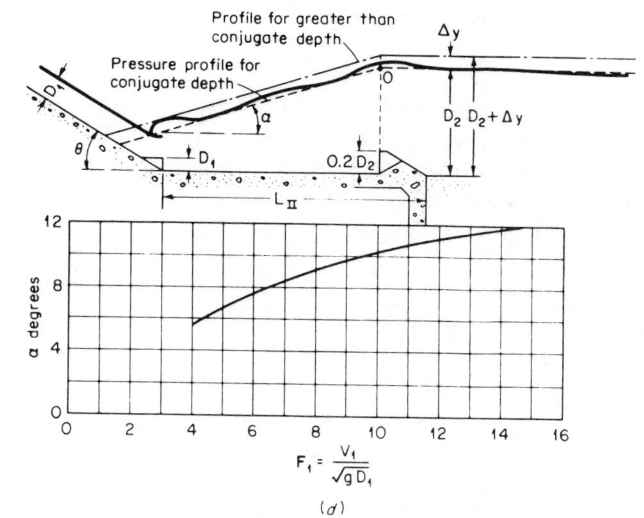

(d)

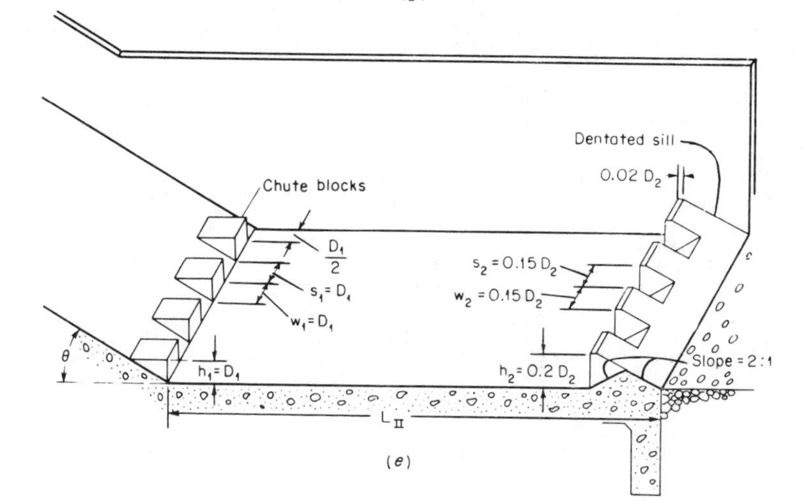

(e)

FIGURE 9.32 *Continued.*

6. It is not necessary to stagger the chute blocks and sill dentates.
7. Apparently the effectiveness of the stilling basin is not a function of the slope of the approach channel. Chow (1959) noted that this is true as long as the velocity distribution and depth of flow in the approach channel are uniform. If the approach channel is long and flat, the flow may become concentrated in one area of the section and an asymmetrical jump with its associated problems may result. When the slope of the approach channel is 1:1 or greater, it is recommended that the usually sharp intersection of the channel and the basin be replaced with a curve of radius R such that

$$R \geq 4y_1$$

The above design rules result in a safe, conservative stilling basin for spillway or falls up to 200 ft (61 m) and for flows of up to 500 (ft^3/s)/ft of stilling basin width [46 (m^3/s)/m] *provided* that the flow enters the basin with a uniform velocity distribution and depth. In situations involving greater falls, higher discharges, or nonuniform velocities and depths, a model study should be conducted.

The USBR Basin IV is designed to control the oscillating jump which is formed in the stilling basin when the Froude number in the approach section lies between 2.5 and 4.5, inclusive. This basin is designed to eliminate the problem at its source by intensifying the roller which appears in the upper portion of the jump with directional jets deflected from large chute blocks (Fig. 9.33). From the viewpoint of hydraulic performance, a minimum of three blocks is required. Better hydraulic performance can be achieved if the blocks are made narrower than shown in Fig. 9.33 and the tailwater depth is 5 to 10 percent deeper than the sequent depth. An optimum chute block width is $0.75y'$. The

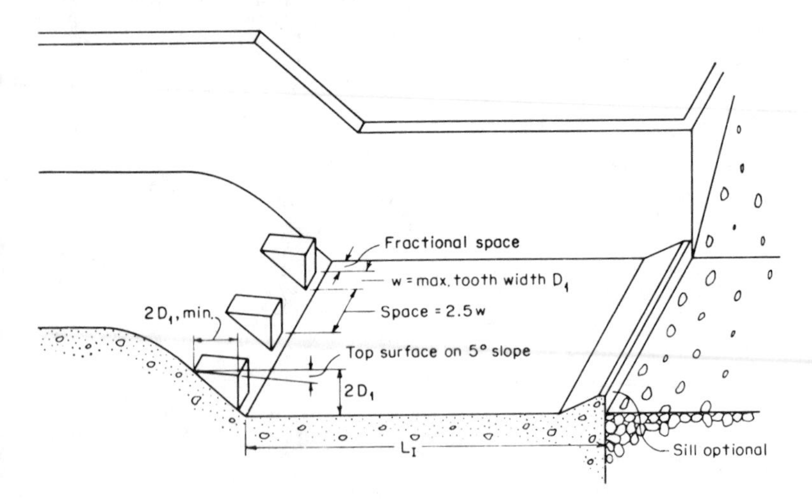

FIGURE 9.33 Proportions of USBR basin IV.

length of the stilling basin is assumed to be equal to the length of a hydraulic jump in a horizontal stilling basin with no appurtenances. Figure 3.8 can be used to estimate this length. It should be noted that USBR Basin IV is applicable only to channels of rectangular cross section.

9.4 DROP SPILLWAYS

The drop spillway is commonly used in small drainage structures to dissipate energy. An aerated, free-falling nappe in a straight-drop spillway will reverse its curvature and result in a supercritical flow on the apron which will, in turn, result in a hydraulic jump (Fig. 9.34). If it is assumed that the depth of flow at the free overfall is critical, then Rand (1955) has demonstrated by analyzing experimental data that

$$\frac{y_1}{\Delta z} = 0.54 \left(\frac{y_c}{\Delta z}\right)^{1.275} \tag{9.4.1}$$

$$\frac{y_1}{y_c} = 0.54 \left(\frac{y_c}{\Delta z}\right)^{0.275} \tag{9.4.2}$$

$$\frac{y_2}{\Delta z} = 1.66 \left(\frac{y_c}{\Delta z}\right)^{0.81} \tag{9.4.3}$$

$$\frac{L_d}{\Delta z} = 4.30 \left(\frac{y_c}{\Delta z}\right)^{0.09} \tag{9.4.4}$$

$$L_j = 6.9\,(y_2 - y_1) \tag{9.4.5}$$

where y_c = critical depth and all other variables are defined in Fig. 9.34. The sill or upward step of $y_2/6$ at the end of the structure serves to locate the jump in the immediate vicinity of the drop structure. Equations (9.4.1) to (9.4.5) and Fig. 9.34 are completely satisfactory for proportioning a simple drop structure. Rand (1955) noted that the equations given above fitted the data with errors of 5 percent or less.

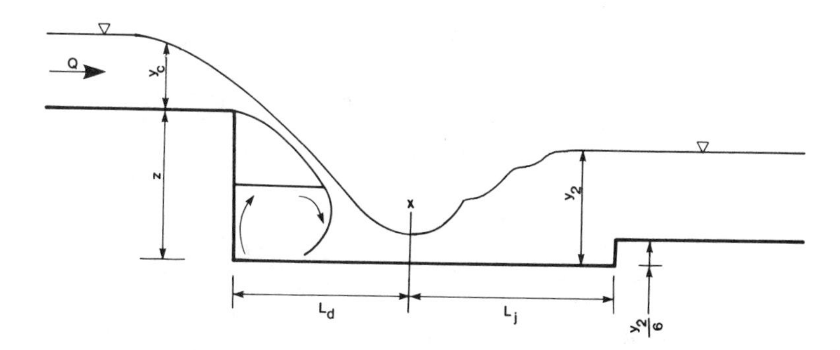

FIGURE 9.34 Schematic of the drop structure.

In practice, many variants of the drop structure given above are used. In Fig. 9.35 a box inlet structure design presented by Blaisdell and Donnelly (1956) is shown. This structure has the advantage of dissipating more energy by making three streams meet at the bottom of the structure. The design rules presented by Blaisdell and Donnelly (1956) for this structure are:

1. The actual depth of flow in the straight section of the structure is

$$y_c = \left[\frac{(Q/b)^2}{g}\right]^{1/3} \tag{9.4.6}$$

where b = width of the straight section (Fig. 9.35) and Q = total discharge.

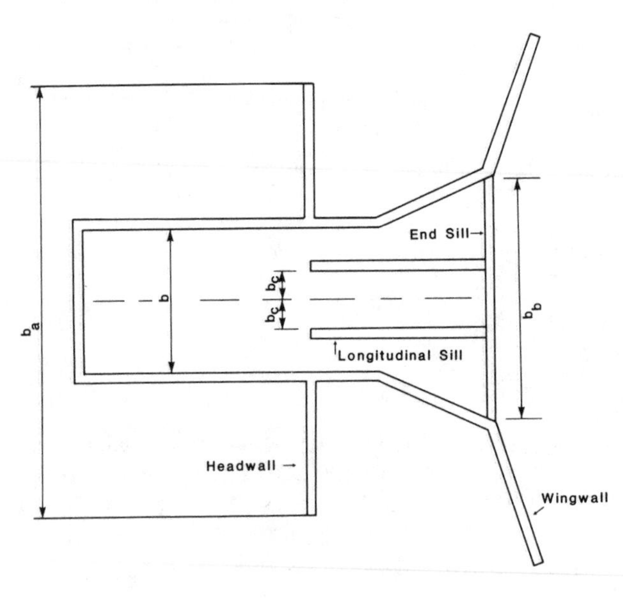

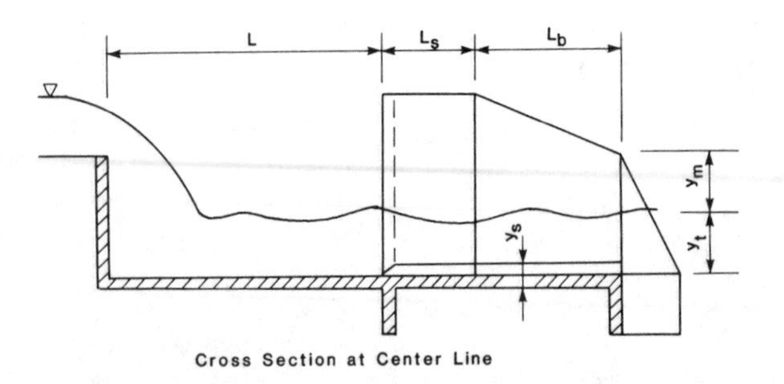

FIGURE 9.35 Definition of variables for drop spillway design. *(Blaisdell and Donnelly, 1956.)*

2. The critical depth of flow at the end of the stilling basin is

$$y_c' = \left[\frac{(Q/b_b)^2}{g}\right]^{1/3} \tag{9.4.7}$$

where b_b = width of the end of the stilling basin.

3. For values of $L/b \geq 0.25$, the minimum length of the straight section is given by

$$L_s = y_c\left(\frac{0.2}{L/b} + 1\right) \tag{9.4.8}$$

where all variables are defined in Fig. 9.35.

4. The flare of the side walls may range from 1 in ∞, that is, parallel extensions of the straight section wall, to 1 in 2.

5. For values of $L/b \geq 0.25$, the minimum length of the stilling basin is given by

$$L_b = \frac{L_c}{2L/b} \tag{9.4.9}$$

where L_c = crest length defined as

$$L_c = 2L + b \tag{9.4.10}$$

(Blaisdell and Donnelly, 1956).

6. When

$$b_b < 11.5y_c'$$

then the minimum tailwater depth over the basin floor is given by

$$y_t = 1.6y_c' \tag{9.4.11}$$

where y_t = tailwater depth. When

$$b_b \geq 11.5y_c'$$

the minimum tailwater depth over the basin floor is given by

$$y_t = y_c' + 0.052b_b \tag{9.4.12}$$

If tailwater depths less than those predicted by Eqs. (9.4.11) and (9.4.12) are used, then increased scour downstream of the structure should be expected.

7. The height of the sill at the end of the channel is given by

$$y_s = \frac{y_t}{6} \tag{9.4.13}$$

where y_s = end sill height.

8. The longitudinal sills shown in Fig. 9.35 improve the flow distribution at the outlet. Blaisdell and Donnelly (1956) recommended that these sills be located as follows:

a. If the stilling basin side walls are parallel, then longitudinal sills are not needed.

b. The center pair of longitudinal sills should extend from the exit of the box inlet through the straight section and stilling basin to the end sill.

c. When

$$b_b < 2.5b$$

only two sills are needed, and these sills should be located at a distance b_c on each side of the centerline where

$$\frac{b}{6} \leq b_c \leq \frac{b}{4}$$

d. When

$$b_b > 2.5b$$

two additional sills located parallel to the outlet centerline and midway between the center sills and the side walls at the exit of the stilling basin are required.

e. The height of the longitudinal sills should be the same as the end sill [Eq. (9.4.13)].

9. The minimum height of the side walls above the water surface at the exit of the stilling basin is given by

$$y_m = \frac{y_t}{3} \tag{9.4.14}$$

Under no circumstances should the side walls be less than the tailwater depth.

10. Wingwalls should be triangular in an elevation view and should have top slopes between 30 and 45°. Further, in plan view, wingwalls should have a flare angle of between 45 and 60°. Wingwalls parallel to the outlet centerline are not recommended (Blaisdell and Donnelly, 1956). It should be noted that Eqs. (9.4.8) and (9.4.14) are empirical equations which are valid only in the English system of units.

9.5 TRANSITION STRUCTURES

The terminology *transition structure* implies a designed channel appurtenance whose purpose is to change the cross-sectional shape of the channel. The function of such a structure is to:

1. Avoid excessive losses of energy

2. Eliminate cross waves, standing waves, and other turbulence
3. Provide safety for both the transition structure and the waterway

The geometric form of a transition structure can vary from a rather simple straight-line design to a complex, streamlined design involving warped surfaces. The common types of transition are inlet and outlet transitions between canal and tunnel, and inlet and outlet transitions between canal and inverted siphon. In the material which follows, the design of transition structures for subcritical flow will be considered.

Transition between Canal and a Flume or Tunnel

In designing a transition between a canal and a flume or tunnel, Chow (1959) noted the following important features:

1. *Proportioning:* The optimum maximum angle between the channel axis and a line connecting the channel sides between entrance and exit sections is 12.5°. Sharp angles in either the water surface or the structure itself which will induce standing waves or turbulence should be avoided.
2. *Losses:* The total energy loss that is incurred in a transition is composed of two parts—a friction loss and a transition loss. The friction loss, which can be estimated with the Manning equation, is usually small, has only a minor effect on the flow profile, and is usually ignored in the preliminary design.

   In the case of inlet transitions, the velocity entering the transition is less than the velocity leaving the transition. In such a situation, the water surface must fall at least the full difference between the velocity heads and a small distance known as the *conversion loss.* The conversion loss for an inlet may be estimated by

$$\Delta y' = \Delta h_u + C_i \Delta h_u = (1 + C_i)\,\Delta h_u \tag{9.5.1}$$

   where $\Delta y'$ = fall in water surface elevation due to inlet loss
   Δh_u = difference in velocity head across the transition
   C_i = inlet loss coefficient

   In the case of outlet transitions, the velocity is reduced and the water surface must rise—the velocity head is recovered—and this rise is accompanied by a conversion loss known as the outlet loss or

$$\Delta y' = \Delta h_u - C_0\,\Delta h_u = (1 - C_0)\,\Delta h_u \tag{9.5.2}$$

   where C_0 = coefficient outlet loss. Values of the coefficients C_i and C_0 are summarized in Table 9.4 for various types of transitions.
3. *Freeboard:* For depths of flow in a transition of less than 12 ft (3.7 m) the freeboard should be designed by the methods described in Chapter 7. When

TABLE 9.4 Average safe design values of C_i and C_o (Chow, 1959)

Type of transition	C_i	C_o
Warped	0.10	0.20
Cylinder-quadrant	0.15	0.25
Simplified straight line	0.20	0.30
Straight line	0.30	0.50
Square-ended	0.30+	0.75

the depth of flow exceeds 12 ft (3.7 m), special attention should be given to the freeboard design within the transition.

For a specified set of entrance conditions, the depth and velocity of flow within the transition are governed by three boundary conditions or

$$b = f_1(x) \tag{9.5.3}$$

$$\Delta z = f_2(x) \tag{9.5.4}$$

$$m = f_3(x) \tag{9.5.5}$$

where b = bottom width within transition
Δz = change in elevation of channel bed within transition
m = reciprocal of side slope of channel
x = longitudinal distance measured from beginning of transition (Fig. 9.36)

The relationship between the depth of flow in the transition and the flow rate is given by the equation of continuity or

$$Q = (b + my)y\bar{u} \tag{9.5.6}$$

where y = depth of flow and $\bar{u}$ = average velocity of flow. A relationship between the depth of flow, velocity of flow, and energy loss can be derived by writing the Bernoulli energy equation between sections i and $i + 1$ or

$$-\Delta z_i + (\Delta z_i + y_f + \Delta y_i) + \frac{\bar{u}_i^2}{2g} = -\Delta z_{i+1} + (\Delta z_{i+1} + y_f + \Delta y_{i+1}) + \frac{\bar{u}_{i+1}^2}{2g} + h_{Li,i+1} \tag{9.5.7}$$

where y_f = depth of flow in the flume or rectangular part of the transition (Fig. 9.36) and $h_{Li,i+1}$ = energy loss between sections i and $i + 1$. In Eq. (9.5.7) it is noted that Δz is considered positive if it is measured downward from the bed of the flume and Δy is considered positive if the water surface in the transition is higher than the water surface in the flume. Equation (9.5.7) can also be written as

$$E_{i+1} = E_i + \Delta z_{i,i+1} - h_{Li,i+1} \tag{9.5.8}$$

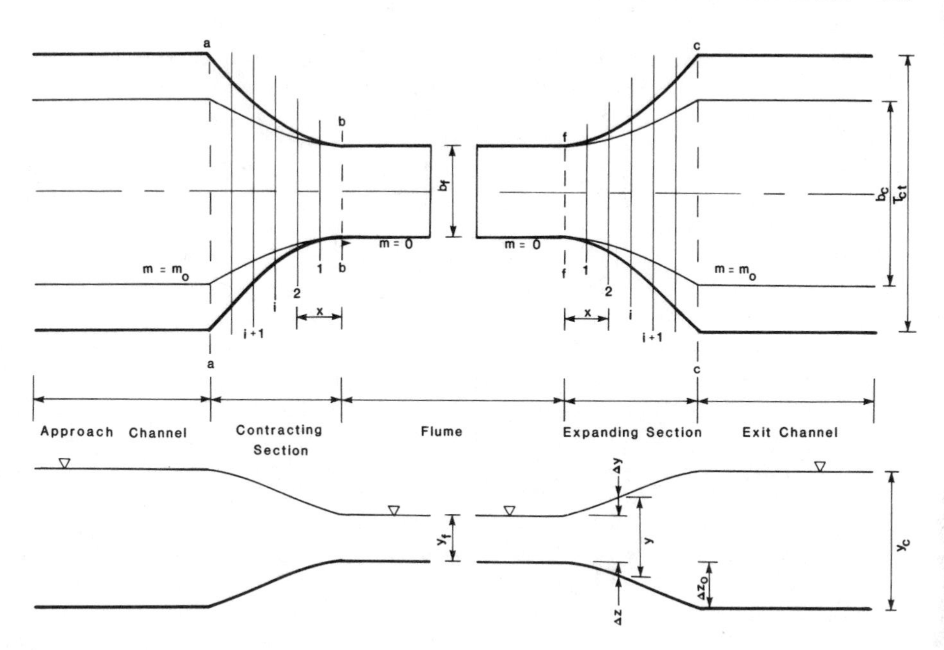

FIGURE 9.36 Transition definition sketch. *(Vittal and Chiranjeevi, 1983.)*

where E = specific energy and

$$\Delta z_{i,i+1} = \Delta z_{i+1} - \Delta z_i \tag{9.5.9}$$

The equation for estimating the headloss through the transition is

$$h_{Li,i+1} = K_H \left(\frac{\overline{u}_i^2 - \overline{u}_{i+1}^2}{2g} \right) = K_H \Delta h_u \tag{9.5.10}$$

(Hinds, 1928).

After the geometric shape of the transition [Eqs. (9.5.3) to (9.5.5)] is defined and the initial conditions at the entrance to the transition are specified, there are a number of techniques which may be used to select the geometry of the transition. Good transition design requires the establishment of the section which yields the least energy loss consistent with the convenience of design and construction.

The technique developed by Hinds (1928) assumes that the water surface profile, i.e.,

$$\Delta y = f_4(x) \tag{9.5.11}$$

in the transition is a compound curve composed of two reverse parabolas which have an inflection point in the center of the transition and which join the water surface at either end of the transition tangentially. The equation of the parabolics is

$$\Delta y = C_1 x^2 \tag{9.5.12}$$

where C_1 = a coefficient and x is measured to the right from section f-f in Fig. 9.36 for the first parabola and measured to the left from section c-c for the second parabola. Hinds (1928) also assumed a linear rise or drop in the channel bed or

$$\Delta z = C_2 x \tag{9.5.13}$$

where C_2 = a constant. The transition is now divided into N subreaches (sections 1-1, 2-2, . . . , in Fig. 9.36) and an arbitrary set of values for m where

$$0 \le m \le m_{0)} \tag{9.5.14}$$

where m_0 = side slope of either the entrance or exit channel. Then, the loss in the total transition section as well as in the subsections of the transition can be estimated. It is noted that Eq. (9.5.10) can be rearranged into an explicit function of the water surface elevation by substituting Eq. (9.5.10) in Eq. (9.5.8) or

$$\Delta h_u = \frac{\Delta y_{i+1} - \Delta y_i}{1 - K_H}$$

and therefore

$$h_{Li,i+1} = \frac{K_H\,(\Delta y_{i+1} - \Delta y_i)}{1 - K_H} \tag{9.5.15}$$

Substitution of Eq. (9.5.15) in Eq. (9.5.7) determines the average velocity of flow in the subsection, and substitution of this value in Eq. (9.5.6) yields a value for the subsection width.

The technique described above begins in the first subreach and proceeds in the downstream direction. If the resulting transition width profile is not smooth and continuous, then the calculations are repeated with a new set of values for m. With regard to the method of Hinds (1928), Vittal and Chiranjeevi (1983) offered the following comments:

1. Since the water surface and bed elevation functions are by definition smooth and continuous, the transition width would also be a smooth and continuous function provided that the values of m are estimated from a smooth and continuous function [Eq. (9.5.5)]. In Table 9.5, a number of relationships for estimating the ratio m/m_0 are summarized. In this table, L = transition length, b_f = bottom width of the flume, b_c = bottom width of the channel, and T_{ct} = water surface width in the exit channel.

2. Even when a smooth and continuous bed with profile is achieved, there is no guarantee that such a design will not result in flow separation, turbulence, and the high headlosses associated with such phenomena.

3. When Eq. (9.5.15) is applied to all N of the subreaches which compose the transition, it demonstrates that in the method of Hinds the energy loss in the transition is a function only of the entry and exit conditions. This is illog-

TABLE 9.5 Relationships for estimating side slope ratios in warped transitions

Relationship	Limits of applicability
$\frac{m}{m_0} = \frac{x}{L}$	$0 \le \frac{x}{L} \le 1$
$\frac{m}{m_0} = \left(\frac{x}{L}\right)^{1/2}$	$0 \le \frac{x}{L} \le 1$
$\frac{m}{m_0} = 1 - \left(1 - \frac{x}{L}\right)^2$	$0 \le \frac{x}{L} \le 1$
$\frac{m}{m_0} = \left(\frac{x}{L}\right)^2$	$0 \le \frac{x}{L} \le 1$
$\frac{m}{m_0} = 0.5 - 2\left(0.5 - \frac{x}{L}\right)^2$	$0 \le \frac{x}{L} < 0.50$
$= 0.5 + 2\left(\frac{x}{L} - 0.5\right)^2$	$0.50 \le \frac{x}{L} \le 1$
$\frac{m}{m_0} = \left(\frac{x}{2L}\right)^{1/2}$	$0 \le \frac{x}{L} < 0.50$
$= 1 - \left[0.5\left(1 - \frac{x}{L}\right)\right]^{1/2}$	$0.50 \le \frac{x}{L} \le 1$
$\frac{m}{m_0} = 0.5 - \left(0.25 - \frac{x}{2L}\right)^{1/2}$	$0 \le \frac{x}{L} < 0.50$
$= 0.5 + \left(\frac{x}{2L} - 0.25\right)^{1/2}$	$0.50 \le \frac{x}{L} \le 1$
$\frac{m}{m_0} = 2\left(\frac{x}{L}\right)^2$	$0 \le \frac{x}{L} < 0.50$
$= 1 - 2\left(1 - \frac{x}{L}\right)^2$	$0.50 \le \frac{x}{L} \le 1$
$\frac{m}{m_0} = \frac{x/L}{[(1 - x/L)(T_{ct}/b_f) + (x/L)]\,[(1 - x/L)(b_c/b_f) + x/L]}$	$0 \le \frac{x}{L} \le 1$

ical and demonstrates one of the theoretical weaknesses of this design technique.

The design of contracting transitions involves an accelerating flow; hence, any transition design that is smooth and continuous is generally acceptable. In the case of an expanding transition, more care must be exercised. Vittal and Chiranjeevi (1983) examined several methods for designing an expanding transition between a flume and canal and developed what they assert to be a superior methodology. On the basis of theoretical and experimental evidence, Vittal and Chiranjeevi (1983) suggested the following design equations.

Transition Bottom Width

$$\frac{b - b_f}{b_c - b_f} = \frac{x}{L}\left[1 - \left(1 - \frac{x}{L}\right)^{\Gamma}\right] \tag{9.5.16}$$

where

$$\Gamma = 0.80 - 0.26\sqrt{m_0} \tag{9.5.17}$$

$$L = 2.35\,(b_c - b_f) + 1.65\,m_0 y_{ch} \tag{9.5.18}$$

b = bottom width of transition and y_{ch} = depth of flow in the channel.

Side-Slope Variation

$$\frac{m}{m_0} = 1 - \left(1 - \frac{x}{L}\right)^{1/2} \tag{9.5.19}$$

Equation (9.5.19) yields variations in the side slope of the transition which vary gradually in the initial length of the transition where the flow velocity is high and rapidly in the latter length of the transition where the flow velocity is low.

Bed-Elevation Profile

$$\Delta z_{i,i+1} = \frac{\Delta z_0}{L}(x_{i+1} - x_i) \tag{9.5.20}$$

where Δz_0 = total change in the bed elevation over the length of the transition.

With regard to this design technique, it should be noted that the underlying theoretical concepts are the minimization of the energy loss and a consideration of flow separation in expanding flow fields. This technique can be used in design schemes based on the concepts of variable depth–variable specific energy and constant specific energy. Further, in contrast to the design procedure of Hinds (1928) which assumes a water surface profile and then defines the boundaries, in this technique the boundaries are defined first and then the water surface profile is determined.

EXAMPLE 9.2

If the following data are specified, design a transition between a flume and trapezoidal channel.

$$Q = 357 \text{ m}^3/\text{s } (12{,}600 \text{ ft}^3/\text{s})$$

$$\Delta z_0 = 0.5 \text{ m (flume higher) } (1.6 \text{ ft})$$

$$b_f = 15.0 \text{ m } (49 \text{ ft})$$

$$b_c = 23.0 \text{ m } (75 \text{ ft})$$

$$m_o = 2.0$$

$$y_{ch} = 6.7 \text{ m } (22 \text{ ft})$$

Note: This example was first given by Vittal and Chiranjeevi (1983).

Solution

In this case, the bed elevation of the flume relative to the channel is predetermined. The specific energy in the flume is

$$E_f = y_f + \frac{Q^2}{2gb_f^2y_f^2} = E_{ch} - \Delta z_0 + h_L$$

where h_L = headloss through the transition and is estimated by either Eq. (9.5.10), Eq. (9.5.15), or another formulation. To estimate h_L, knowledge which is not yet available regarding the flow in the flume would be required; therefore, for the present, E_f will be neglected and, under this assumption,

$$E_{ch} = y_{ch} + \frac{Q^2}{2gA_{ch}^2}$$

where
$$A_{ch} = (b_c + m_0 y_{ch})y_{ch}$$

$$= [23 + 2(6.7)]\ 6.7 = 244 \text{ m}^2 \ (2630 \text{ ft}^2)$$

$$E_{ch} = 6.7 + \frac{(357)^2}{2(9.8)(244)^2} = 6.81 \text{ m } (22.3 \text{ ft})$$

Therefore, an initial estimate of the specific energy in the flume is

$$E_f = E_{ch} - \Delta z_0 = 6.81 - 0.50 = 6.31 \text{ m } (20.7 \text{ ft})$$

The depth of flow in the flume is estimated by solving

$$E_f = 6.31 = y_f + \frac{Q^2}{2g\ b_f^2y_f^2} = y_f + \frac{(357)^2}{2(9.8)(15)^2y_f^2}$$

$$6.31 = y_f + \frac{28.9}{y_f^2}$$

By trial and error

$$y_f = 5.27 \text{ m } (17.3 \text{ ft})$$

and
$$\bar{u}_f = \frac{Q}{A_f} = \frac{357}{b_f y_f} = \frac{357}{(15)(5.27)} = 4.52 \text{ m/s } (14.8 \text{ ft/s})$$

The headloss term h_L can now be evaluated by

$$h_L = K_4\left(\frac{\bar{u}_f^2 - \bar{u}_{ch}^2}{2g}\right)$$

with $K_4 = 0.3$ as recommended by Vittal and Chiranjeevi (1983) for this example.

$$h_L = 0.3 \left[\frac{(4.52)^2 - (357/244)^2}{2(9.8)} \right] = 0.28 \text{ m } (0.92 \text{ ft})$$

At this point, the foregoing calculations for the flow conditions in the flume can be revised to take into account the headloss or

$$E_f = E_{ch} - \Delta z_0 + h_L = 6.81 - 0.50 + 0.28$$

and

$$= 6.59 \text{ m } (21.6 \text{ ft})$$

$$= 6.59 = y_f + \frac{28.9}{y_f^2}$$

Again by trial and error

$$y_f = 5.61 \text{ m } (18.4 \text{ ft})$$

$$\bar{u}_f = \frac{Q}{A_f} = \frac{357}{15(5.61)} = 4.24 \text{ m/s } (13.9 \text{ ft/s})$$

With this revised estimate of the velocity of flow in the flume, the headloss is

$$h_L = 0.3 \left[\frac{(4.24)^2 - (357/244)^2}{2(9.8)} \right] = 0.24 \text{ m } (0.79 \text{ ft})$$

It is concluded that no further trials are required because the headloss estimate has changed very little.

The design length of the transition is estimated by Eq. (9.5.18) or

$$L = 2.35\,(b_c - b_f) + 1.65\, m_0 y_{ch}$$

$$= 2.35\,(23 - 15) + 1.65(2)(6.7) = 40.9 \text{ m} \simeq 41 \text{ m } (134 \text{ ft})$$

Recall that it was previously stated that the maximum angle between the channel axis and a line connecting the channel sides between the entrance and exit sections is 12.5°. In the case considered here the angle is

$$\theta = \tan^{-1} \left[\frac{(b_{ch} - b_f)/2}{L} \right] = \tan^{-1} \left[\frac{(23 - 15)/2}{41} \right] = 5.6°$$

which is significantly less than the maximum permitted.

The transition is then divided into nine subreaches of 4 m (13 ft) each and a tenth subreach of 5 m (16 ft). The additional calculations required for this design are summarized in Table 9.6:

Column 1 Each subreach is assigned an identifying number.

Column 2 The distance from the beginning of the transition is given.

TABLE 9.6 Design computations for Example 9.5 (Vittal and Chiranjeevi, 1983)

Section (1)	x, m (2)	b, m (3)	m (4)	Δz, m (5)	y, m (6)
f-f	0.0	15.00	0.000	0.000	5.64
1	4.0	15.03	0.100	0.049	5.79
2	8.0	15.14	0.206	0.098	5.92
3	12.0	15.32	0.318	0.147	6.04
4	16.0	15.60	0.438	0.196	6.16
5	20.0	15.98	0.569	0.245	6.25
6	24.0	16.48	0.712	0.294	6.36
7	28.0	17.13	0.874	0.343	6.44
8	32.0	18.00	1.063	0.392	6.52
9	36.0	19.18	1.302	0.441	6.60
c-c	41.0	23.00	2.000	0.500	6.70

Column 3 The length of each subreach is specified.

Column 4 By use of Eqs. (9.5.16) to (9.5.18) the bed width of the transition channel is estimated. From Eq. (9.5.16)

$$b = b_f + (b_c - b_f)\left(\frac{x}{L}\right)\left[1 - \left(1 - \frac{x}{L}\right)^{\Gamma}\right]$$

where

$$\Gamma = 0.80 - 0.26\sqrt{m_0}$$
$$= 0.80 - 0.26\sqrt{2} = 0.43$$

Therefore

$$b = 15 + 8\left(\frac{x}{41}\right)\left[1 - \left(1 - \frac{x}{41}\right)^{0.43}\right]$$

For example, at section 5

$$x = 20 \text{ m (65.6 ft)}$$
$$b_5 = 15 + 8\left(\frac{20}{41}\right)\left[1 - \left(1 - \frac{20}{41}\right)^{0.43}\right]$$
$$= 15.98 \text{ m (52.4 ft)}$$

Column 5 From Eq. (9.5.19), the side slope of the transition channel as a function of the distance from the beginning of the channel is

$$m = m_0\left[1 - \left(1 - \frac{x}{L}\right)^{1/2}\right]$$
$$= 2\left[1 - \left(1 - \frac{x}{41}\right)^{1/2}\right]$$

For example, at section 5

$$x = 20.0 \text{ m } (65.6 \text{ ft})$$

and
$$m = 2\left[1 - \left(1 - \frac{20}{41}\right)^{1/2}\right] = 0.569$$

Column 6 The cumulative drop in the bed elevation relative to the bed elevation of the channel is summarized. For example, at section 5

$$x = 20 \text{ m } (65.6 \text{ ft})$$

and
$$\Delta z_5 = \frac{\Delta z_0}{L} x = \frac{0.5}{41}(20) = 0.244 \text{ m } (0.80 \text{ ft})$$

The drop in the bed elevation between sections 4 and 5 is

$$\Delta z_{4,5} = \frac{\Delta z_0}{L}(x_5 - x_4) = \frac{0.5}{41}(20 - 16)$$
$$= 0.049 \text{ m } (0.16 \text{ ft})$$

Column 7 The depth of flow for each subreach is estimated by a trial-and-error solution of the specific energy equation or

$$E_{i+1} = y_{i+1} + \frac{Q^2}{2gA_{i+1}^2} = E_i + \Delta z_{i,i+1} - h_{Li,i+1}$$

where
$$h_{Li,i+1} = K_H\left(\frac{\overline{u}_i^2 - \overline{u}_{i+1}^2}{2g}\right)$$

For example, at section 5

$$x_5 = 20.0 \text{ m } (65.6 \text{ ft})$$
$$m_5 = 0.569$$
$$b_5 = 15.98 \text{ m } (52.4 \text{ ft})$$
$$\Delta z_{4,5} = 0.049 \ (0.16 \text{ ft})$$

and from Table 9.6

$$b_4 = 15.60 \text{ m } (51.2 \text{ ft})$$
$$y_4 = 6.16 \text{ m } (20.2 \text{ ft})$$
$$m_4 = 0.438$$

Therefore,

$$A_4 = (b_4 + m_4 y_4)y_4$$
$$= [15.6 + 0.438\ (6.16)]\ 6.16$$

$$= 113 \text{ m}^2 \ (1220 \text{ ft}^2)$$

$$\overline{u}_4 = \frac{Q}{A_4} = \frac{357}{113} = 3.16 \text{ m/s} \ (10.4 \text{ ft/s})$$

$$E_4 = y_4 + \frac{\overline{u}_4^2}{2g} = 6.16 + \frac{(3.16)^2}{2(9.8)} = 6.67 \text{ m} \ (21.9 \text{ ft})$$

An appropriate value of y_5 is obtained by initially ignoring the parameter $h_{Li,i+1}$ in the specific energy equation or

$$E_5 = y_5 + \frac{Q^2}{2g(b_5 + m_5 y_5)^2 y_5^2} \simeq E_4 + \Delta z_{4,5}$$

$$= y_5 + \frac{(357)^2}{2(9.8)(15.98 + 0.569 y_5)^2 y_5^2} \simeq 6.67 + 0.049$$

$$= y_5 + \frac{(357)^2}{19.6(15.98 + 0.569 y_5)^2 y_5^2} \simeq 6.72$$

Then, by trial and error

$$y_5 = 6.26 \text{ m} \ (20.5 \text{ ft})$$

and

$$\overline{u}_5 = \frac{Q}{A_5} = \frac{357}{[15.98 + 0.569(6.26)]\ 6.26}$$

$$= 2.92 \text{ m/s} \ (9.58 \text{ ft/s})$$

The headloss incurred in this subreach is

$$h_{L4,5} = K_H \left(\frac{\overline{u}_4^2 - \overline{u}_5^2}{2g} \right) = 0.3 \left[\frac{3.16^2 - 2.92^2}{2(9.8)} \right]$$

$$= 0.022 \text{ m} \ (0.072 \text{ ft})$$

A revised estimate of y_5 is then obtained from a trial-and-error solution of the complete energy equation or

$$E_5 = y_5 + \frac{Q^2}{2gA_5^2} = E_4 + \Delta z_{4,5} - h_{L4,5}$$

$$y_5 + \frac{(357)^2}{2(9.8)(15.98 + 0.569 y_5)^2 y_5^2} = 6.67 + 0.049 - 0.022 = 6.70$$

To three significant figures

$$y_5 = 6.25 \text{ m} \ (20.5 \text{ ft})$$

The foregoing calculations illustrate the use of the Vittal and Chiranjeevi (1983) scheme to design an expanding transition under the conditions of variable depth and variable specific energy. The same methodology could be used to design the transition under the conditions of either constant specific energy or constant depth of flow.

In the foregoing paragraphs the analysis and design of gradual transitions have been discussed. In practice, some transitions must be relatively sudden; i.e., the change in the cross-sectional dimensions takes place over a relatively short distance, and rapidly varied flow results. Such transitions may include sudden contractions vertically, horizontally, or both (Fig. 9.37).

As an example of how rapid changes in cross-sectional dimensions are treated, consider the horizontal contraction in Fig. 9.38. Application of the one-dimensional momentum equation to the control volume shown yields

$$\frac{\gamma Q}{g}(\beta_3\overline{u}_3 - \beta_1\overline{u}_1) = P_1 - P_2 - P_3 - F_f$$

or

$$\frac{\gamma Q}{g}(\beta_3\overline{u}_3 - \beta_1\overline{u}_1) = \tfrac{1}{2}\gamma b_1 y_1^2 - \tfrac{1}{2}\gamma (b_1 - b_3)\, y_2^2 - \tfrac{1}{2}\gamma b_3 y_3^2 - F_f \tag{9.5.21}$$

where all variables are defined schematically in Fig. 9.38. With the aid of the continuity equation

$$Q = \overline{u}_1 b_1 y_1 = \overline{u}_3 b_3 y_3$$

and under the assumptions that $F_f = 0$, $\beta_1 = \beta_3 = 1$, and $y_2 = y_3$, Eq. (9.5.21) can be arranged to yield

$$\mathbf{F}_1^2 = \frac{(y_3/y_1)\,[(y_3/y_1)^2 - 1]}{2\,[(y_3/y_1) - (b_1/b_3)]} \tag{9.5.22}$$

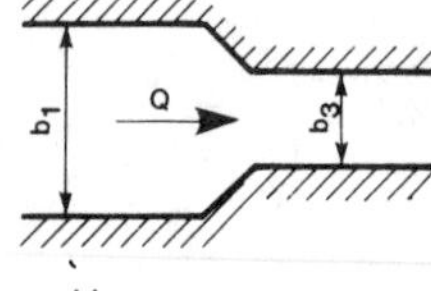

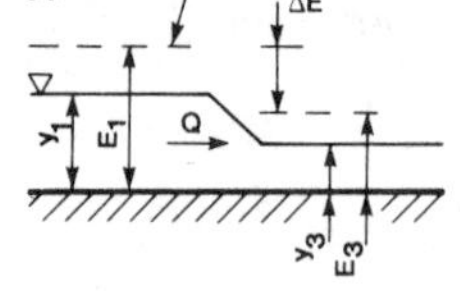

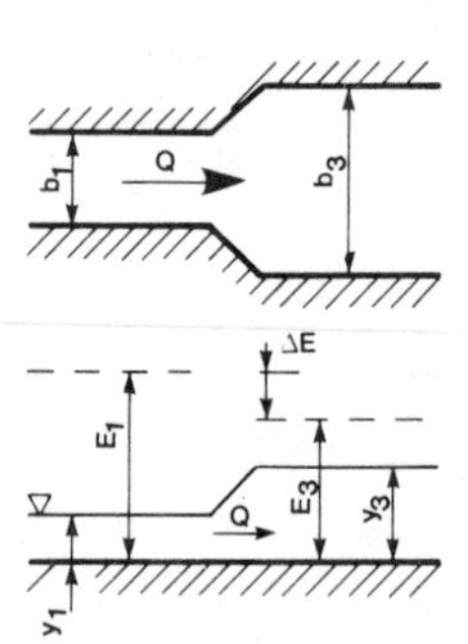

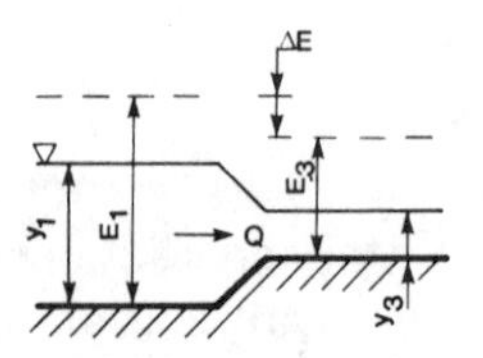

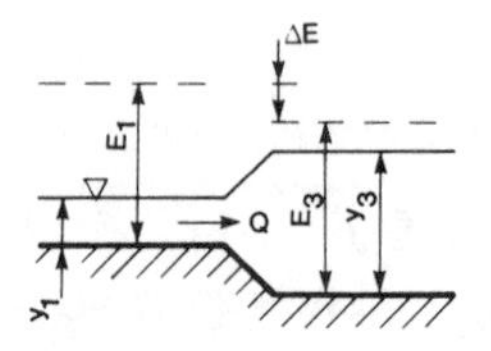

FIGURE 9.37 Sudden horizontal and vertical transitions.

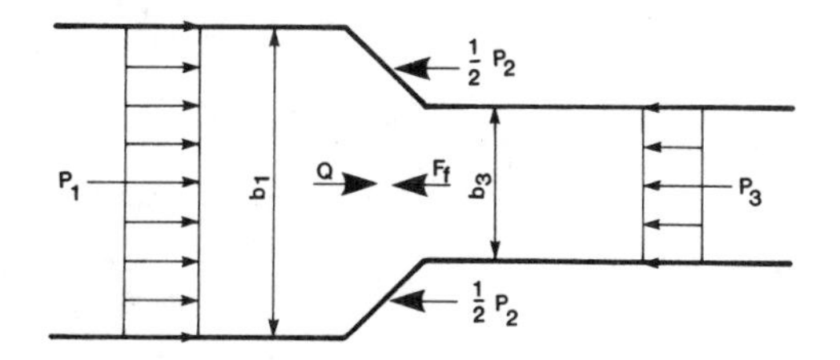

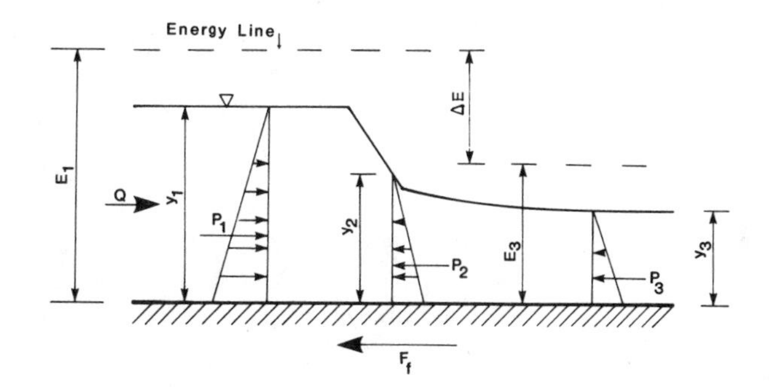

FIGURE 9.38 Sudden horizontal contraction.

where $\mathbf{F}_1$ = Froude number at section 1. This equation is plotted for positive values of $\mathbf{F}_1$ and y_3/y_1 in Fig. 9.39; with regard to this figure, the following observations are relevant:

1. All curves which are members of the family of curves defined by Eq. (9.5.22) pass through the points ($\mathbf{F}_1 = 0$; $y_3/y_1 = 0$) and ($\mathbf{F}_1 = 0$; $y_3/y_1 = 1$) and are asymptotic to the line $y_3/y_1 = b_1/b_3$.

2. The upstream flow is critical or supercritical when $\mathbf{F}_1 \geq 1$. The downstream flow is critical or supercritical when $\mathbf{F}_1^2 \geq (y_3/y_1)^3\ (b_3/b_1)^2$ and subcritical when $\mathbf{F}_1^2 < (y_3/y_1)^3\ (b_3/b_1)^2$. Therefore, in Fig. 9.39 four flow regimes are represented:

   Region 1 Flow is supercritical throughout the transition.

   Region 2 Flow through the transition passes from supercritical to subcritical.

   Region 3 Flow is subcritical throughout the transition.

   Region 4 Flow through the transition passes from subcritical to supercritical.

3. Theoretically some regions of the curves shown in Fig. 9.39 are physically

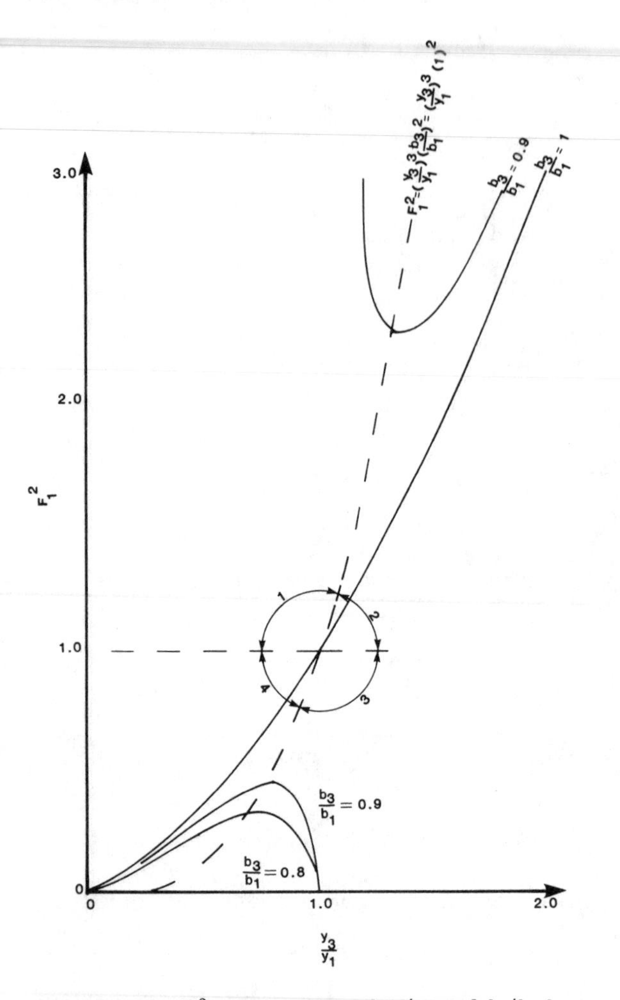

FIGURE 9.39 $\mathbf{F}_1^2$ as a function of y_3/y_1 and b_3/b_1 for rapid horizontal contractions.

impossible because they require an increase in energy. The difference in energy across the transition is given by

$$\Delta E = y_1 + \frac{\overline{u}_1^2}{2g} - y_3 - \frac{\overline{u}_3^2}{2g} \tag{9.5.23}$$

or

$$\frac{\Delta E}{y_1} = 1 + \frac{\mathbf{F}_1^2}{2} - \left[\frac{y_3}{y_1} + \frac{\mathbf{F}_1^2}{2(y_3/y_1)^2\,(b_3/b_1)^2}\right] \tag{9.5.24}$$

With these equations the curves in Fig. 9.39 can be examined, and regions in which $\Delta E > 0$ eliminated from consideration. In practice this theoretical analysis may yield erroneous results because the assumptions ($F_f = 0$, $\beta_1 = \beta_3 = 1$, $y_2 = y_3$) may not be valid.

In a fashion completely analogous to that described for a rapid contraction, the theoretical equation governing rapid expansions is

$$\mathbf{F}_1^2 = \frac{(b_3/b_1)\,(y_3/y_1)\,[1 - (y_3/y_1)^2]}{2\,[(b_1/b_3) - (y_3/y_1)]} \qquad (9.5.25)$$

At this point it is relevant to note that the design of rapid transitions by completely theoretical procedures is usually not satisfactory. The assumptions inherent in the theoretical results limit the usefulness of the results, and in general physical model studies will be required.

BIBLIOGRAPHY

Anonymous, "Hydraulics of Bridge Waterways," U.S. Bureau of Public Roads, Washington, 1960.

Anonymous, "Computation of Water Surface Profiles," *Surface Water Techniques*, book 1, chapter 1, U.S. Geological Survey, Washington, 1964.

Anonymous, "HEC-2 Water Surface Profiles," U.S. Hydrologic Engineering Center, Davis, Calif., August 1979.

Blaisdell, F. W., "Development and Hydraulic Design, Saint Anthony Falls Stilling Basin," *Transactions of the American Society of Civil Engineers*, vol. 113, 1948, pp. 483–520.

Blaisdell, F. W., "The SAF Stilling Basin," *SCS-TP-79*, U.S. Soil Conservation Service, Washington, 1949.

Blaisdell, F. W., and Donnelly, C. A., "The Box Inlet Drop Spillway and Its Outlet," *Transactions of the American Society of Civil Engineers*, vol. 121, 1956, pp. 955–986.

Bradley, J. M., and Peterka, A. J., "The Hydraulic Design of Stilling Basins," *Proceedings of the American Society of Civil Engineers, Journal of the Hydraulics Division*, vol. 83, no. HY5, October 1957.

Chow, V. T., *Open Channel Hydraulics*, McGraw-Hill Book Company, New York, 1959.

Doeringsfeld, H. A., and Barker, C. L., Pressure-Momentum Theory Applied to the Broad-Crested Weir," *Transactions of the American Society of Civil Engineers*, vol. 106, 1941, pp. 934–946.

Eichert, B. S., "Survey of Programs for Water-Surface Profiles," *Proceedings of the American Society of Civil Engineers, Journal of the Hydraulics Division*, vol. 96, no. HY2, February 1970a, pp. 547–563.

Eichert, B. S., and Peters, J., "Computer Determination of Flow through Bridges," *Proceedings of the American Society of Civil Engineers, Journal of the Hydraulics Division*, vol. 96, no. HY7, July 1970b, pp. 1455–1468.

Feldman, A. D., "HEC Models for Water Resources System Simulation: Theory and Experience," *Advances in Hydroscience*, vol. 12, 1981, pp. 297–423.

Forster, J. W., and Skrinde, R. A., "Control of Hydraulic Jump by Sills," *Transactions of the American Society of Civil Engineers,* vol. 115, 1950, pp. 973–987.

Henderson, F. M., *Open Channel Flow,* The Macmillan Company, New York, 1966.

Hinds, J., "The Hydraulic Design of Flume and Syphon Transitions," *Transactions of the American Society of Civil Engineers,* vol. 92, 1928, pp. 1423–1459.

Hsu, E. Y., Discussion of "Control of the Hydraulic Jump by Sills" by J. W. Forster and R. A. Skrinde, *Transactions of the American Society of Civil Engineers,* vol. 115, 1950, pp. 988–991.

Matthai, H. F., "Measurement of Peak Discharge at Width Contractions by Indirect Methods," *Techniques of Water-Resources Investigations of the United States Geological Survey,* U.S. Geological Survey, Washington, 1976.

Rand, W., "Flow Geometry at Straight Drop Spillways," *Proceedings of the American Society of Civil Engineers,* vol. 81, September 1955, pp. 791-1, 791-13.

Vittal, N., and Chiranjeevi, V. V., "Open Channel Transitions: Rational Method of Design," *Proceedings of the American Society of Civil Engineers, Journal of Hydraulic Engineering,* vol. 109, no. 1, January 1983, pp. 99–115.

Yarnell, D. L., "Pile Trestles as Channel Obstructions," Technical Bulletin 429, U.S. Department of Agriculture, Washington, July, 1934a.

Yarnell, D. L., "Bridge Piers as Channel Obstructions," Technical Bulletin 442, U.S. Department of Agriculture, Washington, November 1934b.

TEN

Turbulent Diffusion and Dispersion in Steady Open-Channel Flow

SYNOPSIS

The transport processes known as turbulent diffusion and dispersion are discussed in this chapter. First, the governing equations are developed for both one and two dimensions, and then techniques for estimating the vertical and transverse turbulent diffusion coefficients and the longitudinal dispersion coefficient are discussed. Numerical methods of solving the governing equations are presented, and the problem of numerical dispersion considered. The chapter concludes with a brief discussion of techniques for estimating the vertical turbulent diffusion coefficient in a stratified environment.

10.1 INTRODUCTION

The topic of riverine water quality modeling is almost inseparable from the topic of open-channel hydraulics, with the linkage being established by a common and essential interest in the mechanics of mass transport. In this chapter, turbulent diffusion and dispersion are discussed from the viewpoint of understanding and quantifying these processes rather than from the viewpoint of quantifying the environmental effects of pollutant transport. Thus, no actual pollutants will be mentioned; environmentally neutral tracers will be discussed. This chapter will also be in the form of a summary. A complete treatment of the turbulent diffusion and dispersion transport processes from the viewpoint of hydraulic engineering is available in Fischer et al. (1979) and from an environmental engineering point of view in Krenkel and Novotny (1980).

One of the basic problems in discussing transport processes has been a semantics problem which originated many years ago with imprecise and nonsonsensus definitions of the processes involved. Following Fischer et al. (1979), under the broad category of mixing processes, there are two categories of interest in open-channel flow—turbulent diffusion and dispersion. The terminology *turbulent diffusion* refers to the random scattering of particles in a flow by turbulent motions. *Dispersion* is the scattering of particles by the combined effects of shear and transverse turbulent diffusion. In this chapter, we will discuss advective transport processes, i.e., transport by imposed velocity systems, rather than convective transport which connotes transport due to the hydrodynamic instabilities that are usually associated with temperature gradients. *Shear* refers to the advection of a fluid at different velocities at different positions within the flow. For example, the vertical velocity profile defined by Eq. (1.4.14) results in shear since the fluid particles near the surface are moving faster than those near the bottom.

10.2 GOVERNING EQUATIONS

To begin, consider the control volume defined in Fig. 10.1 and assume that molecular motion is causing mass transfer of a tracer through the control surfaces in the x direction. If it is assumed that the transport of mass through a control surface is proportional to the concentration gradient, then an application of the law of conservation of mass yields

$$\frac{\partial c}{\partial t} = D \frac{\partial^2 c}{\partial x^2} \tag{10.2.1}$$

where c = concentration of tracer and D = molecular diffusion coefficient with the tacit assumption that $D \neq f(x)$. Equation (10.2.1) is a mathematical statement of Fick's first law of diffusion. The fundamental solution of Fick's first law is obtained under the following conditions:

$$c(x, 0) = \delta(x)$$

and

$$c(\pm\infty, t) = 0$$

where $\delta(x)$ designates the Dirac delta function. This function has the properties of being zero everywhere except at $x = 0$ and that

$$\int_{-\infty}^{\infty} \delta(x)\, dx = 1$$

With these initial and boundary conditions, the solution of Eq. (10.2.1) is

$$c(x, t) = \frac{1}{\sqrt{4\pi\, Dt}} \exp\left(-\frac{x^2}{4Dt}\right) \tag{10.2.2}$$

Equation (10.2.2) is a gaussian distribution, and if the mass under the distribution curve is unity, it is known as the normal distribution. In the material which follows, we will often be concerned with the various moments of Eq. (10.2.2) where these moments are defined by

$$M_n \int_{-\infty}^{\infty} x^n c(x, t)\, dx \tag{10.2.3}$$

Thus, the 0 moment of the fundamental solution is given by

$$M_0 = \int_{-\infty}^{\infty} c(x, t)\, dx$$

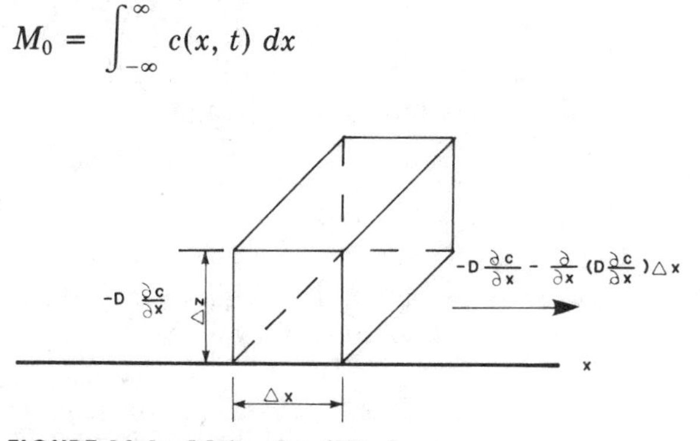

FIGURE 10.1 Molecular diffusion.

The mean μ and the variance σ^2 of the normal or gaussian distribution can be defined in terms of the moments or

$$\mu = \frac{M_1}{M_0} \tag{10.2.4}$$

and

$$\sigma^2 = \frac{M_2}{M_0} - \mu^2 \tag{10.2.5}$$

For a normal distribution,

$$\mu = 0$$

and

$$\sigma^2 = 2Dt$$

At this point, it is noted that

1. The variance of the distribution is a measure of the spread of the distribution about the origin. Recall from statistics (see, for example, Benjamin and Cornell, 1970) that for a gaussian distribution a spread of 4σ includes approximately 95 percent of the total area under the concentration curve.
2. In applied studies, the diffusion coefficient is often determined from

$$\frac{d\sigma^2}{dt} = 2D \tag{10.2.6}$$

3. Fick's first law and its normal boundary conditions belong to a set of differential equations in which the principle of superposition can be applied; i.e., the equation and the usual boundary conditions are linear. This is an important observation since in most practical applications of Eq. (10.2.1) there exist boundaries across which diffusion cannot take place. Suppose, for example, there exist boundaries at $x = \pm L$ across which there is no diffusion (Fig. 10.1). According to Eq. (10.2.1) the appropriate boundary conditions are

$$\frac{\partial c}{\partial x} = 0 \qquad \text{at } x = \pm L$$

To satisfy the boundary condition of $x = -L$, an image source of tracer is added at $x = -2L$; in an analogous fashion, to satisfy the boundary condition at $x = L$, an image source is added at $x = 2L$. The image source at $x = -2L$ makes $\partial c/\partial x = 0$ at $x = -L$ but causes a positive gradient at $x = +L$; therefore, another image source must be added to $x = +4L$. Again, analogous reasoning requires additional image sources at $x = -6L, +8L$, ad infinitum. The solution is then

$$c(x, t) = \frac{1}{\sqrt{4\pi\, Dt}} \sum_{n=\infty}^{\infty} \exp\left(-\frac{x + 2nL}{4Dt}\right) \tag{10.2.7}$$

Now consider a two-dimensional channel flow (Fig. 10.2) defined such that no parameter varies in the direction perpendicular to the plane of paper. At time t, an inert tracer is injected into the flow which is moving from left to right with the velocity distribution shown. As the tracer moves downstream, it spreads out, and the maximum concentration is reduced by dilution. To derive an equation which represents the transport in this situation, define an elemental control volume with dimensions dx by dz and area dA (Fig. 10.2). The principle of conservation of mass can then be applied, and an equation which describes the situation derived.

For this initial development, assume that the spreading of the tracer cloud is due only to molecular diffusion. Two methods of accounting for the movement of the tracer through the control volume suggest themselves. First, the movement of each molecule of the tracer cloud could be described. Although such an approach has theoretical merit, it becomes cumbersome as the movement of each molecule is described. The second method is a continuum mechanics approach similar to that used for the derivation of Fick's first law; i.e., it is assumed that the fluid carries the tracer with it at a rate which depends on both the velocity in the longitudinal direction u and the concentration c. However, the continuum mechanics approach is only approximate since many of the tracer molecules have a velocity which is different from u. Therefore, it is necessary to introduce the molecular diffusion coefficient D to correct for the approximate nature of this solution. The application of the continuum mechanics approach to formulating the required mass balance yields

$$\underbrace{\frac{\partial c}{\partial t}}_{\substack{\text{Time rate}\\ \text{of change of}\\ \text{concentration}\\ \text{at a point}}} + \underbrace{u\frac{\partial c}{\partial x}}_{\substack{\text{advection of}\\ \text{tracer with}\\ \text{fluid}}} = \underbrace{D\frac{\partial^2 c}{\partial x^2} + D\frac{\partial^2 c}{\partial z^2}}_{\substack{\text{molecular}\\ \text{diffusion}}} \tag{10.2.8}$$

which is also known as Fick's second law (Holley, 1969).

Now, assume that the fluid is turbulent. In Fig. 10.3a and b, the velocity and

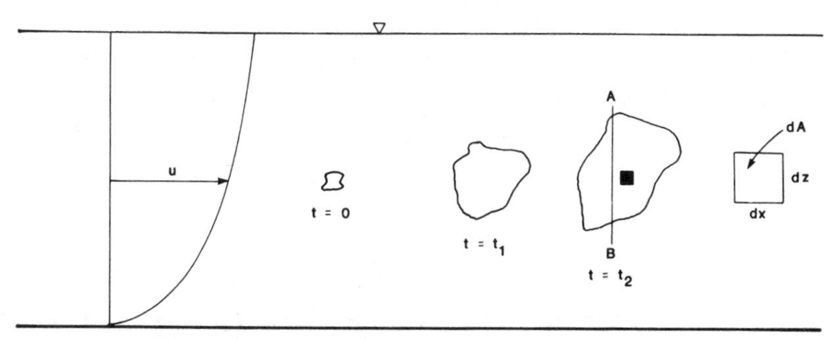

FIGURE 10.2 Transport in two-dimension flow. *(Holley 1969.)*

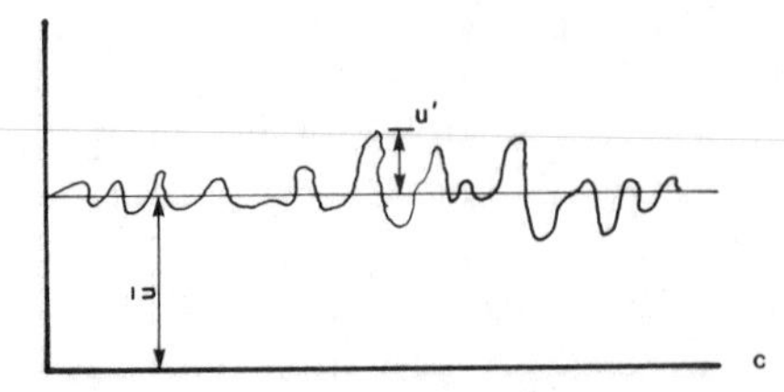

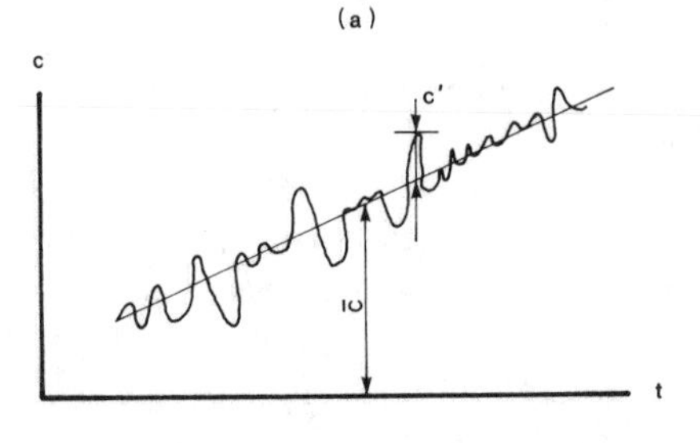

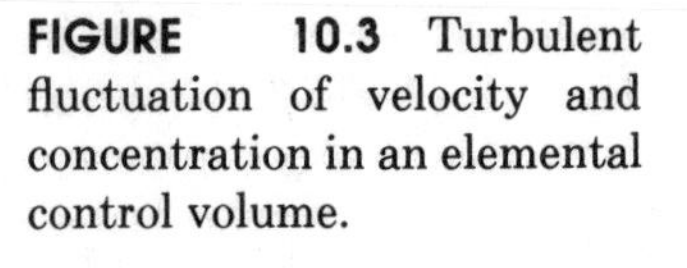

FIGURE 10.3 Turbulent fluctuation of velocity and concentration in an elemental control volume.

tracer concentration in the incremental control volume defined in Fig. 10.2 are shown. In these figures, the dashed lines represent the time-averaged values of u and c. In a manner similar to that used in the case of advective molecular diffusion, the mass balance for the situation is

$$\frac{\partial c}{\partial t} + u\frac{\partial c}{\partial x} + v\frac{\partial c}{\partial z} = D\frac{\partial^2 c}{\partial x^2} + D\frac{\partial^2 c}{\partial z^2} \tag{10.2.9}$$

where u and v are the instantaneous values of the turbulent velocities in the x and z directions, respectively, and c is an instantaneous value of the tracer concentration. It is noted that a term for advection in the z direction is included since, even though the primary direction of flow is in the x direction, there is turbulent motion in the z direction. Equation (10.2.9) takes the effect of transport by turbulent diffusion into account directly because u and v are instantaneous turbulent velocity components. Since instantaneous values of u and v are not usually available, this direct incorporation of the effects of turbulence into the transport equation results in an equation which cannot be solved.

From an applied viewpoint, the effect of turbulent transport of tracer through the incremental control volume dA must be expressed in terms of time-averaged values of u, v, and c or

$$u = \overline{u} + u'$$

$$v = v'$$

and

$$c = \overline{c} + c'$$

where the overbar indicates a time-averaged value of the variable, and the variables with primes are turbulent fluctuations of the variable. Also, recall that there is no average velocity in the z direction by definition. Substitution of these equations for u, v, and c in Eq. (10.2.9) and averaging the resulting equation with respect to time yields

$$\frac{\partial \bar{c}}{\partial t} + \bar{u}\frac{\partial \bar{c}}{\partial x} = D\frac{\partial^2 \bar{c}}{\partial x^2} + D\frac{\partial^2 \bar{c}}{\partial z^2} + \frac{\partial(-\overline{u'c'})}{\partial x} + \frac{\partial(-\overline{v'c'})}{\partial z} \tag{10.2.10}$$

The turbulent diffusion coefficients ϵ_x and ϵ_z are then defined in analogy with molecular diffusion or

$$\overline{u'c'} = -\epsilon_x \frac{\partial c}{\partial x} \tag{10.2.11}$$

and

$$\overline{v'c'} = -\epsilon_z \frac{\partial c}{\partial z} \tag{10.2.12}$$

If it is assumed that $\epsilon_x \neq f(x)$ and $\epsilon_z \neq f(z)$, the substitution of these variables in Eq. (10.2.10) yields

$$\frac{\partial \bar{c}}{\partial t} + \bar{u}\frac{\partial \bar{c}}{\partial x} = (D + \epsilon_x)\frac{\partial^2 \bar{c}}{\partial x^2} + (D + \epsilon_z)\frac{\partial^2 \bar{c}}{\partial z^2} \tag{10.2.13}$$

Equation (10.2.13) represents the transport of the tracer due to both molecular and turbulent diffusion. The terminology *turbulent diffusion* arises, as was the case with molecular diffusion, because the advective transport term written in terms of $\bar{u}$ does not represent the complete advective transport process; therefore, corrective factors are required.

In Fig. 10.4, the tracer cloud is shown progressing through time until time t_4. By this time, turbulent transport has mixed the tracer completely in the vertical direction. After complete mixing in the vertical direction has occurred, the primary variation of the tracer cloud concentration in the two-dimensional problem defined here is in the longitudinal direction. Although Eq. (10.2.13) remains valid, it provides more information than is required since it describes the variation of c in both the x and z directions. For efficiency, a simplification of this equation is sought. Define

$$\bar{u} = U + u'' \tag{10.2.14}$$

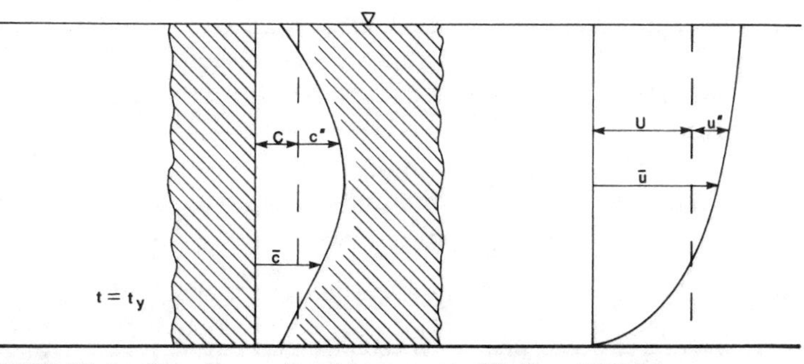

FIGURE 10.4 One-dimensional transport. *(Holley, 1969.)*

and
$$\bar{c} = C + c'' \tag{10.2.15}$$

where U and C are the average values of the velocity and tracer concentration, respectively, *in the cross section.* Substitution of Eqs. (10.2.14) and (10.2.15) in Eq. (10.2.13) yields, after cross-sectional averaging,

$$\frac{\partial C}{\partial t} + U\frac{\partial C}{\partial x} = (D + \epsilon_x)\frac{\partial^2 C}{\partial x^2} + \frac{\partial(-\overline{\overline{u''c''}})}{\partial x} \tag{10.2.16}$$

where the double overbar indicates a cross-sectional average. Taylor (1953, 1954) demonstrated that under certain conditions u'' is proportional to the longitudinal gradient of C or

$$E\frac{\partial C}{\partial x} = -\overline{\overline{u''c''}} + (D + \epsilon_x)\frac{\partial C}{\partial x} \tag{10.2.17}$$

The transport defined by Eq. (10.2.17) is termed *longitudinal dispersion,* and the coefficient E is the coefficient of longitudinal dispersion. Substitution of Eq. (10.2.17) in Eq. (10.2.16) yields the one-dimensional equation of longitudinal dispersion or

$$\frac{\partial C}{\partial t} + U\frac{\partial C}{\partial x} = E\frac{\partial^2 C}{\partial x^2} \tag{10.2.18}$$

With regard to Eqs. (10.2.17) and (10.2.18), the following should be noted:

1. This equation applies whenever both the vertical and transverse mixing are complete, i.e., when the tracer completely fills the cross section and the primary variation in concentration is in the longitudinal direction.

2. Equation (10.2.17) includes both the effect of $\overline{\overline{u''c''}}$ and molecular and turbulent diffusion. After complete mixing is achieved, diffusion acts to smooth the distribution; i.e., after the point of complete mixing is reached, the effect of diffusion is small relative to the effect of the velocity gradient.

3. The role of the velocity gradient in the dispersion process is paramount and is reflected in the term $\overline{\overline{u''c''}}$.

4. Equations (10.2.17) and (10.2.18) are limited since Taylor's assumption was based on the fact that after an initial period of time the rate of advective transport u', either upstream or downstream from a vertical plane which moves at a velocity U, depends only on the tracer concentration gradient and E, which is constant for a uniform flow. In general, Eq. (10.2.18) is valid only after an initial period of time since before the elapse of this presently unspecified time, Taylor's assumption is not valid.

5. Equation (10.2.18) governs a special case in which the flow is steady and uniform, the channel prismatic, and E constant. A much more general case is represented by

$$\frac{\partial}{\partial t}(AC) + \frac{\partial}{\partial x}(UAC) = \frac{\partial}{\partial x}\left(EA\frac{\partial C}{\partial x}\right) \tag{10.2.19}$$

In this situation, U, A, and E can vary with distance and/or time.

At this point it is convenient to note some of the solutions of the partial differential equation

$$\frac{\partial C}{\partial t} + U\frac{\partial C}{\partial x} = \Theta\frac{\partial^2 C}{\partial x} \tag{10.2.20}$$

where both U and Θ are constants. The solution of this equation is dependent on the initial or boundary conditions which are specified. For example, if

$$C(x, 0) = \begin{cases} 0 \text{ for } x > 0 \\ C_0 \text{ for } x < 0 \end{cases}$$

then the solution of Eq. (10.2.20) is

$$C(x, t) = \frac{C_0}{2}\left[1 - \text{erf}\left(\frac{x - Ut}{\sqrt{4\Theta t}}\right)\right] \tag{10.2.21}$$

where erf designates the error function. The above initial condition could, for example, represent the case where one fluid in an open channel is being displaced by a second fluid which has a tracer in it with a concentration C_0 and the fluid moves at a cross-sectional average velocity U.

A much more complex and important case arises when the boundary and initial conditions are

$$C(0, t) = C_0 \qquad \text{for } 0 < t < \infty$$

$$C(x, 0) = 0 \qquad \text{for } 0 < x < \infty$$

In this case, the solution of Eq. (10.2.18) is

$$C(x, t) = \frac{C_0}{2}\left[\text{erfc}\left(\frac{x - Ut}{\sqrt{4\Theta t}}\right) + \text{erfc}\left(\frac{x + Ut}{\sqrt{4\Theta t}}\right)\exp\left(\frac{Ux}{\Theta}\right)\right] \tag{10.2.22}$$

where erfc designates the complementary error function. This set of boundary and initial conditions might represent the situation in which a tracer of concentration C_0 is introduced at the origin of the coordinate system at time zero and continued. It should be obvious that in this case as $t \rightarrow \infty$, $C(x, t) \rightarrow C_0$.

Another problem of interest to the hydraulic engineer is that of the spread of a tracer source in three-dimensional flow. The governing equation for this situation is

$$\frac{\partial C}{\partial t} + U\frac{\partial C}{\partial x} = \Theta_x\frac{\partial^2 C}{\partial x^2} + \Theta_y\frac{\partial^2 C}{\partial y^2} + \Theta_z\frac{\partial^2 C}{\partial z^2}$$

where Θ_x, Θ_y, and Θ_z are the diffusion coefficients in the x, y, and z directions, respectively. This governing equation can be significantly simplified if it is assumed that $\Theta_x = \Theta_y = \Theta_z = \Theta$. The governing equation is then

$$\frac{\partial C}{\partial t} + U\frac{\partial C}{\partial x} = \Theta\left(\frac{\partial^2 C}{\partial x^2} + \frac{\partial^2 C}{\partial y^2} + \frac{\partial^2 C}{\partial z^2}\right) \tag{10.2.23}$$

Although a general solution of Eq. (10.2.23) can be obtained in some situations, it is usually possible to reduce the three-dimensional problem to a two-dimensional problem. For example, in open-channel flow diffusion in the longitudinal direction can usually be neglected (e.g., Fischer et al., 1979). If this is the case, then for an instantaneous injection of tracer of mass M at the origin of the coordinate system, the resulting downstream concentration is

$$C(x, y, z) = \frac{M}{4\pi\Theta x}\exp\left[-\frac{(y^2 + z^2)U}{4x\Theta}\right] \tag{10.2.24}$$

In the case where the tracer injection is not instantaneous but the tracer is added continuously at a mass flow rate of $\dot{M}$, the approximate solution of the problem is given by

$$C(x, y) = \frac{\dot{M}}{U\sqrt{4\pi\Theta x/U}}\exp\left(-\frac{y^2U}{4x\Theta}\right) \tag{10.2.25}$$

It is noted that the solutions provided by Eq. (10.2.24) and (10.2.25) are valid only when the following condition is satisfied:

$$t \gg \frac{2\Theta}{U^2} \tag{10.2.26}$$

In most cases, this condition is satisfied at very small values of t.

10.3 VERTICAL AND TRANSVERSE TURBULENT DIFFUSION AND LONGITUDINAL DISPERSION

When a tracer is injected into a homogeneous channel flow, the complete advective transport process can be conveniently viewed as being composed of three stages. In the first stage, the tracer is diluted by the flow in the channel because of its initial momentum. In the second stage, the tracer is mixed throughout the cross section by turbulent transport processes. And in the third stage longitudinal dispersion tends to erase any longitudinal variations in the tracer concentration. In some cases, the second stage is eliminated because the tracer discharge has a significant amount of initial momentum; however, in many cases the tracer flow is small and has an insignificant amount of momentum associated with it. In this latter case, the first transport stage is eliminated. In this section, only the second and third transport stages will be treated, with the implied assumption that if there is a first stage it can be treated separately.

To develop a quantitative expression for the vertical, turbulent diffusion coefficient, consider a relatively shallow flow in a wide, rectangular channel (Fig. 10.5). From Chapter 1, it is known that the vertical transport of momentum in such a flow is given by

$$\tau = \epsilon_z \rho \frac{du}{dz} \tag{10.3.1}$$

where τ = shear stress distance z above bottom boundary
ρ = fluid density
ϵ_z = vertical, turbulent diffusion coefficient
u = longitudinal velocity

Further, in a two-dimensional flow, such as that defined in Fig. 10.4, the vertical velocity profile can be assumed to be given by

$$u = \frac{u_*}{k}\left(1 + \ln \frac{z}{d}\right) \tag{10.3.2}$$

where u_* = shear velocity
k = von Karman's turbulence coefficient ($k \simeq 0.40$)
d = depth of flow

A shear stress distribution similar to that which prevails in a circular pipe is assumed or

$$\tau = \tau_0\left(1 - \frac{z}{d}\right) \tag{10.3.3}$$

where $\tau_0 = \rho g d S$ = bottom boundary shear stress. Then, as demonstrated by Elder (1959), combining Eq. (10.3.1) to (10.3.3) yields

$$\epsilon_z = k u_* d z'(1 - z') \tag{10.3.4}$$

where $z' = z/d$. The depth-averaged value of ϵ_z is

$$\bar{\epsilon}_z = \frac{1}{d}\int_0^d \epsilon_z \, dz = 0.067 d u_* \tag{10.3.5}$$

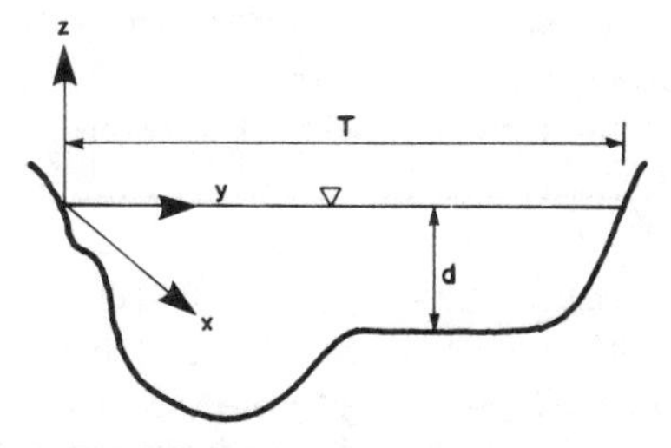

FIGURE 10.5 Coordinate system definition.

In the infinitely wide channel hypothesized for the derivation of Eq. (10.3.4), there does not exist a transverse velocity profile. Thus, a quantitative expression for ϵ_y, the transverse, turbulent diffusion coefficient, cannot be derived in a manner analogous to that used for ϵ_z. Instead, expressions for ϵ_y must be derived from experimental results. A large number of such experiments have been performed and the results of this work are summarized in Fischer et al. (1979) and Lau and Krishnappen (1977). In straight, rectangular channels, an approximate average of the results available is

$$\epsilon_y \simeq 0.15 d u_* \pm 50\% \tag{10.3.6}$$

where ± 50 percent indicates the magnitude of the error incurred in estimating ϵ_y. In natural channels, the value of ϵ_y is significantly higher than the value estimated by Eq. (10.3.6). For channels which can be classified as "slowly meandering" with only moderate boundary irregularities

$$\epsilon_y \simeq 0.60 d u_* \pm 50\% \tag{10.3.7}$$

If the channel has curves of small radii, rapid changes in channel geometry, or severe bank irregularities, e.g., groins, then the value of ϵ_y will be larger than that which is estimated by Eq. (10.3.7). For example, in the case of meanders, Fischer (1969) estimated that

$$\epsilon_y = 25 \frac{\bar{u}^2 d^3}{R_c^2 u_*} \tag{10.3.8}$$

where R_c = curve radius. At present, whether a channel should be considered slowly meandering and Eq. (10.3.7) used to estimate ϵ_y or whether a larger value of ϵ_y should be used is a matter of judgment. Fischer et al. (1979) suggested that a slowly meandering channel is one in which

$$\frac{T\bar{u}}{R_c u_*} < 2 \tag{10.3.9}$$

where T = channel width
$\bar{u}$ = average velocity of flow
R_c = curve radius

At best, Eq. (10.3.9) provides only a crude estimate of channel sinuousity relative to ϵ_y.

As noted in the first paragraph of this section, the complete advective transport process in two-dimensional channel flow can be conveniently viewed as being composed of three stages. In the second stage, the primary transport mechanism is turbulent diffusion, with the diffusion coefficients being defined by Eq. (10.3.5) and (10.3.7), or in some cases Eq. (10.3.8). A comparison of Eqs. (10.3.5) and (10.3.7) demonstrates that

$$\epsilon_y \simeq 10\, \epsilon_z$$

or that the transverse mixing coefficient is roughly 10 times larger than the vertical mixing coefficient. Thus, the rate at which a plume of tracer spreads laterally is an order of magnitude larger than the rate of spread in the vertical direction. However, in almost all cases the channel will be much wider than it is deep; e.g., typical natural channel dimensions might be width 100 ft (30 m) and depth 3.0 ft (1.0 m). An examination of the equation governing the spread of a plume, Eq. (10.2.25) for example, demonstrates that a mixing time can be defined as proportional to the square of the distance divided by the mixing coefficient. Thus, in a typical channel it will take approximately 90 times as long for a plume to spread completely across the channel as it will to mix in the vertical dimension. In most applied problems, it is both convenient and appropriate to begin by assuming that the tracer is uniformly distributed over the vertical.

Recall from Section 10.2 that, in a diffusional process where the tracer is added at a constant mass flow rate $\dot{M}$ to an unbounded channel, the downstream concentration of tracer is given approximately by

$$C = \frac{\dot{M}}{\overline{u}d(4\pi\epsilon_y x/\overline{u})^{1/2}} \exp - \left(\frac{y^2\overline{u}}{4x\epsilon_y}\right) \tag{10.3.10}$$

If the channel has a width T, then the boundary conditions

$$\frac{\partial C}{\partial y} = 0 \qquad \text{at } y = 0$$

and

$$\frac{\partial C}{\partial y} = 0 \qquad \text{at } y = T$$

must be satisfied by the solution. At this point it is convenient to define the following dimensionless variables:

$$\hat{C} = \frac{\dot{M}}{\overline{u}\, dT}$$

$$\hat{x} = \frac{x\epsilon_y}{\overline{u}T^2}$$

and

$$\hat{y} = \frac{y}{T}$$

If the source of the tracer is located at $y = y_0$ where $\hat{y} = \hat{y}_0$, then the downstream concentration is

$$\frac{C}{C_0} = \frac{1}{(4\pi\hat{x})^{1/2}} \sum_{n=-\infty}^{\infty} \left\{ \exp\left[-\frac{(\hat{y} - 2n - \hat{y}_0)^2}{4\hat{x}} \right] + \exp\left[-\frac{(\hat{y} - 2n + \hat{y}_0)}{4\hat{x}} \right] \right\} \tag{10.3.11}$$

where the principle of superposition (Section 10.2) has been used to meet the boundary conditions. In Fig. 10.6, Eq. (10.3.11) is plotted for a centerline injection of tracer, i.e., $\hat{y}_0 = 0.5$, and the relative concentrations of tracer along the centerline of the channel and the side of the channel are shown as functions of $\hat{x}$. In this figure, it is noted that for $\hat{x} \geq 0.1$ the concentration C is within 5 percent of C_0 throughout the channel cross section. This situation is often referred to as *complete mixing*, and for a centerline injection, the distance to the point where complete mixing is achieved is given by

$$L = \frac{0.1\bar{u}T^2}{\epsilon_y} \tag{10.3.12}$$

If the tracer is injected at the side of the channel, the width over which the plume must mix is double that for a centerline injection, but otherwise the boundary conditions are identical and the equations available can be used to obtain a solution.

EXAMPLE 10.1

A tracer is injected at the center of a slowly meandering channel with a flow rate of 0.15 m³/s (5.3 ft³/s). If the concentration of the tracer at the point of injection is 300 mg/L (300 parts per million, ppm), the width of the channel is 300 m (984 ft), the depth of flow is 10 m (33 ft),

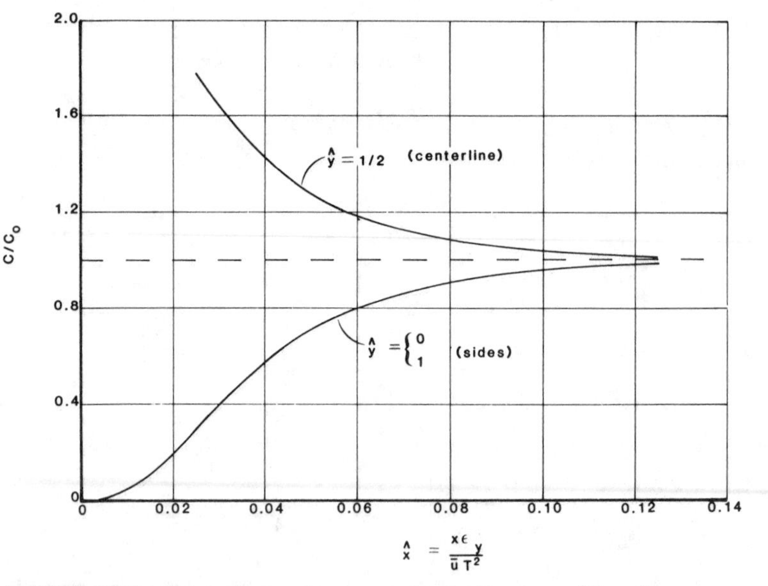

FIGURE 10.6 Downstream tracer concentrations resulting from continuous centerline injection.

the slope of the channel is 0.0001, and the total flow is 1600 m^3/s (57,000 ft^3/s), determine the maximum concentration of tracer 500 m (1600 ft) downstream of the point of injection and estimate the width of the tracer plume at this point.

Solution

It is assumed that the tracer is completely mixed over the vertical dimension at the point of injection. The shear velocity of the flow is then estimated from

$$u_* = \sqrt{gRS} = \sqrt{9.8(10)(0.0001)} = 0.099 \text{ m/s } (0.32 \text{ ft/s})$$

where R, the hydraulic radius, is taken as the depth of flow. From Eq. (10.3.7) the turbulent, transverse diffusion coefficient is

$$\epsilon_y = 0.60du_* = 0.60(10)(0.099) = 0.59 \text{ m}^2/\text{s } (6.4 \text{ ft}^2/\text{s})$$

The mass flow rate of tracer is

$$\dot{M} = Q_T C = 0.15(300) = 45 \text{ (m}^3 - \text{mg)/(s} - \text{L) } (1600 \text{ ft}^3 - \text{ppm/s})$$

where Q_T = the flow rate associated with the tracer. The average velocity of the flow is estimated by assuming that the channel cross section is approximately rectangular in shape; hence

$$\overline{u} = \frac{Q}{A} = \frac{1600}{300(10)} = 0.53 \text{ m/s } (1.7 \text{ ft/s})$$

The width of the plume at x = 500 m (1640 ft) is estimated first to determine if the lateral boundaries of the channel can be ignored in the calculation of the maximum tracer concentration at this distance. The width of the tracer plume is given approximately by

$$b = 4\sigma = 4\sqrt{2\epsilon_y t}$$

where t can be estimated from $t = x/\overline{u}$. Then

$$b = 4\sqrt{2(0.59)\left(\frac{500}{0.53}\right)} = 130 \text{ m } (430 \text{ ft})$$

Therefore, the width of the tracer plume is less than the width of the channel at x = 500 m (1640 ft), and the effect of the lateral channel boundaries can be ignored in computing the maximum tracer concentration at this distance. The maximum tracer concentration in this situation is estimated from Eq. (10.2.25).

$$C_{\max} = \frac{\dot{M}}{\overline{u}\sqrt{4\pi\epsilon_y x/\overline{u}}} \exp\left(-\frac{y^2\overline{u}}{4x\epsilon_y}\right)$$

$$= \frac{45}{0.53\,[4\pi(0.59)(500)/0.53]^{1/2}} \exp\left(\frac{0(0.53)}{4(500)(0.59)}\right)$$
$$= 1.0 \text{ mg/L (1 ppm)}$$

Note: y is taken as zero since an estimate of the maximum concentration is required.

EXAMPLE 10.2

For the data specified in Example 10.1, determine the longitudinal length required to achieve complete mixing in the transverse direction, i.e., the location where the concentration varies no more than 5 percent over the cross section, for (*a*) tracer injection at the channel centerline and (*b*) tracer injection at the side of the channel. How would this distance to the point of complete mixing be affected if the mixing took place as the channel was rounding a bend with a radius of 50 m (160 ft)?

Solution

Again, it is assumed that the tracer is completely mixed over the vertical dimension at the point of injection. Then, the distance to the point of complete mixing can be computed from Eq. (10.3.12) with $\overline{u} = 0.53$ m/s (1.7 ft/s) and $\epsilon_y = 0.59$ m²/s (6.4 ft²/s).

1. When the tracer injection is at the centerline of the channel, the appropriate value of T is 300 m (980 ft), and

$$L = \frac{0.1\overline{u}T^2}{\epsilon_y} = \frac{0.1(0.53)(300)^2}{0.59} = 8100 \text{ m (27,000 ft)}$$

2. When the tracer is injected at the side of the channel, $2T$ is used in Eq. (10.3.12) for the width or

$$L = \frac{0.1\overline{u}(2T)^2}{\epsilon_y} = \frac{0.1(0.53)[2(300)]^2}{0.59} = 32{,}000 \text{ m (105,000 ft)}$$

When the mixing takes place on a curve, the coefficient of turbulent, transverse diffusion is increased, and the distance to the point of complete mixing is decreased. For the data given, ϵ_y is, by Eq. (10.3.8),

$$\epsilon_y = 25\,\frac{\overline{u}^2\,d^3}{R_c^2\overline{u}_*} = 25\left[\frac{(0.53)^2(10)^3}{(50)^2(0.099)}\right] = 28 \text{ m}^2\text{/s (300 ft}^2\text{/s)}$$

Then, for centerline injection

$$L = \frac{0.1\overline{u}T^2}{\epsilon_y} = \frac{0.1(0.53)(300)^2}{28} = 170 \text{ m (560 ft)}$$

and for injection at the side of the channel

$$L = \frac{0.1\overline{u}(2T)^2}{\epsilon_y} = \frac{0.1(0.53)[2(300)]^2}{28} = 680 \text{ m } (2200 \text{ ft})$$

Thus, the distance in both cases required for complete mixing is significantly reduced.

Before a discussion of longitudinal dispersion is initiated, it is appropriate to note that most investigations of turbulent, transverse diffusion have addressed the problem from the viewpoint of channels with a free surface. Although the transverse diffusion process in ice-covered channels has received some attention (see, for example, Engmann, 1977), the results available at the present time must be considered indicative rather than absolute. Apparently, the effect of an ice cover is to cause approximately a 50 percent reduction in the absolute value of ϵ_y when similar hydraulic conditions—constant width, depth, and discharge—are compared. Thus, in an ice-covered channel, a tracer plume will satisfy the complete mix criterion much further downstream than a tracer plume in a comparable channel with a free surface. Engmann's date also indicate that the dimensionless diffusion coefficient, i.e., ϵ_y/Ru_*, for ice-covered channels does not differ significantly from the corresponding coefficient for open-water conditions.

In Section 10.2 the equation of longitudinal dispersion [Eq. (10.2.18)] was developed following the approach used by Elder (1959). At that time, it was noted that this equation is subject to a number of caveats. The most crucial of these is that Eq. (10.2.18) cannot be applied until an initial period of time has elapsed and the rate of advective transport either upstream or downstream of a vertical plane which moves at the cross-sectional average velocity depends only on the tracer concentration and E. This initial period during which Eq. (10.2.18) is not valid in terms of a dimensionless distance is given by

$$0 < x' < 0.4$$

where

$$x' = \frac{x\epsilon_y}{\overline{u}T^2} \tag{10.3.13}$$

with x being the longitudinal distance.

Elder (1959) in his development assumed that the velocity varied only in the vertical dimension and according to Eq. (10.3.2). With this assumption, a theoretical expression for the longitudinal dispersion coefficient can be derived.

$$E = 5.93\, du_* \tag{10.3.14}$$

In Table 10.1, a number of field- and laboratory-determined values of E are summarized along with values of E estimated by Eq. (10.3.14). From this table, it is evident that the Elder equation provides poor estimates of E in both natural and artificial channels. This is the result of Elder's neglecting to include

TABLE 10.1 Summary of dispersion experiments (Fischer et al., 1979)

Source	Channel	Depth, ft	Width, ft	Average velocity, ft/s	Shear velocity, ft/s	E observed, ft^2/s	Eq. (10.3.14)	Eq. (10.3.15)	Eq. (10.3.17)
Thomas (1958)	Chicago Ship Canal	26.5	160	0.88	0.0627	30	10	—	130
State of California, (anon) (1962)	Sacramento River	13.1	—	1.7	0.17	160	13	—	—
Owens et al. (1964)	River Derwent	0.82	—	1.2	0.46	50	2.2	—	—
Glover (1964)	South Platte River	1.5	—	2.2	0.23	170	2.0	—	—
Schuster (1965)	Yuma Mesa A Canal	11.3	—	2.2	1.1	8.2	74	—	—
Fischer (1967)	Laboratory channel	1.1	1.3	0.82	0.066	1.3	0.43	1.4	0.17
		1.5	1.4	1.5	0.12	2.7	1.1	2.7	0.36
		1.1	1.3	1.5	0.12	4.5	0.78	4.0	0.32
		1.1	1.1	1.4	0.11	2.7	0.72	2.7	0.22
		0.69	1.1	1.5	0.11	4.3	0.45	4.8	0.39
		0.69	0.62	1.5	0.13	2.4	0.53	1.8	0.11
Fischer (1968a)	Green-Duwamish River, Washington	3.6	66	—	0.16	70–91	3.4	84	—
Yotsukura et al. (1970)	Missouri River	8.9	660	5.1	0.24	16×10^3	13	—	58×10^3

Godfrey and	Copper	1.6	52	0.89	0.26	220	2.5	64	57
Frederick (1970)	Creek, Va.	2.8	59	2.0	0.33	230	5.5	300	170
[Predicted	(below gage)	1.6	52	0.85	0.26	100	2.5	120	52
values of E from	Clinch	2.8	59	2.0	0.33	150	5.5	160	170
Fischer (1968b)]	River, Tenn.	6.9	200	3.1	0.34	580	14	920	1800
		6.9	170	2.7	0.35	510	14	590	960
	Copper Creek, Va. (above gage)	1.3	62	0.52	0.38	32	2.9	30	23
	Powell River, Tenn.	2.8	110	0.49	0.18	100	3.0	98	63
	Clinch River, Va.	1.9	120	0.69	0.16	87	1.8	320	250
	Coachella Canal, Calif.	5.1	79	2.3	0.14	100	4.2	42	510
McQuivey and	Bayou Anacoco	3.1	85	1.1	0.22	360	4.0	—	141
Keefer (1974)	Colo.	3.0	120	1.3	0.22	420	4.0	—	400
	Nooksack River	2.5	210	2.2	0.89	380	13	—	1100
	Wind/Bighorn	3.6	190	2.9	0.39	450	8.3	—	2400
	Rivers	7.1	230	5.1	0.56	1700	23	—	3800
	John Day	8.1	110	2.7	0.59	700	28	—	200
	River	1.9	82	3.3	0.46	150	5.2	—	920
	Comite River	1.4	52	1.2	0.16	150	1.3	—	190
	Sabine	6.7	340	1.9	0.16	3400	6.3	—	4300
	River	16	420	2.1	0.26	7200	25	—	2100
	Yadkin	7.7	230	1.4	0.33	1200	15	—	450
	River	13	240	2.5	0.43	2800	33	—	710

the effect of the transverse velocity profile in his analysis. Fischer (1967) presented the following equation to estimate E in a natural channel when a cross section has been defined and gaged:

$$E = -\frac{1}{A\partial c/\partial x}\iint_A u'c'\,dA$$

$$= -\frac{1}{A}\int_0^T u'd\int_0^y \frac{1}{\epsilon_y d}\int_0^y u'\,d\,dy\,dy\,dy \qquad (10.3.15)$$

In practice, the integrals of Eq. (10.3.15) are replaced by summations or

$$E = -\frac{1}{A}\sum_{k=2}^{N} q_k'\,\Delta y\left[\sum_{j=2}^{k}\frac{\Delta y}{\epsilon_y d_j}\left(\sum_{i=1}^{j-1} q_i'\,\Delta y\right)\right] \qquad (10.3.16)$$

where $q_i' = [(d_i + d_{i+1})/2]u_i'$
u_i = mean velocity in ith element of cross section
$u_i' = u_i - \overline{u}$
$\overline{u}$ = cross-sectional average velocity of flow
d_i = depth at beginning of ith cross-sectional element
Δy = width of element (constant)
$\epsilon_y = 0.6\,du_*$ = transverse, turbulent diffusion coefficient between (i − 1) and ith cross-sectional elements
N = number of cross-sectional elements

Fischer et al. (1979) stated that N should exceed 20. Equation (10.3.16) is subject to two limitations. First, both the effect of the vertical velocity gradient and the variation of c' in the vertical dimension on the dispersion process are neglected. Second, the development of Eq. (10.3.16) assumes both uniform flow and a channel of constant cross section. In Table 10.1 values of E estimated by this technique are summarized. A comparison of the observed values and the values of E estimated by this method demonstrates reasonable agreement.

EXAMPLE 10.3

For the cross-section defined in Fig. 10.7 and Table 10.2, estimate E by Eq. (10.3.16) if $u_* = 0.100$ m/s (0.33 ft/s).

Solution

The average velocity of flow is determined by summing col. (4) in Table 10.2 and dividing this by the sum of col. (3) in the same table or

$$\overline{u} = \frac{\sum_{i=1}^{N} q_i}{\sum_{i=1}^{N} A_i} = \frac{200}{555} = 0.36 \text{ m/s } (1.2 \text{ ft/s})$$

TABLE 10.2 Channel definition for Example 10.3

Element (1)	Average velocity, m/s (2)	Area, m^2 (3)	Element flow, m^3/s (4)	Depth, left side of element, m (5)
1	0.24	12.5	3.0	0
2	0.30	34.4	10	8.2
3	0.37	49.2	18	14
4	0.40	56.7	23	18
5	0.49	57.1	28	19
6	0.43	54.8	24	18
7	0.43	48.8	21	18
8	0.40	40.9	16	14
9	0.37	35.3	13	12
10	0.34	30.2	10	11
11	0.37	26.5	9.8	9.1
12	0.30	23.7	7.1	8.2
13	0.27	21.4	5.8	7.3
14	0.27	20.0	5.4	6.7
15	0.24	19.0	4.6	6.4
16	0.09	17.2	1.5	6.1
17	0.03	7.90	0.23	5.2

Then values for cols. (3) to (5) of Table 10.3 can be calculated. The values of the transverse, turbulent diffusion coefficient in col. (6) of Table 10.3 are estimated by using the depth of the left-hand side of the element; or for element 2

$$\epsilon_y = 0.6\, du_* = 0.6(8.2)(0.100) = 0.49 \text{ m}^2/\text{s} \ (5.3 \text{ ft}^2/\text{s})$$

The values of col. (7) in Table 10.3 are computed as a running total.

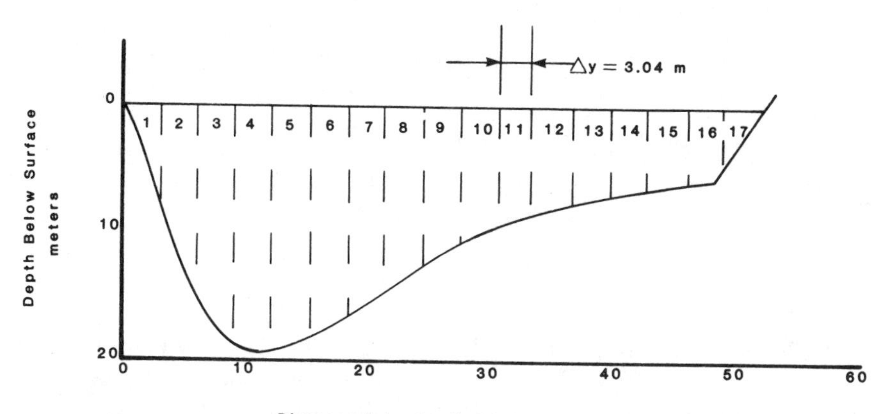

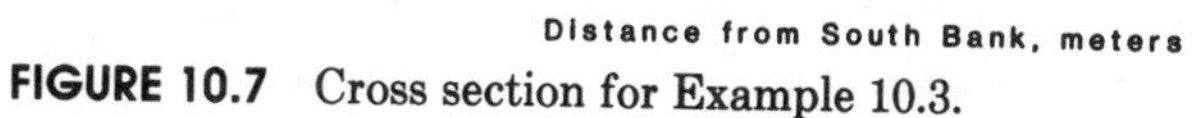
FIGURE 10.7 Cross section for Example 10.3.

TABLE 10.3 Tabular solution of Example 10.3

Element (1)	Depth, left side of element, m (2)	u'_i, m/s (3)	$u'_i A_i = q'_i$, m³/s (4)	$\Sigma q'_i$, m³/s (5)	ϵ_y, m²/s (6)	$\sum_{j=2}^{k} \frac{\Delta y}{\epsilon_y d_j} \sum_i^{j-1} q'_i$, m (7)
1	0	−0.12	−1.5	−1.5	0	0
2	8.2	−0.06	−2.1	−3.6	0.49	−1.1
3	14	0.01	0.49	−3.1	0.84	−2.1
4	18	0.04	2.3	−0.8	1.1	−2.5
5	19	0.13	7.4	6.6	1.1	−2.7
6	18	0.07	3.8	10.4	1.1	−1.6
7	18	0.07	3.4	13.8	1.1	−0.05
8	14	0.04	1.6	15.4	0.84	3.5
9	12	0.01	0.35	15.7	0.72	8.9
10	11	−0.02	−0.60	15.1	0.66	16
11	9.1	0.01	0.26	15.4	0.55	25
12	8.2	−0.06	−1.4	14.0	0.49	36
13	7.3	−0.09	−1.9	12.1	0.44	50
14	6.7	−0.09	−1.8	10.3	0.40	63
15	6.4	−0.12	−2.3	8.0	0.39	76
16	6.1	−0.27	−4.6	3.4	0.37	87
17	5.2	−0.33	−2.6	0.8	0.31	93

The value of this sum is by definition zero for element 1. For element 2, the value of col. (7) is given by

$$\frac{\Delta y}{\epsilon_y d_2} q_1 = \frac{3.0}{0.49(8.2)}(-1.5) = -1.1$$

and for element 3

$$\frac{\Delta y}{\epsilon_y d_3} q_2 + \frac{\Delta y}{\epsilon_y d_2} q_1 = -1.1 + \frac{3.0}{0.84(14)}(-3.6) = -2.0$$

The longitudinal dispersion coefficient is then estimated as

$$E = -\frac{\Sigma(\text{col 4})(\text{col 7})}{\text{flow area}}$$

$$= -\frac{1100}{555} = 2.0 \text{ m}^2/\text{s (22 ft}^2/\text{s)}$$

While the estimates of E yielded by Eq. (10.3.15) are superior to those obtained with Eq. (10.3.14), they are not, in all cases, accurate. The primary reason for the discrepancies which can be noted in Table 10.1 appear to result from the fact that no natural channel completely fulfills the assumptions inherent in the development of Eq. (10.3.15). For example, all natural channels and many artificial channels have bends, changes of cross-sectional shape, pools, and many other irregularities, all of which contribute significantly to the dispersion process. Fischer et al. (1979) presented the following equation which can be used to estimate an approximate value of E:

$$E = \frac{0.011\overline{u}^2 T^2}{du_*} \tag{10.3.17}$$

This equation has the distinct advantage of estimating E from parameters which are usually known. In Table 10.1 values of E estimated by this method can be compared with observed values.

At this point, it is appropriate to note the presence and presumed effect of what has in the literature been termed *dead zones*. This terminology has been used to refer to areas along the boundaries of the channel in which tracer is trapped as the main cloud of tracer passes. The trapped tracer is released from the dead zone slowly and can cause measurable tracer concentrations to be observed downstream long after the main cloud of tracer has passed. The result is downstream concentration versus time plots which are not gaussian and are characterized by a long tail (e.g., Fig. 10.8). At the present time, there is no effective method of quantifying the effect of dead zones either on the dispersion coefficient or on the movement of a tracer plume downstream. One method which has been used by a number of investigators is to treat the tail of the

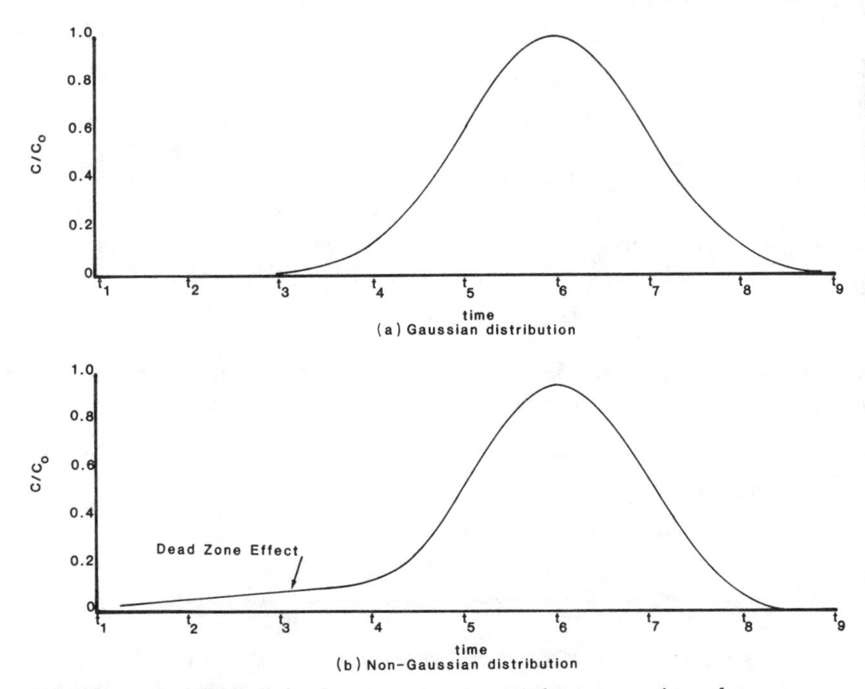

FIGURE 10.8 Effect of dead zones on concentration versus time plots.

concentration versus time plot as if it did not exist with the justification being that the mass of the tracer in this tail is small (see, for example, Fischer, 1968a).

In an examination of the results summarized in Table 10.1, none of the methods which have been suggested for estimating E yields completely satisfactory results. To some extent, the difference between the predicted and observed values of the longitudinal dispersion coefficient are due to the inaccuracy of the method used to determine the observed values. For example, Fischer et al. (1979) have suggested that the observed value is usually accurate only within a factor of 2. Thus, the estimated values of E in Table 10.1 may be better than only a cursory examination would indicate.

There are essentially three methods of calculating a value of E from field data. First, from Section 10.2, by definition, the longitudinal dispersion coefficient measures the rate of change of the variance of the tracer cloud or

$$E = 0.5 \frac{d(\sigma_x^2)}{dt} \tag{10.3.18}$$

where σ_x^2 = the variance of the concentration distribution with respect to distance along the channel. The application of Eq. (10.3.18) to field data is somewhat complicated by the fact that normal field data consist of tabulated tracer concentrations at a specified distance downstream from the point of injection as a function of time. A typical field data base is contained in Table 10.4. Equation (10.3.18) can be transformed to a form which is useful with the typical field data set or

$$E = \frac{\overline{u}^2}{2} \left[\frac{\sigma_{t2}^2 - \sigma_{t1}^2}{\overline{t}_2 - \overline{t}_1} \right] \tag{10.3.19}$$

where σ_{t1}^2 and σ_{t2}^2 = variance of the concentration versus time curves at upstream (station 1) and downstream (station 2), respectively, and $\overline{t}_1$ and $\overline{t}_2$ = mean times of passage of the tracer cloud past the upstream and downstream stations, respectively. Although Eq. (10.3.19) is theoretically correct, a number of practical limitations should be noted. First, a large initial dose of tracer is required to yield measurable downstream concentrations. Second, a long reach of prismatic channel is required. Third, the presence of dead zones can have a significant effect on the value of E which is estimated by this method since the long tail characteristic of dead zones will affect the value of σ^2.

EXAMPLE 10.4

A tracer is injected into a flow occurring in a natural channel, and as the resulting tracer cloud passes two downstream stations, it is sampled. The results of the sampling program are summarized in Table 10.4 and Fig. 10.9. If in the reach between the two sampling stations the average depth, width, and shear velocity are 2.77 ft (0.84 m), 60 ft

TABLE 10.4 Tracer concentration profiles at two stations for dispersion coefficient estimation (Fischer, 1968b)

Time relative to 1st observation, min	Tracer concentration, mg/L
Station 1: 7870 ft (2400 m) from point of injection	
0	0
3	0.26
6	0.67
9	0.95
11	1.09
13	1.13
15	1.10
17	1.04
19	0.95
24	0.72
29	0.50
34	0.31
39	0.21
49	0.08
59	0.02
Station 2: 13,550 ft (4100 m) from point of injection	
32	0
37	0
42	0.07
47	0.22
52	0.40
56	0.50
60	0.58
62	0.59
64	0.59
68	0.54
75	0.44
84	0.27
94	0.14
104	0.06
114	0.03
124	0.025
134	0.02
144	0

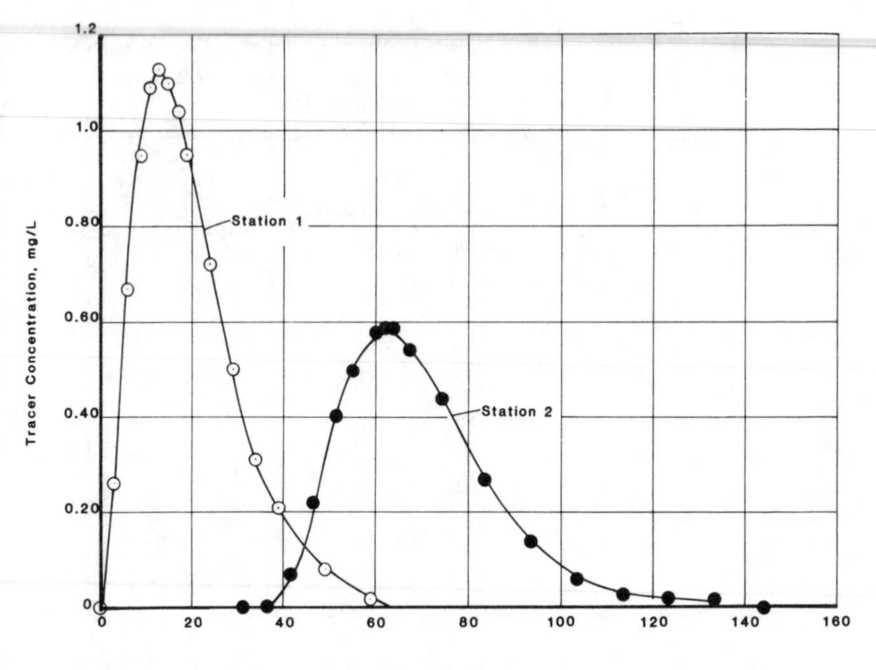

FIGURE 10.9 Tracer concentration at two stations as a function of time, from data in Table 10.4.

(18 m), and 0.33 ft/s (0.10 m/s), respectively, estimate a longitudinal dispersion coefficient for this reach of channel using Eq. (10.3.19). *Note:* The data for this example were first presented by Fischer (1968b).

Solution

The only caveat associated with the use of Eq. (10.3.19) is that both sampling stations must be sufficiently far downstream for Eq. (10.2.18) to apply or

$$x' = \frac{x\epsilon_y}{\overline{u}T^2} > 0.4$$

With reference to Fig. 10.8, the average velocity of flow can be estimated as

$$\overline{u} = \frac{\Delta x}{\Delta t}$$

where Δx = distance between the sampling stations and Δt = travel time between the sampling stations. Δt can be approximated as the time difference between the tracer peaks at the two stations or $\Delta t \simeq$ 49 min. Then

$$\bar{u} = \frac{13{,}550 - 7870}{49} = 116 \text{ ft/min} = 1.94 \text{ ft/s } (0.59 \text{ m/s})$$

and $\quad \epsilon_y = 0.6\, du_* = 0.6(2.77)(0.33) = 0.55 \text{ ft}^2\text{/s } (0.051 \text{ m}^2\text{/s})$

x' is then computed as

$$x' = \frac{7870(0.55)}{1.94(60)^2} = 0.62$$

and, therefore, Eq. (10.3.19) can be used to estimate E. The parameters σ_{ti}^2 can be quickly estimated by

$$\sigma_{ti}^2 = \frac{\sum_{i=1}^{N} C_i t_i^2}{\sum_{i=1}^{N} C_i} - \left(\frac{\sum_{i=1}^{N} C_i t_i}{\sum_{i=1}^{N} C_i} \right)^2$$

where the time base used for the computations is not a factor, but the indicated summations must be performed over uniformly spaced readings. From Fig. 10.9, Table 10.5 is constructed, and we have the following data:

Station 1	Station 2
$\sum_{i=1}^{N} C_i = 9.08$	$\sum_{i=1}^{N} C_i = 4.26$
$\sum_{i=1}^{N} C_i t_i = 181$	$\sum_{i=1}^{N} C_i t_i = 298$
$\sum_{i=1}^{N} C_i t_i^2 = 4776$	$\sum_{i=1}^{N} C_i t_i^2 = 21{,}988$
$\sigma_{t1}^2 = 129$	$\sigma_{t2}^2 = 268$

Although Fischer (1968b) was not precise in his definition of $\bar{t}_1$ and $\bar{t}_2$, Levenspiel and Smith (1957) used the following definition:

$$\bar{t}_i = \frac{\int_0^\infty Ct\, dt}{\int_0^\infty C\, dt}$$

Application of this definition yields

$$\bar{t}_1 = 17.8 \text{ min}$$

TABLE 10.5 Tracer concentration profiles at two stations estimated from data in Fig. 10.9

Station 1		Station 2	
t, min	C, mg/L	t, min	C, mg/L
0	0	37	0
3	0.26	42	0.07
6	0.67	47	0.22
9	0.95	52	0.40
12	1.1	57	0.54
15	1.1	62	0.59
18	1.0	67	0.55
21	0.87	72	0.48
24	0.72	77	0.39
27	0.59	82	0.30
30	0.45	87	0.22
33	0.33	92	0.16
36	0.26	97	0.11
39	0.21	102	0.07
42	0.16	107	0.05
45	0.14	112	0.03
48	0.10	117	0.02
51	0.07	122	0.02
54	0.05	127	0.02
57	0.03	132	0.01
60	0.02	137	0.01
63	0	142	0

and $$\bar{t}_2 = 68.9 \text{ min}$$

Substituting these data in Eq. (10.3.19) yields

$$E = \frac{\bar{u}^2}{2}\left(\frac{\sigma_{t2}^2 - \sigma_{t1}^2}{\bar{t}_2 - \bar{t}_1}\right) = \frac{116^2}{2}\left(\frac{268 - 129}{68.9 - 17.8}\right)$$

$$= 18{,}300 \text{ ft}^2/\text{min } (1700 \text{ m}^2/\text{min})$$

$$= 18{,}300 \frac{\text{ft}^2}{\text{min}}\frac{1 \text{ min}}{60 \text{ s}} = 300 \text{ ft}^2/\text{s } (28 \text{ m}^2/\text{s})$$

The second method of estimating a value of E for a reach of channel uses what is termed in the literature a *routing procedure*. This methodology involves the matching of an observed concentration distribution at a distance x_2 below the point of tracer injection with a distribution predicted by a model which uses a concentration distribution observed at a distance x_1 below the point of injection where $x_1 < x_2$. From Fischer (1968b)

$$C(x_2, t) = \int_{-\infty}^{\infty} \overline{u}C(x, \tau) \frac{\exp\{-[\overline{u}(\overline{t}_2 - \overline{t}_1 - t + \tau)]^2/4E(\overline{t}_2 - \overline{t}_1)\}}{\sqrt{4\pi E(\overline{t}_2 - \overline{t}_1)}} d\tau \tag{10.3.20}$$

where $C(x_1, \tau)$ = observed tracer concentration at a distance x_1 below point of tracer injection at time τ

$C(x_2, t)$ = estimated tracer concentration at distance x_2 below point of tracer injection at time t

τ = time variable of integration

Although Eq. (10.3.20) indicates that the limits of integration are $-\infty < \tau < \infty$, in practice the integration needs only to be performed over the interval $t(1) \leq \tau \leq t(2)$, where $t(1)$ and $t(2)$ are the beginning and ending points of the concentration distribution at x_1 or

For $\tau \leq t(1),\ C(x_1, \tau) = 0$

For $\tau \geq t(2),\ C(x_1, \tau) = 0$

The use of Eq. (10.3.20) to estimate E for a reach of channel involves the following steps:

1. A trial value of E is estimated by Eq. (10.3.19).
2. The trial value of E determined in the previous step is used with Eq. (10.3.20) to estimate a concentration distribution at x_2.
3. The predicted and observed concentration distributions at the downstream station are compared by computing the mean squared concentration difference between the two curves.
4. A search is then made, by trial and error, to determine if there is a value of E which minimizes the mean squared differences between the observed and predicted concentration distribution curves at the downstream station.

EXAMPLE 10.5

For the data given for Example 10.4, use Eq. (10.3.20) to estimate a value of E.

Solution

The most effective method of solving this problem is to develop a simple computer program which will perform the integration indicated in Eq. (10.3.20) to determine $C(x_2, t)$ when t, $\overline{u}$, $\overline{t}_1$, $\overline{t}_2$, E, and a matrix of

$C(x_1, \tau)$ values are specified. From Example 10.4 the following parameter values are available:

$$\bar{u} = 116 \text{ ft/min } (35 \text{ m/min})$$

$$\bar{t}_1 = 17.8 \text{ min}$$

$$\bar{t}_2 = 68.9 \text{ min}$$

The analysis of the data in Table 10.4 presented in Example 10.4 also provides an estimate of the correct value of E or

$$E = 18{,}300 \text{ ft}^2/\text{min } (1700 \text{ m}^2/\text{min})$$

By use of these data, values of $C(x_2, t)$ can be estimated (Table 10.6 and Fig. 10.10). With regard to the data summarized in this table and figure, the following are noted:

1. The estimated peak value arrives at station 2 after the measured peak value; therefore, it might be concluded that the estimated value of $\bar{u}$ is too low. No adjustment can be made to this value without further information.

2. A dispersion coefficient which minimizes some goodness of fit parameter must be searched for; e.g.,

$$D = \frac{[\hat{C}(x_2, t) - C(x_2, t)]^2}{N}$$

TABLE 10.6 Values of $C(x_2, t)$ estimated by Eq. (10.3.20) for E = 18,300 ft²/min

Time, min	Measured value of $C(x_2, t)$, mg/L	Estimated value of $C(x_2, t)$, mg/L
37	0	0.097
42	0.07	0.19
47	0.22	0.31
52	0.40	0.46
56	0.50	0.57
60	0.58	0.65
62	0.59	0.67
64	0.59	0.68
68	0.54	0.68
75	0.44	0.57
84	0.27	0.37
94	0.14	0.18
104	0.06	0.072
114	0.03	0.023
124	0.025	0.0055
134	0.02	0.00088
144	0	0

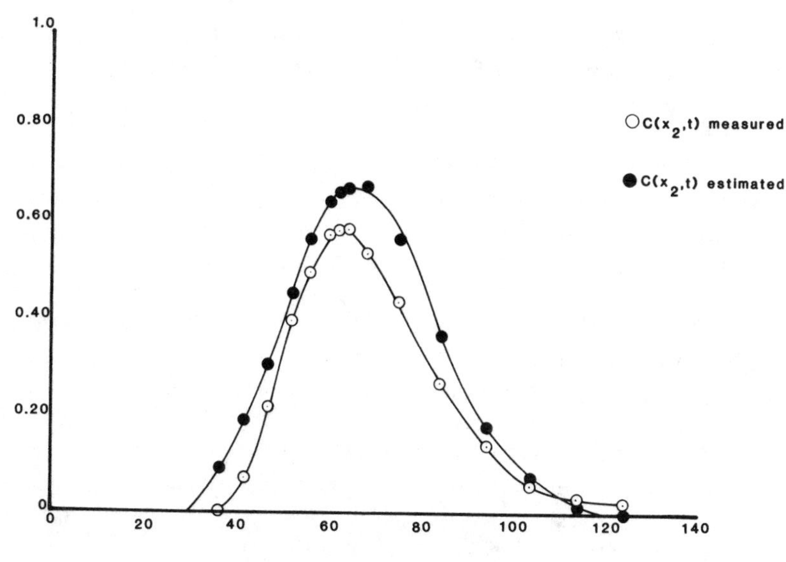

FIGURE 10.10 Measured and estimated values of $C(x_2,t)$ for $E = 19{,}500$ ft^2/min.

where $\hat{C}(x, t)$ = estimated concentration at station 2
$C(x_2, t)$ = measured concentration at station 2
N = number of points considered
For $E = 18{,}300$ ft^2/min (1700 m^2/min)

$$D = \frac{0.1088}{17} = 0.0064$$

Additional trial values of E, both larger and smaller than 18,300 ft^2/min (1700 m^2/min) must be considered.

The third method of estimating E from field data is based on the solution of Eq. (10.2.18) for a pulse input of tracer (Krenkel, 1960) or

$$C = \frac{M}{A\sqrt{4\pi Et}} \exp - \frac{(x - ut)^2}{4Et} \tag{10.3.21}$$

where M = weight of tracer and A = flow area. Rearranging and taking the logarithm of both sides of this equation yields

$$\log (C\sqrt{t}) = \log \left(\frac{M}{A\sqrt{4\pi E}} \right) - \frac{(x - \overline{u}t)^2}{4Et} \log (\mathrm{e})$$

Then, a plot of $\log (C\sqrt{t})$ versus $(x - \overline{u}t)^2/t$ theoretically results in a straight line whose slope is $\log (\mathrm{e})/4E$ from which E can be estimated. Although the use of this method for estimating E requires only a concentration versus time curve from a single downstream station, t is measured from the time of tracer injection. For this reason, the field data for either station in Table 10.4 cannot be used to estimate E with this technique since the time of tracer injection is not known.

At this point, some observations regarding dye studies designed to determine a value of the dispersion coefficient are appropriate. First, it is tacitly assumed that the dye injected into a flow behaves in exactly the same manner as the water. In selecting a dye for a dispersion study, the following characteristics should be considered: (1) detectability, (2) toxicity, (3) solubility, (4) stability, and (5) cost (Kilpatrick et al., 1970). In general, fluorescent dyes have been preferred in such studies (e.g., Kilpatrick et al., 1970, and Wilson, 1968). *Fluorescence* refers to the response of a substance to a light source of a given wavelength. Both rhodamine BA and rhodamine WT have been successfully used in such studies. In general, rhodamine WT dye is preferred even though it costs more than rhodamine BA because rhodamine BA is absorbed quite readily by almost everything it comes in contact with, e.g., aquatic plants, suspended clays, and the channel boundaries. It is also noted that the U.S. Geological Survey recommends that (1) since most dyes have an affinity for plastics, particularly rhodamine WT, glass bottles should be used for sample collection and (2) maximum dye concentration at a water supply intake should be limited to 10 μg/L (0.010 ppm).

Second, several empirical equations which can be used to estimate the quantity of dye required to produce a specified peak concentration at a sampling site are available. The form of these equations is

$$V = \phi \left(\frac{Q_{\max} L}{\overline{u}} \right)^{\Gamma} C_p \qquad (10.3.22)$$

where V = dye volume, L
$Q_{\max}$ = maximum discharge in reach, ft^3/s
L = distance from point of injection to sampling point, miles
$\overline{u}$ = average reach velocity, ft/s
C_p = desired peak concentration at sampling point, μg/L
ϕ and Γ = empirical coefficients

For example, if rhodamine WT 20 percent dye is used, it has been found that

$$\phi = 3.4 \times 10^{-4} \qquad (10.3.23)$$

and

$$\Gamma = 0.93 \qquad (10.3.24)$$

Third, a minimum of two sampling points is required. The first point must be far enough downstream from the point of dye injection that complete lateral and vertical mixing have taken place. The second point of sampling is then at a convenient but sufficient distance downstream of the first sampling point.

Fourth, as noted by Kilpatrick et al. (1970), the schedule for collecting samples at a sampling point is the most uncertain aspect of any study plan. At a specified distance downstream from the point of injection, the time from injection until the peak concentration arrives at the sampling location can be estimated by

$$T_p = 1.47 \frac{L}{\bar{u}} \tag{10.3.25}$$

Then, from Fig. 10.11 the duration, in hours, of the dye cloud passage at that point can be estimated. This figure also recommends the sampling interval which should be used at a downstream station. If the dye cloud is assumed to be symmetrical in shape, then the leading edge of the dye cloud will arrive at the sampling station at

$$T_e = T_p - 0.5T_D \tag{10.3.26}$$

where T_e = expected time of arrival of the leading edge of the dye cloud in hours after the time of injection.

EXAMPLE 10.6

For the situation defined in Fig. 10.12, estimate the following: (1) the volume of rhodamine WT 20% dye which must be injected at mile 0 to

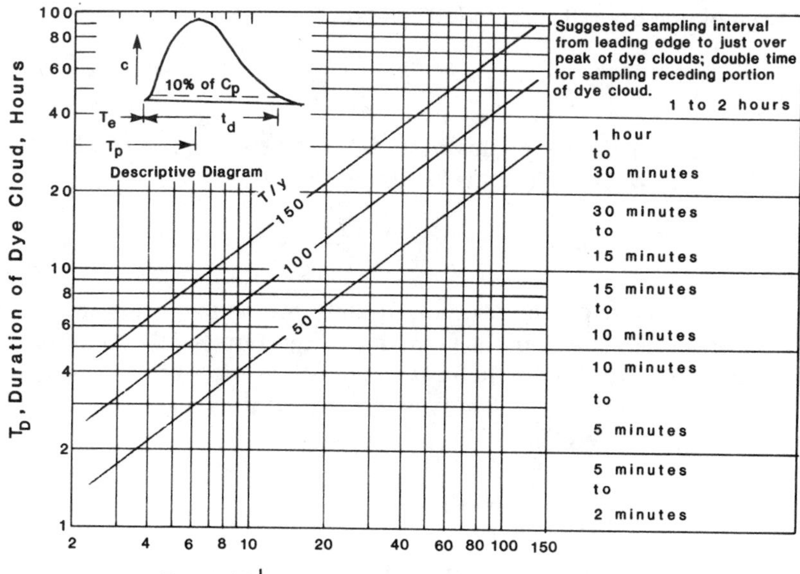

FIGURE 10.11 Duration of dye cloud as a function of travel time to the peak concentration and the average channel width to depth ratio. *(Kilpatrick et al., 1970.)*

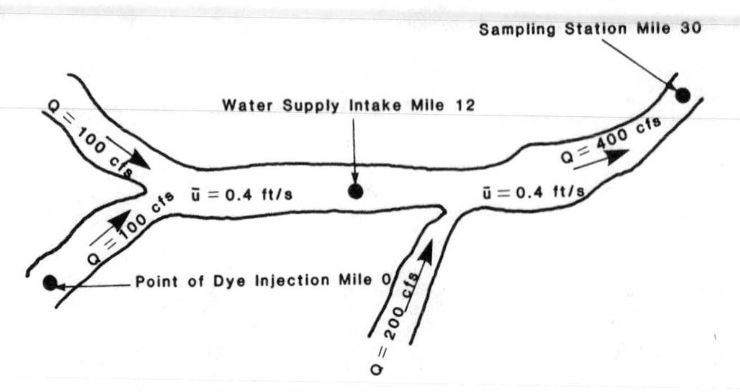

FIGURE 10.12 Example 10.6.

produce a peak concentration of 2 μg/L at mile 30; (2) the peak concentration which can be expected at the water supply inlet located at mile 12; and (3) the time when sampling should begin at mile 30 if the mean width-to-depth ratio for this reach is 100 and the sampling interval which should be used.

Solution

The volume of rhodamine WT 20% dye which must be injected at mile 0 to produce a peak concentration of 2 μg/L at mile 30 is estimated by combining Eq. (10.3.22) to (10.3.24) or

$$V = 3.4 \times 10^{-4} \left(\frac{Q_{max} L}{\bar{u}} \right)^{0.93} C_p$$

The peak discharge for this reach is 400 ft³/s (11 m³/s), and from Fig. 10.12 the average velocity is 0.4 ft/s (0.12 m/s). Then

$$V = 3.4 \times 10^{-4} \left[\frac{400(30)}{0.4} \right]^{0.93} (2) = 9.9 \text{ L}$$

The peak dye concentration at the water supply inlet located 12 miles downstream from the point of dye injection can also be estimated from Eqs. (10.3.22) to (10.3.24) where the average velocity is 0.4 ft/s (0.12 m/s), but $Q_{max} = 200$ ft³/s (5.7 m³/s). Rearranging the equation developed above yields

$$\frac{V}{C_p} = 3.4 \times 10^{-4} \left(\frac{Q_{max} L}{\bar{u}} \right)^{0.93}$$

and

$$\frac{V}{C_p} = 3.4 \times 10^{-4} \left[\frac{200(12)}{0.4} \right]^{0.93} = 1.1$$

If 9.9 L of dye is injected at mile 0, the expected concentration at the water supply diversion would be

$$C_p = \frac{9.9}{1.1} = 9.0\ \mu g/L$$

The time at which sampling should commence 30 miles below the point of injection is estimated by Eq. (10.3.25) and (10.3.26) and Fig. 10.11. From Eq. (10.3.25)

$$T_p = 1.47\frac{L}{\overline{u}} = 1.47\frac{30}{0.4} = 110\ \text{h}$$

From Fig. 10.11, the length of time required for the dye cloud to pass the sampling station is estimated to be, for a width-to-depth ratio (T/y) of 100,

$$T_D = 45\ \text{h}$$

Therefore, it is estimated by Eq. (10.3.26) that the leading edge of the dye cloud will arrive at the sampling station.

$$T_e = T_p - 0.5T_D = 110 - 0.5(45) = 87\ \text{h}$$

after the dye is injected upstream.

Note: At best this is a very crude estimate of the time of arrival. From Fig. 10.11, it is estimated that samples should be obtained at least every hour and perhaps every 30 min.

As noted previously, in the reach of channel defined by $0 < x' < 0.4$ where x' is defined by Eq. (10.3.13), the one-dimensional equation for longitudinal dispersion [Eq. (10.2.18)] is not valid. In many respects, this limitation is crucial since this may be the critical channel reach from the viewpoint of satisfying environmental regulations. Although there are a number of techniques available for modeling the movement of a tracer plume in this reach, space considerations dictate that only two of the many methods be discussed.

Fischer (1968a) developed a very simple model of the spread of a tracer plume in this critical reach of channel. This model has the advantage of requiring that velocities be specified at only one cross section of the reach under consideration, but it also has the disadvantage of requiring that the channel be prismatic and the flow uniform. The steps involved in implementing this model are:

1. The cross section is divided into N subsections of area $A_1, A_2, \ldots, A_N$ (Fig. 10.13).

2. The relative or advective velocity u_j of each channel subsection is determined by

$$u_j' = u_j - \overline{u} \qquad (10.3.27)$$

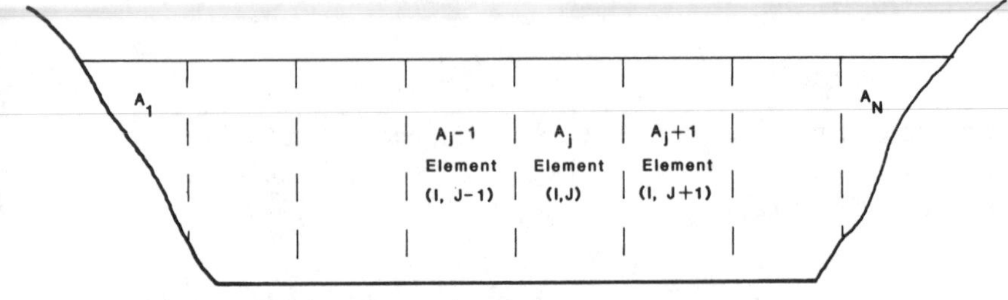

FIGURE 10.13 Definition of notation.

where u_j = measured velocity in the jth subsection and $\overline{u}$ = cross-sectional average velocity. In performing this calculation, care must be used to ensure that the following condition is satisfied:

$$\sum_{j=1}^{N} u_j' A_j = 0$$

3. The longitudinal distance step is defined by

$$\Delta x = |u_j' \,(\max)|\, \Delta t$$

where Δt = selected time step and u_j' (max) = the largest advective velocity without regard to sign [Eq. (10.3.27)]. Thus, the average flow in the subsection where u_j' (max) occurs moves at the rate of one longitudinal grid point per time step.

4. A two-dimensional grid of tracer concentration, $C^k(i, j)$, is defined where the index i refers to the longitudinal distance in a coordinate system which moves at the average velocity of the cross section, and the index j refers to the transverse location of the subsection in the cross section (Fig. 10.13). The superscript k references the current time step.

5. The computation advances in time by performing two sets of computations. First, the tracer is advected and then mixed in the transverse direction.

   The advective movement of the tracer in each subsection either upstream or downstream is computed according to the value of u_j'. In performing the advective computation, it is usually advantageous to convert the advective velocities to units of mesh points per time step or

$$U_j = u_j' \frac{\Delta t}{\Delta x}$$

This operation results in a two-dimensional scratch matrix $D(i, j)$, which contains temporary computation values of the tracer concentration. The $D(i, j)$ matrix is defined by

$$D(i, j) = \begin{cases} \text{if } U_j > 0\text{: } U_j[C^k(i-1, j) - C^k(i, j)] \\ \text{if } U_j < 0\text{: } U_j[C^k(i, j) - C^k(i+1, j)] \end{cases} \qquad (10.3.28)$$

Tracer concentrations which are advected a fraction of a grid space are divided between the grid points involved inversely as the distance from each. It should be noted that use of Eq. (10.3.28) introduces what is termed *numerical dispersion.*

Transverse mixing between the channel subsections is accomplished by a second set of computations. The change in concentration at a grid point (i, j) due to transverse mixing is given by

$$\Delta C(i, j) = \frac{1}{A_j \, \Delta x} (\Delta M_{j,j+1} - \Delta M_{j-1,j})$$

where the variable M designates the mass transport between the subsections. M is calculated under the assumption that the mass transport for the duration of the time step is given by

$$\Delta M_{j,j+1} = h_j \epsilon_j \left(\frac{\Delta C_j}{s_j} \right) \Delta x \, \Delta t$$

where h_j = the area of the plane dividing subsections j and $j + 1$ per unit of downstream length, s_j = distance separating the centroids of the subsections j and $j + 1$, ΔC_j = difference in concentration between subsections j and $j + 1$ or

$$\Delta C_j = C(i, j + 1) - C(i, j)$$

and ϵ_j = turbulent transverse diffusion coefficient given by

$$\epsilon_j = 0.6 d_j u_*$$

6. When the above operations are combined, the tracer concentration at the grid point (i, j) and at time $k + 1$ is

$$C^{k+1}(i, j) = D(i, j) + \left\{ \frac{h_j \epsilon_j}{s_j} [D(i, j + 1) - D(i, j)] - \frac{h_{j-1} \epsilon_{j-1}}{s_{j-1}} [D(i, j) - D(i, j - 1)] \right\} \frac{\Delta t}{A_j} \qquad (10.3.29)$$

Fischer (1968a) noted that the numerical scheme defined by the above operations is stable for all values of j if

$$\frac{\epsilon_j \, \Delta t}{(s_j)^2} < 0.5$$

The obvious limitations of the method described above are that the channel must be prismatic and the flow uniform. Harden and Shen (1979) have developed a method which takes into account channel cross-sectional irregularities but which involves a degree of mathematical sophistication which may not be warranted in many practical situations. If the channel is defined in terms of an orthogonal curvilinear coordinate system oriented so that the longitudinal coordinate surfaces are aligned in the direction of the depth-averaged local velocity vectors, then the depth-averaged two-dimensional advective diffusion equation for steady flow is

$$\frac{\partial C}{\partial t} + \frac{u_x}{m_x}\frac{\partial C}{\partial x} = \frac{1}{m_x m_y d}\frac{\partial}{\partial x}\left(\frac{m_y}{m_x} d\epsilon_x \frac{\partial C}{\partial x}\right) + \frac{u_x}{m_x}\frac{\partial}{\partial q_c}\left(m_x d^2 u_x \epsilon_y \frac{\partial C}{\partial q_c}\right) \tag{10.3.30}$$

where d = local depth of flow
u_x = local depth-averaged velocity in x direction
m_x and m_y = metric coefficients for orthogonal curvilinear coordinate system in x and y directions, respectively
q_c = cumulative discharge = $\int_0^y (m_y d u_x)dy$

In practice, the term in Eq. (10.3.30) describing longitudinal mixing can be ignored, and the governing equation becomes

$$\frac{\partial C}{\partial t} + \frac{u_x}{m_x}\frac{\partial C}{\partial x} = \frac{u_x}{m_x}\frac{\partial}{\partial q_c}\left(m_x d^2 \epsilon_y \frac{\partial C}{\partial q_c}\right) \tag{10.3.31}$$

Equation (10.3.31) can be solved by using a finite difference numerical scheme if values of q, u_x, and d are available. Harden and Shen (1979) noted that in many cases a flow distribution can be synthesized by

$$\frac{q}{\overline{q}} = \Theta\left(\frac{d}{\overline{d}}\right)^{\Gamma} \tag{10.3.32}$$

where q = local flow per unit width
$\overline{q}$ = average flow per unit width
$\overline{d}$ = average depth of flow
Θ and Γ = empirical coefficients

For straight channel reaches with $50 \le T/\overline{d} < 70$, $\Theta = 1$ and $\Gamma = 5/3$, and for $T/d \ge 70$, $\Theta = 0.92$ and $\Gamma = 7/4$. For meandering reaches, Θ varies between 0.95 and 0.5, Γ varies between 2.48 and 1.78, and $T/\overline{d}$ varies between 50 and 100. The advantages of the Harden and Shen (1979) technique are that, first, it predicts the downstream distribution of tracer in channels which are not prismatic and also in the reach of channel where the one-dimensional dispersion equation is not valid. Second, it does not require that a hydrodynamic model be used to describe the flow distribution. The required flow distribution can be either synthesized or measured. The obvious disadvantage is that it is not capable of treating a nonuniform flow.

10.4 NUMERICAL DISPERSION

In the foregoing sections of this chapter, the concepts of turbulent diffusion and dispersion have been summarized from the viewpoint of rather idealized situations. In most of the cases treated, analytic solutions were available; however, in practice, ideal situations where analytic solutions are applicable are rare; in general, a numerical solution of the governing equations is required. The use of finite difference or finite element methods to solve mass transport problems results in a fictitious transport of mass which is termed numerical dispersion, as was noted above. Although a detailed treatment of numerical methods is beyond the scope of this discussion, some consideration is required if the problem of numerical dispersion is to be understood. Additional information regarding finite difference methods is available in Forsythe and Wasow (1960) and Street (1973).

In finite difference schemes, the partial differential equation to be solved is written in terms of a finite difference representation of each derivative. For example, consider the time-averaged one-dimensional dispersion equation for a nonprismatic channel or

$$A\frac{\partial C}{\partial t} + Q\frac{\partial C}{\partial x} = \frac{\partial}{\partial x}\left(EA\frac{\partial C}{\partial x}\right)$$

A grid system for the solution of this equation is shown in Fig. 10.14. In this equation, A, Q, and E are constant in time but variable in space. For the development of a finite difference scheme, it is convenient to represent the value of the tracer concentration at the ith longitudinal node and at the jth node in time as $C_{i,j}$. At time $t = 0$ or $j = 1$, values of C are specified for all values of i, and it is required that $C_{i,j}$ for $j \neq 1$ be determined for all i. In performing this task by finite differences, the terminology of explicit and implicit difference schemes arises. The phrase *explicit difference scheme* refers to a computational scheme

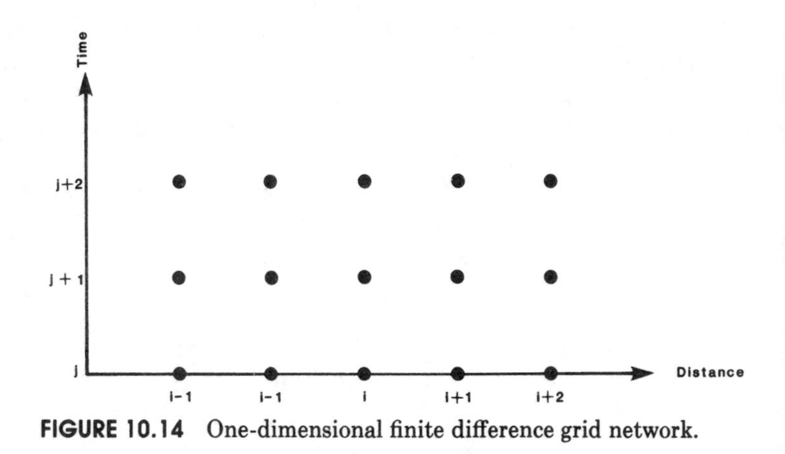

FIGURE 10.14 One-dimensional finite difference grid network.

in which values of $C_{i,j}$, $j \neq 1$ are determined from the solution of explicit equations; i.e., all derivatives are expressed in terms of known values except in the derivative describing the variation of C with time. In explicit difference schemes, the following expressions can be used:

1. *Backward Difference Operator*

$$\frac{\partial C}{\partial x} \simeq \frac{C_{i,j} - C_{i-1,j}}{\Delta x}$$

2. *Forward Difference Operator*

$$\frac{\partial C}{\partial x} \simeq \frac{C_{i+1,j} - C_{i,j}}{\Delta x}$$

3. *Central Difference Operator*

$$\frac{\partial C}{\partial x} \simeq \frac{C_{i+1,j} - C_{i-1,j}}{2\Delta x}$$

The phrase *implicit difference scheme* refers to a computational scheme in which the values of $C_{i,j}$ are determined by solving a system of simultaneous equations; i.e., the expressions for the derivatives are written in terms of unknown values. For example

$$\frac{\partial C}{\partial x} \simeq ½\left(\frac{C_{i+1,j+1} - C_{i-1,j+1}}{2\Delta x} + \frac{C_{i+1,j} - C_{i-1,j}}{2\Delta x}\right)$$

Explicit difference schemes would be preferred because of their computational simplicity if it were not for the fact that they are usually unstable from a numerical viewpoint unless a relatively small time step is used. Implicit schemes, while requiring the solution of a set of simultaneous equations, are stable from a numerical viewpoint, and a much larger time step can be used. The characteristic of both schemes which gives rise to numerical dispersion is that mass concentrations can exist only at nodes within the system (Fig. 10.14). Thus, if a mass is advected and dispersed only part of the distance between two nodes, some mass is returned numerically to the point of origin, and some of the mass is numerically advected and dispersed to the next node.

As an example of the significance of numerical dispersion in a specific situation, consider the following case of pure advection or

$$\frac{\partial C}{\partial t} = -\bar{u}\frac{\partial C}{\partial x}$$

Let the concentration of tracer at node (i, j) be 1, and zero elsewhere. If $\bar{u}\,\Delta t/\Delta x = 0.5$ and a backward difference operator is used, then

$$C_{i,j+1} = C_{i,j} - \frac{\bar{u}\,\Delta t}{\Delta x}(C_{i,j} - C_{i-1,j})$$

$$= 1 - 0.5(1 - 0) = 0.5$$

and
$$C_{i+1,j+1} = C_{i+1,j} - \frac{\overline{u}\,\Delta t}{\Delta x}(C_{i+1,j} - C_{i,j})$$
$$= 0 - 0.5(0 - 1) = 0.5$$

Thus, in this example after one time step, half of the original concentration of tracer remains at node i and half has moved to node $i + 1$. In fact, all the tracer should be located between these two nodes. At the beginning of the computation the variance of the concentration distribution was 0, while at the end of the computation the variance of the concentration distribution is

$$\sigma^2 = \int_{-\infty}^{\infty} (x - \overline{x})\epsilon\, dx = (-0.5\Delta x)^2 0.5 + (0.5\Delta x)^2 0.5 = 0.25(\Delta x)^2$$

If the effect of the numerical procedure on the distribution of the tracer is assumed to be equivalent to a diffusive process, then by Eq. (10.2.6)

$$\frac{d\sigma^2}{dt} = 2E'$$

where E' = a numerical dispersion coefficient and

$$E' = \frac{0.25(\Delta x)^2}{2\Delta t} = \frac{0.125(\Delta x)^2}{\Delta t}$$

This then is what is meant by the term *numerical dispersion.*

A variety of methodologies for dealing with the problem of numerical dispersion have been developed. First, if the numerical dispersion is small compared with the actual dispersion, then it can be ignored. Second, if E' can be accurately estimated, then the value of E can be adjusted downward so that the physically correct result is obtained. Third, numerical schemes which minimize the amount of numerical dispersion in a particular situation can be adapted (e.g., Thatcher and Harleman, 1972).

10.5 VERTICAL, TURBULENT DIFFUSION IN A CONTINUOUSLY STRATIFIED ENVIRONMENT

In Section 10.3 a theoretical expression which has been verified with both laboratory and field data for the vertical, turbulent diffusion coefficient was derived [Eq. (10.3.4)]. In a continuously, stably stratified flow the vertical diffusion of both momentum and mass is inhibited by the stratification, and Eq. (10.3.4) is not valid. In such a situation, the eddy viscosity (that is, the turbulent momentum diffusion coefficient) is larger than the eddy diffusivity (that is, the turbulent diffusion coefficient) of heat and mass. In this section, the problem of estimating values of the eddy viscosity and diffusivity in a continuously stratified flow is briefly treated; however, it must be emphasized that at the present time there does not exist an expression for either the eddy viscosity or the diffusivity which is universally considered valid.

Rossby and Montgomery (1935) proposed an equation relating the stratified vertical eddy viscosity for stratified flow ϵ_z^s to the corresponding value for homogeneous conditions, ϵ_z, of the form

$$\frac{\epsilon_z^s}{\epsilon_z} = (1 + \beta_{RM} Ri)^{-1} \tag{10.5.1}$$

where β_{RM} = a coefficient and Ri = gradient Richardson number, Eq. (1.2.2). In the development of Eq. (10.5.1) it was assumed that the change in kinetic energy per unit mass in going from a neutral or unstratified condition to a stably stratified condition was equal to the potential energy due to displacement over the mixing length from the equilibrium position with a different density. Kent and Pritchard (1957) also used a conservation of energy argument to develop an equation or

$$\frac{\epsilon_z^s}{\epsilon_z} = (1 + \beta_{KP} Ri)^{-2} \tag{10.5.2}$$

where β_{KP} = a coefficient. Holzman (1943) hypothesized a critical value of the gradient Richardson number above which turbulence to accomplish mixing existed, and this hypothesis resulted in

$$\frac{\epsilon_z^s}{\epsilon_z} = (1 - \beta_H Ri) \tag{10.5.3}$$

where β_H = a coefficient. Mamayev, in Anonymous (1974), argued that the ratio of ϵ_z^s to ϵ_z should decrease exponentially with increasing values of Ri or

$$\frac{\epsilon_z^s}{\epsilon_z} = e^{-\beta_M Ri} \tag{10.5.4}$$

where β_M = a coefficient. Munk and Anderson (1948) proposed a generalized form of the Rossby and Montgomery (1935) and Holzman (1943) equations or

$$\frac{\epsilon_z^s}{\epsilon_z} = (1 + \beta_{MA} Ri)^{\alpha_{MA}} \tag{10.5.5}$$

where β_{MA} and α_{MA} = coefficients. Equation (10.5.4) must be considered a completely empirical equation with two free coefficients. French (1979) proposed a second empirical relationship of the form

$$\epsilon_z^s = \beta_F \left(\frac{\epsilon_z}{1 + R_0} \right)^{\alpha_F} \tag{10.5.6}$$

where $\beta_F = \alpha_F$ = coefficients and

$$R_0 = \frac{gD\,(\Delta\rho/\rho)}{u_*^2}$$

Finally, Odd and Rodger (1978) developed a hybrid model based on the original hypothesis of Rossby and Montgomery (1935) or

1. If there is a significant peak in the vertical profile of Ri at a distance $z = z_0$ from the bottom boundary where Θ = peak gradient Richardson number, then

$$\frac{\epsilon_z^s}{\epsilon_z} = (1 + \beta_{OR}\Theta)^{-1} \qquad \text{for } \Theta \leq 1$$

and

$$\frac{\epsilon_z^s}{\epsilon_z} = (1 + \beta_{OR})^{-1} \qquad \text{for } \Theta > 1 \tag{10.5.7}$$

where β_{OR} = a coefficient.

OR

2. If there is no significant peak in the vertical profile of Ri, then

$$\frac{\epsilon_z^s}{\epsilon_z} = (1 + \beta_{OR}Ri)^{-1} \qquad \text{for } Ri \leq 1$$

and

$$\frac{\epsilon_z^s}{\epsilon_z} = (1 + \beta_{OR})^{-1} \qquad \text{for } Ri > 1 \tag{10.5.8}$$

Equations (10.5.7) and (10.5.8) are applied throughout the vertical dimension, but near the boundaries if $\epsilon_t^s > \epsilon_z$, then ϵ_z is used.

The problem with all the above methodologies is that in general they cannot be shown to be universally valid. Values for some of the coefficients used in the above equations are summarized in Table 10.7. With regard to the data summarized in this table, the following should be noted:

TABLE 10.7 Summary of coefficient values for turbulent vertical diffusion of momentum in continuously stratified channel flow

Equation	β	α	Source
10.5.1	2.5	—	Nelson (1972)
	5.0	—	Anonymous (1974)
	30.3	—	French and McCutcheon (1983)
10.5.2	9.8	—	French and McCutcheon (1983)
10.5.3	3.3	—	Nelson (1972)
10.5.4	0.4	—	Anonymous (1974)
10.5.5	10	−0.5	Munk and Anderson (1948)
	30	−0.5	Anonymous (1974)
10.5.6	0.31	0.747	French (1979)
	0.062	0.379	French and McCutcheon (1983)
10.5.7	140–180		Odd and Rodger (1978)

1. Nelson (1972) used published oceanographic, atmospheric, and pipe flow data for his analysis, and the same was true of the analysis by the Delft Hydraulics Laboratory (Anonymous, 1974). Thus, these investigators had no control over the quality of their data.

2. The data used by French (1979) were taken under laboratory conditions, but the flume used for these experiments had a small width-to-depth ratio, and the results may be unduly affected by this fact.

3. Odd and Rodger (1978) used field data from a reach of channel affected by the tide. The Odd and Rodger (1978) data set is perhaps the best data regarding the turbulent vertical diffusion of momentum under stratified conditions presently available.

4. French and McCutcheon (1983) used the Odd and Rodger (1978) data set for their analysis. The coefficient value determined in this work for Eq. (10.5.7) and (10.5.8) differs from that of Odd and Rodger (1978) because of a difference in the definition of goodness of fit.

5. In the past, Eq. (10.5.1) has been the most commonly used method of estimating ϵ_z^s (Nelson, 1972); however, the methods of Odd and Rodger (1978) and French and McCutcheon (1983) or French (1979) may be superior.

6. The Delft Hydraulics Laboratory (Anonymous, 1974) concluded that when $Ri < 0.7$ the scatter of the data available is so great that no best-fit equation can be selected.

A number of models for the eddy diffusivity in stratified flow have also been proposed. One of the most frequently used is

$$\frac{(\epsilon_z^s)^M}{\epsilon_z} = c(1 + \beta' Ri)^\alpha \tag{10.5.9}$$

where $(\epsilon_z^s)^M$ stratified flow, vertical eddy diffusivity, and c, β', and α = coefficients. Munk and Anderson (1948) estimated that $c = 1$, $\alpha = 1.5$, and $\beta' = 3.33$.

At this time it is appropriate to note that stratification apparently also acts to reduce the value of the turbulent transverse diffusion coefficient; however, the results presently available in this area (Sumer, 1976), are not yet extensive enough to warrant inclusion.

BIBLIOGRAPHY

Anonymous, "Sacramento River Water Pollution Survey," Bulletin No. 111, State of California, Department of Water Resources, Sacramento, 1962.

Anonymous, "Momentum and Mass Transfer in Stratified Flows," Report No. R880, Delft Hydraulics Laboratory, Delft, the Netherlands, December 1974.

Benjamin, J. R., and Cornell, C. A., *Probability, Statistics, and Decision for Civil Engineers,* McGraw-Hill Book Company, New York, 1970.

Elder, J. W., "The Dispersion of Marked Fluid in Turbulent Shear Flow," *Journal of Fluid Mechanics,* vol. 5, 1959, pp. 544–560.

Engmann, E. O., "Turbulent Diffusion in Channels with a Surface Cover," *Journal of Hydraulic Research,* International Association for Hydraulic Research, vol. 15, no. 4, 1977, pp. 327–335.

Fischer, H. B., "The Mechanics of Dispersion in Natural Streams," *Proceedings of the American Society of Civil Engineers, Journal of the Hydraulics Division.* vol. 93, no. HY6, November 1967, pp. 187–216.

Fischer, H. B., "Methods for Predicting Dispersion Coefficients in Natural Streams with Application to Lower Reaches of the Green and Duwamish Rivers," Geological Survey Professional Paper 582-A, U.S. Geological Survey, Washington, 1968a.

Fischer, H. B., "Dispersion Predictions in Natural Streams," *Proceedings of the American Society of Civil Engineers, Journal of the Sanitary Engineering Division,* vol. 94, no. SA6, October 1968b, pp. 927–943.

Fischer, H. B., "The Effect of Bends on Dispersion in Streams," *Water Resources Research,* vol. 5, no. 2, 1969, pp. 496–506.

Fischer, H. B., et al., *Mixing in Inland and Coastal Waters,* Academic Press, New York, 1979.

Forsythe, G. E., and Wasow, W. R., *Finite Difference Methods for Partial Differential Equations,* Wiley, New York, 1960.

French, R. H., "Vertical Mixing in Stratified Flows," *Proceedings of the American Society of Civil Engineers, Journal of the Hydraulics Division,* vol. 105, no. HY9, September 1979, pp. 1087–1101.

French, R. H., and McCutcheon, S. C., "Vertical Momentum Transfer in Continuously Stratified Channel Flow," Water Resources Center, Desert Research Institute, Las Vegas, Nev., 1983.

Glover, R. E., "Dispersion of Dissolved or Suspended Materials in Flowing Streams," Geological Survey Professional Paper 433-B, U.S. Geological Survey, Washington, 1964.

Godfrey, R. G., and Frederick, B. J., "Dispersion in Natural Streams," Geological Survey Professional Paper 433-K, U.S. Geological Survey, Washington, 1970.

Harden, T. O., and Shen, H. T., "Numerical Simulation of Mixing in Natural Rivers," *Proceedings of the American Society of Civil Engineers, Journal of the Hydraulics Division,* vol. 105, no. HY4, April 1979, pp. 393–408.

Holley, E. R., "Unified View of Diffusion and Dispersion," *Proceedings of the American Society of Civil Engineers, Journal of the Hydraulics Division,* vol. 95, no. HY2, March 1969, pp. 621–631.

Holzman, B., "The Influence of Stability on Evaporation," *Boundary Layer Problems in the Atmosphere and Ocean,* W. G. Valentine (ed.), vol. XLIV, article 1, 1943, pp. 13–18.

Kent, R. E., and Pritchard, D. W., "A Test of Mixing Length Theories in a Coastal Plain Estuary," *Journal of Marine Research,* vol 1, 1957, pp. 456–466.

Kilpatrick, F. A., Martens, L. A., and Wilson, J. F., Jr., "Measurement of Time of Travel and Dispersion by Dye Tracing," *Techniques of Water-Resources Investigations of the United States Geological Survey,* chapter A9, book 3, U.S. Geological Survey, Washington, 1970.

Krenkel, P. A., "Turbulent Diffusion and the Kinetics of Oxygen Absorption," Ph.D. thesis, University of California, Berkeley, 1960.

Krenkel, P. A., and Novotny, V., *Water Quality Management,* Academic Press, New York, 1980.

Lau, Y. L., and Krishnappen, B. G., "Transverse Dispersion in Rectangular Channels," *Proceedings of the American Society of Civil Engineers, Journal of the Hydraulics Division,* vol. 103, no. HY10, October, 1977, pp. 1173–1189.

Levenspiel, O., and Smith, W. K., "Notes on the Diffusion Type Model for the Longitudinal Mixing of Fluid in Flow," *Chemical Engineering Science,* vol. 6, 1957, pp. 227–233.

McQuivey, R. S., and Keefer, T., "Simple Method for Predicting Dispersion in Streams," *Proceedings of the American Society of Civil Engineers, Journal of the Environmental Engineering Division,* vol. 100, no. EE4, August, 1974, pp. 997–1011.

Munk, W. H., and Anderson, E. R., "Notes on the Theory of the Thermocline," *Journal of Marine Research,* vol. 1, 1948, p. 276.

Nelson, J. E., "Vertical Turbulent Mixing in Stratified Flow—A Comparison of Previous Experiments," Report No. WHM-3, Hydraulic Engineering Laboratory, University of California, Berkeley, December 1972.

Odd, N. V. M., and Rodger, J. G., "Vertical Mixing in Stratified Tidal Flows," *Proceedings of the American Society of Civil Engineers, Journal of the Hydraulics Division,* vol. 104, no. HY3, March 1978, pp. 337–351.

Owens, M., Edwards, R. W., and Gibbs, J. W., "Some Reaeration Studies in Streams," *Air-Water Pollution Institute Journal,* vol. 8, 1964, pp. 469–486.

Rossby, C. G., and Montgomery, R. B., "The Layer of Friction Influence in Wind and Ocean Currents," *Papers in Physical Oceanography and Meteorology,* vol. 3, no. 3, 1935.

Schuster, J. C., "Canal Discharge Measurements with Radioisotopes," *Proceedings of the American Society of Civil Engineers, Journal of the Hydraulics Division,* vol. 91, no. HY2, March 1965, pp. 101–124.

Street, R. L., *Analysis and Solution of Partial Differential Equations,* Brooks/Cole, Monterey, Calif., 1973.

Sumer, S. M., "Transverse Dispersion in Partially Stratified Tidal Flow," Report No.

WHM-20, Hydraulic Engineering Laboratory, University of California, Berkeley, May 1976.

Taylor, G. I., "Dispersion of Soluble Matter in Solvent Flowing Slowly through a Tube," *Proceedings of the Royal Society of London,* series A, vol. 219, August 25, 1953, pp. 186–203.

Taylor, G. I., "The Dispersion of Matter in a Turbulent Flow through a Pipe," *Proceedings of the Royal Society of London,* series A, vol. 223, May 20, 1954, pp. 446–468.

Thatcher, M., and Harleman, D. R. F., "A Mathematical Model for the Prediction of Unsteady Salinity Intrusion in Estuaries," R. M. Parsons Laboratory Report No. 144, Massachusetts Institute of Technology, Cambridge, Mass., 1972.

Thomas, I. E., "Dispersion in Open Channel Flow," Ph.D. thesis, Northwestern University, Evanston, Ill., 1958.

Wilson, J. F., "Fluormetric Procedures for Dye Tracing," *Techniques of Water-Resources Investigations of the U.S. Geological Survey,* chapter A12, book 3, U.S. Geological Survey, Washington, 1968.

Yotsukura, N., Fischer, H. B., and Sayre, W. W., "Measurements of Mixing Characteristics of the Missouri River between Sioux City, Iowa and Portsmouth, Nebraska," Water Supply Paper 1899-G, U.S. Geological Survey, Washington, 1970.

ELEVEN

Turbulent, Buoyant Surface Jets and Associated Phenomena

SYNOPSIS

In this chapter, the topic of turbulent, buoyant surface jets in open-channel flow is treated. In the initial section of the chapter, the fundamental principles of jet analysis such as the scaling of partial differential equations and similarity solutions are introduced as solution techniques. These fundamental solution techniques are demonstrated with applications to the problems of plane or slot jets and axisymmetric jets discharging into an unbounded environment. In subsequent sections, the topics discussed are: entrainment; the length of the zone of flow establishment; and the behavior of a two-dimensional, buoyant surface jet in a cross flow. A numerical scheme for the solution of the two-dimensional jet problem is developed and discussed along with the limited laboratory and field results available for estimating the required coefficient values.

In the final section of the chapter, a one-dimensional analysis of the upstream cooling water wedge problem is discussed. *Note:* This problem usually occurs in conjunction with a buoyant surface jet. Again, theoretical, laboratory, and field results appropriate to the solution of the problem are presented.

11.1 INTRODUCTION

In a dilute form many pollutants are both harmless and ubiquitous. Thus, when a pollutant is discharged into a receiving water there is usually a legal or administrative requirement that significant dilution be achieved as rapidly as possible. One of the many techniques for achieving rapid dilution is the turbulent jet which entrains and mixes large volumes of receiving water with the pollutant discharge.

By definition, the simple jet is a turbulent flow pattern generated by a continuous source of momentum. A closely related phenomenon is the plume; however, the plume has no initial momentum and is generally considered to be a buoyancy-driven phenomenon. Although turbulent jets and plumes can be used to enhance dilution in many different environments, this chapter will consider only turbulent jets discharging into environments similar to rivers and canals.

The behavior of a turbulent jet depends on: (1) jet parameters, (2) environmental parameters, and (3) geometric parameters. The terminology *jet parameters* includes variables such as the initial jet velocity distribution and turbulence level, jet mass and momentum flux, and the density of the jet fluid. Environmental parameters which must be considered are the ambient conditions in the receiving water such as density, density stratification, turbulence level, and the presence of velocity gradients. The geometrical parameters which must be considered include the presence and proximity of boundaries, the orientation of the jet with respect to the boundaries, and the position of the jet in the vertical dimension. In any application, it may be necessary to consider all the above factors; however, this treatment will begin by considering the elementary principles of fluid mechanics which are involved.

11.2 BASIC MECHANICS OF TURBULENT JETS

The turbulent flow phenomena which have been discussed in the foregoing chapters of this book have concerned fluid motion constrained by one or more boundaries. In such cases, the turbulence is termed *wall turbulence*. In the case of an elementary jet there are essentially no boundaries, and the turbulence is called *free turbulence*. In Fig. 11.1 an elementary jet is schematically defined, and with regard to this figure the following assumptions are made:

1. The diameter or width of the jet, $2b$, is small relative to the longitudinal distance x
2. The velocity gradient in the x coordinate direction is assumed to be small relative to the velocity gradient in the y coordinate direction
3. The molecular stresses are assumed small relative to the Reynolds or turbulent stresses

With these assumptions it can be shown (see, for example, Schlichting, 1968) that the equations governing turbulent jet flow are

$$u \frac{\partial u}{\partial x} + v \frac{\partial u}{\partial y} = \frac{1}{\rho} \frac{\partial \tau}{\partial y} \tag{11.2.1}$$

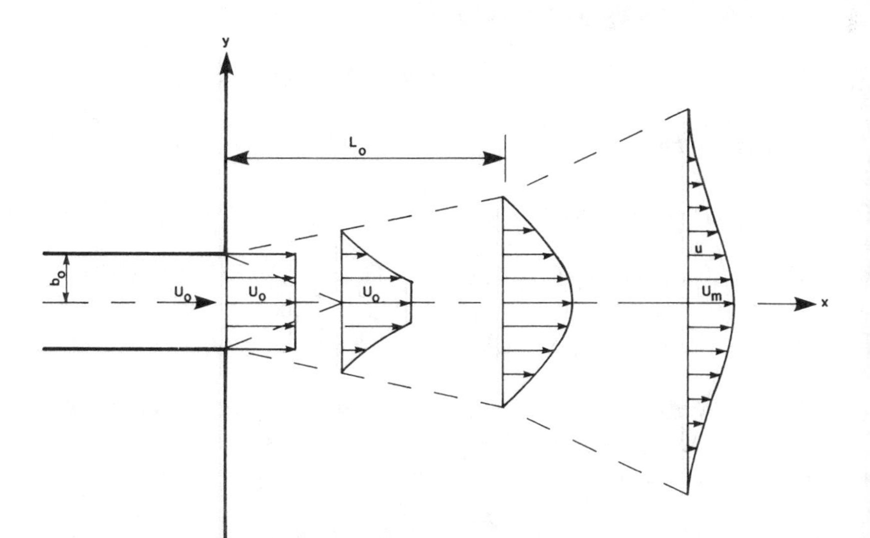

FIGURE 11.1 Schematic for a plane jet.

and

$$\frac{\partial u}{\partial x} + \frac{\partial v}{\partial y} = 0 \tag{11.2.2}$$

where $\tau = \eta\, \partial u/\partial y$ = turbulent shear stress
η = eddy viscosity
ρ = fluid density
u = velocity in x coordinate direction
v = velocity in y coordinate direction

Also in Fig. 11.1, the distance L_0 is the length of the zone of flow establishment or the core length. Within the distance L_0, the jet centerline velocity u_0 is constant. For distances $x > L_0$, the centerline velocity u_m is less than U_0. The dashed lines in Fig. 11.1 define the nominal transverse boundaries of the jet; i.e., these lines define the points at which the jet velocity becomes arbitrarily small.

At this point, assume that the jet is a slot jet with a slot height of $2b_0$. Integrating Eq. (11.2.1) with respect to y yields

$$\underset{(1)}{\rho \int_{-\infty}^{\infty} u \frac{\partial u}{\partial x}\, dy} + \underset{(2)}{\rho \int_{-\infty}^{\infty} v \frac{\partial u}{\partial y}\, dy} = \underset{(3)}{\int_{-\infty}^{\infty} \frac{\partial \tau}{\partial y}\, dy} \tag{11.2.3}$$

The term indicated as (1) in Eq. (11.2.3) can be rearranged to yield

$$\rho \int_{-\infty}^{\infty} u \frac{\partial u}{\partial x}\, dy = \frac{\rho}{2} \frac{\partial}{\partial x} \int_{-\infty}^{\infty} u^2\, dy \tag{11.2.4}$$

With regard to the term indicated as (2) in Eq. (11.2.3), substitution of variables and integration by parts yields

$$\int_{-\infty}^{\infty} v \frac{\partial u}{\partial y}\, dy = uv \Big|_{-\infty}^{\infty} - \int_{-\infty}^{\infty} u \frac{\partial v}{\partial y}\, dy \tag{11.2.5}$$

From Eq. (11.2.2)

$$\frac{\partial v}{\partial y} = -\frac{\partial u}{\partial x} \tag{11.2.6}$$

Then, substitution of Eq. (11.2.4) to (11.2.6) in Eq. (11.2.3) yields

$$\rho \frac{\partial}{\partial x} \int_{-\infty}^{\infty} u^2\, dy + \rho\, uv \Big|_{-\infty}^{\infty} = \tau \Big|_{-\infty}^{\infty} \tag{11.2.7}$$

The boundary conditions for Eq. (11.2.7) which are consistent with the assumptions regarding a simple jet are

$$\text{As } y \to \pm\infty \begin{cases} u \to 0 \\ \dfrac{\partial u}{\partial y} \to 0 \end{cases} \tag{11.2.8}$$

With these boundary conditions, the terms $\tau \mid_{-\infty}^{\infty}$ and $uv \mid_{-\infty}^{\infty}$ in Eq. (11.2.7) become

$$\tau \Big|_{-\infty}^{\infty} = \eta \frac{\partial u}{\partial y} \Big|_{-\infty}^{\infty} = 0 \tag{11.2.9}$$

and

$$uv \Big|_{-\infty}^{\infty} = 0 \tag{11.2.10}$$

Substitution of Eq. (11.2.9) and (11.2.10) in Eq. (11.2.7) yields

$$\rho \frac{\partial}{\partial x} \int_{-\infty}^{\infty} u^2 \, dy = 0$$

or

$$\rho \int_{-\infty}^{\infty} u^2 \, dy = C = \int_{-\infty}^{\infty} (\rho u) u \, dy \tag{11.2.11}$$

where C = a constant of integration. In Eq. (11.2.11) the quantity of $(\rho u) u \, dy$ is the total flux of momentum at any longitudinal section of the jet. Equation (11.2.11) claims that the flux of momentum is constant and independent of the longitudinal coordinate x. This result derives from the tacit assumption in Eq. (11.2.1) that the pressure is constant; hence, the net force on the control volume encompassing the jet is zero. The constant of integration in Eq. (11.2.11) can be evaluated by defining the initial flux of momentum into the system. At $x = 0$

$$\dot{M} = 2 b_0 \rho U_0^2$$

where $\dot{M}$ = momentum flux. Then, since the flux of momentum is constant, Eq. (11.2.11) becomes

$$\int_{-\infty}^{\infty} (\rho u) u \, dy = 2 b_0 \rho U_0^2 \tag{11.2.12}$$

Additional information regarding the behavior of a slot jet discharging into a quiescent environment can be gained by scaling the governing equations (see Chapter 1). Assume that the effective jet width b and the centerline velocity of the jet in the fully developed region, that is, $x > L_0$, can be expressed in terms of the longitudinal coordinate distance or

$$b \sim x^{\theta} \tag{11.2.13}$$

and

$$u_m \sim x^{-\phi} \tag{11.2.14}$$

Given these assumptions, the first term of Eq. (11.2.1) can be evaluated on the basis of order of magnitude as follows:

$$u \frac{\partial u}{\partial y} \sim o\left[\frac{u_m^2}{x}\right] \sim o\left[\frac{x^{-2\phi}}{x}\right] \sim o[x^{-2\phi-1}] \tag{11.2.15}$$

where $o[\ ]$ indicates the order-of-magnitude concept (Chapter 1). From Eq. (11.2.2)

$$v = -\int \frac{\partial u}{\partial x}\,dy \sim o\left[\frac{u_m b}{x}\right]$$

and thus the second term in Eq. (11.2.1) becomes

$$v\frac{\partial u}{\partial y} \sim o\left[\frac{u_m b}{x}\frac{u_m}{b}\right] \sim o\left[\frac{u_m^2}{x}\right] \sim o\,[x^{-2n-1}] \tag{11.2.16}$$

If it is assumed that

$$\frac{\tau}{\rho} \sim o[u_m]^2$$

the third term in Eq. (11.2.1) is

$$\frac{\partial}{\partial y}\left(\frac{\tau}{\rho}\right) \sim o\left[\frac{u_m^2}{b}\right] \sim o[x^{-2\phi-\theta}] \tag{11.2.17}$$

Then, in terms of an order of the magnitude evaluation, Eq. (11.2.1) is

$$o[x^{-2\phi-1}] + o[x^{-2\phi-1}] = o[x^{-2\phi-\theta}] \tag{11.2.18}$$

If each term in Eq. (11.2.18) is equally important and therefore has the same order of magnitude, then

$$-2\phi - 1 = 2\phi - \theta$$

and hence

$$\theta = 1 \tag{11.2.19}$$

From the viewpoint of order of magnitude, then

$$\int \rho u^2\,dy \sim \int \rho u_m^2\,dy \sim o[u_m^2 b] \sim o[x^{-2\phi+\theta}]$$

If the flux of momentum is independent of the longitudinal coordinate distance, then

$$-2\phi + \theta = 0$$

Substituting $\theta = 1$ from Eq. (11.2.19) yields

$$\phi = \frac{\theta}{2} = \frac{1}{2} \tag{11.2.20}$$

The original assumptions regarding u_m and b for a slot jet can be quantified as

$$u_m \sim x^{-\phi} = \frac{1}{\sqrt{x}} \tag{11.2.21}$$

and
$$b \sim x \tag{11.2.22}$$

Although the foregoing discussion provides a qualitative description of jet behavior, it does not provide the equations necessary to describe in a quantitative fashion the size of the jet, the velocity distribution, and the entrainment of ambient fluid by the jet. Assume that the solution of the equations governing the behavior of a jet belong to the class of partial differential equation solutions known as *similarity solutions* (see, for example, Schlichting, 1968), or

$$\frac{u}{u_m} = f\left(\frac{y}{x}\right) = f(\omega)$$

where u = jet velocity at any location y for a specified value of y (Fig. 11.1) and $\omega = y/x$. Solutions of Eqs. (11.2.1) and (11.2.2) are then possible under the assumption of the existence of similarity solutions if a functional form for u/u_m can be determined. Assume

$$\frac{u}{u_m} = f(\omega) = \exp\left(-\frac{y^2}{2C^2x^2}\right) \tag{11.2.23}$$

where C = a constant which must be determined by experiment.

Note: While the velocity distribution assumed in Eq. (11.2.23) is gaussian, any other functional relationship for u/u_m could have been assumed; however, the validity of the solution depends on the validity of the assumed velocity distribution.

Rearrangement of Eq. (11.2.23) and substitution of Eq. (11.2.12) yields

$$\int_{-\infty}^{\infty} \rho u^2\, dy = \int_{-\infty}^{\infty} \rho u_m^2 f^2(\omega)\, dy = 2\rho U_0^2 b_0$$

Substituting in this equation $dy = xd$ yields

$$u_m^2 x I_2 = 2U_0^2 b_0 \tag{11.2.24}$$

where $I_2 = \int_{-\infty}^{\infty} f^2(\omega)\, d\omega$. Rearranging Eq. (11.2.24) yields

$$\frac{u_m}{U_0} = \left(\frac{2b_0}{xI_2}\right)^{1/2} \tag{11.2.25}$$

The length of the core zone or zone of flow establishment can then be determined by noting that within this zone, by definition,

$$\frac{u_m}{U_0} = 1$$

Then from Eq. (11.2.25)

$$L_0 = \frac{2b_0}{I_2} \tag{11.2.26}$$

For $x > L_0$, the discharge per unit width of the jet can be estimated by integrating the local velocity distribution or

$$Q = \int_{-\infty}^{\infty} u\, dy = \int_{-\infty}^{\infty} uf(\omega)x\, d\omega$$

Substituting Eq. (11.2.23) in the above equation and integrating yields

$$Q = U_0 x \left(\frac{2b_0}{xI_2}\right)^{1/2} \int_{-\infty}^{\infty} f(\omega)\, d\omega = U_0 x \left(\frac{2b_0 I_1^2}{xI_2}\right) \tag{11.2.27}$$

where $I_1 = \int_{-\infty}^{\infty} f(\omega)\, d\omega$. Then the ratio of the flow at any longitudinal distance x to the initial flow is

$$\frac{Q}{Q_0} = \frac{U_0[(2b_0 x I_1^2/xI_2)]}{2U_0 b_0} = \left(\frac{xI_1^2}{2b_0 I_2}\right)^{1/2} \tag{11.2.28}$$

In the case of a jet issuing from a circular hole into a quiescent fluid and entraining the ambient fluid, the equations governing the motion of the jet are

$$v_r \frac{\partial v_z}{\partial r} + v_z \frac{\partial v_z}{\partial z} = \frac{1}{\rho r}\frac{\partial(r\tau)}{\partial r} \tag{11.2.29}$$

and

$$\frac{1}{r}\frac{\partial}{\partial r}(rv_r) + \frac{\partial v_z}{\partial z} = 0 \tag{11.2.30}$$

where r = radial coordinate direction (Fig. 11.2)
z = coordinate direction aligned with axis of jet
v_r and v_z = velocities in r and z coordinate directions, respectively

Equations (11.2.29) and (11.2.30) are essentially Eqs. (11.2.1) and (11.2.2) transformed into a radial coordinate system. Also define

$$\tau = \eta \frac{\partial v_z}{\partial r} \tag{11.2.31}$$

and

$$\frac{\eta}{\rho} = \ell^2 \frac{\partial v_z}{\partial r} \tag{11.2.32}$$

where ℓ = the mixing length (see Chapter 1). The boundary conditions are:

As r → ∞:

$$v_z = 0 \tag{11.2.33}$$

and

As r → 0:

$$v_r \to 0 \quad \text{and} \quad \frac{\partial v_z}{\partial r} = 0 \tag{11.2.34}$$

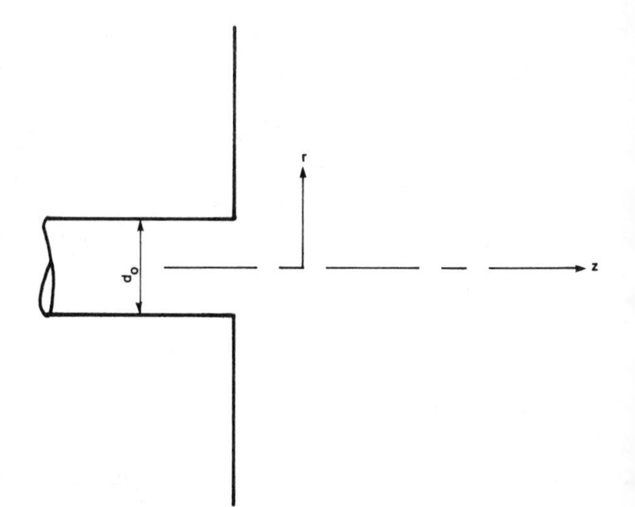

FIGURE 11.2 Coordinate system definition for axisymmetric jet.

It is further assumed that the jet is axisymmetric and can be described at any distance z by the local jet diameter d and the centerline velocity v_{zm}.

As before, assume

$$d \sim z^{\theta} \tag{11.2.35}$$

and

$$v_{zm} \sim z^{-\phi} \tag{11.2.36}$$

An order of magnitude analysis similar to that described for the slot jet would yield

$$\theta = 1$$

and

$$\phi = 1$$

Thus, in the case of an axisymmetric jet issuing from a circular hole into a quiescent fluid

$$d \sim z \tag{11.2.37}$$

and

$$v_{zm} \sim \frac{1}{z} \tag{11.2.38}$$

The conditions implied by Eqs. (11.2.37) and (11.2.38) are first, the Reynolds number is constant throughout the jet; i.e.,

$$\mathbf{R} = \frac{vD}{\nu} = \frac{(1/z)z}{\nu} = \frac{1}{\nu}$$

where $\mathbf{R}$ = Reynolds number and ν = kinematic viscosity, and the second, eddy viscosity defined by Eq. (11.2.32) is constant or

$$\epsilon = \frac{\eta}{\rho} = \ell^2 \frac{\partial v_z}{\partial r} = o\left[d^2 \frac{v_{zm}}{d}\right] = o[z^{m-n}] = o[z^0]$$

where ϵ = kinematic eddy viscosity.

Equations (11.2.29) and (11.2.30) can then be solved by assuming the existence of similarity solutions with the velocity distribution given by

$$\frac{v_z}{v_{zm}} = \left(1 + \frac{v_{zm} r^2}{8\epsilon z}\right) \tag{11.2.39}$$

The velocity distribution defined by Eq. (11.2.39) has been found to be accurate when ϵ is estimated by

$$\epsilon = 0.00196 z v_{zm} \tag{11.2.40}$$

Substitution of Eq. (11.2.40) in Eq. (11.2.39) yields

$$\frac{v_z}{v_{zm}} = \left(1 + \frac{r^2}{0.016z^2}\right)^{-2} \tag{11.2.41}$$

The longitudinal variation of the centerline velocity then given by

$$\frac{v_{zm}}{U_0} = 6.4 \frac{d_0}{z} \tag{11.2.42}$$

where U_0 = initial jet velocity and d_0 = initial jet diameter. It must be noted at this point that the longitudinal distance z is measured from the geometric origin of similarity which is approximately $0.6d_0$ from the actual origin of the jet (Fig. 11.2, and Daily and Harleman, 1973). For

$$z > 7d_0 \tag{11.2.43}$$

the jet is considered fully developed. The volumetric flow rate at any distance z is

$$Q = \int v_z 2\pi r\, dr = \int v_{zm} \left(1 + \frac{r^2}{0.016z^2}\right)^{-2} 2\pi r\, dr \tag{11.2.44}$$

Combining Eqs. (11.2.40) and (11.2.42) yields an expression for the eddy viscosity in terms of the initial jet velocity and diameter or

$$\epsilon = 0.13\, U_0 d_0$$

Substitution of this equation for ϵ in Eq. (11.2.44) yields

$$Q = 8\pi\epsilon z \tag{11.2.45}$$

The ratio of the flow at any longitudinal distance z to the initial flow rate is

$$\frac{Q}{Q_0} = 0.42 \frac{z}{d_0} \tag{11.2.46}$$

At this point a number of theoretical results regarding entrainment can be considered. For the plane or slot jet, Albertson et al. (1950) demonstrated that the velocity distribution specified by Eq. (11.2.23) is well correlated with measured velocity distributions for $C = 0.109$. For this value of C, $I_1 = 0.272$ and $I_2 = 0.192$, and Eq. (11.2.28) becomes

$$\frac{Q}{Q_0} = 0.62 \left(\frac{x}{2b_0}\right)^{1/2}$$

or

$$Q = \frac{0.62 Q_0}{\sqrt{2b_0}} \sqrt{x} \tag{11.2.47}$$

The rate at which the volumetric discharge of the jet occurs with the longitudinal distance is

$$\frac{dQ}{dx} = \frac{1}{2}\left(\frac{0.62Q_0}{\sqrt{2b_0}}\right)\frac{1}{\sqrt{x}} = \frac{0.31\sqrt{2b_0}}{\sqrt{x}} U_0 \tag{11.2.48}$$

For $C = 0.109$ and $x > L_0$, it can be shown that

$$u_m = 2.28\, U_0 \left(\frac{2b_0}{x}\right)^{1/2}$$

and therefore Eq. (11.2.48) can also be written as

$$\frac{dQ}{dx} = 0.136 u_m \tag{11.2.49}$$

Therefore, for the plane or slot jet the rate of increase of volumetric discharge with longitudinal distance is directly proportional to the centerline velocity or inversely proportional to the square root of the longitudinal distance.

In the case of an axisymmetric jet

$$Q = 8\pi\epsilon z = 8\pi(0.00196)v_z z^2$$

or

$$= 0.315 U_0 d_0 z$$

Therefore, the rate of increase of volumetric discharge with longitudinal distance is given by

$$\frac{dQ}{dz} = 0.315 U_0 d_0 \tag{11.2.50}$$

Thus, for an axisymmetric jet, the rate of increase of volumetric discharge with longitudinal distance is constant.

The importance of the foregoing material regarding simple jets in idealized environments is that the analysis of real jets in natural environments closely parallels the analysis of these simple jets.

11.3 BUOYANT SURFACE JETS

In the past three decades, there have been numerous studies of turbulent jets entering a variety of quasinatural environments. For example, Fan (1967), Abraham (1970), Cederwall and Brooks (1971), Stolzenbach and Harleman (1973), Stefan et al. (1975), and Motz and Benedict (1971) have all performed analytical and laboratory studies of the trajectories and dilution of buoyant jets entering flowing environments. Cederwall (1967), Abraham (1967), and Turner (1966) have examined negatively buoyant jets entering stagnant environments, and Anderson et al. (1973) examined negatively buoyant jets in flowing environments. Common to all these investigations are a number of basic assumptions.

1. The fluids involved are incompressible.
2. The fluids are fully turbulent, and there is no Reynolds number dependence. Further, molecular diffusion is negligible relative to turbulent diffusion.
3. Relative to transverse or lateral diffusion, longitudinal diffusion is negligible.
4. The variations in fluid density within the flow field are such that the density effects on the inertial terms in the governing equations can be ignored. However, terms in the governing equations which involve gravity must include density effects. Fan (1967) noted that the assumption of a small density variation within the flow field, also known as the Boussinesq assumption, implies that the conservation of mass flux principle can be approximated by the conservation of volume flux principle.
5. Velocity, relative buoyancy, and tracer concentration profiles are similar in consecutive transverse sections of the jet in the zone of established flow.

In addition to the foregoing assumptions, four additional concepts and definitions are crucial to the analysis and understanding of turbulent jets. First, jets are usually classified as buoyant, indicating that the jet fluid is less dense than the fluid into which the jet is discharging; negatively buoyant, indicating that the jet fluid is denser than the fluid into which the jet is discharging; or neutrally buoyant, indicating that the jet fluid has the same density as the fluid into which it discharges. Jets may further be classified by their geometric point of origin, for example, surface jets.

Second, surface jets may be classifed as either two-or-three-dimensional. A two-dimensional jet is one in which only the width increases along the centerline of the jet, while in a three-dimensional jet both the width and depth of the jet increase along the centerline. As noted by Motz and Benedict (1971) two forces are primarily responsible for the spreading of a jet in the vertical dimension. The difference between the axial velocity of the jet and the ambient velocity of flow creates shear which results in both the lateral and vertical spreading of the jet. When buoyant forces dominate, vertical spreading is suppressed.

Third, Albertson et al. (1950) have demonstrated that a zone of flow establishment must exist beyond the jet efflux section. With reference to Fig. 11.1, the fluid discharged from an opening in the boundary may be assumed to have a rather uniform velocity distribution. When the jet enters the ambient stream, there is by necessity a significant velocity discontinuity between the jet and the surrounding fluid. This velocity discontinuity results in a region of high shear which, in turn, causes lateral mixing. Thus, while the jet gradually decelerates, the surrounding fluid is accelerated and entrained into the jet. By definition, the limit of the zone of flow establishment is reached when the mixing region penetrates the centerline of the jet. Albertson et al. (1950) reported that for a three-dimensional submerged jet discharging into a stagnant environment, the limit of the zone of flow establishment is given by

$$\frac{L}{d_0} = 6.2 \tag{11.3.1}$$

where L = distance from the boundary opening to the limit of the zone of flow establishment and d_0 = initial diameter of the discharging jet. For a jet discharging in a cross flow, the variable L is replaced by the variable S_e which is defined in Fig. 11.3. Fan (1967), using the data of other investigators, developed

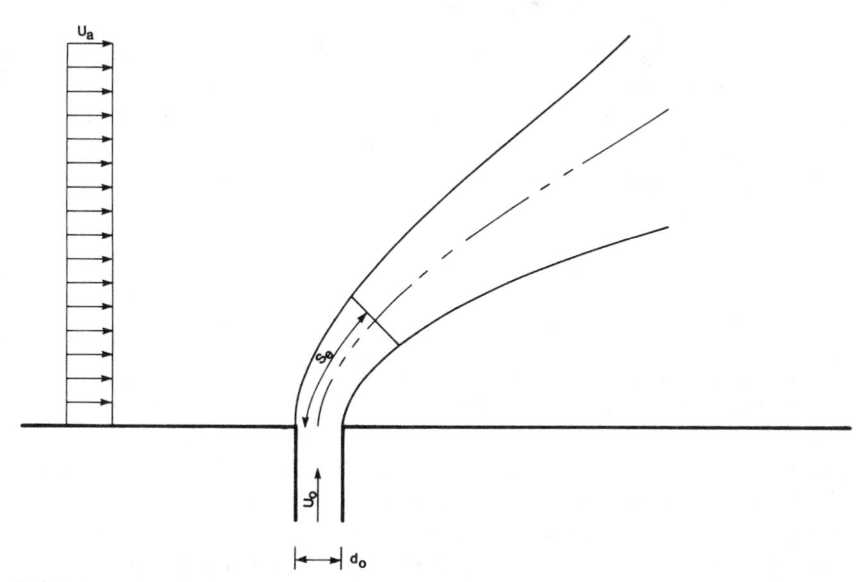

FIGURE 11.3 Schematic definition of S_e.

a plot of S_e/d_0 versus U_0/U_a which defines the zone of flow establishment for a buoyant jet where U_a = ambient fluid velocity and, as before, U_0 = the initial jet velocity. Parker and Krenkel (1969) gave the equation of the line developed by Fan as

$$\frac{S_e}{d_0} = 6.2 \exp\left(-3.22 \frac{U_a}{U_0}\right) \tag{11.3.2}$$

Anderson et al. (1973) performed a least squares fit of all the data available to them in approximately 1973 and concluded

$$\frac{S_e}{d_0} = 5.91 \exp\left(-2.57 \frac{U_a}{U_0}\right) \tag{11.3.3}$$

Equations (11.2.3) and (11.2.2) are very similar, and either can be used to estimate a value of the variable S_e.

Fourth, the concept of entrainment is crucial to most jet analyses. Morton et al. (1956) were perhaps the first researchers to propose an equation to explain the dilution of a jet by entrainment. These investigators asserted that

$$\frac{dQ}{dx} = 2\pi\alpha u b \tag{11.3.4}$$

where dQ/dx = rate of change of jet volume flux in direction of jet flow
u = characteristic jet velocity
b = characteristic jet length usually defined by width of assumed transverse velocity distribution of jet
α = entrainment coefficient

In their analysis, Morton et al. (1956) concluded that α was a constant and had a value of approximately 0.093. Morton (1961) subsequently concluded that the structure of the turbulence within a jet and the rate of entrainment at the edges of the jet depended only on the average density and velocity differences between the jet axis and the ambient fluid. Fan (1967) used this concept in his analysis of a turbulent round jet entering a cross flow and hypothesized

$$\frac{dQ}{dx} = 2\pi\alpha b\,|\Delta\mathbf{u}| \tag{11.3.5}$$

where $|\Delta\mathbf{u}|$ = magnitude of the velocity difference between the jet and the ambient fluid. The boldface symbol is introduced to indicate the vector nature of the velocities involved. Fan (1967) also assumed that α was constant. In a subsequent investigation, Abraham (1965) concluded that the entrainment coefficient α was not a constant and introduced a new parameter which related the rate at which work was done by the turbulent shear per unit time in a layer of incremental thickness at some level per rate of vertical flow. Fan and Brooks (1966), in a discussion of this investigation, concluded that the rate of entrainment must be proportional to the local characteristic velocity and the radius of the jet. Thus, Fan and Brooks (1966) also recognized that the entrainment coefficient was not a constant but was perhaps a function of the local buoyancy of the jet.

Modern electric generating stations require the use of enormous quantities of water for condenser cooling. If the generating station has an overall efficiency in the range of 30 to 49 percent, then the condenser cooling water will carry approximately twice as much energy away from the station as is transmitted away in high-voltage lines. Generating stations with open cooling water systems draw cooling water from rivers, lakes, and oceans; use the water once; and discharge the water, at an elevated temperature, back into the environment. Thus, it is not uncommon to find buoyant surface jets downstream of electric generating stations built on rivers. In general, the analysis of such jets follows the classic, semiempirical jet theory which divides the jet into a region of flow establishment and a fully established flow region. If such a methodology is followed, then the downstream temperature predictions will be in error for the following reasons: (1) The fundamental jet equations upon which the models are based involve assumptions which are in error, e.g., similarity of temperature and velocity profiles; (2) the coefficients used in the fundamental equations are poorly defined; and (3) the correspondence between the idealized discharge situation assumed for the development of their models—usually a long, straight reach of channel with no islands—and the actual situation to be analyzed is usually very poor. Even given the severe limitations of the models currently available for simulating buoyant surface jets, the problem is of such importance that even idealized solutions are valuable.

Zone of Flow Establishment

The variously termed outlet region, core region, or zone of flow establishment can be defined in several different ways:

1. The zone of flow establishment terminates where the time-averaged temperature on the axis of the jet reaches an arbitrary value relative to the outlet temperature. Stefan et al. (1975) asserted that the 90 percent level of time-averaged water temperature measured along the centerline trajectory was a realistic choice to define the end of the zone of flow establishment.

2. It can be arbitrarily decided that the zone of flow establishment terminates where significant changes in the level of turbulence intensity in the profiles of temperature or velocity occur along the axis of the jet.

3. On a log-log plot of the jet centerline surface temperature versus axial distance, a line can be extrapolated backward from the zone of fully established flow to the outlet temperature level to define the zone of flow establishment. Stefan et al. (1975) noted that in the case of buoyant surface jets, the imple-

mentation of this methodology was not trivial because the data do not usually plot as a straight line.

Stefan et al. (1975), using laboratory data developed for a rectangular channel discharging a jet into a deep body of water, developed an empirical equation relating the length of the zone of flow establishment to the excess temperature ratio, the aspect ratio of the discharge channel, the cross flow velocity ratio, and the densimetric Froude number of the outlet or

$$\frac{S_e}{y_o} = (16.0 - 12.8T)^{1.2} A^{(0.85-0.44T)} \exp(-0.9R) \left(1 + \frac{0.5\mathbf{F}_o - 1.5}{\exp(0.4\mathbf{F}_o)}\right) \tag{11.3.6}$$

In this equation with reference to Fig. 11.4, y_o = depth of flow in the rectangular outlet channel, T = excess temperature ratio or

$$T = \frac{T_c - T_a}{T_o - T_a} \tag{11.3.7}$$

T_c = centerline temperature of the jet, T_o = temperature of the fluid in the outlet channel, T_a = ambient temperature of the receiving water, $A = 2b'/y_o$ = channel aspect ratio, $R = U_a/U_o$ = cross flow velocity ratio, U_o = outlet channel velocity, U_a = ambient velocity in the receiving water, $\mathbf{F}_o$ = outlet channel densimetric Froude number.

$$\mathbf{F}_o = \frac{U_o}{\sqrt{(\Delta\rho_o/\rho_o) g y_o}} \tag{11.3.8}$$

where $\Delta\rho_o = \rho_a - \rho_o$
ρ_o = density of fluid in outlet channel
ρ_a = ambient density of receiving water

With regard to Eq. (11.3.6), it is noted that, first, this equation is completely

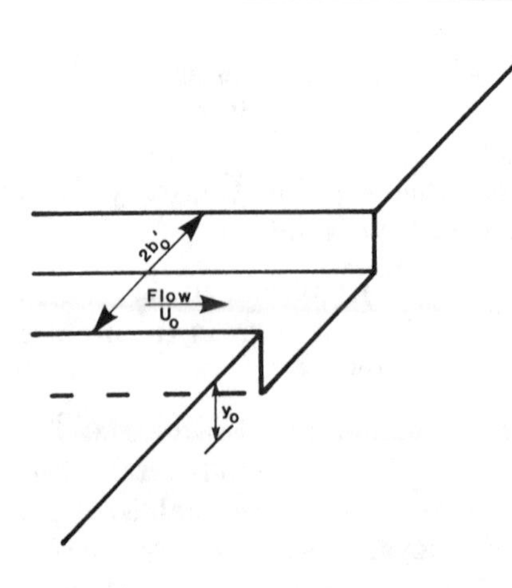

FIGURE 11.4 Definition of variables at jet outlet.

empirical since it is based on data derived from laboratory data. For the experiments from which this equation was derived

$$1 \le A \le 9.6$$
$$0 \le R \le 0.41$$
$$2.0 \le \mathbf{F}_o \le 15$$
$$0.8 \le T \le 0.98$$

Second, the application of this equation to a particular problem requires that a meaningful value of T, the excess temperature ratio, be defined. Stefan et al. (1975) asserted that a value of $T = 0.90$ was a meaningful choice to establish the end of the zone of flow establishment. Third, a comparison of Eqs. (11.3.2) and (11.3.3) with Eq. (11.3.6) demonstrates only one similarity, i.e., the use of the cross flow velocity ratio in an exponential form. Although from a qualitative viewpoint (11.3.6) includes the dimensionless parameters normally believed to be important in this type of problem, it cannot be concluded that this equation is more accurate than Eqs. (11.3.2) and (11.3.3).

Stefan et al. (1975) have also derived from laboratory data an empirical equation for estimating the flow rate within the zone of flow establishment or

$$\frac{Q}{Q_o} = 1 + \left[0.087\left(\frac{x}{2y_o}\right)^{(2.35+0.75A)/(0.90+A)}\right]\left(1 - \frac{0.52}{\mathbf{F}_o^{0.41}}\right) \qquad (11.3.9)$$

where Q = discharge at any section in the zone of flow establishment a distance x from the outlet channel and Q_o = outlet channel discharge. Equation (11.3.9) was derived from laboratory data which had the following parameter ranges:

$$1.8 \le \mathbf{F}_o \le \infty$$
$$1.0 \le A \le 9.6$$

Again, the accuracy of Eq. (11.3.9) in a specific field application cannot be addressed.

It is of theoretical interest to note that the zone of flow establishment can be viewed as being composed of three subregions:

1. A channel region that is characterized by uniform water temperature and a fully developed velocity profile.
2. A core region that is located just beyond the end of the channel region and is characterized by both constant temperatures and velocities along the centerline of the jet.
3. A turbulent transition region that is characterized by fluid temperatures and velocities which fluctuate intermittently with turbulent shear layers growing on the outer edges of the jet. These turbulent shear layers do not penetrate each other.

Beyond the zone of flow establishment is the fully developed jet region that is characterized by a fully developed turbulence field. Within this fully developed jet region, there is a similarity of temperature and velocity distributions in cross sections perpendicular to the jet axis.

Fully Developed Jet Region

If it can be assumed that, in the vicinity of the point of discharge, the heat loss to the atmosphere from a heated jet is negligible, then a rather simple treatment of the jet mechanics is available. In many cases, the assumption of negligible heat loss to the atmosphere is both reasonable and appropriate. For example, Motz and Benedict (1971), using field data, demonstrated that at dimensionless distances of $x/d_o = 40$ from the point of discharge, the observed decrease in centerline jet temperatures due to both lateral mixing and surface heat exchange was at least an order of magnitude larger than the decrease due to surface heat exchange alone.

The model developed by Motz and Benedict (1971) for buoyant surface jets is based on the integration of the governing differential equations for the conservation of volume flux, momentum flux, and heat energy over a two-dimensional momentum jet. In this analysis it was assumed that if

$$Ri = \frac{\Delta\rho g d'}{\rho(u_m - U_a)^2} > 1 \tag{11.3.10}$$

then a two-dimensional jet assumption was valid.

In Eq. (11.3.10) $\Delta\rho$ = density difference between jet and ambient receiving water

u_m = axial velocity of jet

U_a = ambient receiving water velocity

d' = jet depth

ρ = jet density

The assumption, specified by Eq. (11.3.10) and inherent in the Motz and Benedict (1971) analysis, is that the buoyancy forces are dominant and that the vertical spread of the jet is negligible.

The integration of the volumetric continuity equation over the cross section of the jet yields a relation between the rate of change of volume flux along the jet axis and the ambient flow entrainment or

$$\frac{d}{ds}\iint_A u\,dA = \int_c u_e\,dc \tag{11.3.11}$$

where, with reference to Fig. 11.5, u = jet velocity component directed along the s axis, s = a coordinate direction coincident with the jet centerline, dA = differential area, c = circumference through which entrainment takes place, and u_e = entrainment velocity. Motz and Bendict (1971) assumed that the entrainment velocity was proportional to the magnitude of the difference between a velocity characteristic of the jet and the component of the ambient velocity which is parallel to the s axis or

$$u_e = E\,(u_m - U_a \cos\beta) \tag{11.3.12}$$

where u_m = jet centerline velocity
β = angle between jet and ambient velocity of receiving water
u_a = magnitude of ambient velocity in receiving water
E = experimentally determined entrainment coefficient

Substitution of Eq. (11.3.12) in Eq. (11.3.11) yields

$$\frac{d}{ds}\iint_A u\,dA = CE(u_m - U_a \cos\beta) \tag{11.3.13}$$

The momentum equation along the axis of the jet can also be integrated over the cross section of the jet to yield a relationship between the rate of change of jet momentum flux, the rate of entrainment of ambient momentum flux, and the force exerted by the pressure gradient across the jet. The pressure gradient, which is due to the separation of the ambient flow around the jet, is represented as a drag force. At this point, it is convenient to resolve the vector momentum equation into components aligned with the cartesian coordinate direction (Figs. 11.5*a* and *b*).

In the x direction, the rate of change of momentum flux is equal to the rate of entrainment of ambient momentum flux and the x direction component of the drag or

$$\frac{d}{ds}\left(\iint_A u^2\,dA \cos\beta\right) = CE(u_m - U_a \cos\beta)U_a + F_D \sin\beta \tag{11.3.14}$$

where it is hypothesized that

$$F_D = \frac{C_D U_a^2 z_j \sin\beta}{2} \tag{11.3.15}$$

C_D = an experimentally determined drag coefficient and z_j = jet depth.

Since the ambient flow is parallel to the x axis, the rate of change of momentum flux in the y coordinate direction is equal to the y component of the drag force or

$$\frac{d}{ds}\left(\iint_A u^2\,dA \sin\beta\right) = -F_D \sin\beta \tag{11.3.16}$$

Given the small jet surface area in the near field, the temperature excess is by assumption considered conservative or

$$\frac{d}{ds}\iint_A u\phi\,dA = 0 \tag{11.3.17}$$

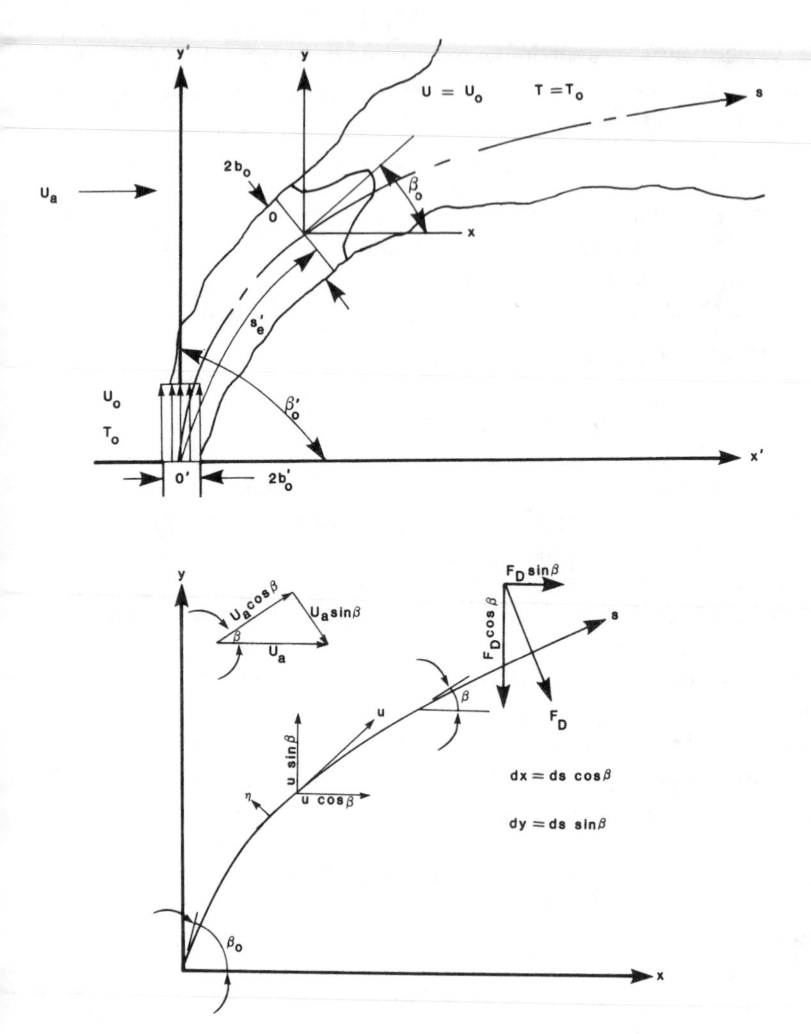

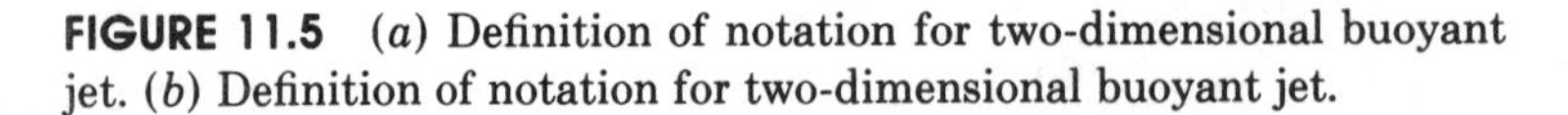

FIGURE 11.5 (*a*) Definition of notation for two-dimensional buoyant jet. (*b*) Definition of notation for two-dimensional buoyant jet.

where $\phi = T_m - T_a$

T_m = jet centerline temperature

T_a = ambient receiving water temperature

Equations (11.3.13), (11.3.14), (11.3.16), and (11.3.17) are the general integral equations for volume, momentum, and excess temperature flux along the s axis.

If the Morton (1961) technique of integral jet analysis is to be applied to the flux equations derived in the previous paragraphs, then assumptions regarding the lateral distributions of velocity and temperature are required. Morton

(1961) asserted that the result of assuming similar profiles of velocity for an integral jet solution is that the details of the lateral structure of the jet are suppressed. Under this assertion, any reasonable transverse profile of velocity can be assumed with only an insignificant effect on the result. In general, a gaussian profile for the lateral distributions of velocity and temperature is a reasonable compromise between mathematical simplicity and accuracy. Motz and Benedict (1971) also assumed that the lateral spread of both heat and momentum occurred at the same rate, although there are data which indicate that heat spreads much more quickly than momentum (see, for example, Rouse et al., 1952).

If the transverse velocity distribution of the jet is approximated as a gaussian profile, then the velocity of the jet at any point in the jet cross section is given by

$$u = u_m \exp\left(-\frac{\eta^2}{2\sigma^2}\right) \tag{11.3.18}$$

where u = jet velocity
σ = standard deviation
η = distance along an axis perpendicular to the s axis

Then

$$\int_{-b}^{b} u_m \exp\left(-\frac{\eta^2}{2\sigma^2}\right) z_j \, d\eta \simeq u_m z_j \int_{-\infty}^{\infty} \exp\left(-\frac{\eta^2}{2\sigma^2}\right) d\eta \tag{11.3.19}$$

where $dA = z_j d\eta$
z_j = jet depth which by assumption of two-dimensional jet is constant
b = jet half-width

Substitution of Eq. (11.3.19) in Eq. (11.3.13) yields

$$\frac{d}{ds}(u_m z_j \sigma \sqrt{2\pi}) - 2zE(u_m - U_a \cos\beta) \tag{11.3.20}$$

where in Eq. (11.3.19)

$$\int_{-\infty}^{\infty} \exp\left(-\frac{\eta^2}{2\sigma^2}\right) d\eta = \sigma\sqrt{2\pi}$$

and $C = 2z_j$ = circumference through which entrainment takes place.

If

$$b = \sigma\sqrt{2} \tag{11.3.21}$$

then Eq. (11.3.20) becomes

$$\frac{d}{ds}(u_m b) = \frac{2E}{\sqrt{\pi}}(u_m - U_a \cos\beta) \tag{11.3.22}$$

where it has been assumed that $z_j \neq f(s)$, which is again the result of the assumption of two-dimensional jet flow.

From Eq. (11.3.18) further assume

$$u^2 = u_m^2 \left[\exp\left(-\frac{\eta^2}{2\sigma^2} \right) \right]^2 \tag{11.3.23}$$

Substitition of this relation and Eqs. (11.3.21) and (11.3.15) in Eq. (11.3.14) yields

$$\frac{d}{ds}(u_m^2 b \cos\beta) = \frac{2\sqrt{2}}{\sqrt{\pi}} E(u_m - U_a \cos\beta)\, U_a + \frac{\sqrt{2}}{2\sqrt{\pi}} C_D U_a^2 \sin^2\beta \tag{11.3.24}$$

In a similar fashion, Eq. (11.3.16) becomes

$$\frac{d}{ds}(u_m^2 b \sin\beta) = -\frac{\sqrt{2}}{2\sqrt{\pi}} C_D U_a \sin\beta \cos\beta \tag{11.3.25}$$

If the transverse temperature distribution is also approximated by a guassian distribution or

$$\phi = \phi \exp\left(-\frac{\eta^2}{2\sigma^2} \right)$$

then Eq. (11.3.17) becomes

$$\frac{d}{ds}(u_m b \phi) = 0 \tag{11.3.26}$$

In addition to Eqs. (11.3.22) and (11.3.24) to (11.3.26), the geometry of the coordinate systems used provides two additional equations or

$$\frac{dx}{ds} = \cos\beta \tag{11.3.27}$$

and

$$\frac{dy}{ds} = \sin\beta \tag{11.3.28}$$

Thus, the behavior of a two-dimensional buoyant surface jet is described by six ordinary differential equations, i.e., Eqs. (11.3.22) and (11.3.24) to (11.3.28) in six unknowns, that is, u_m, ϕ, b, β, x, and y. At this point, it is usually convenient to transform the governing equations to a dimensionless format. For example, define

$$M = \frac{u_m^2 b}{U_0^2 b_0} \tag{11.3.29}$$

where M = dimensionless momentum flux
b_0 = jet half-width at end of zone of flow establishment
U_0 = jet velocity at end of zone of flow establishment

$$V = \frac{u_m b}{U_0 b_0} \tag{11.3.30}$$

where V = a dimensionless volume flux.

$$X = \frac{2E_x}{b_0\sqrt{\pi}} \tag{11.3.31}$$

$$Y = \frac{2E_y}{b_0\sqrt{\pi}} \tag{11.3.32}$$

$$C'_D = \frac{C_D}{4E} \tag{11.3.33}$$

and

$$S = \frac{2Es}{b_0\sqrt{\pi}} \tag{11.3.34}$$

Substitution of these dimensionless variables in the governing equations yields

Continuity

$$\frac{dV}{dS} = \frac{U_a}{U_0}\cos\beta \tag{11.3.35}$$

Momentum Flux, x Component

$$\frac{d(M\cos\beta)}{dS} = \frac{U_a\sqrt{2}}{U_0}\left[\frac{M}{V} - \frac{U_a}{U_0}\cos\beta + \frac{U_a C'_D}{U_0}\sin^2\beta\right] \tag{11.3.36}$$

Momentum Flux, y Component

$$\frac{d(M\sin\beta)}{dS} = -\frac{U_a}{U_0}\sqrt{2}\left(\frac{U_a}{U_0}C'_D\sin\beta\cos\beta\right) \tag{11.3.37}$$

Jet Geometry Equations

$$\frac{dX}{dS} = \cos\beta \tag{11.3.38}$$

$$\frac{dY}{dS} = \sin\beta \tag{11.3.39}$$

Temperature Excess

$$\frac{u_m\phi b}{U_0\phi_0 b_0} = 1 \tag{11.3.40}$$

where $\phi = T_0 - T_a$ and T_0 = jet centerline temperature at the end of the zone of flow establishment.

Thus, the two-dimensional buoyant jet problem has been reduced to six dimensionless ordinary differential equations in six dimensionless variables. If this set of equations is solved for M and V, then the velocity ratio u_m/U_0, the temperature ratio ϕ/ϕ_0, and the jet half-width ratio b/b_0 can be determined. Motz and Benedict (1971) developed a computer program to accomplish this solution, and a listing of this program is presented in their report.

The entrainment coefficient E which appears in the foregoing equations must be estimated from field and laboratory data. Field and laboratory data analyzed by Motz and Benedict (1971) are summarized in Table 11.1. With respect to the laboratory data, Motz and Benedict demonstrated that E was primarily a function of β_0' and further that for $\beta_0' = 90°$, $E = 0.4$, and for $\beta_0' =$ 60 or 45°, $E = 0.2$. For comparable values of β_0', the limited field data available suggest that E is smaller. Motz and Benedict attributed this difference between the laboratory and field results to the fact that in the laboratory experiments the ratio of the ambient flow width to the jet discharge width was 24. At Wid-

TABLE 11.1 Summary of laboratory and field data for E (Motz and Benedict, 1971)

Data source	U_a/U_0	β_0'	β_0	S_e/b_0	E
Motz-Benedict Lab	0.73	90.0	34.5	0.9	0.46
Motz-Benedict Lab	0.44	90.0	47.5	1.4	0.39
Motz-Benedict Lab	0.30	90.0	50.0	1.3	0.31
Motz-Benedict Lab	0.23	90.0	62.0	1.7	0.47
Motz-Benedict Lab	0.20	90.0	71.5	1.9	0.44
Motz-Benedict Lab	0.67	60.0	13.5	0.7	0.24
Motz-Benedict Lab	0.44	60.0	36.5	1.3	0.13
Motz-Benedict Lab	0.30	60.0	40.0	1.5	0.19
Motz-Benedict Lab	0.23	60.0	41.0	2.0	0.25
Motz-Benedict Lab	0.19	60.0	45.0	2.3	0.29
Motz-Benedict Lab	0.66	45.0	25.0	0.7	0.19
Motz-Benedict Lab	0.42	45.0	23.0	1.2	0.13
Motz-Benedict Lab	0.30	45.0	26.5	1.7	0.21
Motz-Benedict Lab	0.23	45.0	33.5	2.1	0.27
Motz-Benedict Lab	0.18	45.0	35.0	2.3	0.22
Motz-Benedict Widows Creek 1	0.50	85.0	—	—	0.16
Motz-Benedict Widows Creek 2	0.67	85.0	—	—	0.16
Motz-Benedict Widows Creek 3	0.57	60.0	—	—	0.16
Motz-Benedict New Johnsonville	0.57	60.0	—	—	0.04

ows Creek this same ratio was 7.5, and at New Johnsonville it was 3.5. Recall that in the derivation of the governing equations it was tacitly assumed that there were no boundaries present to affect the flow field, and in most field situations this assumption is not valid. In addition, the governing equations were derived under the assumption of a uniform ambient velocity field; this assumption is not usually satisfied in a field situation. Hence, in field situations it is found that E is both a function of β_0' and of the ratio of the ambient flow width to the jet discharge width.

The second parameter in this analysis which must be determined from field and laboratory data is the drag coefficient. Motz and Benedict (1971) claimed that C_D was a function of the velocity ratio U_a/U_0 and perhaps β_0'. The laboratory results of these investigations are summarized in Table 11.2 and the field results in Fig. 11.6. The data available regarding C_D must be estimated with extreme caution.

The Motz and Benedict (1971) approach has the distinct advantage of being relatively simple and producing results that are relatively accurate. There are a number of more sophisticated models available, e.g., Stolzenbach and Harleman (1973) and Stefan and Vaidyaraman (1972), which take into account both the three-dimensional nature of most jets and the loss of heat from the water surface to the atmosphere. However, these models also do not account for complex channel geometries and nonuniform ambient velocity distributions. Thus, in many cases, the simpler model presented here may be just as effective in producing useful results as the more complex models.

TABLE 11.2 Partial summary of the Motz and Benedict laboratory results for the drag coefficient C_D

u_a/U_0	β_0'	β_0	C_D
0.73	90.0	34.5	0.7
0.44	90.0	47.5	1.6
0.30	90.0	50.0	3.5
0.23	90.0	62.0	3.8
0.20	90.0	71.5	3.5
0.67	60.0	13.5	0.1
0.44	60.0	36.5	0.8
0.30	60.0	40.0	2.3
0.23	60.0	41.0	2.7
0.19	60.0	45.0	3.9
0.66	45.0	25.0	1.1
0.42	45.0	23.0	0.5
0.30	45.0	26.5	0.8
0.23	45.0	33.5	1.1
0.18	45.0	35.0	1.7

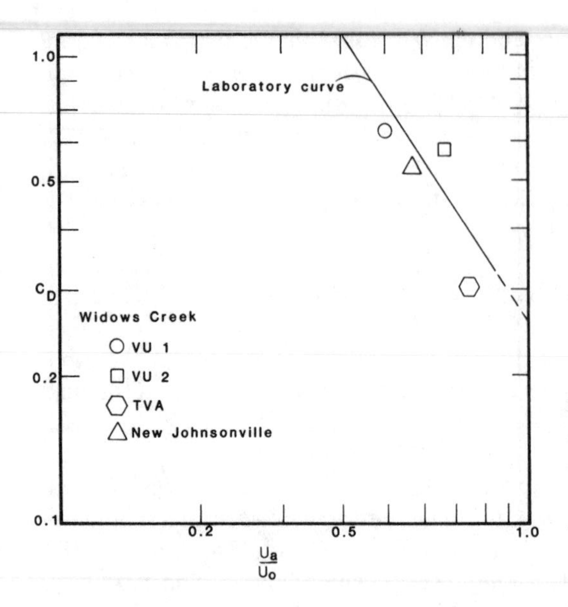

FIGURE 11.6 C_D as a function of u_a/u_0. *(Motz and Benedict, 1971.)*

11.4 UPSTREAM COOLING WATER WEDGES

The upstream cooling water wedge is a phenomenon which is closely associated with the heated buoyant surface jet discussed in the previous sections of this chapter. When cooling water from a power plant is discharged to receiving water in a channel with a weak surface current, an arrested surface density layer may be formed upstream from the point of discharge. In some cases, this wedge of warm water may extend upstream past the cooling water intake (Fig. 11.7). In this latter situation, cooling water recirculation will occur, and the maximum capacity of the generating plant will be reduced. The amount of recirculation depends on the following:

1. The geometry of the intake
2. The design of the outlet
3. The distance between the inlet and outlet
4. The discharge and mean depth of the ambient flow in the channel
5. The relative quantity of cooling water diversion
6. The degree of heating of the diverted cooling water

The complexity of the recirculation problem dictates the separation of the various factors into groups, the effects of which can be studied separately and then compounded for a complete solution. In this treatment, the primary objective is the determination of the shape of the cooling water wedge.

The flow in a channel past a thermal power station (Fig. 11.7) can be conveniently divided into three zones: (1) upstream of the intake, (2) between the intake and outlet, and (3) downstream of the outlet. Before the governing equations for each of these zones are developed, a number of fundamental equations must be developed.

With reference to Fig. 11.8, the one-dimensional equations of motion for the steady, nonuniform flow of a two-layered system are

$$\frac{dy_1}{dx} + \frac{dy_2}{dx} + \frac{\overline{u}_1}{g}\frac{d\overline{u}_1}{dx} + S_{1e} - S_0 = 0 \tag{11.4.1}$$

and

$$\left(1 - \frac{\Delta\rho}{\rho}\right)\frac{dy_1}{dx} + \frac{dy_2}{dx} + \frac{\overline{u}_2}{g}\frac{d\overline{u}_2}{dx} + S_{2e} - S_0 = 0 \tag{11.4.2}$$

where y_1 and y_2 = depths of upper and lower layers, respectively
$\overline{u}_1$ and $\overline{u}_2$ = mean velocities of upper and lower layers, respectively
ρ and $\rho + \Delta\rho$ = densities of upper and lower layers, respectively
S_0 = slope of bottom of channel
S_{1e} and S_{2e} = slopes of energy grade lines for upper and lower layers, respectively

S_{1e} and S_{2e} are defined as follows:

$$S_{1e} = \frac{\tau_i}{\rho g y_1} \tag{11.4.3}$$

and

$$S_{2e} = \frac{\tau_0 - \tau_i}{\rho g y_2} \tag{11.4.4}$$

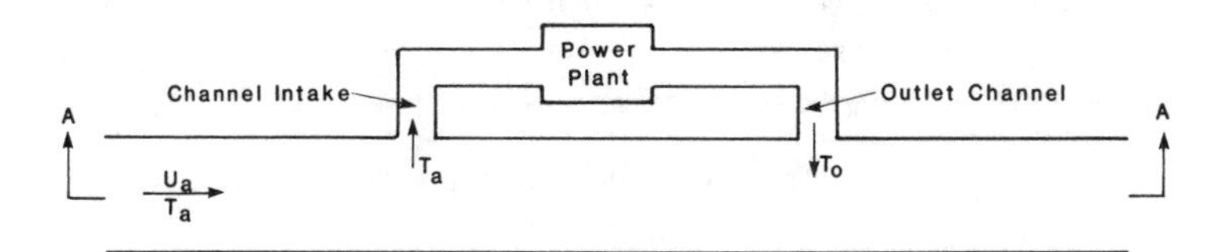

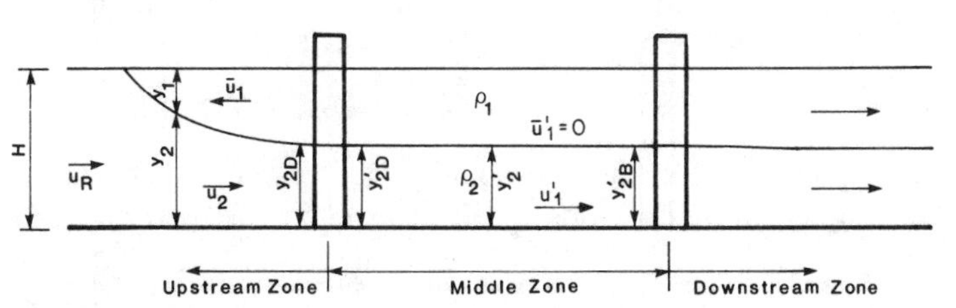

FIGURE 11.7 Schematic of upstream arrested cooling water wedge. *(Harleman, 1969.)*

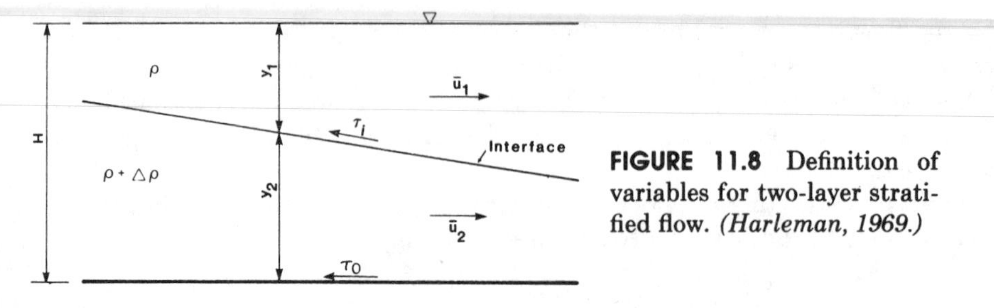

FIGURE 11.8 Definition of variables for two-layer stratified flow. *(Harleman, 1969.)*

where τ_i = shear stress at the interface between the upper and lower layers

$$\tau_i = f_i|\bar{u}_1 - \bar{u}_2|(\bar{u}_1 - \bar{u}_2) \tag{11.4.5}$$

τ_0 = bottom boundary shear stress or

$$\tau_0 = f\frac{\rho}{8}|\bar{u}_2|\bar{u}_2 \tag{11.4.6}$$

f_i = friction factor at the interface between the upper and lower layers, and f = bottom boundary friction factor. The equations of continuity for the upper and lower layers are

$$\bar{u}_1\frac{dy_1}{dx} + y_1\frac{d\bar{u}_1}{dx} = 0 \tag{11.4.7}$$

and

$$\bar{u}_2\frac{dy_2}{dx} + y_2\frac{d\bar{u}_2}{dx} = 0 \tag{11.4.8}$$

The above equations for the steady nonuniform flow of a two-layered system are subject to the following assumptions:

1. The vertical accelerations in the fluids have been neglected.
2. Only the mean velocities of the layers are considered.
3. The standard free surface Froude number based on the total depth and the average velocity of the entire flow is small.

The last assumption implies that changes in the total depth of flow are small relative to changes in the height of the density interface or

$$H = y_1 + y_2 = \text{constant}$$

The equations of motion and continuity for the two layers can be combined into a single differential equation for the slope of density interface (Harleman, 1961) or

$$\frac{dy_2}{dx} = \frac{(\tau_0 - \tau_i/\rho g y_2) - (\tau_i/\rho g y_1)}{(\Delta\rho/\rho)(\mathbf{F}_2^2 + \mathbf{F}_1^2 - 1)} \tag{11.4.9}$$

where
$$\mathbf{F}_1 = \frac{\overline{u}_1}{\sqrt{gy_1(\Delta\rho/\rho)}}$$

and
$$\mathbf{F}_2 = \frac{\overline{u}_2}{\sqrt{gy_2(\Delta\rho/\rho)}}$$

If the expressions for τ_0 and τ_i [Eqs. (11.4.5) and (11.4.6)] are substituted in Eq. (11.4.9), the result is

$$\frac{dy_2}{dx} = \frac{(f/8gy_2)\,|\overline{u}_2|\,\overline{u}_2 - (f_i/8gy_2)[H/(H - y_2)]\,|\overline{u}_1 - \overline{u}_2|\,(\overline{u}_1 - \overline{u}_2)}{(\Delta\rho/\rho)(\mathbf{F}_2^2 + \mathbf{F}_1^2 - 1)} \tag{11.4.10}$$

Outlet Channel

Figure 11.9 is a cross-sectional view of the receiving channel located such that it coincides with the centerline of the outlet channel. The discharge of the outlet channel Q_o and the temperature of this discharge T_o are determined by the design of the thermal power plant, and thus these variables are known. If the mixing between the warm water in the outlet channel and the cold water in the receiving channel is to be minimized, then a tongue of the ambient channel water must be permitted to intrude into the outlet channel (Fig. 11.9). If such a case occurs and the flow is steady, then the average velocity in the intruding tongue of water must be zero. The flowing, upper layer of warm water induces a flow in the lower layer immediately below the interface separating the layers. This induced flow in the lower layer results in a circulation such that the average velocity in the tongue of intruding water is zero. Thus, the interface separating the two layers acts as a moving boundary which prevents the development of large velocity gradients that would result in mixing.

For the situation discussed in the foregoing paragraph, $\overline{u}_{2B} = 0$; therefore,

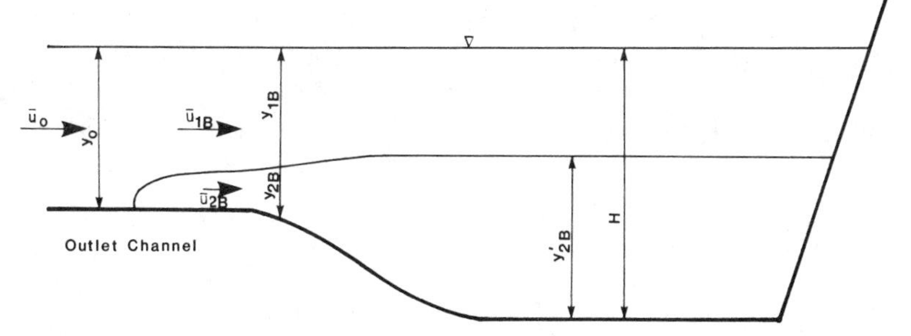

FIGURE 11.9 Cross section along the centerline of the outlet channel. *(Harleman, 1969.)*

$\mathbf{F}_{2B} = 0$. Within the outlet channel, $H = y_o$, and for the outlet channel Eq. (11.4.10) becomes

$$\frac{dy_{2B}}{dx} = \frac{-(f_i/8gy_{2B})\,[y_o/(y_o - y_{2B})]\,\overline{u}_{1B}^2}{(\Delta\rho/\rho)(\mathbf{F}_{1B}^2 - 1)} \tag{11.4.11}$$

where in this case the longitudinal coordinate x is measured in the direction of the upper layer flow. The development of an intrusive tongue of ambient channel water into the outlet channel requires that

$$\frac{dy_{2B}}{dx} > 0$$

and therefore, since the numerator of Eq. (11.4.11) is always negative, the denominator must also be negative or

$$\mathbf{F}_{1B} < 1 \tag{11.4.12}$$

If the outlet channel has a width b_o, which is constant, then the equation of continuity in the upper layer is

$$\overline{u}_o y_o = \overline{u}_{1B} y_{1B} \tag{11.4.13}$$

The local densimetric Froude number for the upper layer is

$$\mathbf{F}_{1B} = \frac{\overline{u}_{1B}}{\sqrt{g(\Delta\rho/\rho)y_{1B}}} \tag{11.4.14}$$

It is noted that $\mathbf{F}_{1B}$ varies as a function of x. Then, using the geometry defined in Fig. 11.9 and Eq. (11.4.13),

$$\mathbf{F}_{1B} = \mathbf{F}_o \left(\frac{y_o}{y_o - y_{2B}}\right)^{3/2} \tag{11.4.15}$$

where

$$\mathbf{F}_o = \frac{\overline{u}_o}{\sqrt{g(\Delta\rho/\rho)y_o}} \tag{11.4.16}$$

Since $\mathbf{F}_{1B}$ must be less than 1, it follows that $\mathbf{F}_o$ must also be less than 1 [Eq. (11.4.16)]. Thus, if mixing in the outlet channel is to be minimized, then $\mathbf{F}_o < 1$ is a design condition which must be satisfied.

Where the outlet channel joins the main channel, the warm upper layer undergoes a rapid expansion, and critical depth occurs in the upper layer or

$$\mathbf{F}_{1B} = 1$$

From Eq. (11.4.15)

$$\frac{y_o - (y_{2B})_c}{y_o} = \mathbf{F}_o^{2/3}$$

or

$$H - y'_{2B} = y_o \mathbf{F}_o^{2/3} \tag{11.4.17}$$

where it has been assumed that since there is no mixing, the upper layer has the same depth in the main channel as at the junction of the two channels.

In practice, the selection of a densimetric Froude number for the outlet channel is limited. As $\mathbf{F}_o$ approaches 1, the intrusive tongue of cold water is expelled from the channel; and as $\mathbf{F}_o$ approaches zero, the length of the intrusive tongue increases. Harleman (1969) suggested that

$$\mathbf{F}_o = 0.5$$

was a reasonable design compromise. Values of $\mathbf{F}_o$ greater than 0.5 result in mixing of the two layers while values of $\mathbf{F}_o$ less than 0.5 yield long discharge channels. Then, since both Q_o and T_o, and hence, the ratio $\Delta\rho/\rho$, are determined by the design of the plant, the only variables which must be determined are the width of the discharge channel b_o and the depth of flow in this channel y_o.

EXAMPLE 11.1

If 57 m^3/s (2000 ft^3/s) and $\Delta\rho/\rho = 0.002$, estimate the width and depth of the discharge or outlet channel.

Solution

Given the foregoing discussion, assume that $\mathbf{F}_o = 0.5$ is a desirable design value for the outlet channel. With this assumption, Eq. (11.4.16) becomes

$$\mathbf{F}_o = 0.5 = \frac{\bar{u}_o}{\sqrt{g y_o(\Delta\rho/\rho)}} = \frac{b_o y_o \bar{u}_o}{b_o y_o^{3/2}\sqrt{g(\Delta\rho/\rho)}} = \frac{Q_o}{b_o y_o^{3/2}\sqrt{g(\Delta\rho/\rho)}}$$

or rearranging

$$y_o = \left[\frac{Q_o}{0.5 b_o \sqrt{g(\Delta\rho/\rho)}}\right]^{2/3} \tag{11.4.18}$$

Values of b_o and y_o which satisfy Eq. (11.4.18) are then determined by assuming a value of b_o and solving for y_o or

b_o, m	y_o, m	$H - y'_{2B}$, m
50	6.4	4.0
100	4.0	2.6
150	3.1	1.9
200	2.6	1.6
250	2.2	1.4
300	1.9	1.2

Any of the above combinations of b_o and y_o will yield a satisfactory design of the outlet channel.

Middle Zone

If there is no recirculation of the cooling water through the inlet, i.e., the length in the upstream direction of the warm water layer is less than the distance between the outlet and inlet, then $\overline{u}_1' = 0$ and $\mathbf{F}_1' = 0$, Fig. 11.7. *Note:* The prime indicates variables and parameters in the middle zones. From Eq. (11.4.10) the slope of the interface separating the two layers is

$$\frac{dy_2'}{dx} = \frac{(\overline{u}_2'/8gy_2')\{f' + f_i'[H/(H - y_2')]\}}{(\Delta\rho/\rho)[(\mathbf{F}_2')^2 - 1]} \tag{11.4.19}$$

with x being positive in the downstream direction. If $\mathbf{F}_2' < 1$, then the slope of the interface is negative and the elevation of the interface would be higher in the vicinity of the inlet than at the outlet. In general, the change of the interface elevation in the middle zone is slight (Harleman, 1961), and it can generally be assumed that the thickness of the upper layer is constant. Wigh [in Harleman (1961)] has shown that there should be no intrusion of cooling water in the middle zone if

$$\frac{Q_R - Q_0}{b_R\sqrt{gH^3(\Delta\rho/\rho)}} > 1 \tag{11.4.20}$$

where b_R = average width of the main channel
Q_R = discharge in the main channel upstream of the cooling water diversion
Q_0 = outlet channel discharge

Harleman (1961) noted that in practice the left-hand side of Eq. (11.4.20) need only be greater than 0.7 to prevent the upstream intrusion of a cooling-water layer.

Intake

Figure 11.10 is a cross-sectional view of the receiving channel located so that it coincides with the centerline of the intake channel. In this figure, a hydraulic appurtenance known as a *skimmer wall* is shown. When a warm cooling-water layer extends upstream to or past the intake channel, recirculation of the cooling water may be controlled with the skimmer wall.

In examining the flow in the vicinity of the inlet channel, the question of whether critical depth occurs in the lower layer must be resolved. If critical depth occurs in the lower layer at the inlet channel, then

$$\frac{\overline{u}_{2D}}{\sqrt{gy_{2D}(\Delta\rho/\rho)}} = 1 \tag{11.4.21}$$

Just upstream of the inlet channel, the equation of continuity requires

$$\overline{u}_R H = \overline{u}_{2D} y_{2D} \tag{11.4.22}$$

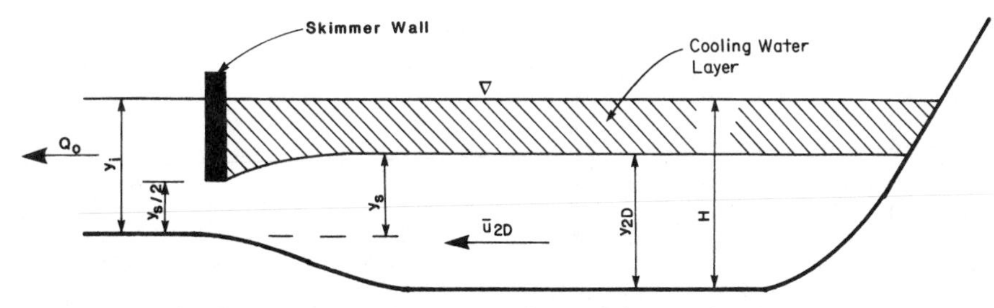

FIGURE 11.10 Cross section along centerline of the intake channel. *(Harleman, 1969.)*

Combining Eqs. (11.4.21) and (11.4.22) yields

$$\frac{(y_{2D})_c}{H} = \mathbf{F}_R^{2/3} \tag{11.4.23}$$

where $(y_{2D})_c$ = critical depth of flow in the lower layer at the inlet and

$$\mathbf{F}_R = \frac{\overline{u}_R}{\sqrt{gH(\Delta\rho/\rho)}} \tag{11.4.24}$$

Then, for a given situation, the determination of whether critical depth occurs in the lower layer at the inlet channel depends on both $\mathbf{F}_R$ and the diversion ratio Γ, where

$$\Gamma = \frac{Q_0}{Q_R} \tag{11.4.25}$$

and Q_R = discharge in the main channel upstream of cooling water diversion. The critical depth question is resolved by writing a momentum equation for the lower layer between sections upstream and downstream of the intake channel (see, for example, Harleman and Elder, 1965). This represents an internal hydraulic jump for the condition that the flow downstream of the inlet channel is reduced by the amount taken as cooling water. Wigh (1967), in Harleman (1969), has demonstrated that the solution of this momentum equation is

$$\Gamma = 1 - \frac{1}{\mathbf{F}_R} \sqrt{\frac{1}{2}\left(\frac{y'_{2D}}{H}\right)\left[3\mathbf{F}_R^{4/3} - \left(\frac{y'_{2D}}{H}\right)^2\right]} \tag{11.4.26}$$

In Fig. 11.11 the ratio y'_{2D}/H is shown as a function of $\mathbf{F}_R$ and Γ. In a design situation, if Γ for the prototype is less than that given by Eq. (11.4.26), then an internal hydraulic jump cannot occur and $y_{2D} = y'_{2D}$ (see notational definition in Fig. 11.7). If the design diversion ratio is greater than that given by Eq. (11.4.26), then an internal hydraulic jump occurs and the value of y_{2D} is estimated by Eq. (11.4.23).

With regard to the skimmer wall shown in Fig. 11.10, Harleman and Elder (1965) demonstrated that the design quantities are the intake flow rate Q_0, the

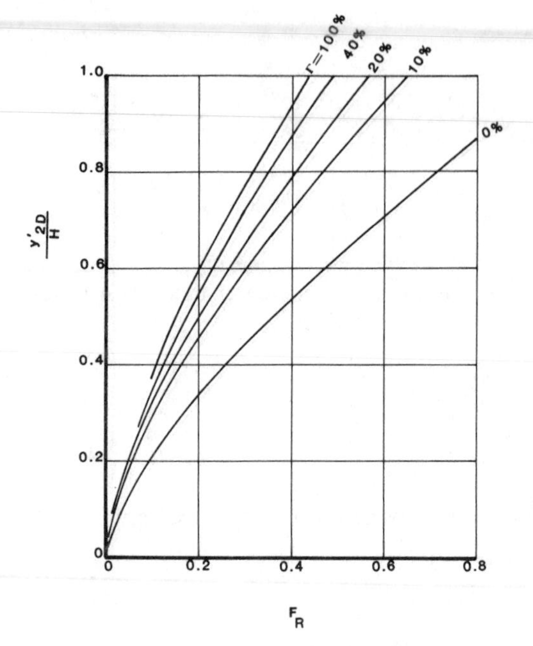

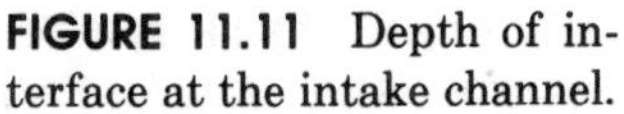

FIGURE 11.11 Depth of interface at the intake channel.

density ratio $\Delta\rho/\rho$, the width of the intake channel b_i, the depth of flow in the intake channel y_i, and the head y_s. If the local densimetric Froude number is equal to 1 at the skimmer wall, then

$$Q_0 = b_i \left[g \frac{\Delta\rho}{\rho} \left(\frac{2}{3} y_s \right)^3 \right]^{1/2} \tag{11.4.27}$$

If the intake flow Q_0 exceeds that given by Eq. (11.4.27), then the interface between the upper and lower layers will be depressed below the top of the skimmer wall and recirculation will occur. In theory, the skimmer wall opening should be less than $\frac{2}{3}y_s$; however, experiments (Harleman and Elder 1965) indicate that the opening should not be greater than $\frac{1}{2}y_s$. From the geometry defined in Fig. 11.10, the relation between y_s and y_i is

$$y_s = y_i - (H - y_{2D}) \tag{11.4.28}$$

where y_{2D} = thickness of the lower layer in the waterway adjacent to the skimmer wall.

Upstream of the Inlet

Bata (1957) combined the equations of motion and continuity for the two layers and solved them to determine the geometry of the intrusion upstream of the inlet or

$$\frac{fx}{H} = \frac{2}{\mathbf{F}_R^2}\eta^4 + \frac{8\theta}{3\mathbf{F}_R^2}\eta^3 + \frac{4\theta(1+\theta)}{\mathbf{F}_R^2}\eta^2 + \frac{8}{\mathbf{F}_R^2}[\theta(1+\theta) - \mathbf{F}_R^2]\eta$$

$$+ \frac{8\theta}{\mathbf{F}_R^2} [(1 + \theta)^3 - \mathbf{F}_R^2] [\ln (1 + \theta - \eta)] + (\text{constant}) \qquad (11.4.29)$$

where

$$\eta = \frac{y_2}{H}$$

$$\theta = \frac{f_i}{f}$$

$\mathbf{F}_R$ is defined by Eq. (11.4.24) and x = longitudinal coordinate measured upstream from the inlet. In using Eq. (11.4.29) to estimate the distance L which the upper layer of flow extends upstream of the intake, two boundary conditions are required. At the upstream limit of the wedge $y_1 = 0$ and, hence, $\eta = 1.0$; it is assumed that at the intake $y_2 = (y_2)_c$, where $(y_2)_c$ = critical depth of flow in the lower layer or

$$(y_2)_c = \left[\frac{u_R^2 H^2}{g(\Delta\rho/\rho)} \right]^{1/3} \qquad (11.4.30)$$

With these boundary conditions, Eq. (11.4.29) becomes

$$\frac{fL}{H} = \frac{2}{\mathbf{F}_R^2} (1 - \mathbf{F}_R^{8/3}) + \frac{8\theta}{3\mathbf{F}_R^2} (1 - \mathbf{F}_R^{6/3}) + \frac{4\theta(1 + \theta)}{\mathbf{F}_R^2} (1 - \mathbf{F}_R^{4/3}) + \frac{8}{\mathbf{F}_R^2} [\theta(1 + \theta) - \mathbf{F}_R^2](1 - \mathbf{F}_R^{2/3}) + \frac{8\theta}{\mathbf{F}_R^2} [(1 + \theta)^3 - \mathbf{F}_R^2] [\ln \theta - \ln (1 + \theta - \mathbf{F}_R^{2/3})] \qquad (11.4.31)$$

In Eq. (11.4.31), if $\mathbf{F}_R < 1$, then the value of L is positive; if $\mathbf{F}_R = 1$, the length of the wedge becomes zero. Keulegan (1957) asserted that the critical value of $\mathbf{F}_R$ is 0.75 rather than 1. If this is the case, then the boundary condition for the length of the wedge upstream of the inlet is

$$\frac{fL}{H} = \frac{2}{\mathbf{F}_R^2} (1 - 2.16\mathbf{F}_R^{8/3}) + \frac{8\theta}{3\mathbf{F}_R^3} (1 - 1.78\mathbf{F}_R^{6/3}) + \frac{4\theta(1 + \theta)}{\mathbf{F}_R^2} (1 - 1.47\mathbf{F}_R^{4/3}) + \frac{8}{\mathbf{F}_R^2} [\theta(1 + \theta)^2 - \mathbf{F}_R^2](1 - 1.21\mathbf{F}_R) + \frac{8\theta}{\mathbf{F}_R^2} [(1 + \theta)^3 - \mathbf{F}_R^2] [\ln \theta - \ln (1 + \theta - 1.21\mathbf{F}_R)] \qquad (11.4.32)$$

Polk et al. (1971) noted that for $\mathbf{F}_R > 0.4$, Eq. (11.4.32) yields significantly smaller estimates of L than does Eq. (11.4.31); however, for $\mathbf{F}_R < 0.4$ the difference in the values of L estimated by Eqs. (11.4.31) and (11.4.32) are insignificant.

Polk et al. (1971) used field data to examine the validity of Eqs. (11.4.31) and (11.4.32) and concluded that:

1. In general, these equations satisfactorily predict the length of cooling wedges upstream of the intake.

2. In cases where there is a long wedge upstream of the intake, it makes little difference whether the critical value of $\mathbf{F}_R$ is assumed to be 0.75 or 1.0.

3. The equations and, hence, the results are sensitive to changes in θ, which is the ratio of the interfacial friction factor to the bottom friction factor. In the four field situations examined by these investigators, bottom friction was rather constant; thus, the observed variations in θ reflected primarily variations in the value of the interfacial friction factor. For the ranges of θ and $\mathbf{F}_R$ encountered in this study, an increase in θ by a factor of 2 resulted in an increase in L/H by a factor of 1.5.

4. The most critical variable in these equations is the value of $\mathbf{F}_R$. Neglecting the initial mixing of the heated discharge prior to wedge development can result in erroneous estimates of L.

At this point it is relevant to note that Polk et al. (1971) estimated the bottom friction factor from

$$S_0 = \frac{f}{4R}\frac{\overline{u}_R^2}{2g}$$

where R = hydraulic radius. The interfacial friction factor f_i was determined by using the Moody diagram for turbulent flow in smooth pipes with the Reynolds number given by

$$\mathbf{R} = \frac{4\overline{u}_2 y_2}{\nu_2}$$

where ν_2 = kinematic viscosity of the lower layer
$\overline{u}_2$ = average velocity of flow in the lower layer
y_2 = average depths of flow in the lower layer

At this point it must be noted that one of the primary difficulties in the foregoing analysis is the accurate estimation of the interfacial friction factor. Although a number of researchers have examined this problem (see, for example, Abraham et al., 1979, and Anonymous, 1974), at this point the results are not conclusive. Field and laboratory results indicate that as the Reynolds number of the lower layer increases, the value of the interfacial friction factor decreases. At large lower-layer Reynolds numbers, f is constant with a value of approximately 15×10^{-4} (Abraham et al., 1979). These researchers also asserted that the value of f_i is a function of the turbulence generated at both the interface and the bottom boundary or

$$\tau_i = f_{i,i}\rho\overline{u}_2^2 + f_{ib}\rho\overline{u}_2^2 \tag{11.4.33}$$

where $f_{i,i}$ = interfacial friction factor expressing the effect of turbulence generated at the interface between the upper and lower layers and f_{ib} = interfacial friction factor expressing the effect of turbulence generated at the bottom boundary. At large lower-layer Reynolds number

$$f_{i,i} \simeq 4 \times 10^{-4}$$

and
$$f_{ib} \simeq 11 - 12 \times 10^{-4}$$

(Abraham et al., 1979). It must be emphasized that additional research in this area is required before a definitive theory will be available.

BIBLIOGRAPHY

Abraham, G., "Horizontal Jets in a Stagnant Fluid of Other Density," *Proceedings of the American Society of Civil Engineers, Journal of the Hydraulics Division,* vol. 91, no. HY4, July 1965, pp. 139–154.

Abraham, G., "Jets with Negative Buoyancy in Homogeneous Fluid," *Journal of Hydraulic Research,* vol. 5, no. 4, 1967, pp. 235–248.

Abraham, G., "The Flow of Round Jets Issuing Vertically in Ambient Fluid Flowing in a Horizontal Direction," *Proceedings of the 5th International Water Pollution Research Conference,* San Francisco, 1970, pp. III 15/1–III 15/7.

Abraham, G., Karelse, M., and Van Os, A. G., "On the Magnitude of Interfacial Shear of Subcritical Stratified Flows in Relation with Interfacial Stability," *Journal of Hydraulic Research,* vol. 17, no. 4, 1979, pp. 273–287.

Albertson, M. L., Dai, Y. B., Jensen, R. A., and Rouse, H., "Diffusion of Submerged Jets," *Transactions of the American Society of Civil Engineers,* vol. 115, 1950, pp. 639–697.

Anderson, J., Parker, F., and Benedict, B., "Negatively Buoyant Jets in a Cross Flow," EPA 660/2-73-012, U.S. Environmental Protection Agency, Washington, October 1973.

Anonymous, "Momentum and Mass Transfer in Stratified Flows," Report R880, Delft Hydraulics Laboratory, Delft, The Netherlands, 1974.

Bata, G., "Recirculation of Cooling Water in Rivers and Canals," *Proceedings of the American Society of Civil Engineers, Journal of the Hydraulics Division,* vol. 83, no. HY3, June 1957, pp. 1265-1–1265-27.

Cederwall, K., "Hydraulics of Marine Waste Disposal," Report No. 42, Hydraulics Division, Chalmers Institute of Technology, Göteborg, Sweden, February 1967.

Cederwall, K., and Brooks, N., "A Buoyant Slot Jet into a Stagnant or Flowing Environment," Report No. KH-R-25, W. M. Keck Laboratory of Hydraulics and Water Resources, California Institute of Technology, Pasadena, March 1971.

Daily, J. W., and Harleman, D. R. F., *Fluid Dynamics,* Addison-Wesley, Reading, Mass., 1973.

Fan, L. N., and Brooks, N. H., Discussion of "Horizontal Jets in a Stratified Fluid of Other Density" by G. Abraham, *Proceedings of the American Society of Civil Engineers, Journal of the Hydraulics Division,* vol. 92, no. HY2, March 1966, pp. 423–429.

Fan, L. N., "Turbulent Buoyant Jets into Stratified or Flowing Ambient Fluids," Technical Report No. KH-R-15, W. M. Keck Laboratory of Hydraulics and Water Resources, California Institute of Technology, Pasadena, June 1967.

Harleman, D. R. F., "Stratified Flow," *Handbook of Fluid Dynamics,* V. L. Streeter (ed.), McGraw-Hill Book Company, New York, 1961.

Harleman, D. R. F., and Elder, R. A., "Withdrawal from Two Layer Stratified Flows," *Proceedings of the American Society of Civil Engineers, Journal of the Hydraulics Division,* vol. 91, no. HY4, July 1965, pp. 43–58.

Harleman, D. R. F., "Mechanics of Condenser-Water Discharge from Thermal-Power Plants," *Engineering Aspects of Thermal Pollution,* F. L. Parker and P. A. Krenkel (eds.), Vanderbilt University Press, Nashville, Tenn., 1969, pp. 144–164.

Keulegan, G. H., "Eleventh Progress Report for Density Currents from Characteristics of Arrested Saline Wedges," National Bureau of Standards Report No. 5482, U.S. Department of Commerce, Washington, October 1957.

Morton, B. R., Taylor, G. I., and Turner, J. S., "Turbulent Gravitational Convection from Maintained and Instantaneous Sources," *Proceedings of the Royal Society of London,* vol. 234A, no. 1196, January 1956, pp. 1–23.

Morton, B. R., "On a Momentum-Mass Flux Diagram for Turbulent Jets, Plumes and Wakes," *Journal of Fluid Mechanics,* vol. 10, 1961, pp. 101–112.

Motz, L., and Benedict, B., "Heated Surface Jet Discharged into a Flowing Ambient Stream," 16130FDQ03/71, U.S. Environmental Protection Agency, Washington, March 1971.

Parker, F. L., and Krenkel, P. A., "Thermal Pollution: Status of the Art," Report No. 3, National Center for Research and Training in Hydrologic and Hydraulic Aspects of Water Pollution Control, Vanderbilt University, Nashville, Tenn., December 1969.

Polk, E. M., Benedict, B. A., and Parker, F. L., "Cooling Water Density Wedges in Streams," *Proceedings of the American Society of Civil Engineers, Journal of the Hydraulics Division,* vol. 97, no. HY10, October 1971, pp. 1639–1652.

Rouse, H., Yih, C. S., and Humphreys, H. W., "Gravitational Convection from a Boundary Source," *Tellus,* vol. 4, 1952, pp. 201–210.

Schlichting, H., *Boundary Layer Theory,* 6th Ed., McGraw-Hill Book Company, New York, 1968.

Stefan, H., and Vaidyaraman, P., "Jet Type Model for the Three Dimensional Thermal Plume in a Crosscurrent and Under Wind," *Water Resources Research,* vol. 8, no. 4, August 1972, pp. 998–1014.

Stefan, H., Bergstedt, L., and Mrosla, E., "Flow Establishment and Initial Entrainment of Heated Water Surface Jets," EPA 660/3-75-014, U.S. Environmental Protection Agency, Corvallis, Oreg., May 1975.

Stolzenbach, K. D., and Harleman, D. R. F., "Three-Dimensional Heated Surface Jets," *Water Resources Research,* vol. 9, no. 1, February 1973, pp. 129–137.

Turner, J. S., "Jets and Plumes with Negative or Reversing Buoyancy," *Journal of Fluid Mechanics,* vol. 26, part 4, 1966, pp. 779–792.

Wigh, R. J., "The Effect of Outlet and Intake Design on Cooling Water Recirculation," S.M. thesis, Department of Civil Engineering, Massachusetts Institute of Technology, Cambridge, 1967.

TWELVE

Gradually Varied, Unsteady Flow

SYNOPSIS

SYNOPSIS

In this chapter, the subject of gradually varied, unsteady flow is treated. In the initial section, both hydraulic and hydrologic solution methodologies are discussed. Subsequent sections emphasize numerical solutions of the St. Venant equations. Although the method of characteristics is briefly considered, four-point implicit difference schemes for channels of arbitrary shape represent the primary solution methodology discussed. Appropriate boundary and initial conditions are presented, and the subjects of calibration and verification of gradually varied, unsteady flow models are discussed.

12.1 INTRODUCTION

Many open-channel flow phenomena of great importance to the hydraulic engineer involve flows which are unsteady; i.e., either the depth of flow and/or the velocity of flow varies with time. Although a limited number of gradually varied, unsteady flow problems can be solved analytically, most problems in this category require a numerical solution of the governing equations and associated boundary conditions. In this chapter, only gradually varied, unsteady flow phenomena are discussed. This terminology refers to flows in which the curvature of the wave profile is mild; the change of depth with time is gradual; the vertical acceleration of the water particles is negligible in comparison with the total acceleration; and the effect of boundary friction must be taken into account. Examples of gradually varied, unsteady flows include flood waves, tidal flows, and waves generated by the slow operation of control structures such as sluice gates and navigation locks.

The mathematical models presently available to treat gradually varied, unsteady flow problems can generally be divided into two categories: (1) hydraulic models which solve the St. Venant equations for gradually varied, unsteady flow and (2) hydrologic models which solve various approximations of the St. Venant equations. Although in this book only the hydraulic models are considered in detail, the practicing engineer should be familiar with the other models which are available and the circumstances under which they can or should be used.

The St. Venant equations which were derived in Chapter 1 consist of the equation of continuity

$$\frac{\partial y}{\partial t} + y\frac{\partial u}{\partial x} + u\frac{\partial y}{\partial x} = 0 \tag{12.1.1}$$

and the one-dimensional conservation of momentum equation

$$\frac{\partial u}{\partial t} + u\frac{\partial u}{\partial x} + g\frac{\partial y}{\partial x} - g(S_x - S_f) = 0 \tag{12.1.2}$$

Alternate but equally valid forms of these equations are

$$T\frac{\partial y}{\partial t} + \frac{\partial(Au)}{\partial x} = 0 \tag{12.1.3}$$

and

$$\frac{1}{g}\frac{\partial u}{\partial t} + \frac{u}{g}\frac{\partial u}{\partial x} + \frac{\partial y}{\partial x} + S_f - S_x = 0 \tag{12.1.4}$$

where u = velocity in longitudinal direction
x = longitudinal coordinate
T = top width
A = flow area
S_x = slope of channel bed in longitudinal direction
S_f = friction slope
g = acceleration of gravity

Equations (12.1.1) and (12.1.2) or (12.1.3) and (12.1.4) compose a group of gradually varied, unsteady flow models which will be termed *complete dynamic models.* Being complete, this group of models can provide accurate results regarding unsteady flow; but, at the same time, they can be very demanding of computer resources. The models in this group are also limited by the assumptions required in the development of the St. Venant equations and the assumptions required to apply them to a specific problem; e.g., assumptions regarding channel irregularities are usually required. From the group of models termed here as complete dynamic models, two groups of simplified models can be derived by making various assumptions regarding the relative importance of various terms in the conservation of momentum equation [Eq. (12.1.3)].

The development and understanding of approximate models can, to some degree, be facilitated by rearranging Eq. (12.1.3) into the form of a rating equation which relates the discharge directly to the depth of flow (Weinmann and Laurenson, 1979). In general, the flow rate is given by

$$Q = \Gamma A R^m \sqrt{S_f} \tag{12.1.5}$$

where Γ = empirical resistance coefficient
R = hydraulic radius
m = empirical exponent

In unsteady flow, S_f varies with both the slope of the wave and the depth of flow. In the case of a steady, uniform flow, the normal discharge is given by

$$Q = Q_N = \Gamma A R^m \sqrt{S_x}$$

or

$$\Gamma A R^m = \frac{Q_N}{\sqrt{S_x}} \tag{12.1.6}$$

Substituting Eq. (12.1.6) in Eq. (12.1.5) yields

$$Q = Q_N \sqrt{\frac{S_f}{S_x}} \tag{12.1.7}$$

Solving Eq. (12.1.4) for S_f

$$S_f = S_x - \frac{1}{g}\frac{\partial u}{\partial t} + \frac{u}{g}\frac{\partial u}{\partial x} + \frac{\partial y}{\partial x}$$

and substituting this expression in Eq. (12.1.7) yields

$$Q = Q_N \left(1 - \frac{1}{S_x}\frac{\partial y}{\partial x} - \frac{u}{S_0 g}\frac{\partial u}{\partial x} - \frac{1}{S_0 g}\frac{\partial u}{\partial t}\right)^{1/2} \tag{12.1.8}$$

Kinematic wave
Diffusion analogy
Complete dynamic

Equation (12.1.8) is termed a *looped rating curve* (Fig. 12.1). In this figure, the points A and B indicate the points of maximum flow and maximum depth, respectively. The width of this loop and, therefore, the order of accuracy achieved by the approximate methods depend on the magnitude of the secondary terms in Eq. (12.1.8).

The group of models known as the diffusion analogy models is based on the continuity equation

$$\frac{\partial A}{\partial t} + \frac{\partial Q}{\partial x} = 0 \tag{12.1.9}$$

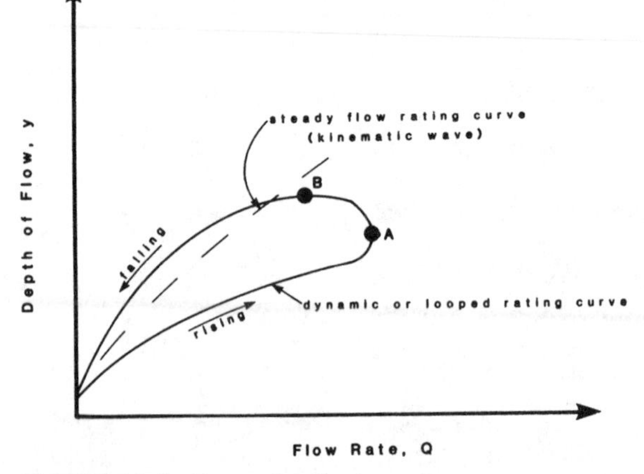

FIGURE 12.1 Looped rating curve.

which is a variant of Eq. (12.1.3) and the indicated terms of Eq. (12.1.8). These two equations can then be combined to yield a single equation

$$\frac{\partial Q}{\partial t} + c\frac{\partial Q}{\partial x} = D\frac{\partial^2 Q}{\partial x^2} \tag{12.1.10}$$

where c = a coefficient responsible for the translation characteristics of the wave and D = the diffusion coefficient which simulates the attenuation of the wave as it passes down the channel. For regular channels

$$c = \frac{1}{T}\frac{dQ}{dy} \tag{12.1.11}$$

and

$$D = \frac{Q}{2TS_0} \tag{12.1.12}$$

The similarity of Eqs. (12.1.10) and (10.2.20) is noted. If D and c are evaluated by fitting them to observed hydrographs, then they can account for the effects of channel irregularities and flood plain storage.

In essence, the diffusion model assumes that in the momentum equation the inertia terms are negligible relative to the pressure, friction, and gravity terms. Ponce et al. (1978), in evaluating the range of applicability of the diffusion type of model, claimed that the diffusion model yields reasonable results in comparison with the complete dynamic model when

$$T_p S_x \left(\frac{g}{y_N}\right)^{1/2} \geq 30 \tag{12.1.13}$$

where T_p = wave period of a sinusoidal perturbation of steady uniform flow and y_N = steady, uniform flow depth. Rearranging Eq. (12.1.13) yields

$$T_p \geq \frac{30}{S_x}\left(\frac{y_N}{g}\right)^{1/2} \tag{12.1.14}$$

If Eq. (12.1.14) is satisfied, a diffusion model will accurately approximate the unsteady flow, and such a model can be used in place of the complete dynamic model.

The kinematic wave group of models assumes that the discharge is always equal to the normal discharge; therefore, the discharge is always a single-valued function of the depth of flow (Fig. 12.1). The St. Venant equations are thus reduced to Eq. (12.1.9) and the indicated terms in Eq. (12.1.8). Combining these equations yields

$$\frac{1}{c}\frac{\partial Q}{\partial t} + \frac{\partial Q}{\partial x} = 0 \tag{12.1.15}$$

where the coefficient c = kinematic wave speed and may, at a particular cross section and for a specified flow rate, be estimated from

$$c = \frac{dQ}{dA} = \frac{1}{T}\frac{dQ}{dy} \tag{12.1.16}$$

The kinematic wave models assume that the inertia and pressure terms are negligible compared with the gravity and friction terms. Thus, the kinematic wave travels without attenuation but with a change of shape at the kinematic wave speed. It should also be noted that there are two subgroups of kinematic models. If c is assumed constant, then the model is linear; however, if c is a variable, then the model is nonlinear.

Ponce et al. (1978) evaluated the range of applicability of kinematic wave models compared with diffusion models. If the kinematic wave model is to be 95 percent as accurate in predicting wave amplitude after one wave propagation period, then

$$T_p \geq \frac{171 y_N}{S_0 u_N} \tag{12.1.17}$$

where u_N = normal flow velocity. An examination of the limit defined by Eq. (12.1.17) demonstrates that the wave period must be very long for the kinematic wave models to be applicable in a channel of mild slope. In general, the steeper the slope of the channel, the shorter the wave period needs to be to satisfy the kinematic wave assumption. Kinematic wave models have been used extensively in investigating overland flow and in simulating slow-rising flood waves which travel unchanged in form.

Before the advent of the high-speed digital computer, necessity required that unsteady flow problems be addressed with approximate models. Even with the advent of the high-speed digital computer, it was still necessary that approximate models be used in many cases because of the limited computer resources available. Today, with the availability of what in essence is almost unlimited computer resources, the need for and the usefulness of approximate models are very limited. Therefore, in this chapter the emphasis is on numerical methods of accurately solving the equations which describe the complete dynamic model of gradually varied, unsteady flow.

12.2 GOVERNING EQUATIONS AND BASIC NUMERICAL TECHNIQUES

From Chapter 1, the equations governing gradually varied, unsteady flow are the equation of continuity and the conservation of momentum equation. In this section, the application and solution of these equations in a rectangular channel for the case of subcritical, unsteady flow are considered.

For a rectangular channel, the equation of continuity is

$$\frac{\partial y}{\partial t} + y\frac{\partial u}{\partial x} + u\frac{\partial y}{\partial x} = 0 \tag{12.2.1}$$

where y = depth of flow
t = time
u = average velocity of flow
x = longitudinal distance

The conservation of momentum equation is

$$\frac{\partial u}{\partial t} + u\frac{\partial u}{\partial x} + g\frac{\partial y}{\partial x} - g(S_x - S_f) = 0 \tag{12.2.2}$$

where S_x = channel slope in the longitudinal direction and S_f = friction slope. Equations (12.2.1) and (12.2.2) are a set of simultaneous equations which can be solved for the two unknowns u and y, given appropriate initial and boundary conditions. The solution of this set of equations is virtually impossible without the aid of a high-speed digital computer.

The most direct method of simultaneously solving Eqs. (12.2.1) and (12.2.2) is an explicit finite difference scheme with a fixed time step. Given the schematic definition of a rectangular finite difference scheme in Fig. 12.2, the derivatives appearing in the governing equations can be approximated by

$$\left(\frac{\partial u}{\partial x}\right)_M = \frac{u_R - u_L}{2\Delta x} \tag{12.2.3}$$

$$\left(\frac{\partial u}{\partial t}\right)_P = \frac{u_P - u_M}{\Delta t} \tag{12.2.4}$$

$$\left(\frac{\partial y}{\partial x}\right)_M = \frac{y_R - y_L}{2\Delta x} \tag{12.2.5}$$

$$\left(\frac{\partial y}{\partial t}\right)_P = \frac{y_P - y_M}{\Delta t} \tag{12.2.6}$$

where x = longitudinal distance between nodes, t = distance in time between

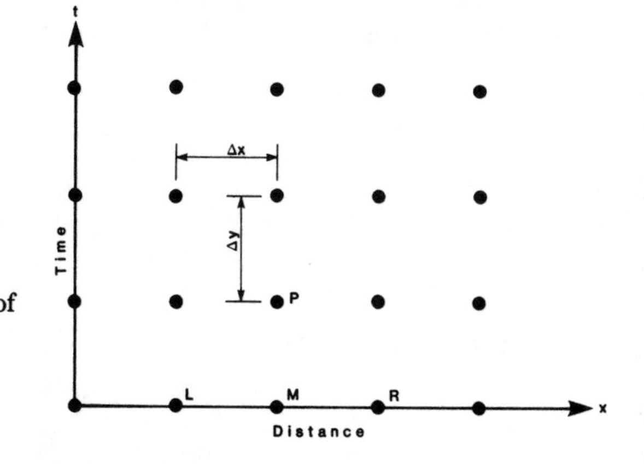

FIGURE 12.2 Definition of finite difference network.

nodes, and the subscript designates the node at which the variable is being evaluated.

Substitution of the above expressions for the derivatives in Eq. (12.2.1) yields

$$\frac{y_P - y_M}{\Delta t} + y_M\left(\frac{u_R - u_L}{2\Delta x}\right) + u_M\left(\frac{y_R - y_L}{2\Delta x}\right) = 0$$

and solving for y_P

$$y_P = y_M + \frac{\Delta t}{2\Delta x}\left[u_M(y_L - y_R) + y_M(u_L - u_R)\right] \tag{12.2.7}$$

Substituting the finite difference approximations for the derivatives in Eq. (12.2.2) yields

$$\frac{u_P - u_M}{\Delta t} + u_M\left(\frac{u_R - u_L}{2\Delta x}\right) + g\left(\frac{y_R - y_L}{2\Delta x}\right) = g(S_x - S_f) \tag{12.2.8}$$

In the computation of unsteady flow, it is usually assumed that the friction slope S_f can be estimated from either the Manning or Chezy resistance equations. Use of the Manning equation yields

$$S_f = \frac{u|u|n^2}{\phi^2 R^{4/3}} \tag{12.2.9}$$

where R = hydraulic radius
n = Manning resistance coefficient
ϕ = coefficient dependent on system of units used (1.49 for English system and 1 for SI system)

It is noted that the absolute value of the velocity of flow is used in combination with a signed value of the velocity to ensure that the frictional resistance always opposes the motion. If the rectangular channel is wide, then $R \simeq y$, and the friction slope can be represented in finite difference form as

$$S_f = \frac{u_P|u_P|n^2}{\phi^2 y_P^{4/3}} \tag{12.2.10}$$

Substitution of Eq. (12.2.10) in Eq. (12.2.8) yields

$$\frac{u_P - u_M}{\Delta t} + u_M\left(\frac{u_R - u_L}{2\Delta x}\right) + g\left(\frac{y_R - y_L}{2\Delta x}\right) = g\left(S_x - \frac{u_P|u_P|n^2}{\phi^2 y_P^{4/3}}\right) \tag{12.2.11}$$

For notational convenience define

$$\Gamma = \frac{\phi^2 y_P^{4/3}}{g\,\Delta t\,n^2}$$

Simplifying and rearranging Eq. (12.2.11) yields

$$u_P|u_P| + \Gamma u_P + \Gamma \left[\frac{u_M \, \Delta t}{2\Delta x} (u_R - u_L) + \frac{g \, \Delta t}{2\Delta x} (y_R - y_L) - g \, \Delta t \, S_x \right] = 0 \quad (12.2.12)$$

Equation (12.2.12) is a quadratic equation in u_p and can be solved by the quadratic formula or

$$u_p = \frac{-\Gamma + (\Gamma^2 - 4\beta)^{1/2}}{2} \quad (12.2.13)$$

where $\beta = \Gamma \left[\frac{u_M \, \Delta t}{2\Delta x} (u_R - u_L) + \frac{g \, \Delta t}{2\Delta x} (y_R - y_L) - g \, \Delta t \, S_x \right]$

with y_P given by Eq. (12.2.7). Thus, the solution of an unsteady flow problem by an explicit finite difference technique with a fixed time step in a wide, rectangular channel proceeds by determining y_p at an advanced time step by Eq. (12.2.7) and using this value in Eq. (12.2.13) to determine u_p.

A second numerical technique which has often been used to solve unsteady flow problems involves the solution of the governing differential equations by the use of characteristics. Again, assuming a wide, rectangular channel and rearranging Eqs. (12.2.1) and (12.2.2) yields

$$H_1 = \frac{\partial y}{\partial t} + y \frac{\partial u}{\partial x} + u \frac{\partial y}{\partial x} = 0 \quad (12.2.14)$$

and

$$H_2 = \frac{\partial u}{\partial t} + u \frac{\partial u}{\partial x} + g \frac{\partial y}{\partial x} - g(S_x - S_f) = 0 \quad (12.2.15)$$

H_1 and H_2 with an unknown multiplier λ can be combined in a linear combination to form a new function H

$$H = \lambda H_1 + H_2 \quad (12.2.16)$$

which, for any two real, distinct values of λ, will produce two equations in u and y that will retain all the attributes of Eqs. (12.2.14) and (12.2.15). Then, combining Eqs. (12.2.14) and (12.2.15) according to Eq. (12.2.16)

$$H = \frac{\partial u}{\partial t} + u \frac{\partial u}{\partial x} + g \frac{\partial y}{\partial x} - g(S_x - S_f) + \lambda \left(\frac{\partial y}{\partial t} + y \frac{\partial u}{\partial x} + u \frac{\partial y}{\partial x} \right)$$

or

$$H = \left[\frac{\partial u}{\partial x} (u + \lambda y) + \frac{\partial u}{\partial t} \right] + \lambda \left[\frac{\partial y}{\partial x} \left(u + \frac{g}{\lambda} \right) + \frac{\partial y}{\partial t} \right] - g(S_x - S_f) \quad (12.2.17)$$

In Eq. (12.2.17) the first and second terms are the total derivatives of the velocity of flow and the depth of flow or

$$\frac{du}{dt} = \frac{\partial u}{\partial x}\frac{dx}{dt} + \frac{\partial u}{\partial t} \qquad \text{if} \qquad \frac{dx}{dt} = u + \lambda y \tag{12.2.18}$$

and

$$\frac{dy}{dt} = \frac{\partial y}{\partial x}\frac{dx}{dt} + \frac{\partial y}{\partial t} \qquad \text{if} \qquad \frac{dx}{dt} = u + \frac{g}{\lambda} \tag{12.2.19}$$

Equation (12.2.17) can then be rewritten as

$$H = \frac{du}{dt} + \lambda\frac{dy}{dt} - g(S_x - S_f) \tag{12.2.20}$$

Equating the expressions for dx/dt in Eqs. (12.2.18) and (12.2.19) yields

$$u + \lambda y = u + \frac{g}{\lambda}$$

and solving yields

$$\lambda = \pm\sqrt{\frac{g}{y}} \tag{12.2.21}$$

The two real, distinct roots for λ specified by Eq. (12.2.21) can be used to transform Eqs. (12.2.14) and (12.2.15) into a pair of ordinary differential equations subject to the restrictions regarding dx/dt specified in Eqs. (12.2.18) and (12.2.19) or

$$du + dy\sqrt{\frac{g}{y}} + g(S_f - S_x)\,dt = 0 \tag{12.2.22}$$

$$dx = (u + \sqrt{gy})\,dt \tag{12.2.23}$$

$$du - dy\sqrt{\frac{g}{y}} + g(S_f - S_x)\,dt = 0 \tag{12.2.24}$$

$$dx = (u - \sqrt{gy})\,dt \tag{12.2.25}$$

At this point, it is noted that the curve defined by Eq. (12.2.23) is termed the *positive characteristic* ($C+$) (curve *LP* in Fig. 12.3). The curve defined by Eq. (12.2.25) is termed the *negative characteristic* ($C-$) (curve *RP* in Fig. 12.3).

The solution of Eqs. (12.2.22) to (12.2.25) must be accomplished by numerical methods. Using a first-order, explicit finite difference technique, Eqs. (12.2.22) to (12.2.25) become

$$u_P - u_L + \sqrt{\frac{g}{y_L}}(y_P - y_L) + (t_P - t_L)(S_{fL} - S_x) = 0 \tag{12.2.26}$$

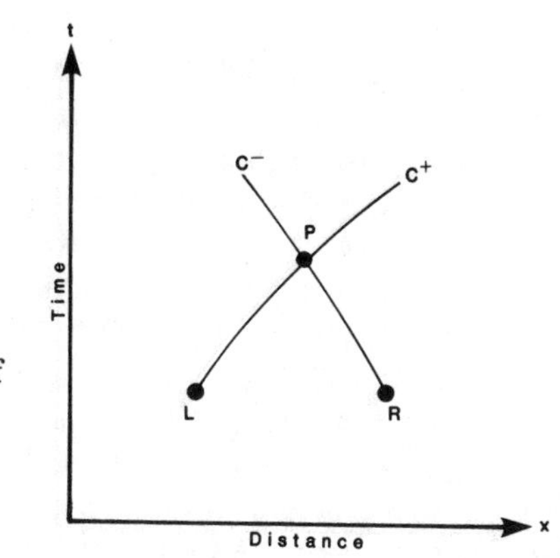

FIGURE 12.3 Definition of characteristic curves.

$$x_P - x_L = (u_L + \sqrt{gy_L})(t_P - t_L) \tag{12.2.27}$$

$$u_P - u_R - \sqrt{\frac{g}{y_R}}\,(y_P - y_R) + (t_P - t_R)(S_{fR} - S_x) = 0 \tag{12.2.28}$$

$$x_P - x_R = (u_R - \sqrt{gy_R})(t_P - t_R) \tag{12.2.29}$$

Unlike the explicit finite difference technique with a fixed time step which was previously discussed, in this technique the length of the time step is determined by subtracting Eq. (12.2.29) from Eq. (12.2.27) or

$$t_P = \frac{x_L - x_R + t_R(u_R - \sqrt{gy_R}) - t_L(u_L - \sqrt{gy_L})}{u_R - u_L - \sqrt{gy_L} - \sqrt{gy_R}} \tag{12.2.30}$$

The distance step in the longitudinal direction is determined from Eq. (12.2.27) or

$$x_P = x_L + (u_L + \sqrt{gy_L})(t_P - t_L) \tag{12.2.31}$$

The depth of flow at node P is determined by subtracting Eq. (12.2.28) from Eq. (12.2.26) and solving for y_P.

$$y_P = \frac{u_L - u_R + y_L\sqrt{\dfrac{g}{y_L}} + y_R\sqrt{\dfrac{g}{y_R}} + (t_P - t_R)(S_{fR} - S_x) - (t_P - t_L)(S_{fL} - S_x)}{\sqrt{\dfrac{g}{y_L}} + \sqrt{\dfrac{g}{y_R}}} \tag{12.2.32}$$

From Eq. (12.2.26) the velocity of flow is given by

$$u_P = u_L - \sqrt{\frac{g}{y_L}}\,(y_P - y_L) - (t_P - t_L)(S_{fL} - S_x) \tag{12.2.33}$$

Before either of the methodologies described in this section can be applied to a problem, initial and boundary conditions must be specified. The term *initial condition* refers to the initial state of flow in the channel, i.e., the depth, velocity, or discharge along the reach under consideration at the time the computation begins. The term *boundary condition* refers to the definition of the depth, velocity, or discharge at the upper and lower ends of the reach for all times after the computation begins. For the general case, the initial condition is usually specified as uniform flow or a gradually varied flow profile (Chapter 6). A typical upstream boundary condition is the specification of a hydrograph, while a downstream condition could be critical depth.

The primary difficulty with explicit finite difference techniques is the problem of numerically unstable solutions. Unstable solutions usually result if Δt is large relative Δx. The Courant stability condition requires

$$\Delta t \leq \frac{\Delta x}{u + c} \tag{12.2.34}$$

where c = wave celerity. However, it has been found that for the type of explicit difference schemes discussed in this section Δt should be approximately 20 percent of the value given by Eq. (12.2.34). Viessman et al. (1972) noted that more stable solutions can be obtained if a diffusing difference approximation is used; i.e., in the foregoing equations, substitute

$$\left(\frac{\partial u}{\partial t}\right)_M = \frac{u_P - 0.5(u_L + u_R)}{\Delta t} \tag{12.2.35}$$

$$\left(\frac{\partial y}{\partial t}\right)_M = \frac{y_P - 0.5(y_L + y_R)}{\Delta t} \tag{12.2.36}$$

$$S_f = \frac{S_{fL} + S_{fR}}{2} \tag{12.2.37}$$

This formulation of the derivatives allows the size of the time increment to be significantly increased; but the Courant stability condition [Eq. (12.2.34)] must still be met. In addition, the diffusing scheme also imposes a friction criteria or

$$\Delta t \leq \frac{\phi^2 R^{4/3}}{g n^2 |u|} \tag{12.2.38}$$

For stability, the lesser value of Δt between Eqs. (12.2.34) and (12.2.38) is used.

12.3 IMPLICIT FOUR-POINT DIFFERENCE SCHEME—CHANNELS OF ARBITRARY SHAPE

As indicated in the foregoing sections, the primary difficulty in analyzing gradually varied, unsteady flow is identifying a numerical scheme which is accurate, fast, and efficient. The explicit finite difference solution of the St. Venant equations discussed in the previous section has the advantage of simplicity, but it is subject to a rather stringent stability condition which imposes a limiting value for the time step in relation to the distance step. Although these stability conditions do not pose a difficulty in the investigation of short-term flows, the solution is both cumbersome and inefficient when treating long-term phenomena such as floods in major rivers. The method of characteristics, also treated in the previous section, is very suitable for treating rapidly varied flows and can be used for flood studies (see, for example, Amein, 1966, and Fletcher and Hamilton, 1967); but in many applications it has the disadvantage of yielding results at nonfixed times and locations. Although a fixed grid can be used, the resulting technique has no significant advantage over an explicit finite difference scheme.

Amein et al. (1968, 1970, 1975) have developed a rapidly convergent and accurate implicit technique which uses a centered finite difference scheme and a Newton iteration method to solve the resulting nonlinear finite difference equations. The equations governing unsteady flow in open channels of arbitrary shape are

$$u\frac{\partial A}{\partial x} + A\frac{\partial u}{\partial x} + \frac{\partial A}{\partial t} - q = 0 \tag{12.3.1}$$

$$\frac{\partial u}{\partial t} + u\frac{\partial u}{\partial x} + g\frac{\partial y}{\partial x} = g(S_x - S_f) - \frac{qu}{A} \tag{12.3.2}$$

where, as before,
u = average velocity
A = flow area
x = longitudinal distance
y = depth of flow
t = time
q = lateral inflow per unit channel length per unit time
S_x = channel bottom slope
S_f = friction slope

The channel bottom slope can be conveniently expressed as

$$S_x = -\frac{dz}{dx} \tag{12.3.3}$$

where z = channel bottom elevation relative to a datum.

The flow area is assumed to be a known function of the depth; therefore, derivatives of A may be expressed in terms of y or

$$\frac{\partial A}{\partial x} = \frac{dA}{dy}\frac{\partial y}{\partial x} = T\frac{\partial y}{\partial x} \tag{12.3.4}$$

and

$$\frac{\partial A}{\partial t} = \frac{dA}{dy}\frac{\partial y}{\partial t} = T\frac{\partial y}{\partial t} \tag{12.3.5}$$

where T = channel top width. At this point it should be noted that if A and T are determined by independent measurements, then measurement errors may cause T to be different from dA/dy. Amein and Fang (1970) noted that for numerical stability, A and T must be compatible. Therefore, if either A or T is determined by measurement, it is essential that the other variable be determined by calculation; see Section 1.2. In this discussion it is assumed that T is determined by

$$T = \frac{dA}{dy} \tag{12.3.6}$$

Equation (12.3.1) becomes

$$\frac{\partial y}{\partial t} + \frac{A}{T}\frac{\partial u}{\partial x} + u\frac{\partial y}{\partial x} - \frac{q}{T} = 0 \tag{12.3.7}$$

Equations (12.3.2) and (12.3.7) are nonlinear, first-order partial differential equations of the hyperbolic type. The numerical solution of this set of governing equations is accomplished in two steps. First, the governing equations are replaced by a set of algebraic finite difference equations. Second, a method of solving the difference equations must be found. The numerical solution of Eqs. (12.3.2) and (12.3.7) will be sought in the (x, t) plane at a discrete number of points arranged to form a rectangular grid (Fig. 12.4). With regard to this network, the following points are noted:

1. The nodes of the network occur at the intersections of straight lines drawn parallel to the x and t axes.
2. The lines parallel to the t axis represent locations along the channel while those drawn parallel to the x axis represent times.
3. The locations lines are drawn with a spacing Δx while the time lines are

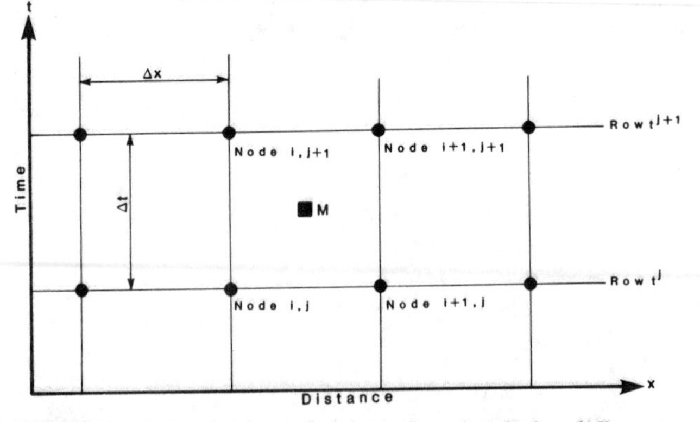

FIGURE 12.4 Grid definition for implicit difference scheme.

drawn with a spacing Δt. Although for convenience in this development Δx and Δt are assumed constant, in practice Δx and Δt may vary in space and time as required.

4. The t axis may be considered the upstream channel boundary location, and the last line drawn parallel to the t axis, termed the Nth line, may be used to represent the downstream channel boundary.

5. Each node in the network is identified by two indices: A subscript designates the x position of the node while the superscript designates the location of the node in time.

6. At time $t = t_0$, it is assumed that values of the velocity and depth of flow are available at all locations.

7. For convenience, it is assumed that at the upstream channel boundary either a stage or discharge hydrograph is available. The downstream channel boundary is assumed to be a control section so that either a stage discharge or velocity-depth relationship is available.

For the development of the necessary set of algebraic finite difference equations, assume that all variables are defined on the row t^j and that it is desired to advance the solution to the row t^{j+1} where $t^{j+1} = t^j + \Delta t$. The equations governing unsteady flow are applied, in a finite difference form, to a point M entirely within a four-point grid (Fig. 12.4). At the point M, the partial derivatives of an arbitrary function are represented as

$$\Gamma\,(M) = \frac{1}{4}\,(\Gamma_i^j + \Gamma_i^{j+1} + \Gamma_i^{j+1} + \Gamma_{i+1}^{j+1}) \tag{12.3.8}$$

$$\frac{\partial \Gamma(M)}{\partial x} = \frac{1}{2\Delta x}\,[(\Gamma_{i+1}^j\,\Gamma_{i+1}^{j+1}) - (\Gamma_i^j + \Gamma_{i+1}^{j+1})] \tag{12.3.9}$$

and

$$\frac{\partial \Gamma(M)}{\partial t} = \frac{1}{2\Delta t}\,[(\Gamma_i^{j+1} + \Gamma_{i+1}^{j+1}) - (\Gamma_i^j + \Gamma_{i+1}^j)] \tag{12.3.10}$$

Since in Eqs. (12.3.8) to (12.3.10) Γ is an arbitrary function, these equations define the variables which appear in the equations governing gradually varied unsteady flow. When y, u, $\partial u/\partial x$, and $\partial u/\partial t$ in Eq. (12.3.7) and (12.3.2) are replaced by their analogies defined in Eq. (12.3.8) to (12.3.10) and two finite difference equations are obtained, the results are

$$\begin{aligned} \frac{1}{2\Delta t}\,[(y_{i+1}^{j+1} + y_i^{j+1}) - (y_{i+1}^j + y_i^j)] + \frac{1}{2\Delta x}\,(u_{i+1/2}^{j+1/2})\,[(y_{i+1}^{j+1} + y_{i+1}^j) \\ - (y_i^{j+1} + y_i^j)] + \frac{1}{2\Delta x}\left(\frac{A}{T}\right)_{i+1/2}^{j+1/2}[(u_{i+1}^j + u_{i+1}^{j+1}) \\ - (u_i^j + u_i^{j+1})] - \left(\frac{q}{T}\right)_{i+1/2}^{j+1/2} = 0 \end{aligned} \tag{12.3.11}$$

and

$$\frac{g}{2\Delta x}[(y_{i+1}^{j+1}+y_{i+1}^{j})-(y_i^{j+1}+y_i^j)]+\frac{1}{2\Delta t}[(u_{i;1}^{j+1}+u_i^{j+1})-(u_{i+1}^{j}+u_i^j)]$$
$$+\frac{1}{2\Delta x}u_{i+1/2}^{j+1/2}[(u_{i+1}^{j+1}+u_{i+1}^{j})-(u_i^j+u_i^{j+1})]+\frac{g}{4}(S_{fi}^j+S_{fi+1}^j$$
$$+S_{fi}^{j+1}+S_{fi+1}^{j+1})+\left(\frac{g}{\Delta x}\right)(z_i^j-z_{i+1}^j)+q\left(\frac{u}{A}\right)_{i+1/2}^{j+1/2}=0 \quad (12.3.12)$$

where

$$u_{i+1/2}^{j+1/2}=¼\,(u_i^j+u_{i+1}^j+u_i^{j+1}+u_{i+1}^{j+1}) \quad (12.3.13)$$

$$\left(\frac{u}{A}\right)_{i+1/2}^{j+1/2}=¼\left(\frac{u_i^j}{A_i^j}+\frac{u_{i+1}^j}{A_{i+1}^j}+\frac{u_i^{j+1}}{A_i^{j+1}}+\frac{u_{i+1}^{j+1}}{A_{i+1}^{j+1}}\right) \quad (12.3.14)$$

$$\left(\frac{q}{T}\right)_{i+1/2}^{j+1/2}=¼\,g\left(\frac{1}{T_i^j}+\frac{1}{T_{i+1}^j}+\frac{1}{T_i^{j+1}}+\frac{1}{T_{i+1}^{j+1}}\right) \quad (12.3.15)$$

$$\left(\frac{A}{T}\right)_{i+1/2}^{j+1/2}=¼\left(\frac{A_i^j}{T_i^j}+\frac{A_{i+1}^j}{T_{i+1}^j}+\frac{A_i^{j+1}}{T_i^{j+1}}+\frac{Q_{i+1}^{j+1}}{T_{i+1}^{j+1}}\right) \quad (12.3.16)$$

$$S_{fi}^j=\frac{(n_i^j)^2\,u_i^j\mid u_i^j\mid(P_i^j)^{4/3}}{\phi^2\,(A_i^j)^{4/3}} \quad (12.3.17)$$

P_i^j = wetted perimeter and ϕ = a coefficient which is dependent on the system of units used (ϕ = 1.49 for the English system and ϕ = 1 for the SI system). In the above equations all variables with the superscript j are known, and all variables with the superscript $j+1$ are unknown. These equations contain only four unknowns: u_i^{j+1}, y_i^{j+1}, u_{i+1}^{j+1}, and y_{i+1}^{j+1}. We note that Eqs. (12.3.11) and (12.3.12) are two nonlinear, algebraic equations in four unknowns. Thus, these equations by themselves are not sufficient to evaluate the four unknowns. However, if these equations are applied to all the nodes in the $j+1$ row, there would result 2 ($N-1$) equations and 2 N unknowns. These equations with two equations expressing the upstream and downstream boundary conditions, respectively, would uniquely define the unknowns. If the stage at the upstream boundary is known as a function of time, then

$$y_1^{j+1}-f_1(t^{j+1})=0 \quad (12.3.18a)$$

is one of the two additional equations required to obtain a solution. If the discharge is known as a function of time at the upstream boundary, then

$$u_1^{j+1}A_1^{j+1}-Q_1^{j+1}=0 \quad (12.3.18b)$$

If, at the downstream boundary, a discharge relationship is known because the section is a control section or the stage is known from a rating curve, then

$$y_N^{j+1} - f_N\,(u_N^{j+1}) = 0 \tag{12.3.19}$$

Then, Eqs. (12.3.11), (12.3.12), (12.3.18*a* or *b*), and (12.3.19) are a determinate system of equations which define the unsteady flow problem.

Amein and Fang (1970) found it convenient to explicitly recognize that the value of the variables at the time step t^j are known and can be treated as constants. Rearranging Eqs. (12.3.18*a*) and (12.3.18*b*) yields

$$G_0\,(y_1) = y_1 - \lambda = 0 \tag{12.3.20a}$$

and

$$G'_0\,(y_1, u_1) = u_1 A_1 - Q_1 = 0 \tag{12.3.20b}$$

where λ = a constant and the superscripts have been dropped because all variables in the above equations and those which follow belong to the row t^{j+1}. Rearrangement of Eq. (12.3.11) yields

$$F_i\,(y_i, u_i, y_{i+1}, u_{i+1}) = (y_{i+1} + y_i) + a + \frac{1}{4}\frac{\Delta t}{\Delta x}\,(y_{i+1} - y_i)\,(u_i + u_{i+1} + b)$$
$$+ c\,(u_i + u_{i+1}) + d + \frac{1}{4}\frac{\Delta t}{\Delta x}\left(\frac{A_i}{T_i} + \frac{A_{i+1}}{T_{i+1}}\right)(u_{i+1} - u_i + e)$$
$$\frac{1}{4}\frac{\Delta t}{\Delta x}\,(hu_{i+1} + mu_i + p) - \frac{\Delta t}{2}\,q\left(\frac{1}{T_i} + \frac{1}{T_{i+1}} + w\right) = 0 \tag{12.3.21}$$

where the constants *a*, *b*, *c*, *d*, *e*, *h*, *m*, *p*, and *w* are defined in Table 12.1. Rearrangement of Eq. (12.3.12) yields

$$G(y_i, u_i, y_{i+1}, u_{i+1}) = (y_{i+1} - y_i - a') + \frac{\Delta x}{g\,\Delta t}\,(u_i + u_{i+1} + b') \tag{12.3.22}$$
$$+ \frac{1}{4g}\,(u_{i+1}^2 + c'\,u_{i+1} + d'\,u_i - u_i^2 + e') + \frac{\Delta x}{2}\,(S_{fi} + S_{fi+1} + h')$$
$$+ \frac{\Delta x\,q}{2g}\left(\frac{u_i}{A_i} + \frac{u_{i+1}}{A_{i+1}} + m'\right) = 0 \tag{12.3.22}$$

where the constants a', b', c', d', e', h', and m' are defined in Table 12.2. In a similar fashion, the downstream boundary condition [Eq. (12.3.19)] can be rearranged to yield

$$F_N\,(y_N, u_N) = y_N - f_N(u_N) = 0 \tag{12.3.23}$$

Equations (12.3.20) to (12.2.23) are then the finite difference equations which approximate the partial differential equations governing unsteady flow and the upstream and downstream boundary conditions. For convenience, the finite difference equations can be assembled as follows:

TABLE 12.1 Definition of constants for Eq. (12.3.21)

Constant	Definition
a	$a = -(y^j_{i+1} + y^j_i)$
b	$b = u^j_i + u^j_{i+1}$
c	$c = y^j_{i+1} - y^j_i$
d	$d = \dfrac{\Delta t}{4\Delta x}(y^j_{i+1} - y^j_i)(u^j_i + u^j_{i+1})$
e	$e = u^j_{i+1} - u^j_i$
h	$h = \dfrac{A^j_i}{T^j_i} + \dfrac{Q^j_{i+1}}{T^j_{i+1}}$
m	$m = -\left(\dfrac{A^j_i}{T^j_i} + \dfrac{A^i_{i+1}}{T^j_{i+1}}\right)$
p	$p = \left(\dfrac{A^j_i}{T^j_i} + \dfrac{A^j_{i+1}}{T^j_{i+1}}\right)\left(u^j_{i+1} - u^j_i\right)$
w	$w = \dfrac{1}{T^j_i} + \dfrac{1}{T^j_{i+1}}$

$$
\begin{aligned}
G_0(y_1, u_1) &= 0 \\
F_1(y_1, u_1, y_1, u_2) &= 0 \\
G_1(y_1, u_1, y_2, u_2) &= 0 \\
&\vdots \\
F_i(y_i, u_i, y_{i+1}, u_{i+1}) &= 0 \\
G_i(y_i, u_i, y_{i+1}, u_{i+1}) &= 0 \\
&\vdots \\
F_{N-1}(y_{N-1}, u_{N-1}, y_N, u_N) &= 0 \\
G_{N-1}(y_{N-1}, u_{N-1}, y_N, u_N) &= 0 \\
F_N(y_N, u_N) &= 0
\end{aligned}
\tag{12.3.24}
$$

Equation (12.3.24) is a system of $2N$ nonlinear algebraic equations in $2N$ unknowns. Thus, the first step of the solution process is complete.

TABLE 12.2 Definition of constants for Eq. (12.3.22)

Constant	Definition
a'	$a' = y_{i+1}^j + y_i^j$
b'	$b' = -(u_{i+1}^j + u_i^j)$
c'	$c' = u_{i+1}^j$
d'	$d' = -2u_i^j$
e'	$e' = u_{i+1}^j - (u_i^j)^2$
h'	$h' = S_{fi+1}^j + S_{fi}^j$
m'	$m' = \dfrac{u_i^j}{A_i^j} + \dfrac{u_{i+1}^j}{A_{i+1}^j}$

Routine methods of solution for systems of nonlinear, algebraic equations are not available. Amein and Fang (1970) and Amein and Chu (1975) recommend solving the system of equations designated as Eq. (12.3.24) by a generalized Newton iteration method (Ralston, 1965). In this technique, trial values are assigned to the unknowns. Substitution of these trial values in a system of equations such as Eq. (12.3.24) usually results in the left-hand side of the equation being nonzero. This nonzero value is termed a *residual,* and the solution is obtained by adjusting the trial variable values until the residuals vanish.

To demonstrate the application of the Newton iterative solution technique to unsteady flow, assume that the values of the unknowns have been approximated through the kth trial. Therefore, it is desired to carry the process through the $k+1$th trial. When the approximated values of the variables through the kth cycle are substituted in Eq. (12.3.24), the right-hand sides of these equations become the residuals at the end of the kth iterative step or

$$
\begin{aligned}
G_0(y_1^k, u_1^k) &= R_{G,0}^k \\
F_1(y_1^k, u_1^k, y_2^k, u_2^k) &= R_{F,1}^k \\
G_1(y_1^k, u_1^k, y_2^k, u_2^k) &= R_{G,1}^k \\
&\vdots \\
F_i(y_i^k, u_i^k, y_{i+1}^k, u_{i+1}^k) &= R_{F,i}^k \\
G_i(y_i^k, u_i^k, y_{i+1}^k, u_{i+1}^k) &= R_{G,i}^k \\
&\vdots
\end{aligned}
\tag{12.3.25}
$$

$$F_{N-1}\,(y_{N-1}^k, u_{N-1}^k, y_N^k, u_N^k) = R_{F,\,N-1}^k$$

$$G_{N-1}\,(y_{N-1}^k, u_{N-1}^k, y_N^k, u_N^k) = R_{G,\,N-1}^k$$

$$F_N\,(y_N^k, u_N^k) = R_{F,\,N}^k$$

where the variable superscript identifies the iteration cycle. Then, according to the generalized Newton iteration technique used by Amein and Fang (1970), the residuals and partial derivatives are related by

$$\frac{\partial G_0}{\partial y_1}\,dy_1 + \frac{\partial G_0}{\partial u_1}\,du_1 = -R_{G,0}^k$$

$$\frac{\partial F_1}{\partial y_1}\,dy_1 + \frac{\partial F_1}{\partial u_1}\,du_1 + \frac{\partial F_1}{\partial y_2}\,dy_2 + \frac{\partial F_1}{\partial u_2}\,du_2 = -R_{F,1}^k$$

$$\frac{\partial G_1}{\partial y_1}\,dy_1 + \frac{\partial G_1}{\partial u_1}\,du_1 + \frac{\partial G_1}{\partial y_2}\,dy_2 + \frac{\partial G_1}{\partial u_2}\,du_2 = -R_{G,1}^k$$

$$\vdots$$

$$\frac{\partial F_i}{\partial y_i}\,dy_i + \frac{\partial F_i}{\partial u_i}\,du_i + \frac{\partial F_i}{\partial y_{i+1}}\,dy_{i+1} + \frac{\partial F_i}{\partial u_{i+1}}\,du_{i+1} = -R_{F,i}^k$$

$$\frac{\partial G_i}{\partial y_i}\,dy_i + \frac{\partial G_i}{\partial u_i}\,du_i + \frac{\partial G_i}{\partial y_{i+1}}\,dy_{i+1} + \frac{\partial G_i}{\partial u_{i+1}}\,du_{i+1} = -R_{G,i}^k \qquad (12.3.26)$$

$$\vdots$$

$$\frac{\partial G_{N-1}}{\partial y_{N-1}}\,dy_{N-1} + \frac{\partial G_{N-1}}{\partial u_{N-1}}\,du_{N-1} + \frac{\partial G_N}{\partial y_N}\,dy_N + \frac{\partial G_N}{\partial u_N}\,du_N = -R_{G,N-1}^k$$

$$\frac{\partial F_N}{\partial y_N}\,dy_N + \frac{\partial F_N}{\partial u_N}\,du_N = -R_{F,N}^k$$

where

$$dy_1 = y_1^{k+1} - y_1^k$$

$$du_1 = u_1^{k+1} - u_1^k$$

$$\vdots$$

$$\begin{aligned} dy_i &= y_i^{k+1} - y_i^k \\ du_i &= u_i^{k+1} - u_i^k \end{aligned} \tag{12.3.27}$$

$$\vdots$$

$$\begin{aligned} dy_N &= y_N^{k+1} - y_N^k \\ du_N &= u_N^{k+1} - u_N^k \end{aligned}$$

With regard to Eq. (12.3.26) the following observations should be noted. First, the partial derivatives specified in this system of equations are evaluated at the kth iteration cycle. Second, this is a set of $2N$ simultaneous linear equations with $2N$ unknowns. Therefore, any of the standard methods for the solution of simultaneous, linear, algebraic equations can be applied, e.g., gaussian elimination or matrix inversion. The solution of this system provides values of y_1^{k+1}, u_1^{k+1}, y_i^{k+1}, u_i^{k+1}, y_N^{k+1}, u_N^{k+1} which are values of the unknowns at the $(k + 1)$ iteration cycle. Third, the matrix of coefficients for Eq. (12.3.26) is sparse, and the nonzero elements are banded about the main diagonal of the matrix. Fourth, the procedure indicated here is repeated until the difference between the value of any unknown in two consecutive iteration cycles falls below a predetermined tolerance level. Fifth, the rate at which this procedure converges depends to a great extent on the choice of the initial trial values. The closer the trial values are to the true values, the faster the convergence. In the case of unsteady open-channel flow, the initial trial values can either be taken as the values from the preceding time step or extrapolated from the preceding time step.

In applying this methodology, it is necessary to evaluate the coefficients of Eq. (12.3.26). The coefficients are the values of the functions F and G at each cycle of iteration. The required partial derivatives can be found by differentiating Eq. (12.3.21) and (12.3.22) or

$$\frac{\partial F_i}{\partial y_i} = 1 - \frac{1}{4}\frac{\Delta t}{\Delta x}(u_i + u_{i+1} + b) + \frac{1}{4}\frac{\Delta t}{\Delta x}(u_{i+1} - u_i + e) \left(1 - \frac{A_i}{T_i^2}\frac{dT_i}{dy_i}\right) - \frac{q}{2}\frac{\Delta t}{T_i^2}\frac{dT_i}{dy_i} \tag{12.3.28}$$

$$\frac{\partial F_i}{\partial y_{i+1}} = 1 + \frac{1}{4}\frac{\Delta t}{\Delta x}(u_{i'} + u_{i+1} + b) + \frac{1}{4}\frac{\Delta t}{\Delta x}(u_{i+1} - u_i + \mathrm{e}) \left(1 - \frac{A_{i+1}}{T_{i+1}^2}\frac{dT_{i+1}}{dy_{i+1}}\right) - \frac{q}{2}\frac{\Delta t}{T_{i+1}^2}\frac{dT_{i+1}}{dy_{i+1}} \tag{12.3.29}$$

$$\frac{\partial F_i}{\partial u_i} = \frac{1}{4}\frac{\Delta t}{\Delta x}(y_{i+1} - y_i + c) - \frac{1}{4}\frac{\Delta t}{\Delta x}\left(\frac{A_i}{T_i} + \frac{A_{i+1}}{T_{i+1}}\right) + \frac{m}{4}\frac{\Delta t}{\Delta x} \tag{12.3.30}$$

$$\frac{\partial F_i}{\partial u_{i+1}} = \frac{1}{4}\frac{\Delta t}{\Delta x}(y_{i+1} - y_i + c) + \frac{1}{4}\frac{\Delta t}{\Delta x}\left(\frac{A_i}{T_i} + \frac{A_{i+1}}{T_{i+1}}\right) + \frac{1}{4}\frac{\Delta t}{\Delta x}h \tag{12.3.31}$$

$$\frac{\partial G_i}{\partial y_i} = -1 - \frac{2}{3}\Delta x\, S_{fi}\left(\frac{1}{P_i}\frac{dP_i}{dy_i} - \frac{T_i}{A_i}\right) - q\frac{\Delta x}{2g}u_i\frac{T_i}{A_i^2} \tag{12.3.32}$$

$$\frac{\partial G_i}{\partial y_{i+1}} = 1 + \frac{2}{3}\Delta x\, S_{fi+1}\left(\frac{1}{P_{i+1}}\frac{dP_{i+1}}{dy_{i+1}} - \frac{T_{i+1}}{A_{i+1}}\right) - q\frac{\Delta x}{2g}u_{i+1}\frac{T_{i+1}}{A_{i+1}^2} \tag{12.3.33}$$

$$\frac{\partial G_i}{\partial u_i} = \frac{1}{g}\frac{\Delta x}{\Delta t} + \frac{1}{4g}(d' - 2u_i) + \Delta x\frac{S_{fi}}{u_i} + \frac{\Delta x}{2g}\frac{q}{A_i}u_i \tag{12.3.34}$$

and

$$\frac{\partial G_i}{\partial u_{i+1}} = \frac{1}{g}\frac{\Delta x}{\Delta t} + \frac{1}{4g}(c' + 2u_{i+1}) + \Delta x\frac{S_{fi+1}}{u_{i+1}} + \frac{\Delta x}{2g}\frac{q}{A_{i+1}}u_{i+1} \tag{12.3.35}$$

At this point, six comments and caveats regarding implicit difference techniques of solving the unsteady flow equations should be noted. First, Amein and Chu (1975) noted that the implicit centered difference technique discussed in detail here is accurate and stable for slowly varying flows. However, this difference scheme produces numerical oscillations under certain transient conditions.

Second, Amein and Chu (1975) also noted that Q rather than u is the preferred dependent variable in the governing equations because Q is, in general, a much smoother function of x and t. For example, between two adjacent sections the area and average velocity may vary significantly while the value of Q, which is the product of the area and average velocity, would change smoothly. If Q is used as the dependent variable, then the governing equations are

$$\frac{\partial Q}{\partial x} + \frac{\partial A}{\partial t} - q = 0 \tag{12.3.36}$$

and

$$\frac{1}{A}\frac{\partial Q}{\partial t} + \frac{1}{A}\frac{\partial}{\partial x}\left(\frac{Q^2}{A}\right) + qu = g(S_x - S_f) - g\frac{\partial y}{\partial x} \tag{12.3.37}$$

Third, Price (1974) compared four numerical methods of flood routing: (*a*) the leap-frog explicit method (Liggett and Woolhiser, 1967), (*b*) the two-step Lax-Wendroff explicit method (e.g., Liggett and Woolhiser, 1967), (*c*) the Amein four-point implicit difference method, and (*d*) the fixed-mesh characteristic method (Baltzer and Lai, 1968). This investigation concluded that the four methods are all accurate and that the four-point difference method of Amein and Fang (1970) is the most efficient method for flood-routing problems. Price (1974) further noted that optimum accuracy is obtained with this method when the finite difference time step is chosen approximately equal to the space step divided by the kinematic wave speed or

$$\Delta t \approx \frac{\Delta x}{\tfrac{5}{3}\, Q/A} \tag{12.3.38}$$

Fourth, implicit difference schemes similiar to the one discussed here are used in many standard unsteady flow models, e.g., the National Weather Service (Fread, 1978) and the USGS model (Schaffranek et al., 1981). Many of these models use what is termed a *weighted* four-point implicit scheme. This numerical technique is favored because it can be readily used with unequal distance steps and its stability-convergence properties can be controlled. In the weighted scheme the continuous $x - t$ plane, time derivatives are specified in a fashion similar to Eq. (12.3.10), but the spatial derivatives are approximated by a finite difference quotient proportioned between two adjacent time lines according to the weighting factors θ and $1 - \theta$ (Fig. 12.5), where $\theta = \Delta t'/\Delta t$. For the arbitrary function

$$\frac{\partial \Gamma}{\partial x} = \frac{\theta}{\Delta x}(\Gamma_{i+1}^{j+1} - \Gamma_i^{j+1}) + \frac{1-\theta}{\Delta x}(\Gamma_{i+1}^{j} - \Gamma_i^{j}) \tag{12.3.39}$$

Variables other than derivatives are approximated in a similar manner or

$$\Gamma = \frac{\theta}{2}(\Gamma_i^{j+1} + \Gamma_{i+1}^{j+1}) + \frac{1-\theta}{2}(\Gamma_i^{j} + \Gamma_{i+1}^{j}) \tag{12.3.40}$$

When $\theta = 1$, a fully implicit scheme is formed which is similar to that used by Amein and Chu (1975). A centered difference scheme similar to the one described in detail in this chapter results if $\theta = 0.5$. If $\theta = 0$, then an explicit difference scheme is the result. The influence of the weighting factor θ on stability and convergence was examined by Fread (1975a), who concluded that the

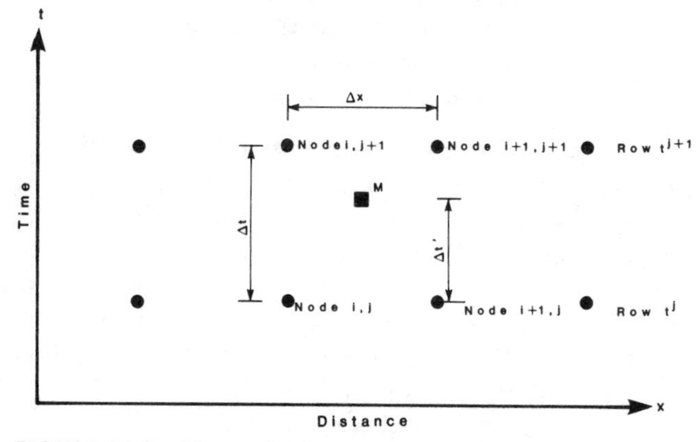

FIGURE 12.5 Network of points on x-t plane.

accuracy of the computation decreases as θ departs from 0.5 and approaches 1.0. This effect on accuracy becomes more pronounced as the time step increases. Fread (1978) recommended a value of $\theta = 0.55$ since such a weighting parameter value minimizes the loss of accuracy and also avoids the instabilities under some transient conditions reported by Amein and Chu (1975), Baltzer and Lai (1968), and Fread (1975b).

Sixth, Price (1974) noted and Fread (1975b) concurred that the implicit method described in this section can be inaccurate when a flood exceeds the channel capacity and propagates along the overbank or flood plain. In such a situation, the rate of flood wave propagation is different for the channel flow relative to the overbank flow owing to the differences in hydraulic properties between the two. In addition, Fread (1975b) noted that when the channel meanders through a flood plain the time of travel of the wave may be different owing to differences in the two reach lengths.

12.4 GENERALIZATION OF THE IMPLICIT FOUR-POINT TECHNIQUE

Of the above comments regarding the implicit four-point difference approximation, the last one regarding overbank flow is crucial from the viewpoint of practice. The modified form of the St. Venant equations developed by Fread (1975b, 1976, 1982) to deal with overbank flow are

$$\frac{\partial Q}{\partial x} + \frac{\partial (A + A_0)}{\partial t} - q = 0 \tag{12.4.1}$$

and

$$\frac{\partial Q}{\partial t} + \frac{\partial (Q^2/A)}{\partial x} + gA\left(\frac{\partial z}{\partial x} + S_f + S_e\right) + L = 0 \tag{12.4.2}$$

where A = active cross-sectional area of flow
A_0 = inactive or off-channel cross-sectional area
S_e = expansion-contraction slope
L = momentum effect of lateral flow assumed here to enter or exit perpendicular to direction of main flow

The term L has the following form: (1) lateral inflow, $L = 0$; (2) lateral outflow by seepage, $L = -0.5\, q\, Q/A$; and (3) bulk lateral outflow, $L = -q\, Q/A$. It is also noted that in this context the terminology *off-channel storage* indicates areas where the flow velocity in the longitudinal direction is negligible relative to the velocity in the main channel section. Examples of dead zones might include embayments, ravines, heavily wooded flood plains, and tributaries which connect to the flow channel but do not pass flow.

As before, the friction slope in Eq. (12.4.2), S_f, is estimated from the Manning uniform flow equation or

$$S_f = \frac{n^2 Q|Q|}{\phi^2 A^2 R^{4/3}} \tag{12.4.3}$$

where R = hydraulic radius of the active cross-sectional area or

$$R = \frac{A}{T}$$

where T = top width of the active cross-sectional area. The term S_e in Eq. (12.4.2) is given by

$$S_e = \frac{k\,\Delta\,(Q/A)^2}{2g\,\Delta x} \tag{12.4.4}$$

where k = an expansion-contraction coefficient (e.g., Morris and Wiggert, 1972) and $\Delta\,(Q/A)^2$ = difference in the average velocity at two adjacent cross sections separated by a distance Δx.

Equations (12.4.1) and (12.4.2) are then modified to account for the differences in flood wave properties of flow occurring simultaneously in the channel and the flood plain section (Fread 1975b, 1976, 1982), or

$$\frac{\partial(K_cQ)}{\partial x_c} + \frac{\partial(K_lQ)}{\partial x_l} + \frac{\partial(K_rQ)}{\partial x_r} + \frac{\partial A}{\partial t} - q = 0 \tag{12.4.5}$$

and

$$\frac{\partial Q}{\partial t} + \frac{\partial(K_c^2Q^2/A_c)}{\partial x_c} + \frac{(K_l^2Q^2/A_l)}{\partial x_l} + \frac{\partial(K_r^2Q^2/A_r)}{\partial x_r} + gA_c\left(\frac{\partial z}{\partial x_c} + S_{fc} + S_e\right) + gA_l\left(\frac{\partial z}{\partial x_l} + S_{fl}\right) + gA_r\left(\frac{\partial z}{\partial x_r} + S_{fr}\right) = 0 \tag{12.4.6}$$

where the subscripts c, l, and r represent the channel, left flood plain, and right flood plain, respectively. The parameters K_c, K_l, and K_r are used to proportion the flow Q into channel flow, left flood plain flow, and right flood plain flow. These parameters have the following definitions:

$$K_c = \frac{1}{1 + k_l + k_r} \tag{12.4.7}$$

$$K_l = \frac{k_l}{1 + k_l + k_r} \tag{12.4.8}$$

$$K_r = \frac{k_r}{1 + k_l + k_r} \tag{12.4.9}$$

where

$$k_l = \frac{Q_l}{Q_c} = \frac{n_c}{n_l}\frac{A_l}{A_c}\left(\frac{R_l}{R_c}\right)^{2/3}\left(\frac{\Delta x_c}{\Delta x_l}\right)^{1/2} \tag{12.4.10}$$

and

$$k_r = \frac{Q_r}{Q_c} = \frac{n_c}{n_r}\frac{A_r}{A_c}\left(\frac{R_r}{R_c}\right)^{2/3}\left(\frac{\Delta x_c}{\Delta x_r}\right)^{1/2} \tag{12.4.11}$$

The friction slope terms in Eq. (12.4.6) are given by

$$S_{fc} = \frac{n_c^2 |K_c Q| K_c Q}{\phi^2 A_c^2 R_c^{4/3}} \tag{12.4.12}$$

$$S_{fl} = \frac{n_l^2 |K_l Q| K_l Q}{\phi^2 A_l^2 R_l^{4/3}} \tag{12.4.13}$$

and

$$S_{fr} = \frac{n_r^2 |K_r Q| K_r Q}{\phi^2 A_r^2 R_r^{4/3}} \tag{12.4.14}$$

In Eq. (12.4.5) the variable A is the total cross-sectional area or

$$A = A_c + A_l + A_r + A_0 \tag{12.4.15}$$

where A_0 = off-channel storage (inactive) area.

Equations (12.4.5) and (12.4.6) constitute a system of partial differential equations of the hyperbolic type which must, in the general case, be solved numerically, e.g., by an implicit four-point difference scheme.

12.5 BOUNDARY AND INITIAL CONDITIONS

As noted previously in this chapter, the solution of the equations governing unsteady flow requires that boundary conditions be specified throughout the simulation period. One set of boundary conditions must be specified at the physical extremities of the system, and additional conditions are required at branch junctions within the system, e.g., at nodes where tributaries join the main channel and at nodes where the main channel divides to pass an island.

Various combinations of boundary conditions can be specified at the physical extremities of a network. External boundary conditions which are commonly specified are zero discharge, discharge as a function of time, stage as a function of time, or a known, unique stage-discharge relationship. Fread (1975b) noted that when there is no significant flow disturbance downstream of the routing reach which can propagate into the reach and influence the flow, a rating equation is always available—the Manning or Chezy uniform flow equation.

The most common internal boundary condition occurs at nodes where the channel is either joined by a tributary or divided by the presence of an island. If the velocity differences and energy losses due to turbulence are neglected at these junctions, then appropriate boundary conditions can be specified (Schaffranek et al., 1981). For a junction composed of N branches, continuity requires

$$\sum_{i=1}^{n} Q_i = W_k \tag{12.5.1}$$

where W_k is a specified external flow at node k. In addition, the stage at a junction must be single-valued. For a junction with N branches

$$y_i = y_{i+1} \qquad \text{for } i = 1, 2, \ldots, (N-1) \tag{12.5.2}$$

Thus, at every internal junction of N branches there is one equation of continuity [(Eq. (12.5.1)] and $(N - 1)$ equations specifying stage compatibility [Eq. (12.5.2)] which must be satisfied. A typical network of nodes for a hypothetical example is shown in Fig. 12.6.

The solution of the equations governing unsteady flow and the specified boundary conditions also requires that initial values of the unknown variables be specified. Initial values may be obtained from measurements or estimated

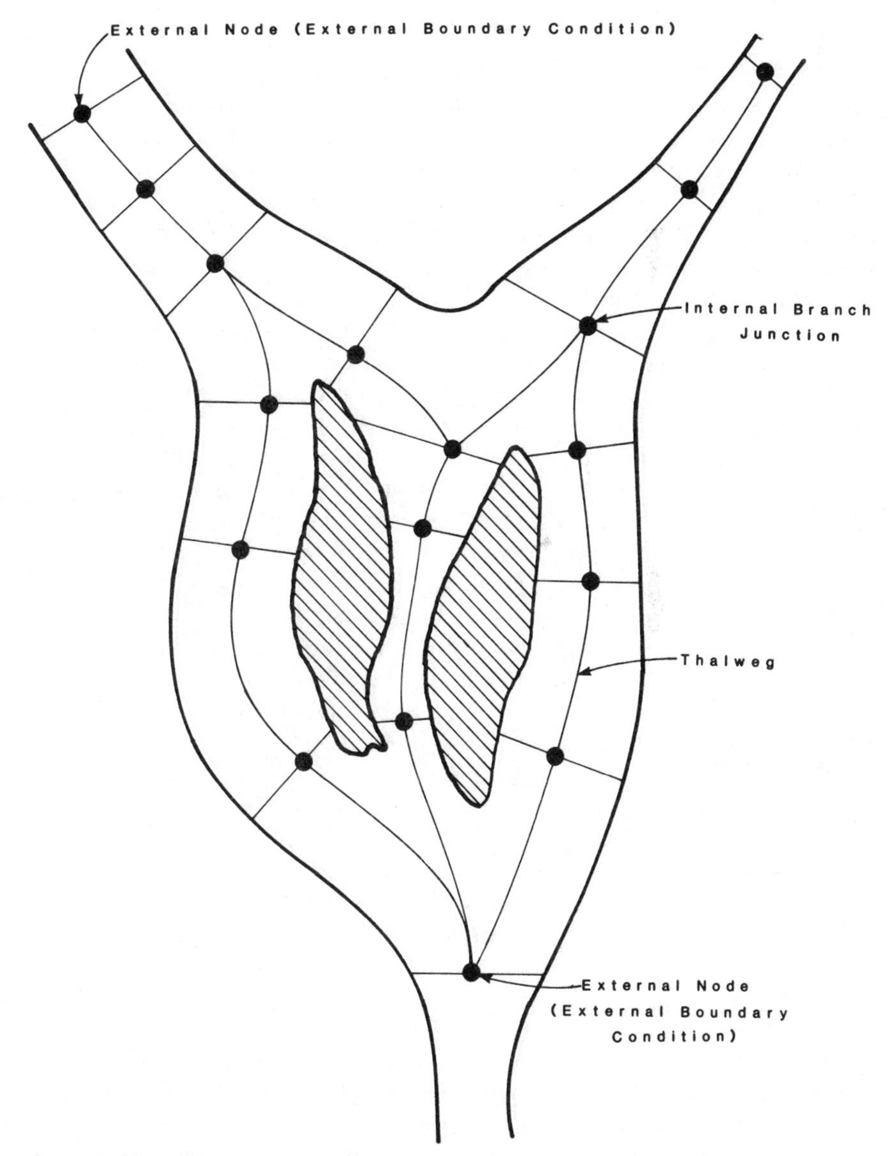

FIGURE 12.6 Node network for unsteady flow.

by computation, e.g., uniform flow or gradually varied flow approximations or previous unsteady flow simulations. While reasonable initial conditions are desirable, estimates can be used if a sufficient period of time is provided for the unsteady flow model to dissipate the errors and converge to the true solution. Models of flow systems having high rates of dissipation will converge more rapidly than models of systems having low rates of energy dissipation (Lai, 1965a, 1965b).

12.6 CALIBRATION AND VERIFICATION

As is the case with all models, unsteady flow models must be calibrated and verified. These are two processes which require accurate prototype data. As noted by French and Krenkel (1980), the calibration and verification processes require two independent and statistically reliable sets of prototype data. One data set is used to establish the optimum values of the "free" coefficients, and the second data set is used to verify that the calibration process established optimum coefficient values.

In the calibration process, prototype data are used to refine the values of the least quantifiable variables, e.g., the flow resistance coefficient. The objective of this process is to adjust these variables so that the prototype system for a range of flow conditions is accurately replicated. Although the reproduction of the water surface elevations by the model is of value, the replication of the prototype discharge hydrograph in both time and amplitude is much more important. If a high level of calibration can be achieved and verified, then it may be possible to extend the application of the model beyond the limits of the data used in the calibration and verification processes. In general, the data for the calibration and verification of an unsteady flow model consists of two time series of measured discharges with simultaneously measured water surface elevations. The goal of the calibration and verification processes is to satisfy some goodness of fit criterion, e.g., minimizing the sum of the squares of the differences between the measured and simulated values.

While in theory all aspects and variables in a model may be subject to adjustment and refinement, in practice there are variables which cannot be determined directly. For example, in the case of unsteady flow models, channel geometry data are not usually adjusted since such data can be measured with reasonable accuracy. In such models, attention is usually centered on the value of the resistance coefficient, a variable which is not normally measured directly.

The difficulty in determining a theoretically accurate value of the flow resistance coefficient results primarily from the fact that in applied hydraulics the energy dissipation relation used is the result of empiricism. In addition, the dissipation formulation used in unsteady flow models is an approximation borrowed from an empirical or semiempirical formula developed for steady, uniform flow. Actually, little factual information regarding the effect of perimeter resistance under unsteady flow conditions is available.

Schaffranek et al. (1981) have suggested that the calibration of an unsteady flow model should begin with a flow data set gathered during steady or nearly steady flow conditions. If such a data set is available, then the resistance coefficient can be either determined from the Manning or Chezy uniform flow equations or simply estimated. At this point a number of comments regarding the calibration process will be made:

1. The use of a resistance coefficient which is too small reduces the flow resistance and increases the momentum or inertia. The simulations which result will have peaks which are too high and minimum points which are too low. In general, there is also a phase shift which may or may not be desirable. The use of a resistance coefficient which is too large has the opposite effect on the simulation results.
2. The recorded water surface elevations may be in error due to datum errors, surveying inaccuracies, or vertical displacement of the gaging structure. The result of an error of this type is an increase or decrease in the water surface slope throughout the reach in which the error occurs. An increased water surface slope yields increased flow in the downstream direction while a decreased water surface slope has the opposite effect.
3. The use of too large a cross-sectional area results in magnified peaks and troughs while using a cross-sectional area which is too small has the opposite effect.
4. The weighting factor θ [Eq. (12.3.39)] is viewed by some as being a calibration variable (see, for example, Schaffranek et al., 1981). Appropriate values of θ appear to be in the range $0.6 \leq \theta \leq 1.0$. Values of θ less than 0.6 may result in instabilities. *Note:* This problem has been mentioned previously in Section 12.3, and Fread (1978) recommends $\theta = 0.55$. Schaffranek et al. (1981) further noted that $\theta = 1.0$, while yielding the greatest computational stability, also results in a damping of the computed wave.

Schaffranek et al. (1981) noted that the calibration of unsteady flow models may range from simple to difficult depending on the complexity of the flow regime being simulated and the interconnection of the channels within the prototype. Channel networks in which the flow has only one path of travel between any two locations are generally easier to calibrate than channel networks where it is possible for the flow to use more than one route between locations. In this latter situation, erroneous circulations may appear internally and render the model useless.

As can be discerned from the foregoing discussion, the determination of an optimum value of the flow resistance coefficient is the primary task in the calibration of a one-dimensional unsteady flow model. An obvious and widely used method of accomplishing this is a trial-and-error technique in which the governing equations are repeatedly solved for different values of the flow resistance coefficient. The value of this coefficient which produces the best fit, as measured by some goodness-of-fit criterion, between the measured and simulated values is the optimum value. This trial-and-error search for optimum values can be tedious, difficult, and expensive. In addition, in the case of a model for a river system, Fread and Smith (1978) have noted that modification of the resistance coefficient value in the tributary can affect the flow both upstream and downstream of the tributary in the main stem of the river as well as the flow in the other tributaries. This is particularly true in large rivers with major tributaries and/or when the bottom slope of the river is mild, e.g., 2 ft/mile (0.38 m/km).

In general, the determination of parameter values in a set of partial differential equations for a specified set of boundary and initial conditions is termed the *inverse problem*. Fread and Smith (1978) have developed an optimization methodology for determining the value of the flow resistance coefficient in one-dimensional unsteady flow models. In this technique, the resistance coefficient is assumed to vary with the stage and discharge in the channel but is not a function of longitudinal distance. The goodness-of-fit criterion used by Fread and Smith (1978) is the minimization of the absolute value of the sum of the differences between the observed and simulated values of either the stage or discharge.

In their development, Fread and Smith (1978) assumed that the flow was governed by Eqs. (12.4.1) and (12.4.2) with the friction slope given by Eq. (12.4.3). These investigators further assumed that these equations were solved by a weighted, nonlinear, four-point difference scheme and that the following conditions were specified:

1. The initial condition was the specification of the stages and discharges at all computational nodes in the reach *A-B* (Fig. 12.7).
2. The upstream boundary condition is a known discharge hydrograph or

$$Q_A = Q'_A(t) \tag{12.6.1}$$

3. The downstream boundary condition is a known stage hydrograph or

$$y_B = y'_B(t) \tag{12.6.2}$$

4. All lateral inflows or outflows are known.

It is noted that by specifying a known discharge hydrograph as the upstream boundary condition all flow perturbations occurring upstream of the reach *A-B* which could affect the flow in reach *A-B* are taken into account. By specifying the downstream boundary condition, all flow pertubations downstream of the reach which might affect the flow within the reach are also taken into account. Finally, as noted above, the flow resistance coefficient n does not vary as a function of distance within the reach *A-B* but is a function of the stage of

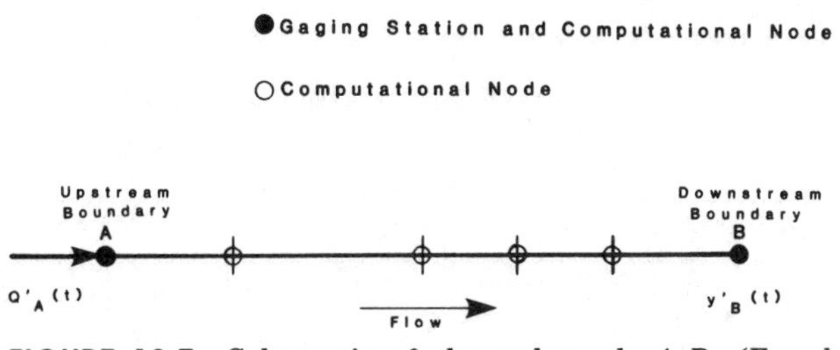

FIGURE 12.7 Schematic of channel reach *A-B*. *(Fread and Smith, 1978.)*

discharge. Fread and Smith (1978) assumed that the variation of n with the discharge is described by a piecewise, continuous linear function of the average discharge through the reach (Fig. 12.8). In this figure, the total range of possible discharges is divided into a number of strata ($j = 1, 2, \ldots, J$), and each stratum is associated with an (n, Q) breakpoint in the $n(Q)$ piecewise continuous linear function.

In this analysis, an optimal $n(Q)$ value is sought which minimizes the absolute value of the sum of the difference between the estimated stages $\hat{y}_A$ and the measured stages y_A at the upstream boundary. Under this definition, the objective function for miminimization is

$$\min\,(S_T) = \left|\sum_{j=1}^{J} S_j\right| = \left|\sum_{j=1}^{J} \frac{1}{M_J} \sum_{i=1}^{M_J} \left(\hat{y}_A - y_A^i\right)\right| \tag{12.6.3}$$

where M_J = total number of stages associated with discharges in the jth stratum. Within each stratum it is convenient to minimize the objective function for that stratum S_j where

$$S_j = \frac{1}{M_J} \sum_{i=1}^{M_J} \left(\hat{y}_A - y_A^i\right) \tag{12.6.4}$$

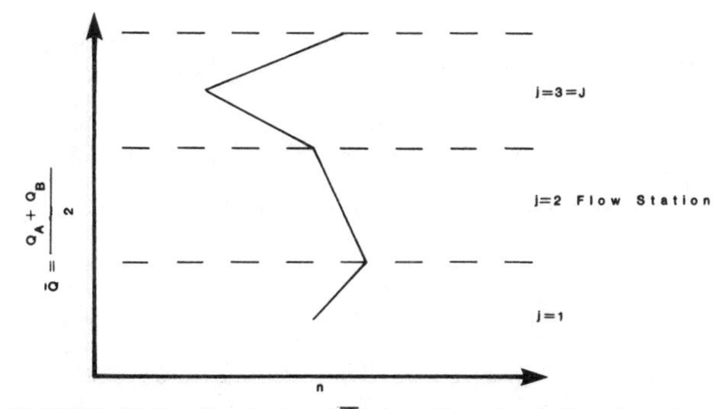

FIGURE 12.8 Typical $n\,(\overline{Q})$ functional relationship for each *A-B* (Fig. 12.7). *(Fread and Smith, 1978.)*

It is noted that since

$$\left| \sum_{j=1}^{J} S_j \right| \leq \sum_{j=1}^{J} \left| S_j \right| \tag{12.6.5}$$

the minimization of S_j for each stratum also results in a minimum S_T.

The objective function for the jth stratum can be expressed in the following functional form:

$$\min S_j \{y_{Aj} [n_j (\overline{Q}_j)]\}; \qquad j = 1, 2, \ldots, J \tag{12.6.6}$$

where S_j = a function of the estimated and measured stages at the upstream boundary. According to Fread and Smith (1978) an equivalent form of Eq. (12.6.6) is

$$S_j \{y_{Aj} [n_j (\overline{Q}_j)]\} = 0; \qquad j = 1, 2, \ldots, J \tag{12.6.7}$$

By expressing either Eq. (12.6.4) or (12.6.6) in the form of Eq. (12.6.7), a gradient-type Newton-Raphson algorithm can be used to improve the estimate of $n(\overline{Q})$ such that the objective function is minimized. With these definitions, the solution proceeds as follows:

1. Initial values of the n $(\overline{Q})$ function are estimated. These initial values can be either sophisticated guesses or the result of the application of the Manning equation to the boundry condition data; e.g.,

$$n_j^1 = \frac{1}{M_j} \sum_{i=1}^{Mj} \frac{\phi A R^{2/3}}{(Q_A + q\,\Delta x)} \left(\frac{y_{Ai} - y_{Bi}}{\Delta x} \right)^{1/2} \tag{12.6.8}$$

   where the superscript of the Manning resistance coefficient denotes the number of iterations and Δx is the distance between stations A and B. If the beginning estimates deviate significantly from the optimum values, then later steps in this computational scheme may not converge.

2. The governing equations, i.e., Eqs. (12.4.1), (12.4.2), (12.6.1), and (12.6.2), are solved.

3. The objective function, Eq. (12.6.3), is then evaluated for each discharge stratum.

4. Improved values of the n $(\overline{Q})$ function are then estimated. If $k = 1$, then

$$n_j^{k+1} = n_j^1 \, \frac{1.00 - 0.01\, S_j^1}{|\, S_j^1 \,|} \qquad \text{for } j = 1, 2, \ldots, J \tag{12.6.9}$$

   If $k > 1$, then

$$n_j^{k+1} = n_j^k - \frac{S_j^k \left(n_j^k - n_j^{k-1} \right)}{S_j^k - S_j^{k-1}} \qquad \text{for } j = 1, 2, \ldots, J \tag{12.6.10}$$

5. At this point, the objective function is evaluated, and the following inequalities are used to determine whether additional iterations are required. If

$$S_T^k = \frac{1}{J}\sum_{j=1}^{J} S_j^k < \epsilon \qquad (12.6.11)$$

or

$$S_T^k \geq S_T^{k-1} \qquad (12.6.12)$$

where ϵ is a convergence criterion, then an optimal $n(\overline{Q})$ function has been found.

The technique of determining an optimal $n(\overline{Q})$ function can be applied to river systems consisting of multiple reaches and/or tributaries. Fread and Smith (1978) noted that when river systems are treated, discharge measurements are required only at the upstream extremity of the main stem of the river and at each tributary; however, stage measurements are required at all gaging stations. Such a set of required measurements is well suited to large, relatively flat river systems where gradually varied flow profiles may make discharge measurements very difficult.

BIBLIOGRAPHY

Amein, M., "Streamflow Routing on Computer by Characteristics," *Water Resources Research,* vol. 2, no. 1, 1966, pp. 123–130.

Amein, M., "An Implicit Method for Numerical Flood Routing," *Water Resources Research,* vol. 4, no. 4, 1968, pp. 719–726.

Amein, M., and Fang, C. S., "Implicit Flood Routing in Natural Channels," *Proceedings of the American Society of Civil Engineers, Journal of the Hydraulics Division,* vol. 96, no. HY12, December 1970, pp. 2481–2500.

Amein, M., and Chu, H. L., "Implicit Numerical Modeling of Unsteady Flows," *Proceedings of the American Society of Civil Engineers, Journal of the Hydraulics Division,* vol. 101, no. HY6, June 1975, pp. 717–731.

Baltzer, R. A., and Lai, C., "Computer Simulations of Unsteady Flow in Waterways," *Proceedings of the American Society of Civil Engineers, Journal of the Hydraulics Division,* vol. 94, no. HY 4, July 1968, pp. 1083–1117.

Fletcher, A. G., and Hamilton, W. S., "Flood Routing in an Irregular Channel," *Proceedings of the American Society of Civil Engineers, Journal of the Engineering Mechanics Division,* vol. 93, no. EM3, June 1967, pp. 45–62.

Fread, D. L., "Numerical Properties of Implicit Four Point Finite Difference Equations of Unsteady Flow," *NOAA Technical Memorandum NWS HYDRO,* 18, NOAA, 1975a.

Fread, D. L., Discussion of "Comparison of Four Numerical Method for Flood Routing," by R. K. Price, *Proceedings of the American Society of Civil Engineers, Journal of the Hydraulics Division,* vol. 101, no. HY3, 1975b, pp. 565–567.

Fread, D. L., "Flood Routing in Meandering Rivers with Flood Plains," *Proceedings, Rivers '76,* Third Annual Symposium of Waterways, Harbors and Coastal Engineering Division, American Society of Civil Engineers, vol. 1, 1976, pp. 16–35.

Fread, D. L., "National Weather Service Operational Dynamic Wave Model," National Weather Service, NOAA, Silver Spring, Md., 1978.

Fread, D. L., "DAMBRK: The NWS Dam Break Flood Forecasting Model," Office of Hydrology, National Weather Service, Silver Spring, Md., 1982.

Fread, D. L., and Smith, G. F., "Calibration Technique for 1-D Unsteady Flow Models," *Proceedings of the American Society of Civil Engineers, Journal of the Hydraulics Division,* vol. 104, no. HY7, July 1978, pp. 1027–1044.

French, R. H., and Krenkel, P. A., "Effectiveness of River Models," *Progress in Water Technology,* no. 13, no. 99, 1980.

Lai, C., "Flows of Homogeneous Density in Tidal Reaches: Solution by Method of Characteristics," U.S. Geological Survey Open File Report, Washington, 1965a.

Lai, C., "Flows of Homogeneous Density in Tidal Reaches: Solution by Implicit Method," U.S. Geological Survey Open File Report, Washington, 1965b.

Liggett, J. A., and Woolhiser, D. A., "Difference Solution of Shallow-Water Equations," *Proceedings of the American Society of Civil Engineers, Journal of the Engineering Mechanics Division,* vol. 93, no. EM2, April 1967, pp. 39–71.

Morris, H. M., and Wiggert, J. M., *Applied Hydraulics in Engineering,* Ronald Press, New York, 1972, pp. 570–573.

Ponce, V. M., Li, R. M., and Simons, D. B., "Applicability of Kinematic and Diffusion Models," *Proceedings of the American Society of Civil Engineers, Journal of the Hydraulics Division,* vol. 104, no. HY3, March 1978, pp. 353–360.

Price, R. K., "Comparison of Four Numerical Methods for Flood Routing," *Proceedings of the American Society of Civil Engineers, Journal of the Hydraulics Division,* vol. 100, no. HY7, July 1974, pp. 879–899.

Ralston, A., *A First Course in Numerical Analysis,* McGraw-Hill Book Company, New York, 1965.

Schaffranek, R. W., Baltzer, R. A., and Goldberg, D. E., "A Model for Simulation of Flow in Singular and Interconnected Channels," chapter C3, *Techniques of Water Resources Investigations of the United States Geological Survey,* Washington, 1981.

Viessman, W., Jr., Knapp, J. W., Lewis, G. L., and Harbaugh, T. E., *Introduction to Hydrology,* 2d ed., Harper & Row, New York, 1972.

Weinmann, D. E., and Laurenson, E. M., "Approximate Flood Routing Methods: A Review," *Proceedings of the American Society of Civil Engineers, Journal of the Hydraulics Division,* vol. 105, no. HY12, December 1979, pp. 1521–1536.

THIRTEEN

Rapidly Varied, Unsteady Flow

SYNOPSIS

SYNOPSIS

In this chapter, flows classified as rapidly varied and unsteady are discussed. This type of flow can result from the rapid operation of flow control structures, tidal action, the catastrophic failure of a dam, or supercritical flow in a steep channel at normal depth (roll waves). Beginning with an analytical treatment of elementary surges, the chapter includes a detailed treatment of the dam breach problem with emphasis on techniques amenable to hand estimations of discharge and downstream depth of flow, numerical methods for evaluating surge problems, techniques for determining the conditions under which roll waves will form, methods of estimating where the roll waves will form, and a technique for estimating the height of the resulting roll waves.

13.1 INTRODUCTION

As noted in the previous chapter, many open-channel flow phenomena of great importance to the hydraulic engineer involve flows which are unsteady. The term *rapidly varied unsteady flow* refers to flows in which the curvature of the wave profile is large; the change of the depth of flow with time is rapid; the vertical acceleration of the water particles is significant relative to the total acceleration; and the effect of boundary friction can be neglected. Examples of rapidly varied, unsteady flows include the catastrophic failure of a dam, tidal bores, and surges which result from the quick operation of control structures such as sluice gates.

13.2 ELEMENTARY SURGES

By definition, a surge or surge wave is a moving wave front which results in an abrupt change of the depth of flow. If the surge is the result of tidal action, e.g., the River Severn in England and the Chien Tang River in the Peoples Republic of China, the surge is termed a *tidal bore*. As an elementary example, consider a positive surge which results from the rapid, partial closure of a sluice gate in a rectangular, level, frictionless channel (Fig. 13.1*a*). In this figure it is assumed that the observer is stationary and watches the surge pass; clearly, this is by definition an unsteady phenomenon since the observer sees a change in the depth of flow as a function of time. In Fig. 13.1*b* the coordinate system has been transformed to a system which is translating in the direction the surge is moving at the velocity of the surge. In this figure, it is as if the observer were moving with the surge, and to the observer the phenomenon appears steady. The steady continuity equation applied to Fig. 13.1*b* yields

$$(u_1 + c)y_1 = (u_2 + c)y_2 \tag{13.2.1}$$

The momentum equation applied to Fig. 13.1*b* yields

$$\gamma \frac{y_1^2}{2} - \gamma \frac{y_2^2}{2} = \frac{\gamma}{g} y_1(u_1 + c)(u_2 + c - u_1 - c)$$

or

$$\frac{\gamma}{2}(y_1^2 - y_2^2) = \frac{\gamma}{g} y_1(u_1 + c)(u_2 - u_1) \tag{13.2.2}$$

where boundary friction has been neglected. Eliminating u_2 in Eq. (13.2.2) by the manipulation of Eq. (13.2.1) yields

$$u_1 + c = \sqrt{gy_1}\left[\frac{y_2}{2y_1}\left(1 + \frac{y_2}{y_1}\right)\right]^{1/2} \tag{13.2.3}$$

With regard to Eq. (13.2.3), the following observations should be noted:

1. If y_1 approaches y_2, i.e., the wave height becomes incremental, then Eq. (13.2.3) becomes

$$u_1 + c = \sqrt{gy_1}$$

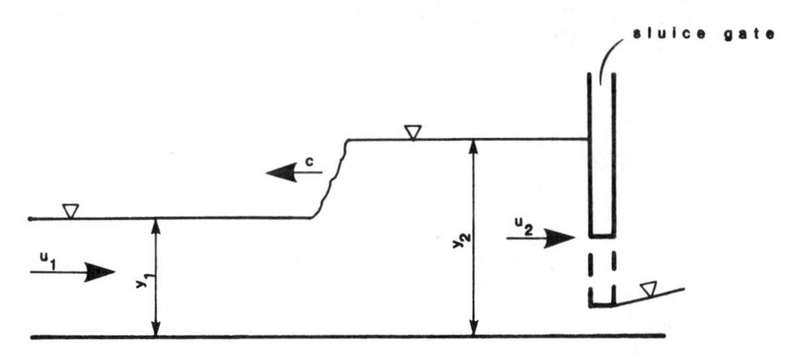

(a)

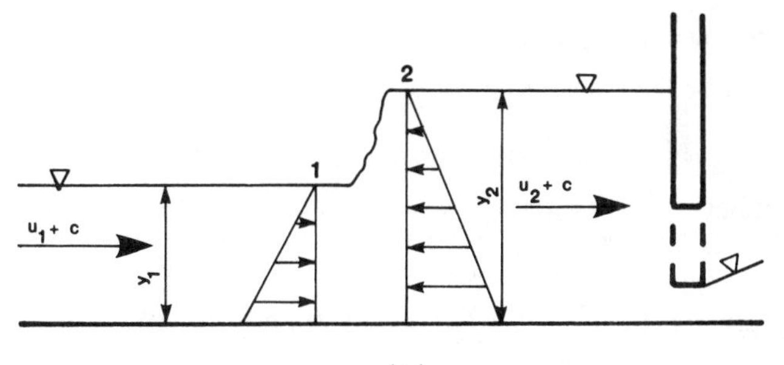

(b)

FIGURE 13.1 (*a*) Positive surge wave in a rectangular channel; observer stationary. (*b*) Positive surge wave in a rectangular channel; observer moving with surge.

Further, if the water is still, then

$$c = \sqrt{gy_1}$$

or

$$1 = \frac{c}{\sqrt{gy_1}}$$

which is the definition of the Froude number.

2. If $c = 0$, then Eq. (13.2.3) is analogous to those derived for the hydraulic jump in Chapter 3, i.e., Eqs. (3.2.4) and (3.2.5). Thus, a positive surge may be viewed as a moving hydraulic jump.

EXAMPLE 13.1

A level, frictionless rectangular channel 3.0 m (9.8 ft) wide conveys a flow of 18 m^3/s (640 ft^3/s) at a depth of 2.0 m (6.6 ft). If the flow rate is suddenly reduced to 12 m^3/s (420 ft^3/s), estimate the height and speed of the resulting surge wave. This example was first presented by Streeter and Wylie (1975).

Solution

According to the problem statement

$$y_1 = 2.0 \text{ m (6.6 ft)}$$

and

$$u_1 = \frac{Q_1}{A_1} = \frac{18}{(2.0)(3.0)} = 3.0 \text{ m/s (9.8 ft/s)}$$

Further,

$$q_2 = u_2y_2 = \frac{Q_2}{b} = \frac{12}{3.0} = 4.0 \text{ (m}^3\text{/s)/m [43 (ft}^3\text{/s)/ft]}$$

Expanding Eq. (13.2.1)

$$u_1y_1 + y_1c = u_2y_2 + y_2c$$

$$u_1y_1 = u_2y_2 + c(y_2 - y_1)$$

and substituting yields

$$(3.0)(2.0) = 4.0 + c(y_2 - 2.0)$$

$$6.0 = 4.0 + c(y_2 - 2.0)$$

Solving the above equation for the celerity of the wave yields

$$c = \frac{6.0 - 4.0}{y_2 - 2.0} = \frac{2.0}{y_2 - 2.0} \tag{13.2.4}$$

Rearranging Eq. (13.2.2) yields

$$y_2^2 - y_1^2 = \left(\frac{2y_1}{g}\right)(u_1 + c)(u_1 - u_2)$$

and substituting gives

$$y_2^2 - (2.0)^2 = 2\left(\frac{2.0}{9.8}\right)(c + 3.0)(3.0 - u_2)$$

$$y_2^2 - (2.0)^2 = 0.41(c + 3.0)(3.0 - u_2)$$

Substituting Eq. (13.2.4) in the above equation for c and $u_2 = 4.0/y_2$ yields

$$y_2^2 - 4.0 = 0.41\left(\frac{2.0}{y_2 - 2.0} + 3.0\right)\left(3.0 - \frac{4.0}{y_2}\right)$$

This equation is solved by trial and error, and it is found that

$$y_2 = 2.8 \text{ m } (9.2 \text{ ft})$$

and

$$u_2 = \frac{4.0}{y_2} = \frac{4.0}{2.8} = 1.4 \text{ m/s } (4.6 \text{ ft/s})$$

Therefore, the height of the surge is 0.80 m (2.6 ft) and the celerity or speed of the wave is

$$c = \frac{2}{y_2 - 2.0} = \frac{2}{2.8 - 2.0} = 2.5 \text{ m/s } (8.2 \text{ ft/s})$$

The positive surge is usually the result of a downstream decrease in the flow rate, while a negative surge may result from an increase in the downstream flow rate. In Fig. 13.2*a* a negative surge is shown from the viewpoint of a stationary system of coordinates, and Fig. 13.2*b* is the same situation from the viewpoint of a system of coordinates translating at the velocity of the surge. Application of the equation of continuity between points 1 and 2 in Fig. 13.2*b* yields

$$(u - \delta u - c)(y - \delta y) = (u - c)y$$

and expanding

$$-u\,\delta y + c\,\delta y - y\,\delta u - yc + yu + \delta y\,\delta u = uy - cy$$

If the term $\delta y\,\delta u$ is neglected, the above equation can be simplified to yield

$$(c - u)\,\delta y = y\,\delta u \tag{13.2.5}$$

or

$$\frac{\delta u}{\delta y} = \frac{c - u}{y} \tag{13.2.6}$$

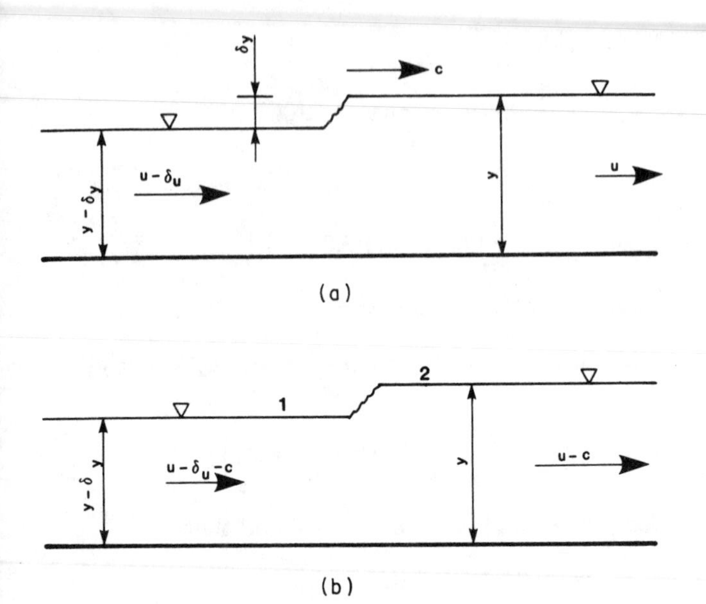

FIGURE 13.2 (*a*) Negative surge wave in a rectangular channel; stationary coordinate system. (*b*) Negative surge wave in a rectangular channel; translating coordinate system.

The momentum equation for the control volume defined in Fig. 13.2*b* is

$$\gamma \frac{(y - \delta y)^2}{2} - \gamma \frac{y^2}{2} = \frac{\gamma}{g} [u - c - (u - \delta u - c)](u - c)\, y$$

Simplifying this equation and neglecting the products of all small quantities yields

$$\delta y = \frac{c - u}{g} \delta u$$

or

$$\frac{\delta u}{\delta y} = \frac{g}{c - u} \tag{13.2.7}$$

Equating Eqs. (13.2.6) and (13.2.7) and solving for c yields

$$\frac{c - u}{y} = \frac{g}{c - u}$$

$$(c - u)^2 = gy$$

or

$$c = u \pm \sqrt{gy} \tag{13.2.8}$$

Equation (13.2.8) shows that the velocity of an elementary surge in still water is $\sqrt{gy}$; in the case of an elementary negative surge its celerity is $\sqrt{gy}$ relative to the flow.

If c is eliminated from Eqs. (13.2.6) and (13.2.7), the result is

$$\frac{du}{dy} = \pm \sqrt{\frac{g}{y}} \tag{13.2.9}$$

Then, given appropriate boundary conditions, the problem can be solved.

EXAMPLE 13.2

An elementary negative surge wave forms downstream of a sluice gate after it is partially closed. If the channel is rectangular and frictionless, determine the celerity of the surge and the velocity of flow just behind the surge.

Solution

With reference to Fig. 13.3, before the gate was lowered, the depth of flow was y_0 and the velocity of flow u_0. After the gate is lowered, the depth of flow is y_1 and the velocity u_1. Since the surge is moving in the $+x$ coordinate direction, Eq. (13.2.8) is

$$\frac{du}{dy} = \sqrt{\frac{g}{y}}$$

Integrating this equation and eliminating the constant of integration by applying the initial condition yields

$$u = u_0 - 2\sqrt{g}(\sqrt{y_0} - \sqrt{y}) \tag{13.2.10}$$

The celerity of the surge is given by Eq. (13.2.8) as

$$c = u + \sqrt{gy}$$

or substituting for u

$$c = u_0 - 2\sqrt{gy_0} + 3\sqrt{gy} \tag{13.2.11}$$

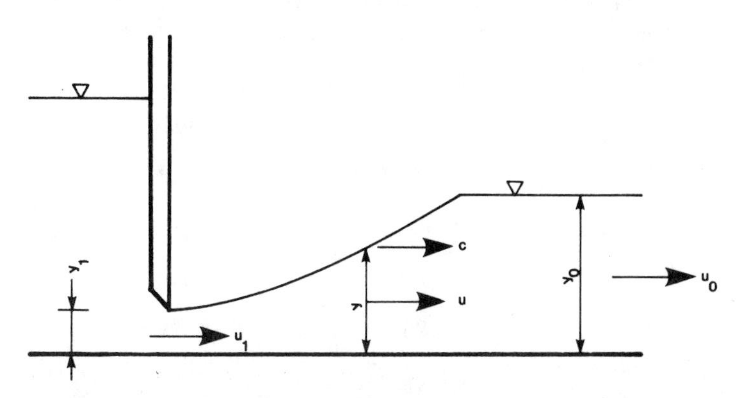

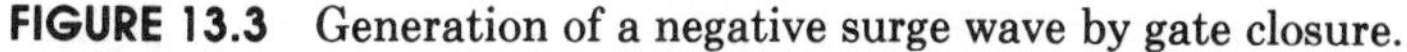
FIGURE 13.3 Generation of a negative surge wave by gate closure.

If the gate is lowered at time $t = 0$, then the position of the surge is given by

$$x = ct$$

or

$$x = (u_0 - 2\sqrt{gy_0} + 3\sqrt{gy})t \tag{13.2.12}$$

If y is eliminated from Eqs. (13.2.11) and (13.1.12), the result is

$$u = \frac{u_0}{3} + \frac{2}{3}\frac{x}{t} - \frac{2}{3}\sqrt{gy_0}$$

EXAMPLE 13.3

In Fig. 13.3 the gate is moved at time $t = 0$ in such a manner that the flow is decreased by 50 percent. If initially, $u_0 = 20$ ft/s (6.1 m/s) and $y_0 = 10$ ft (3.0 m), determine u_1, y_1, and the water surface profile.

Solution

At time $t = 0$, the following continuity relationship is valid:

$$q = u_0 y_0 = 20(10) = 200 \text{ (ft}^3\text{/s)/ft [18.6 (m}^3\text{/s)/m]}$$

where q is the discharge per unit width. After the gate is lowered

$$q = 0.5(u_0 y_0) = 0.5(200) = 100 = u_1 y_1 \tag{13.2.13}$$

By Eq. (13.2.10)

$$u_1 = u_0 - 2\sqrt{g}(\sqrt{y_0} - \sqrt{y_1})$$

or

$$= 20 - 2\sqrt{32.2}(\sqrt{10} - \sqrt{y_1}) \tag{13.2.14}$$

$$= 20 - 11.3(3.2 - \sqrt{y_1})$$

Equations (13.2.13) and (13.2.14) must be solved simultaneously by trial and error, or some other technique, to determine the correct values of y_1 and u_1. Or

$$y_1 = 7.0 \text{ ft (2.1 m)}$$

and

$$u_1 = 14 \text{ ft/s (4.3 m/s)}$$

The shape of the water surface is given by Eq. (13.2.12) or

$$x = (u_0 - 2\sqrt{gy_0} + \sqrt{gy})t$$

and

$$= [20 - 2\sqrt{32.2(10)} + 3\sqrt{32.2y}]t$$

$$= (17\sqrt{y} - 16)t$$

One of the most critical and interesting problems in the area of rapidly varied, unsteady flow is that of the catastrophic failure of a dam. An analytical solution to this problem can be obtained when it is assumed that a frictionless, horizontal, rectangular channel with water of depth y_0 on one side of a gate and no water on the other side exists (Fig. 13.4*a*). The gate is then suddenly removed. If vertical accelerations are neglected, then $u_0 = 0$, y varies from y_0 to 0, and the velocity of flow at any section is given by Eq. (13.2.10) or

$$u = -2\sqrt{g}(\sqrt{y_0} - \sqrt{y}) \tag{13.2.15}$$

and the shape of the water surface is given by Eq. (13.2.12) or

$$x = (3\sqrt{gy} - 2\sqrt{gy_0})t \tag{13.2.16}$$

y_0

x = 0

(a)

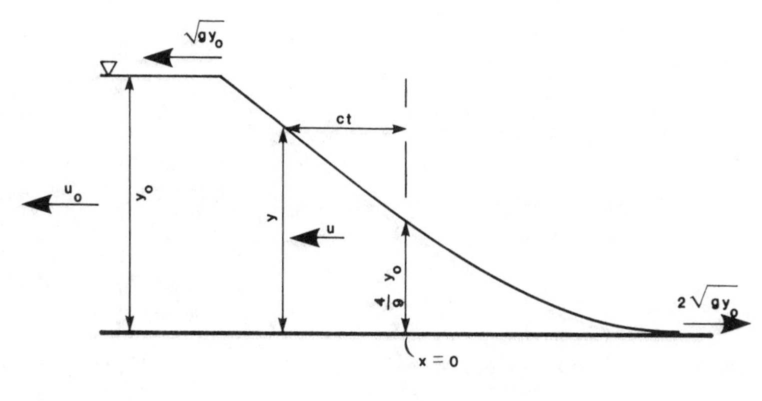

(b)

FIGURE 13.4 (*a*) Dambreak problem before gate removal. (*b*) Dambreak problem after gate removal.

From Eq. (13.2.16), it is determined that at $x = 0$ for all t

$$u = \tfrac{4}{9} y_0 \tag{13.2.17}$$

Substitution of this relation in Eq. (13.2.15) yields

$$u = -\tfrac{2}{3} \sqrt{g y_0} \tag{13.2.18}$$

which demonstrates that the velocity of this flow is independent of time. In this equation, the negative sign indicates that the surge progresses upstream. The shape of the water surface is a parabola with its vertex at the leading edge concave up (Fig. 13.4*b*). In the case of an actual dambreak, this solution can provide a crude estimate of the actual situation in which both friction and vertical accelerations are important considerations.

When two surges traveling in opposite directions meet (Fig. 13.5*a*), the result is asserted to be two new surges traveling in reversed directions (Fig. 13.5*b*). The solution is obtained by applying the principles of continuity and momentum to the situation shown in Fig. 13.5*c*.

Continuity

$$(u^L + c^L)y^L = (u + c^L)y \tag{13.2.19}$$

$$(u^R + c^R)y^R = (u - c^R)y \tag{13.2.20}$$

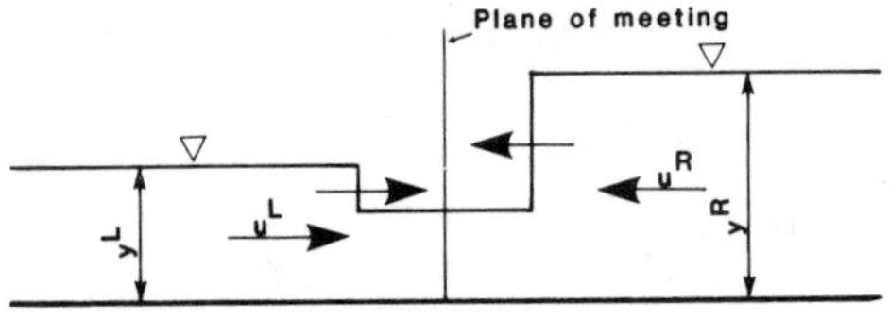

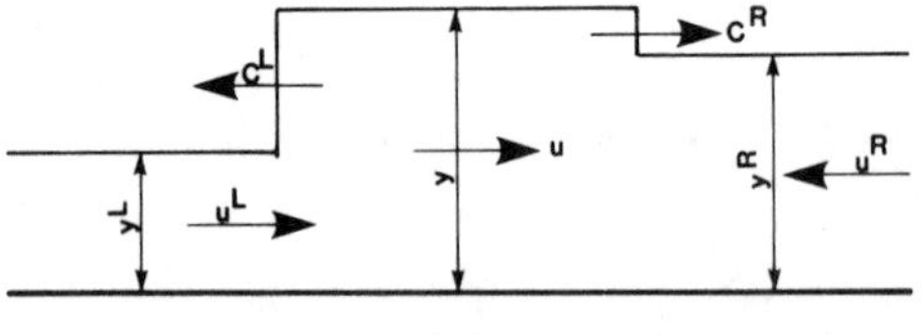

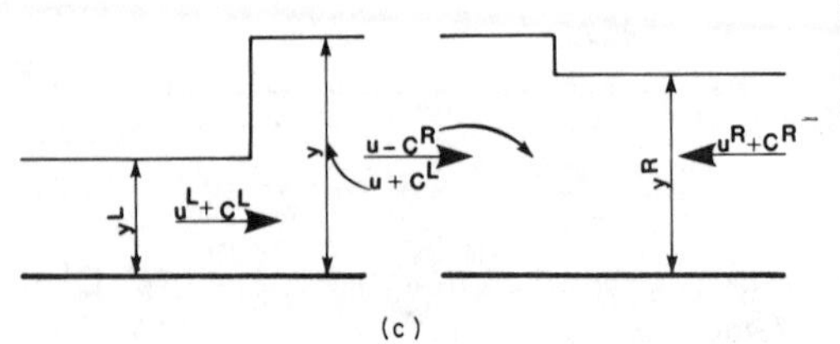

FIGURE 13.5 (*a*) Meeting of two surges; before meeting. (*b*) Meeting of two surges; after meeting. (*c*) Meeting of two surges reduced to steady state.

Momentum

$$[(y^L)^2 - y^2] = \frac{2y^L}{g}(u^L + c^L)(u^L - u) \tag{13.2.21}$$

$$[y^2 - (y^R)^2] = \frac{2y^R}{g}(u^R + c^R)(u - u^R) \tag{13.2.22}$$

In Eqs. (13.2.19) to (13.2.22), u^R, y^R, u^L, y^L, and g are known while y, c^R, c^L, and u are unknowns. Thus, there are four equations and four unknowns, and y, c^R, c^L, and u can be determined by a simultaneous solution of the equations.

The presence of a step in a channel also results in a rather complex situation when a surge passes over it.

> *Note:* A step may be either real or fictitious. A fictitious step may, for example, be used to simulate the effect of a channel slope. In the previous portions of this chapter, the slope of the channel has been assumed negligible; if this is not the case, then a simplified surge analysis may be performed by treating the effect of the slope as a series of steps. If such an analysis is performed, the channel length is divided into a number of reaches, and each reach is treated as if it were horizontal. The actual drop in bed elevation is represented by a step whose height F is equal to the slope of the channel multiplied by the length of the reach (Fig. 13.6*a*).

When a surge arrives at a step, the result is two new surges, one traveling upstream and one traveling downstream (Fig. 13.6*b*). Again, the problem can be solved by applying the principles of continuity and momentum (Fig. 13.6*c*) or

Continuity

$$(u^L + c^L)y^L = (u + c^L)y \tag{13.2.23}$$

$$(u^R + c^R)y^R = (u^L - c^R)(y + F) \tag{13.2.24}$$

Momentum

$$(y^L)^2 - y^2 = \frac{2}{g}(u^L + c^L)y^L(u^L - u) \tag{13.2.25}$$

$$(y + F)^2 - (y^R)^2 = \frac{2y^R}{g}(u^R + c^R)(u^L + u^R) \tag{13.2.26}$$

In the above equations u, c^L, u^L, c^R, and y are unknown. Thus, there are four equations in five unknowns. The relationship required for the solution of the problem is

$$uy = u^L(y + F) \tag{13.2.27}$$

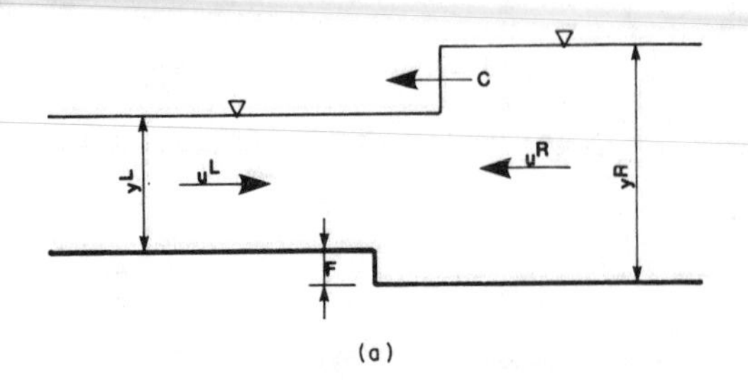

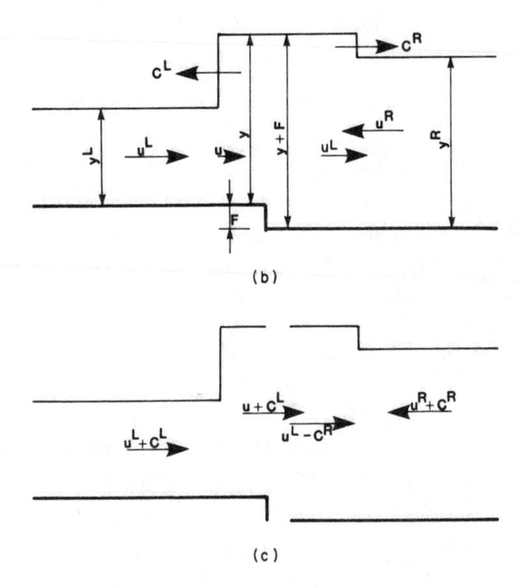

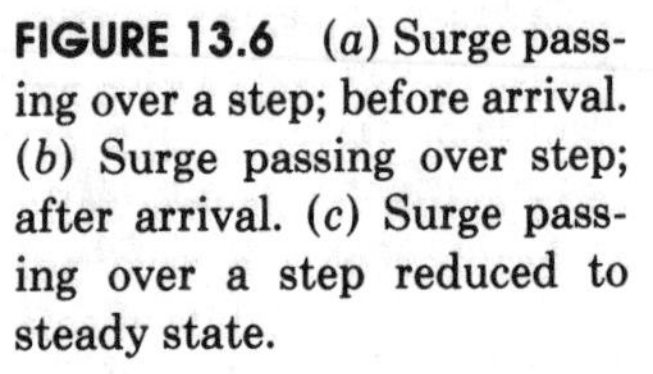

FIGURE 13.6 (*a*) Surge passing over a step; before arrival. (*b*) Surge passing over step; after arrival. (*c*) Surge passing over a step reduced to steady state.

13.3 DAMBREAK

A significant and critically important example of rapidly varied, unsteady flow is the failure of a dam. When a dam fails, the impounded water is released, and catastrophic flooding may occur in the downstream valley. The U.S. Army Corps of Engineers (Anonymous, 1975) has estimated that within the United States there are approximately 50,000 dams with heights greater than 25 ft (7.6 m) or storage volumes in excess of 50 acre·ft (62,000 m^3). This report further classified 20,000 of these dams as being located such that the failure of the dam would result in loss of human life and/or significant property damage.

Dam failures are often caused by overtopping of the structure due to inadequate spillway capacity, seepage or piping through the dam or along internal conduits, slope failure, earthquake damage, liquefaction of earthen dams by earthquakes, or waves generated within the reservoir by landslides. Middlebrooks (1952) describes earthen dam failures occurring within the United States previous to 1951; Johnson and Illes (1976) have summarized dam failures throughout the world.

In the previous section, a simplified solution of the dambreak problem was presented and discussed. Recall that in this analysis it was assumed that the entire structure failed instantaneously. Although, in general, instantaneous failure of the structure is not a realistic assumption, a number of investigators have used it (see, for example, Stoker, 1957, Su and Barnes, 1970, and Sakkas and Strelkoff, 1973). Other investigators (e.g., the Corps of Engineers, Anonymous, 1960, 1961, and Fread, 1982) have recognized that the formation of a dam breach is a time-dependent phenomenon. In the discussion which follows, the term *breach* will be used to refer to the opening in the dam which forms as the dam fails.

Constructed earth dams, which significantly outnumber all other types of dams, do not fail completely nor do they fail instantaneously. Fread (1982) noted that a fully formed breach in an earthen dam tends to have an average width $\bar{b}$ which is generally in the following range

$$h_D < \bar{b} < 3h_D \tag{13.3.1}$$

where h_D = height of the dam. For earthen dams, the width of the breach is less than the total width of the dam. Further, the formation of the breach requires a finite amount of time which may range from a few minutes to a few hours, depending on the height of the dam, the construction materials used, the degree of compaction, and the magnitude and duration of the escaping water. Failure by piping takes place at some point below the top of the dam and is the result of an internal channel formed in the dam by escaping water. As the erosion process proceeds, a larger and larger breach is formed, and the size of this breach is greatly increased when the top of the dam eventually falls into the cavity which is created. It should be noted at this point that poorly constructed earthen dams and embankments which act as dams under extreme conditions, e.g., road and railroad embankments and piles of waste materials, may fail within minutes and may have breach widths which are at the upper end of the range defined by Eq. (13.3.1) or even exceed the maximum width defined by this equation. It should also be noted that debris dams formed by landslides will undoubtedly exhibit failure characteristics which may differ significantly from those noted above for constructed earthen dams.

Concrete gravity dams also tend to fail by partial breachings as one or more of the monolithic concrete sections formed during construction are forced apart by the escaping water. The time required for breach formation in this situation is in the range of a few minutes.

In the case of concrete arch type dams, the assumption of complete and instantaneous failure may closely approximate the actual situation.

At this point, the dam break model developed by Fread (1982) will be discussed. Although there are a number of other models available (e.g., Brevard and Theurer, 1979, Balloffet et al., 1974, Rajar, 1978, and Gundlach and Thomas, 1977), this model represents a state-of-the-art treatment and has a fair degree of flexibility.

Reservoir Outflow Hydrograph

As noted by Fread (1981), a significant source of error in estimating the outflow hydrograph is the quantification of the breach size, shape, and time of formation. Of these three considerations, the shape is the least important, and the time of breach formation becomes increasingly insignificant as the reservoir volume becomes large. At the time of failure, i.e., at any time after a breach is formed, the total reservoir outflow consists of the flow through designed outlets and the flow through the breach which can be approximated as a flow over a broad-crested weir or

$$Q = Q_D + Q_b \tag{13.3.2}$$

where Q = total outflow
Q_D = flow through designed exits
Q_b = flow through the breach

With reference to Fig. 13.7, the outflow through the breach in the computer model DAMBRK (Fread, 1982) is estimated by

$$Q_b = C_1(h - h_b)^{1.5} + C_2(h - h_b)^{2.5} \tag{13.3.3}$$

where

$$C_1 = 3.1 b_i C_v K_s \tag{13.3.4}$$

$$C_2 = 2.45 z C_v K_s \tag{13.3.5}$$

$$h_b = h_d - (h_d - h_{bm})\frac{t_b}{\tau} \qquad \text{for } t_b \le \tau \tag{13.3.6}$$

$$= h_{bm} \qquad \text{for } t_b > \tau \tag{13.3.7}$$

$$b_i = \frac{b t_b}{\tau} \qquad \text{for } t_b \le \tau \tag{13.3.8}$$

$$C_v = 1.0 + \frac{0.023 Q^2}{B_d^2 (h - h_{bm})^2 (h - h_b)} \tag{13.3.9}$$

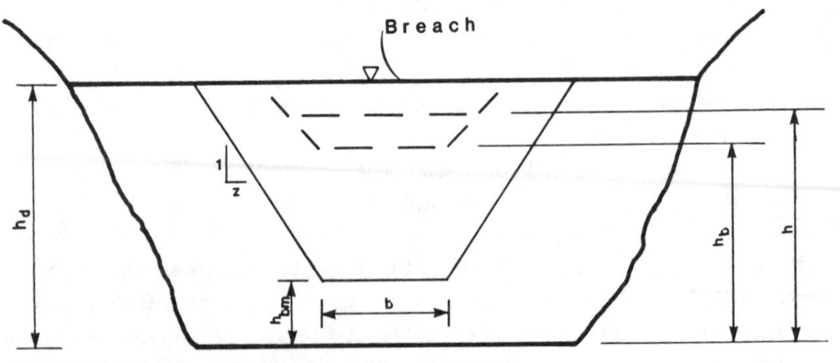

FIGURE 13.7 Breach of a dam. *(Fread, 1982.)*

$$K_s = 1.0 \qquad \text{for } \frac{h_t - h_b}{h - h_b} \leq 0.67 \tag{13.3.10}$$

$$= 1.0 - 27.8\left(\frac{h_t - h_b}{h - h_b} - 0.67\right) \qquad \text{for } \frac{h_t - h_b}{h - h_b} > 0.67 \tag{13.3.11}$$

where h = impoundment water surface elevation
h_b = instantaneous elevation of breach bottom
h_d = dam height
b_i = instantaneous bottom width of breach
t_b = time interval since breach began forming
h_t = tailwater depth
K_s = corrective factor for effect of tailwater effects on weir discharge
C_v = corrective factor for velocity of approach
τ = time from initiation of breach until "complete failure"
h_{bm} = final elevation of breach bottom
z = breach side slope (Fig. 13.7)
b = final breach width

The tailwater elevation h_t is estimated from Manning's equation or

$$Q = \frac{\phi}{n} AR^{2/3}\sqrt{S}$$

where each term in this equation applies to the channel immediately downstream of the dam. Since both A and R are functions of h_t, the Manning equation in this situation is implicit.

If the breach is formed by piping, then the equations stated above, which assume broad-crested weir flow, must be replaced by the equation for orifice flow or

$$Q_b = 4.8A_p(h - \overline{h})^{0.5} \tag{13.3.12}$$

where

$$A_p = [2b_i + 4z(h_f - h_b)](h_f - h_b) \tag{13.3.13}$$

$$\overline{h} = h_f \qquad \text{for } h_t \leq (2h_f - h_b) \tag{13.3.14}$$

$$= h_t \qquad \text{for } h_t > (2h_f - h_b) \tag{13.3.15}$$

$$h_b = h_f - (h_f - h_{bm})\frac{t_b}{\tau} \qquad \text{for } t_b \leq \tau \tag{13.3.16}$$

where h_f = height of the water surface when piping begins. If $\overline{h} = h_f$ and

$$h - h_b < 2.2(h_f - h_b) \tag{13.3.17}$$

then the flow ceases to be orifice flow and becomes a broad-crested weir flow.

All the equations governing the discharge from the breached impoundment depend on the water surface elevation of the impoundment. Obviously, outflow from the reservoir causes a decrease in h which results in a decrease in Q; any flow into the reservoir causes an increase in h which results in an increase in Q. Thus, the determination of an instantaneous value of Q requires a consideration of outflow, inflow, and the reservoir storage characteristics. The model DAMBRK uses a hydrologic routing technique based on the principle of conservation of mass or

$$\frac{dS}{dt} = I - Q \tag{13.3.18}$$

where dS/dt = time rate of change of reservoir storage volume
I = inflow
Q = total outflow

In finite difference form Eq. (13.3.18) becomes

$$\frac{\Delta S}{\Delta t} = \frac{I(t + \Delta t) - I(t)}{2} - \frac{Q(t + \Delta t) - Q(t)}{2} \tag{13.3.19}$$

and

$$\Delta S = \frac{[A_s(t + \Delta t) + A_s(t)][h(t + \Delta t) - h(t)]}{2} \tag{13.3.20}$$

where A_s = reservoir surface area corresponding to a water surface elevation of h. Combining Eqs. (13.3.2), (13.3.3) or (13.3.12), (13.3.19), and (13.3.20) yields a function to estimate the discharge. In the resulting equation, only A_s and h are unknown and A_s is a function of h. Thus, the equation can be solved by an iterative technique to determine h, and Q can then be estimated. The time step used in the calculation should be sufficiently small to minimize the error incurred by numerical integration, e.g., $\tau/50$.

Note: The use of a hydrologic routing technique implies that the surface of the reservoir is level, which is a good approximation only if the dam breach occurs slowly. However, if the dam fails instantaneously and results in a negative wave in the reservoir, or if the magnitude of the inflow is such that a positive wave is formed, then a more sophisticated hydraulic routing technique should be used.

The foregoing method of estimating an outflow hydrograph for a breached dam is a component of the model DAMBRK and requires access to a rather large and fast digital computer for solution. In many cases, the use of a sophisticated model such as DAMBRK may not be feasible because the warning-response time may be limited and/or adequate computing facilities may not be available. For these reasons, Wetmore and Fread (1981) developed a simplified technique for estimating both the peak outflow through a breach and the maximum stage at a specified distance downstream of the breached dam. The original model developed has been modified (Anonymous, 1982).

In the simplified dam break model, it is assumed that the breach is rectangular in shape (Fig. 13.8). Then, the instantaneous flow through the breach is given by the broad-crested weir equation or

$$Q_b = 3.1 b_r h^{(1.5)} \tag{13.3.21}$$

where b_r = breach width, ft
h = head on breach, ft
Q_b = flow through breach, ft^3/s

If the breach forms in a finite time interval τ, the volume of water discharged from the reservoir is given by the integral of the instantaneous flow rate. This outflow volume must also be equal to the product of the reservoir surface area A_s multiplied by the integral of the instantaneous drawdown y_d over the total change in reservoir pool level y_f or

$$3.1b_r \int_0^{\tau} h^{1.5}\, dt = A_s \int_0^{y_f} dy_d \tag{13.3.22}$$

The development of Eq. (13.3.22) tacitly assumes that the reservoir surface area is constant during the time interval τ during which the breach is formed.

The instantaneous head on the weir can be expressed in terms of the instantaneous drawdown or

$$h = (H - y_b) - y_d \tag{13.3.23}$$

where y_b = instantaneous height of the bottom of the breach (Fig. 13.8). Substitution of Eq. (13.3.23) in Eq. (13.3.22) yields an expression which cannot be integrated analytically. Wetmore and Fread (1981) assumed that Eq. (13.3.23) could be approximated by

$$h = \frac{1}{\Gamma}(H - y_d) \tag{13.3.24}$$

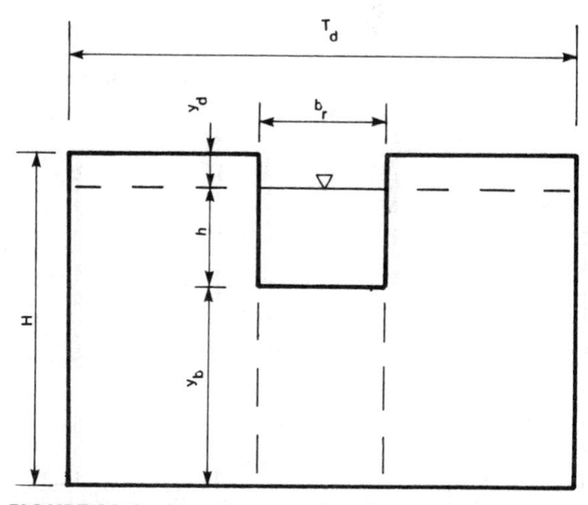

FIGURE 13.8 Instantaneous break geometry for simplified discharge calculation with $0 \le t \le \gamma$. *(Wetmore and Fread, 1981.)*

where Γ = an empirical coefficient which corrects for the assumption inherent in this equation. Substitution of Eq. (13.3.24) in Eq. (13.3.22) yields

$$\frac{3.1b_r}{\Gamma}\int_0^{\tau}(H - y_d)^{1.5}\,dt = A_s\int_0^{y_f} dy_d$$

Rearranging this equation

$$\int_0^{\tau} dt = \frac{\Gamma A_s}{3.1b_r}\int_0^{y_f}\frac{dy_d}{(H - y_d)^{3/2}} \tag{13.3.25}$$

and solving and evaluating at the limits of integration yield

$$\tau = \frac{2A_s\Gamma}{3.1b_r}\left(\frac{1}{\sqrt{H - y_f}} - \frac{1}{\sqrt{H}}\right) \tag{13.3.26}$$

where A_s = reservoir surface area, ft^2
b_r = breach width, ft
τ = time of failure, s

With regard to these equations, it should be noted that the assumption that Eq. (13.3.23) could be approximated with Eq. (13.3.24) was verified by comparing flows computed with this assumption with those estimated by the DAMBRK model. From these comparisons, it was also determined that

$$\Gamma = 3$$

Substitution of this value for Γ in Eq. (13.3.26) yields

$$\tau = \frac{1.94A_s}{b_r}\left(\frac{1}{\sqrt{H - y_f}} - \frac{1}{\sqrt{H}}\right) \tag{13.3.27}$$

If τ is large and the reservoir storage volume is small, then the assumptions made in the above development result in estimates of the peak breach outflow which are much larger than the flow rates that will actually be observed.

Rearrangement of Eq. (13.3.27) yields an expression for the maximum head on the weir or

$$H - y_f = h(\max) = \left\{\frac{1.94A_s/b_r}{\tau + [1.94A_s/(b_r\sqrt{H})]}\right\}^2 \tag{13.3.28}$$

Then, from Eq. (13.3.21) the maximum breach discharge is

$$Q(\max) = 3.1b_r[h(\max)]^{1.5} = 3.1b_r\left\{\frac{1.94A_s/b_r}{\tau + [1.94A_s/(b_r\sqrt{H})]}\right\}^3 \tag{13.3.29}$$

As noted previously, on rare occasions a dam will fail rapidly, and a negative wave which can significantly affect the outflow through the breach will be formed. In such a case, Eq. (13.3.29) will not provide an accurate estimate of the peak outflow. If

$$\tau < 3.6H$$

then

$$Q(\max) = 3.1 b_r (I_v I_n H)^{1.5} \tag{13.3.30}$$

where
$$I_v = 1.0 + 0.148 \left(\frac{b_r}{T_d}\right)^2 (\beta + 1)^2 - 0.083 \left(\frac{b_r}{T_d}\right)^3 (\beta + 1)^{5/2}$$

$$I_n = 1.0 - 0.5467 \left(\frac{b_r}{T_d}\right) (\beta + 1)^{1/2} + 0.2989 \left(\frac{b_r}{T_d}\right)^2 (\beta + 1)^{1/4}$$
$$- 0.1634 \left(\frac{b_r}{T_d}\right)^3 (\beta + 1)^{1/8} + 0.0893 \left(\frac{b_r}{T_d}\right)^4 (\beta + 1)^{1/16}$$
$$- 0.0486 \left(\frac{b_r}{T_d}\right)^5 (\beta + 1)^{1/32}$$

T_d = valley top width at the crest of the dam and β = a channel-fitting coefficient which is defined below (Anonymous, 1982).

At this point in the analysis, the depth of flow immediately downstream of the breached dam for $Q(\max)$ must be estimated to determine if the maximum discharge must be corrected for outflow submergence. Although a method of interpolating the required geometric channel variables is presented in Chapter 1, the method of interpolation used by Wetmore and Fread (1981) will be used here. The channel downstream of the dam is assumed to be a prismatic channel which can be defined by a single distance weighted cross section with the top width of the channel being a power law function of the depth. Sakkas and Strelkoff (1973) also describe a method similar to this one for defining the downstream channel. The approximation of the downstream channel as a prismatic channel is performed in three steps. First, top width versus depth data are obtained from either field surveys or topographic maps. Then, for each depth y_i a distance-weighted top width $\overline{T}_i$ is defined by

$$\overline{T}_i = \sum_{j=2}^{N} \left[\frac{(X_j - X_{j-1})\,(T_{i,j-1} + T_{i,j})/2}{X_N - X_1} \right] \tag{13.3.31}$$

where
y_i = ith depth ($i = 1, 2, \ldots, M$)
$T_{i,j}$ = ith top width corresponding to depth y_i
$\overline{T}_i$ = distance-weighted ith top width
X_j = downstream distance to jth cross section ($j = 1, 2, \ldots, N$)

The recursive use of Eq. (13.3.31) yields a table of $\overline{T}_i$ versus y_i values which may be used to develop a regression equation for the distance-weighted top width of the form

$$T = \theta y^{\beta} \tag{13.3.32}$$

which describes a prismatic channel geometry. The parameters β and θ are estimated by

$$\beta = \frac{\sum_{i=1}^{M} [(\log y_i)(\log \overline{T}_i)] - \left[\sum_{i=1}^{M} (\log y_i) \sum_{i=1}^{M} (\log \overline{T}_i) \right] \Big/ M}{\sum_{i=1}^{M} (\log y_i)^2 - \left(\sum_{i=1}^{M} \log y_i \right)^2 \Big/ M} \tag{13.3.33}$$

and

$$\log \theta = \frac{\sum_{i=1}^{M} \log \overline{T}_i}{M} - \beta \left(\frac{\sum_{i=1}^{M} \log y_i}{M} \right) \tag{13.3.34}$$

or

$$\theta = 10^{\log \theta} \tag{13.3.35}$$

In the case of a downstream channel with steep side walls or very dense vegetation in close proximity to the channel, it may be necessary to define a depth y_s beyond which Eq. (13.3.32) is no longer valid (Fig. 13.9).

Given this empirical definition of the downstream channel and the maximum discharge from the breached dam, the maximum depth of flow immedi-

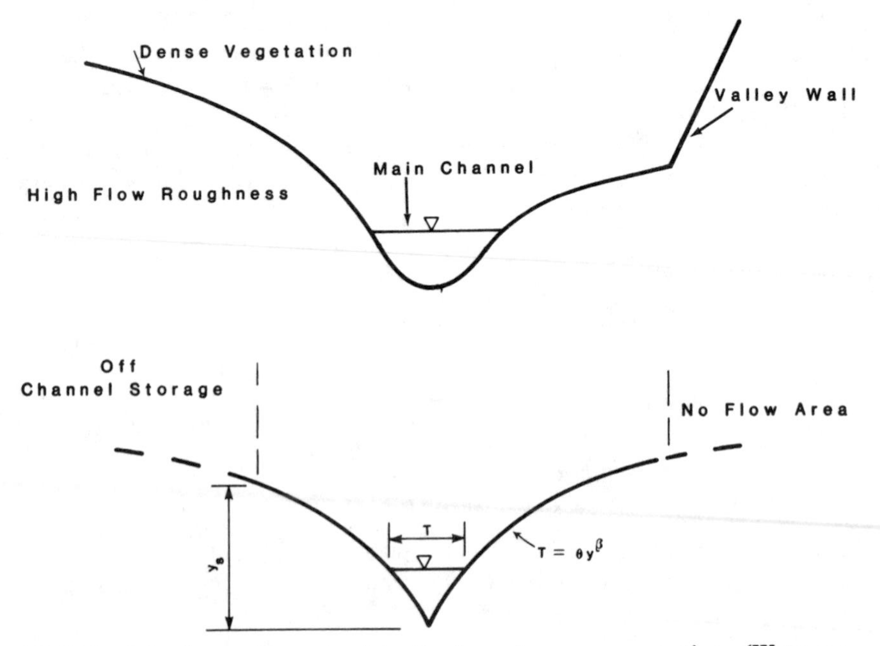

FIGURE 13.9 Approximated prismatic downstream cross section. *(Wetmore and Fread, 1981.)*

ately downstream of the dam can be estimated. To start, the flow which corresponds to a normal depth of flow y_s is calculated or

$$Q_s = \frac{\phi}{n}\sqrt{S}A_s\left(\frac{A_s}{T_s}\right)^{2/3} \tag{13.3.36}$$

Given the assumption of a prismatic channel, T_s is given by Eq. (13.3.32) with $y = y_s$, and A_s is estimated by

$$A_s = \frac{\theta y_s^{\beta+1}}{\beta+1} \tag{13.3.37}$$

Substitution of Eqs. (13.3.32) and (13.3.37) in Eq. (13.3.36) yields

$$Q_s = \frac{\phi}{n}\sqrt{S}\left[\frac{\theta}{(\beta+1)^{5/3}}\right]y_s^{\beta+5/3} \tag{13.3.38}$$

If the flow rate given by Eq. (13.3.38) is greater than the estimated peak discharge from the breached reservoir, i.e.,

$$Q_s > Q(\max)$$

then the maximum depth of flow immediately below the dam is given by

$$y(\max) = \left[\frac{Q(\max)}{\zeta}\right]^{1/(\beta+5/3)} \tag{13.3.39}$$

where

$$\zeta = \frac{\phi}{n}\sqrt{S}\frac{\theta}{(\beta+1)^{5/3}}$$

Recall that ϕ is a coefficient whose value is determined by the system of units being used; that is, $\phi = 1$ for SI units and $\phi = 1.49$ for English units. If the flow rate given by Eq. (13.3.38) is less than the peak discharge from the breached reservoir, i.e.,

$$Q_s < Q(\max)$$

then the maximum depth of flow immediately below the dam is given by

$$y(\max) = [Q(\max)]^{3/5}\left[\frac{1}{\zeta(\beta+1)^{5/3}y_s^{\beta}}\right]^{3/5} + \frac{\beta}{\beta+1}y_s \tag{13.3.40}$$

Once $y(\max)$ has been estimated by either Eq. (13.3.39) or (13.3.40), this depth of flow must be compared with the head on the breach to determine if a submergence correction for tailwater effects is required.

Note: In the case of instantaneous failure, a submergence correction is not required.

If

$$\frac{y(\max)}{h(\max)} > 0.67 \tag{13.3.41}$$

where $h(\max)$ is given by Eq. (13.3.28), then a tailwater correction factor must be estimated according to

$$K_s^* = 1 - 27.8\left[\frac{y(\max)}{h(\max)} - 0.67\right]^3 \tag{13.3.42}$$

This value of K_s^* is then used in Eq. (13.3.43) to obtain an averaged submergence correction factor or

$$K_s^k = \frac{K_s^* + K_s^{k-1}}{2} \tag{13.3.43}$$

where the superscript k is an iteration counter ($k = 1, 2, \ldots$) and $K_s^0 = 1$. At this point, it is noted that in some cases K_s^* may be estimated to be less than 0; in such cases, it should be set to 0 to prevent overcompensating for submergence. A corrected breach outflow may now be estimated by

$$Q^k(\max) = K_s^k Q^{k-1}(\max) \tag{13.3.44}$$

$Q^k(\max)$ is then compared with Q_s and a new outflow stage $y^k(\max)$ is estimated from either Eq. (13.3.39) or (13.3.40). Also, since there is decreased flow through the breach, a corrected level on the breach must be estimated or

$$h^k(\max) = h^{k-1}(\max) + [Q^{k-1}(\max) - Q^k(\max)]\frac{\tau}{2A_s} \tag{13.3.45}$$

where τ is in seconds and A_s is in square feet. The ratio $y^k(\max)/h^k(\max)$ is computed and used in Eq. (13.3.42) to estimate a new value of K_s^*. If the new averaged submergence computed by Eq. (13.3.43) is significantly different from that computed in the previous iteration, then this procedure must be repeated. Generally, two or three iterations are sufficient to obtain an accurate estimate of the peak discharge through the breach.

EXAMPLE 13.4

For the reservoir and channel data summarized in Tables 13.1 and 13.2, estimate the peak flow and maximum stage just below the dam. The Manning resistance coefficient for the downstream channel is assumed to be 0.045. This example was first presented in Anonymous (1982).

TABLE 13.1 Reservoir—breach data

Parameter description	Parameter	Value
Reservoir surface area at the maximum pool level	A_s	350 acre
Height of dam	H	50 ft
Final breach width	b_r	100 ft
Time of failure	τ	0.75 h
Downstream slope	S	0.0008

TABLE 13.2 Downstream channel description

Mile 0.0		Mile 12.3		Mile 40.5	
Depth, ft	Top width, ft	Depth, ft	Top width, ft	Depth, ft	Top width, ft
0	0	0	0	0	0
8	480	8	437	8	456
18	900	18	826	18	858
28	1300	28	1337	28	1338
38	1350	38	1407	38	1456

Solution

The first step in the solution of this problem is the establishment of a single distance-weighted cross section. For $y = 8$ ft (2.4 m), the data in Table 13.2 and Eq. (13.3.31) yield

$$\overline{T} = \sum_{j=2}^{N} \left[\frac{(X_j - X_{j-1})(T_{i,j-1} + T_{i,j})/2}{X_N - X_1} \right]$$

and

$$\overline{T}_8 = \frac{(12.3 - 0)(480 + 437)/2 + (40.5 - 12.3)(437 + 456)/2}{40.5 - 0}$$

$$= 450.1 \text{ ft } (137 \text{ m})$$

For $y = 18$ ft (5.5 m)

$$\overline{T}_{18} = \frac{(12.3 - 0)(1300 + 1337)/2 + (40.5 - 12.3)(826 + 858)/2}{40.5 - 0}$$

$$= 848.4 \text{ ft } (259 \text{ m})$$

For $y = 28$ ft (8.5 m)

$$\overline{T}_{28} = \frac{(12.3 - 0)(1300 + 1337)/2 + (40.5 - 12.3)(1337 + 1338)/2}{40.5 - 0}$$

$$= 1331.7 \text{ ft } (406 \text{ m})$$

For $y = 38$ ft (12 m)

$$\overline{T}_{38} = \frac{(12.3 - 0)(1350 + 1407)/2 + (40.5 - 12.3)(1407 + 1456)/2}{40.5 - 0}$$

$$= 1415.4 \text{ ft } (431 \text{ m})$$

An examination of the data in Table 13.2 and the above calculations indicate that above a depth of 28 ft (8.5 m) the top widths increase very slowly. Therefore, a reasonable value of y_s is approximately 30 ft (9.1 m). The distance-weighted top width for $y_s = 30$ ft (9.1 m) is determined by first interpolating appropriate top widths in Table 13.2 and then using Eq. (13.3.31) or

$$\overline{T}_{30} = \frac{(12.3 - 0)(1310 + 1351)/2 + (40.5 - 12.3)(1351 + 1362)/2}{40.5 - 0}$$

$$= 1348.6 \text{ ft } (411 \text{ m})$$

The coefficient β for Eq. (13.3.32) is estimated by Eq. (13.3.33) or

$$\beta = \frac{\sum_{i=1}^{M} [(\log y_i)(\log \overline{T}_i)] - \left[\left(\sum_{i=1}^{M} \log y_i\right)\left(\sum_{i=1}^{M} \log \overline{T}_i\right)\right] \Big/ M}{\sum_{i=1}^{M} (\log y_i)^2 - \left(\sum_{i=1}^{M} \log y_i\right)^2 \Big/ M}$$

The data for this computation are summarized in Table 13.3.

$$\beta = \frac{15.22 - [(5.09)(11.83)]/4}{6.67 - (5.09)^2/4} = 0.86$$

The coefficient θ for Eq. (13.3.32) is estimated by Eqs. (13.3.34) and (13.3.35) or

$$\log \theta = \frac{\sum_{i=1}^{M} \log \overline{T}_i}{M} - \beta \left(\frac{\sum_{i-1}^{M} \log y_i}{M} \right)$$

By use of data from Table 13.3

$$\log \theta = \frac{11.83}{4} - 0.86 \left(\frac{5.09}{4} \right) = 1.86$$

and
$$\theta = 10^{\log \theta} = 10^{1.86} = 73.0$$

Therefore, the distance-weighted top width of the channel below the reservoir is given by

$$\overline{T} = 73.0 y^{0.86} \qquad \text{for } 0 \le y \le 30 \text{ ft}$$

TABLE 13.3 Summary of data for the calculation of β and θ

y_i, ft	$\log y_i$	$(\log y_i)^2$	$\overline{T}_i$, ft	$\log \overline{T}_i$	$\log \overline{T}_i \log y_i$
8	0.90	0.82	450	2.65	2.39
18	1.26	1.58	848	2.93	3.69
28	1.45	2.09	1332	3.12	4.52
30	1.48	2.18	1349	3.13	4.62
	$\Sigma (\log y_i) = 5.09$	$\Sigma (\log y_i)^2 = 6.67$		$\Sigma (\log \overline{T}_i) = 11.83$	$\Sigma (\log \overline{T}_i)(\log y_i) = 15.22)$

The maximum breach outflow is given by Eq. (13.3.29) or

$$Q(\max) = 3.1 b_r \left\{ \frac{1.94 A_s / b_r}{\tau + [1.94 A_s / (b_r \sqrt{H})]} \right\}^3$$

$$= 3.1\,(100) \left\{ \frac{1.94(350)(43{,}560)/100}{0.75(60)(60) + [1.94(350)(43{,}560)/(100\sqrt{50})]} \right\}^3$$

$$= 90{,}800 \text{ ft}^3/\text{s } (2570 \text{ m}^3/\text{s})$$

where by definition there are 43,560 ft^2/acre. The downstream flow corresponding to a depth y_s = 30 ft (9.1 m) is given by Eq. (13.3.38) or

$$Q_s = \frac{1.49}{n} \sqrt{S} \left[\frac{\theta}{(\beta + 1)^{5/3}} \right] y_s^{\beta + 5/3}$$

$$= \frac{1.49}{0.045} \sqrt{0.0008} \left[\frac{73}{(0.86 + 1)^{5/3}} \right] (30)^{0.86 + 5/3}$$

$$= 131{,}000 \text{ ft}^3/\text{s } (3700 \text{ m}^3/\text{s})$$

Since $Q_s > Q(\max)$, Eq. (13.3.39) must be used to determine $y(\max)$ or

$$\zeta = \frac{\phi}{n} \sqrt{S} \frac{\theta}{(\beta + 1)^{5/3}}$$

$$= \frac{1.49}{0.045} \sqrt{0.0008} \left[\frac{73}{(0.86 + 1)^{5/3}} \right] = 24.3$$

and

$$y(\max) = \left[\frac{Q(\max)}{\zeta} \right]^{1/(\beta + 5/3)} = \left(\frac{90{,}000}{24.3} \right)^{1/(0.86 + 5/3)}$$

$$= 25.8 \text{ ft } (7.86 \text{ m})$$

To check if submergence must be corrected for, the head on the weir must be estimated by Eq. (13.3.28).

$$h(\max) = \left\{ \frac{1.94 A_s / b_r}{\tau + [1.94 A_s / (b_r \sqrt{H})]} \right\}^2$$

$$= \left\{ \frac{1.94(350)(43{,}560)/100}{0.75(60)(60) + [1.94(350)(43{,}560)/(100\sqrt{50})]} \right\}^2$$

$$= 44.1 \text{ ft } (13.4 \text{ m})$$

The ratio of $y(\max)$ to $h(\max)$ is then checked [Eq. (13.3.41)] or

$$\frac{y(\max)}{h(\max)} = \frac{25.8}{44.1} = 0.58$$

Since this ratio is less than 0.67, the tailwater does not affect the flow through the breach.

Downstream Routing

After the outflow hydrograph for a breached reservoir has been estimated, the occurrence of flooding in downstream areas can be predicted by any one of a number of routing techniques. A distinguishing feature of flood waves emanating from a breached dam is their great magnitude compared with ordinary flood waves and their rather short time of duration. The time from the beginning of rise to the peak discharge is generally of the same order of magnitude as τ and, thus, may range from a few minutes to a few hours. Given the short time of duration and the large magnitude of the peak flow, characteristics which combine to yield acceleration components far in excess of those encountered in runoff-generated flood waves, a dynamic wave method is usually required to accurately route the flow downstream (Fread, 1982). Such a method uses the St. Venant equations (Chapter 12), and the applicability of these equations to the case of a breached dam has been demonstrated by Terzidis and Strelkoff (1970) and Martin and Zovne (1971).

Wetmore and Fread (1981) and Anonymous (1982) have also developed a technique of routing the dam break flood wave through the downstream valley by hand. The routing curves used in this technique were developed using data obtained from numerous executions of the NWS DAMBRK model and are grouped into families based on the Froude number associated with the flood wave peak. The use of these curves requires the definition of several routing parameters.

The distance parameter X_c is calculated from either Eq. (13.3.46) or (13.3.47). If the height of the dam H is less than y_s (Fig. 13.9), then

$$X_c = \frac{\beta + 1}{\theta}\left(\frac{V}{H^{\beta+1}}\right)\left[\frac{6}{1 + 4(0.5)^{\beta+1}}\right] \tag{13.3.46}$$

where X_c = distance parameter, ft
H = height of dam, ft
V = reservoir volume, ft^3

If the height of the dam is greater then y_s, then

$$X_c = \frac{6V}{\theta y_s^{\beta}\{3H - 5[\beta y_s/(\beta + 1)]\}} \tag{13.3.47}$$

Within the distance X_c downstream of the breached dam, the flood wave attenuates such that at the point X_c the stage is y_x (Fig. 13.10), which is a function of the maximum stage in the reach $y(\max)$. The average stage in the reach is, by definition,

$$\bar{y} = \frac{y(\max) + y_x}{2} = \Gamma y(\max) \tag{13.3.48}$$

where $\bar{y}$ = average stage and Γ = a weighting factor that is determined iteratively. The initial estimate of Γ is 0.95.

The average hydraulic depth in the reach is estimated by

$$\text{If } Q(\max) < Q_s, \text{ then } D = \frac{\Gamma y(\max)}{\beta + 1} \tag{13.3.49}$$

$$\text{If } Q(\max) > Q_s, \text{ then } D = \Gamma y(\max) - y_s + \frac{y_s}{\beta + 1} \tag{13.3.50}$$

With the hydraulic depth determined, the average velocity of flow in the reach can be estimated from the Manning equation

$$\bar{u} = \frac{\phi}{n} \sqrt{S}\, D^{2/3} \tag{13.3.51}$$

where S = average slope of the entire downstream routing reach. The average Froude number for the reach under consideration is

$$\mathbf{F} = \frac{\bar{u}}{\sqrt{gD}} \tag{13.3.52}$$

A dimensionless volume parameter V^* is then defined as the ratio of the reservoir volume to the flow volume in the reach defined by the distance X_c or

$$V^* = \frac{V}{A_x X_c} \tag{13.3.53}$$

where

$$\text{If } Q(\max) < Q_s, \text{ then } A_x = \theta[\Gamma y(\max)]^{\beta} D \tag{13.3.54}$$

$$\text{If } Q(\max) > Q_s, \text{ then } A_x = \theta y_s^{\beta} D \tag{13.3.55}$$

Table 13.4 summarizes the equations used to compute the downstream routing parameters.

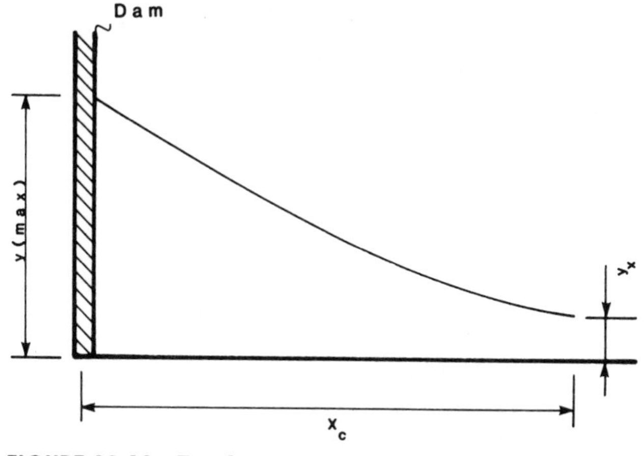

FIGURE 13.10 Dambreak flood wave attenuation.

TABLE 13.4 Summary of routing parameter equations

Parameter	Case I		Case II	
	Condition	Equation	Condition	Equation
X_c	$H < y_s$	$X_c = \frac{\beta + 1}{\theta}\left(\frac{V}{H^{\beta+1}}\right)\left[\frac{6}{1 + 4(0.5)^{\beta+1}}\right]$	$H > y_s$	$X_c = \frac{6V}{\theta y_s^{\beta}\{3H - 5[\beta y_s/(\beta + 1)]\}}$
Γ	$\Gamma = 0.95$ (initial estimate)			
D, ft	$Q(\max) < Q_s$	$D = \frac{\Gamma y(\max)}{\beta + 1}$	$Q(\max) > Q_s$	$D = \Gamma y(\max) - y_s + \frac{y_s}{\beta + 1}$
$\overline{u}$, ft/s	$\overline{u} = \frac{1.49}{n} \sqrt{S}\, (D)^{2/3}$			
$\mathbf{F}$	$\mathbf{F} = \frac{\overline{u}}{\sqrt{gD}}$			
A_x, ft^2	$Q(\max) < Q_s$	$A_x = \theta[\Gamma y(\max)]^{\beta} D$	$Q(\max) > Q_s$	$A_x = \theta y_s^{\beta} D$
V^*	$V^* = \frac{V}{A_x X_c}$			

With values of **F** and V^* known, the correct routing curve in Figs. 13.11 to 13.13 can be identified, and the original estimate of the weighting factor used in Eq. (13.3.48) checked. That is, the ordinate of the appropriate routing curve at $X^* = 1$ is the ratio of the peak flow Q_x at $x = X_c$ to the peak flow at the breach $Q(\max)$. With Q_x known, the stage at $x = X_c$ can be estimated from

If $Q_s > Q_x$

$$y_x = \left[\frac{Q_x}{\zeta}\right]^{1/(\beta+5/3)} \tag{13.3.56}$$

where

$$\zeta = \frac{\phi}{n}\sqrt{S}\,\frac{\theta}{(\beta+1)^{5/3}}$$

If $Q_s < Q_x$

$$y_x = (Q_x)^{3/5}\left[\frac{1}{\zeta(\beta+1)^{5/3}y_s^{\beta}}\right]^{3/5} + \frac{\beta}{\beta+1}\,y_s \tag{13.3.57}$$

The value of Γ in Eq. (13.3.48) can be checked by

$$\Gamma = \frac{y(\max) + y_x}{2y(\max)} \tag{13.3.58}$$

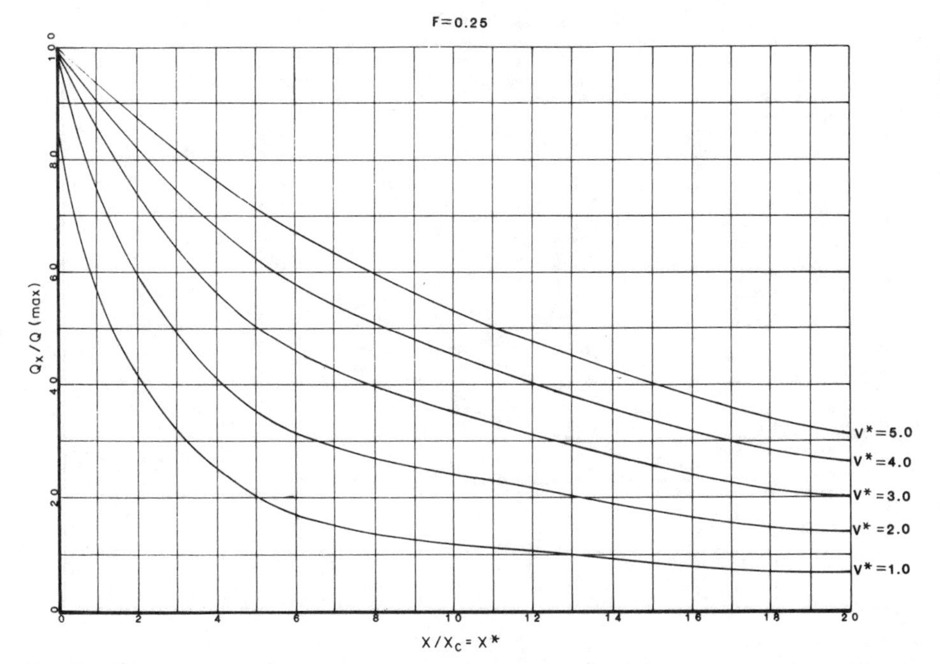

FIGURE 13.11 Simplified dambreak routing curves, $FC = 0.25$. *(Anonymous, 1982.)*

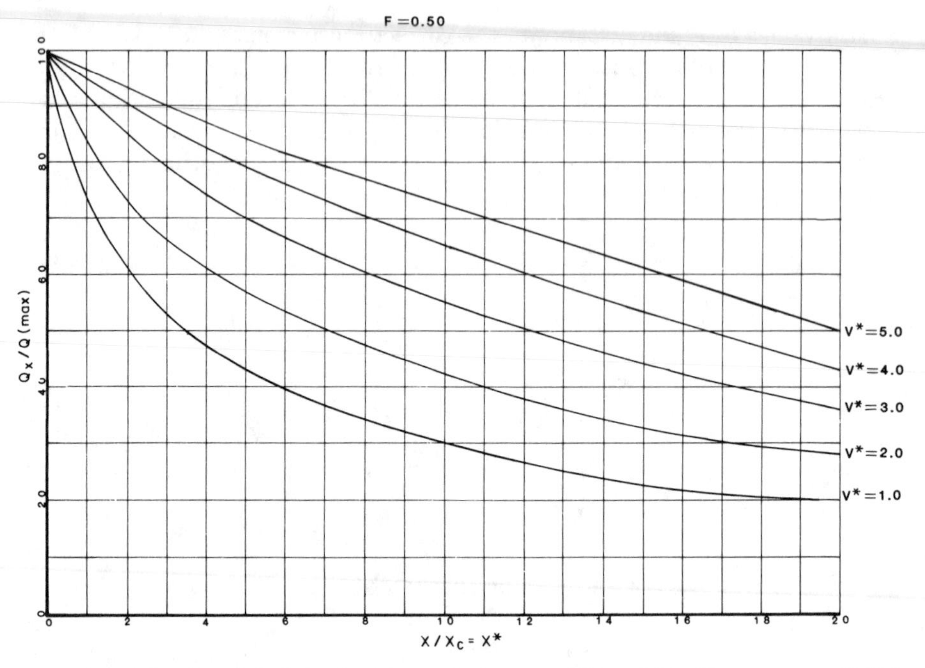

FIGURE 13.12 Simplified dambreak routing curves, $FC = 0.50$. *(Anonymous, 1982.)*

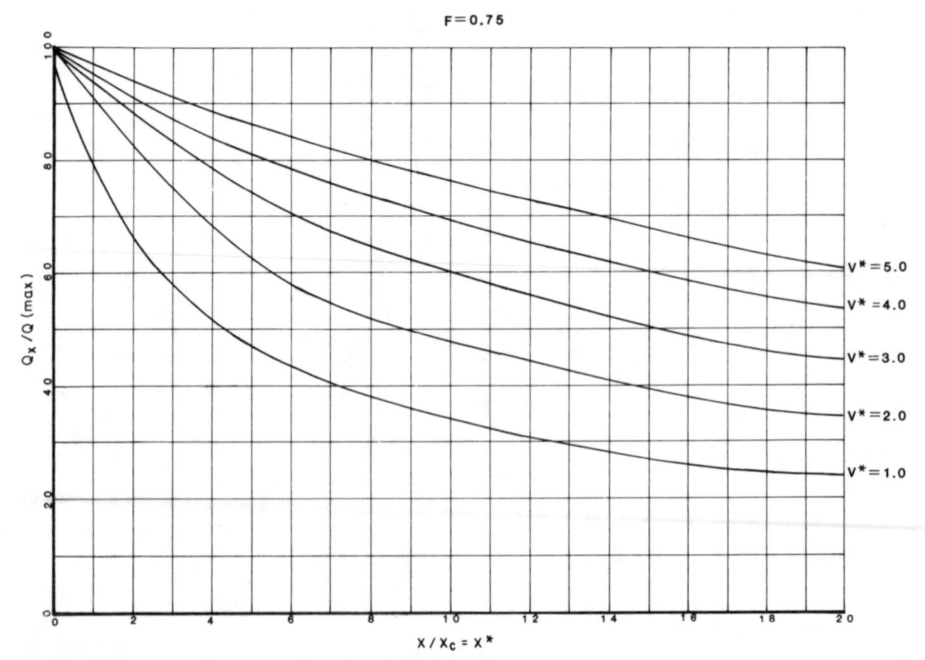

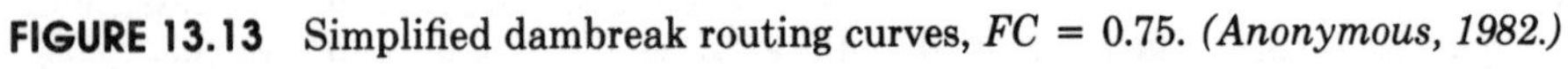

FIGURE 13.13 Simplified dambreak routing curves, $FC = 0.75$. *(Anonymous, 1982.)*

If there is a significant difference between the value of Γ estimated by Eq. (13.3.58) and the original estimate of Γ, then the calculations indicated in Eqs. (13.3.49) to (13.3.57) must be repeated.

Once the correct routing parameters have been calculated, the correct routing curve can be identified and the downstream distances to the required peak flow estimate points are nondimensionalized by

$$X_i^* = \frac{x_i}{X_c} \tag{13.3.59}$$

where x_i $(i = 1, 2, \ldots)$ = the downstream distance to the ith forecast point. Then, to find the peak flow and stage at a downstream point, the appropriate routing curve in Figs. 13.11 to 13.13 is consulted and the ordinate corresponding to V^* at X_i^* is determined. Multiplying the value of this ordinate by $Q(\max)$ yields an estimate of the peak flow at a distance x_i below the breached dam. The peak stage at this location can be estimated from either Eq. (13.3.56) or (13.3.57).

The time of travel for the flood wave to the point x_i can be estimated by first calculating the average velocity of flow at the midpoint between the dam and the point x_i. In this calculation, the routing curves are again used to determine the peak flow at $X^* = 0.5$, and the peak stage corresponding to this flow is determined by applying Eq. (13.3.56) or (13.3.57). For this stage and flow rate the hydraulic depth is

$$\text{If } y_x < y_s, \text{ then } D_x = \frac{y_x}{\beta + 1} \tag{13.3.60}$$

$$\text{If } y_x > y_s, \text{ then } D_x = y_x - y_s + \frac{y_s}{\beta + 1} \tag{13.3.61}$$

and the average velocity of flow is

$$\overline{u} = \frac{\phi}{n} \sqrt{S}\,(D_x)^{2/3} \tag{13.3.62}$$

The speed of the flood wave is then given by the wave celerity equation or

$$c = \overline{u}\left[\frac{5}{3} - \frac{2}{3}\left(\frac{\beta}{\beta + 1}\right)\right] \tag{13.3.63}$$

where c = wave celerity in feet per second. Then, the time until peak discharge and stage at a position x_i feet below the breached dam is

$$t_a = \tau + \frac{x_i}{c} \tag{13.3.64}$$

where t_a = time of arrival in seconds if τ is specified in seconds, x_i in feet, and c in feet per second.

A user of this technique may also wish to estimate the time at which flooding begins or ceases at a station at a specified distance downstream. First, the flow rate Q_f which corresponds to the flood stage in the cross section must be specified or computed by

$$Q_f = \theta y^{\beta} \tag{13.3.65}$$

then

$$t_f = t_a - \frac{Q_x - Q_f}{Q_x}\tau \tag{13.3.66}$$

where t_f = time to flooding in seconds. An estimate of the time when flooding ceases is given by

$$t_c = t_f + \left(\frac{2V}{Q_x} - \tau\right)\left(\frac{Q_x - Q_f}{Q_x}\right) \tag{13.3.67}$$

where t_c = time at which flooding ceases at a downstream station, in seconds.

EXAMPLE 13.5

For the data presented and estimated in Example 13.4, determine (1) the peak discharge, (2) the maximum stage, and (3) the time of flood wave arrival at stations located 12.3 and 40.5 miles below the breached dam. For convenience, the given and calculated parameter values are summarized in Table 13.5. This example was first presented in Anonymous (1982).

Solution

From Example 13.4, the height of the dam exceeds y_s; therefore, Eq. (13.3.47) is used to determine the routing parameter or

$$X_c = \frac{6V}{\theta y_s^{\beta}\{3H - 5[\beta y_s/(\beta + 1)]\}}$$

$$= \frac{6(8750)(43{,}560)}{73(30)^{0.86}\{3(50) - 5[0.86(30)/(0.86 + 1)]\}}$$

$$= 20{,}800 \text{ ft } (6340 \text{ m})$$

where by definition there are 43,560 ft²/acre. Assuming $\Gamma = 0.95$, the average stage in the reach below the dam is

$$\bar{y} = \Gamma y(\max) = 0.95(25.8) = 24.5 \text{ ft } (7.47 \text{ m})$$

Since $Q(\max) < Q_s$, Eq. (13.3.49) is used to estimate the downstream average hydraulic depth or

$$D = \frac{\Gamma y(\max)}{\beta + 1} = \frac{0.95(25.8)}{0.86 + 1} = 13.2 \text{ ft } (4.0 \text{ m})$$

TABLE 13.5 Summary of variable values for Example 13.5

Parameter description	Parameter	Value
Reservoir volume	V	8750 acre·ft
Height of dam	H	50 ft
Maximum breach outflow	$Q(\max)$	90,800 ft^3/s
Maximum head on weir	$h(\max)$	44.1 ft
Height of valley wall	y_s	30 ft
Channel "fitting" coefficients	θ	73.0
	β	0.86
Downstream channel slope	S	0.0008
Manning's roughness coefficient	n	0.045
Maximum stage below dam	$y(\max)$	25.8 ft

By use of Eq. (13.3.51), the average downstream velocity is

$$\bar{u} = \frac{1.49}{n}\sqrt{S}D^{2/3} = \frac{1.49}{0.045}\sqrt{0.0008}(13.2)^{2/3} = 5.2 \text{ ft/s } (1.6 \text{ m/s})$$

By Eq. (13.3.52), the Froude number for this reach is

$$\mathbf{F} = \frac{\bar{u}}{\sqrt{gD}} = \frac{5.2}{\sqrt{32.2(13.2)}} = 0.25$$

The flow area corresponding to this estimated hydraulic depth is given by Eq. (13.3.54) since $Q(\max) < Q_s$ or

$$A_x = \theta[\Gamma y(\max)]^{\beta}D = 73[0.95(25.8)]^{0.86}(13.2)$$
$$= 15{,}100 \text{ ft}^2 \ (1400 \text{ m}^2)$$

The dimensionless routing parameter V^* is given by Eq. (13.3.53).

$$V^* = \frac{V}{A_x X_c} = \frac{8750(43{,}560)}{15{,}100(20{,}800)} = 1.21$$

Then with $\mathbf{F} = 0.25$ and $V^* = 1.2$, Fig. 13.12 and 13.13 can be used to check the assumed value of Γ. From Fig. 13.12

$$\text{for } \mathbf{F} = 0.25,\ x/X_c = 1, \text{ and } V^* = 1.2\text{: } Q_x/Q(\max) = 0.60$$

Thus, the peak flow at $x = X_c$ is

$$Q_x = 0.60Q(\max) = 0.60(90{,}800) = 54{,}500 \text{ ft}^3/\text{s } (1540 \text{ m}^3/\text{s})$$

The maximum depth of flow at $x = X_c$ is estimated by Eq. (13.3.56) since $Q_s > Q_x$ or

$$y_x = \left(\frac{Q_x}{\zeta}\right)^{1/(\beta+5/3)}$$

where

$$\zeta = \frac{\phi}{n}\sqrt{S}\frac{\theta}{(\beta + 1)^{5/3}} = \frac{1.49}{0.045}\sqrt{0.0008}\frac{73}{(0.86 + 1)^{5/3}} = 24.3$$

and
$$y_x = \left(\frac{54{,}500}{24.3}\right)^{1/(0.86+5/3)} = 21.2 \text{ ft } (6.5 \text{ m})$$

A new value of the weighting factor Γ is then estimated from Eq. (13.3.58).

$$\Gamma = \frac{y(\max) + y_x}{2y(\max)} = \frac{25.8 + 21.2}{2(25.8)} = 0.91$$

The percentage difference between the value of Γ which was assumed and the value estimated above is

$$\Delta\% = \frac{0.95 - 0.91}{0.95} 100 = 4.2\%$$

The dimensionless distances X_i^* to the forecast points at miles 12.3 and 40.5 are

$$X_1^* = \frac{x}{X_c} = \frac{12.3(5280)}{20{,}800} = 3.1$$

and
$$X_2^* = \frac{x}{X_c} = \frac{40.5(5280)}{20{,}800} = 10.3$$

Forecast at Mile 12.3

From Fig. 3.11

$$\text{for } \mathbf{F} = 0.25,\ X_1^* = 3.1, \text{ and } V^* = 1.2\text{: } \frac{Q_x}{Q(\max)} = 0.34$$

Thus, the peak flow at this station is

$$Q_x = 0.34Q(\max) = 0.34(90{,}800) = 30{,}900 \text{ ft}^3\text{/s } (875 \text{ m}^3\text{/s})$$

The peak stage is estimated from Eq. (13.3.56) or

$$y_x = \left(\frac{Q_x}{\zeta}\right)^{1/(\beta+5/3)} = \left(\frac{30{,}900}{24.3}\right)^{1/(0.86+5/3)} = 16.9 \text{ ft } (5.2 \text{ m})$$

To estimate the time of arrival of the flood wave at this station, the velocity of flow at the midpoint between the dam and mile 12.3 must be determined. The dimensionless distance parameter for the midpoint is

$$\frac{x}{X_c} = \frac{12.3(5280)}{2(20{,}800)} = 1.6$$

From Fig. 13.11

$$\text{for } \mathbf{F} = 0.25,\ X^* = 1.6, \text{ and } V^* = 1.2: \frac{Q_{x/2}}{Q(\max)} = 0.5$$

and $Q_{x/2} = 0.5Q(\max) = 0.5(90{,}800) = 45{,}400 \text{ ft}^3/\text{s } (1290 \text{ m}^3/\text{s})$

The average peak stage in the reach is estimated by Eq. (13.3.56) since $Q_s > Q_{x/2}$ or

$$y_{x/2} = \left(\frac{Q_{x/2}}{\zeta}\right)^{1/(\beta+5/3)} = \left(\frac{45{,}400}{24.3}\right)^{1/(0.86+5/3)} = 19.7 \text{ ft } (6.0 \text{ m})$$

The hydraulic depth at the midpoint is estimated by Eq. (13.3.60) since $y_{x/2} < y_s$ or

$$D_{x/2} = \frac{y_{x/2}}{\beta + 1} = \frac{19.7}{0.86 + 1} = 10.6 \text{ ft } (3.2 \text{ m})$$

The average velocity of flow is by Eq. (13.3.62)

$$\bar{u} = \frac{1.49}{n}\sqrt{S}D_{x/2}^{2/3} = \frac{1.49}{0.045}\sqrt{0.0008}(10.6)^{2/3} = 4.5 \text{ ft/s } (1.4 \text{ m/s})$$

The celerity of the flood wave is estimated by Eq. (13.3.63)

$$c = \bar{u}\left[\frac{5}{3} - \frac{2}{3}\left(\frac{\beta}{\beta + 1}\right)\right] = 4.5\left[\frac{5}{3} - \frac{2}{3}\left(\frac{0.86}{0.86 + 1}\right)\right]$$
$$= 6.1 \text{ ft/s } (1.9 \text{ m/s})$$

Then the time of arrival of the peak discharge and stage at mile 12.3 is given by Eq. (13.3.64) as

$$t_a = \tau + \frac{x_i}{c} = 0.75(60)(60) + \frac{12.3(5280)}{6.1} = 13{,}347 \text{ s}$$

$$= \frac{13{,}347}{60(60)} = 3.7 \text{ h}$$

Forecast at Mile 40.5

The procedure is repeated to find the peak flow and stage and time of arrival at mile 40.5 with the following results.

$$X_2^* = \frac{x}{X_c} = \frac{40.5(5280)}{20{,}800} = 10.3$$

$$\frac{Q_x}{Q(\max)} = 0.14$$

$$Q_x(X_2^* = 10.2) = 12{,}700 \text{ ft}^3/\text{s } (360 \text{ m}^3/\text{s})$$

$$y_x = 11.9 \text{ ft } (3.6 \text{ m})$$

$$Q_{x/2} = 0.20(90{,}800) = 18{,}200 \text{ ft}^3/\text{s } (515 \text{ m}^3/\text{s})$$

$$y_{x/2} = 13.7 \text{ ft } (4.2 \text{ m})$$

$$D_{x/2} = 7.37 \text{ ft } (2.2 \text{ m})$$

$$\overline{u} = 3.55 \text{ ft/s } (1.1 \text{ m/s})$$

In concluding this section, a number of comments should be made. First, although the use of a sophisticated model such as DAMBRK is preferred from the viewpoint of accuracy in estimating the downstream flow rate and stage, in many cases the time available for a forecast may preclude the use of such a complex model. In such cases, reasonable estimates can be made with the simplified technique developed by Wetmore and Fread (1981) and Anonymous (1982). Second, when the downstream river valley widens rapidly, the simplified dam break methodology is not applicable. For example, in the case of the Teton Dam failure, hand calculations could not be used past a station located 8.5 miles below the dam. Third, the most time-consuming task in using either a digital or simplified model is the acquisition of data. Wetmore and Fread (1981) have presented a set of default data values (Table 13.6).

Fread (1981) has also examined the effect of errors in various parameters on

TABLE 13.6 Default input data values (Wetmore and Fread, 1981)

Variable	Units	Default	Description
b_r	ft	2 (breach depth)	Breach width
τ	h	0.1–2	Time of failure
y_s	ft	$H/2$	Elevation of valley, wall base, or thick vegetation
n		0.06	Downstream Manning's n
S		Dam height/reservoir length	Bottom slope of downstream channel

downstream estimates of flow rate and the depth of flow. The conclusions of this study are:

1. As noted earlier, a potentially significant source of error is the prediction of the breach size, shape, and time of formation. Of these three characteristics, the shape of the breach is the least important and the time of breach formation becomes increasingly important as the reservoir volume becomes large. One method of evaluating the effect of uncertainty in breach width and time of formation is to consider two possibilities: first, maximum probable breach width and the minimum time of breach formation; second, minimum probable breach width and the maximum probable time of breach formation. The first possibility produces an estimate of the maximum possible downstream flood, while the second possibility produces an estimate of the minimum downstream flood.
2. To some extent, errors are always present in the description of a channel cross section. In general, errors in the estimate of flow area result in significantly smaller errors in the depth of flow.
3. Errors in the estimate of the flow resistance coefficient are the result of factors such as (*a*) large dam breach flows inundate areas which have not been previously flooded; (*b*) the inundated area often contains roughness elements which are not typical of normal floods, for example, buildings and trees; (*c*) the flow over a flood plain may short-circuit the flow past features such as meanders; (*c*) debris uprooted and transported by the flood waters may cause flow blockage if the debris is trapped against permanent structures; and (*e*) sediment trapped behind the dam may result in reduced flow resistance if it is deposited on the flood plain. The effect of features (*a*) and (*b*) on the estimated downstream depth of flow and the celerity of the flood wave can be analyzed analytically. Fread (1981) demonstrated that the effect of errors in the flow resistance coefficient due to these factors was damped in estimating both the depth of flow and the wave celerity.

It should also be noted at this point that, in dam breach floods where high flows inundate areas not normally covered with water, measurable losses of flow volume may occur. Such flow losses are difficult to predict and must normally be neglected.

13.4 SURGES IN OPEN CHANNELS

There are a number of engineered structures typically found in irrigation and hydroelectric systems which, under certain conditions of operation, can result in a surge being formed either upstream or downstream of the structure. For example, a surge may develop in a hydroelectric power canal when there is a sudden load rejection; surges may develop downstream of reservoirs whose outflows are primarily regulated by the need for hydroelectric power; and surges may develop in irrigation channels as the result of the rapid opening or closing of control gates. These types of flow situations generally have gradually varied, unsteady flow occurring on both sides of the surge. In the past, the method of characteristics (Section 12.2) has been used to estimate the gradually varied, unsteady portions of the flow; the surge equations have been used to estimate the flow conditions just upstream and downstream of the surge (see, for example, Streeter and Wylie, 1967). Such a solution technique is both rigorous and accurate, but it is very difficult to implement since it requires that first the inception of the surge must be detected and then the surge tracked as it moves through the system. A number of investigators (see, for example, Martin and DeFazio, 1969, Martin and Zovne, 1971, and Terzidis and Strelkoff, 1970) have demonstrated that, when the surge equations are neglected and the flow is treated as if it were unsteady and gradually varied throughout the system, the results compare favorably with both the rigorous method of solution and the available laboratory and field data.

In Chapter 12, the equations governing gradually varied, unsteady flow were presented and solved by what was termed a weighted four-point implicit difference scheme (Sections 12.3 and 12.4). It was further noted in Sections 12.3 and 12.4 that the weighting factor θ should lie in the following range:

$$0.5 \leq \theta \leq 1.0$$

with a recommended value of approximately 0.55. Chaudhry and Contractor (1973) have noted that in the case of surge flow the optimum value of θ may be different than that recommended for gradually varied, unsteady flow. Chaudhry and Contractor (1973) reached the following conclusions:

1. The use of $\theta = 0.5$ for surge flows results in a steep wave front but also a significant numerical instability after the passage of the surge.
2. The use of $\theta = 1$ results in a diffused surge front but no numerical instability.
3. Values of θ in the range of 0.5 to 1.0 yield solutions which have characteristics between the extremes noted above.
4. The use of $\theta = 0.6$ results in solutions which had little or no numerical instability and also exhibit a steep surge front.
5. Values of θ less than 0.6 may be appropriate in channels with boundary roughness coefficients larger than those studied by Chaudhry and Contractor (1973).

Note: In the cases studied by these investigators, the Manning roughness coefficient was in the range of 0.01 to 0.03. It was further noted that the optimum value of θ may also be a function of the steepness of the surge.

Thus, it is concluded that with proper calibration and verification the numerical methods presented in Chapter 12 for treating gradually varied, unsteady flow may also be used to estimate surge flow in open channels.

13.5 PULSATING FLOW; ROLL WAVES

When the slope of a channel becomes very steep, the resulting supercritical flow at normal depth may develop into a series of shallow-water waves known as *roll waves*. As these waves progress downstream, they eventually break and form hydraulic bores or shock waves. Figure 13.14 schematically shows the development of a pulsating flow (roll waves) in a channel of steep slope. When this type of flow occurs, the increased depth of flow requires increased freeboard, and the concentrated mass of the wave fronts may require additional structural factors of safety. In considering the phenomena of pulsating flow, it is first necessary to define the types of channel in which it may occur and then estimate the resulting wave height.

Escoffier (1950) and Escoffier and Boyd (1962) considered the theoretical conditions under which a uniform flow must be considered unstable. In the case of a nonuniform distribution of velocity in the vertical direction, i.e., two-dimensional flow, Escoffier (1950) demonstrated that when

$$\sqrt{\frac{A}{gT}}\frac{du}{dy} > 1 \tag{13.5.1}$$

or

$$\sqrt{\frac{A}{gT}}\frac{du}{dy} < -1 \tag{13.5.2}$$

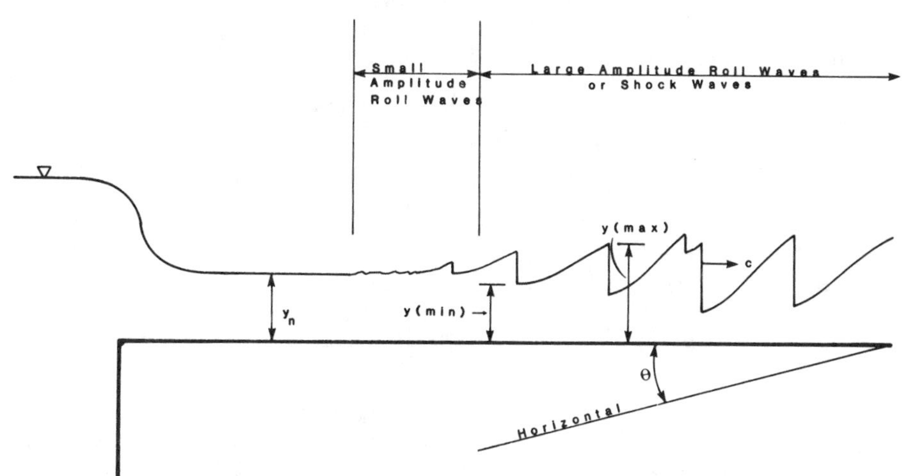

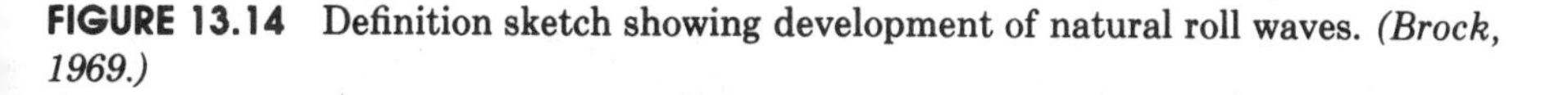

FIGURE 13.14 Definition sketch showing development of natural roll waves. *(Brock, 1969.)*

the flow must be considered unstable because small waves that are generated have a tendency to increase. In these equations, A = flow area, T = top width, u = velocity in the longitudinal direction, and y = vertical coordinate. Montuori (1963) demonstrated that the conditions specified by Eqs. (13.5.1) and (13.5.2) are equivalent to the Vedernikov number where

$$\mathbf{V} = \beta\theta\mathbf{F} \tag{13.5.3}$$

where β = exponent associated with the hydraulic radius in the general uniform flow equation (β = 2 for laminar flow; β = ½ for turbulent flow if the Chezy equation is used; and β = ⅔ for turbulent flow if the Manning equation is used) and θ = a channel shape factor defined as

$$\theta = 1 - R\frac{dP}{dA} \tag{13.5.4}$$

where R = hydraulic radius
P = wetted perimeter
A = flow area
$\mathbf{F}$ = Froude number (see, for example, Chow 1959)

If $\mathbf{V} < 1$, then stable flow prevails; however, if $\mathbf{V} > 1$, then unstable flow will exist. In a subsequent paper Escoffier and Boyd (1962) developed an analogous relationship for the transition point between stable and unstable flow for one-dimensional flow, i.e., a uniform velocity in the vertical. In this development, if

$$J > \left| 5.723 \frac{\rho^{1/3}}{\delta^{1/2}} \frac{d\zeta}{d\rho} \right| \tag{13.5.5}$$

then the flow is unstable. In Eq. (13.5.5)

$$\begin{aligned} \zeta &= \frac{y}{b} \\ \rho &= \frac{R}{b} \\ \delta &= \frac{D}{b} \\ J &= \frac{b^{1/6}\sqrt{S}}{n} \end{aligned} \tag{13.5.6}$$

where b = bottom width of channel for rectangular and trapezoidal sections and diameter for circular sections
D = hydraulic depth
S = channel slope
n = Manning's roughness coefficient

EXAMPLE 13.5

If

$$S_0 = 0.10$$
$$y_N = 1.0 \text{ ft } (0.30 \text{ m})$$
$$\overline{u} = 32 \text{ ft/s } (9.8 \text{ m/s})$$
$$b = 10 \text{ ft } (3.0 \text{ m})$$

and the channel is rectangular in shape, determine if the resulting flow is stable or unstable.

Solution

Escoffier and Boyd (1962) asserted that the transition point between stable and unstable flow defined by Eq. (13.5.5) provides a more conservative estimate than Eqs. (13.5.1) and (13.5.2). The parameters on the right-hand side of Eq. (13.5.5) are

$$P = b + 2y = 10 + 2(1) = 12 \text{ ft } (3.7 \text{ m})$$
$$A = by = 10(1) = 10 \text{ ft}^2 \ (0.93 \text{ m}^2)$$
$$D = \frac{A}{T} = \frac{A}{b} = \frac{10}{10} = 1 \text{ ft } (0.30 \text{ m})$$
$$R = \frac{A}{P} = \frac{10}{12} = 0.83 \text{ ft } (0.25 \text{ m})$$

and

$$\rho = \frac{R}{b} = \frac{0.83}{10} = 0.083$$
$$\delta = \frac{D}{b} = \frac{1}{10} = 0.10$$
$$\zeta = \frac{y}{b} = \frac{1}{10} = 0.10$$

The derivative $d\zeta/d\rho$ which appears on the right-hand side of Eq. (13.5.5) is estimated numerically or

y	R	$\zeta = y/b$	$\rho = R/b$	$d\zeta/d\rho$
0.5	0.45	0.05	0.045	
				1.3
1.0	0.83	0.10	0.083	
				1.4
1.5	1.15	0.15	0.12	

In the range of interest

$$\frac{d\zeta}{d\rho} = 1.35$$

Then $$\left| 5.723 \frac{\rho^{1/3}}{\sqrt{\delta}} \frac{d\zeta}{d\rho} \right| = \left| 5.723 \frac{(0.083)^{1/3}}{\sqrt{0.10}} (1.35) \right| = 11$$

The left-hand side of Eq. (13.5.5) is given by Eq. (13.5.6) or

$$J = \frac{b^{1/6}\sqrt{S}}{n}$$

n is estimated from the Manning equation or

$$n = \frac{1.49}{\bar{u}} R^{2/3}\sqrt{S} = \frac{1.49}{32} (0.83)^{0.667}\sqrt{0.1}$$
$$= 0.013$$

Therefore

$$J = \frac{(10)^{1/6}\sqrt{0.10}}{0.013} = 36$$

Since the left-hand side of Eq. (13.5.5) is greater than the right-hand side, it is concluded that a pulsating flow will occur in this channel.

The conclusion reached above, using the Escoffier and Boyd (1962) results, can be checked by estimating the Vedernikov number or

$$\mathbf{V} = \beta\theta\mathbf{F}$$

where $$\beta = 2/3 \qquad \text{(Manning equation)}$$

$$\mathbf{F} = \frac{\bar{u}}{\sqrt{gy}} = \frac{32}{\sqrt{32.2(1)}} = 5.6$$

and $$\theta = 1 - R\frac{dP}{dA}$$

The derivative dP/dA is estimated by a numerical scheme or

y	P	A	dP/dA
0.5	11	5	
			0.20
1.0	12	10	
			0.20
1.5	13	15	

and $dP/dA = 0.20$. Therefore

$$\theta = 1 - R\frac{dP}{dA} = 1 - 0.83(0.20) = 0.83$$

and
$$\mathbf{V} = (2/3)(0.83)(5.6) = 3.1$$

Since $\mathbf{V} \geq 1$, this technique also predicts unstable flow.

Most investigations of pulsating flow or roll waves have focused on establishing the criteria that must be satisfied before roll wave trains can be established. Although such criteria are useful in alerting the designer to a potential problem, they provide no information regarding maximum and minimum depths of flow which may result. Neither do these criteria provide information regarding the properties of the roll wave train as a function of distance along the channel. Brock (1969) in a series of experiments examined four properties of roll wave trains: $y(\max)$ = the maximum depth of flow downstream (Fig. 13.14), $y(\min)$ = the minimum depth of flow downstream (Fig. 13.14), c = wave velocity, and T = wave period. For each of the three channel slopes used in these experiments (S = 0.05, 0.08, and 0.12), the Froude number varied only slightly ($3.5 < \mathbf{F} < 5.6$), and the normal depth of flow was changed by a factor of 2. The conclusions of this study were:

1. For S fixed and $\mathbf{F}$ varying only slightly $\overline{y}'(\max)$, $\overline{T}'$, $\overline{y}'(\min)$, $\sigma'_{y(\max)}$, and σ'_T could be expressed as functions of x' where $\overline{y}'(\max) = \overline{y}(\max)/y_N$ = dimensionless maximum depth, $\overline{T}' = S\overline{T}/\sqrt{gy_N}$ = dimensionless mean wave period, $\overline{y}'(\min) = \overline{y}(\min)/y_N$ = dimensionless mean minimum depth, $\sigma'_{y(\max)} = \sigma_{y(\max)}/y_N$ = standard deviation of $y(\max)$, $\sigma'_T = S\sigma_T/\sqrt{gy_N}$ = dimensionless standard deviation of T, and $x' = x/y_N$ = dimensionless longitudinal distance.
2. As S and $\mathbf{F}$ were increased, roll waves of significant amplitude appeared at smaller values of x'. (*Note:* Significant amplitude was defined as $\overline{y}'(\max)$ = 1.1.)
3. The rate of increase in the amplitude of the roll waves with distance, $\partial y(\max)/\partial x$, for a specified value of $\overline{y}'(\max)$ increased as $\mathbf{F}$ was increased by increasing S.
4. Two phases of roll wave development were identified. In the initial development phase, roll waves are formed but they do not overtake and combine with other waves; hence, $\overline{T}'$ is constant in this phase. In the final development phase, waves overtake each other and combine. In this phase $\overline{T}'$ increases in an almost linear fashion with x'.
5. The frequency distributions of $y(\max)$, T, and c for a given flow at a given station can be approximated by a gaussian distribution. The distribution of $y(\min)$ values was skewed toward small values.

6. The nature of the channel bottom in the vicinity of the inlet apparently has a significant influence on the distance from the inlet where roll waves are observed and can be measured. If the channel bottom near the inlet is smooth, then roll waves begin to develop further upstream than if the bottom in the vicinity of the inlet is rough. Brock (1969) hypothesized that these results could be explained in terms of boundary layer development.

Given the above observations and his experimental data, Brock (1969) described the following applications. First, in a wide, rectangular channel, the data summarized in Table 13.7 can be used to predict roll wave development. (*Note:* In terms of this development, roll waves begin to develop when $\bar{y}'(\max) = 1.1$.)

The effect of the channel slope on the development of roll waves could not be determined from the laboratory results. Brock (1969) suggested that as a first approximation the values of x' in Table 13.7 should be viewed as a function of **F**. For example, for a flow in which $\mathbf{F} = 3.75$, roll waves will begin to appear at $x' = 2100$ and shock waves—the waves formed by the overtaking process—at $x' = 2600$. Second, for wide, rectangular channels in which values of **F** and S are comparable with those studied by Brock, Figs. 13.15 and 13.16 can be used in conjunction with

$$y(\max)_{\max} = \bar{y}(\max) + 2.58\sigma_{y(\max)} \tag{13.5.7}$$

to estimate the expected maximum downstream depth of flow.

Thorsky and Haggman (1970) in a discussion noted that the results and theory presented by Brock (1969) appear to agree well with previous literature regarding roll waves. However, it is also clear that additional work in this area is required. For example, Thorsky and Haggman (1970) note that the shape of the channel can have a decisive influence on the generation of roll waves, and this is an observation which is not addressed in the literature currently available. Further, waveless channel shapes are possible, and this subject has not been adequately addressed.

EXAMPLE 13.6

If

$$S = 0.10$$
$$y_N = 0.30 \text{ m } (1.0 \text{ ft})$$
$$\bar{u} = 9.8 \text{ m/s } (32 \text{ ft/s})$$
$$b = 3.0 \text{ m } (9.8 \text{ ft})$$

and the channel is rectangular in shape, estimate the distance to the beginning of roll wave formation, the distance to the beginning of shock wave formation, and maximum downstream depth of flow.

TABLE 13.7 Experimental values of x' for the formation of roll and shock waves (Brock, 1969)

Wave condition	Values of x'		
	$S = 0.05011$ $3.71 < \mathbf{F} < 3.81$	$S = 0.08429$ $4.96 < \mathbf{F} < 5.06$	$S = 0.1192$ $5.60 < \mathbf{F} < 5.98$
$\bar{y}(\text{max}) = 1.1$	2100	1300–1650	1150–1250
Shock waves	2600	2150	1800–2250

Solution

From the given data

$$\mathbf{F} = \frac{\bar{u}}{\sqrt{gy_N}} = \frac{9.8}{\sqrt{9.8(0.3)}} = 5.7$$

The specified channel slope and the computed Froude number indicate that the data in the fourth column of Table 13.7 should be used to estimate the distance to roll wave formation and shock wave formation. The distance to roll wave formation is estimated by

$$x' = \frac{x}{y_N} \simeq 1200$$

$$x \simeq 1200y_n = 1200(0.3) = 360 \text{ m } (1200 \text{ ft})$$

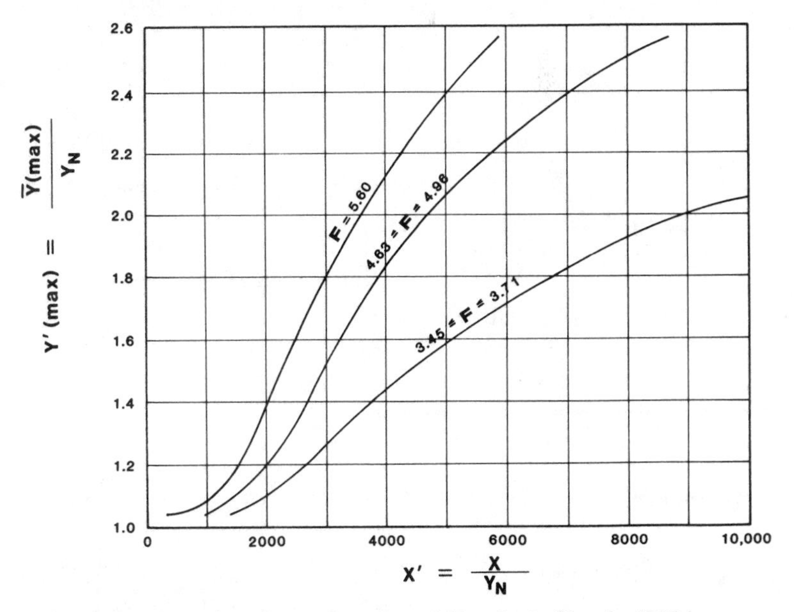

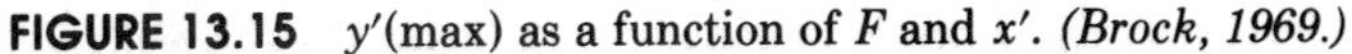
FIGURE 13.15 $y'(\text{max})$ as a function of F and x'. *(Brock, 1969.)*

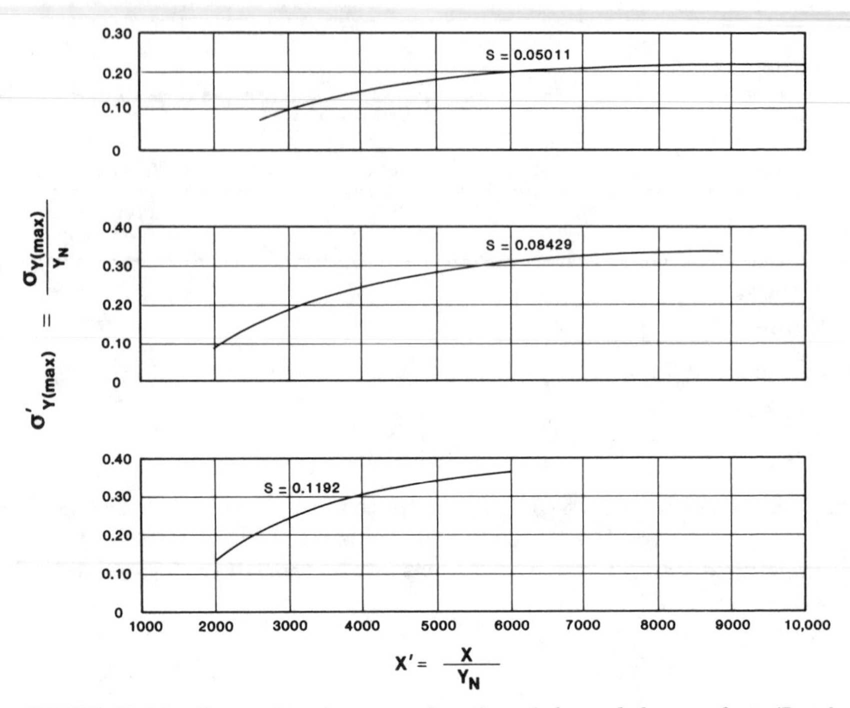

FIGURE 13.16 Curves for $\sigma'_{y(\max)}$ as a function of channel slope and x'. *(Brock, 1969.)*

The distance to shock wave formation is estimated by

$$x' = \frac{x}{y_N} \simeq 2000$$

$$x \simeq 2000y_N = 2000(0.3) = 600 \text{ m (2000 ft)}$$

An estimate of the expected maximum downstream depth of flow is obtained from Eq. (13.5.7) or

$$y(\max)_{\max} = \overline{y}(\max) + 2.58\sigma_{y(\max)}$$

From Fig. 13.15 with $x' = 2000$

$$\overline{y}'(\max) = \frac{\overline{y}(\max)}{y_N} \simeq 1.4$$

$$\overline{y}(\max) \simeq 1.4y_n = 1.4(0.3) = 0.42 \text{ m (1.4 ft)}$$

and from Fig. 13.16 with $x' = 2000$

$$\sigma'_{y(\max)} = \frac{\sigma_{y(\max)}}{y_N} = 0.15$$

$$= \simeq 0.15y_N = 0.15(0.3) = 0.045 \text{ m (0.15 ft)}$$

Therefore

$$y(\text{max})_{\text{max}} = 0.42 + 2.58(0.045) = 0.54 \text{ m } (1.8 \text{ ft})$$

BIBLIOGRAPHY

Anonymous, "Floods Resulting from Suddenly Breached Dams—Conditions of Minimum Resistance, Hydraulic Model Investigation," Misc. Paper 2-374, Report 1, U.S. Army Corps of Engineers, Waterways Experiment Station, Vicksburg, Miss., February 1960.

Anonymous, "Floods Resulting from Suddenly Breached Dams—Conditions of High Resistance, Hydraulic Model Investigation," Misc. Paper 2-374, Report 2, U.S. Army Corps of Engineers, Waterways Experiment Station, Vicksburg, Miss., November 1961.

Anonymous, "National Program of Inspection of Dams," Department of the Army, Office of Chief of Engineers, Washington, 1975.

Anonymous, "Simplified Dam-Break Model Users Manual," U.S. National Weather Service, Silver Spring, Md., 1982.

Balloffet, A., Cole, E., and Balloffet, A. F., "Dam Collapse Wave in a River," *Proceedings of the American Society of Civil Engineers, Journal of the Hydraulics Division,* vol. 100, no. HY5, May 1974, pp. 645–665.

Brevard, J. A., and Theurer, F. D., "Simplified Dam-Break Routing Procedure," *Technical Release Number 66,* U.S. Soil Conservation Service, Washington, 1979.

Brock, R. R., "Development of Roll-Wave Trains in Open Channels," *Proceedings of the American Society of Civil Engineers, Journal of the Hydraulics Division,* vol. 95, no. HY4, July 1969, pp. 1401–1427.

Chaudhry, Y. M., and Contractor, D. N., "Application of the Implicit Method to Surges in Open Channels," *Water Resources Research,* vol. 9, no. 6, December 1973, pp. 1605–1612.

Chow, V. T., *Open Channel Hydraulics,* McGraw-Hill Book Company, New York, 1959.

Escoffier, F. F., "A Graphical Method for Investigating the Stability of Flow in Open Channels or in Closed Conduits Flowing Full," *Transactions of the American Geophysical Union,* vol. 31, no. 4, August 1950.

Escoffier, F. F. and Boyd, M. B., "Stability Aspects of Flow in Open Channels," *Proceedings of the American Society of Civil Engineers, Journal of the Hydraulics Division,* vol. 88, no. HY6, November 1962, pp. 145–166.

Fread, D. L., "Some Limitations of Dam-Break Flood Routing Models," Preprint, *American Society of Civil Engineers,* Fall Convention, St. Louis, Mo., October 1981.

Fread, D. L., "DAMBRK: The NWS Dam-Break Flood Forecasting Model," National Weather Service, Office of Hydrology, Silver Spring, Md., January 1982.

Gundlach, D. L., and Thomas, W. A., "Guidelines for Calculating and Routing a Dam-Break Flood," *Research Note No. 5,* U.S. Army Corps of Engineers, Hydrologic Engineering Center, Davis, Calif., 1977.

Johnson, F. A., and Illes, P., "A Classification of Dam Failures," *Water Power and Dam Construction,* December 1976, pp. 43–45.

Martin, C. S., and DeFazio, F. G., "Open Channel Surge Simulation by Digital Computer," *Proceedings of the American Society of Civil Engineers, Journal of the Hydraulics Division,* vol. 95, no. HY6, June 1969, pp. 2049–2070.

Martin, C. S., and Zovne, J. J., "Finite Difference Simulation of Bore Propagation," *Proceedings of the American Society of Civil Engineers, Journal of the Hydraulics Division,* vol. 97, no. HY7, July 1971, pp. 993–1010.

Middlebrooks, T. A., "Earth Dam Practice in the United States," *Centennial Transactions,* American Society of Civil Engineers, 1952, pp. 697–722.

Montuori, C., Discussion of "Stability Aspects of Flow in Open Channels," by F. F. Escoffier, *Proceedings of the American Society of Civil Engineers, Journal of the Hydraulics Division,* vol. 89, no. HY4, July 1963, pp. 264–273.

Rajar, R., "Mathematical Simulation of Dam Break Flow," *Proceedings of the American Society of Civil Engineers, Journal of the Hydraulics Division,* vol. 104, no. HY7, July 1978, pp. 1011–1026.

Sakkas, I. G., and Strelkoff, T., "Dam Break Flood in a Prismatic Dry Channel," *Proceedings of the American Society of Civil Engineers, Journal of the Hydraulics Division,* vol. 99, no. HY12, December 1973, pp. 2195–2216.

Stoker, J. J., *Water Waves,* Interscience Publishers, New York, 1957.

Streeter, V. L., and Wylie, E. B., *Hydraulic Transients,* McGraw-Hill Book Company, New York, 1967.

Streeter, V. L., and Wylie, E. B., *Fluid Mechanics,* 6th ed., McGraw-Hill Book Company, New York, 1975.

Su, S. T., and Barnes, A. H., "Geometric and Frictional Effects on Sudden Releases," *Proceedings of the American Society of Civil Engineers, Journal of the Hydraulics Division,* vol. 96, no. HY11, November 1970, pp. 2185–2200.

Terzidis, G., and Strelkoff, T., "Computation of Open Channel Surges and Shocks," *Proceedings of the American Society of Civil Engineers, Journal of the Hydraulics Division,* vol. 96, no. HY12, December 1970, pp. 2581–2610.

Thorsky, G. N., and Haggman, D. C., Discussion of "Development of Roll Wave Trains in Open Channels," *Proceedings of the American Society of Civil Engineers, Journal of the Hydraulics Division,* vol. 96, no. HY4, April 1970, pp. 1069–1072.

Wetmore, J. N., and Fread, D. L., "The NWS Simplified Dam Break Flood Forecasting Model," *Proceedings of the Fifth Canadian Hydrotechnical Conference,* Fredericton, New Brunswick, Canada, May 1981.

FOURTEEN

Hydraulic Models

SYNOPSIS

In this chapter, the design, construction, and use of physical models to examine open-channel flow phenomena are discussed. Among the types of models considered are geometrically distorted and undistorted, fixed and movable bed, and ice.

In the case of fixed-bed models, special attention is given to the problem of simulating dispersion in a geometrically distorted model. It is demonstrated that in the general case, dispersion cannot be accurately modeled with a distorted physical model.

In the case of movable-bed models both empirical and theoretical techniques of model design are discussed.

Consideration is also given to the materials and methods which are used to construct physical models. The dual problems of calibration and verification are also discussed.

In the final section of the chapter, physical models of ice problems in open channels are discussed. Two types of models are considered. In the first type both hydrodynamics and hydroelasticity are modeled. Such models must be used when, for example, the forces developed by an ice sheet in contact with a structure are studied. In the second type of model only the hydrodynamics must be simulated correctly, for example, models to study drifting-ice problems.

14.1 INTRODUCTION

Unlike many other fields of scientific endeavor, modern hydraulic engineering has been and remains to a large extent based on experiment. A study of the history of hydraulics (see, for example, Rouse and Ince, 1963) indicates that periods of experimental investigation have alternated with periods of analysis. For example, as noted by the American Society of Civil Engineers' (ASCE) Hydraulics Division Committee (Anonymous, 1942), it was not until the eighteenth century that sufficient experimental data had been accumulated to permit the foretelling of future progress, and another 100 years passed before the results marking the achievement of this progress were available.

During the twentieth century, hydraulic engineering greatly benefited from the developments of boundary layer and turbulence theory in the field of fluid mechanics. Before these developments and the advent of the high-speed digital computer, many problems in the field of hydraulics could be resolved only with model studies. Today, the dual approach of theoretical and model studies is available to the engineer.

The use of models in the laboratory environment to solve hydraulic engineering problems requires a clear and accurate understanding of the principles of similitude. With regard to similarity, previously discussed in Chapter 1, there are three distinct viewpoints.

1. *Geometric Similarity:* Two objects are said to be geometrically similar if the ratios of all corresponding dimensions are equal. Thus, the *geometric similarity* refers only to similarity in form.
2. *Kinematic Similarity:* Two motions are said to be kinematically similar if (*a*) the patterns or paths of motion are geometrically similar and (*b*) the ratios of the velocities of the particles involved in the two motions are equal.
3. *Dynamic Similarity:* Two motions are said to be dynamically similar if (*a*) the ratio of the masses of the objects involved are equal and (*b*) the ratios of the forces which affect the motion are equal.

At this point, it is noted that while geometric and kinematic similarity can be achieved in most modeling situations, complete dynamic similarity is an ideal which can seldom, if ever, be achieved in practice (Anonymous, 1942).

As noted in Section 1.5, one approach to developing appropriate parameters to ensure dynamic similarity is the scaling of the governing equations (Section 1.4). A second method of achieving this objective is the use of the concept of dimensionless analysis and the Buckingham Π theorem (see, for example, Streeter and Wylie, 1975).

Experience dictates that in most models it is necessary to confine similarity to a single force. Since in most open-channel flow problems the force of gravity is the primary force, this chapter focuses on models designed for similitude with respect to the force of gravity. Further, the modeler is almost invariably constrained to use water as the model fluid. It should be mentioned that there are, of course, exceptions to these statements when special circumstances require exceptions (see, for example, Walski, 1979). In addition, in most models of interest to hydraulic engineers, the flow should be turbulent rather than laminar. Recall that turbulent flow generally exists when the Reynolds number, based on the hydraulic radius, exceeds 2000. Flow is laminar when the Reynolds number is less than 500. Between these limits lies the transition zone which should be avoided in designing models.

In addition to theoretical considerations, only a few of which have been discussed in the foregoing paragraphs, the modeler must be concerned with the following practical limitations: (1) time, (2) money, (3) laboratory space, and (4) the available water supply. These four practical considerations may have a crucial and significant impact on any modeling effort.

The foregoing theoretical and practical considerations combine to produce model scale ratios in the following ranges:

1. Models of spillways, conduits, and other appurtenances, the prototypes of which have relatively smooth surfaces and have scale ratios that range from 1:50 (model distance prototype distance) to 1:15. Such models should never be distorted.

2. Models of rivers, harbors, estuaries, and reservoirs, the prototypes of which have relatively rough surfaces and have horizontal-scale ratios that range from 1:2000 to 1:100 and vertical-scale ratios that range from 1:150 to 1:50. Models in this category are almost always distorted.

Models with distorted-scale ratios depart from strict dynamic similarity, and such a departure is acceptable only when the requirement of strict similitude can be waived.

The need for distorted-scale models, or what is more accurately termed geometrically distorted models, is usually found only in the study of open-channel phenomena and arises from two sets of circumstances.

1. The area required for an undistorted model would be so large that space and economic considerations dictate that the horizontal scale of the model must be made small. In such a situation, if the vertical scale were equal to the horizontal scale, the depth of flow would be so small that it could not be measured satisfactorily. Given these circumstances, it is usually considered desirable to increase the vertical scale of the model relative to the horizontal scale.
2. In the case of a movable-bed model whose purpose is the simulation of the movement of bed material, the slopes and fluid velocities encountered in an undistorted model would usually be too small to move any of the materials typically available for use as bed materials. As before, it is usually considered desirable to use a vertical scale which is larger than the horizontal scale.

Although in the subsequent sections of this chapter the primary principle used will be the Froude law, it is appropriate to consider, at this point, not only the implication of the Froude law but also a number of other specialized modeling laws.

Froude law models assert that the primary force causing fluid motion is gravity and that all other forces such as fluid friction and surface tension can be neglected. As noted in Section 1.5 if only Froude number similarity is required, then

$$\mathbf{F}_M = \mathbf{F}_P \tag{14.1.1}$$

where $\mathbf{F}$ = Froude number and the subscripts M and P designate the model and prototype Froude numbers, respectively. Equation (14.1.1) can be solved to yield

$$U_R = \frac{U_M}{U_P} = \left(\frac{g_M}{g_P}\frac{L_M}{L_P}\right)^{1/2} = \sqrt{g_R L_R} \tag{14.1.2}$$

where R = subscript indicating ratio of model to prototype variables
U_R = velocity ratio
L_R = length scale ratio
g_R = gravity ratio

Since from a practical viewpoint the acceleration of gravity cannot be altered

$$g_R = 1$$

and Eq. (14.1.2) becomes

$$U_R = \sqrt{L_R} \tag{14.1.3}$$

Since the velocity of flow can be expressed in terms of distance and time, the time-scale ratio can be derived from Eq. (14.1.3)

$$\frac{U_M}{U_P} = \frac{L_M/T_M}{L_P/T_P} = \frac{T_P L_M}{T_M L_P} = \frac{L_R}{T_R}$$

Substitution of Equation (14.1.3) in this result yields

$$T_R = \sqrt{L_R} \tag{14.1.4}$$

where T_R = time-scale ratio. Previously in this section it was noted that in some cases Froude models with distorted scales must be used. In such cases, the distorted model should be designed so that

$$U_R = \sqrt{Y_R} \tag{14.1.5}$$

where Y_R = vertical-scale ratio. The time scale is

$$T_R = \frac{L_R}{\sqrt{Y_R}} \tag{14.1.6}$$

where L_R = horizontal-scale ratio. Although Eqs. (14.1.5) and (14.1.6) are recommended formulations for distorted models (see, for example, Anonymous 1942), many distorted-scale models depart from these recommendations.

Reynolds law models assert that viscosity is the primary force that must be considered and, hence, that gravitational and surface tension forces can be safely neglected. In models based on the Reynolds law

$$\mathbf{R}_R = 1 \tag{14.1.7}$$

where $\mathbf{R}$ = Reynolds number; therefore

$$U_R = \frac{\mu_R}{L_R \rho_R} \tag{14.1.8}$$

and

$$T_R = \frac{L_R \rho_R}{\mu_R} \tag{14.1.9}$$

where ρ_R = density ratio and μ_R = dynamic viscosity ratio.

Weber law models ensure similitude between the model and prototype with regard to surface tension effects. Although this modeling law is of paramount importance in the study of droplet formation, it can also be a crucial consideration in some open-channel models. For example, surface tension effects are noticeable in flow over weirs when the head on the weir is small. In addition, some wave phenomena noticeable in models are due to surface tension. The Weber number is defined by

$$\mathbf{W} = \frac{\rho U^2 L}{\sigma} \tag{14.1.10}$$

where U = characteristic velocity
L = characteristic length
ρ = fluid density
σ = surface tension force

Equating the Weber numbers for a model and prototype yields the following relationships for the velocity and time ratios:

$$U_R = \left(\frac{\sigma_R}{L_R \rho_R}\right)^{1/2} \tag{14.1.11}$$

and

$$T_R = \left(\frac{L_R^3 \rho_R}{\sigma_R}\right)^{1/2} \tag{14.1.12}$$

Table 14.1 summarizes the relations that must exist between the several properties of the fluids used in the model and prototype if perfect similitude is to be obtained under the various modeling laws discussed. Each expression in Table 14.1 is a combination of ratios, model to prototype, between an arbitrary length and force characteristic of each law. Time is generally considered the dependent ratio and becomes fixed once the length ratio and fluids for the model and prototype are selected.

In open-channel flow, the presence of the free surface ensures that gravitational forces are always dominant; therefore, the Froude law is the primary principle on which almost all open-channel models are built. Surface tension effects are important only when the radius of curvature of the free surface or the distances from solid boundaries are very small. Thus, surface tension effects are negligible in most prototypes, and care must be taken to ensure that they are also negligible in the models. From a practical viewpoint, if model depths of flow are greater than 2 in (0.05 m), surface tension effects can usually be ignored. Viscosity is much more important and can exert its influence in many different situations. The only method of keeping viscous effects exactly the same in an open-channel model and prototype is to keep both the Reynolds and Froude numbers equal in both the model and prototype. From a practical viewpoint, this requirement cannot be satisifed. However, if the flow in the model is turbulent, then the form drag will be accurately modeled, but surface drag will not be accurately modeled. Thus, the modeler must at all times be concerned with what are termed *scale effects*, i.e., distortions introduced by forces other than the dominant force on which the model is based.

TABLE 14.1 Fluid property scales (Anonymous, 1942)

Characteristic	Dimension	Scale ratio for the laws of			
		Froude	Reynolds	Weber	Cauchy
Force characteristics		Gravity	Viscosity	Surface tension	Elasticity
Length	L	L_R	L_R	L_R	L_R
Area	L^2	$(L_R)^2$	$(L_R)^2$	$(L_R)^2$	$(L_R)^2$
Volume	L^3	$(L_R)^3$	$(L_R)^3$	$(L_R)^3$	$(L_R)^3$
Kinematic properties					
Time	T	$[(L\rho/\gamma)_R]^{1/2}$	$(L^2\rho/\mu)_R$	$[(L^3\rho/\sigma)_R]^{1/2}$	$[L(\rho/K)^{1/2}]_R$
Velocity	L/T	$(L\gamma/\rho)_R$	$(\mu/L\rho)_R$	$[(\sigma/L\rho)_R]^{1/2}$	$[(K/\rho)^{1/2}]_R$
Acceleration	L/T^2	$(\gamma/\rho)_R$	$(\mu^2/\rho^2L^3)_R$	$(\sigma/L^2\rho)_R$	$(K/L\rho)_R$
Discharge	L^3/T	$[L^{5/2}(\gamma/\rho)^{1/2}]_R$	$(L\mu/\rho)_R$	$[L^{3/2}(\sigma/\rho)^{1/2}]_R$	$[L^2(K/\rho)^{1/2}]_R$
Kinematic viscosity	L^2/T	$[L^{3/2}(\gamma/\rho)^{1/2}]_R$	$(\mu/\rho)_R$	$[(L\sigma/\rho)_R]^{1/2}$	$[L(K/\rho)^{1/2}]_R$
Dynamic properties					
Mass	M	$(L^3\rho)_R$	$(L^3\rho)_R$	$(L^3\rho)_R$	$(L^3\rho)_R$
Force	ML/T^2	$(L^3\gamma)_R$	$(\mu^2/\rho)_R$	$(L\sigma)_R$	$(L^2K)_R$
Density	M/L^3	ρ_R	ρ_R	ρ_R	ρ_R
Specific weight	M/L^2T^2	γ_R	$(\mu^2/L^3\rho)_R$	$(\sigma/L^2)_R$	$(K/L)_R$
Dynamic viscosity	M/LT	$[L^{3/2}(\rho\gamma)^{1/2}]_R$	μ_R	$[(L\rho\sigma)_R]^{1/2}$	$[L(K\rho)^{1/2}]_R$
Surface tension	M/T^2	$(L^2\gamma)_R$	$(\mu^2/L\rho)_R$	σ_R	$(LK)_R$
Volume elasticity	M/LT^2	$(L\gamma)_R$	$(\mu^2/L^2\rho)_R$	$(\sigma/L)_R$	K_R
Pressure intensity	M/LT^2	$(L\gamma)_R$	$(\mu^2/L^2\rho)_R$	$(\sigma/L)_R$	K_R
Momentum impulse	ML/T	$[L^{7/2}(\rho\gamma)^{1/2}]_R$	$(L^2\mu)_R$	$[L^{5/2}(\rho\sigma)^{1/2}]_R$	$[L^3(K\rho)^{1/2}]_R$
Energy and work	ML^2/T^2	$(L^4\gamma)_R$	$(L\mu^2/\rho)_R$	$(L^2\sigma)_R$	$(L^3K)_R$
Power	ML^2/T^3	$(L^{7/2}\gamma^{3/2}/\rho^{1/2})_R$	$(\mu^3/L\rho^2)_R$	$[\sigma^{3/2}(L/\rho)^{1/2}]_R$	$(L^2K^{3/2}/\rho^{1/2})_R$

14.2 FIXED-BED RIVER OR CHANNEL MODELS

For river or channel studies in which the motion of the bed is unimportant, either an undistorted or a distorted fixed-bed model can be used, depending on the characteristics of the flow which must be represented. An undistorted model is recommended if the study involves the reproduction of supercritical flow, transitions, wave patterns, or water surface profiles. If the model is to reproduce channel capacity or channel storage capacity, then a distorted-scale model is satisfactory.

The use of an undistorted model presents the modeler with a minimum of design and analytical problems. In a distorted-scale model, the primary difficulty is ensuring that the model is sufficiently rough so that conversions of kinetic to potential energy and vice versa are not distorted. To demonstrate this difficulty, consider a distorted-scale model in which the vertical-scale ratio is Y_R and the horizontal-scale ratio is L_R. The average velocities of flow in the prototype and model are given by the Manning equation or

$$\overline{u}_P = \frac{\phi R_P^{2/3}\sqrt{S_P}}{n_P} \tag{14.2.1}$$

and

$$\overline{u}_M = \frac{\phi R_M^{2/3}\sqrt{S_M}}{n_M} \tag{14.2.2}$$

where R = hydraulic radius
S = channel bottom slope
n = Manning resistance coefficient
ϕ = coefficient whose value depends on system of units used (1 for SI system, 1.49 for English system)

Then, combining these equations to form the velocity scale ratio yields

$$U_R = \frac{\overline{u}_M}{\overline{u}_P} = \frac{(R_R)^{2/3}\sqrt{S_R}}{n_R}$$

and since $S_R = Y_R/L_R$

$$U_R = \frac{(R_R)^{2/3}\sqrt{Y_R}}{n_R\sqrt{L_R}} \tag{14.2.3}$$

The volumetric discharge ratio can be obtained by noting that

$$Q_R = (L_R Y_R U_R)$$

or

$$Q_R = \frac{(R_R)^{2/3}\sqrt{L_R}\,(Y_R)^{3/2}}{n_R} \tag{14.2.4}$$

A time scale can then be defined as follows

$$T_R = \frac{\text{volume scale}}{\text{discharge scale}} = \frac{L_R^2 Y_R}{Q_R} \tag{14.2.5}$$

Note: In a distorted hydraulic model, the hydraulic radius cannot be estimated analytically but must be computed on a section-by-section basis.

EXAMPLE 14.1

If

$$L_R = 1/200$$
$$Y_R = 1/80$$
$$Q_R = 1/128{,}000$$
$$n_P = 0.024$$

and corresponding channel sections of the prototype and model are as shown in Fig. 14.1, determine n_M and T_R. This example was first presented in Anonymous (1942).

Solution

The slope ratio is given by

$$S_R = \frac{Y_R}{L_R} = \frac{1/80}{1/200} = \frac{200}{80} = 2.5$$

From the information in Fig. 14.1

$$R_R = \frac{R_M}{R_P} = \frac{0.245}{23.9} = \frac{1}{97.5} = 0.0103$$

From Eq. (14.2.4)

$$Q_R = \frac{1}{n_R}(R_R)^{2/3}(Y_R)^{3/2}\sqrt{L_R} = \frac{1}{n_R}\left(\frac{1}{97.5}\right)^{2/3}\left(\frac{1}{80}\right)^{3/2}\sqrt{\frac{1}{200}}$$

or

$$= \frac{1}{128{,}000} = \frac{1}{n_R}4.66 \times 10^{-6}$$

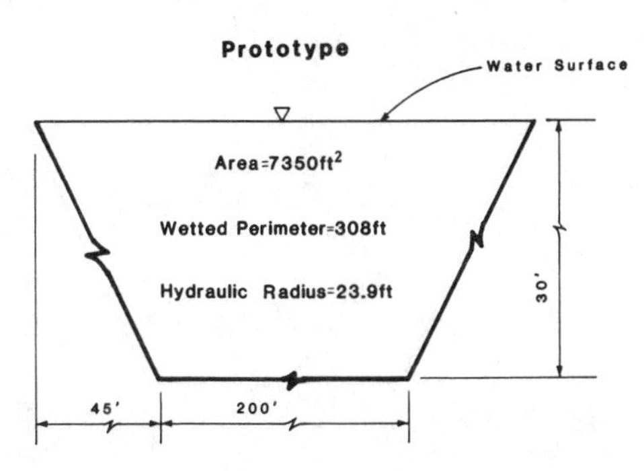

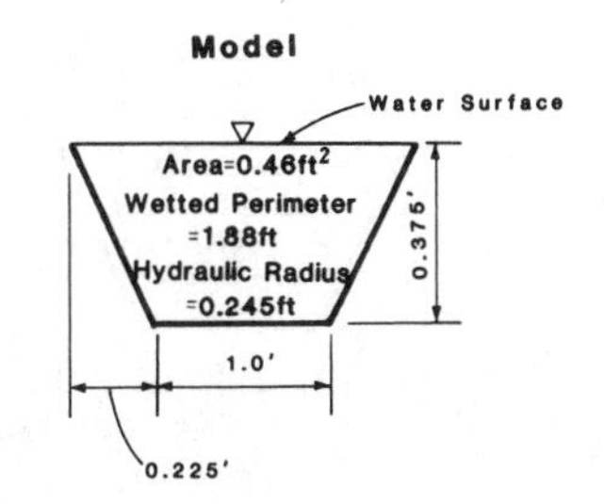

FIGURE 14.1 Schematic for Example 14.1.

and $$n_R = 4.66 \times 10^{-6}(128{,}000) = 0.596$$

The required value of n for the model is

$$n_M = 0.596 n_P = 0.596(0.024) = 0.014$$

The time-scale ratio for this set of circumstances is estimated by Eq. (14.2.5) or

$$T_R = \frac{L_R^2 Y_R}{Q_R} = \frac{(1/200)^2(1/80)}{(1/128{,}000)} = 0.04 = \frac{1}{25}$$

Thus, 1 h of prototype time corresponds to 2.4 min in the model.

In the preceding example, a close examination of the computations demonstrates that under some circumstances it is conceivable that the model boundaries would be rougher than the prototype boundaries. This would be the case if $Q_R < 1/214{,}590$. Such situations are often encountered in practice and are resolved in some cases by setting pebbles in the model boundaries or by placing folded metal strips in the boundary to increase the roughness.

At this point, it is constructive to consider the model of the Mississippi River which has been built by the U.S. Army Corps of Engineers at Vicksburg, Mississippi. The Mississippi River basin extends approximately 2000 miles (3200 km) in each direction. Therefore, a horizontal scale of 1:2000 was chosen, and the resulting model covers an area of approximately 1 mile^2 (2.6 km^2). Even at this rather large scale, the model river is only 2 to 3 ft wide (0.61 to 0.91 m) where the prototype river is 1 mile wide (1.6 km). In the vertical dimension, a scale of 1:2000 would result in extremely shallow depths of flow with the attendant problems of surface tension effects and laminar motion; therefore, in the model a vertical-length scale ratio of 1:100 was chosen. This length scale results in flow depths of at least 1 in (0.025 m), eliminates surface tension effects, and makes the flow turbulent. The scale distortion does not seriously distort the gross flow pattern and in most cases yields satisfactory results.

There are a number of hydraulic processes which cannot be modeled with distorted models, for example, the study of waste dispersion in a river or estuary. Recall from Section 10.2 that dispersion in a one-dimensional open channel is governed by

$$\frac{\partial}{\partial t}(Ac) + \frac{\partial}{\partial x}(\overline{u}Ac) = \frac{\partial}{\partial x}\left(EA\frac{\partial c}{\partial x}\right) \tag{14.2.6}$$

where c = concentration of conservative dissolved substance
t = time
$\overline{u}$ = average velocity in longitudinal direction
A = cross-sectional flow area
E = longitudinal dispersion coefficient

Equation (14.2.6) uses only quantities which are averaged over the cross section and assumes that all mass transport other than that caused by advection is proportional to the concentration gradient and the dispersion coefficient. In Chapter 10, it was also asserted that E is a function of the velocity gradients in both the vertical and transverse directions. Given these observations, it might be expected that a distorted hydraulic model, especially if exaggerated roughness elements are used, will not correctly model dispersive processes.

By definition, the correct modeling of pollutant dispersion is a simulation in which the relative concentration at each point in the model is equal to the relative concentration at each geometrically corresponding point in the prototype. The relative concentration is the actual concentration divided by the mass input per unit volume of the system. For the correct modeling of a pollutant cloud, a necessary, though not a sufficient condition, is

$$(\sigma_x)_R = L_R \tag{14.2.7}$$

where σ_x = longitudinal variance of the dispersing cloud, and, as before, L_R = horizontal-length scale ratio. Then, by Eq. (10.2.6)

$$E = \frac{1}{2}\frac{d}{dt}(\sigma_x^2)$$

Combining Eq. (10.2.6) with Eqs. (14.1.6) and (14.2.7) yields a necessary condition for correct modeling or

$$E_R = \frac{L_R^2}{T_R} = \frac{L_R}{\sqrt{Y_R}} \tag{14.2.8}$$

If Eq. (14.2.8) is satisfied, then the concentration profiles are simply transposed between the prototype and the model.

Fischer and Holley (1971) examined the problem of correctly reproducing a dispersing cloud in a distorted-scale hydraulic model by noting that the overall dispersion coefficient E can be viewed as having two components, namely, (1) E_v, which is the dispersion coefficient component associated with the vertical velocity distribution, and (2) E_t, which is the dispersion coefficient component associated with the transverse velocity distribution. In a steady flow dominated by vertical velocity gradients, the dispersion coefficient can be estimated by

$$E_v = 5.9 y u_* \tag{14.2.9}$$

where $u_* = \sqrt{gRS}$
y = depth of flow
R = hydraulic radius
S = longitudinal channel slope

Fischer and Holley (1971) assumed that

$$(u_*)_R = (g_R Y_R S_R)^{1/2} = \frac{Y_R}{\sqrt{L_R}} \tag{14.2.10}$$

where $g_R = 1$. Combining Eqs. (14.2.9) and (14.2.10) yields

$$(E_v)_R = \frac{Y_R^2}{\sqrt{L_R}} \tag{14.2.11}$$

A comparison of Eqs. (14.2.8) and (14.2.11) demonstrates that even in this elementary case a distorted hydraulic model will not correctly model longitudinal dispersion. For flows dominated by transverse velocity gradients, Fischer (1967) demonstrated that

$$E_t = \frac{0.3\langle u''^2\rangle b^2}{R u_*} \tag{14.2.12}$$

where b = distance on the water surface from the thread of maximum velocity to the most distant bank, u'' = difference between the point velocity and the cross-sectional mean velocity, and $\langle\ \rangle$ indicates a cross-sectional average of the variable enclosed. If it is assumed that

$$\frac{\langle u_R''\rangle}{\bar{u}_R^2} = 1$$

and Eq. (14.2.12) is combined with this assumption (Fischer and Holley, 1971), then

$$(E_t)_R = \frac{L_R^{5/2}}{Y_R} \tag{14.2.13}$$

Equation (14.2.13) indicates that a distorted-scale hydraulic model will not model the dispersion of a pollutant cloud correctly when the flow is dominated by transverse velocity gradients. The addition of roughness strips to most distorted-scale models increases the discrepancy that will be noted between model and prototype. Based on the results presented above and additional work, Fischer and Holley (1971) concluded that:

1. In steady flow in a distorted hydraulic model, the dispersive effects of the vertical velocity distribution are magnified while the dispersive effects of the transverse gradients of velocity are diminished. Although it is conceivable that in specific cases these effects may cancel and yield a correct simulation, the chances of this occurring must be considered minimal.

2. In oscillatory flows such as those found in tidal estuaries, the dispersive effects of the vertical velocity distribution are magnified while the effects of the transverse velocity distribution may be magnified, diminished, or correctly modeled, depending on the value of the dimensionless time scale for the prototype.

3. In very wide tidal estuaries, the dispersive effects of both the transverse and vertical velocity gradients are magnified; therefore, dispersion occurs much more quickly in the model than in the prototype.

Other investigators (see, for example, Moretti and McLaughlin, 1977, and Roberts and Street, 1980) have also reached these conclusions. Moretti and McLaughlin (1977) concluded that similarity criteria could be used to develop models which modeled either transverse or vertical mixing correctly, but that a model which simulated both processes simultaneously was not feasible with a distorted-scale model. Roberts et al. (1979) described an initial series of experiments in what they termed a variable distortion facility (see also Roberts et al., 1979) and reached the conclusion that at present the details of the mixing processes in a turbulent flow and how they vary with geometric-scale distortion are unknown. Thus, the general conclusion is that distorted hydraulic models are not appropriate to the study of phenomena in which the dispersion of a pollutant is a critical issue.

14.3 MOVABLE-BED MODELS

When the movement of the materials which compose the sides and bed of a channel is of paramount importance, a movable-bed model is used. Such models can be used to address the following problems:

1. General river morphology: Changes in river grades, cross-sectional shape, meanders, erosion, sedimentation, and changes in discharge and sediment yield associated with upstream hydraulic structures or land use changes
2. River training: Elimination of bends or meanders, relocation of the main channel, and the optimum location and design of groins
3. Flood plain development
4. Bridge pier location and design
5. Scour below dams
6. Pipeline crossings

In comparison with the fixed-bed models described in the foregoing section of this chapter, the design and operation of a movable-bed model are much more complex. Two major difficulties are:

1. The boundary roughness of the model is not controlled by design but is a dynamic variable controlled by the motion of the sediment and the bed forms which are developed.
2. The model must correctly simulate not only the prototype water movement but also the prototype sediment movement.

There are three methods of designing a movable-bed model which range from the purely empirical trial and error to the essentially theoretical method developed by Einstein and Chien (1956b).

Trial-and-Error Design

The earliest known movable-bed hydraulic model study was conducted in 1875 by Louis Jerome Fargue for improvements in the Garonne River at Bordeaux (Zwamborn, 1967). In 1885, Osborne Reynolds constructed a physical model with a movable bed to study the estuary of the River Mersey in England. These early models were constructed and effectively used many decades before the development of modern theoretical modeling principles. The basis for design in these early models was the hypothesis that if a model can be adjusted to reproduce events that *have* occurred in the prototype, then the model should also reproduce events which *will* occur in the prototype. The accuracy of the results obtained by Fargue and Reynolds proves the validity of this hypothesis, which basically requires the development of a model by trial and error coupled with verification.

The trial-and-error approach to the design of movable-bed hydraulic models involves the selection of a bed material. The velocity of flow required to move the selected material and the resistance coefficient which characterizes the material are established. Then, the vertical scale of the model required to scour the bed material is estimated. The model is then constructed, and the model variables are adjusted as required to achieve agreement between the results in the model and prototype for a verification event or a series of verification events.

The verification event or events should satisfy the following criteria:

1. The event or events from the prototype which are used to verify the model should involve phenomena pertinent to the goals and objectives of the proposed model study. For example, a model which is to be used to study bed load movement should not be verified with flood flow data.
2. The event or events on which the verification is to be based should be continuous and of reasonable duration.
3. The event or events on which the verification is based should not be extreme unless the study of such events is the purpose of the modeling program.
4. The greater the departure of the conditions which are to be modeled from the conditions under which the model was verified, the greater the uncertainty in the modeling results.

As indicated above, the development of a movable-bed hydraulic model by this design process, i.e., comparison of model and prototype data, is a trial-and-error process which is repeated until satisfactory results are achieved. Among the model variables which can be adjusted and manipulated are:

1. Discharge scale
2. Water surface slope

3. Type of bed material
4. Magnitude of the time scale
5. Roughness of the fixed boundary

At this point a number of miscellaneous comments regarding movable-bed models can be made:

1. In practice, it has been found that the low velocities of flow typically found in a model are not sufficient to scour many bed materials. Therefore, a model bed material which is significantly less dense than natural sand (specific gravity $\simeq$ 2.65) is required. Bituminous materials, e.g., coal dust (specific gravity 1.3 or less), or plastics (specific gravity $\simeq$ 1.2 or less), are commonly used.
2. While in fixed-bed models geometric distortion can be used in many cases without creating unsolvable problems, in a movable-bed model geometric-scale distortion may result in channel side slopes that exceed the angle of repose of the model bed material. If the side slopes are stable in the prototype, then the side slopes in the model can be constructed of rigid materials.
3. Geometric-scale distortion also results in an increase in the longitudinal slope of the model and requires an increase in the boundary roughness. However, the boundary roughness is not only a function of the material selected but also a function of the movement of the bed itself. Thus, in movable-bed models, geometric-scale distortion should be either avoided or minimized.

Theoretical Methods

Although movable-bed hydraulic models can be constructed by trial and error, there are explicit design techniques available which are based on reasonably well-established theoretical and semitheoretical concepts. For example, consider Fig. 7.13, which delineates the threshold of particle movement as a function of two dimensionless parameters $\mathbf{R}_*$ and F_s where

$$\mathbf{R}_* = \frac{u_* d}{\nu} \tag{14.3.1}$$

$$F_s = \frac{\tau_0}{(S_s - 1)d} = \frac{u_*^2}{(S_s - 1)gd} \tag{14.3.2}$$

where R_* = Reynolds number based on shear velocity and particle size
u_* = shear velocity
ν = fluid kinematic viscosity
S_s = specific gravity of particles composing perimeter
d = diameter of particles composing perimeter of channel

As suggested earlier in this chapter, the state of the channel bed is governed by Fig. 7.13 and the location of the $\mathbf{R}_* - F_s$ line in this figure. Therefore, it is asserted that if F_s and $\mathbf{R}_*$ are the same in the model and prototype, then the state of the bed will be the same in both cases. In addition, it can be assumed that the equivalent roughness–to–particle-size ratio will be the same in both the prototype and the model.

In a flow which transports sediment, the sediment suspended in the flow reduces the flow resistance, but this effect is small in comparison with the effect of the formation of dunes in the bed of the channel. Einstein and Barbarossa (1956a) assumed that the shear stress on the bed of the channel was composed of two components: τ_0' = a shear stress due to the intrinsic roughness of the sediment composing the bed and τ_0'' = a shear stress due to the bed or form roughness. Einstein and Barbarossa (1956a) further assumed that the total flow area A and the total wetted perimeter P could also be divided into components; and therefore,

$$\tau_0' = \gamma R'S \tag{14.3.3}$$

$$\tau_0'' = \gamma R''S \tag{14.3.4}$$

and
$$\tau_0 = \tau_0' + \tau_0'' = \gamma(R' + R'')S \tag{14.3.5}$$

where R' and R'' are the component hydraulic radii associated with τ_0' and τ_0'', respectively. In 1923 Strickler (see, for example, Chow, 1959, or Henderson, 1966) or Section 4.3 asserted that

$$n = 0.034d^{1/6} \tag{14.3.6}$$

where n = Manning's n and d = diameter of the sediment. Then, if as stated in the previous paragraph, the equivalent roughness-to-particle-size ratio is the same in both the model and prototype

$$n_R = d_R^{1/6} \tag{14.3.7}$$

and
$$\left(\frac{R'}{R''}\right)_R = 1 \tag{14.3.8}$$

From Eq. (14.2.3)

$$U_R = \frac{(R_R)^{2/3}\sqrt{Y_R}}{n_R\sqrt{L_R}}$$
$$n_R = \frac{(R_R)^{2/3}}{\sqrt{L_R}} \tag{14.3.9}$$

where $U_R = \sqrt{Y_R}$. Combining Eqs. (14.3.7) and (14.3.9) yields

$$d_R^{1/6} = \frac{(R_R)^{2/3}}{\sqrt{L_R}} \tag{14.3.10}$$

The implication of Eq. (14.3.8) is

$$\tau_0 = \gamma_R R_R S_R = \frac{\gamma_R R_R Y_R}{L_R} \tag{14.3.11}$$

It then follows from Eqs. (14.3.1) and (14.3.2) that

$$\frac{R_R Y_R}{\alpha_R L_R d_R} = 1 \tag{14.3.12}$$

and

$$\frac{R_R Y_R d_R^2}{L_R \nu_R^2} = 1 \tag{14.3.13}$$

where $\alpha = S_s - 1$. R_R is a function of Y_R and L_R, and if $\nu_R = 1$, then Eqs. (14.3.10) to (14.3.13) are a set of three equations in four unknowns. Thus, in designing a movable-bed model by this method, one model ratio is selected and the other three model ratios are determined from Eqs. (14.3.10), (14.3.12), and (14.3.13).

With regard to Eqs. (14.3.10), (14.3.12), and (14.3.13) the following implications are noted:

1. If $Y_R = L_R$, that is, the model is undistorted, then a scale model cannot be constructed, and any model will have to have the same scale as the prototype.

2. If the specific gravity ratio is selected in advance, then from Eqs. (14.3.12) and (14.3.13)

$$d_R = \alpha_R^{-(1/3)} \tag{14.3.14}$$

if $\nu_R = 1$. Equation (14.3.14) implies that if the sediment material used in the model has a specific gravity less than the material present in the prototype, then the size of the material used in the model should be larger.

3. If R_R cannot be assumed equal to Y_R, then the scale ratios L_R and Y_R must be found by trial and error. However, if $R_R = Y_R$, then

$$L_R = \alpha_R^{5/3} \tag{14.3.15}$$

and

$$Y_R = \alpha_R^{7/6} \tag{14.3.16}$$

4. It can be shown that the ratio of the rate of sediment transport per unit width to the flow per unit width must be larger in the model than in the prototype.

5. In most cases of the design of movable-bed models, the Froude number model law is not considered an absolute rule. For example, if the Froude number is low, then gravitational effects are not pronounced; therefore, slight changes in the Froude number between the model and prototype can be tolerated.

6. The particle size ratio referred to in Eq. (14.3.10) characterizes the roughness of the bed of the channel, while the particle size ratio used in Eqs. (14.3.12) and (14.3.13) is characteristic of the sediment transport properties of the particles. It should be recognized that these ratios which characterize the particle size distributions in the model and prototype are not necessarily the same.

EXAMPLE 14.2

A channel reach that is approximately parabolic in section has a surface width of 60m (197 ft) when the depth of flow is 1.5 m (4.9 ft). In the prototype, the quartzite sand which composes the bed has a specific gravity of 2.65. If a movable-bed model of this reach is to be constructed of material which has a specific gravity of 1.2, determine the appropriate particle size in the model and the length-scale ratios L_R and Y_R.

Solution

If the Einstein and Chien (1956b) method of analysis is used and $U_R = 1$, then the particle size ratio (model to prototype) is given by Eq. (14.3.14) or

$$d_R = \alpha_R^{-1/3} = \left(\frac{1.2 - 1}{2.65 - 1}\right)^{-1/3} = 2.0$$

Therefore, the size of the particles used in the model must be approximately twice the size of the particles in the prototype.

The scale ratios L_R and Y_R are then determined by the simultaneous solution of Eqs. (14.3.12) and (14.3.13). Since $R_R = f(L_R, Y_R)$, the solution of these equations must be by trial and error. Initial trial values for L_R and Y_R are determined by assuming $R_R = Y_R$; hence, by Eqs. (14.3.15) and (14.3.16)

$$L_R = \alpha_R^{5/3} = \left(\frac{1.2 - 1}{2.65 - 1}\right)^{5/3} = 0.030$$

and

$$Y_R = \alpha_R^{7/6} = \left(\frac{1.2 - 1}{2.65 - 1}\right)^{7/6} = 0.085$$

For the parabolic channel section defined, the hydraulic radius in the prototype is (see Table 1.1)

$$R_P = \frac{2T^2 y}{3T^2 + 8y^2} = \frac{2(60)^2(1.5)}{3(60)^2 + 8(1.5)^2} = 1.00 \text{ m (3.28 ft)}$$

The hydraulic radius in the model for the assumed geometric scale is

$$R_M = \frac{(2TL_R)^2 yY_R}{(3TL_R)^2 + 8(yY_R)^2} = \frac{2[60(0.03)]^2(1.5)(0.085)}{3[(60(0.03)]^2 + 8[1.5(0.085)]^2}$$

$$= 0.084 \text{ m } (0.28 \text{ ft})$$

Therefore,

$$R_R = \frac{R_M}{R_P} = \frac{0.084}{1.0} = 0.084$$

At this point, it must be demonstrated that the assumed geometric-scale ratios satisfy Eqs. (14.3.12) and (14.3.13). Equation (14.3.12) requires

$$\frac{R_R Y_R}{\alpha_R L_R d_R} = 1$$

and

$$= \frac{0.084(0.085)}{[(1.2 - 1)/(2.65 - 1)](0.030)(2.0)} = 0.98$$

Equation (14.3.13) requires

$$\frac{R_R Y_R d_R^2}{L_R \nu_R^2} = 1$$

$$= \frac{0.084(0.085)(2.0)^2}{0.03(1.0)^2} = 0.95$$

Since Eqs. (14.3.12) and (14.3.13) are satisfied almost exactly, it is concluded that

$$L_R = 0.03$$

and

$$Y_R = 0.085$$

A few investigators [see, for example, Bogardi 1959, or Bogardi (in Henderson, 1966), and Raudkivi, 1976] have asserted that the requirement that F_s and $\mathbf{R}_*$ have exactly the same value in the model and prototype can and should be relaxed somewhat. When $\mathbf{R}_* \geq 100$, the flow around the sediment grains is fully turbulent, and, as is the case with other models, it is not necessary for $(\mathbf{R}_*)_M = (\mathbf{R}_*)_P$ as long as $(\mathbf{R}_*)_M \geq 100$. It is also asserted that the bed formation is a function only of the parameter β which is defined by

$$\frac{1}{\beta} = d^{0.88}\left(\frac{u_*^2}{gd}\right) \tag{14.3.17}$$

where the unit of length used is the centimeter. Henderson (1966) noted that all the evidence quoted in support of Eq. (14.3.17) results from grain water systems in which $S_s = 2.65$ and $\nu = 0.011$ cm^2/s (1.18×10^{-5} ft^2/s). Therefore, Eq. (14.3.2) can be rewritten as

$$F_s = \frac{u_*^2}{(S_s - 1)gd} = \frac{456}{\beta^{1.41}}\left(\frac{u_* d}{\nu}\right)^{-5/6} \tag{14.3.18}$$

The values of β, in centimeters, given by Bogardi for the various bed formations, are summarized in Table 14.2. These values, when used in Eq. (14.3.18), define a series of bands in the F_s − $\mathbf{R}_*$ plane (Fig. 14.2) which graphically define the bed forms. In Fig. 14.2 the lines defining the bed form bands all terminate at $F_s = 0.05$, which is the threshold of motion. Figure 14.2 also demonstrates a contradiction in the theory of Bogardi. Specifically, it was claimed earlier that if $\mathbf{R}_* \geq 100$, then the bed forms in the model and prototype would be similar; however, Fig. 14.2 demonstrates that even if $\mathbf{R}_* \geq 100$, the bed forms in the model and prototype could be different. The conclusion is that, to achieve similarity between the model and prototype, the values of F_s and $\mathbf{R}_*$ do not have

TABLE 14.2 Bed form type as a function of β (Henderson, 1966)

Bed form	β, cm
Threshold of motion	550
Ripples form	322
Dunes develop	66
Transition occurs	24
Antidunes develop	18.5

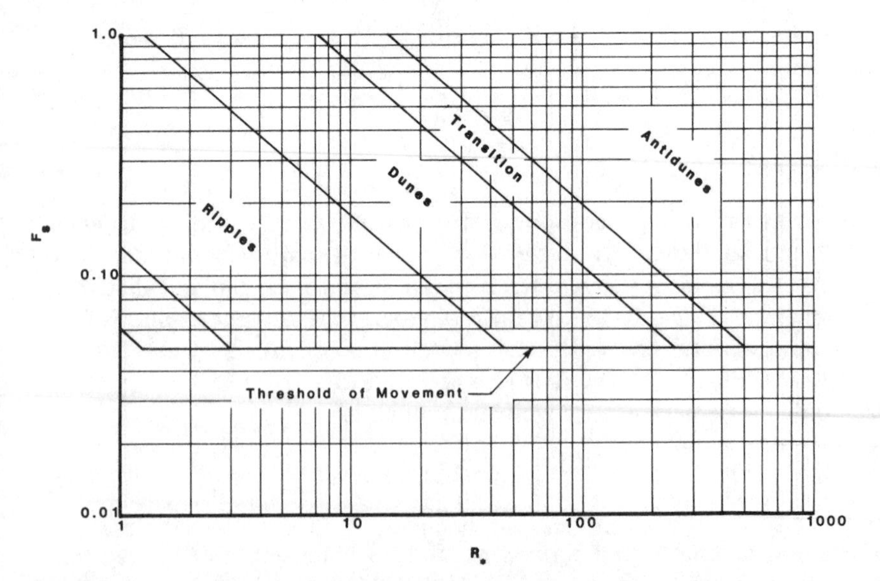

FIGURE 14.2 Relation of bed formation to position on the F_s–$\mathbf{R}_*$ plane.

to be exactly equal, but the points representing the model and prototype should be within the same bed form band of Fig. 14.2.

Zwamborn (1966, 1967, 1969) has developed a set of similarity criteria for movable-bed models which have been validated by model-prototype correlations. The similarity criteria are:

1. Flow in natural rivers is always fully turbulent. To ensure that the flow in the model is also fully turbulent, the Reynolds number for the model should exceed 600.

2. Dynamic similarity between the model and prototype is achieved when the model-to-prototype ratios of the inertial, gravitational, and frictional forces are all equal. Similarity is achieved when:

   a. The Froude numbers for the model and prototype are equal.

   b. The friction criteria are satisfied when three conditions are met. First, the product of the grain roughness coefficient and the inverse square root of the slope are equal in the model and prototype or

$$C'_R = \frac{1}{\sqrt{S_R}} \tag{14.3.19}$$

where C' = grain roughness coefficient

$$C' = 5.75\sqrt{g}\log\left(\frac{12R}{d_{90}}\right) \tag{14.3.20}$$

d_{90} = 90 percent smaller grain diameter, and R = hydraulic radius. Second, the ratio of the shear to the settling velocity in the model and prototype are equal or

$$\left(\frac{u_*}{u_s}\right)_R = 1 \tag{14.3.21}$$

where $u_* = \sqrt{gRS}$ = shear velocity and u_s = settling velocity. The settling velocities of selected model sediment materials are summarized in Fig. 14.3 as a function of grain size and water temperature. Third, the Reynolds number based on the sediment grain diameter in the model should be about one-tenth of the grain Reynolds number for the prototype or

$$(\mathbf{R}_*)_R < 1/10 \tag{14.3.22}$$

where $R_* = ud/\nu$ = grain Reynolds number
u = velocity
ν = kinematic viscosity

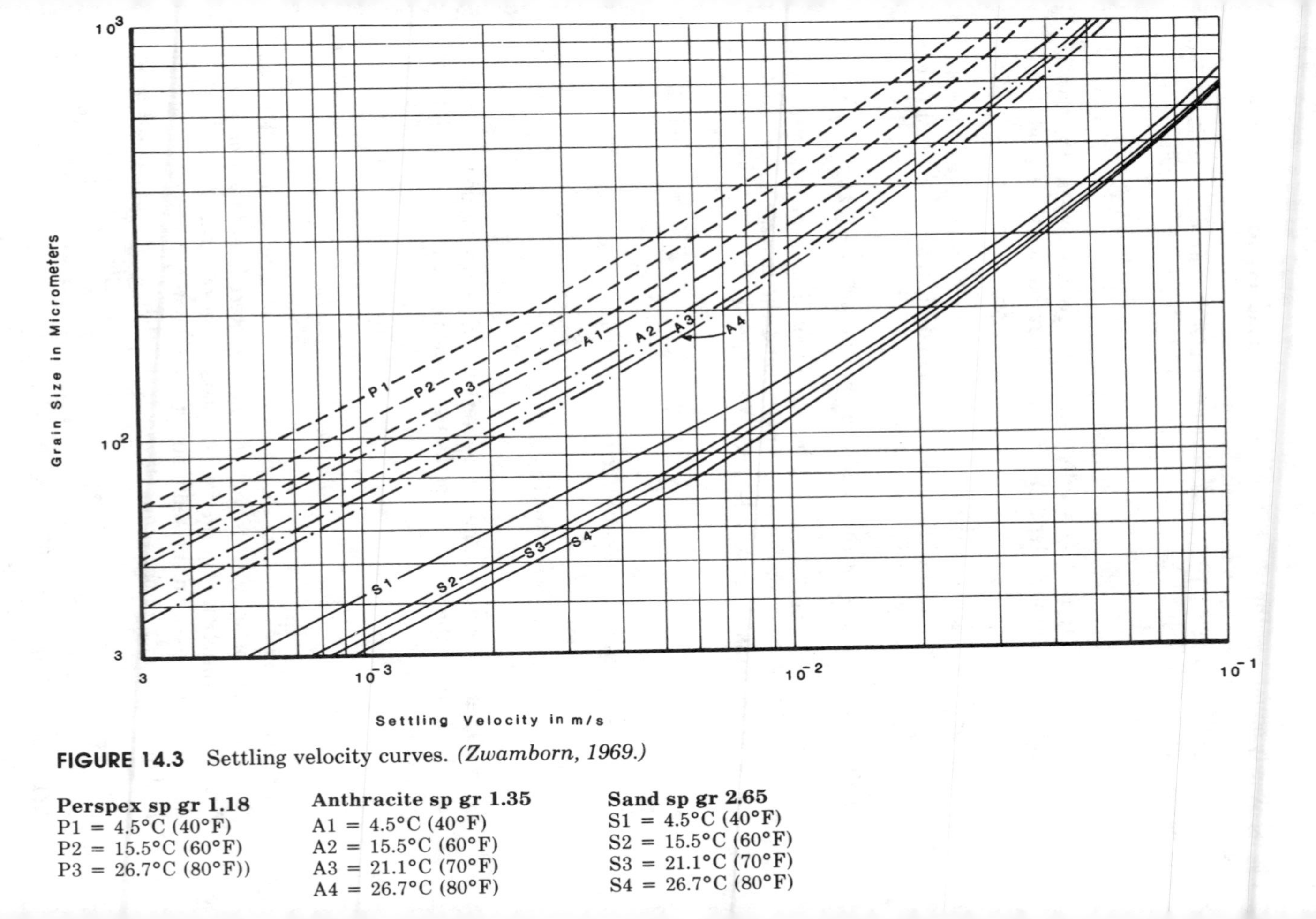

FIGURE 14.3 Settling velocity curves. *(Zwamborn, 1969.)*

Perspex sp gr 1.18	**Anthracite sp gr 1.35**	**Sand sp gr 2.65**
P1 = 4.5°C (40°F)	A1 = 4.5°C (40°F)	S1 = 4.5°C (40°F)
P2 = 15.5°C (60°F)	A2 = 15.5°C (60°F)	S2 = 15.5°C (60°F)
P3 = 26.7°C (80°F))	A3 = 21.1°C (70°F)	S3 = 21.1°C (70°F)
	A4 = 26.7°C (80°F)	S4 = 26.7°C (80°F)

c. The geometric-scale distortion should not be large; i.e.,

$$\frac{Y_R}{L_R} < \frac{1}{4} \tag{14.3.23}$$

3. Sediment motion is closely related to the bed form type which is, in turn, a function of u_*/u_s, $\mathbf{R}_*$, and $\mathbf{F}$. For specified values of u_*/u_s and $\mathbf{F}$, similar bed forms and consequently similar models of bed movement are ensured when the value of $\mathbf{R}_*$ for the model falls within the range as determined by u_*/u_s and the appropriate range of $\mathbf{F}$ defining a type of bed form (Fig. 14.4).

4. Given the similarity criteria defined above, the following scale factors can be derived.

Flow Rate

$$Q_R = L_R Y_R^{1.5} \tag{14.3.24}$$

Hydraulic Time

$$T_R = \frac{L_R}{U_R} \tag{14.3.25}$$

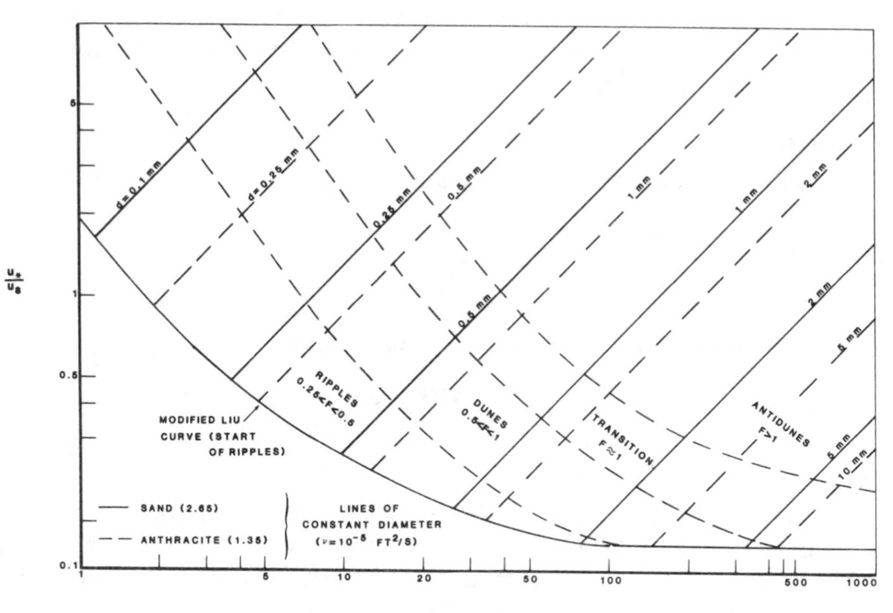

FIGURE 14.4 Bed shape criteria. *(Zwamborn, 1969.)*

Sedimentological Time

$$(T_s)_R = \frac{L_R Y_R \Delta_R}{(q_s)_R} \tag{14.3.26}$$

where T_s = sediment time scale
Δ = relative submerged sediment density
q_s = sediment bed load per unit width

From Eq. (14.3.26) it is evident that the determination of the time scale for the sediment requires that the bed load in both the model and the prototype be known. Although the bed load of the prototype can be estimated (see, for example, Rouse, 1958), Zwamborn (1966, 1967, and 1969) gave the following approximate relationship:

$$(T_s)_R \simeq 10 T_R \tag{14.3.27}$$

The validity of the similarity criteria stated above has been confirmed by comparison of the results achieved in models with those observed in prototypes (see, for example, Zwamborn, 1966, 1967, and 1969).

EXAMPLE 14.3

In the prototype $u_*/u_s = 5.5$, $\mathbf{F} = 0.5$, and the bed shape is dunes. Using the criteria of Zwamborn (1966, 1967, and 1969) which are summarized in Fig. 14.4, determine the appropriate range of grain Reynolds numbers for the model.

Solution

The Froude numbers for the model and prototype are equal, and Eq. (14.3.21) requires

$$\left(\frac{u_*}{u_s}\right)_R = 1$$

Therefore from Fig. 14.4 if the bed form is dunes

$$2 < (\mathbf{R}_*)_M < 5$$

Regime Method

A channel is said to be in regime when it has adjusted its slope and shape to an equilibrium condition. The implication of a channel being in regime is that although the bed of the channel is in motion it is a stable movement since the rate of transport equals the rate of sediment supply. The regime hypothesis is a completely empirical approach which incorporates no theoretical considerations. This hypothesis was developed in India from field measurements taken in canals of human origin which were rather uniform in character but had widely varying sizes. Three equations were developed which can be rearranged into a form that implies that any one of three variables—average depth, average velocity, and channel width—is a unique function of discharge. These equations were developed by plotting the available data on log-log graphs and determining the best-fit straight lines. As noted by Einstein and Chien (1956b) in reply to a discussion by Blench (1956), because the field data were obtained from channels which were either empty or running at their design discharge, the data were regular.

In particular, the regime hypothesis asserts that

$$T \propto \sqrt{Q} \tag{14.3.27}$$

$$Y \propto Q^{1/3} \tag{14.3.28}$$

$$S \propto Q^{-1/6} \tag{14.3.29}$$

From Eqs. (14.3.27) to (14.3.29), appropriate model scale ratios can be derived by induction or

$$L_R = \sqrt{Q_R} \tag{14.3.30}$$

and

$$Y_R = Q_R^{1/3} \tag{14.3.31}$$

As a result of Eqs. (14.3.30) and (14.3.31), the following modeling ratios can be derived:

$$S_R = \frac{Y_R}{L_R} = \frac{Q_R^{1/3}}{\sqrt{Q_R}} = Q_R^{-1/6} \tag{14.3.32}$$

$$Y_R = L_R^{2/3} \tag{14.3.33}$$

and

$$Q_R = L_R Y_R^{3/2} \tag{14.3.34}$$

With regard to this method of movable-bed model design, the following observations can be made:

1. From Eq. (14.3.33), it is obvious that the model has distorted geometric scales.
2. Equation (14.3.34) agrees with the Froude modeling law.
3. If this design method is used, the bed material in the model is tacitly assumed to be the same as that found in the prototype.

Thus, it is concluded that the regime hypothesis yields results which are surprisingly in accordance with the requirements of similarity. Two objections to this design process must be noted. First, as asserted by Einstein and Chien (1956b), theories or hypotheses which satisfactorily represent flow in designed channels may be unsatisfactory when applied to natural rivers. Einstein and

Chien (1956b), using data from U.S. rivers, clearly demonstrated that the application of Eqs. (14.3.27) to (14.3.29) to natural rivers can lead to significant errors. Second, if the bed material present in the prototype is used in the model, then Eqs. (14.3.15) and (14.3.16) predict that $X_R = Y_R$, and hence the model is the prototype.

14.4 MODEL MATERIALS AND CONSTRUCTION

In constructing a physical, hydraulic model, the modeler must be aware that the model must:

1. Be an accurate, scaled geometric replica of the prototype
2. Retain its geometric consistency and accuracy during operation
3. Contain the appurtenances necessary to control and measure the flow
4. Be amenable to quick and easy changes in detail
5. Be consistent with the purpose of the study and its cost allocations

Models are generally constructed from materials which are readily available, e.g., wood, concrete, metal, wax, paraffin, plastic, sand, and coal. The shop facilities required to prepare and shape these materials must also be available. In the following paragraphs of this section, a select number of topics will be discussed from the viewpoint of materials and methods of model construction. In many cases, most of the time required to conduct a model study is used in the design and construction processes. It is noted that additional information regarding materials and construction techniques can be found in Anonymous (1942) and Sharp (1981).

Materials

Wood The best-suited types of wood for the construction of hydraulic models are sugar pine, cypress, redwood, mahogany, and marine grade plywood. In general, since wood of all types is to a greater or lesser degree affected by water, the wood should be waterproofed in some manner; e.g., an initial coat of linseed oil followed by one or more coats of varnish is usually satisfactory. Wood has the advantages of being relatively inexpensive and easily shaped; in the case of plywood, it can be bent or warped slightly. It has the disadvantages of being affected dimensionally by water content and warping and being opaque.

Sheet Metal Sheet metal is also relatively inexpensive and can be easily shaped. It may be cut to any shape and can be attached to wood with countersunk screws. The disadvantages of sheet metal are that it tends to buckle under unequal stress and under thermal stress; it cannot be shaped in warped surfaces; its surfaces must be protected from corrosion by either painting or galvanizing; and it is opaque.

Plastics Plastics such as lucite and plexiglass are transparent and can be molded with heat, machined, or cut and glued into many shapes. They are particularly useful when one of the goals of the modeling program is flow visualization. Plastics have the disadvantage of being affected by humidity, temperature, and sunlight.

Wax Wax has the advantages of being very workable and unaffected by water, and this material may also be formed and shaped into a wide variety of geometric configurations. Wax is particularly well suited for constructing curved surfaces which are required to be very accurate. The primary disadvantages of wax are that it is fragile, devoid of structural strength, opaque, and easily damaged by temperature changes.

Cement Cement has the advantages of being workable when wet, unaffected by water, and unaffected by temperature changes. With templates, it can be molded into many shapes and can be made very smooth by working with a trowel until the initial set occurs. Experience has shown that the best mix is one part cement to one to three parts sand (Anonymous, 1942). This material is well suited to models involving warped surfaces or large areas. In general, even after the final set, cement can be easily cut and patched. The primary disadvantage of cement is that it is opaque. The designer should also be aware that in a newly constructed cement model, salts will be dissolved from the cement, and instrumentation systems which measure the conductivity of water may be affected by this dissolution process.

Fiberglass and Resins When complex shapes must be constructed, they may be constructed from fiberglass or cast from resins. It should be noted that the continuous exposure of technicians to epoxy resin fumes may result in significant health problems (Sharp, 1981).

Polyurethane Foam In recent years, high-density polyurethane foam (6 to 12 lb/ft^3 or 940 to 1900 N/m^3) has been used to model small hydraulic structures. Although this is a rather expensive construction material, the labor costs are reduced by the ease with which this material can be cut and formed to the correct shape. Building models with polyurethane foam usually involves gluing the sheets of foam together with epoxy resin, forming them into the proper shape, and finally sealing the model. Although various sealers may be used, metal paint primer appears to yield satisfactory results (Sharp, 1981). The primary disadvantage of this material is that when glues are used, separation at the joints may occur. This problem is apparently overcome when a resin, for example, fiberglass, is used. Some separation may occur even if fiberglass resin is used when the foam sheets are less than 1 in (2.5 cm) in thickness.

As the above discussion indicates, there are many materials which can be used in the construction of a model. Each material has certain advantages and also a number of disadvantages which preclude its use as a universal model material.

In the actual construction of a hydraulic model, reliable and accurate horizontal and vertical controls must be established. The method of horizontal control is usually strongly dependent on the size and shape of the model. For example, if the model is long and narrow, then horizontal control may be established with a traverse which approximates the centerline of the model. In the case of a model which is both long and wide, horizontal control may be established with a traverse around the perimeter of the model. If the model is very large, then a perimeter traverse supplemented with a cartesian grid system is used.

The establishment of vertical control is usually based on the location of a benchmark located in the vicinity of the model. If the model is large, then supplementary benchmarks may be required.

Once horizontal and vertical control has been established, the topography must be established. There are three commonly used methods of accomplishing this:

1. *Female Templates:* Female templates represent prototype cross sections of the channel which have been reduced to an appropriate size by the geometric-scale ratios during the design process (Fig. 14.5). These templates are plotted and cut with precision and then are securely anchored in models using the horizontal and vertical control nets to establish their location. The surface of the model is then molded in accordance with the surface established by the templates.

   The materials used to make the templates include sheet metal, masonite, and wood. It is recommended that the templates be spaced no farther apart than 2 ft (0.6 m) and no closer than 6 in (0.15 m).

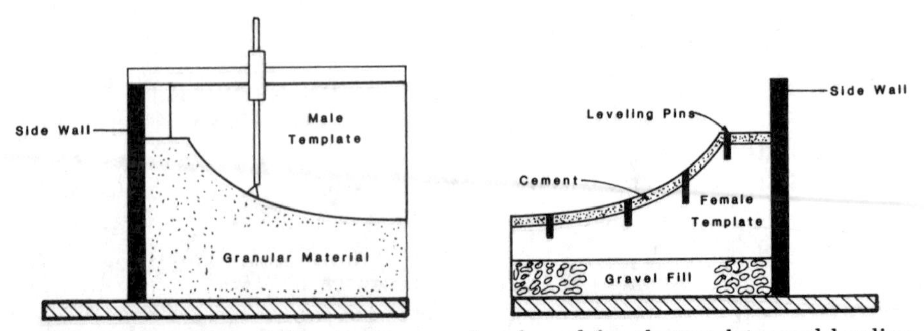

FIGURE 14.5 Schematic illustrating use of male and female templates and leveling pegs in model construction.

2. *Male Templates:* The channel topography may also be established with the aid of male templates (Fig. 14.5). In this case the templates sit on graded rails and are removed when the channel surfaces have been shaped.
3. *Pegs:* The topography of an area may also be established by pegs which are placed in the model such that their tops reflect the appropriate elevation. The surface is then molded in accordance with these peg elevations.
4. *Horizontal Templates:* Although the use of male or female templates, which compose the group termed *vertical templates,* can be used in most situations, in some cases horizontal templates cut to the shape of the contour lines are appropriate. If the thickness of the template is identical to the scaled vertical distance between the contours, then the templates can be laid one on top of another and a solid model formed. Although plywood and polyurethane have been used successfully in this manner, it is more common to use thin templates separated by vertical spacers to ensure correct elevations. If this latter methodology is used, then the templates must be fairly rigid plywood or sheet metal and covered with a material such as fiberglass.

In considering the model construction techniques discussed above, it must be realized that none of the techniques will result in a model which is geometrically exact. Although in all the methods there is an emphasis on horizontal and vertical control between the points, lines, and planes of control, the topography must be interpreted by the human hand and eye. At this point, the following three points are noted. First, the female and male template construction techniques are usually considered optimum when large areas with considerable topographic relief must be covered. Second, when the male or female template method is used, the volume between the templates is usually filled with an inexpensive filler material; e.g., a 50/50 mixture of clay and fine sand has been found to be satisfactory. This material may then be covered with a thin layer of cement. Third, the horizontal template technique has been found to be very useful when the horizontal extent of the model is large and the relief slight.

The techniques used to construct movable-bed models are slightly different from those used for fixed-bed models. The construction of movable-bed models generally requires positioned guides or guide rails along the sides of the model. These rails or guides may be the model tank walls if they have been accurately constructed. The male template method of construction may then be used; i.e., the templates are suspended from the guides or guide rails and the proper topography formed. Depending on the size of the model basin and the presence and distribution of bedrock in the prototype, the bottom of the basin can be filled with inexpensive filler materials as was suggested in the case of fixed-bed models. In the top layers of a movable-bed model, the model sediment is placed. In the placement of this material, care must be taken to ensure that the depth of the mobile model sediment is greater than the greatest feasible depth of scour.

14.5 PHYSICAL MODEL CALIBRATION AND VERIFICATION

As is the case with numerical models, a model has no value unless it can be used to predict prototype behavior. Thus, after a physical hydraulic model has been constructed, it must be calibrated and verified; i.e., a determination must be made whether prototype events are accurately reproduced in the model. In numerical models, prototype model agreement is usually achieved by adjusting the coefficients used in the model until adequate agreement between the prototype and model is achieved. In a physical model, prototype model agreement is attained by adjusting physical characteristics such as the bed roughness, the discharge, and/or the water levels.

The successful calibration of either a numerical or a physical model requires accurate prototype data regarding channel geometry, water surface elevations, discharge, sediment transport, and velocities. It is noted that the collection of a data base suitable for calibrating a model can be an expensive and time-consuming undertaking. With regard to the calibration process, the following should be noted:

1. In the case of models in which the conditions for exact similarity are closely met, very little calibration will be required. For example, the gross flow characteristics at a hydraulic structure can be modeled at a relatively large scale and with very little, if any, geometric distortion. Thus, the model will reproduce prototype behavior with very little calibration.

2. When frictional resistance is important, e.g., in channel flow, the problem of calibration becomes much more complex. At a minimum, water surface elevations between the model and prototype must be compared for discharges close to the operating discharge for the model. If the model is to simulate discharges in the prototype over a wide range, then the model should be calibrated for a wide range of discharges.

3. Velocity distribution measurements are also valuable even though in models with geometric-scale distortions it may be unrealistic to expect the model to reproduce these distributions accurately. However, the comparison of velocity distributions provide valuable information regarding the overall accuracy of the model.

4. If the model is to simulate point velocities in a reach, then accurate field data regarding the velocity of flow at these points is required.

5. If the model is to simulate unsteady flow events, then the movement of the unsteady event through the reach should be reproduced, at the appropriate spatial and temporal scales, in the model.

6. If the model is to simulate sediment transport, then the calibration process requires accurate data regarding sediment transport and bed forms in the prototype. Sharp (1981) noted that although steady-state flow information

regarding sediment transport is useful, measurements of sediment transport under unsteady conditions for model calibration are particularly valuable.

After reasonable agreement is obtained for the calibration event or events, the verification process begins. Verification requires data independent of the calibration data and seeks to confirm that the model has been correctly calibrated. In many cases, a rigorous verification process is not performed because of the high cost involved in obtaining adequate and reliable field data. The calibration and verification processes are essential if the modeling program is to yield useful and reliable information.

14.6 SPECIAL-PURPOSE MODELS

Ice

As has been discussed in Section 5.4, ice can be an important and critical problem in the design and analysis of open channels in northern climates. Given the importance of ice problems in many areas of the world, it is crucial that physical models of ice problems be considered. Two different types of models have been developed. When the internal strength of an ice sheet is important, e.g., in considering the forces developed when an ice sheet comes in contact with bridge piers, not only must the hydrodynamics of the flow be correctly modeled, but there must also be hydroelastic similarity between the prototype and the model. On the other hand, ice jams which result from the accumulation of drifting ice can in many cases be adequately addressed without considering hydroelastic similarity.

In the case where the internal strength of the ice must be considered, the manner in which the ice load is applied to the structure (see, for example, Neill, 1976) is important. For example, when ice pushes against a vertical pier, it fails by crushing (Fig. 14.6). If it is assumed that the flow is turbulent and that grav-

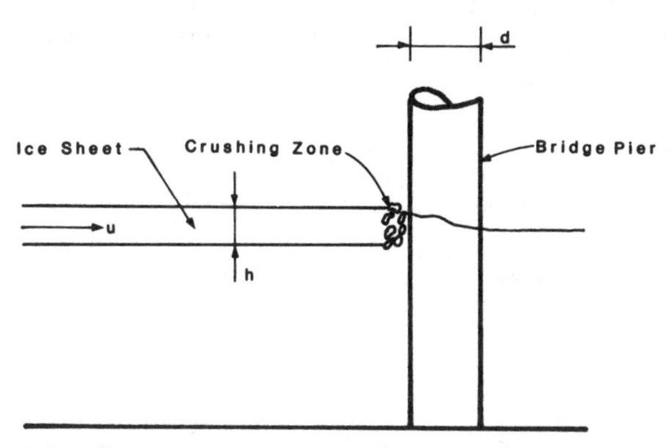

FIGURE 14.6 Ice failure at a vertical bridge pier.

ity does not enter into the problem because the ice sheet neither bends nor rides up on the pier, then

$$F = g(h, d, u, \sigma_c, \epsilon, E) \qquad (14.6.1)$$

where F = force on pier
h = ice sheet thickness
d = characteristic size of pier
u = ice sheet velocity
σ_c = compressive strength of ice at strain rate ϵ
E = Young's modulus of elasticity

Application of the Buckingham Π theorem to the functional formulation assumed in Eq. (14.6.1) yields

$$\frac{F}{\sigma_c dh} = g\left(\frac{d}{h}, \frac{u}{\epsilon h}, \frac{E}{\sigma_c}\right) \qquad (14.6.2)$$

Sharp (1981) noted that this formulation also assumes that the ice crystal size is small relative to the width of the pier; the ice is broken into small fragments; and ϵ must be included because of the known variation of the compressive strength of ice with the rate of deformation.

In the case of ice pushing against an inclined pier (Fig. 14.7), the ice can be assumed to fail by a combination of crushing, shear, and bending. In this case, gravitational forces must be considered because the ice rides up on the pier face. Neill (1976) has suggested that in this case

$$\frac{F}{\sigma_f h^2} = g\left(\frac{d}{h}, \Gamma, \frac{u^2}{gh}, \frac{E}{\sigma_f}\right) \qquad (14.6.3)$$

where σ_f = bending strength of ice
Γ = pier force slope angle
g = acceleration of gravity

The primary difficulty associated with modeling ice problems in which the internal strength of ice must be considered is the development of a model ice

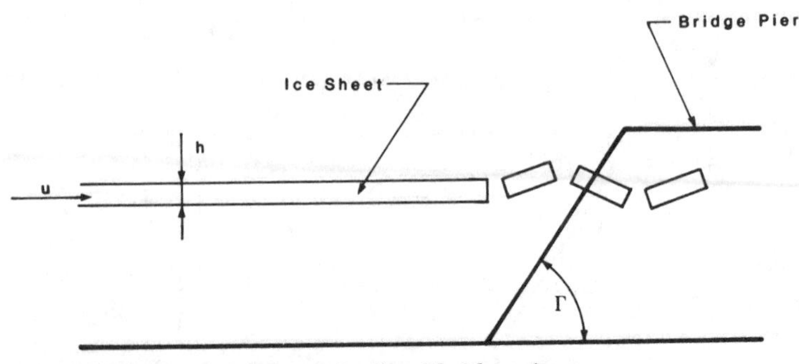

FIGURE 14.7 Ice failure at inclined bridge pier.

in which all the mechanical properties of prototype ice are properly scaled. A number of techniques have been developed to form model ice, and it has been demonstrated that the flexural strength of the ice sheet depends on: (1) the salinity of the water in which the ice forms; (2) the rate of growth of the ice sheet; and (3) the temperature at which the ice sheet is held after formation. If these variables are suitably controlled, then it is possible to develop ice sheets with bending strengths ranging between prototype values and 0.01 of prototype values. However, the primary problem is in developing model ice sheets with the proper elastic modulus. With naturally formed prototype ice the ratio E/σ_f has the range

$$2000 < \frac{E}{\sigma_f} < 8000$$

Whereas in many models E/σ_f may be as low as 500, low values of E/σ_f yield ice which deflects an inordinate amount before breaking (Schwarz, 1977). Given the importance of this characteristic ratio to ice modeling, a few comments regarding the measurement of σ_f and E are appropriate. The flexural strength of ice both in the laboratory and field is usually determined by cutting a cantilever beam in the ice and testing it to failure. σ_f is then given by

$$\sigma_f = \frac{6P_L L}{bh^2} \tag{14.6.4}$$

where P_L = applied load
L = length of beam
b = beam width
h = beam thickness

Although the elastic modulus is difficult to measure in the field environment, it can be conveniently measured in the laboratory by a combination of load and deflection measurements. For an ice sheet which is continuous and with boundaries sufficiently distant from the point of loading so that the ice sheet can be considered infinite, the deflection Δ for a load P_L is given by

$$\Delta = \frac{P_L}{8S_w l^2} \tag{14.6.5}$$

where S_w = specific gravity of water and l = a measure of the length of the infinite ice sheet over which the stresses are distributed. l may be estimated from the theory of infinite slabs supported by an elastic foundation or

$$l^4 = \frac{Eh^3}{12S_w(1 - n_p^2)} \tag{14.6.6}$$

where n_p^2 = Poisson's ratio.

An alternative to using real ice generated by refrigeration in a model is to use artificial ice which is composed of various chemicals bound in a plaster or

TABLE 14.3 Properties of freshwater ice, sea ice, and Mod-Ice (Kotras et al., 1977)

Property	Freshwater ice	Seawater ice	Mod-Ice scale = $1/30$
Ice thickness, ft (m)	0–2.0 (0–0.6)	0.1–11 (0–3.5)	0.3–20 (0.1–6.0)
Ice flexural strength, lb/ft^2 (kPa)	10,000–21,000 (500–1000)	5200–16,000 (250–750)	8400–21,000 (400–1000)
Crushing strength/flexural strength	2.0–5.0	2.0–5.0	0.7–7.0
Elastic modulus/flexural strength	1000–5000	1000–8000	500–11,000
Darcy Weisbach friction factor	0.3–0.7	—	0.3–0.8
Coefficient of sliding friction	0.1–0.4	0.1–0.4	0.1–0.55

wax matrix. Mod-Ice used by Arctec of Columbia, Maryland, employs five separate chemicals in a mixture, which is proprietary, to achieve appropriate and controllable values of bending and crushing strength, elastic modulus, density, and roughness (Kotras et al., 1977). In Table 14.3 the relevant properties of freshwater ice, seawater ice, and Mod-Ice are summarized. In models using Mod-Ice, the model scales are usually limited by the minimum value of the bending strength—approximately 2.5 lb/in^2 (13.8 kN/m^2); however, this does not appear to be a significant limitation (Sharp, 1981). In addition to the proprietary formulations for ice, such as the one noted above, there are also formulations described and discussed in the open literature (see, for example, Timco, 1979).

If the model is designed to study ice drift, ice control structures, or ice jams, the ice is assumed to be perfectly rigid and thermal effects can be neglected; then the model study can usually be conducted using polyethylene, polypropylene, wood, or paraffin to simulate the ice blocks. In such a case, fluid motion is the primary concern; however, the study should be considered qualitative rather than quantitative. The difficulty in obtaining field data for the calibration of these models is a major problem.

BIBLIOGRAPHY

Anonymous, "Hydraulic Models," *The American Society of Civil Engineers Manual of Practice,* No. 25, American Society of Civil Engineers, New York, 1942.

Blench, T., "Blench on Distorted Models," *Transactions of the American Society of Civil Engineers,* vol. 121, 1956, pp. 458–459.

Bogardi, J., "Hydraulic Similarity of River Models with Movable Bed," *Acta Tech. Acad. Sci. Hung.*, vol. 14, 1959, pp. 417–445.

Chow, V. T., *Open Channel Hydraulics,* McGraw-Hill Book Company, New York, 1959.

Einstein, H. A., and Barbarossa, N. L., "River Channel Roughness," *Transactions of the American Society of Civil Engineers,* vol. 177, 1956a, pp. 440–457.

Einstein, H. A., and Chien, N., "Einstein-Chien on Distorted Models," *Transactions of the American Society of Civil Engineers*, vol. 121, 1956b, pp. 459–462.

Fischer, H. B., "The Mechanics of Dispersion in Natural Streams," Proceedings of the *American Society of Civil Engineers, Journal of the Hydraulics Division,* vol. 93, no. HY6, November 1967, pp. 187–216.

Fischer, H. B., and Holley, E. R., "Analysis of the Use of Distorted Hydraulic Models for Dispersion Studies," *Water Resources Research,* vol. 7, no. 1, February 1971, pp. 46–51.

Henderson, F. M., *Open Channel Flow,* Macmillan, New York, 1966.

Kotras, T., Lewis, J., and Etzel, R., "Hydraulic Modeling of Ice-Covered Waters," *Proceedings of the 4th International Conference on POAC,* vol. 1, Memorial University of Newfoundland, St. John's, Newfoundland, Canada, 1977, pp. 453–463.

Moretti, P. M., and McLaughlin, D. K., "Hydraulic Modeling in Stratified Lakes," *Proceedings of the American Society of Civil Engineers, Journal of the Hydraulics Division,* vol. 103, no. HY4, April 1977, pp. 368–380.

Neill, C. R., "Dynamic Ice Forces on Piers and Piles: An Assessment of Design Guidelines in the Light of Recent Research," *Canadian Journal of Civil Engineering,* vol. 3, 1976, pp. 305–341.

Raudkivi, A. J., *Loose Boundary Hydraulics,* 2d ed., Pergamon Press, New York, 1976.

Roberts, B. R., Findikakis, A. N., and Street, R. L., "A Program of Research on Turbulent Mixing in Distorted Hydraulic Models," The American Society of Civil Engineers, *Proceedings of the Specialty Conference on Conservation and Utilization of Water and Energy Resources,* 1979, pp. 57–64.

Roberts, B. R., and Street, R. L., "Impact of Vertical Scale Distortion on Turbulent Mixing in Physical Models," The American Society of Civil Engineers, *Proceedings of the Specialty Conference on Computer and Physical Modeling in Hydraulic Engineering,* 1980, pp. 261–271.

Rouse, H., *Engineering Hydraulics,* Wiley, New York, 1958.

Rouse, H., and Ince, S., *History of Hydraulics,* Dover Publications, New York, 1963.

Schwarz, J., "New Developments in Modelling Ice Problems," *Proceedings of the 4th International Conference on POAC,* vol. 1, Memorial University of Newfoundland, St. John's, Newfoundland, Canada, 1977, pp. 45–61.

Sharp, J. J., *Hydraulic Modelling,* Butterworth, London, 1981.

Streeter, V. L., and Wylie, E. B., *Fluid Mechanics,* 6th ed., McGraw-Hill Book Company, New York, 1975.

Timco, G. W., "The Mechanical and Morphological Properties of Doped Ice," *Proceedings of the 5th International Conference on POAC,* Norwegian Institute of Technology, Trondheim, Norway, 1979, pp. 719–739.

Walski, T. M., "Properties of Steady Viscosity-Stratified Flow to a Line Sink," Technical Rept. EL-79-6, U.S. Army Corps of Engineers, Waterways Experiment Station, Vicksburg, Miss., July 1979.

Zwamborn, J. A., "Reproducibility in Hydraulic Models of Prototype River Morphology," *La Houille Blanche,* no. 3, 1966, pp. 291–298.

Zwamborn, J. A., "Solution of River Problems with Moveable-Bed Hydraulic Models," MEG 597, Council for Scientific and Industrial Research, Pretoria, South Africa, 1967.

Zwamborn, J. A., "Hydraulic Models," MEG 795, Council for Scientific and Industrial Research, Stellenbosch, South Africa, 1969.

PROBLEMS

Chapter 1: Concepts of Fluid Flow

1. Steady flow occurs when
   a. Conditions at all points of interest steadily change with time
   b. Conditions are the same at adjacent points at any instant
   c. Conditions do not change with time at any point

2. Turbulent flow occurs in situations involving
   a. Very viscous fluids
   b. Very small velocities of flow
   c. Capillary tubes
   d. None of the above

3. Eddy viscosity, defined in Eq. (1.3.36), is
   a. A physical property of the fluid
   b. The viscosity divided by the density of the fluid
   c. Dependent on the flow and density
   d. Independent of the nature of the flow

4. Viscous forces are weak relative to inertial forces in
   a. Laminar flows
   b. Turbulent flows

5. The Reynolds number may be defined as the ratio of
   a. Viscous forces to inertial forces
   b. Viscous forces to gravity forces
   c. Gravity forces to inertial forces
   d. Pressure forces to elastic forces
   e. Inertial forces to gravity forces
   f. None of the above

6. The Froude number may be defined as the ratio of
   a. Viscous forces to inertial forces
   b. Viscous forces to gravity forces
   c. Gravity forces to inertial forces
   d. Pressure forces to elastic forces
   e. Inertial forces to gravity forces
   f. None of the above

7. A canoe is floating in a lake of constant volume. In the canoe is a 1-ft^3 block (0.028 m^3) with a specific gravity of 2. If the concrete block is taken out of the canoe and dropped in the lake, does the average depth of the lake
   a. Increase?
   b. Decrease?
   c. Stay the same?

   Justify your answer with calculations.

8. Classify the following flow situations as either steady or unsteady from the viewpoint of the specified observer:

Case	*Observer*
I. A boat crossing a reservoir at a constant velocity	a. Standing on boat b. Standing on shore
II. Flow of a river around bridge piers	a. Standing on bridge b. In a boat drifting with the current
III. A ship moving upstream at a constant velocity in a river	a. Standing on the bank b. In a boat drifting with the current c. Standing on the ship
IV. Movement of a flood down a river valley after the collapse of a dam	a. Standing on the river bank at the dam site b. Running downstream on the river bank at the velocity of the dambreak surge

9. Water has an average velocity of 10 ft/s (3.0 m/s) through a 24-in (0.61-m) pipe flowing full. Estimate the discharge through the pipe in cubic feet per second.

10. Given the schematic figure below of a constant-width rectangular channel, answer the following questions:
   a. Estimate the net force on the sluice gate if all losses are neglected.
   b. Sketch the pressure distribution on the surface *AB* (note that the pressures at points *A* and *B* are zero gage pressure).
   c. Is the pressure distribution sketched in part (b) a hydrostatic pressure distribution?
   d. How is the pressure distribution sketched in part (b) related to the force estimated in part (a)?

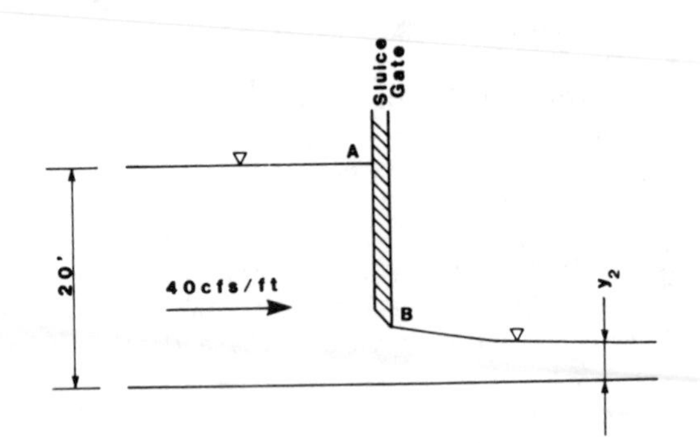

11. A bridge has its piers at a center distance of 7 m (23 ft). A short distance upstream of the bridge, the water depth is 3 m (9.8 ft) and the velocity is 4 m/s (13 ft/s); when the flow has gone far enough downstream to even out again after the disturbance caused by the piers, the depth of flow is 2.7 m (8.9 ft). Neglecting the bed slope and boundary resistance, and assuming that the pressure variation is given by the hydrostatic law, find the thrust (force) on *each* pier.

12. A cylinder of infinite length is placed in a uniform parallel flow stream of water as shown in the figure below. Using the information provided in the figure, calculate the force per unit length on the cylinder.

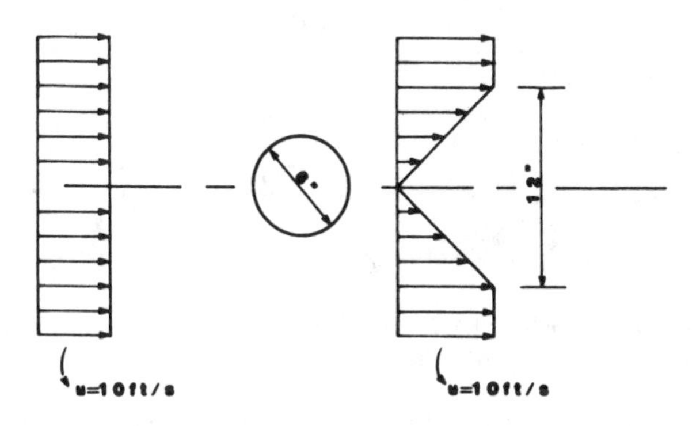

13. Water is flowing in a rectangular channel of unit width as shown in the figure below. Neglecting all losses, determine the possible depths of flow at section *B*.

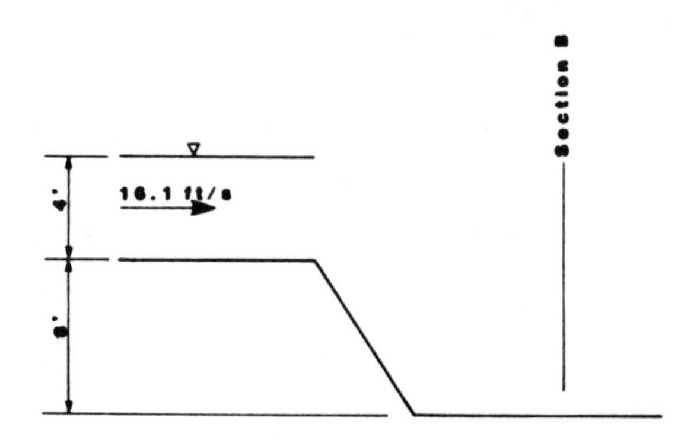

14. Given the data regarding a natural channel on page 170, estimate

a. The top width (English units)

b. The flow area (English units)

c. The centroid of the flow area (English units)

when the river stage is 11.5 m (38 ft).

15. Given the data provided in the table below, estimate the top width, the flow area, and the centroid of the flow area when the depth of flow is 45 ft (13.7 m):

Depth, ft (1)	Top width, ft (2)	Area, ft^2 (3)
0	93	0
10	110	1015
20	208	2065
30	319	5240
40	410	8885
50	494	13405
60	575	18750

16. Given the data regarding a natural channel in Table 10.2 (page 481), estimate

a. The kinetic energy correction factor

b. The momentum correction factor

17. Given the data in the figure below, and knowing

$$u_1 = 0.5 \text{ ft/s } (0.15 \text{ m/s})$$

$$u_2 = 10 \text{ ft/s } (3.0 \text{ m/s})$$

$$u_3 = 1 \text{ ft/s } (0.30 \text{ m/s})$$

estimate the momentum correction factor and the kinetic energy correction factor using Eqs. (1.3.40) and (1.3.41).

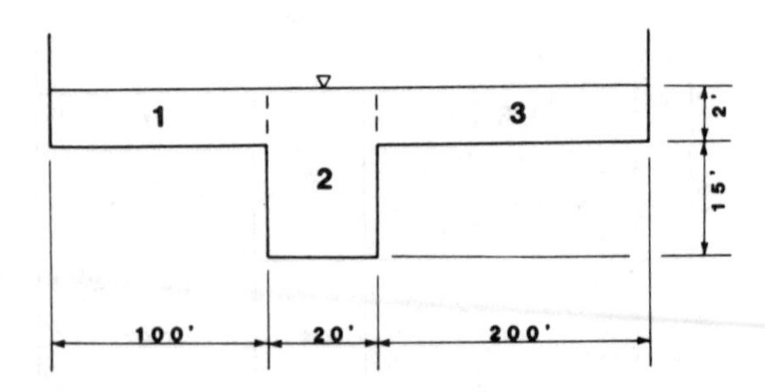

18. Using Eq. (1.4.14), prove that the average velocity of flow $\overline{u}$ can be estimated as follows:

a. Measurements at 0.2 and 0.8 of the depth of flow with $\overline{u}$, the average velocity, given by

$$\overline{u} = \frac{u_{0.2D} + u_{0.8D}}{2}$$

where $u_{0.2D}$ is the measured velocity at 0.2 of the depth and $u_{0.8D}$, is the measured velocity at 0.8 of the depth.

b. A single measurement at 0.6 of depth of flow. Is the distance measured from the bottom of the channel up or from the water surface down?

19. A laboratory channel 0.5 m (1.6 ft) wide conveys a uniform flow of 0.01 m^3/s (0.35 ft^3/s) at a depth of 0.05 m (0.16 ft). If the longitudinal slope is 0.0009 m/m and the roughness height k_s of the channel bed is 7.6×10^{-4} m (0.0025 ft), estimate the length of channel required for the full development of the boundary layer if the boundary layer is turbulent at the channel entrance. Also, plot the profile of the boundary layer as a function of distance along the channel bottom.

20. A 0.5-in (1.27-cm) snail is cleaning the bottom of a hydraulically smooth chute that has a roughened entrance section and conveys sewage from a flow equalization pool to the sewage treatment plant.

a. If the free-stream velocity entering this chute is 17 ft/s (5.2 m/s), how close can the snail crawl to the entrance if the snail is not to encounter any of the free-stream flow? Assume the kinematic viscosity of the sewage is 1.2×10^{-5} ft^2/s (1.1×10^{-6} m^2/s).

b. How will this situation be affected if (1) the free-stream velocity is increased, or (2) the sewage temperature increases?

21. Give two *good* reasons why manhole covers are round.

22. Why are sanitary sewers generally designed as open channels?

23. Given the results stated in Eq. (1.5.4) and assuming that the fluid in the prototype is water, what is a common fluid that should be used in the model?

PROBLEMS

Chapter 2: Energy Principle

1. For $Q = 100\ ft^3/s$ ($2.8\ m^3/s$) flowing in a rectangular channel, plot specific energy as a function of the depth of flow for channel widths of 10, 15, 20, and 25 ft (3.0, 4.6, 6.1, and 7.6 m, respectively).

2. Calculate the critical depth for a discharge of $10\ m^3/s$ ($350\ ft^3/s$) in the following channels:
   a. Rectangular channel; $b = 3.0$ m (9.8 ft)
   b. Triangular channel; $z = 0.5$ m (1.6 ft)
   c. Trapezoidal channel; $b = 3.0$ m (9.8 ft), $z = 1.5$
   d. Circular channel; $D = 3.0$ m (9.8 ft)

3. Calculate the bottom width of a channel that is required to convey a discharge of $20\ m^3/s$ ($710\ ft^3/s$) as a critical flow at a depth of 1.5 m (4.9 ft) if the channel section is
   a. Rectangular
   b. Trapezoidal; $z = 1.5$

4. Water is flowing at a velocity of 4.6 m/s (15 ft/s) and a depth of 0.75 m (2.5 ft) in a channel of rectangular section. Find the change in the absolute depth of flow produced by
   a. A smooth upward step of 0.15 m (0.49 ft)
   b. A smooth downward step of 0.15 m (0.49 ft)
   c. The maximum allowable size of an upward step for the flow to be possible upstream as described

5. Water is flowing at a velocity of 11 ft/s (3.4 m/s) and a depth of 11 ft (3.4 m) in a channel of rectangular section with a width of 11 ft (3.4 m). Find the changes in depth and absolute water level produced by
   a. A smooth contraction to a width of 10 ft (3.0 m)
   b. A smooth expansion to a width of 14 ft (4.3 m)
   c. The greatest allowable contraction for the flow to be possible upstream as described

6. Water is flowing at a velocity of 10 ft/s (3.0 m/s) and a depth of 10 ft (3.0 m) in a channel of rectangular section 10 ft (3.0 m) wide. There is a smooth upward step of 4 ft (1.2 m) in the channel bed. What expansion in width must take place for the upstream flow to be possible as specified?

7. Water flows from a lake into a steep rectangular channel 5 m (16 ft) wide. If the lake level is 8 m (26 ft) above the channel bed at the outfall, estimate the discharge from the lake to the channel.

8. A trapezoidal channel with a base width of 20 ft (6.1 m) and side slopes of 2 (horizontal): 1 (vertical) carries a flow of 2000 ft^3/s (57 m^3/s) at a depth of 8 ft (2.4 m). There is a smooth transition to a rectangular section 20 ft (6.1 m) wide accompanied by a gradual lowering of the channel bed by 2 ft (0.61 m).
   a. Find the depth of water within the rectangular section and the change in water surface elevation.
   b. What is the minimum amount by which the bed must be lowered for the upstream flow to be possible as specified?

9. For the situation shown in the figure below, adapt the viewpoint of an observer who is moving with the elementary surge. Neglecting all squares of Δy and Δv, use the continuity and momentum equations to show that

$$c^2 = gy$$

Then, by using continuity and the energy equations, show that the same result can be achieved. In both cases, list all important assumptions and justify them. Is one of these two methods of proof preferable? Why?

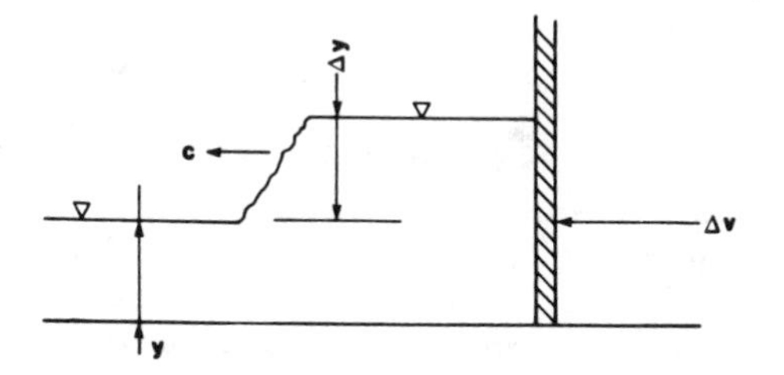

10. For the situation shown in the figure below, prove that

$$\Gamma_L = \left(\frac{b_2}{b_1}\right)^2 = \frac{27(\mathbf{F}_1)^2}{[2 + (\mathbf{F}_1)^2]^3}$$

assuming that $E_1 = E_2$. Γ_L is known as the limiting contraction ratio and can be used to distinguish Yarnell's class A or unchoked flow and class B choked flow; see Chapter 9.

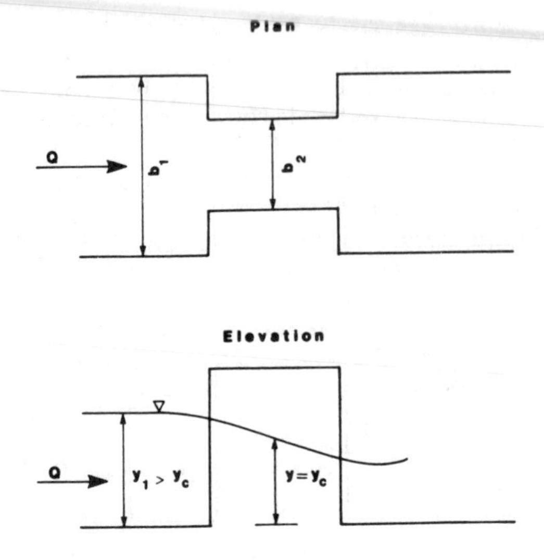

11. A rectangular channel 14 ft (4.3 m) wide conveys 756 ft^3/s (21 m^3/s) at a uniform depth of flow of 9 ft (2.7 m). This flow must pass under a road in a single barrel box culvert without creating a backwater effect and with the water surface a minimum of 1.5 ft (0.46 m) below the top of the culvert. Assuming that the culvert slope is sufficient to convey the flow uniformly, determine the minimum required culvert size.

12. Flow at the rate of 4800 ft^3/s (136 m^3/s) occurs in a rectangular channel 40 ft (12 m) wide and at a depth of 12 ft (3.7 m). Estimate the maximum width a single bridge pier may be if it is placed at midstream and is not to cause an increase in the depth of flow upstream. Compare this result with that obtained from the Yarnell equation in Prob. 10.

13. For the upstream flow data specified in Prob. 12, assume that it was decided not to put the bridge pier in midchannel but to protrude the abutments into the channel. How far could each abutment protrude into the channel if no losses are incurred and no increase in the upstream depth of flow is caused?

14. For the upstream flow specified in Prob. 12, assume that the bridge abutments are placed protruding into the channel such that the span between their faces is 24 ft (7.3 m). Also assume that the energy loss through this section is 1.0 ft·lb/lb of flow (0.30 m·N/N of flow).

a. Will the flow pass through these abutments without causing an increase in the upstream depth of flow?
b. If the flow cannot pass through the abutments without causing an increase in the upstream depth of flow, then estimate the height of the water surface immediately upstream.

15. A lake discharges freely into a steep channel. Develop a theoretical equation for estimating discharge and list all assumptions.

16. The effects of the channel slope have been ignored in this chapter under the assumption that the channel slope is small.
a. Show, schematically, total energy in a situation where the slope of the channel is not small.
b. Specific energy for a channel where the slope is small is given by Eq. (2.1.3). Using the results from part (a), derive a specific energy equation for the situation where the slope of the channel cannot be ignored.
c. If you are in a boat and measured the depth of flow with a lead line, what depth, with reference to part (a) of this problem, do you measure?

17. Consider a channel of irregular and varying cross section but with the low point of the section remaining at the same level along the channel. Neglect all losses, and show that

$$\left(\frac{dy}{dx}\right)(1 - \mathbf{F}^2) - \mathbf{F}^2\left(\frac{y}{B}\right)\left(\frac{db}{dx}\right) = 0$$

The channel is shown in the figure below.

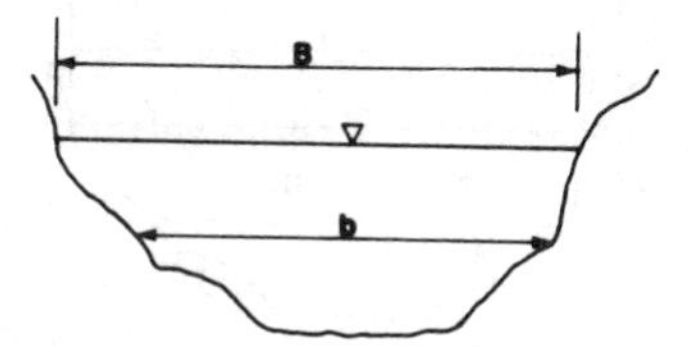

PROBLEMS

Chapter 3: The Momentum Principle

1. If the channel in the figure below is rectangular in shape and 10 ft (3 m) wide, calculate
   a. The force on the gate.
   b. The Froude number at sections 1 and 2.
   c. The condition required downstream at section 3 for a hydraulic jump to form.
   d. Assuming a hydraulic jump is formed downstream, estimate the efficiency and length of the jump.

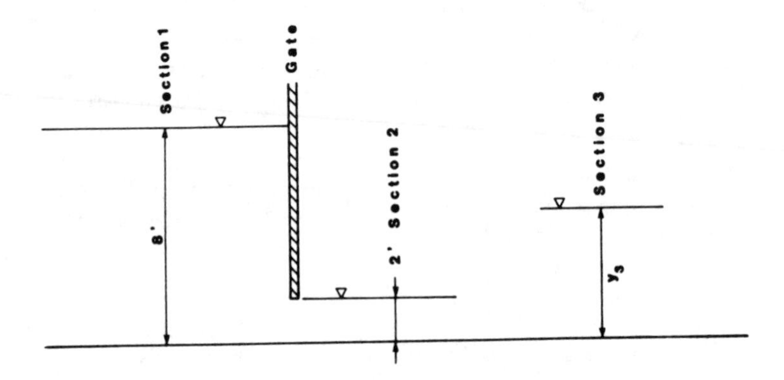

2. Apply the momentum and continuity equations to the submerged hydraulic jump that occurs at a sluice gate outlet in a rectangular channel (see the figure below) to prove that

$$\frac{y_s}{y_2} = \left[1 + 2(\mathbf{F}_2)^2\left(1 - \frac{y_2}{y_1}\right)\right]^{1/2}$$

where all variables are defined in the figure below. Neglect the channel bed friction F_f.

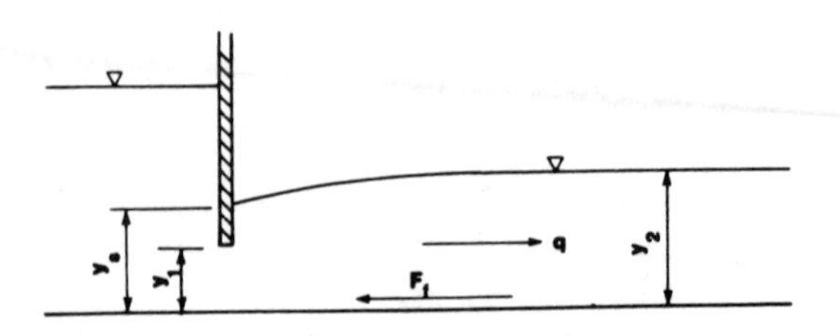

3. Prove that the energy loss in a hydraulic jump occurring in a horizontal rectangular channel is given by Eq. (3.2.24) or

$$\Delta E = \frac{(y_2 - y_1)^3}{4y_1y_2}$$

4. A rectangular channel 8 ft (2.4 m) wide carries 100 ft^3/s (2.8 m^3/s) at a depth of 0.5 ft (0.15 m). This channel is connected by a 50-ft- (15-m) long straight wall transition to a rectangular channel 10 ft (3.0 m) wide. Determine whether a hydraulic jump occurs in this transition if the head loss within the transition is h_f ft. (Assume that h_f is uniformly distributed through the transition.) If a hydraulic jump occurs, what is the approximate location of the jump relative to the beginning of the transition section?
   a. Downstream depth 4 ft (1.2 m) and $h_f = 0$ ft.
   b. Downstream depth of 4 ft (1.2 m) and $h_f = 1$ ft (0.30 m).
   c. Compare the answers of parts (a) and (b) and discuss any differences.
   d. Downstream depth 3.6 ft (1.1 m) and $h_f = 1$ ft (0.30 m).

5. Water flows at a velocity of 25 ft/s (7.6 m/s) and a depth of 3.5 ft (1.1 m) in a rectangular channel 16 ft (4.9 m) wide. Calculate the downstream depth necessary to form a jump, the head loss, and the horsepower dissipated in the jump.

6. A hydraulic jump is to be caused in a circular culvert 3 m (9.8 ft) in diameter and conveying a flow of 12 m^3/s (420 ft^3/s). If the depth of flow downstream of the jump cannot exceed 60 percent of the diameter of the pipe, what is the limiting upstream depth of flow? Is the limiting value found a minimum or maximum value? If assumptions are made in obtaining a solution, then confirm that they are valid.

7. For the situation shown below, prove that

$$\Gamma_L = \left(2 + \frac{1}{\Gamma_L}\right)^3 \frac{\mathbf{F}_3^4}{(1 + 2\mathbf{F}_3^2)^3}$$

assuming that $M_2 = M_3$ and that the channel is rectangular in shape. Γ_L is known as the limiting contraction ratio and can be used to distinguish between Yarnell's class A (unchoked) and class B (choked) flow; see Chapter 9.

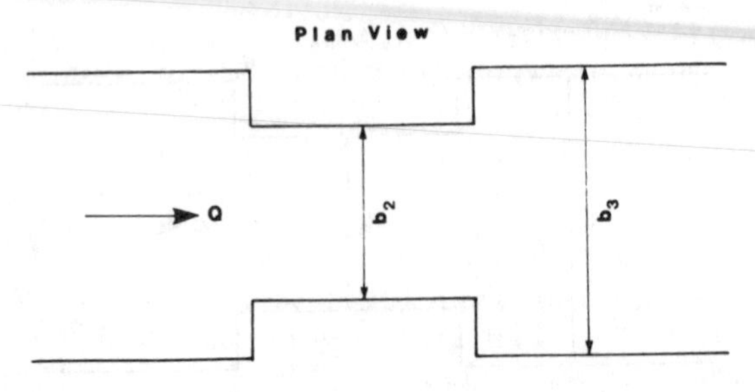

8. A bridge has piers 1 m (3.3 ft) wide at 7-m (23-ft) centers. Some distance upstream of the bridge, the depth and velocity of flow are 1.5 m (4.9 ft) and 1 m/s (3.3 ft/s), respectively. Neglecting the slope of the channel and bed friction, estimate the downstream depth of flow if the drag of the piers P_f is given by

$$P_f = \tfrac{1}{2} C_D A \rho u^2$$

where C_D = drag coefficient, A = area of pier projected on a plane normal to the flow, ρ = fluid density, and u = upstream velocity. For this problem, take C_D = 1.5. Use the results from Prob. 7 to determine if the flow choked.

9. A spillway discharges a flow of 8 (m^3/s)/m (86 (ft^3/s)/ft). A model study demonstrated that on the downstream horizontal apron the depth of flow was 0.6 m (2.0 ft). Estimate the tail water depth that is required for a hydraulic jump to form. Assuming that a hydraulic jump is formed,
   a. Classify the jump type.
   b. Estimate the jump length.
   c. Estimate the head loss incurred through the jump.

10. A trapezoidal channel is 20-ft (6.1-m) wide at the bottom and has side slopes of 1.5 (horizontal) to 1 (vertical). For a flow of 500 ft^3/s (14 m^3/s), calculate the depth sequent to a supercritical depth of 2.0 ft (0.61 m).

11. A rectangular channel has a longitudinal slope of 0.15 ft/ft (0.15 m/m). This sloping channel ends in a horizontal channel section, Fig. 3.9. When a discharge of 120 (ft^3/s)/ft (11 (m^3/s)/m) occurs in the channel at a depth of 2.0 ft (0.61 m), a hydraulic jump occurs when the sloping channel meets the horizontal channel. Calculate
   a. The length of the jump
   b. The energy lost in the jump

12. A sluice gate similar to that shown in Fig. 3.5 is used to control flow in an irrigation system, and the flow from this gate discharges into a rectangular channel of the same width as the gate. Under normal circumstances, the flow immediately downstream of the gate has a depth of 4 ft (1.2 m) and a velocity of 60 ft/s (18 m/s). Given these circumstances, estimate

a. The sequent depth of the tail water

b. The length of lined channel required to protect the natural soil from supercritical flow and the hydraulic jump

c. The efficiency of the jump with the downstream sequent depth conditions

d. The type of jump that is to be expected if sequent depth conditions are met

13. For the data provided in Prob. 10, describe qualitatively the result of having a tail-water depth of 32 ft (9.8 m). Can such a situation be described quantitatively? If it can, list all assumptions.

14. A wave is moving down a wide rectangular channel in the direction of flow. Upstream of the wave, the water velocity is 5 ft/s (1.5 m/s) with a depth of 5 ft (1.5 m). Downstream of the wave, the water velocity is 4 ft/s (1.2 m/s) with a depth of 7 ft (2.1 m). What is the velocity of the wave?

15. A rectangular irrigation channel 10 ft (3.0 m) wide conveys a flow of 500 ft^3/s (14 m^3/s) at a depth of 10 ft (3.0 m). If the upstream discharge is suddenly increased to 700 ft^3/s (20 m^3/s), estimate the height and velocity of the resulting surge wave.

16. A river is flowing at a depth of 10 m (33 ft) and a velocity of 3 m/s (9.8 ft/s). A tidal bore is moving up the river, and at the point where the river meets the bore, the depth abruptly increases to 15 m (49 ft). Estimate the velocity of the tidal bore up the river. What is the velocity and direction of flow behind the bore?

PROBLEMS

Chapter 5: Computation of Uniform Flow

1. Manning's n for a channel is estimated as $n = 0.015$. If the channel is rectangular in shape with a bottom width of 10 ft (3.0 m) and a longitudinal slope of 0.001 ft/ft,
   a. Estimate the discharge of this channel by both the Manning and Chezy equations when the depth of flow is 2 ft (0.61 m).
   b. Estimate the discharge of this channel by both the Manning and Chezy equations when the depth of flow is 20 ft (6.1 m).
   c. Discuss the results of parts (a) and (b).

2. The following survey field notes describe a channel reach. Using these field notes, estimate a value of Manning's n for the channel reach.

   Course: straight, 661-ft (200-m) long

   Cross section: very little variation in shape

   Side slopes: fairly regular

   Soil: lower part, yellowish gray clay
   upper part, light gray silty clay loam

   Condition: side slopes covered with heavy growth of poplar trees 2 to 3 in (5.1 to 7.6 cm) in diameter, large willows, and climbing vines; thick growth of water weed on bottom.

3. The following survey notes describe a channel reach. Using these field notes, estimate a value of Manning's n for the channel reach.

   Course: straight, 1710-ft (520-m) long

   Cross section: not much variation

   Side slopes: somewhat irregular

   Bottom: rather irregular and uneven

   Soil: heavy silty clay

   Condition: practically no vegetation or obstructions; appears to be newly dredged.

4. Determine the normal discharges for $y_N = 6.0$ ft (1.8 m), $n = 0.015$, and $S = 0.0020$ ft/ft in channels having the following geometric characteristics:
   a. A rectangular section having a width of 20 ft (6.1 m).
   b. A triangular section with an internal bottom angle equal to 60°.
   c. A trapezoidal section with a bottom width of 20 ft (6.1 m) and a side slope of 1 (vertical): 2 (horizontal).

5. Prove that in English units

$$Q = \frac{0.47y^{8/3}S^{1/2}f(z)}{n}$$

where

$$f(z) = \frac{z^{5/3}}{[1 + (1 + z^2)^{1/2}]^{2/3}}$$

is the equation for discharge in a triangular highway gutter having one vertical side, one side sloped 1 on "z," Manning's n, depth of flow y, and longitudinal slope S. The figure below defines the problem; note the break in side slope and change of n when the gutter overflows.

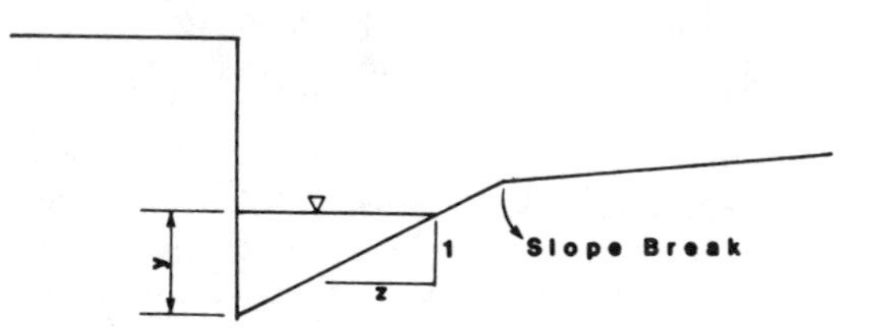

6. For the previous problem, estimate the flow in the gutter if

$$z = 24$$
$$n = 0.017$$
$$y = 0.22 \text{ ft } (0.07 \text{ m})$$
$$S = 0.30 \text{ ft/ft}$$

7. Prove for a circular pipe of diameter D that $R = D/4$. Show in doing this that the Darcy-Weisbach equation for pipe flow is equivalent to a special case of the Chezy equation provided that

$$C = \left(\frac{8g}{f}\right)^{1/2}$$

8. A canal of rectangular section and $n = 0.014$ is to be laid on a slope of 0.001 ft/ft and is to carry a discharge of 1000 ft^3/s (28 m^3/s). Calculate the flow area required for the following values of the ratio b/y: 1, 1.5, 2, 2.5, 3, 3.5, and plot the values of flow area as a function of b/y.

9. Estimate the normal depth of flow and velocity for a trapezoidal channel with the following characteristics: $b = 6.1$ m (20 ft), $z = 2$, $Q = 11$ m^3/s (390 ft^3/s), $S = 0.0016$ m/m, and $n = 0.025$.

10. A critical uniform flow at a specified normal depth may be produced in a channel of fixed geometry and roughness by varying the slope and discharge. For a wide rectangular channel of width b, develop a theoretical expression for the critical slope in terms of n and q.

11. A lined rectangular channel ($n = 0.011$) is to be built to convey flows that range from 400 ft^3/s (11 m^3/s) to 8000 ft^3/s (230 m^3/s). If other considerations dictate that the channel must be 40 ft (12 m) wide, what is the range of slopes for this channel such that the flow will be subcritical?

12. Estimate the normal discharge in channels having the following sections when $y = 2.0$ m (6.6 ft), $n = 0.015$, and $S = 0.0020$ m/m:

a. A triangular section with an internal bottom angle of 45°

b. A circular section 4.5 m (15 ft) in diameter

c. A parabolic section having a top width of 5 m (16 ft) at a depth of 1.2 m (3.9 ft)

13. Based on field observations of historical floods on alluvial fans, it has been suggested that natural alluvial fan channels stabilize at the point where a decrease in depth would result in a 200-fold increase in top width or

$$\frac{dy}{dT} = -0.005$$

(See, for example, French, R.H., *Hydraulic Processes on Alluvial Fans*, Elsevier Science Publishers, Amsterdam, 1987.) For a wide rectangular channel and specified values of Q, n, and S, develop equations for

a. The channel top width

b. The depth of flow

c. The average velocity

In doing this, the resulting equations should include a factor that accounts for either English or SI units.

14. A number of investigators (see, for example, Jarrett, R.D., "Hydraulics of High Gradient Streams," *Journal of Hydraulic Engineering*, Vol. 110, No. 11, 1984, pp. 1519–1539) have recognized that the available guidelines for estimating n in high gradient streams ($S > 0.002$) are based on very limited data. An empirical expression for n in such cases is

$$n = 0.39S^{0.38}R^{-0.16}$$

Assume in natural channels on alluvial fans that n is given by the above equation and that the channel is subject to the condition of Prob. 13.

Then for a wide rectangular channel

a. Estimate T in terms of Q and S.

b. Estimate the depth of flow in terms of Q and S.

c. Estimate u in terms of Q and S.

15. For $Q = 28$ m^3/s (990 ft^3/s), $n = 0.0115$ and $S = 0.05$ m/m, use the results from both Probs. 13 and 14 to

a. Estimate T.

b. Estimate y.

c. Estimate u.

d. Estimate F.

e. Discuss and compare the results.

16. In a laboratory experiment, the bed of a rectangular flume is artificially roughened, but the side walls are smooth. If $Q = 0.024$ m^3/s (0.85 ft^3/s), $y = 0.073$ m (0.24 ft), $b = 0.85$ m (2.8 ft), $S = 0.00278$ m/m, and the water temperature is 25.5°C, estimate f_b and u_{*b}.

17. A trapezoidal irrigation channel is constructed through alluvial materials ($n = 0.02$) and has the following dimensions: $b = 15$ ft (4.6 m), $z = 1.5$, $S = 0.0003$ ft/ft and a total safe depth of 4 ft (1.2 m). After it is built, it is found that there is excessive seepage and two proposals are advanced to solve this problem: (1) line only the sides (alternative 1) and (2) line the bottom (alternative 2). If the proposed lining material is concrete ($n = 0.012$), determine the following:

a. The normal discharge in the unlined channel if 1 ft (0.30 m) of freeboard is required.

b. The equivalent roughness and normal discharge [allowing 1 ft (0.30 m) of freeboard] for alternative 1.

c. The equivalent roughness and normal discharge [allowing 1 ft (0.30 m) of freeboard] for alternative 2.

18. A circular storm water drainage pipe 6 ft (1.8 m) in diameter is laid on a slope of 0.0004 ft/ft. If $n = 0.015$ and $Q = 75$ ft^3/s (2.1 m^3/s), what is the depth of flow in the drainage pipe?

19. In a circular open channel, at what depth does the maximum normal discharge take place? How much larger is the maximum discharge than the just-full pipe discharge?

20. A trapezoidal channel excavated in earth with a straight alignment and uniform cross section has a bottom width of 2 ft (0.61 m), side slopes of 1:1, and a longitudinal slope of 0.003 ft/ft. If the clean channel is 10 years old and the normal depth is 1.0 ft (0.30 m), estimate the velocity of flow and the discharge.

21. For the channel in the accompanying figure with a slope of 0.005 ft/ft, estimate the following:

a. For $n_A = n_B = 0.02$, what is the discharge? Is there more than one estimate of this discharge?

b. For $n_A = 0.02$ and $n_B = 0.05$, what is the discharge? Is there more than one discharge estimate?

c. For the situation defined in the figure below, what is the meaning of average velocity?

d. Compare and comment on the results from parts (a) and (b).

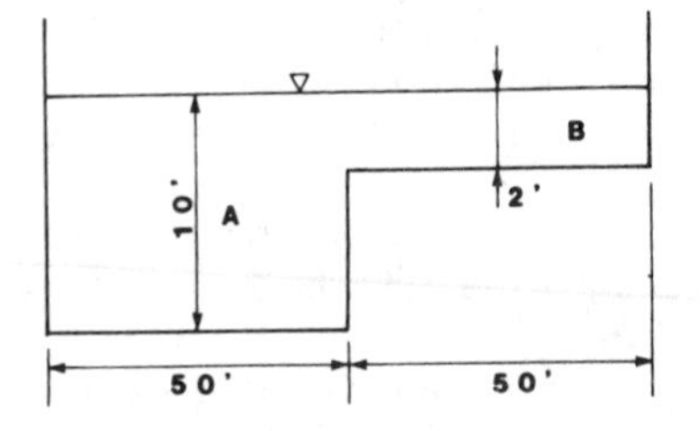

22. A trapezoidal channel parallels a highway and has a bottom width of 4 ft (1.2 m), side slopes 3:1, and a grade of 1 percent. At the lower end of the channel, the flow must be turned 90° and passed under the roadway. The designer can line the channel with concrete ($n = 0.015$) or grass ($n = 0.04$). The design flow is 300 ft^3/s (8.5 m^3/s).

a. From the viewpoint of hydraulic engineering only, which lining would you choose? Why?

b. Discuss other factors that would influence the choice made in part (a).

23. A trapezoidal channel has a bottom width of 11 m (36 ft) and $z = 1.5$. If this channel is lined with smooth concrete and $S = 0.0003$ m/m, estimate the mean velocity and discharge for a normal depth of flow of 3.5 m (11 ft). Considering that Manning's n is an approximate coefficient, estimate the maximum and minimum values of discharge. Discuss the results.

24. A trapezoidal channel with a bottom width of 5 m (16 ft), $n = 0.015$, $z = 2$, and $S = 0.0015$ m/m, conveys a flow of 25 m^3/s (880 ft^3/s). Estimate the normal depth of flow for this situation.

25. A triangular channel with an apex angle of 50° conveys a flow of 50 ft^3/s (1.4 m^3/s) at a depth of 1 ft (0.30 m). If the longitudinal slope of the channel is 0.009 ft/ft, estimate the roughness coefficient of the channel. From the viewpoint of engineering design and construction, is this an attainable roughness coefficient value?

26. A trapezoidal channel has a bottom width of 11 m (36 ft) and $z = 1.5$. If the channel is lined with smooth concrete, estimate the longitudinal channel slope required to convey a flow of 60 m^3/s (2100 ft^3/s) at a depth of 3.5 m (11 ft).

27. A concrete lined trapezoidal channel ($n = 0.015$) with $z = 1.0$ and a longitudinal slope of 0.004 m/m must be designed to convey a flow of 150 m^3/s (5300 ft^3/s) at a normal depth of 3.0 m (9.8 m).

a. Estimate the bottom width of the channel required to accomplish this.

b. Is the flow in this channel subcritical or supercritical?

28. Estimate the slope required to convey 200 ft^3/s (5.7 m^3/s) in a rectangular channel 10 ft (3.0 m) wide and lined with poorly troweled concrete so that the Froude number of the flow is 2.

PROBLEMS

Chapter 6: Theory and Analysis of Gradually and Spatially Varied Flow

1. The very wide rectangular channel shown in the figure below has a unit discharge of 25 $(ft^3/s)/ft$ (2.3 $(m^3/s)/m$) and $n = 0.02$. For these conditions,
   a. Calculate the normal and critical depths of flow for each slope. Draw lines representing these depths of flow on the figure and label them as either NDL (normal depth of flow) or CDL (critical depth of flow).
   b. On the figure, draw all possible flow profiles and classify and label them; for example, M1, S3, etc.

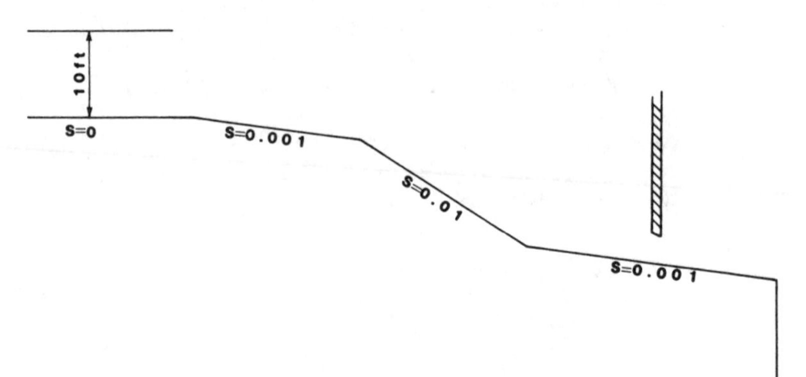

2. A rectangular channel 2.4 m (7.9 ft) wide with a longitudinal slope of 0.008 m/m conveys a discharge of 4.2 m^3/s (150 ft^3/s). If $n = 0.014$ and the depth of flow is 1.5 m (4.9 ft), what is the flow profile classification? What is the upstream boundary condition?

3. A rectangular channel 25 ft (7.6 m) wide has a longitudinal slope of 0.002 ft/ft and conveys a discharge of 400 ft^3/s (11 m^3/s). If $n = 0.03$, $\alpha = 1.15$ and the downstream depth of flow at a control point is 7 ft (2.1 m), classify the flow profile upstream of the control point and estimate the length of the flow profile. What is the upstream boundary condition?

4. A trapezoidal channel with a longitudinal slope of 0.001 m/m carrying a discharge of 30 m^3/s (1060 ft^3/s), a Manning's n of 0.025, a bottom width of 6 m (19.7 ft), and side slopes of 1.5 (horizontal):1 (vertical) terminates in a weir. What is the upstream boundary condition for this problem? Define and estimate the gradually varied flow profile that results from this situation.

5. A trapezoidal channel with a longitudinal slope of 0.001 ft/ft, carrying a discharge of 1000 ft^3/s (28 m^3/s), a Manning's n of 0.025, a bottom width

of 20 ft (6.1 m), and side slopes of 1.5 (horizontal):1 (vertical) terminates in a sluice gate structure. If the depth of water at the sluice gate is 10 ft (3.0 m), use the step method with a depth increment of 0.5 ft (0.15 m) to estimate the water surface profile behind this sluice gate. What is the upstream boundary condition for this problem?

6. A trapezoidal channel with a longitudinal slope of 0.0169 ft/ft, carrying a discharge of 400 ft^3/s (11 m^3/s), a Manning's n of 0.025, a bottom width of 20 ft (6.1 m), and side slopes of 2 (horizontal):1 (vertical) terminates at a dam where the depth of water is 5 ft (1.5 m). If the upstream boundary condition is critical flow, estimate the water surface profile behind the dam using the step method with a depth increment of 0.5 ft (0.15 m).

7. Determine the hydraulic exponents M and N for a trapezoidal channel with a bottom width b and side slopes z.

8. A flow of 5000 ft^3/s (140 m^3/s) is carried in a rectangular channel 40 ft (12 m) wide. At one point, this channel constricts to a width of 25 ft (7.6 m). It is known from model tests that the energy loss in this constriction is approximately 2.5 ft (0.76 m). This channel has a longitudinal slope of 0.002 ft/ft, and Manning's n has been estimated to be 0.16. Given this information,

a. Is the constriction sufficient to cause a backwater curve?

b. If the answer to part (a) is yes, then estimate the depth of flow immediately upstream of the constriction.

c. If the answer to part (a) is yes, what is the depth of flow 2000 ft (610 m) upstream of the constriction?

9. A trapezoidal irrigation channel has the following characteristics: $b = 15$ ft (4.6 m), $z = 0.02$, $S = 0.0004$ ft/ft, and a normal depth of 10 ft (3.0 m). If this channel empties into a pond at the downstream end and the pond elevation is 4 ft (1.2 m) higher than the channel elevation at the downstream end, calculate and plot the gradually varied flow profile that results from this situation.

10. A 6-ft- (1.8-m) diameter sewer pipe ($n = 0.015$) has a slope of 0.001 ft/ft and carries a discharge of 100 ft^3/s (2.8 m^3/s). If this pipe ends in a free overfall, estimate the resulting gradually varied flow profile.

11. A small river has a cross section that can be approximated as a trapezoid. If $n = 0.015$, $Q = 3500$ ft^3/s (99 m^3/s), and the cross-sectional properties of three typical cross sections are given in the following table, find the water surface elevations at sections II and III if the water surface elevation at section I is 345 ft (105 m).

Section	Distance from mouth, mi (km)	Bed elevation above MSL, ft (m)	Bed width, ft (m)	Side slope, z
I	100	328	46	2
	(161)	(99.97)	(14)	
II	101.5	331	41	1.5
	(163)	(100.9)	(12)	
III	102.5	333	33	3
	(165)	(101.5)	(10)	

12. In the accompanying table, values of the wetted perimeter P and the cross-sectional area A are given for three channel sections as a function of water surface elevation or stage. The distance from section 1 to section 2 is 600 m (1970 ft), and the distance from section 2 to section 3 is 700 m (2300 ft). In $n = 0.035$, $Q = 200\ m^3/s$ ($7060\ ft^3/s$), and the stage at section 1 corresponding to this flow rate is 508 m (1670 ft), estimate the water surface elevation at sections 2 and 3.

	Section 1		Section 2		Section 3	
Stage, m (ft)	P, m (ft)	A, m^2 (ft^2)	P, m (ft)	A, m^2 (ft^2)	P, m (ft)	A, m^2 (ft^2)
505	80	110	70	37	—	—
(1657)	(262)	(1184)	(230)	(398)	—	—
506	91	130	100	64	0	0
(1660)	(299)	(1399)	(328)	(689)	(0)	(0)
507	100	160	120	100	30	6.5
(1663)	(328)	(1722)	(394)	(1076)	(98)	(70)
508	110	200	130	135	44	18
(1667)	(361)	(2153)	(426)	(1453)	(144)	(194)
509			135	175	52	32
(1670)			(443)	(1884)	(171)	(344)
510			142	220	58	50
(1673)			(466)	(2368)	(190)	(538)
511			150	260	65	70
(1677)			(492)	(2799)	(213)	(753)
512			154	310	70	88
(1680)			(505)	(3337)	(230)	(947)
513					76	110
(1683)					(249)	(1184)
514					81	135
(1686)					(266)	(1453)
515					86	160
(1690)					(282)	(1722)
516					91	190
(1693)					(299)	(2045)

13. A trapezoidal channel with a base width of 30 ft (9.1 m) and side slopes of 2 (horizontal):1 (vertical) has a longitudinal slope of 0.001 ft/ft, a Manning's n of 0.027, and conveys a flow of 2500 ft^3/s (71 m^3/s) to a reservoir whose level varies as a function of time. At the point where the channel enters the lake, the invert of the channel is 10 ft (3.0 m) above a datum. Estimate the elevation of the water surface 1 mile upstream of the point where the channel enters the lake when

a. The elevation of the lake is 14 ft (4.3 m) above datum.

b. The elevation of the lake is 21 ft (6.4 m) above datum.

PROBLEMS

Chapter 7: Design of Channels

1. Demonstrate that the best hydraulic section for
   a. A rectangular channel is half a square
   b. A trapezoidal channel is half a hexagon

2. Calculate the dimensions of the most economical trowel-finished concrete-lined trapezoidal channel to convey 250 m^3/s (8800 ft^3/s) with a slope of 0.0003 m/m.

3. A trapezoidal channel with b = 20 ft (6.1 m), z = 2 on a slope of 0.001 ft/ft conveys a uniform flow at a normal depth of 6 ft (1.8 m). Given the above, calculate
   a. The *average* unit tractive force and shear velocity on the channel bottom.
   b. The maximum tractive force and shear velocity on the channel side slopes.
   c. The maximum tractive force and shear velocity on the channel bottom.

4. A trapezoidal channel has been designed to convey 8.5 m^3/s (300 ft^3/s) at a slope of 0.0015 m/m in noncolloidal coarse gravels with z = 2 and b = 6.1 m (20 ft). As the senior design engineer, you are to examine the appropriateness of this design. What is your conclusion?

5. Use the method of maximum permissible velocity to design an unlined trapezoidal channel in fine gravel that will convey a discharge of 600 ft^3/s (17 m^3/s). The channel will have a longitudinal slope of 0.0016 ft/ft. Compare the designs developed assuming clear water and water-transporting colloidal silts. What are the freeboard requirements for each design?

6. Design a nonerodible (lined) channel to carry 200 ft^3/s (5.7 m^3/s) with n = 0.020 and S = 0.0020 ft/ft. Use engineering judgment and list all assumptions. Do not forget to estimate the lining height and the freeboard required. Show the final design in a sketch.

7. The All-American canal is designed to divert 15,155 ft^3/s (430 m^3/s) of desilted water from the Colorado River to irrigate the Imperial Valley in Southern California. This canal is 80 miles (130 km) long. The typical maximum section has a bottom width of 160 ft (49 m), a top width at the water surface of 232 ft (71 m), a water depth of 20.6 ft (6.3 m), a minimum freeboard of 6 ft (1.8 m), and a bank width of 27 to 30 ft (8.2 to 9.1 m). The terminal capacity of this canal is 2600 ft^3/s (74 m^3/s). The canal is excavated primarily in alluvial materials, ranging from light sandy or silty

loams to adobe and having an average particle size of 0.0025 in (0.063 mm). Review the hydraulic design of this section.

8. Determine the cross section and discharge of the stable hydraulic section of a channel excavated in noncohesive material with $\tau_0 = 0.10$ lb/ft^2, $S = 0.0004$ ft/ft, $\theta = 31°$, and $n = 0.020$.

9. For the data given in Prob. 8, determine the modified channel section if the channel is to carry
- **a.** 75 ft^3/s (2.1 m^3/s)
- **b.** 300 ft^3/s (8.5 m^3/s)

10. Design a waterway lined with Bermuda grass on erosion-resistant soil to carry 200 ft^3/s (5.7 m^3/s). The average slope of the channel is 3 percent. Use the curve for moderate vegetal retardance.

11. The channel section shown below is to convey 2000 ft^3/s (57 m^3/s) of heavily silted water. It is excavated on a slope of 0.001 ft/ft through country composed of moderately rounded alluvium. The canal will be armored with selected rock averaging 2 in (51 mm) in diameter. Do you foresee any problems with this design?

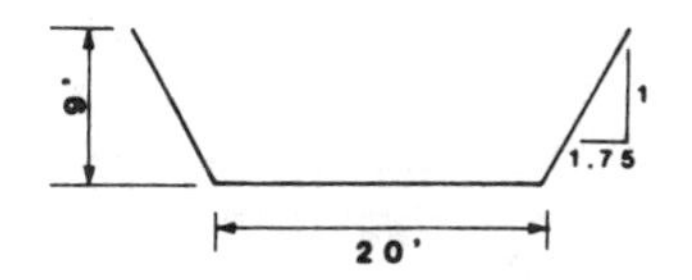

12. A trapezoidal channel with a bottom width of 4 ft (1.2 m); side slopes of 4:1; and a longitudinal slope of one percent is lined with Bermuda grass in easily erodible soil. If the grass is kept mowed to a height of 2 to 4 in (51 to 102 mm) and the design flow is 100 ft^3/s (2.8 m^3/s), evaluate the adequacy of the design.

13. A trapezoidal channel with a bottom width of 4 ft (1.2 m), side slopes of 3:1, and a longitudinal slope of 0.005 ft/ft must convey a continuous flow of 30 ft^3/s (0.85 m^3/s) and a flood flow of 170 ft^3/s (4.8 m^3/s). If the channel is to be built in sandy loam and the flood water conveys fine silts, evaluate the following lining options if supercritical flow is not acceptable:
- **a.** No lining
- **b.** Complete concrete lining
- **c.** Combination of mowed Bermuda grass and concrete lining as shown at the top of the next page

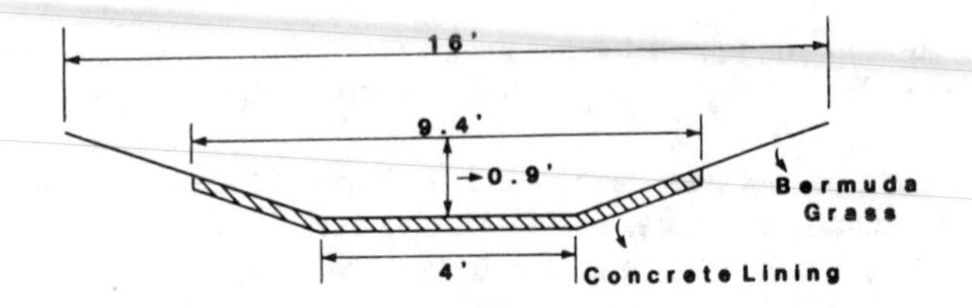

d. If none of the above designs are acceptable, recommend and justify a new design using a combination of concrete and mowed Bermuda grass linings.

PROBLEMS

Chapter 8: Flow Measurement

1. The field data in the accompanying table are stream gaging data resulting from a preliminary study. Estimate the discharge at this station and make recommendations for the next gaging at this station.

Distance from bank to current meter, m	Depth, m	Meter depth (m) from free surface	Velocity, m/s
0.80	0.30	0.18	0.20
1.50	1.20	0.96	0.35
		0.24	0.10
2.30	1.75	1.40	0.51
		0.35	0.12
3.40	2.10	1.68	0.61
		0.42	0.15
4.20	1.50	1.20	0.44
		0.30	0.11
5.0	0.75	0.45	0.16
5.70	0.30	0.18	0.07
6.50	0	—	—

2. For the field data provided in Prob. 1, estimate the kinetic energy and momentum correction factors.

3. The field data in the accompanying table are stream gaging data on the Cumberland River in the vicinity of Cumberland City, Tennessee. Estimate the discharge at this station and the average velocity of flow. Comment on the data provided.

Distance from South Bank, ft	Depth, ft	Velocity at 0.8 depth, ft/s	Velocity at 0.2 depth, ft/s
0 South Bank	0	0	0
101	8	0.17	0.12
148	19	0.25	0.18
220	32	0.45	0.40
308	40	0.58	0.49
330	40	0.28	0.35
414	38	0.53	0.57
447	35	0.30	0.29
480	33	0.47	0.59
589	26	0.38	0.63
642	23	0.26	0.27
700	20	0.39	0.57
759	18	0.51	0.58

Distance from South Bank, ft	Depth, ft	Velocity at 0.8 depth, ft/s	Velocity at 0.2 depth, ft/s
836	16	0.30	0.36
914	13	0.28	0.28
983	11	0.38	0.31
1070	11	0.35	0.35
1230 North Bank	0	0	

4. For the field data provided in Prob. 3, estimate the kinetic energy and momentum correction factors.

5. A lake freely discharges over a spillway crest into a rectangular open channel. Derive a discharge equation for this situation in terms of the elevation of the lake surface.

6. A rectangular channel 6 ft (1.8 m) wide has a discharge of 14 ft^3/s (0.40 m^3/s). Find the height of a rectangular weir spanning the complete width of the channel that will pass this discharge while maintaining an upstream depth of 2.75 ft (0.84 m).

7. For the weir designed in Prob. 6, what is the discharge when the upstream depth is 3.5 ft (1.1 m)?

8. For the weir designed in Prob. 6, if the flow rate is 50 ft^3/s (1.4 m^3/s), then
   a. What is the depth of water immediately upstream of the weir?
   b. If the downstream channel is rectangular, 6 ft (1.8 m) wide, concrete lined, and has a slope of 0.004 ft/ft, what is the depth of flow downstream?
   c. What is the total air demand of this weir? Further, if this air demand is provided by a steel pipe 15 ft (4.6 m) long with three right-angle elbows and a sharp-cornered entrance, determine the diameter of pipe required.

PROBLEMS

Chapter 10: Turbulent Diffusion and Dispersion in Steady Open-Channel Flow

1. A facility discharges 5 ft^3/s (0.14 m^3/s) of wastewater containing 100 ppm (100 mg/L) of a conservative substance into the center of a very wide meandering channel. The stream is 25 ft (7.6 m) deep, the mean velocity of flow is 1.75 ft/s (0.53 m/s), and the shear velocity is 0.15 ft/s (0.05 m/s). Assuming this discharge is well mixed in the vertical, estimate the maximum concentration of the substance 500 ft (152 m), 1000 ft (305 m), and 1500 ft (457 m) downstream.

2. A facility discharges a conservative substance at the side of a straight rectangular channel. If the channel is 610 m (2000 ft) wide, n = 0.02, S = 0.0002 m/m, and conveys a flow of 560 m^3/s (19,800 ft^3/s), estimate the length of channel required for complete mixing.

3. An industry discharges 4 million gallons per day (0.174 m^3/s) of effluent containing 400 ppm (400 mg/L) of a conservative substance at the center of a slowly meandering channel 200 ft (61 m) wide and 20 ft (6.1 m) deep. If the shear velocity is 0.2 ft/s (0.06 m/s) and the average velocity is 1 ft/s (0.30 m/s), then
   a. What is the width of the plume and maximum concentration 1000 ft (305 m) downstream?
   b. What is the distance to the point of complete mixing?

4. A conservative substance is discharged at the bottom and in the center of a straight rectangular channel in which

   Average velocity of flow = 0.61 m/s (2 ft/s)

   Depth of flow = 1.5 m (4.9 ft)

   Slope = 0.0002 m/m

   Width = 60 m (197 ft)

   a. If the substance is completely mixed in the transverse direction, what is the length of channel required for complete vertical mixing?
   b. If the substance is completely mixed in the vertical direction, what is the length of channel required for complete transverse mixing?

5. For the cross section and field data summarized below, for a meandering channel estimate the value of the dispersion coefficient if Δy = 2.5 ft and u_* = 0.33 ft/s.

Element (1)	Distance from left bank to start of element, ft (2)	Mean velocity in element, ft/s (3)	Depth at left side of element, ft (4)
1	0	0.15	0
2	2.5	0.25	1.00
3	5.0	0.35	2.15
4	7.5	0.42	2.61
5	10.0	0.65	3.30
6	12.5	1.35	3.41
7	15.0	1.80	3.32
8	17.5	2.30	3.14
9	20.0	2.35	3.00
10	22.5	2.40	3.01
11	25.0	2.50	3.10
12	27.5	2.55	2.99
13	30.0	2.70	2.78
14	32.5	2.75	2.64
15	35.0	2.65	2.59
16	37.5	2.45	2.66
17	40.0	2.30	2.90
18	42.5	2.15	2.98
19	45.0	1.85	2.84
20	47.5	1.50	2.78
21	50.0	1.10	2.72
22	52.5	0.70	2.39
23	55.0	0.40	1.82
24	57.5	0.20	0.84
25	60.0	0.10	0.34

6. For the cross section and field data summarized below, estimate the value of the dispersion coefficient if S = 0.001 ft/ft and the channel meanders. Is there any problem apparent with the data provided?

Element (1)	Element average, ft/s (2)	Element area, ft² (3)	Element flow, ft³/s (4)	Depth, left side of element, ft (5)
1	0.08	400	32	0
2	0.19	630	120	8
3	0.32	1800	580	19
4	0.48	3200	1500	32
5	0.43	880	380	40
6	0.43	3300	1400	40
7	0.42	1200	500	38
8	0.42	1100	460	35
9	0.52	3200	1700	33

Element (1)	Element average, ft/s (2)	Element area, ft^2 (3)	Element flow, ft^3/s (4)	Depth, left side of element, ft (5)
10	0.38	1300	490	26
11	0.38	1200	460	23
12	0.52	1100	570	20
13	0.44	1300	570	18
14	0.30	1100	330	16
15	0.32	800	260	13
16	0.35	960	340	11
17	0.18	880	160	11

7. For the cross section defined in the table below, estimate the dispersion coefficient if Δy = 15 ft (4.6 m) and u_* = 0.4 ft/s (0.12 m/s). Compare the estimated values obtained by using Eq. (10.3.17).

Element (1)	Average velocity, ft/s (2)	Area, ft^2 (3)	Element flow, ft^3/s (4)	Depth, left side of element, ft (5)
1	0.80	202	162	0
2	1.0	548	548	27
3	1.2	788	946	46
4	1.3	907	1180	59
5	1.6	907	1450	62
6	1.4	885	1240	59
7	1.4	788	1100	59
8	1.3	638	829	46
9	1.2	562	674	39
10	1.1	495	544	36
11	1.2	428	514	30
12	1.0	382	382	27
13	0.90	345	310	24
14	0.90	322	290	22
15	0.79	308	243	21
16	0.30	278	83	20
17	0.10	128	12.8	17

8. A facility accidentally discharges a slug of pollutant into a river with a natural flow section that is approximated as a rectangle 200 ft (61 m) wide and 5 ft (1.5 m) deep. If the velocity of flow is 2 ft/s (0.61 m/s) and the longitudinal slope is 0.002 ft/ft, estimate the distance downstream before a dispersion analysis can be used. Estimate an appropriate value of the dispersion coefficient for this problem.

9. For the stream reach and flow conditions shown in the figure below, estimate the volume of Rhodamine WT 20% dye that should be injected at Mile 0 to produce a peak concentration of 5 μg/L at Mile 30.

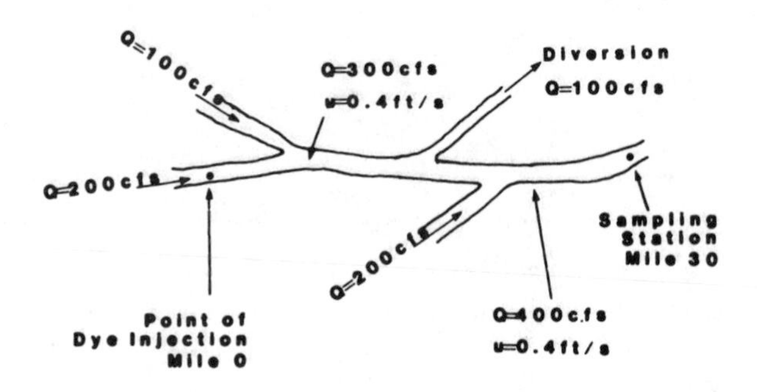

PROBLEMS

Chapter 14: Hydraulic Models

1. The Reynolds number may be defined as the ratio of
   a. Viscous forces to inertial forces
   b. Viscous forces to gravity forces
   c. Inertial forces to viscous forces
   d. Gravity forces to inertial forces
   e. None of the above

2. Select the situations in which inertial forces are not important:
   a. Flow over a spillway crest
   b. Flow through an open channel transition
   c. Flow through a long capillary tube
   d. Flow through a half-open valve

3. A model of a spillway is to be constructed. If $n_P = 0.013$ and the minimum roughness achievable in the model is $n_M = 0.009$, discuss the implications from a laboratory viewpoint.

4. A fixed bed model is to be constructed of a channel reach that is rectangular in section and 100 m (328 ft) wide. The scale ratios to be used are $L_R = 1/100$ and $Y_R = 1/10$. Manning's n for the prototype is 0.025 and the slope is 0.001 m/m. Calculate the discharges in the model when the prototype depth of flow is 1.0 m (3.28 ft) and 2 m (6.56 ft). What are the required values of n in the model for these depths of flow?

5. A fixed bed model is to be constructed of a channel reach that is triangular in section. In the prototype, $z = 2$ and $n = 0.030$. If $L_R = 0.005$ and $Y_R = 0.05$, find the required values of n in the model when the prototype depths of flow are 5 ft (1.5 m) and 10 ft (3.0 m).

6. A fixed bed model is to be constructed of a channel reach that is approximately parabolic in section. When the depth of flow in the prototype channel is 4 ft (1.2 m), the top width is 100 ft (30.5 m). If $L_R = 1/150$ and $Y_R = 1/20$, estimate the required values of n in the model when the prototype depth of flow is 3.5 ft (1.1 m) if $n_P = 0.035$.

7. For the data provided in Prob. 6, estimate the value of n required in the model when the depth of flow in the prototype is 6 ft (1.83 m).

8. A fixed bed model of a wide channel reach is to be constructed. Space in the laboratory is such that $L_R = 1/200$ and the maximum flow to the laboratory is 0.04 m^3/s (1.4 ft^3/s). In the prototype, $n = 0.04$, and laboratory materials available indicate that in the model n can be no less than 0.014. If the maximum discharge in the prototype is 2000 m^3/s (70,600 ft^3/s), cal-

culate the range of Y_R and the corresponding values of n and maximum discharge in the model.

9. A movable bed model of a rectangular channel reach 60 m (197 ft) wide is to be constructed. The specific gravity of the material composing the prototype bed is 2.65 and that composing the model bed is 1.2. Use Eqs. (14.3.10) through (14.3.13) to determine suitable values of d_R, L_R, and Y_R.

APPENDIX I

Sample computer program for determining the normal depth of flow in rectangular, triangular, trapezoidal, circular, and natural channels

```
      common ych(7),y(10),ar(10)
      dimension s(20), index(20)
      write(6,100)
      read(5,*) arm
      write(6,101)
      read(5,102) id
      if( id .eq. 40 .or. id .eq. 30) write(6,103)
      if( id .eq. 40 .or. id .eq. 30) read(5,*) z
      if(id .eq. 20 .or. id .eq. 40) write(6,104)
      if( id .eq. 20 .or. id .eq. 40) read(5,*) b
      if(id .eq. 50)write(6,106)
      if(id .eq. 50)read(5,*)d
      if(id .eq. 50)go to 1
      if(id .eq. 60)go to 1
      i=1
      ych(i)=0.05
      i=i+1
      ych(i) = 0.5
      i=i+1
      ych(i) = 1.0
      i=i+1
      ych(i) = 5.0
      i=i+1
      ych(i) = 10.0
      i=i+1
      ych(i) = 15.0
      i=i+1
      ych(i) = 20.0
      err = 0.0
      if(id .eq. 20) call rect(arm,b,err)
      if(id .eq. 30)b=0.
      if(id .eq. 30 .or. id .eq. 40)call trap(arm,b,z,err)
1     if(id .eq. 50)call circ(arm,d,err)
      if(id .eq. 60)call natdat(arm,err,m)
      if(err .eq. 1.) go to 500
      if(id .ne. 60)n=10
      if(id .eq. 60)n=m
      call spcoef(n,ar,y,s,index)
      call spline(n,ar,y,s,index,arm,ycomp)
      if(id .ne. 60)write(6,105)ycomp
      if(id .eq. 60)write(6,107)ycomp
100   format(1x,"input computed value of ar**0.667")
```

```
101   format(1x,"identify channel type: 20 rectangle;",/
     c,24x,"30 triangle;",/,24x,"40 trapezoid;",/
     c,24x,"50 circular;",/,24x,"60 natural")
102   format(i2)
103   format(1x,"input z")
104   format(1x,"input bottom width")
105   format(1x,"depth of flow = ", f10.2)
106   format(1x,"input diameter")
107   format(1x,"normal stage of flow =",f10.2)
500   stop
      end
      subroutine rect(arm,b,err)
      common ych(7),y(10),ar(10)
      dimension arch(6)
      do 1 i=1,6
      a=b*(ych(i))
      p=b+2.*ych(i)
      arch(i)=a*(a/p)**0.667
1     continue
      do 2 i=1,5
      if(arm .ge. arch(i) .and. arm .lt. arch(i+1)) go to 3
2     continue
      write(6,100)
100   format(1x,"ar out of range")
      err = 1.
      go to 4
3     y1=ych(i)
      y2=ych(i+1)
      dy=(ych(i+1)-ych(i))/10.
      j=1
      y(j)=ych(i)
      ar(j)= arch(i)
      do 5 i=2,10
      y(i)= y(i-1) + dy
      a = b*y(i)
      p = b+2.*y(i)
      ar(i) = a*(a/p)**0.667
5     continue
4     continue
      return
      end
      subroutine tria(arm,z,err)
      common ych(7),y(10),ar(10)
      dimension arch(6)
      do 1 i=1,6
      a=z*ych(i)**2
      p= 2.*ych(i)*sqrt(1+z**2)
      arch(i)= a*(a/p)**0.667
1     continue
      do 2 i=1,5
      if(arm .ge. arch(i) .and. arm .lt. arch(i+1)) go to 3
2     continue
      write(6,100)
100   format(1x,"ar out of range")
      err = 1.
      go to 4
3     y1=ych(i)
      y2=ych(i+1)
      dy=(y2-y1)/10.
      j=1
      y(j)= y1
      ar(j)=arch(i)
      do 5 i=2,10
      y(i)=y(i-1)+dy
      a=z*y(i)**2
      p=2.*y(i)*sqrt(1.+z**2)
      ar(i)=a*(a/p)**0.667
5     continue
```

```
 4      continue
        return
        end
        subroutine trap(arm,b,z,err)
        common ych(7),y(10),ar(10)
        dimension arch(6)
        do 1 i=1,6
        a = ych(i)*(b+z*ych(i))
        p= b+2.*ych(i)*sqrt(1.  + z**2)
        arch(i)=a*(a/p)**0.667
 1      continue
        do 2 i=1,5
        if(arm .ge. arch(i) .and. arm .lt. arch(i+1)) go to 3
 2      continue
        write(6,100)
 100    format(1x,"ar out of range")
        err=1.
        go to 4
 3      y1=ych(i)
        y2=ych(i+1)
        dy=(y2-y1)/10.
        j=1
        y(j)=y1
        ar(j)=arch(i)
        do 5 i=2,10
        y(i)=y(i-1)+dy
        a=y(i)*(b+z*y(i))
        p=b+2.*y(i)*sqrt(1.+z**2)
        ar(i)= a*(a/p)**0.667
 5      continue
 4      continue
        return
        end
        subroutine spcoef(n,xn,fn,s,index)
        dimension xn(n),fn(n),s(n),index(n)
        dimension rho(50),tau(50)
        nm1=n-1
        do 23017 i=1,n
        index(i)=i
23017   continue
        do 23019 i=1,nm1
        ip1=i+1
        do 23021 j=ip1,n
        ii= index(i)
        ij= index(j)
        if(.not.(xn(ii) .gt.  xn(ij))) go to 23023
        itemp = index(i)
        index(i) = index(j)
        index(j) = itemp
23023   continue
23021   continue
23019   continue
        nm2= n-2
        rho(2)= 0.
        tau(2)= 0.
        do 23025 i=2,nm1
        iim1 = index(i-1)
        ii = index(i)
        iip1= index(i+1)
        him1= xn(ii)-xn(iim1)
        hi= xn(iip1)-xn(ii)
        temp =(him1/hi)*(rho(i)+2.) + 2.
        rho(i+1)= 1./temp
        d = 6.  *((fn(iip1)-fn(ii))/hi-(fn(ii)-fn(iim1))/him1)/hi
        tau(i+1) = (d-him1*tau(i)/hi)/temp
23025   continue
        s(i) = 0.
        s(n) = 0.
        do 23027 i=1,nm2
```

```
      ib= n-1
      s(ib)= rho(ib+1)*s(ib+1)+tau(ib+1)
23027 continue
      return
      end
      subroutine spline(n,xn,fn,s,index,x,sp)
      dimension xn(n),fn(n),s(n),index(n)
      i1=index(1)
      if(.not.(x.lt.xn(i1))) go to 23029
      i2= index(2)
      h1= xn(i2)-xn(i1)
      sp= fn(i1) +(x-xn(i1))*((fn(i2)-fn(i1))/h1-h1*s(2)/6.)
      go to 23030
23029 continue
      in = index(n)
      if(.not.(x .gt.  xn(in))) go to 23031
      inm1 = index(n-1)
      hnm1 = xn(in)-xn(inm1)
      sp= fn(in)+(x-xn(in))*((fn(in)-fn(inm1))/hnm1+hnm1*s(n-1)/6.)
      go to 23032
23031 continue
      do 23033 i=2,n
      ii = index(i)
      if(.not.(x .le.  xn(ii))) go to 23035
      go to 23034
23035 continue
23033 continue
23034 continue
      l=i-1
      il= index(l)
      ilp1= index(l+1)
      a = xn(ilp1)-x
      b = x-xn(il)
      hl= xn(ilp1)- xn(il)
      sp= a*s(l)*(a**2/hl-hl)/6.+b*s(l+1)*(b**2/hl-hl)/6.+(a*fn(il)+b*
     + fn(ilp1))/hl
23032 continue
23030 continue
      return
      end
      subroutine circ(arm,d,err)
      common ych(7),y(10),ar(10)
      dimension arch(6)
      dd=0.15
      do 1 i=1,6
      ych(i)=d*dd*i
      z=ych(i)-d/2.
      r=d/2.
      a=(3.14*r*r/2.)+z*sqrt(r*r-z*z)+r*r*asin(z/r)
      if(z .ge. 0.)go to 6
      zz=abs(z)
      theta=2.*acos(zz/r)
      go to 7
6     theta=asin(z/r)
      theta=2.*theta+3.14
7     p=r*theta
      arch(i)=a*(a/p)**0.667
1     continue
      do 2 i=1,5
      if(arm .ge. arch(i) .and. arm .lt. arch(i+1))go to 3
2     continue
      write(6,100)
100   format(1x,"ar out of range")
      err=1.
      go to 4
3     y1=ych(i)
      y2=ych(i+1)
      dy=(ych(i+1)-ych(i))/10.
```

```
      j=1
      y(j)=y1
      ar(j)=arch(i)
      do 5 i=2,10
      y(i)=y(i-1)+dy
      z=y(i)-r
      a=(3.14*r*r/2.)+z*sqrt(r*r-z*z)+r*r*asin(z/r)
      if(z .ge. 0.)go to 8
      zz=abs(z)
      theta=2.*acos(zz/r)
      go to 9
8     theta=asin(z/r)
      theta=2.*theta+3.14
9     p=r*theta
      ar(i)=a*(a/p)**0.667
5     continue
4     continue
      return
      end
      subroutine natdat(arm,err,n)
      common ych(7),y(10),ar(10)
      dimension a(20),p(20),s(20),x1(20),x2(20)
      write(6,100)
      write(6,101)
      read(5,102)n
      do 1 i=1,n
      write(6,103)i
      read(5,*)s(i),x1(i),x2(i)
      if(i .eq. 1)go to 1
      dy=s(i)-s(i-1)
      dx1=x1(i-1)-x1(i)
      dx2=x2(i)-x2(i-1)
      if(i .gt. 2)go to 2
      p(i-1)=sqrt(dy**2+dx1**2)+sqrt(dy**2+dx2**2)
      a(i-1)=0.5*dy*(x2(i)-x1(i))
      go to 3
2     bt=x2(i)-x1(i)
      bb=x2(i-1)-x1(i-1)
      a(i-1)=a(i-2)+0.5*dy*(bt+bb)
      p(i-1)=p(i-2)+sqrt(dy**2+dx1**2)+sqrt(dy**2+dx2**2)
 3    ar(i-1)=a(i-1)*(a(i-1)/p(i-1))**0.667
      y(i-1)=s(i)
1      continue
      do 4 i=2,n
      if(ar(i-1) .le. arm .and. ar(i) .gt. arm)go to 5
 4    continue
      err=1.
      write(6,104)
 5    continue
100   format(1x,"this is an interactive data program",/
     c,1x,"define the channel perimeter in terms of a",/
     c,1x,"stage above a datum and two intersection points")
101   format(1x,"input number of triplets to be read i2")
102   format(i2)
103   format(1x,"input triplet number",2x,i2)
104   format(1x,"ar out of range")
       return
       end
```

APPENDIX II

Table for determining $F(u,N)$ for positive slopes*

u \ N	2.2	2.4	2.6	2.8	3.0	3.2	3.4	3.6	3.8	4.0
0.00	0.000	0.000	0.000	0.000	0.000	0.000	0.000	0.000	0.000	0.000
0.02	0.020	0.020	0.020	0.020	0.020	0.020	0.020	0.020	0.020	0.020
0.04	0.040	0.040	0.040	0.040	0.040	0.040	0.040	0.040	0.040	0.040
0.06	0.060	0.060	0.060	0.060	0.060	0.060	0.060	0.060	0.060	0.060
0.08	0.080	0.080	0.080	0.080	0.080	0.080	0.080	0.080	0.080	0.080
0.10	0.100	0.100	0.100	0.100	0.100	0.100	0.100	0.100	0.100	0.100
0.12	0.120	0.120	0.120	0.120	0.120	0.120	0.120	0.120	0.120	0.120
0.14	0.140	0.140	0.140	0.140	0.140	0.140	0.140	0.140	0.140	0.140
0.16	0.161	0.161	0.160	0.160	0.160	0.160	0.160	0.160	0.160	0.160
0.18	0.181	0.181	0.181	0.180	0.180	0.180	0.180	0.180	0.180	0.180
0.20	0.202	0.201	0.201	0.201	0.200	0.200	0.200	0.200	0.200	0.200
0.22	0.223	0.222	0.221	0.221	0.221	0.220	0.220	0.220	0.220	0.220
0.24	0.244	0.243	0.242	0.241	0.241	0.241	0.240	0.240	0.240	0.240
0.26	0.265	0.263	0.262	0.262	0.261	0.261	0.261	0.260	0.260	0.260
0.28	0.286	0.284	0.283	0.282	0.282	0.281	0.281	0.281	0.280	0.280
0.30	0.307	0.305	0.304	0.303	0.302	0.302	0.301	0.301	0.301	0.300
0.32	0.329	0.326	0.325	0.324	0.323	0.322	0.322	0.321	0.321	0.321
0.34	0.351	0.348	0.346	0.344	0.343	0.343	0.342	0.342	0.341	0.341
0.36	0.372	0.369	0.367	0.366	0.364	0.363	0.363	0.362	0.362	0.361
0.38	0.395	0.392	0.389	0.387	0.385	0.384	0.383	0.383	0.382	0.382

*This table is reproduced from V. T. Chow, "Integrating the Equation of Gradually Varied Flow," *Proceedings of the American Society of Civil Engineers*, vol. 81, November 1955, pp. 1–32.

u \ N	2.2	2.4	2.6	2.8	3.0	3.2	3.4	3.6	3.8	4.0
0.40	0.418	0.414	0.411	0.408	0.407	0.405	0.404	0.403	0.403	0.402
0.42	0.442	0.437	0.433	0.430	0.428	0.426	0.425	0.424	0.423	0.423
0.44	0.465	0.460	0.456	0.452	0.450	0.448	0.446	0.445	0.444	0.443
0.46	0.480	0.483	0.479	0.475	0.472	0.470	0.468	0.466	0.465	0.464
0.48	0.514	0.507	0.502	0.497	0.494	0.492	0.489	0.488	0.486	0.485
0.50	0.539	0.531	0.525	0.521	0.517	0.514	0.511	0.509	0.508	0.506
0.52	0.565	0.557	0.550	0.544	0.540	0.536	0.534	0.531	0.529	0.528
0.54	0.592	0.582	0.574	0.568	0.563	0.559	0.556	0.554	0.551	0.550
0.56	0.619	0.608	0.599	0.593	0.587	0.583	0.579	0.576	0.574	0.572
0.58	0.648	0.635	0.626	0.618	0.612	0.607	0.603	0.599	0.596	0.594
0.60	0.676	0.663	0.653	0.644	0.637	0.631	0.627	0.623	0.620	0.617
0.61	0.691	0.678	0.667	0.657	0.650	0.644	0.639	0.635	0.631	0.628
0.62	0.706	0.692	0.680	0.671	0.663	0.657	0.651	0.647	0.643	0.640
0.63	0.722	0.707	0.694	0.684	0.676	0.669	0.664	0.659	0.655	0.652
0.64	0.738	0.722	0.709	0.698	0.690	0.683	0.677	0.672	0.667	0.664
0.65	0.754	0.737	0.724	0.712	0.703	0.696	0.689	0.684	0.680	0.676
0.66	0.771	0.753	0.738	0.727	0.717	0.709	0.703	0.697	0.692	0.688
0.67	0.787	0.769	0.754	0.742	0.731	0.723	0.716	0.710	0.705	0.701
0.68	0.804	0.785	0.769	0.757	0.746	0.737	0.729	0.723	0.718	0.713
0.69	0.822	0.804	0.785	0.772	0.761	0.751	0.743	0.737	0.731	0.726
0.70	0.840	0.819	0.802	0.787	0.776	0.766	0.757	0.750	0.744	0.739
0.71	0.858	0.836	0.819	0.804	0.791	0.781	0.772	0.764	0.758	0.752
0.72	0.878	0.855	0.836	0.820	0.807	0.796	0.786	0.779	0.772	0.766
0.73	0.898	0.874	0.854	0.837	0.823	0.811	0.802	0.793	0.786	0.780
0.74	0.918	0.892	0.868	0.854	0.840	0.827	0.817	0.808	0.800	0.794
0.75	0.940	0.913	0.890	0.872	0.857	0.844	0.833	0.823	0.815	0.808
0.76	0.961	0.933	0.909	0.890	0.874	0.861	0.849	0.839	0.830	0.823
0.77	0.985	0.954	0.930	0.909	0.892	0.878	0.866	0.855	0.846	0.838
0.78	1.007	0.976	0.950	0.929	0.911	0.896	0.883	0.872	0.862	0.854
0.79	1.031	0.998	0.971	0.940	0.930	0.914	0.901	0.889	0.879	0.870
0.80	1.056	1.022	0.994	0.970	0.950	0.934	0.919	0.907	0.896	0.887
0.81	1.083	1.046	1.017	0.992	0.971	0.954	0.938	0.925	0.914	0.904
0.82	1.110	1.072	1.041	1.015	0.993	0.974	0.958	0.945	0.932	0.922
0.83	1.139	1.099	1.067	1.039	1.016	0.996	0.979	0.965	0.952	0.940
0.84	1.171	1.129	1.094	1.064	1.040	1.019	1.001	0.985	0.972	0.960
0.85	1.201	1.157	1.121	1.091	1.065	1.043	1.024	1.007	0.993	0.980
0.86	1.238	1.192	1.153	1.119	1.092	1.068	1.048	1.031	1.015	1.002
0.87	1.272	1.223	1.182	1.149	1.120	1.095	1.074	1.055	1.039	1.025
0.88	1.314	1.262	1.228	1.181	1.151	1.124	1.101	1.081	1.064	1.049
0.89	1.357	1.302	1.255	1.216	1.183	1.155	1.131	1.110	1.091	1.075

u \ N	2.2	2.4	2.6	2.8	3.0	3.2	3.4	3.6	3.8	4.0
0.90	1.401	1.343	1.294	1.253	1.218	1.189	1.163	1.140	1.120	1.103
0.91	1.452	1.389	1.338	1.294	1.257	1.225	1.197	1.173	1.152	1.133
0.92	1.505	1.438	1.351	1.340	1.300	1.266	1.236	1.210	1.187	1.166
0.93	1.564	1.493	1.435	1.391	1.348	1.311	1.279	1.251	1.226	1.204
0.94	1.645	1.568	1.504	1.449	1.403	1.363	1.328	1.297	1.270	1.246
0.950	1.737	1.652	1.582	1.518	1.467	1.423	1.385	1.352	1.322	1.296
0.960	1.833	1.741	1.665	1.601	1.545	1.497	1.454	1.417	1.385	1.355
0.970	1.969	1.866	1.780	1.707	1.644	1.590	1.543	1.501	1.464	1.431
0.975	2.055	1.945	1.853	1.773	1.707	1.649	1.598	1.554	1.514	1.479
0.980	2.164	2.045	1.946	1.855	1.783	1.720	1.666	1.617	1.575	1.536
0.985	2.294	2.165	2.056	1.959	1.880	1.812	1.752	1.699	1.652	1.610
0.990	2.477	2.333	2.212	2.106	2.017	1.940	1.873	1.814	1.761	1.714
0.995	2.792	2.621	2.478	2.355	2.250	2.159	2.070	2.008	1.945	1.889
0.999	3.523	3.292	3.097	2.931	2.788	2.663	2.554	2.457	2.370	2.293
1.000	∞	∞	∞	∞	∞	∞	∞	∞	∞	∞
1.001	3.317	2.931	2.640	2.399	2.184	2.008	1.856	1.725	1.610	1.508
1.005	2.587	2.266	2.022	1.818	1.679	1.506	1.384	1.279	1.188	1.107
1.010	2.273	1.977	1.757	1.572	1.419	1.291	1.182	1.089	1.007	0.936
1.015	2.090	1.807	1.602	1.428	1.286	1.166	1.065	0.978	0.902	0.836
1.020	1.961	1.711	1.493	1.327	1.191	1.078	0.982	0.900	0.828	0.766
1.03	1.779	1.531	1.340	1.186	1.060	0.955	0.866	0.790	0.725	0.668
1.04	1.651	1.410	1.232	1.086	0.967	0.868	0.785	0.714	0.653	0.600
1.05	1.552	1.334	1.150	1.010	0.896	0.802	0.723	0.656	0.598	0.548
1.06	1.472	1.250	1.082	0.948	0.838	0.748	0.672	0.608	0.553	0.506
1.07	1.404	1.195	1.026	0.896	0.790	0.703	0.630	0.569	0.516	0.471
1.08	1.346	1.139	0.978	0.851	0.749	0.665	0.595	0.535	0.485	0.441
1.09	1.295	1.089	0.935	0.812	0.713	0.631	0.563	0.506	0.457	0.415
1.10	1.250	1.050	0.897	0.777	0.681	0.601	0.536	0.480	0.433	0.392
1.11	1.209	1.014	0.864	0.746	0.652	0.575	0.511	0.457	0.411	0.372
1.12	1.172	0.981	0.833	0.718	0.626	0.551	0.488	0.436	0.392	0.354
1.13	1.138	0.950	0.805	0.692	0.602	0.529	0.468	0.417	0.374	0.337
1.14	1.107	0.921	0.780	0.669	0.581	0.509	0.450	0.400	0.358	0.322
1.15	1.078	0.892	0.756	0.647	0.561	0.490	0.432	0.384	0.343	0.308
1.16	1.052	0.870	0.734	0.627	0.542	0.473	0.417	0.369	0.329	0.295
1.17	1.027	0.850	0.713	0.608	0.525	0.458	0.402	0.356	0.317	0.283
1.18	1.003	0.825	0.694	0.591	0.509	0.443	0.388	0.343	0.305	0.272
1.19	0.981	0.810	0.676	0.574	0.494	0.429	0.375	0.331	0.294	0.262
1.20	0.960	0.787	0.659	0.559	0.480	0.416	0.363	0.320	0.283	0.252
1.22	0.922	0.755	0.628	0.531	0.454	0.392	0.341	0.299	0.264	0.235
1.24	0.887	0.725	0.600	0.505	0.431	0.371	0.322	0.281	0.248	0.219

u \ N	2.2	2.4	2.6	2.8	3.0	3.2	3.4	3.6	3.8	4.0
1.26	0.855	0.692	0.574	0.482	0.410	0.351	0.304	0.265	0.233	0.205
1.28	0.827	0.666	0.551	0.461	0.391	0.334	0.288	0.250	0.219	0.193
1.30	0.800	0.644	0.530	0.442	0.373	0.318	0.274	0.237	0.207	0.181
1.32	0.775	0.625	0.510	0.424	0.357	0.304	0.260	0.225	0.196	0.171
1.34	0.752	0.605	0.492	0.408	0.342	0.290	0.248	0.214	0.185	0.162
1.36	0.731	0.588	0.475	0.393	0.329	0.278	0.237	0.204	0.176	0.153
1.38	0.711	0.567	0.459	0.378	0.316	0.266	0.226	0.194	0.167	0.145
1.40	0.692	0.548	0.444	0.365	0.304	0.256	0.217	0.185	0.159	0.138
1.42	0.674	0.533	0.431	0.353	0.293	0.246	0.208	0.177	0.152	0.131
1.44	0.658	0.517	0.417	0.341	0.282	0.236	0.199	0.169	0.145	0.125
1.46	0.642	0.505	0.405	0.330	0.273	0.227	0.191	0.162	0.139	0.119
1.48	0.627	0.493	0.394	0.320	0.263	0.219	0.184	0.156	0.133	0.113
1.50	0.613	0.480	0.383	0.310	0.255	0.211	0.177	0.149	0.127	0.108
1.55	0.580	0.451	0.358	0.288	0.235	0.194	0.161	0.135	0.114	0.097
1.60	0.551	0.425	0.335	0.269	0.218	0.179	0.148	0.123	0.103	0.087
1.65	0.525	0.402	0.316	0.251	0.203	0.165	0.136	0.113	0.094	0.079
1.70	0.501	0.381	0.298	0.236	0.189	0.153	0.125	0.103	0.086	0.072
1.75	0.480	0.362	0.282	0.222	0.177	0.143	0.116	0.095	0.079	0.065
1.80	0.460	0.349	0.267	0.209	0.166	0.133	0.108	0.088	0.072	0.060
1.85	0.442	0.332	0.254	0.198	0.156	0.125	0.100	0.082	0.067	0.055
1.90	0.425	0.315	0.242	0.188	0.147	0.117	0.094	0.076	0.062	0.050
1.95	0.409	0.304	0.231	0.178	0.139	0.110	0.088	0.070	0.057	0.046
2.00	0.395	0.292	0.221	0.169	0.132	0.104	0.082	0.066	0.053	0.043
2.10	0.369	0.273	0.202	0.154	0.119	0.092	0.073	0.058	0.046	0.037
2.20	0.346	0.253	0.186	0.141	0.107	0.083	0.065	0.051	0.040	0.032
2.3	0.326	0.235	0.173	0.129	0.098	0.075	0.058	0.045	0.035	0.028
2.4	0.308	0.220	0.160	0.119	0.089	0.068	0.052	0.040	0.031	0.024
2.5	0.292	0.207	0.150	0.110	0.082	0.062	0.047	0.036	0.028	0.022
2.6	0.277	0.197	0.140	0.102	0.076	0.057	0.043	0.033	0.025	0.019
2.7	0.264	0.188	0.131	0.095	0.070	0.052	0.039	0.029	0.022	0.017
2.8	0.252	0.176	0.124	0.089	0.065	0.048	0.036	0.027	0.020	0.015
2.9	0.241	0.166	0.117	0.083	0.060	0.044	0.033	0.024	0.018	0.014
3.0	0.230	0.159	0.110	0.078	0.056	0.041	0.030	0.022	0.017	0.012
3.5	0.190	0.126	0.085	0.059	0.041	0.029	0.021	0.015	0.011	0.008
4.0	0.161	0.104	0.069	0.046	0.031	0.022	0.015	0.010	0.007	0.005
4.5	0.139	0.087	0.057	0.037	0.025	0.017	0.011	0.008	0.005	0.004
5.0	0.122	0.076	0.048	0.031	0.020	0.013	0.009	0.006	0.004	0.003
6.0	0.098	0.060	0.036	0.022	0.014	0.009	0.006	0.004	0.002	0.002
7.0	0.081	0.048	0.028	0.017	0.010	0.006	0.004	0.002	0.002	0.001
8.0	0.069	0.040	0.022	0.013	0.008	0.005	0.003	0.002	0.001	0.001
9.0	0.060	0.034	0.019	0.011	0.006	0.004	0.002	0.001	0.001	0.000
10.0	0.053	0.028	0.016	0.009	0.005	0.003	0.002	0.001	0.001	0.000
20.0	0.023	0.018	0.011	0.006	0.002	0.001	0.001	0.000	0.000	0.000

u \ N	4.2	4.6	5.0	5.4	5.8	6.2	6.6	7.0	7.4	7.8
0.00	0.000	0.000	0.000	0.000	0.000	0.000	0.000	0.000	0.000	0.000
0.02	0.020	0.020	0.020	0.020	0.020	0.020	0.020	0.020	0.020	0.020
0.04	0.040	0.040	0.040	0.040	0.040	0.040	0.040	0.040	0.040	0.040
0.06	0.060	0.060	0.060	0.060	0.060	0.060	0.060	0.060	0.060	0.060
0.08	0.080	0.080	0.080	0.080	0.080	0.080	0.080	0.080	0.080	0.080
0.10	0.100	0.100	0.100	0.100	0.100	0.100	0.100	0.100	0.100	0.100
0.12	0.120	0.120	0.120	0.120	0.120	0.120	0.120	0.120	0.120	0.120
0.14	0.140	0.140	0.140	0.140	0.140	0.140	0.140	0.140	0.140	0.140
0.16	0.160	0.160	0.160	0.160	0.160	0.160	0.160	0.160	0.160	0.160
0.18	0.180	0.180	0.180	0.180	0.180	0.180	0.180	0.180	0.180	0.180
0.20	0.200	0.200	0.200	0.200	0.200	0.200	0.200	0.200	0.200	0.200
0.22	0.220	0.220	0.220	0.220	0.220	0.220	0.220	0.220	0.220	0.220
0.24	0.240	0.240	0.240	0.240	0.240	0.240	0.240	0.240	0.240	0.240
0.26	0.260	0.260	0.260	0.260	0.260	0.260	0.260	0.260	0.260	0.260
0.28	0.280	0.280	0.280	0.280	0.280	0.280	0.280	0.280	0.280	0.280
0.30	0.300	0.300	0.300	0.300	0.300	0.300	0.300	0.300	0.300	0.300
0.32	0.321	0.320	0.320	0.320	0.320	0.320	0.320	0.320	0.320	0.320
0.34	0.341	0.340	0.340	0.340	0.340	0.340	0.340	0.340	0.340	0.340
0.36	0.361	0.361	0.360	0.360	0.360	0.360	0.360	0.360	0.360	0.360
0.38	0.381	0.381	0.381	0.380	0.380	0.380	0.380	0.380	0.380	0.380
0.40	0.402	0.401	0.401	0.400	0.400	0.400	0.400	0.400	0.400	0.400
0.42	0.422	0.421	0.421	0.421	0.420	0.420	0.420	0.420	0.420	0.420
0.44	0.443	0.442	0.441	0.441	0.441	0.441	0.440	0.440	0.440	0.440
0.46	0.463	0.462	0.462	0.461	0.461	0.461	0.460	0.460	0.460	0.460
0.48	0.484	0.483	0.482	0.481	0.481	0.481	0.480	0.480	0.480	0.480
0.50	0.505	0.504	0.503	0.502	0.501	0.501	0.501	0.500	0.500	0.500
0.52	0.527	0.525	0.523	0.522	0.522	0.521	0.521	0.521	0.520	0.520
0.54	0.548	0.546	0.544	0.543	0.542	0.542	0.541	0.541	0.541	0.541
0.56	0.570	0.567	0.565	0.564	0.563	0.562	0.562	0.561	0.561	0.561
0.58	0.592	0.589	0.587	0.585	0.583	0.583	0.582	0.582	0.581	0.581
0.60	0.614	0.611	0.608	0.606	0.605	0.604	0.603	0.602	0.602	0.601
0.61	0.626	0.622	0.619	0.617	0.615	0.614	0.613	0.612	0.612	0.611
0.62	0.637	0.633	0.630	0.628	0.626	0.625	0.624	0.623	0.622	0.622
0.63	0.649	0.644	0.641	0.638	0.636	0.635	0.634	0.633	0.632	0.632
0.64	0.661	0.656	0.652	0.649	0.647	0.646	0.645	0.644	0.643	0.642
0.65	0.673	0.667	0.663	0.660	0.658	0.656	0.655	0.654	0.653	0.653
0.66	0.685	0.679	0.675	0.672	0.669	0.667	0.666	0.665	0.664	0.663
0.67	0.697	0.691	0.686	0.683	0.680	0.678	0.676	0.675	0.674	0.673
0.68	0.709	0.703	0.698	0.694	0.691	0.689	0.687	0.686	0.685	0.684
0.69	0.722	0.715	0.710	0.706	0.703	0.700	0.698	0.696	0.695	0.694

u \ N	4.2	4.6	5.0	5.4	5.8	6.2	6.6	7.0	7.4	7.8
0.70	0.735	0.727	0.722	0.717	0.714	0.712	0.710	0.708	0.706	0.705
0.71	0.748	0.740	0.734	0.729	0.726	0.723	0.721	0.719	0.717	0.716
0.72	0.761	0.752	0.746	0.741	0.737	0.734	0.732	0.730	0.728	0.727
0.73	0.774	0.765	0.759	0.753	0.749	0.746	0.743	0.741	0.739	0.737
0.74	0.788	0.779	0.771	0.766	0.761	0.757	0.754	0.752	0.750	0.748
0.75	0.802	0.792	0.784	0.778	0.773	0.769	0.766	0.763	0.761	0.759
0.76	0.817	0.806	0.798	0.791	0.786	0.782	0.778	0.775	0.773	0.771
0.77	0.831	0.820	0.811	0.804	0.798	0.794	0.790	0.787	0.784	0.782
0.78	0.847	0.834	0.825	0.817	0.811	0.806	0.802	0.799	0.796	0.794
0.79	0.862	0.849	0.839	0.831	0.824	0.819	0.815	0.811	0.808	0.805
0.80	0.878	0.865	0.854	0.845	0.838	0.832	0.828	0.823	0.820	0.818
0.81	0.895	0.881	0.869	0.860	0.852	0.846	0.841	0.836	0.833	0.830
0.82	0.913	0.897	0.885	0.875	0.866	0.860	0.854	0.850	0.846	0.842
0.83	0.931	0.914	0.901	0.890	0.881	0.874	0.868	0.863	0.859	0.855
0.84	0.949	0.932	0.918	0.906	0.897	0.889	0.882	0.877	0.872	0.868
0.85	0.969	0.950	0.935	0.923	0.912	0.905	0.898	0.891	0.887	0.882
0.86	0.990	0.970	0.954	0.940	0.930	0.921	0.913	0.906	0.901	0.896
0.87	1.012	0.990	0.973	0.959	0.947	0.937	0.929	0.922	0.916	0.911
0.88	1.035	1.012	0.994	0.978	0.966	0.955	0.946	0.938	0.932	0.927
0.89	1.060	1.035	1.015	0.999	0.986	0.974	0.964	0.956	0.949	0.943
0.90	1.087	1.060	1.039	1.021	1.007	0.994	0.984	0.974	0.967	0.960
0.91	1.116	1.088	1.064	1.045	1.029	1.016	1.003	0.995	0.986	0.979
0.92	1.148	1.117	1.092	1.072	1.054	1.039	1.027	1.016	1.006	0.999
0.93	1.184	1.151	1.123	1.101	1.081	1.065	1.050	1.040	1.029	1.021
0.94	1.225	1.188	1.158	1.134	1.113	1.095	1.080	1.066	1.054	1.044
0.950	1.272	1.232	1.199	1.172	1.148	1.128	1.111	1.097	1.084	1.073
0.960	1.329	1.285	1.248	1.217	1.188	1.167	1.149	1.133	1.119	1.106
0.970	1.402	1.351	1.310	1.275	1.246	1.319	1.197	1.179	1.162	1.148
0.975	1.447	1.393	1.348	1.311	1.280	1.250	1.227	1.207	1.190	1.173
0.980	1.502	1.443	1.395	1.354	1.339	1.288	1.262	1.241	1.221	1.204
0.985	1.573	1.508	1.454	1.409	1.372	1.337	1.309	1.284	1.263	1.243
0.990	1.671	1.598	1.537	1.487	1.444	1.404	1.373	1.344	1.319	1.297
0.995	1.838	1.751	1.678	1.617	1.565	1.519	1.479	1.451	1.416	1.388
0.999	2.223	2.102	2.002	1.917	1.845	1.780	1.725	1.678	1.635	1.596
1.000	∞	∞	∞	∞	∞	∞	∞	∞	∞	∞
1.001	1.417	1.264	1.138	1.033	0.951	0.870	0.803	0.746	0.697	0.651
1.005	1.036	0.915	0.817	0.737	0.669	0.612	0.553	0.526	0.481	0.447
1.010	0.873	0.766	0.681	0.610	0.551	0.502	0.459	0.422	0.389	0.360
1.015	0.778	0.680	0.602	0.537	0.483	0.440	0.399	0.366	0.336	0.310
1.02	0.711	0.620	0.546	0.486	0.436	0.394	0.358	0.327	0.300	0.276

u \ N	4.2	4.6	5.0	5.4	5.8	6.2	6.6	7.0	7.4	7.8
1.03	0.618	0.535	0.469	0.415	0.370	0.333	0.300	0.272	0.249	0.228
1.04	0.554	0.477	0.415	0.365	0.324	0.290	0.262	0.236	0.214	0.195
1.05	0.504	0.432	0.374	0.328	0.289	0.259	0.231	0.208	0.189	0.174
1.06	0.464	0.396	0.342	0.298	0.262	0.233	0.209	0.187	0.170	0.154
1.07	0.431	0.366	0.315	0.273	0.239	0.212	0.191	0.168	0.151	0.136
1.08	0.403	0.341	0.292	0.252	0.220	0.194	0.172	0.153	0.137	0.123
1.09	0.379	0.319	0.272	0.234	0.204	0.179	0.158	0.140	0.125	0.112
1.10	0.357	0.299	0.254	0.218	0.189	0.165	0.146	0.129	0.114	0.102
1.11	0.338	0.282	0.239	0.204	0.176	0.154	0.135	0.119	0.105	0.094
1.12	0.321	0.267	0.225	0.192	0.165	0.143	0.125	0.110	0.097	0.086
1.13	0.305	0.253	0.212	0.181	0.155	0.135	0.117	0.102	0.090	0.080
1.14	0.291	0.240	0.201	0.170	0.146	0.126	0.109	0.095	0.084	0.074
1.15	0.278	0.229	0.191	0.161	0.137	0.118	0.102	0.089	0.078	0.068
1.16	0.266	0.218	0.181	0.153	0.130	0.111	0.096	0.084	0.072	0.064
1.17	0.255	0.208	0.173	0.145	0.123	0.105	0.090	0.078	0.068	0.060
1.18	0.244	0.199	0.165	0.138	0.116	0.099	0.085	0.073	0.063	0.055
1.19	0.235	0.191	0.157	0.131	0.110	0.094	0.080	0.068	0.059	0.051
1.20	0.226	0.183	0.150	0.215	0.105	0.088	0.076	0.064	0.056	0.048
1.22	0.209	0.168	0.138	0.114	0.095	0.080	0.068	0.057	0.049	0.042
1.24	0.195	0.156	0.127	0.104	0.086	0.072	0.060	0.051	0.044	0.038
1.26	0.182	0.145	0.117	0.095	0.079	0.065	0.055	0.046	0.039	0.033
1.28	0.170	0.135	0.108	0.088	0.072	0.060	0.050	0.041	0.035	0.030
1.30	0.160	0.126	0.100	0.081	0.066	0.054	0.045	0.037	0.031	0.026
1.32	0.150	0.118	0.093	0.075	0.061	0.050	0.041	0.034	0.028	0.024
1.34	0.142	0.110	0.087	0.069	0.056	0.045	0.037	0.030	0.025	0.021
1.36	0.134	0.103	0.081	0.064	0.052	0.042	0.034	0.028	0.023	0.019
1.38	0.127	0.097	0.076	0.060	0.048	0.038	0.032	0.026	0.021	0.017
1.40	0.120	0.092	0.071	0.056	0.044	0.036	0.028	0.023	0.019	0.016
1.42	0.114	0.087	0.067	0.052	0.041	0.033	0.026	0.021	0.017	0.014
1.44	0.108	0.082	0.063	0.049	0.038	0.030	0.024	0.019	0.016	0.013
1.46	0.103	0.077	0.059	0.046	0.036	0.028	0.022	0.018	0.014	0.012
1.48	0.098	0.073	0.056	0.043	0.033	0.026	0.021	0.017	0.013	0.010
1.50	0.093	0.069	0.053	0.040	0.031	0.024	0.020	0.015	0.012	0.009
1.55	0.083	0.061	0.046	0.035	0.026	0.020	0.016	0.012	0.010	0.008
1.60	0.074	0.054	0.040	0.030	0.023	0.017	0.013	0.010	0.008	0.006
1.65	0.067	0.048	0.035	0.026	0.019	0.014	0.011	0.008	0.006	0.005
1.70	0.060	0.043	0.031	0.023	0.016	0.012	0.009	0.007	0.005	0.004
1.75	0.054	0.038	0.027	0.020	0.014	0.010	0.008	0.006	0.004	0.003
1.80	0.049	0.034	0.024	0.017	0.012	0.009	0.007	0.005	0.004	0.003
1.85	0.045	0.031	0.022	0.015	0.011	0.008	0.006	0.004	0.003	0.002

u \ N	4.2	4.6	5.0	5.4	5.8	6.2	6.6	7.0	7.4	7.8
1.90	0.041	0.028	0.020	0.014	0.010	0.007	0.005	0.004	0.003	0.002
1.95	0.038	0.026	0.018	0.012	0.008	0.006	0.004	0.003	0.002	0.002
2.00	0.035	0.023	0.016	0.011	0.007	0.005	0.004	0.003	0.002	0.001
2.10	0.030	0.019	0.013	0.009	0.006	0.004	0.003	0.002	0.001	0.001
2.20	0.025	0.016	0.011	0.007	0.005	0.004	0.002	0.001	0.001	0.001
2.3	0.022	0.014	0.009	0.006	0.004	0.003	0.002	0.001	0.001	0.001
2.4	0.019	0.012	0.008	0.005	0.003	0.002	0.001	0.001	0.001	0.001
2.5	0.017	0.010	0.006	0.004	0.003	0.002	0.001	0.001	0.000	0.001
2.6	0.015	0.009	0.005	0.003	0.002	0.001	0.001	0.001	0.000	0.000
2.7	0.013	0.008	0.005	0.003	0.002	0.001	0.001	0.000	0.000	0.000
2.8	0.012	0.007	0.004	0.002	0.001	0.001	0.001	0.000	0.000	0.000
2.9	0.010	0.006	0.004	0.002	0.001	0.001	0.000	0.000	0.000	0.000
3.0	0.009	0.005	0.003	0.002	0.001	0.001	0.000	0.000	0.000	0.000
3.5	0.006	0.003	0.002	0.001	0.001	0.000	0.000	0.000	0.000	0.000
4.0	0.004	0.002	0.001	0.000	0.000	0.000	0.000	0.000	0.000	0.000
4.5	0.003	0.001	0.001	0.000	0.000	0.000	0.000	0.000	0.000	0.000
5.0	0.002	0.001	0.000	0.000	0.000	0.000	0.000	0.000	0.000	0.000
6.0	0.001	0.000	0.000	0.000	0.000	0.000	0.000	0.000	0.000	0.000
7.0	0.001	0.000	0.000	0.000	0.000	0.000	0.000	0.000	0.000	0.000
8.0	0.000	0.000	0.000	0.000	0.000	0.000	0.000	0.000	0.000	0.000
9.0	0.000	0.000	0.000	0.000	0.000	0.000	0.000	0.000	0.000	0.000
10.0	0.000	0.000	0.000	0.000	0.000	0.000	0.000	0.000	0.000	0.000
20.0	0.000	0.000	0.000	0.000	0.000	0.000	0.000	0.000	0.000	0.000

u \ N	8.2	8.6	9.0	9.4	9.8
0.00	0.000	0.000	0.000	0.000	0.000
0.02	0.020	0.020	0.020	0.020	0.020
0.04	0.040	0.040	0.040	0.040	0.040
0.06	0.060	0.060	0.060	0.060	0.060
0.08	0.080	0.080	0.080	0.080	0.080
0.10	0.100	0.100	0.100	0.100	0.100
0.12	0.120	0.120	0.120	0.120	0.120
0.14	0.140	0.140	0.140	0.140	0.140
0.16	0.160	0.160	0.160	0.160	0.160
0.18	0.180	0.180	0.180	0.180	0.180
0.20	0.200	0.200	0.200	0.200	0.200
0.22	0.220	0.220	0.220	0.200	0.220
0.24	0.240	0.240	0.240	0.240	0.240
0.26	0.260	0.260	0.260	0.260	0.260
0.28	0.280	0.280	0.280	0.280	0.280
0.30	0.300	0.300	0.300	0.300	0.300
0.32	0.320	0.320	0.320	0.320	0.320
0.34	0.340	0.340	0.340	0.340	0.340
0.36	0.360	0.360	0.360	0.360	0.360
0.38	0.380	0.380	0.380	0.380	0.380
0.40	0.400	0.400	0.400	0.400	0.400
0.42	0.420	0.420	0.420	0.420	0.420
0.44	0.440	0.440	0.440	0.440	0.440
0.46	0.460	0.460	0.460	0.460	0.460
0.48	0.480	0.480	0.480	0.480	0.480
0.50	0.500	0.500	0.500	0.500	0.500
0.52	0.520	0.520	0.520	0.520	0.520
0.54	0.540	0.540	0.540	0.540	0.540
0.56	0.561	0.560	0.560	0.560	0.560
0.58	0.581	0.581	0.580	0.580	0.580
0.60	0.601	0.601	0.601	0.600	0.600
0.61	0.611	0.611	0.611	0.611	0.610
0.62	0.621	0.621	0.621	0.621	0.621
0.63	0.632	0.631	0.631	0.631	0.631
0.64	0.642	0.641	0.641	0.641	0.641
0.65	0.652	0.652	0.651	0.651	0.651
0.66	0.662	0.662	0.662	0.661	0.661
0.67	0.673	0.672	0.672	0.672	0.671
0.68	0.683	0.683	0.682	0.682	0.681
0.69	0.694	0.693	0.692	0.692	0.692

u \ N	8.2	8.6	9.0	9.4	9.8
0.70	0.704	0.704	0.703	0.702	0.702
0.71	0.715	0.714	0.713	0.713	0.712
0.72	0.726	0.725	0.724	0.723	0.723
0.73	0.736	0.735	0.734	0.734	0.733
0.74	0.747	0.746	0.745	0.744	0.744
0.75	0.758	0.757	0.756	0.755	0.754
0.76	0.769	0.768	0.767	0.766	0.765
0.77	0.780	0.779	0.778	0.777	0.776
0.78	0.792	0.790	0.789	0.788	0.787
0.79	0.804	0.802	0.800	0.799	0.798
0.80	0.815	0.813	0.811	0.810	0.809
0.81	0.827	0.825	0.823	0.822	0.820
0.82	0.839	0.837	0.835	0.833	0.831
0.83	0.852	0.849	0.847	0.845	0.844
0.84	0.856	0.862	0.860	0.858	0.856
0.85	0.878	0.875	0.873	0.870	0.868
0.86	0.892	0.889	0.886	0.883	0.881
0.87	0.907	0.903	0.900	0.897	0.894
0.88	0.921	0.918	0.914	0.911	0.908
0.89	0.937	0.933	0.929	0.925	0.922
0.90	0.954	0.949	0.944	0.940	0.937
0.91	0.972	0.967	0.961	0.957	0.953
0.92	0.991	0.986	0.980	0.975	0.970
0.93	1.012	1.006	0.999	0.994	0.989
0.94	1.036	1.029	1.022	1.016	1.010
0.950	1.062	1.055	1.047	1.040	1.033
0.960	1.097	1.085	1.074	1.063	1.053
0.970	1.136	1.124	1.112	1.100	1.087
0.975	1.157	1.147	1.134	1.122	1.108
0.980	1.187	1.175	1.160	1.150	1.132
0.985	1.224	1.210	1.196	1.183	1.165
0.990	1.275	1.260	1.243	1.228	1.208
0.995	1.363	1.342	1.320	1.302	1.280
0.999	1.560	1.530	1.500	1.476	1.447
1.000	∞	∞	∞	∞	∞
1.001	0.614	0.577	0.546	0.519	0.494
1.005	0.420	0.391	0.368	0.350	0.331
1.010	0.337	0.313	0.294	0.278	0.262
1.015	0.289	0.269	0.255	0.237	0.223
1.020	0.257	0.237	0.221	0.209	0.196

u \ N	8.2	8.6	9.0	9.4	9.8
1.03	0.212	0.195	0.181	0.170	0.159
1.04	0.173	0.165	0.152	0.143	0.134
1.05	0.158	0.143	0.132	0.124	0.115
1.06	0.140	0.127	0.116	0.106	0.098
1.07	0.123	0.112	0.102	0.094	0.086
1.08	0.111	0.101	0.092	0.084	0.077
1.09	0.101	0.091	0.082	0.075	0.069
1.10	0.092	0.083	0.074	0.067	0.062
1.11	0.084	0.075	0.067	0.060	0.055
1.12	0.077	0.069	0.062	0.055	0.050
1.13	0.071	0.063	0.056	0.050	0.045
1.14	0.065	0.058	0.052	0.046	0.041
1.15	0.061	0.054	0.048	0.043	0.038
1.16	0.056	0.050	0.045	0.040	0.035
1.17	0.052	0.046	0.041	0.036	0.032
1.18	0.048	0.042	0.037	0.033	0.029
1.19	0.045	0.039	0.034	0.030	0.027
1.20	0.043	0.037	0.032	0.028	0.025
1.22	0.037	0.032	0.028	0.024	0.021
1.24	0.032	0.028	0.024	0.021	0.018
1.26	0.028	0.024	0.021	0.018	0.016
1.28	0.025	0.021	0.018	0.016	0.014
1.30	0.022	0.019	0.016	0.014	0.012
1.32	0.020	0.017	0.014	0.012	0.010
1.34	0.018	0.015	0.012	0.010	0.009
1.36	0.016	0.013	0.011	0.009	0.008
1.38	0.014	0.012	0.010	0.008	0.007
1.40	0.013	0.011	0.009	0.007	0.006
1.42	0.011	0.009	0.008	0.006	0.005
1.44	0.010	0.008	0.007	0.006	0.005
1.46	0.009	0.008	0.006	0.005	0.004
1.48	0.009	0.007	0.005	0.004	0.004
1.50	0.008	0.006	0.005	0.004	0.003
1.55	0.006	0.005	0.004	0.003	0.003
1.60	0.005	0.004	0.003	0.002	0.002
1.65	0.004	0.003	0.002	0.002	0.001
1.70	0.003	0.002	0.002	0.001	0.001
1.75	0.002	0.002	0.002	0.001	0.001
1.80	0.002	0.001	0.001	0.001	0.001
1.85	0.002	0.001	0.001	0.001	0.001

u \ N	8.2	8.6	9.0	9.4	9.8
1.90	0.001	0.001	0.001	0.001	0.000
1.95	0.001	0.001	0.001	0.000	0.000
2.00	0.001	0.001	0.000	0.000	0.000
2.10	0.001	0.000	0.000	0.000	0.000
2.20	0.000	0.000	0.000	0.000	0.000
2.3	0.000	0.000	0.000	0.000	0.000
2.4	0.000	0.000	0.000	0.000	0.000
2.5	0.000	0.000	0.000	0.000	0.000
2.6	0.000	0.000	0.000	0.000	0.000
2.7	0.000	0.000	0.000	0.000	0.000
2.8	0.000	0.000	0.000	0.000	0.000
2.9	0.000	0.000	0.000	0.000	0.000
3.0	0.000	0.000	0.000	0.000	0.000
3.5	0.000	0.000	0.000	0.000	0.000
4.0	0.000	0.000	0.000	0.000	0.000
4.5	0.000	0.000	0.000	0.000	0.000
5.0	0.000	0.000	0.000	0.000	0.000
6.0	0.000	0.000	0.000	0.000	0.000
7.0	0.000	0.000	0.000	0.000	0.000
8.0	0.000	0.000	0.000	0.000	0.000
9.0	0.000	0.000	0.000	0.000	0.000
10.0	0.000	0.000	0.000	0.000	0.000
20.0	0.000	0.000	0.000	0.000	0.000

APPENDIX III

Table for determining $F(u,N)$ for negative slopes*

u \ N	2.0	2.2	2.4	2.6	2.8	3.0	3.2	3.4	3.6	3.8
0.00	0.000	0.000	0.000	0.000	0.000	0.000	0.000	0.000	0.000	0.000
0.02	0.020	0.020	0.020	0.020	0.020	0.020	0.020	0.020	0.020	0.020
0.04	0.040	0.040	0.040	0.040	0.040	0.040	0.040	0.040	0.040	0.040
0.06	0.060	0.060	0.060	0.060	0.060	0.060	0.060	0.060	0.060	0.060
0.08	0.080	0.080	0.080	0.080	0.080	0.080	0.080	0.080	0.080	0.080
0.10	0.099	0.100	0.100	0.100	0.100	0.100	0.100	0.100	0.100	0.100
0.12	0.199	0.119	0.120	0.120	0.120	0.120	0.120	0.120	0.120	0.120
0.14	0.139	0.139	0.140	0.140	0.140	0.140	0.140	0.140	0.140	0.140
0.16	0.158	0.159	0.159	0.160	0.160	0.160	0.160	0.160	0.160	0.160
0.18	0.178	0.179	0.179	0.180	0.180	0.180	0.180	0.180	0.180	0.180
0.20	0.197	0.198	0.199	0.199	0.200	0.200	0.200	0.200	0.200	0.200
0.22	0.216	0.217	0.218	0.219	0.219	0.220	0.220	0.220	0.220	0.220
0.24	0.234	0.236	0.237	0.238	0.239	0.240	0.240	0.240	0.240	0.240
0.26	0.253	0.255	0.256	0.257	0.258	0.259	0.259	0.260	0.260	0.260
0.28	0.272	0.274	0.275	0.276	0.277	0.278	0.278	0.279	0.280	0.280
0.30	0.291	0.293	0.294	0.295	0.296	0.297	0.298	0.298	0.299	0.299
0.32	0.308	0.311	0.313	0.314	0.316	0.317	0.318	0.318	0.319	0.319
0.34	0.326	0.329	0.331	0.333	0.335	0.337	0.338	0.338	0.339	0.339
0.36	0.344	0.347	0.350	0.352	0.354	0.356	0.357	0.357	0.358	0.358
0.38	0.362	0.355	0.368	0.371	0.373	0.374	0.375	0.376	0.377	0.377

*This table is reproduced from V. T. Chow, Closure to "Integrating the Equation of Gradually Varied Flow," *Proceedings of the American Society of Civil Engineers, Journal of the Hydraulics Division*, vol. 83, no. HY 1, February 1957, pp. 9–22.

u \ N	2.0	2.2	2.4	2.6	2.8	3.0	3.2	3.4	3.6	3.8
0.40	0.380	0.384	0.387	0.390	0.392	0.393	0.394	0.395	0.396	0.396
0.42	0.397	0.401	0.405	0.407	0.409	0.411	0.412	0.413	0.414	0.414
0.44	0.414	0.419	0.423	0.426	0.429	0.430	0.432	0.433	0.434	0.435
0.46	0.431	0.437	0.440	0.444	0.447	0.449	0.451	0.452	0.453	0.454
0.48	0.447	0.453	0.458	0.461	0.464	0.467	0.469	0.471	0.472	0.473
0.50	0.463	0.470	0.475	0.479	0.482	0.485	0.487	0.489	0.491	0.492
0.52	0.479	0.485	0.491	0.494	0.499	0.502	0.505	0.507	0.509	0.511
0.54	0.494	0.501	0.507	0.512	0.516	0.520	0.522	0.525	0.527	0.529
0.56	0.509	0.517	0.523	0.528	0.533	0.537	0.540	0.543	0.545	0.547
0.58	0.524	0.533	0.539	0.545	0.550	0.554	0.558	0.561	0.563	0.567
0.60	0.540	0.548	0.555	0.561	0.566	0.571	0.575	0.578	0.581	0.583
0.61	0.547	0.556	0.563	0.569	0.575	0.579	0.583	0.587	0.589	0.592
0.62	0.554	0.563	0.571	0.578	0.583	0.578	0.591	0.595	0.598	0.600
0.63	0.562	0.571	0.579	0.585	0.590	0.595	0.599	0.603	0.607	0.609
0.64	0.569	0.579	0.586	0.592	0.598	0.602	0.607	0.611	0.615	0.618
0.65	0.576	0.585	0.592	0.599	0.606	0.610	0.615	0.619	0.623	0.626
0.66	0.583	0.593	0.600	0.607	0.613	0.618	0.622	0.626	0.630	0.634
0.67	0.590	0.599	0.607	0.614	0.621	0.626	0.631	0.635	0.639	0.643
0.68	0.597	0.607	0.615	0.622	0.628	0.634	0.639	0.643	0.647	0.651
0.69	0.603	0.613	0.621	0.629	0.635	0.641	0.646	0.651	0.655	0.659
0.70	0.610	0.620	0.629	0.637	0.644	0.649	0.654	0.659	0.663	0.667
0.71	0.617	0.627	0.636	0.644	0.651	0.657	0.661	0.666	0.671	0.674
0.72	0.624	0.634	0.643	0.651	0.658	0.664	0.669	0.674	0.679	0.682
0.73	0.630	0.641	0.650	0.659	0.665	0.672	0.677	0.682	0.687	0.691
0.74	0.637	0.648	0.657	0.665	0.672	0.679	0.684	0.689	0.694	0.698
0.75	0.643	0.655	0.664	0.671	0.679	0.686	0.691	0.696	0.701	0.705
0.76	0.649	0.661	0.670	0.679	0.687	0.693	0.699	0.704	0.709	0.713
0.77	0.656	0.667	0.677	0.685	0.693	0.700	0.705	0.711	0.715	0.719
0.78	0.662	0.673	0.683	0.692	0.700	0.707	0.713	0.718	0.723	0.727
0.79	0.668	0.680	0.689	0.698	0.705	0.713	0.719	0.724	0.729	0.733
0.80	0.674	0.685	0.695	0.703	0.712	0.720	0.726	0.732	0.737	0.741
0.81	0.680	0.691	0.701	0.710	0.719	0.727	0.733	0.739	0.744	0.749
0.82	0.686	0.698	0.707	0.717	0.725	0.733	0.740	0.745	0.751	0.755
0.83	0.692	0.703	0.713	0.722	0.731	0.740	0.746	0.752	0.757	0.762
0.84	0.698	0.709	0.719	0.729	0.737	0.746	0.752	0.758	0.764	0.769
0.85	0.704	0.715	0.725	0.735	0.744	0.752	0.759	0.765	0.770	0.775
0.86	0.710	0.721	0.731	0.741	0.750	0.758	0.765	0.771	0.777	0.782
0.87	0.715	0.727	0.738	0.747	0.756	0.764	0.771	0.777	0.783	0.788
0.88	0.721	0.733	0.743	0.753	0.762	0.770	0.777	0.783	0.789	0.794
0.89	0.727	0.739	0.749	0.758	0.767	0.776	0.783	0.789	0.795	0.800

u \ N	2.0	2.2	2.4	2.6	2.8	3.0	3.2	3.4	3.6	3.8
0.90	0.732	0.744	0.754	0.764	0.773	0.781	0.789	0.795	0.801	0.807
0.91	0.738	0.750	0.760	0.770	0.779	0.787	0.795	0.801	0.807	0.812
0.92	0.743	0.754	0.766	0.776	0.785	0.793	0.800	0.807	0.813	0.818
0.93	0.749	0.761	0.772	0.782	0.791	0.799	0.807	0.812	0.818	0.823
0.94	0.754	0.767	0.777	0.787	0.795	0.804	0.813	0.813	0.824	0.829
0.950	0.759	0.772	0.783	0.793	0.801	0.809	0.819	0.823	0.829	0.835
0.960	0.764	0.777	0.788	0.798	0.807	0.815	0.824	0.829	0.835	0.841
0.970	0.770	0.782	0.793	0.803	0.812	0.820	0.826	0.834	0.840	0.846
0.975	0.772	0.785	0.796	0.805	0.814	0.822	0.828	0.836	0.843	0.848
0.980	0.775	0.787	0.798	0.808	0.818	0.825	0.830	0.839	0.845	0.851
0.985	0.777	0.790	0.801	0.811	0.820	0.827	0.833	0.841	0.847	0.853
0.990	0.780	0.793	0.804	0.814	0.822	0.830	0.837	0.844	0.850	0.856
0.995	0.782	0.795	0.806	0.816	0.824	0.832	0.840	0.847	0.753	0.859
1.000	0.785	0.797	0.808	0.818	0.826	0.834	0.842	0.849	0.856	0.862
1.005	0.788	0.799	0.810	0.820	0.829	0.837	0.845	0.852	0.858	0.864
1.010	0.790	0.801	0.812	0.822	0.831	0.840	0.847	0.855	0.861	0.867
1.015	0.793	0.804	0.815	0.824	0.833	0.843	0.850	0.858	0.864	0.870
1.020	0.795	0.807	0.818	0.828	0.837	0.845	0.853	0.860	0.866	0.872
1.03	0.800	0.811	0.822	0.832	0.841	0.850	0.857	0.864	0.871	0.877
1.04	0.805	0.816	0.829	0.837	0.846	0.855	0.862	0.870	0.877	0.883
1.05	0.810	0.821	0.831	0.841	0.851	0.859	0.867	0.874	0.881	0.887
1.06	0.815	0.826	0.837	0.846	0.855	0.864	0.871	0.879	0.885	0.891
1.07	0.819	0.831	0.841	0.851	0.860	0.869	0.876	0.883	0.889	0.896
1.08	0.824	0.836	0.846	0.856	0.865	0.873	0.880	0.887	0.893	0.900
1.09	0.828	0.840	0.851	0.860	0.870	0.877	0.885	0.892	0.898	0.904
1.10	0.833	0.845	0.855	0.865	0.874	0.881	0.890	0.897	0.903	0.908
1.11	0.837	0.849	0.860	0.870	0.878	0.886	0.894	0.900	0.907	0.912
1.12	0.842	0.854	0.864	0.873	0.882	0.891	0.897	0.904	0.910	0.916
1.13	0.846	0.858	0.868	0.878	0.886	0.895	0.902	0.908	0.914	0.919
1.14	0.851	0.861	0.872	0.881	0.890	0.899	0.905	0.912	0.918	0.923
1.15	0.855	0.866	0.876	0.886	0.895	0.903	0.910	0.916	0.922	0.928
1.16	0.859	0.870	0.880	0.890	0.899	0.907	0.914	0.920	0.926	0.931
1.17	0.864	0.874	0.884	0.893	0.902	0.911	0.917	0.923	0.930	0.934
1.18	0.868	0.878	0.888	0.897	0.906	0.915	0.921	0.927	0.933	0.939
1.19	0.872	0.882	0.892	0.901	0.910	0.918	0.925	0.931	0.937	0.942
1.20	0.876	0.886	0.896	0.904	0.913	0.921	0.928	0.934	0.940	0.945
1.22	0.880	0.891	0.900	0.909	0.917	0.929	0.932	0.938	0.944	0.949
1.24	0.888	0.898	0.908	0.917	0.925	0.935	0.940	0.945	0.950	0.955
1.26	0.900	0.910	0.919	0.927	0.935	0.942	0.948	0.954	0.960	0.964
1.28	0.908	0.917	0.926	0.934	0.945	0.948	0.954	0.960	0.965	0.970

u \ N	2.0	2.2	2.4	2.6	2.8	3.0	3.2	3.4	3.6	3.8
1.30	0.915	0.925	0.933	0.941	0.948	0.955	0.961	0.966	0.981	0.975
1.32	0.922	0.931	0.940	0.948	0.955	0.961	0.967	0.972	0.976	0.980
1.34	0.930	0.939	0.948	0.955	0.962	0.967	0.973	0.978	0.982	0.986
1.36	0.937	0.946	0.954	0.961	0.968	0.973	0.979	0.983	0.987	0.991
1.38	0.944	0.952	0.960	0.967	0.974	0.979	0.985	0.989	0.993	0.996
1.40	0.951	0.959	0.966	0.973	0.979	0.984	0.989	0.993	0.997	1.000
1.42	0.957	0.965	0.972	0.979	0.984	0.989	0.995	0.998	1.001	1.004
1.44	0.964	0.972	0.979	0.984	0.990	0.995	1.000	1.003	1.006	1.009
1.46	0.970	0.977	0.983	0.989	0.995	1.000	1.004	1.007	1.010	1.012
1.48	0.977	0.983	0.989	0.994	0.999	1.005	1.008	1.011	1.014	1.016
1.50	0.983	0.990	0.996	1.001	1.005	1.009	1.012	1.015	1.017	1.019
1.55	0.997	1.002	1.007	1.012	1.016	1.020	1.022	1.024	1.026	1.028
1.60	1.012	1.017	1.020	1.024	1.027	1.030	1.032	1.034	1.035	1.035
1.65	1.026	1.029	1.032	1.035	1.037	1.039	1.041	1.041	1.042	1.042
1.70	1.039	1.042	1.044	1.045	1.047	1.048	1.049	1.049	1.049	1.048
1.75	1.052	1.053	1.054	1.055	1.056	1.057	1.056	1.056	1.055	1.053
1.80	1.064	1.064	1.064	1.064	1.065	1.065	1.064	1.062	1.060	1.058
1.85	1.075	1.074	1.074	1.073	1.072	1.071	1.069	1.067	1.066	1.063
1.90	1.086	1.085	1.084	1.082	1.081	1.079	1.077	1.074	1.071	1.066
1.95	1.097	1.095	1.092	1.090	1.087	1.085	1.081	1.079	1.075	1.071
2.00	1.107	1.103	1.100	1.096	1.093	1.090	1.085	1.082	1.078	1.075
2.10	1.126	1.120	1.115	1.110	1.104	1.100	1.094	1.089	1.085	1.080
2.20	1.144	1.136	1.129	1.122	1.115	1.109	1.102	1.096	1.090	1.085
2.3	1.161	1.150	1.141	1.133	1.124	1.117	1.110	1.103	1.097	1.090
2.4	1.176	1.163	1.152	1.142	1.133	1.124	1.116	1.109	1.101	1.094
2.5	1.190	1.175	1.162	1.150	1.140	1.131	1.121	1.113	1.105	1.098
2.6	1.204	1.187	1.172	1.159	1.147	1.137	1.126	1.117	1.106	1.000
2.7	1.216	1.196	1.180	1.166	1.153	1.142	1.130	1.120	1.110	1.102
2.8	1.228	1.208	1.189	1.173	1.158	1.146	1.132	1.122	1.112	1.103
2.9	1.239	1.216	1.196	1.178	1.162	1.150	1.137	1.125	1.115	1.106
3.0	1.249	1.224	1.203	1.184	1.168	1.154	1.140	1.128	1.117	1.107
3.5	1.292	1.260	1.232	1.206	1.185	1.167	1.151	1.138	1.125	1.113
4.0	1.326	1.286	1.251	1.223	1.198	1.176	1.158	1.142	1.129	1.117
4.5	1.352	1.308	1.270	1.235	1.205	1.183	1.162	1.146	1.131	1.119
5.0	1.374	1.325	1.283	1.245	1.212	1.188	1.166	1.149	1.134	1.121
6.0	1.406	1.342	1.292	1.252	1.221	1.195	1.171	1.152	1.136	1.122
7.0	1.430	1.360	1.303	1.260	1.225	1.199	1.174	1.153	1.136	1.122
8.0	1.447	1.373	1.313	1.266	1.229	1.201	1.175	1.154	1.137	1.122
9.0	1.461	1.384	1.319	1.269	1.231	1.203	1.176	1.156	1.137	1.122
10.0	1.471	1.394	1.324	1.272	1.233	1.203	1.176	1.156	1.137	1.122

u \ N	4.0	4.2	4.5	5.0	5.5
0.00	0.000	0.000	0.000	0.000	0.000
0.02	0.020	0.020	0.020	0.020	0.020
0.04	0.040	0.040	0.040	0.040	0.040
0.06	0.060	0.060	0.060	0.060	0.060
0.08	0.080	0.080	0.080	0.080	0.080
0.10	0.100	0.100	0.100	0.100	0.100
0.12	0.120	0.120	0.120	0.120	0.120
0.14	0.140	0.140	0.140	0.140	0.140
0.16	0.160	0.160	0.160	0.160	0.160
0.18	0.180	0.180	0.180	0.180	0.180
0.20	0.200	0.200	0.200	0.200	0.200
0.22	0.220	0.220	0.220	0.220	0.220
0.24	0.240	0.240	0.240	0.240	0.240
0.26	0.260	0.260	0.260	0.260	0.260
0.28	0.280	0.280	0.280	0.280	0.280
0.30	0.300	0.300	0.300	0.300	0.300
0.32	0.320	0.320	0.320	0.320	0.320
0.34	0.339	0.340	0.340	0.340	0.340
0.36	0.359	0.360	0.360	0.360	0.360
0.38	0.378	0.379	0.380	0.380	0.380
0.40	0.397	0.398	0.398	0.400	0.400
0.42	0.417	0.418	0.418	0.419	0.420
0.44	0.436	0.437	0.437	0.439	0.440
0.46	0.455	0.456	0.457	0.458	0.459
0.48	0.474	0.475	0.476	0.478	0.479
0.50	0.493	0.494	0.495	0.497	0.498
0.52	0.512	0.513	0.515	0.517	0.518
0.54	0.531	0.532	0.533	0.536	0.537
0.56	0.549	0.550	0.552	0.555	0.558
0.58	0.567	0.569	0.570	0.574	0.576
0.60	0.585	0.587	0.589	0.593	0.595
0.61	0.594	0.596	0.598	0.602	0.604
0.62	0.603	0.605	0.607	0.611	0.613
0.63	0.612	0.615	0.616	0.620	0.622
0.64	0.620	0.623	0.625	0.629	0.631
0.65	0.629	0.632	0.634	0.638	0.640
0.66	0.637	0.640	0.643	0.647	0.650
0.67	0.646	0.649	0.652	0.656	0.659
0.68	0.654	0.657	0.660	0.665	0.668
0.69	0.662	0.665	0.668	0.674	0.677

u \ N	4.0	4.2	4.5	5.0	5.5
0.70	0.670	0.673	0.677	0.682	0.686
0.71	0.678	0.681	0.685	0.690	0.694
0.72	0.686	0.689	0.694	0.699	0.703
0.73	0.694	0.698	0.702	0.707	0.712
0.74	0.702	0.705	0.710	0.716	0.720
0.75	0.709	0.712	0.717	0.724	0.728
0.76	0.717	0.720	0.725	0.731	0.736
0.77	0.724	0.727	0.733	0.739	0.744
0.78	0.731	0.735	0.740	0.747	0.752
0.79	0.738	0.742	0.748	0.754	0.760
0.80	0.746	0.750	0.755	0.762	0.768
0.81	0.753	0.757	0.762	0.770	0.776
0.82	0.760	0.764	0.769	0.777	0.783
0.83	0.766	0.771	0.776	0.784	0.790
0.84	0.773	0.778	0.783	0.791	0.798
0.85	0.780	0.784	0.790	0.798	0.805
0.86	0.786	0.791	0.797	0.804	0.812
0.87	0.793	0.797	0.803	0.811	0.819
0.88	0.799	0.803	0.810	0.818	0.826
0.89	0.805	0.810	0.816	0.825	0.832
0.90	0.811	0.816	0.822	0.831	0.839
0.91	0.817	0.821	0.828	0.837	0.845
0.92	0.823	0.828	0.834	0.844	0.851
0.93	0.829	0.833	0.840	0.850	0.857
0.94	0.835	0.840	0.846	0.856	0.864
0.950	0.840	0.845	0.852	0.861	0.869
0.960	0.846	0.861	0.857	0.867	0.875
0.970	0.851	0.866	0.863	0.972	0.881
0.975	0.854	0.859	0.866	0.875	0.883
0.980	0.857	0.861	0.868	0.878	0.886
0.985	0.859	0.863	0.870	0.880	0.889
0.990	0.861	0.867	0.873	0.883	0.891
0.995	0.864	0.869	0.876	0.885	0.894
1.000	0.867	0.873	0.879	0.887	0.897
1.005	0.870	0.874	0.881	0.890	0.899
1.010	0.873	0.878	0.884	0.893	0.902
1.015	0.875	0.880	0.886	0.896	0.904
1.020	0.877	0.883	0.889	0.898	0.907
1.03	0.882	0.887	0.893	0.902	0.911
1.04	0.888	0.893	0.898	0.907	0.916

u \ N	4.0	4.2	4.5	5.0	5.5
1.05	0.892	0.897	0.903	0.911	0.920
1.06	0.896	0.901	0.907	0.915	0.924
1.07	0.901	0.906	0.911	0.919	0.928
1.08	0.905	0.910	0.916	0.923	0.932
1.09	0.909	0.914	0.920	0.927	0.936
1.10	0.913	0.918	0.923	0.931	0.940
1.11	0.917	0.921	0.927	0.935	0.944
1.12	0.921	0.926	0.931	0.939	0.948
1.13	0.925	0.929	0.935	0.943	0.951
1.14	0.928	0.933	0.938	0.947	0.954
1.15	0.932	0.936	0.942	0.950	0.957
1.16	0.936	0.941	0.945	0.953	0.960
1.17	0.939	0.944	0.948	0.957	0.963
1.18	0.943	0.947	0.951	0.960	0.965
1.19	0.947	0.950	0.954	0.963	0.968
1.20	0.950	0.953	0.958	0.966	0.970
1.22	0.956	0.957	0.964	0.972	0.976
1.24	0.962	0.962	0.970	0.977	0.981
1.26	0.968	0.971	0.975	0.982	0.986
1.28	0.974	0.977	0.981	0.987	0.990
1.30	0.979	0.978	0.985	0.991	0.994
1.32	0.985	0.986	0.990	0.995	0.997
1.34	0.990	0.992	0.995	0.999	1.001
1.36	0.994	0.996	0.999	1.002	1.005
1.38	0.998	1.000	1.003	1.006	1.008
1.40	1.001	1.004	1.006	1.009	1.011
1.42	1.005	1.008	1.010	1.012	1.014
1.44	1.009	1.013	1.014	1.016	1.016
1.46	1.014	1.016	1.017	1.018	1.018
1.48	1.016	1.019	1.020	1.020	1.020
1.50	1.020	1.021	1.022	1.022	1.022
1.55	1.029	1.029	1.029	1.028	1.028
1.60	1.035	1.035	1.034	1.032	1.030
1.65	1.041	1.040	1.039	1.036	1.034
1.70	1.047	1.046	1.043	1.039	1.037
1.75	1.052	1.051	1.047	1.042	1.039
1.80	1.057	1.055	1.051	1.045	1.041
1.85	1.061	1.059	1.054	1.047	1.043
1.90	1.065	1.060	1.057	1.049	1.045
1.95	1.068	1.064	1.059	1.051	1.046

u \ N	4.0	4.2	4.5	5.0	5.5
2.00	1.071	1.068	1.062	1.053	1.047
2.10	1.076	1.071	1.065	1.056	1.049
2.20	1.080	1.073	1.068	1.058	1.050
2.3	1.084	1.079	1.071	1.060	1.051
2.4	1.087	1.081	1.073	1.061	1.052
2.5	1.090	1.083	1.075	1.062	1.053
2.6	1.092	1.985	1.076	1.063	1.054
2.7	1.094	1.087	1.077	1.063	1.054
2.8	1.096	1.088	1.078	1.064	1.054
2.9	1.098	1.089	1.079	1.065	1.055
3.0	1.099	1.090	1.080	1.065	1.055
3.5	1.103	1.093	1.082	1.066	1.055
4.0	1.106	1.097	1.084	1.067	1.056
4.5	1.108	1.098	1.085	1.067	1.056
5.0	1.110	1.099	1.085	1.068	1.056
6.0	1.111	1.100	1.085	1.068	1.056
7.0	1.111	1.100	1.086	1.068	1.056
8.0	1.111	1.100	1.086	1.068	1.056
9.0	1.111	1.100	1.086	1.068	1.056
10.0	1.111	1.100	1.086	1.068	1.056

Author Index

SUBJECT INDEX

About the Author

Richard H. French received his B.S.C.E. and M.S. in Civil Engineering from the Ohio State University and continued his graduate studies at the University of California, Berkeley. He was the recipient of the Boris Bakhmeteff Research Fellowship in Fluid Mechanics from Columbia University in 1973–1974. After receiving his Ph.D. in Civil Engineering from the University of California in 1975, he was appointed an Assistant Professor of Environmental and Water Resources Engineering at Vanderbilt University. In 1979, French joined the staff of the Water Resources Center of the Desert Research Institute, University of Nevada System in Las Vegas, Nevada where he is currently a Research Professor and the Associate Director. He has also served as a consultant on water quality and water resources issues to the Attorney General of Nevada and the state legislature.

In the fall of 1983 as a Visiting Professor of Civil Engineering, French presented an invited series of lectures at the Japanese National Defense Academy in Yokosuka, Japan.

French is a registered Professional Engineer in Nevada and in California, and is a member of the American Society of Civil Engineers, the International Association for Hydraulic Research, the American Water Resources Association, and several honorary fraternities. His previous works include: *Principles of Hydraulics* and *Salinity in Watercourses and Reservoirs.*